W0112966

Soil and Climate

Advances in Soil Science

Series Editors: Rattan Lal and B. A. Stewart

Published Titles

Soil Processes and the Carbon Cycle
R. Lal, J. M. Kimble, R. F. Follett, and B. A. Stewart

Global Climate Change and Cold Regions Ecosystems
R. Lal, J. M. Kimble, and B. A. Stewart

Assessment Methods for Soil Carbon
R. Lal, J. M. Kimble, R. F. Follett, and B. A. Stewart

Soil Erosion and Carbon Dynamics
E.J. Roose, R. Lal, C. Feller, B. Barthès, and B. A. Stewart

Soil Quality and Biofuel Production
R. Lal and B. A. Stewart

Food Security and Soil Quality
R. Lal and B. A. Stewart

World Soil Resources and Food Security
R. Lal and B. A. Stewart

Soil Water and Agronomic Productivity
R. Lal and B. A. Stewart

Principles of Sustainable Soil Management in Agroecosystems
R. Lal and B. A. Stewart

Soil Management of Smallholder Agriculture
R. Lal and B. A. Stewart

Soil-Specific Farming: Precision Agriculture
R. Lal and B. A. Stewart

Soil Phosphorus
R. Lal and B. A. Stewart

Urban Soils
R. Lal and B. A. Stewart

Soil Nitrogen Uses and Environmental Impacts
R. Lal and B. A. Stewart

For more information about this series, please visit:
https://www.crcpress.com/Advances-in-Soil-Science/book-series/CRCADVSOILSCI

Soil and Climate

Edited by
Rattan Lal
B. A. Stewart

CRC Press
Taylor & Francis Group
Boca Raton London New York

CRC Press is an imprint of the
Taylor & Francis Group, an **informa** business

CRC Press
Taylor & Francis Group
6000 Broken Sound Parkway NW, Suite 300
Boca Raton, FL 33487-2742

First issued in paperback 2021

© 2019 by Taylor & Francis Group, LLC
CRC Press is an imprint of Taylor & Francis Group, an Informa business

No claim to original U.S. Government works

ISBN 13: 978-1-03-209476-2 (pbk)
ISBN 13: 978-1-4987-8365-1 (hbk)

This book contains information obtained from authentic and highly regarded sources. Reasonable efforts have been made to publish reliable data and information, but the author and publisher cannot assume responsibility for the validity of all materials or the consequences of their use. The authors and publishers have attempted to trace the copyright holders of all material reproduced in this publication and apologize to copyright holders if permission to publish in this form has not been obtained. If any copyright material has not been acknowledged please write and let us know so we may rectify in any future reprint.

Except as permitted under U.S. Copyright Law, no part of this book may be reprinted, reproduced, transmitted, or utilized in any form by any electronic, mechanical, or other means, now known or hereafter invented, including photocopying, microfilming, and recording, or in any information storage or retrieval system, without written permission from the publishers.

For permission to photocopy or use material electronically from this work, please access www.copyright.com (http://www.copyright.com/) or contact the Copyright Clearance Center, Inc. (CCC), 222 Rosewood Drive, Danvers, MA 01923, 978-750-8400. CCC is a not-for-profit organization that provides licenses and registration for a variety of users. For organizations that have been granted a photocopy license by the CCC, a separate system of payment has been arranged.

Trademark Notice: Product or corporate names may be trademarks or registered trademarks, and are used only for identification and explanation without intent to infringe.

Library of Congress Cataloging-in-Publication Data

Names: Lal, R., editor. | Stewart, B. A. (Bobby Alton), 1932- editor.
Title: Soil and climate / editors: Rattan Lal and B. A. Stewart.
Other titles: Advances in soil science (Boca Raton, Fla.)
Description: Boca Raton, FL : CRC Press, Taylor & Francis Group, 2018. | Series: Advances in soil science
Identifiers: LCCN 2018006065 | ISBN 9781498783651 (hardback)
Subjects: LCSH: Soils and climate. | Soils--Environmental aspects. | Soil management. |
Climate change mitigation.
Classification: LCC S596.3 S64 2018 | DDC 631.4--dc23
LC record available at https://lccn.loc.gov/2018006065

Visit the Taylor & Francis Web site at
http://www.taylorandfrancis.com

and the CRC Press Web site at
http://www.crcpress.com

Publisher's Note
The publisher has gone to great lengths to ensure the quality of this reprint but points out that some imperfections in the original copies may be apparent.

Contents

Preface

The year 2015 was historic in terms of two key proclamations of global significance: The Paris Climate Summit and U.N. Sustainable Development Goals. The Paris COP21 Summit (November 30 to December 12, 2015) proposed to limit the global temperature to 1.5°C above the pre-industrial levels. For the first time ever, all U.N. member nations agreed to curb emissions, strengthen resilience, and join in the common pursuit to implement common climate action. Critical clauses of the historic agreement include: (1) *mitigation-reducing emission* fast enough to achieve the temperature goal; (2) *a complete transparency* of the global stock accounting system for any climate action plan, (3) *adaptation*—strengthening ability of countries to mitigate climate change; (4) *loss and damage*—strengthening ability to recover from climate impacts; and (5) *support*, including finance, for nations to rebuild a clean, resilient future.

On September 25, 2015, the U.N. adopted 17 Sustainable Development Goals (SDGs). Notable among these with reference to soil are (1) no poverty, (2) zero hunger, (3) good health and wellbeing, (5) gender equality, (6) clear water and sanitation, (7) affordable and clean energy, (8) decent work and economic growth, (12) responsible consumption and production, (13) climate action, and (15) life on land. SDG-Targets 2.4 and 15.3 are specifically focused on land and soil. For example, SDG-Target 2.4 states, "By 2030, ensure sustainable food production systems and implement resilient agriculture practices that increase productivity and production, that help maintain ecosystems that strengthen capacity for adaptation to climate change, extreme weather, drought, flooding and other disasters, and that progressively improve land and soil quality." Similarly, SDG-Target 15.3 states, "By 2020, combat desertification, restore degraded land and soil, including land by desertification, drought and floods, and strive to achieve a land degradation neutral world."

Soils are critical to achieving the objectives outlined in the COP21 and SDGs, especially with reference to climate change. Soils affect and are affected by the climate. Weathering and rate of soil formation are affected by climate, being an active factor of soil formation. In turn, soils affect climate as sinks and sources of atmospheric carbon dioxide (CO_2) and other greenhouse gases (GHGs), namely methane (CH_4) and nitrous oxide (N_2O). Soils are the largest terrestrial reservoir of carbon, estimated at 3000 Pg of soil organic carbon (SOC) to 3-m depth. In addition, there are 940 Pg of soil inorganic C. Frozen soils (permafrost) are a large reservoir of C, estimated to contain ~1500 Pg to 3-m depth. These are vast reservoirs compared with about 820 Pg contained in the atmosphere and 620 Pg in live and dead phytomass. Small changes in the soil C pool can strongly impact atmospheric concentrations of CO_2 and CH_4. Sustainable management of soils, aimed at creating a positive soil C budget, can create a notable drawdown of atmospheric CO_2. It is in this context that the objectives of COP21 and SDGs of the U.N. are in accord with the restoration and sustainable management of world soils.

There is also a strong soil–water–food–climate nexus. Anthropogenic or natural changes in soil strongly impact water quality and renewability, food and nutritional security, and adaptation and mitigation of climate change. The French Minister of Agriculture, M Stephane Le Foll, proposed at COP21 that C sequestration in world soils is an important strategy to adapt and mitigate climate change, while also advancing the global food and nutritional security. Therefore, the specific objectives of the proposed volume entitled *Soil and Climate* are to describe state-of-the-knowledge regarding the climate–soil nexus in relation to:

1. *Soil Processes*: weathering, decomposition of organic matter, erosion, leaching, salinization, biochemical, transformations, gaseous flux, and elemental cycling,
2. *Soil Properties*: physical, chemical, biological, and ecological,
3. *Atmospheric Chemistry*: gaseous concentrations of CO_2, CH_4, N_2O, water vapors, soot, dust, and particulate matter,

4. *Mitigation and Adaptation*: source and sink of GHGs (CO_2, CH_4, N_2O), land use and soil management, soil C sink capacity, permafrost,
5. *Soil Management*: sequestration of organic and inorganic C, nutrient requirements, water demands, coupled cycling of H_2O, N, P, S, and
6. *Policy and Outreach*: carbon farming, payments for ecosystem services, COP21, SDGs, land degradation neutrality.

Rattan Lal
Bobby A. Stewart

Editors

Rattan Lal, PhD, is a distinguished university professor of soil science and director of the Carbon Management and Sequestration Center, The Ohio State University, and an adjunct professor at the University of Iceland. His current research focus is on climate-resilient agriculture, soil carbon sequestration, sustainable intensification, efficient use of agroecosystems, and sustainable management of soil resources. He received honorary degrees of Doctor of Science from Punjab Agricultural University (2001); the Norwegian University of Life Sciences, Aas (2005); Alecu Russo Balti State University, Moldova (2010); Technical University of Dresden, Germany (2015); and University of Lleida, Spain (2017). He was president of the World Association of the Soil and Water Conservation (1987–1990), the International Soil Tillage Research Organization (1988–1991), the Soil Science Society of America (2005–2007), and is the president of the International Union of Soil Science (2017–2018). He was a member of the Federal Advisory Committee on U.S. National Assessment of Climate Change—NCADAC (2010–2013); member of the SERDP Scientific Advisory Board of the US–DOE (2011–); Senior Science Advisor to the Global Soil Forum of the Institute for Advanced Sustainability Studies, Potsdam, Germany (2010–2016); member of the Advisory Board of Joint Program Initiative of Agriculture, Food Security and Climate Change (FACCE-JPI) of the European Union (2013–2016); and chair of the Advisory Board of Institute for Integrated Management of Material Fluxes and Resources of the United Nations University (UNU-FLORES), Dresden, Germany (2014–2019). Professor Lal was a lead author of the IPCC (1998–2000). He has mentored 110 graduate students and 54 postdoctoral researchers and hosted 169 visiting scholars. He has authored/co-authored 868 refereed journal articles, and has written 20 and edited/co-edited 68 books. For 3 years (2014, 2015, 2016), Thomson Reuters listed him among the world's most influential scientific minds and having citations of publications among the top 1% of scientists in agricultural sciences.

B. A. Stewart is Director of the Dryland Agriculture Institute and a distinguished professor of Agriculture at West Texas A&M University, Canyon, Texas. He is a former director of the USDA Conservation and Production Laboratory at Bushland, Texas; past president of the Soil Science Society of America; and a member of the 1990–1993 Committee on Long-Range Soil and Water Policy, National Research Council, National Academy of Sciences. He is a fellow of the Soil Science Society of America, American Society of Agronomy, Soil and Water Conservation Society, a recipient of the USDA Superior Service Award, a recipient of the Hugh Hammond Bennett Award of the Soil and Water Conservation Society, and was an honorary member of the International Union of Soil Sciences in 2008. In 2009, Dr. Stewart was inducted into the USDA Agriculture Research Service Science Hall of Fame. Dr. Stewart is very supportive of education and research on dryland agriculture. The B.A. and Jane Ann Stewart Dryland Agriculture Scholarship Fund was established in West Texas A&M University in 1994 to provide scholarships for undergraduate and graduate students with a demonstrated interest in dryland agriculture.

Contributors

Muhammad Shakeel Arshad
Department of Agronomy
University of Agriculture
Faisalabad, Pakistan

Frank Baumann
Department of Geosciences, Soil Science
    and Geomorphology
University of Tübingen
Tübingen, Germany

Asmeret Asefaw Berhe
Life and Environmental Sciences
University of California
Merced, California

Jens Boy
Institute of Soil Science
Leibniz University
Hannover, Germany

Constanze Buhk
Institute of Environmental Sciences,
    Geoecology & Physical Geography
University Koblenz–Landau
Landau, Germany

Maria Luz Cayuela
Department of Soil and Water Conservation
    and Waste Management
CEBAS-CSIC, Espinardo
Murcia, Spain

Annette Cowie
University of New England
New South Wales Department
    of Primary Industries
Armidale, New South Wales, Australia

Eugenio Díaz-Pinés
Institute of Soil Research
University of Natural Resources
    and Life Sciences Vienna
Vienna, Austria

Jennifer A.J. Dungait
Sustainable Agriculture Sciences
Rothamsted Research
Okehampton, United Kingdom

Stefan Erasmi
Institute of Geography, Cartography, GIS and
    Remote Sensing
University of Göttingen
Göttingen, Germany

Muhammad Farooq
Department of Agronomy
University of Agriculture
Faisalabad, Pakistan

and

The UWA Institute of Agriculture
The University of Western Australia
Perth, Western Australia

and

Department of Crop Sciences
College of Agricultural and Marine Sciences
Sultan Qaboos University
Muscat, Oman

Nirmali Gogoi
Department of Environmental Science
Tezpur University
Assam, India

Andrew S. Gregory
Rothamsted Research
Harpenden, United Kingdom

Thomas Guillaume
Laboratory of Ecological Systems
Swiss Federal Institute for Forest, Snow
    and Landscape Research
Lausanne, Switzerland

Jin-Sheng He
Department of Ecology
College of Urban and Environmental Sciences
Peking University
Beijing, China

Jessica Henkner
Department of Geosciences, Soil Science
    and Geomorphology
University of Tübingen
Tübingen, Germany

Katrin Hofmann
Institute of Soil Research
University of Natural Resources
    and Life Sciences Vienna
Vienna, Austria

David W. Hopkins
The Royal Agricultural University
Cirencester, Gloucestershire, United Kingdom

Thomas Horvath
Faculty of Natural and Environmental Sciences
University Koblenz–Landau
Landau, Germany

H. H. Janzen
Agriculture and Agri-Food Canada
Lethbridge, Alberta, Canada

Hermann F. Jungkunst
Institute of Environmental Sciences,
    Geoecology & Physical Geography
University Koblenz–Landau
Landau, Germany

Thomas Kätterer
Department of Ecology
Swedish University of Agricultural Sciences
Uppsala, Sweden

Kathi J. Kemper
College of Medicine
The Ohio State University
Columbus, Ohio

Peter Kühn
Department of Geosciences, Soil Science
    and Geomorphology
University of Tübingen
Tübingen, Germany

Jan Paul Krüger
UDATA GmbH
Neustadt an der Weinstrasse, Germany

Jeffery Lakritz
Veterinary Clinical Medicine
The Ohio State University
Columbus, Ohio

Rattan Lal
Carbon Management and Sequestration Center
School of Environment and Natural
    Resources
The Ohio State University
Columbus, Ohio

Johannes Lehmann
School of Integrative Plant Sciences
Atkinson Center for a Sustainable Future
Cornell University
Ithaca, New York

Ram Swaroop Meena
Carbon Management and Sequestration Center
Ohio State University
Columbus, Ohio

Katharina H.M. Meurer
Department of Ecology
Swedish University of Agricultural Sciences
Uppsala, Sweden

Umakant Mishra
Argonne National Laboratory
Lemont, Illinois

Tarik Mitran
Carbon Management and Sequestration Center
The Ohio State University
Columbus, Ohio

Faisal Nadeem
Department of Agronomy
University of Agriculture
Faisalabad, Pakistan

T. Ravisankar
National Remote Sensing Centre
Balanagar, Hyderabad, India

D. C. Reicosky
Retired
USDA-ARS
Morris, Minnesota

Sabine Reinsch
Centre for Ecology and Hydrology
Bangor, United Kingdom

Muhammad Sanaullah
Institute of Soil and Environmental Sciences
University of Agriculture
Faisalabad, Pakistan

Per-Marten Schleuss
Bayreuth Centre of Ecology and Environmental
    Research
University of Bayreuth
Bayreuth, Germany

Jörg Schnecker
Department of Microbiology and Ecosystem
    Science
University of Vienna
Vienna, Austria

Julia Schneider
Institute of Environmental Sciences,
    Geoecology & Physical Geography
University Koblenz–Landau
Landau, Germany

Thomas Scholten
Department of Geosciences, Soil Science
    and Geomorphology
University of Tübingen
Tübingen, Germany

Klaus Schützenmeister
Institute of Environmental Sciences,
    Geoecology & Physical Geography
University Koblenz–Landau
Landau, Germany

Darryl D. Siemer
ISU Nuclear Engineering
Idaho Falls, Idaho

Saran Sohi
School of Geosciences
Edinburgh University
Edinburgh, United Kingodom

Caroline Spann
Institute of Soil Research
University of Natural Resources
    and Life Sciences Vienna
Vienna, Austria

K. Sreenivas
National Remote Sensing Centre
Balanagar, Hyderabad, India

B. A. Stewart
Retired
Department of Agricultural Sciences
West Texas A&M University
Canyon, Texas

David A.N. Ussiri
Carbon Management and Sequestration Center
The Ohio State University
Columbus, Ohio

Thea Whitman
Department of Soil Science
University of Wisconsin–Madison
Madison, Wisconsin

Dominic Woolf
School of Integrative Plant Sciences
Cornell University
Ithaca, New York

Sophie Zechmeister-Boltenstern
Institute of Soil Research
University of Natural Resources
    and Life Sciences Vienna
Vienna, Austria

1 Soil and Climate

Rattan Lal

CONTENTS

1.1 CLIMATE–SOIL INTERACTION

The climate–soil interaction goes back to the origin of Earth and the solar system. The initial atmosphere, 4.5–5 billion years (Ga) ago, consisted of H_2 and He, and these light gases escaped into space. The primary atmosphere was formed during the first 500 million years. These gases were replaced by a secondary atmosphere consisting of a mixture of gases attributed to outgassing and accretion (Kasting 1993). Outgassing implies a release of gases by volcanism, which releases water, carbon monoxide (CO), carbon dioxide (CO_2), methane (CH_4), ammonia (NH_3), nitrogen (N_2), and other gases (e.g., SO_2, S_2, Cl_2). Differences in atmosphere chemistry of Earth from those of Mars and Venus (Prinn and Fegley 1987; NASA 2017) are due to the presence of life, especially that of plants on Earth.

There was no O_2 in the primary atmosphere. Cooling of the Earth condensed H_2O vapor into liquid forms and led to the formation of oceans and the hydrosphere about 4 Ga ago. O_2 was created by the origin of green plants and a combination of CO_2 with water by absorption of ultraviolet rays leading to photosynthesis:

$$6\,CO_2 + 12\,H_2O + UV\ radiation \rightarrow C_6H_{12}O_6 + 6\,O_2 + 6\,H_2O \qquad (1.1)$$

It was the interaction of CO_2 with silicate rocks and their weathering which absorbed CO_2 and formed carbonates according to the Urey reactions:

$$CaSiO_3 + H_2CO_3 = CaCO_3 + SiO_2 + H_2O \qquad (1.2)$$

The Urey reactions, over geologic timescale, remove CO_2 from the atmosphere and its burial in the marine sediments. Thus, over time CO_2 has been removed and O_2 concentration has been increased. The slow silicate rock weathering has balanced atmospheric CO_2 over a millennial timescale. Chemical weathering is faster in the humid tropics than in temperate climates. Weathering of parent rock material and new soil formation is strongly impacted by the atmosphere, especially by its chemistry (e.g., gaseous composition).

The rate of chemical weathering of silicates, eventually leading to the formation of new soil through action with plants and other biota (Jenny 1943), is strongly dependent on climate, especially temperature and precipitation. The rate of silicate weathering is doubled with every 10°C increase in temperature (Vont Hoff Rule). Similarly, an increase in precipitation increases the rate of weathering through an increase in the hydrolysis. There exists a close link between the temperature and precipitation. Thus, climate (atmosphere, temperature, and precipitation) has a direct impact on soil (Jenny 1943).

1.2   SOIL LIFE AND THE ATMOSPHERE

As the Earth cooled and formed a crust, water began to condense, leading to the formation of the hydrosphere. Atmosphere, comprising of gases from volcanic activities (outgoing) and escape of H_2 and He (degassing), concentration of CO_2, CH_4, and H_2O vapors increased in the atmosphere. Concentration of CO_2 peaked during the Archean era (Figure 1.1) at ~15% due to volcanic activity. Dissolution of CO_2 and NH_3 in the water led to the formation of H_2CO_3 and NH_4^+ ions, which reacted with the rocks according to the Urey reactions:

$$\left.\begin{aligned} CO_2 + H_2O &\rightarrow H_2CO_3 \\ H_2CO_3 + CaSiO_3 &\rightarrow CaCO_3 + SiO_2 + H_2O \\ CO_2 + CaSiO_3 &\rightarrow CaCO_3 + SiO_2 \\ CO_2 + MgSiO_3 &\rightarrow MgCO_2 + SiO_2 \end{aligned}\right\} \qquad (1.3)$$

These reactions absorbed CO_2 from the atmosphere and led to the formation of carbonates and increased concentration of C into the rocks and geological strata.

The solar energy received from the Earth's surface is partly absorbed and partly reflected back as albedo. The amount of solar energy retained in the Earth's atmosphere depends on the concentration

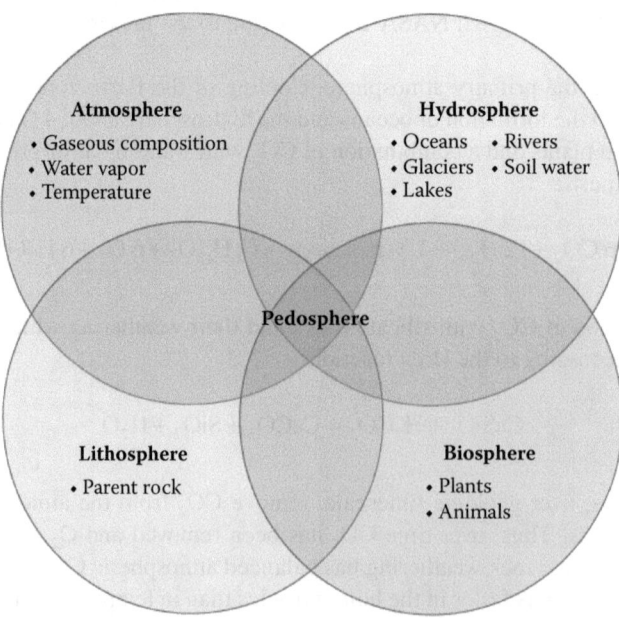

FIGURE 1.1   The co-evolution of the atmosphere, oceans, rocks, plants, and soil.

and type of radiatively-active gases (CO_2, CH_4, N_2O, O_3, H_2O). With the weathering of rocks, the formation of soil, and the gradual increase in soil organic carbon (SOC) concentration, soil became a sink of atmospheric CO_2 and oxidation of CH_4. Over billions of years, soil has become the largest reservoir of terrestrial C pool.

1.3 IMPACTS OF CLIMATE ON SOIL

Climatic parameters with a strong effect on soil are temperature, water, gaseous composition, and nutrients. Climate impacts on soil may be direct and indirect (Figure 1.1). Directly, climate impacts soil and air temperature, rate and type of soil erosion, rate of weathering, salinization, and desertification. Indirectly, climate impacts soil through its effects on net primary productivity (NPP), hydrological cycles, energy balance, and making soil a source or sink of C (Figure 1.2). Through its impact on the amount of water availability, climate affects the rate of soil formation through the weathering of rocks, and the transportation and distribution of the weathered material over the landscape. The rate of new soil formation through weathering of the parent material is also influenced by the temperature regime. Thus, hydrothermal regimes have a strong impact on soil properties. Furthermore, warm and moist climates (humid tropics) have a higher NPP than colder and drier climates, and the soil:biomass ratio of the C stock varies with the moisture and temperature regimes. In addition to the rate of weathering, the climate also impacts the rate of accumulation of SOC and leaching of carbonates. Soils of wet climates are strongly leached of the basic cations and are acidic with low pH and high concentrations of Al, Mn, Fe, etc. In comparison, soils of drier climates have higher concentrations of carbonates. Climate also impacts SOC dynamics through moderation of soil biota (e.g., micro-, meso-, and macro-organisms).

The CO_2 fertilization effect, due to the enrichment of CO_2 in the atmosphere, can enhance the NPP provided that N and water are adequate. The mean resident time of C in soil may range from 100 to 10,000 years (Parton et al. 1995) with a strong impact on atmospheric chemistry. Similar to temperature, changes in the moisture regime by climate fluctuations (D'Odorico et al. 2000) may also impact SOC stocks and pedosphere processes. Therefore, the projected climate change may impact soil organic carbon (SOC) stocks. An increase in temperature may also impact the soil-water supply, NPP, and the magnitude of soil degradation caused by accelerated erosion and salinization. Because of their strong impact on the hydrological cycle and vegetation type, a combination of moisture and temperature regimes along with that of vegetation affects the eco-regions or biomes, ranging from tundra to the humid tropics.

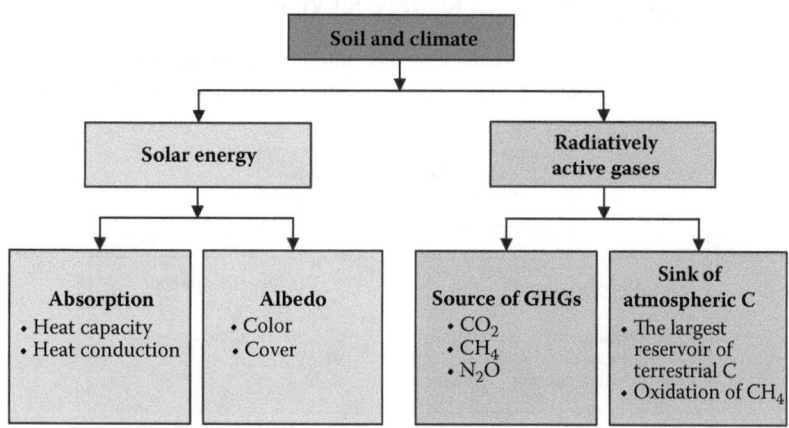

FIGURE 1.2 Soil affects and is affected by climate. Both are intricately interlinked.

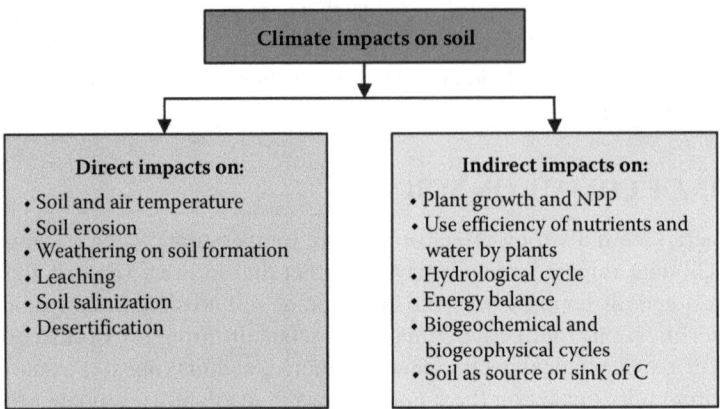

FIGURE 1.3 Direct and indirect impacts of climate on soil.

1.4 IMPACTS OF SOIL ON CLIMATE

Soil also impacts climate, through moderation of the atmospheric chemistry by coupled cycling of H_2O, C, N, P, and S (Lal 2010a), and changes in vapor pressure of water through a soil moisture regime. Soil moderates climates at the micro, meso, and macro scales, and over short and long time horizons (Figure 1.3). Notable changes in soil moisture can have strong feedback. Changes in the soil moisture regime can also impact the temperature (both air and soil) through changes in evaporative cooling (Seneviratne et al. 2013). The soil moisture regime is also the key link in the soil–climate–vegetation nexus (Rodriquez-Iturbe 2000), and is an important factor moderating the climate. Soil can also impact climate by being a source or sink of atmospheric CO_2, CH_4, and N_2O. Soil degradation and desertification can deplete the SOC stock and emit GHGs into the atmosphere (Lal 2003). In contrast, restoration of the soil and ecosystems can restore the SOC stock, offset anthropogenic and natural emissions, and mitigate climate change. The annual global CO_2 flux from soil to the atmosphere may be as much as 68 ± 4 Pg C/yr (Raich and Schlesinger 1992) and has impacts on vegetation and climate. The SOC stock also impacts soil color and thus the albedo. Soil color differs among climates. Soils in high latitudes, with relatively high SOC stock, have darker soil and low albedo. In contrast, soil in arid climates, with relatively low SOC stock, have a high albedo. Soils of the tropics (humid and sub-humid), because of a high concentration of sesquioxides, can have red or orange color and differ in albedo. Saline soils, with high salt content and an accumulation of carbonates in the surface, may have whitish coloration and a high albedo.

1.5 THE SOIL–CLIMATE–WATER–ENERGY NEXUS

The schematics in Figures 1.1–1.4 indicate a strong interconnectivity between soil, climate, and vegetation. The discussion presented also indicates other ramifications of climate: energy and

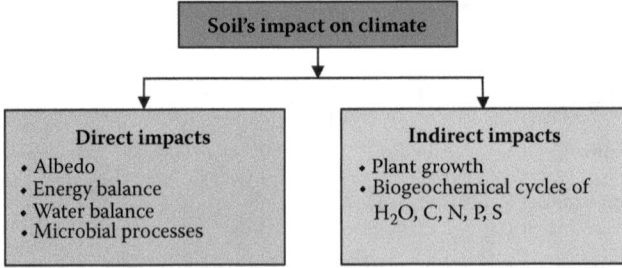

FIGURE 1.4 Soil in moderate climates at micro, meso, and macro scales by direct and indirect impacts over short and long timescale. Thus, soil can be analyzed to understand past climate and predict future changes in climate.

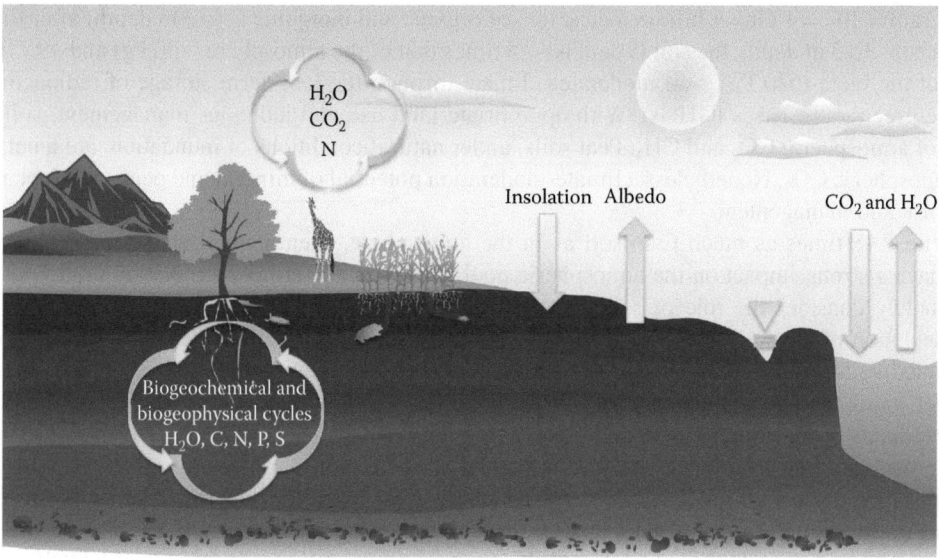

FIGURE 1.5 The soil–climate–energy–water nexus (SCEW).

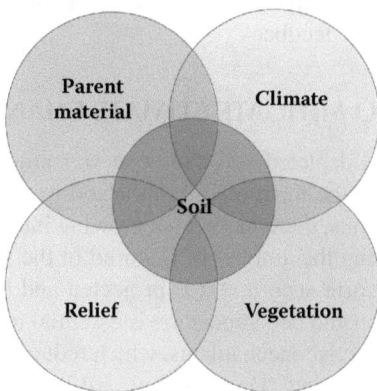

FIGURE 1.6 Interaction of climate and soil also affects vegetation and relief, and is affected by the parent material.

water. Indeed, water and energy balances are closely inter-linked. Therefore, understanding the interdependence of water and energy budgets, especially for some critical regions such as the Arctic (Semmler et al. 2005), is important to understanding the global circulation. While water vapors in the atmosphere absorb the incoming radiation, evaporation and convection transfer 25% and 5% of incoming solar energy from the surface to the atmosphere (Lindsey 2009) thus moderating the Earth's energy budget. In addition, the water–energy budget is also affected by the complex soil–climate–water–energy nexus or the SCEW nexus (Figure 1.5). Soil and vegetation strongly alter and moderate the water and energy budgets, and impact climate (Figure 1.6).

1.6 THE SOIL–CLIMATE ECOSYSTEM SERVICES

As a natural capital, soil is a source of natural ecosystem services including those related to moderation of climate through its impact on atmospheric chemistry (Costanza et al. 1997; Daily 1997; Baveye et al. 2016). Soil is the second largest reservoir of C (storage capacity of >6000 Pg [Pg =

petagram = 10^{15} = 1 Gt = 1 billion metric ton] of organic and inorganic C to 3 m depth) after that of the ocean. To 3 m depth, the soil C pool is ~7.5 times that of the atmosphere (800 Pg) and ~9.7 times that of the biota (620 Pg). Soil moderates climate through the long-term storage of radioactively-active greenhouse gases (GHGs). With appropriate land use and judicious management, soil is a sink of atmospheric CO_2 and CH_4. Peat soils, under natural conditions of inundation, are a net sink of atmospheric CO_2. Nonetheless, climate-moderation potential of mineral and peat soils depend on land use and management.

With 7.45 times as much C in soil as in the atmosphere, even a small change in soil C pool can have a strong impact on the atmospheric pool. However, the climate prediction models do not adequately consider the role of soil. Changes in land use and soil management that exacerbate the depletion of the soil C reserves can have a positive feedback to climate change. Restoration of degraded soils and conversion to a restorative land use, which create a positive soil/ecosystem C budget, would lead to a negative feedback. Thus, soil and its C storage and dynamics are important determinants of the climate system.

Cryosols, frozen soils of the arctic, tundra, and alpine regions, are a major reservoir of organic C. Presently, these soils are a net sink of atmospheric CO_2. With the thawing of permafrost, soil C can be emitted into the atmosphere as CO_2 through enhanced decomposition and as CH_4 through methanogenesis. In addition, soil N can be emitted as N_2O through O_2-induced nitrification/denitrification processes. Thus, rising temperatures with the projected global warming can exacerbate the loss of soil C to the atmosphere, creating a massive positive soil C climate feedback, and accelerate the global warming. The temperature sensitivity of soil C decomposition (Davidson and Janssens 2006) governs the positive feedback.

1.7   MANAGING SOILS TO MITIGATE CLIMATE CHANGES

The restoration of degraded and depleted soil can sequester atmospheric CO_2 and improve soil health. The strategy is to increase the input of C into the soil to exceed the losses through decomposition, erosion, and leaching. Thus, there has been a growing interest in reducing the atmospheric concentration of CO_2 by increasing the amount of C stored in the soil as organic and inorganic C. The objective is to store C in a form such that it is projected and has a long mean residence time (MRT). Protection against erosion and decomposition is essential to long-term storage of SOC. The MRT is determined more by protective mechanisms, which reduce accessibility, than by the molecular structures of the organic substances (Dungait et al. 2012). There are several protective mechanisms (Six et al. 2002) including physical (stable micro-aggregates), chemical (organo-mineral complexes), biological (recalcitrance), and ecological (deep translocation into sub-soil). Thus, SOC storage can be enhanced by creating a positive soil C budget and enhancing protective mechanisms that increase the MRT.

There are also options of enhancing the storage of soil inorganic C as secondary carbonates and through the leaching of bicarbonates (Monger et al. 2015). However, storage of soil inorganic carbon (SIC) to 3 m depth, the mechanisms of enhancing the formation of secondary carbonates, and the leaching of bicarbonates are not widely studied.

These are numerous co-benefits of increasing soil C stock. In addition to mitigating climate change, improvement in SOC also enhances soil health (Lal 2016a) and productivity (Lal 2016b). Important other benefits include water filtration and purification, denaturing of pollutants, and improvement of biodiversity.

1.8   SOIL HEALTH AND CLIMATE CHANGE

The strong link between soil health and climate change (Singh et al. 2011) is receiving global attention. The Paris Climate Summit COP21 initiated the program "4 per Thousand" to sequester C in soil for climate change and food (Lal 2016c; Chambers et al. 2016). A strong decline of soil health by degradation processes (e.g., erosion, compaction, nutrient depletion, salinization, acidification) can deplete the SOC stock and emit GHG from the soil into the atmosphere. Accelerated soil erosion preferentially removes the SOC faction, as is indicated by a high enrichment ratio. The preferential removal of SOC is caused by low density (0.6–0.8 Mg/m^3) and the fact that it is concentrated in vicinity of the surface layer. The SOC being transported by alluvial and aeolian processes and redistributed over the landscape is prone to accelerated decomposition because of the breakdown of aggregates and changes in soil temperature and moisture regimes (Lal 2003). Thus, the restoration of eroded and degraded soils can sequester atmospheric CO_2 and offset anthropogenic emissions (Lal 2010a,b, 2004). It is precisely in this context that desertification control has a large potential to sequester C and mitigate the greenhouse effect (Lal 2001, 2013). The concept of land degradation neutrality (Lal et al. 2012), similar to that of "4 per Thousand," is also an aspirational goal of restoring degraded and desertified lands to improve the environment. Target 15.3 of Sustainable Development Goals (SDG) states "By 2030, combat desertification, restore degraded lands and soil, including land affected by desertification, drought and floods, and strive to achieve land degradation-neutral world." This Target is an integral component of SDG 15, which states "Protect, restore and promote sustainable use of terrestrial ecosystems, sustainably manage forests, combat desertification, and halt and reverse land degradation and halt biodiversity loss."

There are multiple benefits of restoring the health of degraded/desertified soils. Over and above the vast potential to mitigate climate change, soil restoration also advances food and nutritional security while improving the environment. Furthermore, there are high economic costs of no action (Nkonya et al. 2016).

Despite multiple benefits, there are numerous challenges to implementing global programs on desertification control (Glenn et al. 1998; Zhang et al. 2016; Ambalam 2014). The real challenges to implementing global programs are key benchmark locations and involving the local community. It is important to move away from conferences and meeting toward a concrete and meaningful program to restore degraded soils (Stringer 2008).

1.9 TRANSLATING SCIENCE INTO ACTION

The strategy is to translate knowledge of "soil and climate" into action by identifying and implanting soil-specific soil use, management, and restoration strategies. There is a strong scientific knowledge base in soil properties and processes which impact and are impacted by climate change. The scientific knowledge must be integrated with the real-world and site-specific needs for adaptation and mitigation of climate change in diverse ecoregions. It is important to translate generic soil management/adaptation principles into soil-specific management options for validation and fine-tuning. Climate change information must be integrated into management plans and appropriate activities to facilitate translating knowledge of soil science into action. Decision-makers must know how the higher temperatures and extreme events will impact the community and local businesses, and be prepared to monitor changes as they occur. Short-term climate forecasts can be useful to linking science and action (Buizer et al. 2016).

COP21 (Paris 2015), COP22 (Marrakesh 2016), and COP23 (Berlin 2017) have linked advances in soil science to action for adaptation and mitigation of climate change. Specifically, the "4 per Thousand" program was adopted by COP21, "Adapting African Agriculture (AAA)" by COP22, and developing specific programs to implement "4 per Thousand" by COP23. Thus, the enhanced awareness about the importance of soil to mitigate climate change has mobilized the global political agenda (Janowiak et al. 2014; UNCCD 2016).

The political agenda can be made effective by involving agro-industry in soil C sequestration to improve soil health and accrue benefits through provisioning of numerous ecosystem services. Strategies of reducing net emissions by industries (e.g., machinery, fertilizers, seed production, agro-chemicals, food-processing, and value addition) through soil C sequestration and measures to enhance use efficiency can be win–win scenarios. Industries, more so than government and political organizations, can also motivate land managers/farmers to adopt best management practices (BMPs). Payments, cash or kind, for ecosystem services (i.e., C credits) are essential to promote adoption of BMPs, which lead to soil restoration, improvement of the environment, and meeting the growing demands of humanity.

1.10 CONCLUSIONS

Soil properties and processes affect and are affected by climate. Principal soil properties that affect climate, the same as those that also affect soil health, are physical (clay content and mineralogy, structure, plant available water, infiltration rate, color, albedo, bulk density), chemical (acidity, charge properties, nutrient reserves and dynamics, elemental balance), biological (soil organic carbon concentration and attributes, microbial biomass C, respiration quotient, enzyme activity, species diversity), and ecological (erosion, decomposition, leaching, compaction, aeration, energy balance, water balance). The relative importance of these properties and processes vary among ecoregions.

Climate also affects soil properties and processes. For example, the topsoil depth with high SOC concentration varies with the climate. Climate determines the egetation, and the latter controls the depth distribution and the magnitude of SOC stock. The soil orders vary with climate. Predominant soil orders from the North Pole to the Equator are: Gelisols, Crysols, Inceptisols, Alfisols, Mollisols, Aridisols, Ultisols, Oxisols. Indeed, soil types and vegetation change with change in climate.

REFERENCES

Ambalam, K. 2014. Challenges of compliance with multilateral environmental agreements: The case of UNCCD in Africa. *Journal of Sustainable Development Studies.* 5:145–168.

Baveye, P., Baveye, J., Gowdy, J. 2016. Soil ecosystem services and natural capital: Critical appraisal of research or uncertain ground. *Frontiers in Environmental Science.* doi.org/10.3389/fenvs.2016.00041.

Buizer, J., Jacobs, K., Cash, D. 2016. Making short-term climate forecasts useful: Linking science and action. *Proceedings of the National Academy of Sciences of the United States of America.* 113(17):4597–4602. Available doi 10.1073/pnas.0900518107.

Chambers, A., Lal, R., Paustian, K. 2016. Soil carbon sequestration potential of US croplands and grasslands: Implementing the 4 per Thousand Initiative. *Journal of Soil and Water Conservation.* 71(3):68A–76A.

Costanza, R., d'Arge, R., de'Groot, R., Farber, S., Grasso, M., Hannon, B., Limburg, K., Naeem, S., ONeill, R., Paruelo, J. et al. 1997. The value of the world's ecosystem services and natural capital. *Nature.* 387(6630):253-260. Available doi 10.1038/387253a0.

D'Odorico, P., Ridolfi, L., Porporato, A., Rodriguez-Iturbe, I. 2000. Preferential states of seasonal soil moisture: The impact of climate fluctuations. *Water Resources Research.* 36(8):2209–2219. Available doi 10.1029/2000WR900103.

Daily, G.C. (ed). 1997. *Nature's Services: Societal Dependence on Natural Ecosystem.* Island Press, Washington, 392 pp.

Davidson, E., Janssens, I. 2006. Temperature sensitivity of soil carbon decomposition and feedbacks to climate change. *Nature.* 440(7081):165–173. Available doi 10.1038/nature04514.

Dungait, J., Hopkins, D., Gregory, A., Whitmore, A. 2012. Soil organic matter turnover is governed by accessibility not recalcitrance. *Global Change Biology.* 18(6):1781–1796. Available doi 10.1111 /j.1365-2486.2012.02665.x.

Glenn, E., Stafford Smith, M., Squires, V. 1998. On our failure to control desertification: Implication of global change issues, and a research agenda for the future. *Environmental Science & Policy.* 1:71–78.

Janowiak, M., Swanston, C., Nagel, L., Brandt, L., Butler, P., Handler, S., Shannon, P., Iverson, L., Matthews, S., Prasad, A. et al. 2014. A practical approach for translating climate change adaptation principles into forest management actions. *Journal of Forestry.* 112(5):424–433. Available doi 10.5849/jof.13-094.

Jenny, H. 1943. *Factors of Soil Formation.* McGraw Hill, N.Y., 281 pp.

Kasting, J.F. 1993. Earth's early atmosphere. *Science.* 259:920–926.

Lal, R. 2001. Potential of desertification control to sequester carbon and mitigate the greenhouse effect. *Climatic Change.* 15:35–72.

Lal, R. 2003. Soil erosion and the global carbon budget. *Environment International.* 29:437–450.

Lal, R. 2004. Soil carbon sequestration impacts on global climate change and food security. *Science.* 304:1623–1627.

Lal, R. 2010a. Managing soils and ecosystems for mitigating anthropogenic carbon emissions and advancing global food security. *BioScience.* 60(9):708–721.

Lal, R. 2010b. Carbon sequestration in saline soils. *Journal of Soil Salinity and Water Quality.* 1(1&2):30–40.

Lal, R. 2013. Sustainable soil management under changing climate and desertification. *Annals of Arid Zone.* 50:279–296.

Lal, R. 2016a. Soil health and carbon management. *Food and Energy Security.* 5(4):212–222.

Lal, R. 2016b. Feeding 11 billion on 0.5 hectare of cropland. *Food and Energy Security Journal.* 5(4):239–251.

Lal, R. 2016c. Beyond COP21: Potential and challenges of the "4 per Thousand" initiative. *Journal of Soil and Water Conservation.* 71:20A–25A

Lal, R. U. Safriel and B. Boer. 2012. Zero Net Land Degradation. UNCCD Position Paper for Rio+20. Bonn, Germany.

Lindsey, R. 2009. Climate and Earth's budgets. https://earthobservatory.nasa.gov/Features/EnergyBalance/.

Monger, H., Kraimer, R., Khresat, S., Cole, D., Wang, X., Wang, J. 2015. Sequestration of inorganic carbon in soil and groundwater. *Geology.* 43(5):375–378. Available doi 10.1130/G36449.1.

NASA. 2017. Planetary fact sheet - ratio to Earth values. Retrieved from https://nssdc.gsfc.nasa.gov/planetary/factsheet/planet_table_ratio.html.

Nkonya, E., Mirzabaev, A., Von Braun, J. (Eds). 2016. Economics of Land Degradation and Improvement—A Global Assessment for Sustainable Development. Springer, Dordrecht, Holland, 686 pp.

Parton, W., Scurlock, J., Ojima, D., Schimel, D., Hall, D., Coughenour, M., Garcia Moya, E., Gilmanov, T., Kamnalrut, A., Kinyamario, J. et al. 1995. Impact of climate-change on grassland production and soil carbon worldwide. *Global Change Biology.* 1(1):13–22. Available doi 10.1111/j.1365-2486.1995.tb00002.x.

Prinn, R.G., Fegley, jr. B. 1987. The atmospheres of Venus, Earth and Mars: A critical comparison. *Annual Review of Earth and Planetary Sciences.* 15:171–212.

Raich, J., Schlesinger, W. 1992. The global carbon-dioxide flux in soil respiration and its relationship to vegetation and climate. *Tellus Series B-Chemical and Physical Meteorology.* 44(2):81–99. Available doi 10.1034/j.1600-0889.1992.t01-1-00001.x.

Rodriquez-Iturbe, I. 2000. Ecohydrology: A hydrologic perspective of climate-soil-vegetation dynamics. *Water Resources Research.* 36:3–9.

Semmler, T., Jacob, D., Schlunzen, K., Podzun, R. 2005. The water and energy budget of the Arctic atmosphere. *Journal of Climate.* 18(13):2515–2530. Available doi 10.1175/JCLI3414.1.

Seneviratne, S., Wilhelm, M., Stanelle, T., van den Hurk, B., Hagemann, S., Berg, A., Cheruy, F., Higgins, M., Meier, A., Brovkin, V. et al. 2013. Impact of soil moisture-climate feedbacks on CMIP5 projections: First results from the GLACE-CMIP5 experiment. *Geophysical Research Letters.* 40(19):5212–5217. Available doi 10.1002/grl.50956.

Singh, B.P., Cowie, A.L., Chan, K.Y. (Eds). 2011. *Soil Health and Climate Change.* Springer, New York, USA, 399 pp.

Six, J., Conant, R., Paul, E., Paustian, K. 2002. Stabilization mechanisms of soil organic matter: Implications for C-saturation of soils. *Plant and Soil.* 241(2):155–176. Available doi 10.1023/A:1016125726789.

Stringer, L.C. 2008. Reviewing the International Year of Deserts and Desertification (IYD 2006): What contribution towards combating global desertification and implementing the United Nations Convention to Combat Desertification. *Journal of Arid Environments.* 72:2065–2074.

UNCCD. 2016. Towards a land degradation in neutral world: A sustainable development priority. UNCCD, Bonn.

Zhang, J., Dai, M., Wang, L., Zeng, C., Su, W. 2016. The challenge and future of rocky desertification control in karst areas in southwest China. *Solid Earth.* 7(1):83–91. Available doi 10.5194/se-7-83-2016.

2 Soil—The Hidden Part of Climate

Microbial Processes Regulating Soil–Atmosphere Exchange of Greenhouse Gases

Sophie Zechmeister-Boltenstern, Eugenio Díaz-Pinés,
Caroline Spann, Katrin Hofmann, Jörg Schnecker,
and Sabine Reinsch

CONTENTS

2.1   INTRODUCTION

Climate change has, and will continue to have, fundamental impacts on the natural environment and on human well-being (IPCC 2007). Human activities are altering the chemical composition of the atmosphere, which will lead to a predicted global warming of between 0.3 and 4.8°C by the end of this century (IPCC 2013). Furthermore, state-of-the-art climate models project global intensification of heavy precipitation events (IPCC 2013; Fischer and Knutti 2014). Climate change can fundamentally alter the temporal and spatial variability of temperature and precipitation (Seneviratne et al. 2012). Human activities lead not only to an increase in global atmospheric greenhouse gas (GHG) concentrations and air temperature and changes in precipitation regimes, but also to rising atmospheric nitrogen (N) deposition and land use change (Galloway et al. 2008). These factors modify terrestrial ecosystems by strongly influencing carbon (C) (Fischlin et al. 2007) and N cycling (Fowler et al. 2015) and thus land–atmosphere exchange processes of GHGs (Gritsch et al. 2016).

More C is present in soils than in plant biomass and the atmosphere combined (Jobbágy and Jackson 2000), mostly in the form of soil organic matter (SOM). SOM accumulates over time due to an excess of primary production over both ecosystem respiration (CO_2) and losses of dissolved organic matter (DOM) via leaching. A shift in the balance between the primary production and these losses could trigger significant CO_2 uptake by, or release of CO_2 from soils (Trumbore 1996; Davidson and Janssens 2006). Since N limits primary production in many terrestrial ecosystems (Vitousek and Howarth 1991), responses of N processes in soils (nitrification and denitrification) to global environmental change are of particular importance. This is because these processes can lead to ecosystem N losses, for example by NO_3^- leaching, or by the release of N-containing gases to the atmosphere (Wrage et al. 2001; Smith 2010; Gritsch et al. 2016).

Investigating the response of the C and N cycles to global environmental changes like temperature, precipitation, land use, and N supply is an important prerequisite to developing reasonable global climate change mitigation and adaptation policies (Dick et al. 2006). Long-term measurements of GHG emission exchange rates between the biosphere and atmosphere in different ecosystems and management scenarios is necessary to develop robust emission baselines. Understanding the microbial processes that are responsible for GHG emissions is critical as well if we are to anticipate ecosystem responses to changing environmental conditions. Experimental data are needed to develop process-based emission parameterization of GHG fluxes from natural and agricultural ecosystems in response to climate and N deposition for incorporation into atmosphere–biosphere circulation models (Gritsch et al. 2016).

2.1.1   Sources and Sinks of Greenhouse Gases

The main contributors to the increasing atmospheric CO_2 concentrations are fossil fuel emissions (9.3 Pg C a^{-1}, or 91%) and land use change (~1 Pg C a^{-1}, or 9%), both due to anthropogenic activities. However, only 4.5 Pg C (or 44%) of those emissions accumulate annually in the atmosphere, while the land and ocean absorb 3.1 (31%) and 2.6 Pg C a^{-1} (26%) of the anthropogenic emissions, respectively (Le Quéré et al. 2016). However, natural gross fluxes in terrestrial ecosystems are up to one order of magnitude higher than anthropogenic emissions: gross primary production takes up about 123 Pg C annually, while 118 Pg C are released annually to the atmosphere in the form of above- and belowground respiration and fire (Ciais et al. 2013), largely due to plant and microbial activities.

Microbial denitrification and nitrification processes are responsible for 87% of the annual global N_2O budget (Syakila and Kroeze 2011), with roughly equal contributions from natural soils (35%), agriculture (27%), and oceans (25%). Non-biological sources are responsible for the remaining 13% (fuel combustion, biomass burning and industrial processes). Recent measurements of isotopic N_2O composition in the atmosphere are consistent with the assumption (Park et al. 2012) that increased N-fertilizer use is responsible for rising N_2O concentrations (from 270 ppb to over 319 ppb) since the industrial revolution (Davidson and Kanter 2014).

Atmospheric CH_4 concentrations have more than doubled during the last 200 years, mostly due to human activities. Atmospheric CH_4 sinks include reactions with OH radicals and CH_4 oxidation in soils, while atmospheric sources are agriculture and waste, including enteric CH_4 production by ruminants (34%), wetlands (30%), fossil fuel production and use (18%), biomass burning (6%), natural emissions from the other sources (e.g., geological, termites, permafrost, 11%) (Saunois et al. 2016).

2.1.2 MICROBIAL PROCESSES RESPONSIBLE FOR GREENHOUSE GAS EMISSIONS

Various processes lead to GHG emissions from soil and litter to the atmosphere. The flux of C from soil to the atmosphere occurs primarily in the form of CO_2, which is the result of soil respiration and constitutes the largest terrestrial source of CO_2 to the atmosphere (Raich and Tufekciogul 2000). The production of CO_2 in non-calcareous soils originates almost entirely from autotrophic (respiration from root, mycorrhizal fungi and microbes living on exudates) and heterotrophic respiration, i.e., the respiration of heterotrophic microorganisms decomposing SOM (Högberg et al. 2001), with the relative contribution from each respiration component largely varying from 10 to 90% of the total soil respiration (Hanson et al. 2000).

Nitrous oxide emissions strongly depend on the activity of the microbiological N turnover processes of nitrification and denitrification during which N_2O is produced as a facultative or obligate intermediate (Firestone and Davidson 1989; Butterbach-Bahl et al. 2004), as demonstrated by experiments using inhibitors, ^{15}N-tracers, and other techniques (Tiedje et al. 1984; Anderson and Levine 1986; Papen et al. 1989; Li et al. 2000). While nitrification occurs under aerobic conditions, the denitrification process occurs under anaerobic conditions and leads to N_2O and N_2 emissions. The ratio of these products depends on localized environmental conditions in the soil (Davidson 1992). Microbial denitrification occurs in soils when environmental conditions become unfavorable for aerobic degradation of organic matter. Denitrification can be performed by archaea, bacteria, and fungi, using oxidized N compounds (nitrate or nitrite) as an alternative electron acceptor in the absence of oxygen (Butterbach-Bahl et al. 2013), and this is a highly temporally and spatially distributed process. This pathway requires oxygen depletion, availability of N oxides, and availability of easily degradable C substrates (Butterbach-Bahl et al. 2012). Significant losses of N_2O due to denitrification from fertilized croplands have been reported when soils become water-saturated, due to heavy rainfall or irrigation, which results in a decrease of molecular oxygen (O_2) diffusion in soils (Fowler et al. 2015; Gritsch et al. 2016).

Nitrification in soils is also responsible for the production and consumption of N oxides (NO and N_2O). During this process, ammonium (NO_4^+) is oxidized to nitrate (NO_3^-), with hydroxylamine and NO_2^- as essential intermediates. Nitrification in soils involves a wide range of microorganisms. NO and N_2O are by-products of the process and thought to operate when conditions are suboptimal for further oxidation to NO_3^- (Conrad 1996; Baggs 2008). Generally, NO emissions are associated with nitrification processes, and N_2O emissions with denitrification processes. Chemodenitrification is the chemical decomposition of NO_2 and another important source of NO in soils with pH below 4 (Stevenson et al. 1970; Zumft 1997; Gritsch et al. 2016).

Methane is produced by microbial methanogenesis under anaerobic conditions, while the net oxidation of atmospheric CH_4 by bacteria often takes place in aerated soils (Saari et al. 1997), thus counteracting the increase in atmospheric CH_4 concentration and global warming (Born et al. 1990; Dörr et al. 1993; Saunois, Bousquet et al. 2016). Generally, CH_4 is formed by methanogenic archaea in soils if oxygen is lacking and sufficient SOM is available as energy source (Cappenberg 1974; Gambrell and Patrick Jr 1978). They are obligate anaerobes and require highly reduced conditions for growth (Cicerone and Oremland 1988). Therefore, water-saturated ecosystems are major sources of terrestrial CH_4 emissions (Saunois, Bousquet et al. 2016). Net CH_4 emissions occur at low redox potentials, when production of CH_4 by methanogenesis outweighs consumption by methanotrophy (Le Mer and Roger 2001; Gritsch et al. 2016).

2.1.3 FACTORS AFFECTING C AND N FLUXES FROM SOIL TO ATMOSPHERE

2.1.3.1 Climate

Climate change refers to the change in the primary climate drivers, temperature and precipitation, due to human activities which have modified the chemical composition of the atmosphere (Sutton et al. 2013). The processes that regulate gaseous transfers between the atmosphere and terrestrial reservoirs are in turn sensitive to climate due to its effect on soil environmental conditions. Transfers are primarily mediated by biological processes, especially microbiological transformations, which require concomitant measurements to investigate the potential range of effects (Fowler et al. 2009; Monks et al. 2009; Gritsch et al. 2016).

Rates of biological processes, including microbiological ones, generally increase with temperature (Meixner and Yang 2006) until a characteristic temperature optimum, beyond which biological activity progressively decreases (Saad and Conrad 1993; Tuomi et al. 2008). There is generally a positive correlation between microbial respiration (CO_2) and temperature, as long as other factors like moisture content or N input are not limiting (Lloyd and Taylor 1994; Davidson et al. 1998; Kirschbaum 2006).

Warming generally increases net or gross N mineralization and immobilization rates (Rustad et al. 2001; Shaw and Harte 2001; Liu et al. 2017). If global and regional temperatures continue to increase, there is a potential for nitrification and denitrification rates to also increase. Nitrification processes are highly temperature dependent (Matson and Vitousek 1981; Slemr and Seiler 1991) and increase with increasing temperatures (Ingwersen et al. 1999) to an optimum (Li et al. 2000). The response of denitrification to rising temperature is debated in the literature. Some authors have found a positive effect (Skiba et al. 1998; Schindlbacher et al. 2004) but others no significant effect of temperature on N_2O emissions (McHale et al. 1998; Kiese and Butterbach-Bahl 2002; Tang et al. 2006).

Changes in soil moisture driven by changes in precipitation regimes can significantly alter rates of C and N cycling processes (Barnard et al. 2006; Dijkstra et al. 2010; Larsen et al. 2011). In particular, very low and very high soil moisture contents decrease microbial respiration, while increased soil moisture can result in enhanced N mineralization and immobilization. Water-limiting conditions enhance nitrification rates (Jamieson et al. 1999; Avrahami and Bohannan 2007), while water-saturated conditions result in increased denitrification rates (Barnard et al. 2006). In general, changes in soil moisture will very likely dominate the overall climate change response and overwhelm any direct temperature effects on denitrification, nitrification and NO and N_2O emissions (Fowler et al. 2015).

2.1.3.2 Land Use

Land use change refers to changes in vegetation cover, land use and management, substrate supply to soil microorganisms, as well as modifications in soil and catchment hydrology (MacLeod et al. 2010).

The potential impacts of changes in land use on the global C and N cycles are considerable in both the range and magnitude of effects (Fowler et al. 2015).

During the past few decades, European forests have acted as a C sink (Wallberg et al. 2014). Forests play a pertinent role in the climate system by absorbing CH_4 (Schaufler et al. 2010) and storing large C pools in tree biomass and soils (Schlamadinger and Marland 1996). Grasslands are generally dominated by high microbial activity associated with high root density, which lead to high plant production rates (i.e., atmospheric CO_2 fixation). At the same time, presence of ruminant for grazing and management may favor net N_2O and CH_4 production in grassland soils (Hörtnagl and Wohlfahrt 2014; Merbold et al. 2014). Peatlands constitute unique ecosystems that contribute to water purification and biodiversity conservation (Kimmel and Mander 2010). Additionally, peatlands play an important role in the regulation of climate change through the sequestration of a major proportion of atmospheric C (Millennium Ecosystem Assessment 2005). Agricultural soils are a large source of various N-containing trace gases, especially in response to fertilization (Sutton et al. 2015). Globally, agriculture is directly responsible for approximately 14% of anthropogenic GHG emissions, while indirect emissions due to the conversion of natural landscapes to agricultural systems may contribute an additional 17% (Vermeulen et al. 2012). Expected climate change might lead to additional changes in land use and land management. For example, the area of irrigated agricultural land is expanding quickly due to water scarcity (Trost et al. 2013), which may trigger increases in N_2O emissions caused by increased denitrification activity in such soils (Liu et al. 2010).

2.1.3.3 N Input

Atmospheric N deposition is known to increase gross and potential nitrification and denitrification rates, through increases in soil inorganic N availability (Barnard et al. 2005). A wide range of atmospheric N compounds are emitted by, and/or dry-deposited to the earth's surface (vegetation, soils, water bodies, built-up areas). N deposition mainly appears in the form of reduced N (gaseous ammonia (NH_3), and wet ammonium (NH_4^+)), and oxidized N (wet nitrate (NO_3^-)) (Flechard et al. 2011). European ecosystems have been, and will continue to be, threatened by dry and wet N deposition from animal farming, fertilization treatments in agriculture, and fossil fuel combustion. Additionally, a warmer climate will increase NH_3 emissions from sources such as animal manure (Sutton et al. 2013). Nitrogen inputs to ecosystems can also affect yields and can lead to pollution of watersheds in heavily fertilized regions. The resulting increase in soil N availability is hypothesized to increase rates of primary production and C accumulation (Schimel et al. 1994; Asner et al. 1997; Medlyn et al. 2000). There is debate about whether N limitation of plant growth will notably affect future ecosystem C storage. Arguably, climate effects of N_2O emissions are of more concern than N-interactions with the C cycle (Sutton et al. 2015; Gritsch et al. 2016). To this end it is important to scrutinize the microbial, as well as the global, drivers of N_2O emission production and consumption processes.

2.2 N₂O EMISSIONS AND THE NITROGEN CYCLE

Nitrogen oxide is a potent GHG, with a global warming potential 298 times higher than CO_2. Further, N_2O reacts with stratospheric ozone causing severe environmental problems (Ravishankara et al. 2009). Since the Industrial Revolution, anthropogenic sources of N_2O have increased substantially, enhancing its atmospheric concentrations from about 260 ppb in 1850 to about 320 ppb at the present time, thus contributing to global warming (Forster et al. 2007). Harter et al. (2014) attribute this mainly to the expansion of farming activities followed by the intensification of fertilizer use. It has been suggested that 56–70% of N_2O emissions can be traced back to agricultural emissions (Butterbach-Bahl et al. 2013). Next to anthropogenic emissions, soils and oceans are the main sources and sinks of natural N_2O emissions.

The global N budget illustrates the importance of soils, which annually cycle around 300 Tg N (Figure 2.1). Atmospheric N enters the soil through both natural and human-driven process.

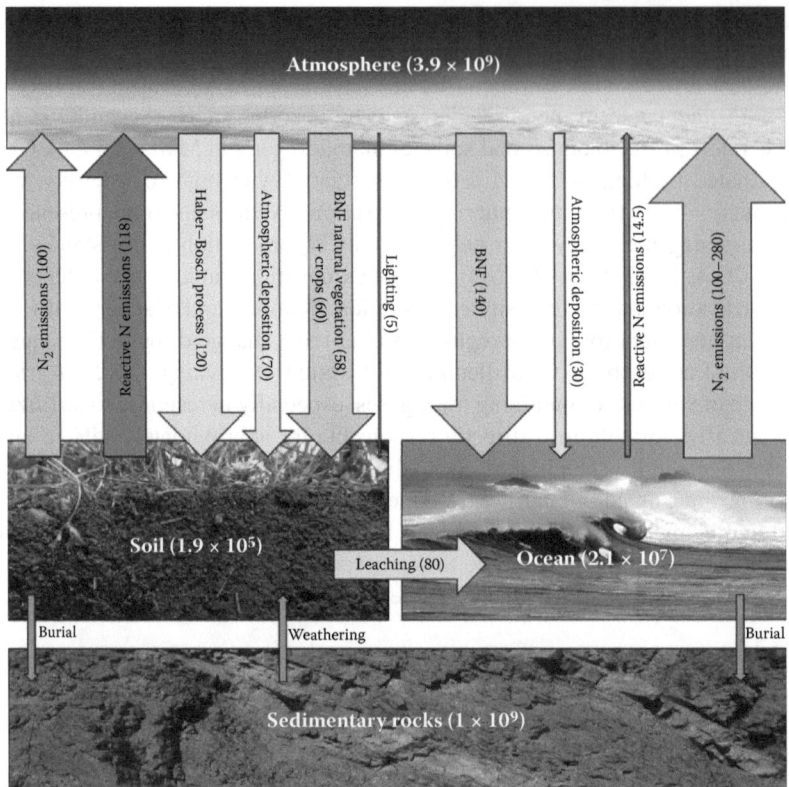

FIGURE 2.1 Global N cycle (fluxes in Tg N a^{-1}; pools in Tg N). Width of the flux arrows proportional to the magnitude of the flux. Pools not to scale. (Fluxes source: From Fowler, D., Steadman, C.E., Stevenson, D., Coyle, M., Rees, R.M., Skiba, U.M. et al. *Atmospheric Chem. Phys.* **15**: 13849–13893, 2015, and references therein.) (Pools source: From Galloway, J.N. *Treatise on Geochemistry*, Elsevier Ltd., Amsterdam, 2003.) Photo sources (all creative commons): Land, NRCS Soil Health; Ocean, Brocken inaglory; Atmosphere, Andres Rueda; sedimentary rocks, The paleobear.

Biological N fixation in natural vegetation is the main non-anthropogenic N input into the soil (58 Tg N a^{-1}), whereas the remaining N pathways into the soil have a direct or indirect human influence. The so-called Haber–Bosch process, by which N_2 is transformed into NH_3 for fertilization purposes, contributes about 100 Tg N a^{-1} (Galloway et al. 2008), along with 20 Tg N a^{-1} from other industrial processes. Biological N fixation via crops is responsible for about 60 Tg N a^{-1}, with the remaining 75 Tg N a^{-1} entering the soil through the deposition of atmospheric reactive N. The fate of the N entering terrestrial ecosystems is to be taken up by plants, or transformed by microbes, or both, before leaving the system, either to the atmosphere (as N_2, about 100 Tg N a^{-1}, or as reactive N, about 120 Tg N a^{-1}, mainly as NH_3 and NO_x, but also as N_2O (Fowler et al. 2013)), or to the ocean via leaching and run-off (Figure 2.1).

The local N cycle and the site specific N flux between soil and atmosphere are strongly determined by microbial production and consumption processes (Butterbach-Bahl et al. 2013). Robertson and Groffman (2015) identified the main microbial activities involved in the N cycle in soil as N mineralization, N immobilization, nitrification, and denitrification.

During N mineralization (also known as ammonification), organic N is transformed to inorganic N, usually involving depolymerization of N-containing polymers into monomers by microbial extracellular enzymes as a previous step (Schimel and Bennett 2004). During the decomposition of SOM and detritus by microorganisms, ammonium (NO_4^+) is produced, which can be assimilated (or "immobilized") by soil microorganisms or taken up by plants. Both N ammonification and

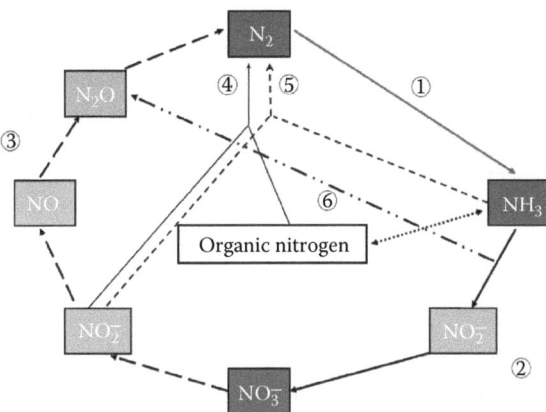

FIGURE 2.2 Main microbiological processes of the N cycle. (1) Nitrogen fixation; (2) bacterial nitrification, archaeal nitrification, and heterotrophic nitrification; (3) aerobic and anaerobic bacterial denitrification, nitrifier denitrification, fungal denitrification, and archaeal denitrification; (4) and (5) co-denitrification and archaeal denitrification; (4) and (5) co-denitrification (by fungi); (5) anammox; and (6) N_2O production during nitrification (ammonia oxidation) (modified after Hayatsu et al. 2008, © Japanese Society of Soil Science and Plant Nutrition, reprinted by permission of Taylor & Francis Ltd, www.tandfonline.com on behalf of Japanese Society of Soil Science and Plant Nutrition).

immobilization can occur simultaneously as different microorganisms undertake diverse processes in the SOM breakdown (Robertson and Groffman 2015).

Along with nitrification and denitrification, newly discovered biotic pathways such as commamox (COMplete AMMonia OXidiser (Daims et al. 2015)), anammox (Anaerobic AMMonia Oxidation [Mulder et al. 1995]), or dissimilatory nitrate reduction to ammonium (DNRA [Silver et al. 2001]), and a number of abiotic N transformation processes (e.g., feammox, chemodenitrification) contribute to N transformation processes. This exemplifies the enormous complexity of N cycling in the soil (Figure 2.2) and highlights the need for increasing our current process understanding from the microsite to the field level to explore feedbacks of N emissions to changes in environmental conditions and land use (Butterbach-Bahl et al. 2013).

2.2.1 NITRIFICATION

Nitrification refers to the microbial conversion of NO_4^+ (or NH_3) to NO_3^-, involving hydroxylamine (NH_2OH) and nitrite (NO_2^-) as obligatory intermediate products (Firestone and Davidson 1989:

$$NH_4^+ + 2\,O_2 \rightarrow NO_3^- + H_2O + 2\,H^+ \qquad (2.1)$$

Nitrification can be split further into the two steps ammonia oxidation and nitrite oxidation. Ammonia monooxygenase and hydroxylamine oxidoreductase enzymes are involved in the ammonia oxidation process, while the nitrite oxidoreductase enzyme catalyses the conversion of NO_2^- to NO_3^- (Hirsch and Mauchline 2015). Nitrification rates are primarily determined by the availability of NO_4^+ for microbial transformation, as well as by environmental conditions, especially the level of O_2 in the soil (Butterbach-Bahl et al. 2012; Robertson and Groffman 2015). Nitrification is favored by aerobic soil conditions (e.g., Schindlbacher et al. 2004; Pilegaard 2013), optimum soil temperatures, pH values around neutrality (6–7), and high NO_4^+ concentrations. Autotrophic nitrification is recognized as being the dominant process of nitrification in most systems, although heterotrophic microbes and archaea have also been found to be capable of conducting nitrification (Robertson and Groffman).

Nitrification activity is considered to be the limiting step in N_2O production, since it provides the necessary substrate (i.e., NO_3^-) for subsequent denitrification (Ambus et al. 2006). In a case study,

Gödde and Conrad (1999) even showed that nitrification per se may contribute up to 80% to total soil N_2O emissions. A special form of denitrification is also performed by autotrophic nitrifiers (Kool et al. 2010) through which NH_3 is oxidized to NO_2^- and subsequently reduced to NO and N_2O by the same organism, when soil N availability is high but both O_2 and organic C levels are low (Butterbach-Bahl et al. 2013; Robertson and Groffman 2015). There are a number of studies suggesting a rather large contribution of nitrifier denitrification to overall N_2O production (e.g., Kool et al. 2010), suggesting that the role of bacterial ammonia oxidation may be overestimated if nitrifier denitrification is not taken into account.

2.2.2 Denitrification

During denitrification, NO_3^- is reduced to the N gases NO, N_2O and N_2 (Equation 2.2). Denitrification takes place in soils, sediments, and fresh and marine waters (Galloway et al. 2008), and it is a crucial mechanism for the global N cycle as it returns N to the atmosphere as N_2 (Robertson and Groffman 2015). The process takes place only when O_2 concentrations are very low (microaerobic conditions) or zero (anaerobic conditions), because O_2 is a more efficient terminal electron acceptor during respiration than NO_3^- (Robertson and Groffman 2015). In addition to anaerobic conditions, denitrification rates are usually fostered by increasing concentrations of both N substrates (especially NO_3^-) and organic C as energy supply (Butterbach-Bahl et al. 2012):

$$4\,NO_3^- + 4\,H^+ + 5\,C_{org} \rightarrow 5\,CO_2 + 2\,N_2 + 2\,H_2O \qquad (2.2)$$

There are four enzymatic steps involved in dissimilatory denitrification: 1) the conversion of NO_3^- to NO_2^-, which involves a NO_3^- reductase gene (NarG); 2) the conversion of NO_2^- to NO, facilitated by two NO_2^- reductase genes (NirK or NirS); 3) the transformation of NO to N_2O, with the mediation of two NO reductase genes (NorB or NorC); and 4) the reduction of N_2O to N_2 by the N_2O reductase gene (NosZ) (Hirsch and Mauchline 2015). Depending on denitrifying organisms and environmental conditions, not all enzymatic steps may take place. For instance, N_2O is not always subsequently converted to N_2, and N_2O can become the end product of the process chain if the N_2O reductase is not expressed, for example in soils with low pH. On the other hand, some microorganisms lack genes of the earlier pathway steps but can reduce N_2O to N_2 (Hirsch and Mauchline 2015).

2.2.3 Microorganisms Involved in N_2O Emissions

All major pathways of soil N_2O emissions are microbially mediated, including ammonia oxidation, heterotrophic denitrification, and nitrifier denitrification (Figure 2.3). It has been shown that microorganisms belonging not only to Bacteria but also to Eukarya and Archaea are involved in the denitrification and nitrification processes. Many heterotrophic microorganisms contribute to the nitrification process, including both ammonia oxidation and nitrite oxidation. The ability to perform aerobic denitrification has been detected in various bacterial genera. Modern molecular techniques and genome analyses are providing new insights into the ecology of microorganisms responsible for conventional bacterial nitrification and denitrification. Recently, novel microbiological pathways have been discovered, and new light has been shed on the specific role of certain microbial groups for N turnover processes. For example, it has been shown that NO_4^+ and NO_2^- can be converted to N_2 under anaerobic conditions through anammox, a process performed by bacteria belonging to the Planctomycetes group (Mulder et al. 1995). Some fungi are able to produce N_2O and N_2 through both fungal denitrification and co-denitrification pathways (Shoun et al. 2012). Further, recent studies have shown definitive evidence that Archaea mediate nitrification (Könneke et al. 2005). Thus, researchers are beginning to recognize that a greater diversity of microorganisms is involved in the N cycle than was previously understood (Hayatsu et al. 2008).

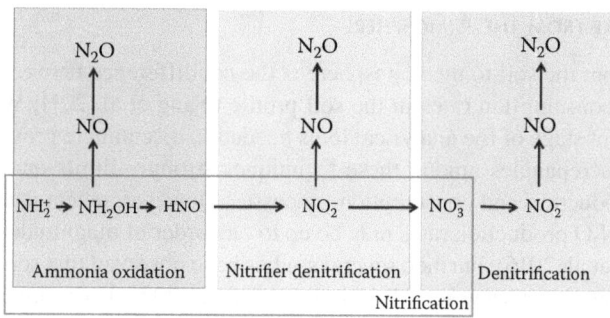

FIGURE 2.3 Predominant microbial pathways of soil N_2O production. (Adapted by permission from Macmillan Publishers Ltd: Scientific Reports, Huang et al, Ammonia-oxidation as an engine to generate nitrous oxide in an intensively managed calcareous Fluvo-aquic soil. *Scientific Reports* © 2014.)

The nitrification process is performed by nitrifying autotrophic bacteria, by heterotrophic bacteria and fungi, as well as by NH_3 oxidizing archaea, all ubiquitous microorganisms in soil (Ussiri and Lal 2013). The most abundant NH_3 oxidizing archaeon is from the genus *Nitrosphaera* (Zhalnina et al. 2013). Regarding the autotrophic bacteria involved in nitrification, the so-called *Nitrosobacteria* perform the first part of nitrification by oxidizing NO_4^+ to NO_2^-, while the subsequent NO_2^- oxidation to NO_3^- is performed by bacteria, including the genera *Nitrobacteria* (Pilegaard 2013) and *Nitrospira* (Hirsch and Mauchline 2015).

Many bacteria and certain fungi are able to produce N_2O when generating energy (Philippot et al. 2007; Thomson et al. 2012). Over 50 genera with over 125 denitrifying species have been identified, with a high predominance of heterotrophic, facultative anaerobes (Robertson and Groffman 2015). Denitrification by bacteria usually takes place under anaerobic or micro-aerophilic conditions, and bacterial denitrification is usually dominated by *Pseudomonas* and *Alcaligenes* genera. However, there are several bacteria that are able to produce N_2O in the presence of O_2, for example, *Paracoccus denitrificans* (Hayatsu et al. 2008). In some ammonia-oxidizing organisms and denitrifying bacteria the respiratory reductase (NOR) can be found, an enzyme that contributes to biological N_2O production in many environments (Spiro 2012; Thomson et al. 2012).

Fungi can utilize dissimilatory and assimilatory nitrate reductase to denitrify NO_3^-. Usually, the final product of fungal denitrification is N_2O because the fungal denitrification chain is generally truncated at the formation of N_2O (Butterbach-Bahl et al. 2013). This means that the composition of the soil microbial community strongly influences the total N_2O efflux from the soil (Hayatsu et al. 2008; Thomson et al. 2012). Examples for denitrifying fungi are Ascomycota like *Cylindrocarpon tonkinense* and *Gibberella fujikuroii*, or Basidiomycota like *Trichosporon cutaneum* (Shoun et al. 2012). Also, some ectomycorrhizal fungi possess the ability to produce N_2O, and they could play a yet unexplored role when it comes to N_2O emissions in forest ecosystems (Prendergast-Miller et al. 2011). Studies working on the inhibition of fungi in the soil N cycle suggested a significant importance of fungi in the N_2O production process. However, such results need to be treated with care as the inhibition of an essential member of the soil microbial community might provoke cascading perturbations in the soil food web. Furthermore, Keuschnig (2016) showed that abiotic NO release from sterile NO_2^- medium (NO_2^- is unstable under anoxic conditions) can lead to the detection of false positives when screening for denitrifying N_2O-producing fungi. This may result in distinctly different patterns of N_2O release in comparison to true fungal denitrifiers such as *Fusarium* spp. The consequence may be that only few of the fugal strains, which have been shown to release N_2O from NO_2^- media in the laboratory so far, do actually denitrify under field conditions.

2.2.4 N$_2$O Uptake from the Atmosphere

The N$_2$O released from the soil to the atmosphere is the net difference between gross N$_2$O production and gross N$_2$O consumption rates in the soil profile (Yang et al. 2011; Wen et al. 2016). The still early development stage of the analytical tools needed to disentangle gross N$_2$O turnover rates, and the observed discrepancies among these techniques strongly limits our current understanding of gross N$_2$O production and consumption processes. However, recent studies suggest that the magnitude of gross N$_2$O production rates may be up to one order of magnitude higher than net N$_2$O emission rates (Wen et al. 2016). Further, several studies have observed that soils can function as net sinks of N$_2$O (for a detailed review see Chapuis-Lardy et al. 2007). There are confronting opinions about the interpretation of these observations, and some data analyses suggest that the majority of the observed N$_2$O uptake rates are due to limits in the detection of particular fluxes (Erickson et al. 2002; Pinto et al. 2006; Cowan et al. 2014). However, the increasing number of observations across different ecosystems, including forests (Butterbach-Bahl et al. 2002; Rosenkranz et al. 2006), grasslands (Flechard et al. 2005) and savannahs (Donoso et al. 1993), together with new studies that use very sensitive modern measurement techniques (Savage et al. 2014) strongly suggest that net soil N$_2$O uptake is not solely the result of instrumental artifacts.

The only pathway known to consume N$_2$O in the soil is through the activity of the enzyme respiratory N$_2$O reductase (N$_2$OR), which is found in denitrifying bacteria (Spiro 2012; Thomson et al. 2012). In the light of this, net N$_2$O uptake may be confined that net N$_2$O uptake may be confined to ecosystems limited in N, especially in NO$_3^-$ (Glatzel and Stahr 2001; Rosenkranz et al. 2006). This argumentation is further reinforced by observations of negative soil N$_2$O fluxes during water-limited periods (Donoso et al. 1993; Rosenkranz et al. 2006), which likely reduced or suppressed soil nitrification. Under those specific conditions, the demand for N$_2$O by denitrifying organisms may outweigh the low NO$_3^-$ supply and, therefore, atmospheric N$_2$O instead of nitrification- or denitrification-derived N$_2$O may be used and reduced to N$_2$. The extent to which a soil may act as a sink for N$_2$O depends on its potential for the reduction to N$_2$, the ease of N$_2$O diffusion through soil and its dissolution in soil water, as well as the soil microbial community composition (Chapuis-Lardy et al. 2007). The rate of N$_2$O uptake seems to depend further on the presence of labile organic C and O$_2$, water content, soil temperature, and pH (Chapuis-Lardy et al. 2007).

Net N$_2$O uptake by soils has not been studied systematically so far, and there are still great uncertainties about why and when soils may act as net sinks for N$_2$O. We consider that net uptake rates should always be documented and not necessarily considered as measurement errors. This, along with specifically targeted studies, may help to disentangle the underlying mechanisms and identify the conditions under which N$_2$O uptake from the atmosphere into the soil takes place.

2.2.5 Factors Promoting N$_2$O Emissions

2.2.5.1 Climate

Several environmental parameters influence the microbial production and consumption of N$_2$O in the soil, including soil temperature and especially moisture (Figure 2.4), as the latter strongly influences O$_2$ availability and soil redox potential (Butterbach-Bahl et al. 2013). Further, soil physical properties such as water retention potential and soil aggregation (Quin et al. 2014), soil chemical properties (e.g., pH, organic N, dissolved organic C), and soil biological properties (e.g., microbial biomass content, N cycling enzymes, macro fauna, etc.) can affect production rates of N$_2$O (e.g., Schindlbacher et al. 2004; Ambus et al. 2006; Pilegaard et al. 2006; Butterbach-Bahl et al. 2012, 2013). Further, the relative contributions of different microbial pathways to total N$_2$O release vary along with environmental conditions. Since all driving parameters for N$_2$O exchange vary with rapid temporal dynamics and at different spatial scales due to changing environmental conditions (e.g., episodic rainfalls) or by management actions (e.g., fertilization), local N$_2$O emissions from soils are usually dominated by "hot spots" and "hot moments" (Wolf et al. 2010; Zona et al. 2013; Savage et al. 2014).

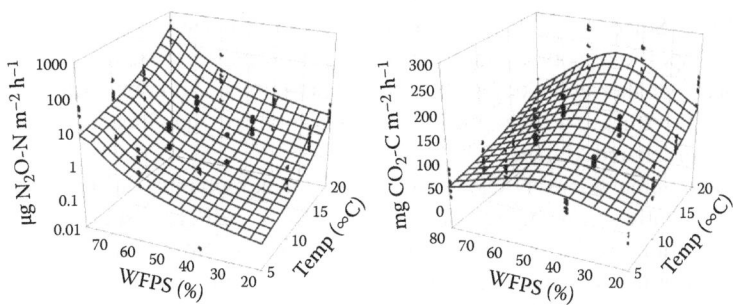

FIGURE 2.4 Soil temperature and moisture (depicted as water filled pore space [WFPS]) as drivers of N_2O and CO_2 emissions. (Data from Schaufler, G., Kitzler, B., Schindlbacher, A., Skiba, U., Sutton, M.A., and Zechmeister-Boltenstern, S., *Eur. J. Soil Sci.* 61, 683–696, 2010, calculated and drawn by Barbara Kitzler, BFW, Vienna, Austria.)

The temperature sensitivity of soil N_2O emissions is usually much higher than what has been observed for soil CO_2 fluxes (Schaufler et al. 2010; Díaz-Pinés et al. 2014), probably because the temperature effect multiplies with the several possible processes of N_2O production. Besides the direct effect of temperature on enzymatic processes involved in N_2O production, increases in temperature can lead to indirect effects as a consequence of enhanced respiratory activity and the subsequent depletion of O_2 in the soil (Butterbach-Bahl et al. 2013). On the other hand, increases in temperature have also been associated with a decrease in electron acceptors under anaerobic conditions, which results in the expression of the full enzymatic chain involved in the denitrification processes, thereby yielding N_2 as end-product of denitrification at the extent of N_2O release (Butterbach-Bahl and Dannenmann 2011).

Soil moisture strongly modulates both the absolute rates of nitrification and denitrification in the soil and the relative amount of N_2O produced in the course of these microbial processes (Pilegaard et al. 2006; Butterbach-Bahl et al. 2013; Pilegaard 2013). Whereas nitrification rates are usually high when soil water-filled pore space (WFPS) is low, the predominant product of nitrification under these environmental conditions is NO, whereas N_2O production is negligible (Pilegaard 2013). With increasing water content, aeration deteriorates and denitrifiers dominate N_2O formation, with optimum rates for N_2O production around 70%–80% WFPS (Davidson et al. 2000), although a specific threshold may vary depending on soil types and climatic conditions (Barton et al. 1999). When soil becomes fully anaerobic, denitrification rates further increase, but virtually all the N_2O is reduced to N_2 (Zou et al. 2007), resulting in diminished N_2O emissions in totally flooded ecosystems.

In view of climate change, soil temperature and soil water regimes are changing and will continue to change in the future, with potentially strong implications for the production of N_2O by soil microorganisms. Many model scenarios agree in predicting warmer temperatures and wetter soils for large parts of the globe (IPCC 2013), which strongly suggests that soil N_2O emissions will increase in the future (Butterbach-Bahl et al. 2012). In addition to the changes in annual temperature and annual precipitation, the spatial heterogeneity and temporal distribution of precipitation is predicted to increase, with both higher frequency and intensity of extreme weather events such as drought spells and episodic heavy rainfalls. Rewetting of soils after drought periods usually provokes accelerated N turnover rates by microorganisms, resulting in over-proportionated pulses of N_2O compared to the expected and/or observed rates in constantly moist soils (Borken and Matzner 2009). Whereas the exact underlying mechanisms are not completely known, it has been hypothesized that there is an accumulation of labile substrate originating from microbial necromass and the osmotic equilibration during drought periods. This substrate becomes available for the remaining living microorganisms once the adequate environmental conditions are reestablished (i.e., after the first rainfalls), thereby resulting in a rapid release of both C and N (Butterbach-Bahl et al. 2013;

Gelfand et al. 2015). Since the prediction of future weather patterns is associated with high uncertainties, and the response of soils to abrupt changes in soil moisture status is not yet fully understood, it is unclear whether enhanced post-wetting N_2O pulses will compensate for reduced N_2O emissions or even the shift to a temporary N_2O sink (Goldberg and Gebauer 2009) during water-limited periods (Borken and Matzner 2009).

In temperate and boreal zones, N_2O can also be emitted at high rates during freeze–thaw events, contributing significantly to annual N_2O budgets (Wolf et al. 2010; Luo et al. 2012). Global change (IPCC 2013) in addition to regional air circulation pattern shifts (Zeeman et al. 2017) are likely to reduce the snow-cover period length, and "colder soils in a warmer world" are expected (Groffman et al. 2001), probably increasing the frequency of freeze-thaw events and the associated release of N_2O from the soil.

In some case studies, variations in soil moisture and soil temperature have usually shown adequate explanatory power at the site level (Pilegaard et al. 2006), with up to 95% of the temporal variation of N_2O emissions being explained by changes in moisture and temperature (Kitzler et al. 2006). However, this is more the exception than the rule because of the usually extremely high temporal and spatial variation of soil N_2O fluxes at different scales (Butterbach-Bahl et al. 2013). Therefore, it has not been possible yet to derive reliable estimates of the effects of climate change on future soil N_2O emissions at a global level. A conceptual approach performed by Butterbach-Bahl et al. (2012) suggests that denitrification will be positively affected by climate change, primarily due to decreased evapotranspiration and enhanced soil respiration and soil moisture, all pointing towards lower O_2 availability in the soil. On the other hand, nitrification will likely decrease in the view of climate change. In any case, it is clear that we are still far from gaining a full understanding of 1) the effects of environmental parameters affecting soil microbial communities and their activities; 2) the complex interactions between those parameters; and 3) the indirect effects associated with changes in landscape hydrology, land use and land management as a consequence of climate change.

2.2.5.2   Land Use

Land use type and land management have a strong effect on the amount of inputs received by soils (e.g., amendments, fertilization), as well as on the outputs (e.g., harvesting, grazing). It further influences soil environmental conditions by predefining specific vegetation covers or by management actions (e.g., watering or drainage), which eventually can lead to changes in the physical properties of the soil (e.g., tillage). As a consequence, land use has a great influence on the extent of N_2O emitted by soils (e.g., Schaufler et al. 2010).

Wetland soils cover only 2.6% of the Earth's terrestrial surface, but they deliver important ecosystem services (Latham et al. 2014) and are highly sensitive to changes in environmental conditions. In wet soils, gas diffusivity and O_2 availability is low, which creates adequate conditions for denitrification and leads to substantial N_2O production rates. The effect of water probably outweighs the influence of climate, resulting in overall high N_2O emissions from wetlands regardless of climatic regions. Thus, significant emissions of N_2O have been observed in wetlands under tropical (Allen et al. 2007), mediterranean (Meijide et al. 2017), temperate (Søvik et al. 2006) and boreal (Repo et al. 2009) climatic conditions.

Permanent grasslands cover roughly one third of the global land surface (Latham et al. 2014). Managed grasslands usually involve the presence of grazing animals, which constitute a source of N to the soil via excreta, resulting in substantial soil N_2O emissions (Velthof and Oenema 1995). At the same time, intense grazing can interactively influence the effects of freeze–thaw events on N_2O pulses, resulting in strongly reduced N_2O emissions (Wolf et al. 2010). The type of grazing animals also play a role, with sheep-grazed grasslands usually emitting less N_2O compared to cattle-grazed ones (Saggar et al. 2008; Oertel et al. 2016). Nitrogen fertilization of grasslands is also a common feature intended to enhance plant performance and sustain soil productivity. Fertilization usually involves the addition of manure, which does not only add N to the soil but also a substantial amount of labile C, usually leading to high denitrification rates and significant pulses of N_2O (Merbold et al. 2014).

Forests are supposed to undergo a closed N cycle as they are natural or close to natural systems (Schulze 2000) and therefore low soil N_2O fluxes are expected in these ecosystems. However, high atmospheric N deposition rates from industrial sources and the close vicinity of forest stands to heavily fertilized agricultural fields have drastically increased N inputs into forest ecosystems, which has further provoked that forests have become a substantial source of N_2O (Butterbach-Bahl and Kiese 2005). Forests on sites with high atmospheric N deposition show higher N_2O emissions compared to sites with low atmospheric N input (Bowden et al. 1991; Castro et al. 1992; Luo et al. 2012; Oertel et al. 2016).

Agricultural lands, including arable soils, cover around 12.6% of the Earth's land surface (Latham et al. 2014) and the vast majority of humanity depends on agriculture to fulfill their nutritional needs. Globally, agriculture contributes to more than two thirds of the total anthropogenic N_2O emissions, with a substantial amount of those emissions coming from microbial activity in soils after fertilizers are used to increase food production (Butterbach-Bahl and Dannenmann 2011). Recent studies have summarized the current knowledge to provide effective mitigation measures, which aim at minimizing the N_2O footprint of agriculture without compromising food production. Since agricultural activities take place under a broad range of climatic, environmental, social, and economic contexts, no general patterns can be derived, but current research agrees on focusing on optimized fertilization strategies (timing, dose, and type of fertilizer) and appropriate water management in order to improve the N use efficiency of agricultural systems (Snyder et al. 2014; Paustian et al. 2016; Sanz-Cobena et al. 2017).

Consequences of land use change on the net balance of N_2O emitted from soils have been receiving increasing scientific attention in the last years. However, variations are largely dependent on soil and environmental conditions, direction of the land use change, and the temporal scale under investigation. The latter aspect is crucial since a stabilization phase may take place after a disturbance until a new equilibrium is reached, similarly to what occurs with relation to soil C sequestration after land use change (Poeplau et al. 2011). For example, former grasslands showed enhanced N_2O emission rates one year after they were afforested, but this was related to pulses associated with the soil tillage activities intended to prepare the soil (Mishurov and Kiely 2010), as has also been observed in tilled grasslands **not** subjected to land use change (Cowan et al. 2016).

2.2.5.3 N Input

Nitrogen input to the soil is a major controlling factor of N_2O emissions, since it increases N availability for plant uptake but also provides substrate for microorganisms that may produce N_2O during N transformation processes. Large N inputs take place in form of fertilization as well as unintended atmospheric N deposition caused by anthropogenic activities in urban, industrial, and agricultural areas (Lovett and Rueth 1999). Agricultural fields receiving high loads of N in the form of fertilizer tend to be short-term "hot spots" for denitrification, showing large N_2O pulses after fertilizer application, but also substantial NO_3^- losses through leaching (e.g., Ussiri and Lal 2013). Organic fertilizers often promote denitrification more than mineral ones, probably because of the additional amount of readily available organic C present in the manure. However, N can be released more slowly from organic fertilizers, especially if they are composted (opposed to fresh animal slurry) prior to their application; this allows for a more efficient plant N uptake and reduced N_2O losses compared to mineral fertilizers (Priemé and Christensen 2001; Philippot et al. 2007).

Several anthropogenic activities (industry, transport, agriculture) can lead to the release of reactive N to the atmosphere (NO_x, NH_3), which is dispersed with the wind and can be deposited elsewhere. As a consequence, unintentional N addition can take place, increasing N availability in the soil. While this may have positive effects on vegetation growth for selected plant species (Hurkuck et al. 2015), it can also lead to N saturation processes, with natural systems becoming stronger sources of N_2O (Bühlmann et al. 2015) in addition to associated problems such as biodiversity loss (Bleeker et al. 2011) or enhanced water eutrophication.

Population growth strongly enhances the demand for food, fiber, and biofuel, calling for highly productive lands. In this context, fertilizers provide a unique opportunity to deliver these increasingly needed goods. On the other hand, a substantial amount of N is lost annually in form of reactive N due to low N efficiency of agricultural production systems and N losses during the trading of food and feed (Lassaletta et al. 2014). Since changing environmental conditions due to global change are likely to further trigger N losses in general and soil N_2O emissions in particular, there is a crucial need to revert this loop. The only plausible way is probably an optimization of agroecosystems based on our current knowledge on ecosystem functioning, aiming at minimizing the N leakages along all the steps of production, distribution and commercialization of food, so that the needs of our societies can be met sustainably without irreversibly endangering the environment.

2.3   SOIL CH_4 FLUXES FROM TERRESTRIAL ECOSYSTEMS

To date, CH_4 is the second most abundant greenhouse gas in the Earth's atmosphere after CO_2 and it contributes substantially to global change (Denman et al. 2007; Saunois, Jackson et al. 2016). Ice-core data indicate rather constant concentrations of CH_4 of about 0.7 ppm for approximately 2000 years before industrialization began in the 19th century. Following the industrial revolution anthropogenic activities have more than doubled the atmospheric concentration of CH_4 to the current level of about 1.8 ppm (Crutzen and Lelieveld 2001). During the 1980s, a yearly increase of 10 ppb CH_4 in the atmosphere was recorded. However, the average yearly growth rate of atmospheric CH_4 declined to 5 ppb during the 1990s and even grew close to zero between 1999 and 2005. Around 2008 atmospheric CH_4 concentrations began to increase again and they continue to rise. Current models predict that atmospheric CH_4 concentrations will reach 2.55 ppm by the year 2050 (Lelieveld et al. 1998). Several potential explanations for the lack of increase in the late 1990s and early 2000s have been proposed, including (1) reduced gas losses from pipelines (estimated 29 to 50 Tg CH_4 y^{-1}), which had been leaking after the collapse of the Soviet Union, (2) the depletion of fossil fuels in Antarctica and Greenland, and (3) the reduction of rice agriculture (Aronson et al. 2013). The high number of possible explanations shows that there is still a lack of knowledge concerning the CH_4 cycle and its interplay with biotic and abiotic factors and that more research is urgently needed in order to develop strategies to reduce CH_4 emissions.

Despite the fact that CH_4 is currently about 200 times less abundant in the atmosphere than CO_2, its global warming potential (GWP) is about 25 to 30 times higher calculated over a 100-year horizon compared with CO_2 because of its efficiency in absorbing infrared radiation (Crutzen and Lelieveld 2001). Furthermore, reemission of trapped radiation is known as a crucial driver of the destruction of the Earth's ozone layer, hence promoting climate warming.

Methane is emitted from both natural and anthropogenic sources. Currently, total annual CH_4 emissions, which are for the most part of microbiological origin (Conrad 2009), are estimated to range between 503 and 610 Tg CH_4 y^{-1} (Denman et al. 2007). CH_4 is emitted naturally from sources such as wetlands, it is produced in the ocean's water columns and sediments, through fermentation in termite guts and by geological sources (e.g., seepage, gas hydrates). The largest individual source of CH_4 is wetlands, which account for $\geq 70\%$ of natural CH_4 emissions. CH_4 is produced naturally by methanogenic archaea during the decomposition of organic matter under O_2-free conditions, as long as no alternative electron acceptors (e.g., NO_3^-, sulfate, ferric iron) are present (Conrad 2009). Recently, plant leaves have been recognized as an additional—though abiogenic—source of CH_4 (Keppler et al. 2006). To date, it is believed that UV radiation is cleaving methyl-groups from the pectin of plant tissues by a photochemical reaction (McLeod et al. 2008) and/or that the emissions are due to diffusion of dissolved CH_4 from soils into plant leaves (Nisbet et al. 2009). However, the contribution of this source to the total CH_4 budget is estimated to be rather small.

The annual emissions of CH_4 become, at least partly, neutralized by CH_4 sinks that are of similar magnitude. Still, for decades CH_4 sinks have been slightly smaller compared with sources, leading to the observed steady growth of atmospheric CH_4 concentrations. Chemical destruction of CH_4 in the troposphere is primarily responsible for the removal of CH_4 from the atmosphere and accounts for ≥87% of the total CH_4 sink (Lelieveld et al. 1998). This process is mainly driven by photochemical oxidation of CH_4 through OH radicals. Another 7% of the atmospheric CH_4 is removed in the stratosphere. Additionally, methanotrophic bacteria inhabiting soils are able to use CH_4 as C and energy source (Hanson and Hanson 1996), representing the only biological sink for CH_4. However, the exact magnitude of the soil sink is still under debate. One of the classic reviews on methanotrophs estimates the biological CH_4 sink to range between 40 and 60 Tg y^{-1} (Hanson and Hanson 1996), whereas the latest climate report published by the IPCC (Intergovernmental Panel on Climate Change) assumes it to be 30 ± 15 Tg y^{-1} (Denman et al. 2007), which would account for 2.5 to 7.5% of the total sink. Recent research suggests that approximately 29 Tg of the annually emitted CH_4 is re-oxidized to CO_2 within soils (Smith et al. 2000). However, the calculated uncertainty range is rather wide (between 7 Tg y^{-1} and ca. 100 Tg y^{-1}) (Smith et al. 2000), because the vast majority of studies focusing on CH_4 oxidation in soils were performed in North America (Adamsen and King 1993; Castro et al. 1994, 1995; Whalen and Reeburgh 1996) and Northern Europe (Bárcena et al. 2011), while data on other biomes are rare (Smith et al. 2000).

2.3.1 METHANE PRODUCTION

2.3.1.1 Methanogenesis

Methanogenesis represents the last step of the anoxic food chain in many anaerobic environments, in which methanogenic archaea generally convert acetate (acetotrophic methanogens, equation 2.3) or H_2 and CO_2 (hydrogenotrophic methanogens, equation 2.4) to CH_4 (Liu and Whitman 2008) (Figure 2.5):

$$CH_3COO^- + H_2O \rightarrow CH_4 + HCO_3^- \quad (2.3)$$

$$CO_2 + 4H_2 \rightarrow CH_4 + 2\,H_2O \quad (2.4)$$

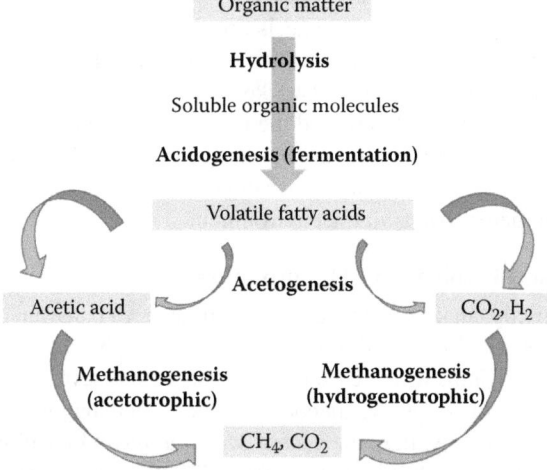

FIGURE 2.5 The two most common pathways leading to CH_4 formation. (Adapted from Cheng, H., and Wang, L. In *Biomass Mow–Sustainable Growth and Use*, ed. M.D. Matovic, 347–388. Rijeka. Intech. CC BY 3.0.)

In spite of their phylogenetic diversity, methanogens can only use a small number of substrates to supply themselves with energy. Besides CO_2 and acetate this restricted pool of substrates includes simple methylated compounds (such as formate, methanol, methylated amines, and sulfides). Most organic compounds that occur in nature, such as fatty acids and carbohydrates, cannot be used directly by methanogenic archaea. Instead, such complex compounds need to be degraded by a cooperation of different anaerobic microorganisms in the anaerobic food chain. Initially, organic polymers are broken down to simple sugars, volatile fatty acids, lactate, and alcohols by different groups of heterotrophic anaerobes. These breakdown products are then used by syntrophic fermenters, which produce acetate, formate, H_2, and CO_2 that can be metabolized by methanogens in the terminal step of the anaerobic degradation of organic matter (Liu and Whitman 2008). However, this conversion step that results in substrates for methanogenesis is only favorable at H_2 concentrations below 102 Pa (Zinder 1993). Methanogens are the key to solving this problem, since they are capable to remove excess H_2 very quickly and maintain H_2 concentrations around 10 Pa (Hedderich and Whitman 2006). This interplay between both groups is known as interspecies hydrogen transfer. Generally, methanogenic archaea prefer habitats in which alternative electron acceptors (O_2, NO_3^-, Fe^{3+}, SO_4^{2-}) are rare or absent. Because there are three main substrates for methanogenesis (CO_2, CH_3COOH, and methylated compounds), there are also three slightly different biochemical pathways (hydrogenotrophic, acetoclastic, and methylotrophic pathway). However, all pathways share the same unique coenzymes: ferredoxin (Fd), methanofuran (MF), tetrahydromethanopterin (H4MPT), coenzyme F420 (F420), coenzyme M (CoM), and coenzyme B (CoB).

Most methanogens are hydrogenotrophic and reduce CO_2 to CH_4 with H_2 as the electron donor. Few methanogens can also use formate as donor of electrons by oxidation of four molecules of formate to three molecules of CO_2. This step is carried out by the enzyme formate dehydrogenase. The produced H_2 is then used to reduce one molecule of CO_2 (Figure 2.5).

Some methanogens are capable of using methylated compounds (e.g., methanol) as a substrate (methylotrophic pathway). This ability is, however, restricted (pathway not shown in Figure 2.5). In a disproportionation (i.e., a specific type of redox reaction in which an element from a reaction undergoes both oxidation and reduction to form two different products), one methyl group has to be oxidized to CO_2 in order to allow the reduction of three other methyl groups to CH_4.

In the third pathway of methanogenesis, CH_4 is produced by reduction of acetate (acetoclastic pathway, Figure 2.5). Although only two genera are capable of utilizing acetate, about two thirds of the produced CH_4 in nature is derived from acetoclastic methanogenesis (Garcia et al. 2000). Compared to hydrogenotrophic methanogens, acetoclastic ones have to first oxidize the carboxyl-group of CH_3COOH to CO_2 before reducing it to CH_4.

Although the three pathways of methanogenesis are different at first sight, they share common features. These common steps include the transfer of the methyl-group to CoM and the reduction of methyl-CoM to CH_4 (Thauer 1998). Two key enzymes are involved in these reactions: a membrane-bound methyltransferase (MTR) and the methyl-coenzyme A reductase (MCR) (Thauer 1998). The genes which encode for the latter are homologous and well-conserved among the methanogenic archaea. Therefore, the mcrA gene is nowadays a widely used marker gene for the culture-independent detection of methanogens in the environment.

2.3.1.2 Microorganisms Involved in CH_4 Formation

All methanogens are strictly anaerobic archaea that produce CH_4 as an end product of anaerobic respiration (Garcia et al. 2000). All known methanogens belong to a monophyletic group within the Euryarchaeota phylum, which currently consists of the seven orders: Methanosarcinales, Methanobacteriales, Methanococcales, Methanomicrobiales, Methanopyrales, and Methanocellales (Nazaries et al. 2013). Additionally, the new genera *Methanolinea* (Imachi et al. 2008; Sakai et al. 2012), *Methanosphaerula* (Cadillo-Quiroz et al. 2009), and *Methanoregula* (Bräuer et al. 2011; Yashiro et al. 2011) were recently isolated and characterized. Each of these strains utilizes H_2/CO_2 and formate for growth and CH_4 production, except for *Methanonlinea mesophila* (Sakai et al. 2012), which did not grow on formate. *Methanolinea tarda* was shown to optimally grow under thermophilic conditions (50°C) (Imachi et al. 2008), which is a rare feature among methanogens. Indeed, the vast majority of the cultivated methanogenic strains are mesophilic. Most cultivable methanogens are neutrophils, optimally growing at pH values ranging from 6.0 to 8.0 (Garcia et al. 2000). Although methanogens may be exposed to physiological stress in acidic soils, adaptation was shown to be possible (Le Mer and Roger 2001; Bräuer, Yashiro et al. 2006). In acidic soils, enrichment culture studies have demonstrated that acid-tolerant methanogens are physiologically active at low pH (Sizova et al. 2003; Bräuer, Yashiro et al. 2006). Additionally, several acid-tolerant and acidophilic strains have been isolated in recent years (Bräuer, Cadillo-Quiroz et al. 2006; Kotsyurbenko et al. 2007; Cadillo-Quiroz et al. 2009, 2014). The recently discovered genera were shown to be closely affiliated with the order Methanomicrobiales.

2.3.2 Methane Consumption

2.3.2.1 Methanotrophy

Methanotrophic bacteria are usually found at the oxic–anoxic interfaces of environments such as geothermal habitats, landfills, wetland soils, peat bogs, and aquatic habitats. Here, these microbes feed on the CH_4 produced by methanogens on site, therefore, preventing a large fraction of the produced CH_4 from escaping into the atmosphere. According to estimations, about 42% of the total amount of CH_4 produced in these environments is consumed by microbes and, hence, never reaches the atmosphere (Reeburgh 2006). The responsible microorganisms are so-called low-affinity (Type I) methanotrophs, which can only consume CH_4 at high concentrations (>100 ppm) due to the low substrate-affinity of their methane monooxygenases. However, CH_4-consuming microbes also inhabit upland soils, which represent the only known biological sink for atmospheric CH_4 (Conrad 2009). CH_4 at atmospheric concentrations (about 1.8 ppm) is a quite meager source of carbon and energy and, therefore, atmospheric CH_4-oxidizers need to be adapted to such low concentrations of CH_4. To mediate this uptake, these high-affinity methanotrophs (Type II) are equipped with enzymes that have high affinities for CH_4 (in the nanomolar range) (Bender and Conrad 1992). Due to emerging molecular methods, it was possible to detect high-affinity methanotrophs by cultivation-independent approaches using the pmoA gene sequences as marker a few years later (Holmes et al. 1999; Henckel et al. 2000; Knief et al. 2003). However, their exact phylogenetic affiliation is still unknown and all attempts to cultivate them have only been partly successful so far (Kolb 2009).

Methane monooxygenase (MMO) is the crucial enzyme, which is responsible for the conversion of CH_4 to methanol (Hanson and Hanson 1996). Generally, two types of MMOs have been found in methanotrophic bacteria—a soluble type (sMMO) as well as a membrane-bound particulate type (pMMO) (Prior and Dalton 1985). Despite different enzyme structures and encoding gene operons, both enzymes catalyze the same reaction but require different electron donors. The presence of O_2 is a prerequisite for the activity of the MMOs (Lieberman and Rosenzweig 2004):

$$CH_4 + 2\,O_2 \rightarrow CO_2 + 2\,H_2O \qquad (2.5)$$

As mentioned before, two pathways for carbon assimilation are available in methanotrophs, the ribulose monophosphate (RuMP) pathway (Type I methanotrophs) as well as the serine pathway (Type II methanotrophs) (Trotsenko and Murrell 2008) (Figure 2.6). The RuMP pathway starts with the combination of formaldehyde and ribulose-5-phosphate to hexulose-6-phosphate via the enzyme hexulose-6-phosphate synthase (HPS). The initial step of the serine pathway results in the formation of the amino acid serine, which is generated by the reaction of formaldehyde and glycine.

Two enzymes that are unique to methanotrophs, hydroxypyruvate reductase (HPR) and glycerate kinase (GK), are responsible for the reduction of hydroxypyruvate to glycerate and the addition of a phosphate group from ATP to the latter, forming 2-phosphoglycerate, which is further processed as shown in Figure 2.6.

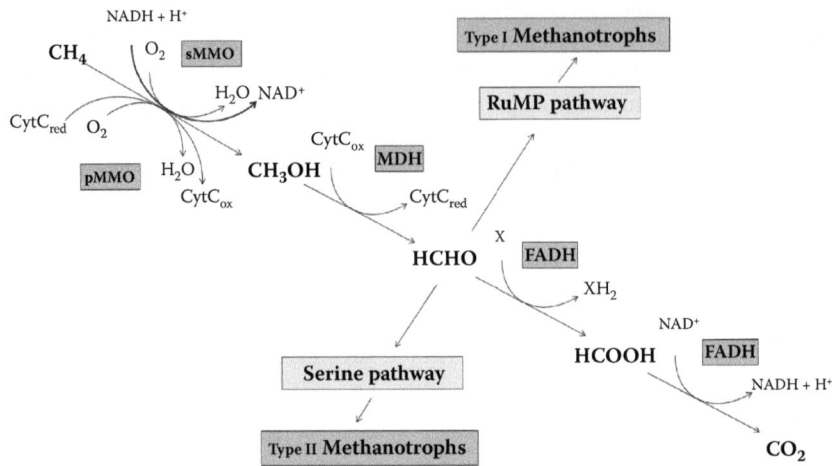

FIGURE 2.6 Pathways for the oxidation of CH_4 and the assimilation of formaldehyde. *CytC*, cytochrome c; *FADH*, formaldehyde dehydrogenase; *FDH*, formate dehydrogenase. (Adapted from Hanson, R.S., and Hanson, T.E., *Microbial. Rev.* **60**, 439–471, 1996.)

2.3.2.2 Microorganisms Involved in CH_4 Consumption

Methanotrophs are aerobic, Gram-negative bacteria that are able to cover their need of C and energy via oxidation of CH_4 to CO_2 (Hanson and Hanson 1996). So far, there are 20 genera, which either belong to the Gammaproteobacteria (15 different genera) or to the Alphaproteobacteria (5 different genera) classes of the Proteobacteria phylum. The Alphaproteobacteria class contains genera belonging to the family Methylocystaceae (*Methylocystis* and *Methylosinus*) and the family Beijerinckiaceae (*Methylocapsa*, *Methylocella* and *Methyloferula*), while the remaining 15 genera are exclusively assembled within the Methylococcaceae family of the Gammaproteobacteria. As described in the previous section, methanotrophs have originally been referred to as Type I and Type II methanotrophs according on their pathway of carbon fixation (Type I methanotrophs use the ribulose monophosphate pathway (RuMP), whereas Type II methanotrophs use the serine pathway) (Trotsenko and Murrell 2008). Another distinguishing feature is the structure of the intracytoplasmic membrane (ICM) (Trotsenko and Murrell 2008). Moreover, the predominant membrane phospholipid fatty acids (PLFAs) of Type I methanotrophs differ from those found in Type II methanotrophs (16C vs. 18C) (Hanson and Hanson 1996).

Recently, novel methanotrophic bacteria belonging to the Verrucomicrobia phylum have been isolated (Dunfield et al. 2007; Pol et al. 2007; Islam et al. 2008). The detection of these novel thermoacidophilic methane-oxidizers led to dramatic changes to the phylogeny of aerobic methanotrophs. Most cultivable methanotrophs are neutrophiles growing at pH values ranging between 6 and 8 and mesophilic (Hanson and Hanson 1996). However, there has already been evidence that some methanotrophs can withstand and grow in hot and extremely acidic habitats. These data also indicate hardly any activity losses, since CH_4 oxidation rates measured at pH 1.8 and 70°C were comparable with those of mesophilic soils (Castaldi and Tedesco 2005). Other methanotrophs that do not fit into the Type I and Type II scheme are *Crenothrix polyspora* and *Clonothrix fusca*, due to their filamentous morphology. Both strains have been suggested to be CH_4-oxidizers based on DNA sequencing, which was proved later by physiological experiments (Stoecker et al. 2006; Vigliotta et al. 2007).

2.3.3   ENVIRONMENTAL FACTORS PROMOTING CH$_4$ RELEASE

Soil methanotrophy and methanogenesis are influenced by physicochemical soil properties and climatic factors (Le Mer and Roger 2001). Especially methanogens are highly sensitive toward variations of **soil pH** and the growth optimum of most cultivated strains ranges from neutral to slightly alkaline conditions (Garcia et al. 2000). Although methanogens may be exposed to physiological stress in acidic soils, adaptation was shown to be possible (Le Mer and Roger 2001; Bräuer, Yashiro et al. 2006). In acidic soils, enrichment culture studies have demonstrated that acid-tolerant methanogens are physiologically active at low pH (Sizova et al. 2003; Bräuer, Yashiro et al. 2006). Additionally, several acid-tolerant and acidophilic strains have been isolated in recent years (Bräuer, Cadillo-Quiroz et al. 2006; Kotsyurbenko et al. 2007; Cadillo-Quiroz et al. 2009; Bräuer, Cadillo-Quiroz et al. 2011; Cadillo-Quiroz et al. 2014). Nevertheless, methanotrophs seem to be adapted to a wider range of soil pH than methanogens, since acidophilic, neutrophilic as well as alkaliphilic strains have been isolated. Moreover, CH$_4$ oxidation rates of temperate soils remain at comparable levels at pH values ranging from values as low as 3.5 up to 8.0 (Born et al. 1990). However, the effect of soil pH on net CH$_4$ flux between soil and atmosphere is more pronounced. Alkalinity and acidity, for instance, reduced the overall net CH$_4$ fluxes in forest and grassland soils (Amaral et al. 1998; Reay et al. 2001). A meta-analysis of methanotrophic community structure in forest soils has revealed some influence of pH, especially on atmospheric methane oxidizers. In acidic forest soils, alphaproteobacterial methanotrophs were predominant, whereas in pH-neutral forest soils mainly gammaproteobacterial methanotrophs were responsible for CH$_4$ oxidation (Kolb 2009). The recent discovery of the genus *Methylacidiphilum* and its ability to grow at pH 1 (Op den Camp et al. 2009) strengthened the view that methanotrophy in soils is only slightly affected by pH changes. Taken together, this indicates that pH influences the net CH$_4$ flux of soils by alteration of the rates of methanogenesis rather than those of methanotrophy.

2.3.3.1   Climate

Increasing soil temperature has been previously shown to lead to greater net CH$_4$ uptake rates, for instance in temperate forest soil (Castro et al. 1995). However, the enzymes responsible for methanotrophy optimally work at an average temperature of 25°C (Hanson and Hanson 1996), whereas the enzymes involved in methanogenesis optimally work at temperatures ranging from 30 to 40°C (Le Mer and Roger 2001). This discrepancy leads to the suggestion that temperature changes might differentially affect the activity of methanogens and methanotrophs. Although temperature affects both the enzyme activities of the resident methanotrophs as well as the diffusion of CH$_4$ into the soils, the reported effect is rather small with Q_{10} values (relative increase of activity per 10°C of temperature increase) of around 1.4 (Smith et al. 2003). The reason for this might be the relatively broad temperature range of methanotrophic bacteria, since the described strains are able to grow at psychrophilic, mesophilic, and thermophilic conditions (Nazaries et al. 2013). On the other hand, temperature exerts a great influence on CH$_4$ production rates. CH$_4$ emissions increased in wetland soil samples (peat, swamp) by approximately 7 times when the temperature was increased from 10 to 23°C. More recent studies showed that CH$_4$ production in neutral and acidic peat soil samples from Northern wetlands increased with temperatures ranging from 1 to 25°C (Metje and Frenzel 2005; Høj et al. 2008). At temperatures above 25°C production rates decreased again (Metje and Frenzel 2005). The effect of temperature on methanogenesis in well-aerated upland soils seems to be contrasting compared to wetlands. Laboratory incubations of alpine soil sampled in a glacier foreland pointed to the presence of a thermophilic methanogenic community, since CH$_4$ production was highest at 50°C (Hofmann et al. 2013). A similar response could be detected in grazed pasture soils of the subalpine altitudinal belt (Prem et al. 2014). By contrast, optimal CH$_4$ production rates were found to occur at 37°C in a subalpine fallow soil (Praeg et al. 2014). So far, our knowledge on the effect of increasing temperature on CH$_4$ production in upland soils is still limited. Nevertheless, this activity is expected to increase with elevated temperatures due to a physiological response of methanogens. When the global temperature rises by 3.4°C, the global CH_4 emissions from wetlands are estimated to increase by 78% (Shindell et al. 2004).

Temperature changes are also known to affect community structures of microorganisms. When forest soils capable of atmospheric CH_4 removal were incubated at temperatures ranging from 5 to 45°C, the methanotrophic community composition changed. Individual species responded differently to this increase, which suggests that the community consisted of methanotrophs with different temperature optima (Mohanty et al. 2007). A major drawback of this study is that the high CH_4 concentrations used for the soil incubations (ca. 10 000 ppm) together with the high incubation temperatures did not mimic representative atmospheric conditions. Hence, the response of the community to temperature increases under in situ conditions remains unclear. In arctic wetland soils both acetoclastic and hydrogenotrophic methanogens (including species belonging to the order Methanomicrobiales and the genera *Methanobacterium, Methanosarcina,* and *Methanosaeta*) were found (Høj et al. 2005, 2008). On the one hand, seasonal temperature changes were shown to distinctly affect the resident community (Høj et al. 2005). On the other hand, overall temperature increase boosted the relative abundances of the dominant methanogens and the diversity of the entire community (assessed by DGGE profiling of the methanogens) (Høj et al. 2008). However, there is hardly any study addressing the responses of the methanogenic community and their population sizes in upland soils.

As mentioned before, methanogens are restricted to anaerobic habitats characterized by low oxydo-reduction potentials (E_h <–200mV) and low concentrations of available SO_4^{3-} and NO_3^- as these anions are more favorable electron acceptors compared to CO_2 (Le Mer and Roger 2001). Therefore, the nature and availability of organic matter but also soil moisture content that regulates O_2 availability play a role for methane production in soils. However, CH_4 production has also been shown to occur in well-aerated soils, although the overall conditions were thought to be inhibitory (low soil moisture, high concentrations of SO_4^{3-} and NO_3^-). This could be explained by the complex spatial structure of the soil, which could lead to the creation of O_2-free micro-sites in which methanogens could be active. Improving these conditions from a methanogenic point of view by rising soil moisture content (and thereby decreasing O_2 availability) usually increases both abundance and activity of methanogens (Chen et al. 2011). Accordingly, it could be observed that Methanomicrobiales, Methanobacteriaceae, and *Methanosaeta* clearly prefer sites with high water contents, while other methanogens tolerate less reduced conditions (Høj et al. 2006). Temporal water saturation of normally well-aerated upland soils can also enhance CH_4 production. This was demonstrated in grasslands and arable soils when soil moisture increased during heavy storms or snow-melt (Wang and Bettany 1995). Similarly, when well-drained glacier foreland soils were incubated under water saturation in order to simulate spring snow-melt, they showed significantly more CH_4 release compared with field-moist controls (Hofmann et al. 2013).

The relationship between CH_4 consumption and soil moisture content is more complicated. There are many investigations showing that soil CH_4 consumption decreased when water content increased. This relationship was shown for many ecosystems, including forests and grasslands located in temperate, boreal, and tropical regions, and in arable and tundra ecosystems (Adamsen and King 1993; Castro et al. 1994; West and Schmidt 1999). However, other studies indicate that this generalization is not justified because increases in soil moisture do not always lead to decreases in CH_4 consumption (Castro et al. 1992). It was shown, for instance, that CH_4 consumption was actually promoted by increasing soil moisture during the rainy season (Singh et al. 1997). During the dry season, however, the same activity was hampered by rising soil moisture contents. Generally, CH_4 uptake was recognized to increase until water contents close to field capacity are reached, but to decrease above this value (Le Mer and Roger 2001). For example, in temperate spruce-fir forest soils CH_4 uptake was reduced when 60 to 100% of the soil pores were filled with water. However, CH_4 uptake increased between 22 and 60% WFPS (Castro et al. 1992).

2.3.3.2 Land Use

For decades, methanogens were thought to be restricted to highly reduced environments. This assumption was mainly based on physiological aspects since O_2 causes damage to the F420–hydrogenase complex, an important enzyme involved in methanogenesis (Schönheit et al. 1981). Pure-culture studies further revealed that reactive oxygen species (ROS), which are formed in the presence of O_2, can damage cell membranes and cell walls of methanogens (Storz et al. 1990). Nevertheless, during the last decades there has been increasing evidence that methanogens are not as sensitive to oxic conditions as previously thought. It has been possible to activate methanogenesis in desert, forest, meadow, and savannah soils after incubation as slurries under anoxic conditions (Peters and Conrad 1995). Further research efforts demonstrated that under favorable conditions even alpine soils can be CH_4 sources (West and Schmidt 2002). Recent studies along a Central European alpine glacier foreland and subalpine meadow, fallow, and abandoned soil revealed that the soil CH_4 production potential depends on soil age and moisture conditions as well as on the degree of cattle-grazing (Hofmann et al. 2013; Prem et al. 2014).

Additionally, it has been shown that upland forest soils contain methanogens (Angel et al. 2012; Christiansen et al. 2016), and that CH_4 production in these soils can be induced by increasing the soil moisture content (Christiansen et al. 2017). In a study across a natural soil moisture gradient, going from upland to wetland forest types, Christiansen et al. (2016) showed that variations in CH_4 fluxes were related to soil hydrology in both upland and wet soils. In upland forest soils, CH_4 uptake rates increased with decreasing soil moisture content, whereas CH_4 emissions were inversely related to water table depth in wet soils.

The occurrence of methanogenic sequences has recently been demonstrated in arid soil biocrusts (Angel et al. 2012) and in high-altitudinal cold deserts (Aschenbach et al. 2013). This indicates that methanogens are tolerant toward extreme aeration caused by low soil moisture, and that they can also withstand cold climatic conditions. The majority of the detected sequences were assigned to the genera *Methanosarcina* and *Methanocella*, while *Methanobacterium* spp. was only detected to a limited extent. Accordingly, it was hypothesized that these methanogens could be autochthonous members of the methanogenic community in aerated soils worldwide (Angel et al. 2011, 2012; Aschenbach et al. 2013). Apart from this, sequences affiliated with the classes Methanobacteria, Methanomicrobia, and Methanococci were retrieved from high-altitudinal cold soils of the Tibetan plateau (Wang et al. 2015). *Methanobrevibacter* spp., which is a typical member of the gut microbiome of cows, was recently detected in a subalpine fallow soil from the Austrian Central Alps. This methanogenic archaeon was very likely introduced into the soil through cattle manure and surprisingly survived for at least six months in soil microcosms under oxic conditions, which again highlights the ability of methanogens to survive exposure to O_2 (Praeg et al. 2014). Other methanogenic groups, besides *Methanosarcina* spp. and *Methanocella* spp., might also be members of the microbial communities in non-wetland soils and contribute to CH_4 emissions. However, our knowledge about the abundance and composition of the methanogenic community in upland soils is limited to a few reports from soils of arid and semi-arid regions (Angel et al. 2011, 2012; Aschenbach et al. 2013).

2.3.3.3 N Input

Studies on the response of soil CH_4 flux to N fertilization (urea, NH_4^+, NO_3^-) led to contradictory results, possibly due to contrasting responses of methanogenesis and methanotrophy and/or differences in microbial community composition at individual sites. These conflicting studies show either no influence, inhibition, or even stimulation of CH_4 oxidation in soils when N fertilizers are added (Bodelier and Laanbroek 2004). A recently published meta-analysis provides evidence that the extent to which CH_4 consumption is reduced depends on fertilizer type, dose, biome, as well as on the land use history of the respective site (Aronson and Helliker 2010). The initial CH_4 consumption rates of croplands and pastures are not only generally smaller compared to those of natural forests and grasslands, but they are also distinctly more reduced upon fertilizer addition (Aronson and Helliker 2010). At the enzyme level, competitive inhibition of MMO has been proposed to be the reason for the observed reduction in CH_4 oxidation. Since the responsible genes for both CH_4 oxidation and NH_4^+ oxidation are functionally and evolutionary related, some methanotrophs are able to change from CH_4 to NH_4^+ oxidation when N is added (Dunfield and Knowles 1995), which would consequently reduce CH_4 uptake by soils. However, low N applications (<100 kg N ha^{-1} y^{-1}) stimulated CH_4 uptake, and the observed stimulation resulted in a reduction of total CH_4 emissions. When soils were fertilized with >100 kg N ha^{-1} y^{-1}, CH_4 uptake decreased and total emissions increased (Aronson and Helliker 2010). A possible explanation for the N limitation of methanotrophy in upland (Aronson et al. 2012) and wetland soils (Bodelier 2011) might be N fixation, which is predominantly found in type II and some type X methanotrophs. In N-limited environments, methanotrophs are forced to use reducing equivalents for N fixation instead of regeneration of the methane monooxygenase, which consequently reduces CH_4 oxidation rates (Bodelier and Laanbroek 2004). However, type II methanotrophs have been observed to reduce growth in order to carry out both N fixation and CH_4 oxidation (Steenbergh et al. 2010). Therefore, the stimulation of the overall rates of CH_4 consumption cannot be due to Type II methanotrophs. More likely, these effects are caused by the stimulation of type I methanotrophs, which are only seldom capable of atmospheric N fixation (Auman et al. 2001).

The response of methanogens and CH_4 production to N depends on fertilizer type, dosage, and time of application (Bodelier 2011). For instance, NO_3^- amendment was shown to inhibit CH_4 production in rice field soil. Two distinct mechanisms (direct inhibition by toxic compounds, indirect inhibition by competition) have been proposed to explain the observed suppression. Firstly, methanogenesis could be directly suppressed by toxic intermediates of denitrification processes such as NO_2^-, NO, and N_2O (Roy and Conrad 1999). Secondly, others observed that the application of NO_3^- and N_2O to rice field soils inhibited CH_4 production, because both compounds led to an increase of the electron acceptors SO_4^{2-} and Fe^{3+}, possibly through the oxidation of sulfur species and reduced iron (Klüber and Conrad 1998). Subsequently, methanogens were outcompeted by sulfate- and iron-reducing bacteria, which consumed H_2 (Klüber and Conrad 1998). However, detection of inhibitory effects of NO_3^- on methanogenesis remains restricted to laboratory experiments. Under field conditions no such effect has been reported, and it has been proposed that the reason for this might be that the amended NO_3^- was assimilated by plants or denitrifying microorganisms before it could inhibit methanogenesis (Nazaries et al. 2013). However, across a range of natural wetlands soil N concentration was positively related to the abundance of methanogens. This result points to a possible N limitation of methanogenic archaea (Liu et al. 2011). Indeed, methanogens depending on N fixation would benefit from additional N supply, which would enable them to invest more energy into growth.

Deducing from the physiology of the involved methanogenic archaea and methanotrophic bacteria and from field studies it seems logical that net CH_4 release is a delicate balance between CH_4 production and uptake rates. Recent research on methanogenesis and methanotrophy along an altitudinal gradient (Hofmann et al. 2016) provides evidence that increasing temperatures due to climate change may lead to an increase in the abundance of methanogenic microorganisms and consequently to increased CH_4 production rates. As methanogens are N-limited to some extent whereas methanotrophs are inhibited by high N dosage (>100 kg N ha^{-1} y^{-1}), a global acceleration of the N-cycle may feed back positively on climate change not only via increased N_2O emissions, but also via increased CH_4 emissions. This holds true under the premise that there is no significant change in soil moisture, which remains the most important driver of non-CO_2 greenhouse gas fluxes.

2.4 CO$_2$ EMISSIONS AND THE TERRESTRIAL CARBON CYCLE

Soils are major contributors to the atmospheric pool of CO_2 (IPCC 2013). Recent estimates show that about 94.3 ± 17.9 Pg C are released by soils at the global scale annually (Xu and Shang 2016).

This CO_2 is mainly originated during either plant or microbial respiration. Soil fauna respire too, but contribute only 5% (up to 10%) of CO_2 to the total amount of the CO_2 released by the soil (Petersen and Luxton 1982; Sørensen et al. 2006). Additionally, soil respiration differs across ecosystems (Carey et al. 2016). Plants shape the microbial community (e.g., De Deyn et al. 2008) and abiotic factors (Moyano et al. 2013) set the boundaries for the microbial community and the magnitude of plant-microbe interaction. Abiotic factors such as soil structure and soil moisture change the soil microbial environment by modulating gas transport (e.g., oxygen), solute transport (e.g., nitrogen), the metabolic cost and predation (Moyano et al. 2013), and thereby affect the microbial processes in which C is respired (Bardgett et al. 2008).

The size and the temporal dynamics of the soil C stock depend on the net sum of the ecosystem and human-driven processes that affect soil C gains and losses (Figure 2.7). Factors such as climate, land use, and N deposition can change ecosystem processes related to ecosystem C (and nutrient) stores, and hence modulate the soil–atmosphere feedback.

Terrestrial C cycling starts with the assimilation of atmospheric CO_2 via plant photosynthesis (Figure 2.8). This C is further respired to sustain plant energetic demands or it is used to build up biomass in plant components aboveground or allocated to roots belowground (Bowling et al. 2008; Brüggemann et al. 2011). Carbon compounds in roots are either used for root processes, released as CO_2 due to respiration, or released into the rooting zone in form of labile C (root exudates), which can stimulate soil microbial activity (Kuzyakov et al. 2001; Paterson et al. 2007). Additionally, when plants or parts of them die, their necromass is incorporated into the soil, both aboveground (litterfall) and belowground (root decay). Microbes use plant-derived C compounds to grow and invest C in the production of exoenzymes for SOM and litter decomposition and, as a result, increase soil nutrient availability (Fontaine et al. 2011). Soil microbes can receive C from i) the breakdown of organic matter (e.g., litter and roots), ii) symbiotic association (mycorrhizal fungi) with plants roots, iii) plant root rhizodeposition, and iv) the breakdown of SOM (Figure 2.9).

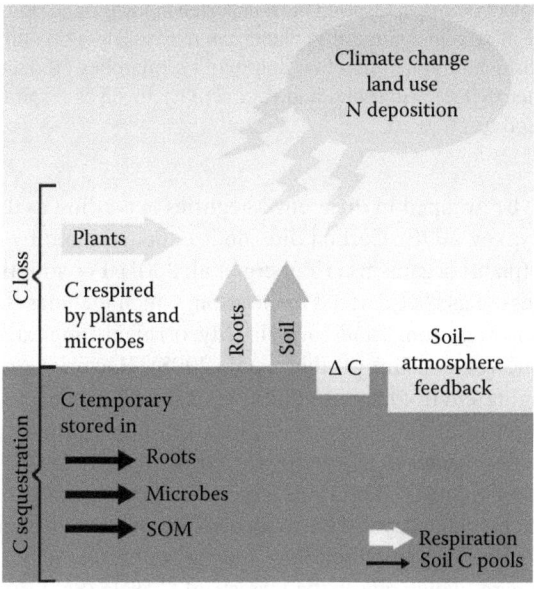

FIGURE 2.7 The terrestrial C stock is a dynamic pool that is modulated by C loss via respiration and soil C sequestration. Carbon is respired by aboveground and belowground plant structures, soil microbes (bacteria, fungi and archaea) and soil fauna. The main soil C stocks are roots and their associated mycorrhizal network, microbes and soil organic matter (SOM). External factors such as climate, land use, and N deposition affect the ratio of C loss to C sequestration and hence may change the soil–atmosphere C feedback.

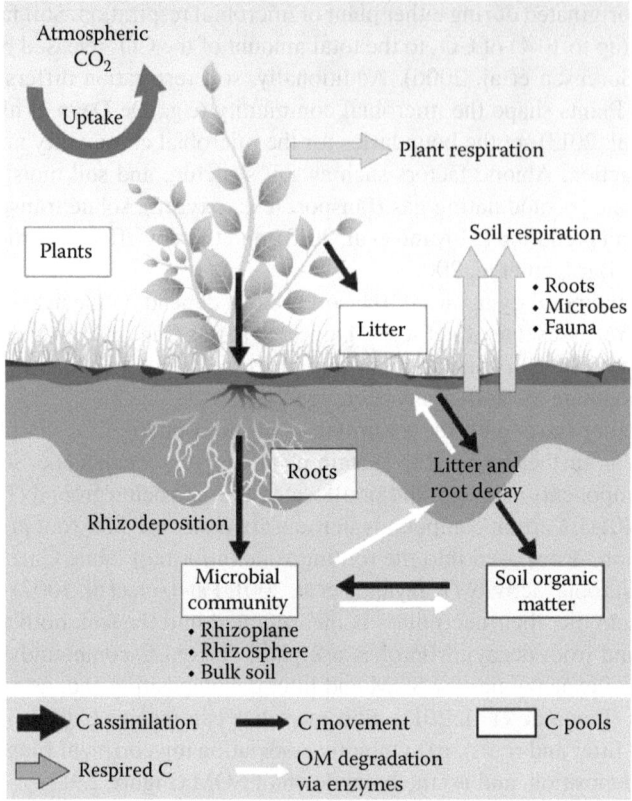

FIGURE 2.8 The terrestrial C cycle with focus on soil processes: atmospheric CO_2 is assimilated by plants, C is stored in aboveground plant biomass, respired or transported belowground into roots; roots can be directly colonized by the rhizoplane microbial community; plants can release labile C compounds (rhizodeposits) into the soil matrix that are consumed by the microbial community; microbes produce enzymes that mineralize litter and soil organic matter (SOM). Microbial and root activity involves respiration processes; respired C eventually leaves the soil (soil respiration).

Soil respired CO_2 can be grouped in different categories according to the processes from which the CO_2 originates (Kuzyakov 2006). Carbon dioxide is either respired by heterotrophic (microbes and fauna) or autotropic (plant) organisms (Högberg et al. 2001). For simplification, the autotrophic component is often addressed as root-derived respiration, which includes the respiration from roots and root-associated mycorrhizal fungi and bacteria. Mycorrhizal fungi are closely associated with plant roots in a symbiotic relationship (Talbot et al. 2008). Bacterial groups in the rhizosphere rapidly respond to changing environmental conditions such as exudation of labile C compounds by plant roots (Fontaine et al. 2003; Reinsch et al. 2014). Saprophytic fungi and actinomycetes (a sub-group of Gram-positive bacteria) decompose SOM. The actinomycetes group has a fungal-like life cycle and appearance and has the ability to break down complex C compounds (Lacey 1997). Carbon that is assimilated by plants, cycled between ecosystem compartments, and allocated below-ground is eventually respired by roots, microbes and the soil fauna and diffuses through the soil pores back to the atmosphere, which closes the terrestrial C cycle (Schlesinger and Andrews 2000; Amundson 2001) (Figure 2.8).

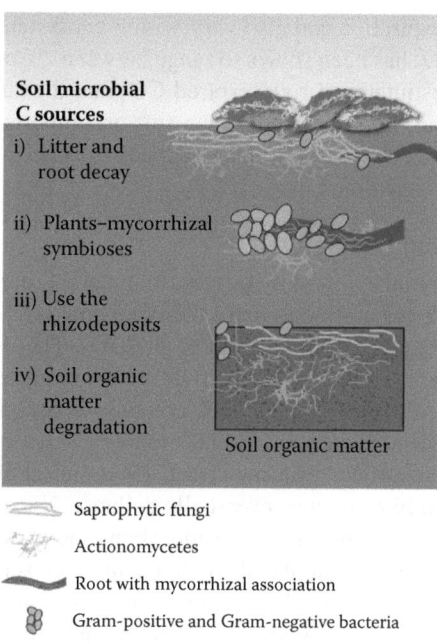

Saprophytic fungi

Actionomycetes

Root with mycorrhizal association

Gram-positive and Gram-negative bacteria

FIGURE 2.9   Soil C sources are accessed by different components of the soil microbial community: i) actinomycetes (filamentous Gram-positive bacteria) and saprophytic fungi break down organic material such as litter and roots, making C available to the soil bacterial community; ii) mycorrhizal association with plants supplies C to mycorrhizal fungi in exchange for nutrients; iii) plant roots exude labile C compounds called rhizodeposits (process = rhizodeposition) into the soil matrix which can be utilized by the rhizosphere microbial community (Gram-positive and Gram-negative bacteria); iv) soil organic matter in the bulk soil can be accessed by saprophytic fungi and bulk soil bacteria after a physical breakdown of the soil structure.

2.4.1   Soil Respiration

2.4.1.1   Roots and Plant–Mycorrhizal Symbiosis

Active roots exploring the soil matrix for water and nutrients contribute to soil respiration and soil C loss (Raich and Tufekciogul 2000). The complex interaction of plant roots and the rhizosphere microbes has caught increasing interest in recent decades (Singh et al. 2004). It has long been known that plants form symbiotic relationships between their roots and mycorrhizal fungi where both organisms are tightly interlinked to exchange plant C for nutrients (Paul and Kucey 1981; Harley and Smith 1983). The root–mycorrhizal network hosts a highly diverse microbial community that is indirectly powered by decaying plant material and microbial turnover (Wiant 1967; Nannipieri et al. 2003). The connection between plant C assimilates and roots and thus mycorrhizal fungi has been demonstrated in a tree-girdling experiment (Högberg et al. 2001): stripping off the bark of pine trees stopped the transport of C assimilates to the roots and their associated mycorrhizal fungi without damaging the root–mycorrhizal network, thereby reducing the autotrophic respiration by over 50% (Högberg et al. 2001). The measurement of the contribution of the autotrophic respiration to total soil respiration, however, remains challenging (e.g., Kuzyakov and Larionova 2005; Slavíková et al. 2017). The contribution of root respired CO_2 to the total soil CO_2 efflux was estimated to be roughly half of the total amount of soil respired CO_2 (for a review on soil respiration see Ryan and Law 2005) but is highly variable and ranges from 10% to 90% across ecosystems

(Hanson et al. 2000). Root respiration can also vary within ecosystems (as shown by Subke et al. 2006), where root-respired CO_2 has been shown to range between 2% and 97% in coniferous forests alone. The discrepancies in estimates of root-respired CO_2 and the autotrophic component of soil respiration show that roots and root-associated organisms are important drivers of soil respiration with potential relevance for the ecosystem-climate feedback (Mohan et al. 2014), but the magnitude of importance remains to be determined.

2.4.1.2 Rhizodeposition and Priming

Plants stimulate the soil microbial community by releasing labile C compounds into the soil, which leads to increased rhizosphere respiration to drive the breakdown of SOM and make soil N available (Blagodatskaya and Kuzyakov 2008; Dijkstra et al. 2013). The use of labile C compounds is more energy-efficient than the use of more stable SOM which requires the production of extracellular enzymes (Xu and Shang 2016). When labile C compounds are present, the fast-growing part of the microbial community is activated, microbial biomass increases and thus, CO_2 production increases and constitutes the so-called "priming effect" (Fontaine et al. 2003). After the activation of soil microbes, the microbial biomass may eventually collapse. Nitrogen that was formerly bound in microbial biomass now becomes available to plants (Fontaine et al. 2003; Dijkstra et al. 2013). However, SOM priming is not only stimulated by plant rhizodeposition (Kuzyakov et al. 2000). Any change in easily available C, N and phosphorus or in environmental conditions can lead to a change in SOM mineralization rates. If SOM mineralization rates increase, soil C loss is enhanced (positive priming); if SOM mineralization rates decrease, soil C loss is decreased (negative priming) (reviewed in Kuzyakov et al. 2000).

Rhizosphere "hotspots" of labile C substrates especially increase the activity of Gram-negative bacteria (Kuzyakov and Blagodatskaya 2015), which are usually the quickest to use recent plant-assimilated C (Reinsch et al. 2014) and prefer rhizodeposits over SOM-derived C (Kramer and Gleixner 2008). Consequently, factors such as grazing or drought that influence plant C assimilation and C allocation within the plant will affect the magnitude of plant rhizodeposition and this will translate into changes in soil C respired by Gram-negative bacteria.

2.4.1.3 Soil Organic Matter Degradation

Carbon stored in SOM can be protected by physical and geochemical forces, preventing the microbial breakdown of SOM (Kögel-Knabner et al. 2008; Schmidt et al. 2011; Doetterl et al. 2015). Main mechanisms of SOM persistence include spatial separation, sorption to minerals or metal ions, and aggregation (von Lützow et al. 2006; Schmidt et al. 2011). Soil microorganisms only occupy about one percent of the soil volume (Ekschmitt et al. 2008). They are restricted in their movements and have to rely on diffusion and transport of substrate through a network of soil pores. This setup and the potential for disconnection of microbes and their substrate is one of the reasons for the persistence of SOM in deep soils (Schmidt et al. 2011).

Another stabilization mechanism of SOM is the formation of organo-mineral complexes. Dissolved substances, either directly leached from plant litter or solubilized by microbial enzyme activities can interact with clay particles or metal ions in soils and become unavailable for immediate uptake by microbes. The type of bond between organic matter and the mineral surface or metal ions (e.g., ligand exchange, polyvalent cation bridges, hydrophobic interactions or Van Der Waals forces) determines the strength of the bonds (von Lützow et al. 2006) and how organic matter can again be released from the mineral surfaces. Further, the formation of aggregates increases the persistence of organic matter in soils (Six et al. 2002). Microbial cell compounds and exudates as well as rhizodeposits and mucus can adhere to each other and form microaggregates which can, together with clay minerals and particulate plant detritus, be enmeshed by fungal hyphae to form macro-aggregates (von Lützow et al. 2006). Organic material inside aggregates is protected by reduced access for soil microbes and enzymes but also by the reduced oxygen contend in the aggregates due to low diffusion. However, a change in soil physical properties such as SOM breakdown by the soil fauna (e.g., Bocock 1964), changes in precipitation (e.g., Robinson et al. 2016), and changes in land use (Bronick and Lal 2005) can make formerly protected SOM accessible to the soil microbial community. Saprophytic fungi and actinomycetes produce extracellular enzymes that can break down complex organic matter (Makoi and Ndakidemi 2008). The microbial network of saprophytes and actinomycetes in association with bulk soil bacteria can break down SOM into labile C compounds. The preference of Gram-positive bacteria to use SOM-derived C compounds over plant-derived C (Kramer and Gleixner 2008; also in Kuzyakov and Blagodatskaya 2015) make them, together with saprophytic fungi and actinomycetes, a significant source of SOM-derived respired CO_2.

Decades of research on soil respiration and its many contributing sources has advanced our understanding of the importance of the microbial community in terrestrial C turnover. This has also been recognized by the modeling community suggesting that information on soil microbial physiology is needed to improve Earth system models (Wieder et al. 2013; Xu et al. 2014).

2.4.2 Soil Carbon Sequestration by Microbes

Increasing global temperatures and changing weather patterns resulting from rising atmospheric CO_2 and other GHG concentrations demand effective mitigation strategies. One strategy is the sequestration of C into persistent soil C pools (Lal 2008), in which soil microbes play an important part.

In recent years, the perception of how SOM is formed and stabilized has changed. Classically, the persistence of chemically recalcitrant compounds was emphasized; now the persistence of SOM is believed to be due to SOM sorption to minerals or metal ions, spatial separation, and aggregation (von Lützow et al. 2006; Schmidt et al. 2011). Concurrently, the view that chemically recalcitrant plant components will accumulate in soils has been challenged by findings that microbial compounds make up the majority of SOM (Grandy and Neff 2008; Schmidt et al. 2011; Miltner et al. 2012; Throckmorton et al. 2012; Creamer et al. 2016; Kallenbach et al. 2016). In fact, at least 40% of SOM is microbially transformed (Miltner et al. 2012). Furthermore, microbes that were fed only with simple sugar compounds could, over time, build chemically diverse organic matter that was similar in its chemistry to SOM found in nature (Kallenbach et al. 2016). The C accumulation in soils depends more on microbial transformation than on the chemistry of C inputs, which is further supported by the inconsistent effects of input of chemically recalcitrant plant components in agricultural soils (Leifeld and Fuhrer 2010; Pittelkow et al. 2015). Inputs that promote microbial biomass production are thus considered to be helpful in terms of C accumulation in soils (Cotrufo et al. 2013). Efficient incorporation of C into microbial biomass might be the first step for the effective stabilization of SOM.

2.4.3 Microbial Carbon Use Efficiency

Microbial C use efficiency (CUE) describes the investment of C into microbial biomass production relative to total microbial uptake of C (Sinsabaugh et al. 2013). With a high CUE, microbes transform plant products and SOM more efficiently into microbial compounds and exhibit little loss of C as CO_2 to the atmosphere. Microbial compounds, released after cell death, can then be adsorbed to clay minerals (von Lützow et al. 2006; Throckmorton et al. 2012) or incorporated into soil aggregates, which are key mechanisms for the persistence of C in the soil system (Schmidt et al. 2011). High CUE can thus translate into high C persistence in the soil if a great abundance of microbial compounds is effectively generated (Bradford et al. 2013) and is introduced into the mineral-associated organic matter (MaOM) pool (Creamer et al. 2016; Kallenbach et al. 2016). Besides this crucial first step, a high CUE will also reduce C losses during the recycling of microbial necromass (Geyer et al. 2016). Microbial CUE is thus a proximal agent of microbial biomass accumulation, with a higher CUE resulting in greater soil C sequestration. Indeed, Kallenbach et al. (2015) have related CUE to higher SOM content in an organic agricultural system compared to a conventional one, despite lower inputs and greater tillage pressure in the organic system. Soil C models that explicitly incorporate microbial CUE have demonstrated that changes in CUE might have big effects for soil C storage and C sequestration (Frey et al. 2013; Hagerty et al. 2014; Wieder et al. 2014).

While the stabilization of organic matter on the mineral component of the soil matrix also strongly depends on the nature of available minerals and metal ions, microbial CUE determines the proportion of plant input that is available for interactions with clay minerals. How the interactions of microbial CUE, mineral surfaces and the consequences of these interactions for the loss of C as CO_2 to the atmosphere play out can be explained based on the Microbial Efficiency-Matrix Stabilization (MEMS) framework (Cotrufo et al. 2013; Figure 2.10). Following this conceptual approach, CUE depends on external parameters such as substrate quality and complexity, nutrient availability and environmental factors such as temperature. A high CUE will result in a lower proportion of C lost as CO_2 to the atmosphere and a higher abundance of microbial products that can, depending on the soil matrix, be stabilized through organo-mineral interactions. Since matrix stabilization also controls the availability of substrate for microbial recycling, it further feeds back to CUE and CO_2 loss.

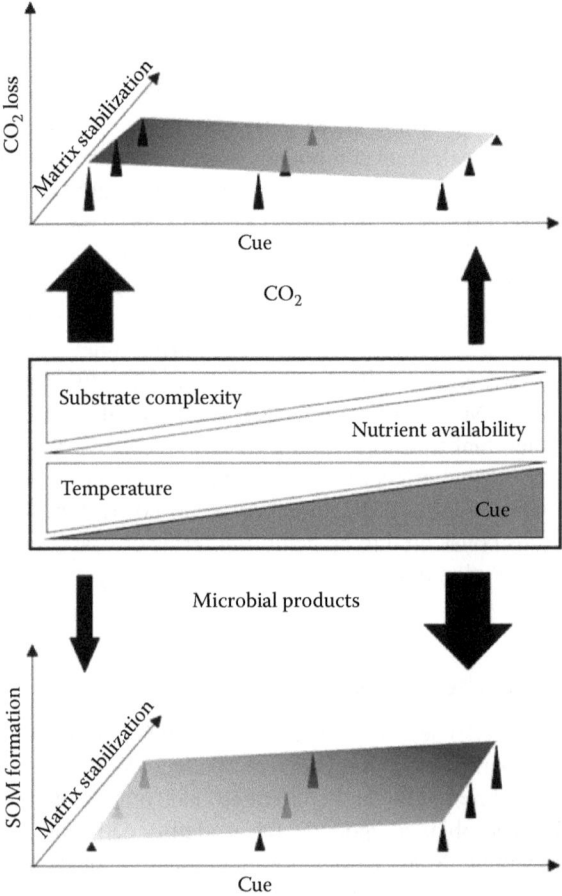

FIGURE 2.10 Microbial carbon use efficiency (CUE) depends on substrate chemistry and complexity, nutrient availability, and environmental factors such as temperature. When CUE is high, little CO_2 is lost to the atmosphere and more microbial biomass is produced. Microbial products can be stabilized on the mineral soil matrix. How much stable soil organic matter (SOM) will accumulate further depends on the magnitude of the soil matrix stabilization (e.g., high content of clay minerals or a high abundance of metal ions). The concept is based on the Microbial Efficiency-Matrix Stabilization (MEMS) framework. (From Cotrufo, M.F., Wallenstein, M.D., Boot, C.M., Denef, K., and Paul, E., *Glob. Change Biol.* **19**, 988–995, 2013.)

Carbon use efficiency explicitly encompasses microbial C uptake and growth and is dependent on environmental conditions, substrate chemistry and nutrient availability. These dependencies make CUE highly sensitive to global change (e.g., climate change, land use change, and N deposition).

Climate change: An increase in temperature is generally believed to increase microbial activity. This requires more energy, and therefore microbes have to invest more C into respiration to fulfill their physiological needs (Marr et al. 1963). If growth is not stimulated to the same extent, CUE will decrease (Manzoni et al. 2012). The effects of moisture on CUE are less straightforward. Drought might decrease the connectivity between microbes and their substrate, force microbes to invest in osmotically active substances and induce stress reactions, all of which will decrease CUE (Manzoni et al. 2012). Under water-saturated conditions, microbes might switch to alternative, less energy-efficient metabolic pathways, which will also affect CUE (Manzoni et al. 2012).

Land use change will affect the chemical composition of plant inputs. Woody plant debris is generally higher in complex polymers such as lignin and cellulose than is litter from herbaceous plants (Taylor et al. 1989). To break up complex substances for microbial uptake, microbes need to produce extracellular enzymes. This process requires C, N, and energy (Sinsabaugh et al. 2013). A higher proportion of complex structural carbohydrates thus reduces overall CUE. On the other hand, a high proportion of C monomers that are readily assimilable will increase CUE (Cotrufo et al. 2013). Addition of chemically simple substrates to the soil usually leads to a more efficient production of biomass, incorporation into the mineral-associated SOM pool, and to the accumulation of stable C in soils (Cotrufo et al. 2013; Creamer et al. 2016). Besides the investment in extracellular enzymes to take up substrates or to counteract shortage in C or nutrients, substrate chemistry itself can affect CUE. Different monomeric substances yield varying amounts of energy for microbes (LaRowe and Van Cappellen 2011), and different substrates have been shown to be preferentially used for energy or growth (Manzoni et al. 2012). The relative abundances of individual substrates will affect the overall soil microbial community CUE (Geyer et al. 2016).

N input: Along with land use change, the nutrient availability for microbes in soils will change. Herbaceous plants have lower C:N ratios than woody species (Taylor et al. 1989). Additionally, conversion of forests into managed grasslands and agricultural systems often includes the introduction of fertilizer. On top of that, atmospheric N-deposition is increasing, especially in industrialized areas, and this changes nutrient availability in soils (Galloway et al. 2008).

Nutrient availability (e.g., N and phosphorus) has been shown to influence CUE directly (Spohn et al. 2016). Microbes need C and nutrients to grow and to sustain their physiological needs. There is, however, a discrepancy between microbial C:nutrient ratios and the C:nutrient ratio of the microbial substrate in soils (Zechmeister-Boltenstern et al. 2015). Globally, soil microorganisms have an average C:N ratio of 7 while their main substrates have much higher C:N ratios. Leaf litter for instance has a C:N ratio of 71 and SOM of 17 (Mooshammer et al. 2014). This results in an excess of C compared to N for microbial growth. Various mechanisms have been proposed to explain how microbes deal with the imbalance between microbial biomass C:N and soil C:N (Mooshammer et al. 2014; Zechmeister-Boltenstern et al. 2015). Many of these mechanisms directly or indirectly affect CUE. In the case of high C:N ratios, microbes can respire the excess C (overflow respiration), which decreases CUE (Manzoni et al. 2012), or they can invest in N acquiring enzymes, which is a C-, N- and energy-intensive process which also reduces CUE (Sinsabaugh et al. 2013). Microbes can also increase the speed of N cycling from microbial necromass, which decreases the apparent C:N ratio of the remaining substrate pool (Kaiser et al. 2014). In soils where N is scarce, additional N can increase microbial growth, which also increases CUE. Excess N will have fewer effects on C dynamics, but an increase in N mineralization can lead to increased losses of N from the system. Each one of these potential changes by itself might have strong effects on the CUE of a given soil microbial community. These changes, however, are expected to happen in concert. Changes in temperature, nutrient availability and substrate chemistry can thus, depending on the system, resonate or counteract each other.

Adaptations of the microbial community: Changes in environmental parameters, substrate chemistry and nutrient availability can be buffered or completely counteracted by physiological adjustments of the microbial activity or by changes in the microbial community composition (Kaiser et al. 2014). As discussed before, microbes can invest in extracellular enzymes to counteract shortages of specific nutrients (Sinsabaugh et al. 2013). Different microbial groups may also have different biomass C:N ratios (e.g., bacteria vs. fungi), which results in altered C:N demand (Keiblinger et al. 2010; Manzoni et al. 2012). Greater C:N ratio and longer residence time for C associated with fungi compared to bacterial biomass has been related to high CUE of fungal communities compared to bacterial biomass, despite the fact that available information is not yet sufficient to derive robust conclusions (Strickland and Rousk 2010). Shifts in the microbial community composition in favor of one of these groups might shift the absolute nutrient demand and the CUE of the microbial community. Different communities may also have different specific substrate use efficiencies, and some microbial groups might be better adapted to complex substrates (Keiblinger et al. 2010; Kallenbach et al. 2016).

The microbial communities in soil respond to changes in substrate chemistry and N availability by shifts in community composition, and also adaptations of their extracellular enzymes (Kaiser et al. 2010) and physiology (Koranda et al. 2013). These adjustments could result in a stabilization of CUE. Interestingly, most studies measuring CUE in response to soil warming have shown either no adaptation (Schindlbacher et al. 2015), or late adaptation after several years of continuous warming. These adaptations are suspected to be connected to changes in substrate availability or chemistry (Frey et al. 2013).

Currently, our knowledge about in situ CUE dynamics and how the combined controls, their interactions, and the responses of the microbial community influence CUE is rather limited. This makes it hard to predict the dynamics of CUE under a future climate. However, as a central parameter of microbial physiology that describes the efficient transformation of plant compounds into microbial biomass, a high CUE is the first step to efficiently accumulate stable SOM.

2.4.4 Environmental Factors Promoting CO_2 Release

The accumulation of stable SOM is influenced by the CUE of soil organisms, which also have a strong impact on the net release of CO_2 from soils. Changes in climatic conditions are known to affect the C balance of soils, although the magnitude of ecosystem responses to higher atmospheric CO_2 concentration, continuous warming, and drought or flood extremes are difficult to predict (Heimann and Reichstein 2008). Plants and soil microbes are adaptable to changes in environmental conditions, and microbes in particular possess the ability to overcome unfavorable conditions (Evans and Wallenstein 2014; Reinsch et al. 2014). In addition to changes in climate, land use changes are highly disruptive and pose strong disturbances to the soil system: they may completely change the soil physical and chemical environment (Bronick and Lal 2005) to which plants and soil microbes then have to adapt. Additionally, changes in N input in the form of fertilizer or by means of enhanced atmospheric N deposition alter ecosystem nutrient availability and potentially decrease microbial biomass (Treseder 2008).

2.4.4.1 Climate

Several scenarios predict increasing atmospheric temperatures, rising atmospheric CO_2 concentrations, changes in precipitation patterns, and more frequent occurrence of extreme weather events in the future (IPCC 2014). Consequently, changing climatic conditions will alter processes involved in the terrestrial C cycling and may result in changed ecosystem C fluxes. CO_2 enrichment further accelerates inputs of CO_2 into the atmosphere (Hungate et al. 1998; Selsted et al. 2012).

Elevated atmospheric CO_2 concentrations increase C turnover rates by raising CO_2 assimilation efficiency, increasing aboveground and belowground plant biomass, and stimulating respiratory processes (King et al. 2004; Wan et al. 2007; Selsted et al. 2012). Carney et al. (2007) observed a decrease in soil C stocks after four years of elevated CO_2 exposure in a shrub-oak forest, which is in agreement with earlier modeling results (Amundson 2001). However, the effects of increased atmospheric CO_2 concentrations are still controversial, and elevated atmospheric CO_2 concentrations do not necessarily result in soil C loss (Norby and Zak 2011).

Increased temperatures generally accelerate C turnover by extending plant growing seasons, which are closely linked with belowground C allocation processes, especially in cold ecosystems (Luo 2007; Heimann and Reichstein 2008). Recent findings show that warming may decrease the topsoil C stocks in high-latitude areas by ~13% per 1°C warming (Crowther et al. 2016), likely due to the high temperature sensitivity of microbial processes at lower temperature ranges (Karhu et al. 2014; Alster et al. 2016).

Future precipitation patterns are likely to lead to more heavy rain events alternating with longer drought periods (IPCC 2014). Whereas as warming generally has an accelerating effect on the C cycling, drought periods or heat waves can decrease total terrestrial C turnover (Ciais et al. 2005; Luo 2007; Heimann and Reichstein 2008). Extended drought periods decrease belowground C turnover (e.g., Reinsch et al. 2014), decrease soil respiration (e.g., Selsted et al. 2012; Wang et al. 2014), decrease soil water availability (Goebel et al. 2011; Robinson et al. 2016), and can change soil properties (e.g., Robinson et al. 2016). In contrast, heavy rain events can lead to flooding, resulting in anoxic soil conditions and reduced gas transport within the soil profile. This may result in the death of roots and associated soil microbes and animals, leading to lower microbial C use (Keiluweit et al. 2016). However, the effects of drought or flooding, as well as warming, on soil microbial processes depend on baseline ecosystem properties (Moyano et al. 2013) and ecosystem integrity (Schmidt et al. 2011).

2.4.4.2 Land Use

Different land use types such as forests, grasslands, and agricultural land foster different microbial communities (e.g., Drenovsky et al. 2010; Lauber et al. 2013), which may perform different functions. The conversion from forest to grasslands to agricultural fields will change the microbial community due to changes in soil structure and organic matter input. Drenovsky et al. (2010) observed that dry soils favor Gram-negative bacteria and fungi whereas wet soils were dominated by Gram-positive, anaerobic and sulphate-reducing bacteria. Microbial communities in agricultural soils are more sensitive to soil moisture and temperature than those in grassland soils (Lauber et al. 2013), potentially due to the disruption of fungal hyphae by agricultural practice (Drenovsky et al. 2010). Soil tillage also disrupts the soil structure, breaking down bigger soil aggregates and thereby releasing formerly protected SOM for microbial utilization (Six et al. 1998; Islam and Weil 2000). Further organic matter loss from agricultural soils occurs with the removal of litter and plant harvest, which decreases microbial biomass C and SOM content (Schlesinger and Andrews 2000; Geraei et al. 2016). Interestingly, Jangid et al. (2011) suggest that land use history (not land use per se) is a stronger determinant of soil microbial communities than vegetation and soil properties. However, the sequence of land use types determines soil properties, which in turn can be good predictors of a soil's microbial community (e.g., Lauber et al. 2008; Kuramae et al. 2012).

2.4.4.3 N Input

Terrestrial N input, either in the form of fertilization or atmospheric N deposition, changes the nutrient pool of ecosystems. Plant production is increased with N fertilization (if other nutrients are not limiting), which consequently affects the soil microbial community (Farrer and Suding 2016). Especially the abundance of mycorrhizal fungi is often reduced by N fertilization (Egerton-Warburton and Allen 2000; Boot et al. 2016), reducing the plant-mycorrhizal exchange of C and nutrients. This could lead to an increase of SOM priming and a reduction of CUE, thereby increasing the plants' C investment in exchange for soil nutrients. Zhang et al. (2014) observed different responses to simulated N deposition, with soil microbial respiration being more sensitive to

N additions than root respiration. Additionally, the competition of plants and microbes for N may increase with increased plant biomass production due to fertilization (Farrer and Suding 2016), which can lead to a decrease in microbial biomass (e.g., Boot et al. 2016) and a reduction of soil respiration by microbes (Lagomarsino et al. 2006). Treseder (2008) showed in meta-analysis across biomes that microbial biomass decreased by ~15% with increased N input. In agreement, a more recent meta-analysis suggests that mineral N additions decrease microbial biomass by 12% in grasslands, while increasing microbial biomass by 13.6% in annual cropping systems (Geisseler et al. 2016).

Increased soil N concentrations can reduce plant–microbial competition for N, which would lead to a decrease in SOM priming, and reduced degradation of SOM and thus a reduction in soil respiration per unit microbial biomass, which would lead to higher SOM stocks. However, this is only true if N and no other nutrients (e.g., phosphorus) or water are limiting. Excessive N can, in turn, reduce the soil pH and may have a negative effect on the soil microbial community, thereby decreasing soil respiration (e.g., Bowden et al. 2004).

2.5   SUMMARY AND CONCLUSIONS

Rising temperatures and irregular weather conditions are threatening human health, crop production and natural ecosystems. Climate is influenced by CO_2 emissions, not only in the course of fossil fuel combustion for energy and vehicles, but also by land use and management. Changes in climate affect vegetation performance and dynamics, with direct and indirect effects on soil processes. Soil is the living environment for a multitude of microorganisms which thrive belowground hidden from sight. Soil microorganisms are constantly processing organic matter, thereby releasing nutrients essential for plant growth. As a side effect of a multitude of soil microbial processes the major greenhouse gases (GHG) responsible for climate change, namely N_2O, CH_4 and CO_2 can be produced and released to the atmosphere. In our text we reveal the microbial processes which may be responsible for the global increase in GHG concentrations and we highlight the soil systems holding environmental conditions conducive to significant net GHG release (Table 2.1). Further, we have summarized existing information at the global scale in an exercise to show the overall importance of soils as emitters of GHG at the terrestrial scale by comparing them with anthropogenic and other natural sources (mainly oceanic systems). Roughly one third of the CH_4 emissions, and half of the total CO_2, and and N_2O emitted worldwide, respectively, is soil-related, which translates into nearly 50% of all the global warming potential produced across the globe (including anthropogenic sources and natural sources in terrestrial and oceanic systems) having its origin in the soils (Table 2.2). However, soils are the basis of plant growth, and hence sustain photosynthesis, which is the most important global C sink. Furthermore, given specific environmental conditions, some soil microorganisms are also able to take up GHG and convert them to climate-neutral products (Table 2.1). Often, production and uptake occur simultaneously and hence the exchange of gas fluxes between soil and atmosphere is the net result of a variety of microbial activities, thus while being massive GHG emitters, the soils may be also holding the key for strongly reducing the GHG footprint of terrestrial ecosystems.

The overall picture is even more complicated if we consider that there is usually a trade-off between the net balance of three main GHGs, a phenomenon usually addressed as "pollution swapping." For example, mitigation measures aiming at reducing soil CH_4 emissions by disrupting anaerobic conditions may dramatically trigger soil N_2O emissions due to incomplete denitrification and/or deplete soil C stocks due to aerobic decomposition of SOM. In the same way, addition of N to enhance soil C sequestration may have the undesired consequence of enhancing soil N_2O fluxes due to increasing availability of N for microbial nitrification and denitrification. We therefore suggest that climate change mitigation measures should carefully take into account the consequences of the management actions on the overall balance of the three GHG and their interactions.

TABLE 2.1

Processes That Lead to Release or Uptake of Greenhouse Gases (GHGs) and Examples of Soil Systems, Which Are Hot Spots for the Respective Processes

GHG	Release	Uptake
CO_2	Autotrophic and heterotrophic soil respiration (recently cultivated soils)	Stabilization of carbon in soil organic matter (SOM) (set aside soils)
CH_4	Methanogenesis (wetlands, paddy fields)	Methanotrophy (upland soils, especially forests)
N_2O	Denitrification and nitrification (drained and fertilized peatland, cropland, fertilized/ grazed grassland)	N_2O reduction (waterlogged soils, very dry soils)

TABLE 2.2

Terrestrial Sources of Greenhouse Gases and the Explicit Contribution of Soils

Greenhouse Gas	Soils	Anthropogenic Sources (Excluding Soils*)	Natural Sources (Excluding Soils)	TOTAL	Soil Contribution to Terrestrial GHG Emissions (%)	References
CO_2 (Pg C a^{-1})	98	11.2	100.5	210	47	a, b, c
CH_4 (Tg C a^{-1})	191	221	97	509	38	d
N_2O (Tg N a^{-1})	9.3**	5.5	4.0	18.8	49	e
Total GWP*** (Pg CO_2-eq a^{-1})	379	53	377	809	47	

Estimates sources: (a) Bond-Lamberty and Thomson (2010); (b) Le Quéré et al. (2016); (c) Ciais et al. (2013); (d) Saunois et al. (2016); (e) (Syakila and Kroeze (2011).

* CH_4 emissions from rice cultivation and N_2O emissions from agricultural soils were excluded from anthropogenic sources.
** Includes indirect N_2O emissions due to fertilization.
*** Assuming a global warming potential of 34 and 298 for CH_4 and N_2O, respectively (Myhre et al. 2013).

If climate change results in warmer and wetter soil conditions—as predicted for large parts of the globe—stimulation of microbial activities may increase the strength of soils as source of GHG emissions. Hence soil organisms may not only be affected by climate change, but may also trigger a positive feedback to climate change. However, the overall impact strongly depends on human management of soil, especially land use and application of fertilizer. The best way to minimize losses of hazardous C and N compounds to the atmosphere is to promote plant growth to the same extent as microbial decomposer activity and to recycle plant material to the soil, so that the global soil pools of C and N are constantly replenished. In this way, the capacity of soils as GHG sinks will be strengthened while its capacity as GHG source will decrease. If human society manages to increase soil C and decrease soil GHG emissions in a careful way, soil microbes could—in addition to reductions of GHG emissions from fossil fuels—help to reduce GHG concentrations in the atmosphere and stabilize the climate.

ACKNOWLEDGMENTS

We thank Rattan Lal for his encouragement to write this manuscript and his patience and support. Christine Gritsch and Jutta Grabenhofer contributed significantly to this book chapter by agreeing that parts of their thesis introductions could be adapted and included. Sonja Leitner worked on layout and referencing and together with Sue Grayston provided an internal review of the manuscript. We thank Johanna Kohl, BFW, Vienna, for designing Figures 2.8–2.10 after a concept of Sabine Reinsch. The writing was supported by work and material from the research projects DRAIN (ACRP KR13AC6K11008), AXA (2012-Doc-University of Natural Resources and Life Sciences-LEITNER S), NitroAustria (ACRP KR14AC7K11916), INFOSOM (FWF P25438), ECLAIRE (FP7-Env.2011.1.1.2-1), and ExtremeGrass (ACRP KR15AC8K12624).

REFERENCES

Adamsen, A.P.S. and King, G.M. (1993) Methane consumption in temperate and subarctic forest soils: Rates, vertical zonation, and responses to water and nitrogen. *Appl. Environ. Microbiol.* **59**: 485–490.

Allen, D.E., Dalal, R.C., Rennenberg, H., Meyer, R.L., Reeves, S., and Schmidt, S. (2007) Spatial and temporal variation of nitrous oxide and methane flux between subtropical mangrove sediments and the atmosphere. *Soil Biol. Biochem.* **39**: 622–631.

Alster, C.J., Koyama, A., Johnson, N.G., Wallenstein, M.D., and von Fischer, J.C. (2016) Temperature sensitivity of soil microbial communities: An application of macromolecular rate theory to microbial respiration. *J. Geophys. Res. Biogeosciences.* **121**: 1420–1433.

Amaral, J.A., Ren, T., and Knowles, R. (1998) Atmospheric methane consumption by forest soils and extracted bacteria at different pH values. *Appl. Environ. Microbiol.* **64**: 2397–2402.

Ambus, P., Zechmeister-Boltenstern, S., and Butterbach-Bahl, K. (2006) Sources of nitrous oxide emitted from European forest soils. *Biogeosciences.* **3**: 135–145.

Amundson, R. (2001) The carbon budget in soils. *Annu. Rev. Earth Planet. Sci.* **29**: 535–562.

Anderson, I.C. and Levine, J.S. (1986) Relative rates of nitric oxide and nitrous oxide production by nitrifiers, denitrifiers, and nitrate respirers. *Appl. Environ. Microbiol.* **51**: 938–945.

Angel, R., Claus, P., and Conrad, R. (2012) Methanogenic archaea are globally ubiquitous in aerated soils and become active under wet anoxic conditions. *ISME J.* **6**: 847–862.

Angel, R., Matthies, D., and Conrad, R. (2011) Activation of methanogenesis in arid biological soil crusts despite the presence of oxygen. *PLoS One* **6**: e20453.

Aronson, E.L. and Helliker, B.R. (2010) Methane flux in non-wetland soils in response to nitrogen addition: A meta-analysis. *Ecology.* **91**: 3242–3251.

Aronson, E.L., Allison, S.D., and Helliker, B.R. (2013) Environmental impacts on the diversity of methane-cycling microbes and their resultant function. *Front. Microbiol.* **4**: 225.

Aronson, E.L., Vann, D.R., and Helliker, B.R. (2012) Methane flux response to nitrogen amendment in an upland pine forest soil and riparian zone. *J. Geophys. Res. Biogeosciences.* **117**: G03012.

Aschenbach, K., Conrad, R., Řeháková, K., Doležal, J., Janatková, K., and Angel, R. (2013) Methanogens at the top of the world: Occurrence and potential activity of methanogens in newly deglaciated soils in high-altitude cold deserts in the Western Himalayas. *Front. Microbiol.* **4**: 359.

Asner, G.P., Seastedt, T.R., and Townsend, A.R. (1997) The decoupling of terrestrial carbon and nitrogen cycles. *BioScience.* **47**: 226–234.

Auman, A.J., Speake, C.C., and Lidstrom, M.E. (2001) nifH sequences and nitrogen fixation in type I and type II methanotrophs. *Appl. Environ. Microbiol.* **67**: 4009–4016.

Avrahami, S. and Bohannan, B.J.M. (2007) Response of Nitrosospira sp. strain AF-like ammonia oxidizers to changes in temperature, soil moisture content, and fertilizer concentration. *Appl. Environ. Microbiol.* **73**: 1166–1173.

Baggs, E.M. (2008) A review of stable isotope techniques for N$_2$O source partitioning in soils: Recent progress, remaining challenges and future considerations. *Rapid Commun. Mass Spectrom.* **22**: 1664–1672.

Bárcena, T.G., Finster, K.W., and Yde, J.C. (2011) Spatial patterns of soil development, methane oxidation, and methanotrophic diversity along a receding glacier forefield, Southeast Greenland. *Arct. Antarct. Alp. Res.* **43**: 178–188.

Bardgett, R.D., Freeman, C., and Ostle, N.J. (2008) Microbial contributions to climate change through carbon cycle feedbacks. *ISME J.* **2**: 805–814.

Barnard, R., Le Roux, X., Hungate, B.A., Cleland, E.E., Blankinship, J.C., Barthes, L., and Leadley, P.W. (2006) Several components of global change alter nitrifying and denitrifying activities in an annual grassland. *Funct. Ecol.* **20**: 557–564.

Barnard, R., Leadley, P.W., and Hungate, B.A. (2005) Global change, nitrification, and denitrification: A review. *Glob. Biogeochem. Cycles.* **19**: GB1007.

Barton, L., McLay, C.D.A., Schipper, L.A., and Smith, C.T. (1999) Annual denitrification rates in agricultural and forest soils: A review. *Aust. J. Soil Res.* **37**: 1073–1094.

Bender, M. and Conrad, R. (1992) Kinetics of CH_4 oxidation in oxic soils exposed to ambient air or high CH4 mixing ratios. *FEMS Microbiol. Lett.* **101**: 261–269.

Blagodatskaya, E. and Kuzyakov, Y. (2008) Mechanisms of real and apparent priming effects and their dependence on soil microbial biomass and community structure: Critical review. *Biol. Fertil. Soils.* **45**: 115–131.

Bleeker, A., Hicks, W.K., Dentener, F., Galloway, J., and Erisman, J.W. (2011) N deposition as a threat to the world's protected areas under the Convention on Biological Diversity. *Environ. Pollut.* **159**: 2280–2288.

Bocock, K.L. (1964) Changes in the amounts of dry matter, nitrogen, carbon and energy in decomposing woodland leaf litter in relation to the activities of the soil fauna. *J. Ecol.* **52**: 273–284.

Bodelier, P.L. (2011) Interactions between nitrogenous fertilizers and methane cycling in wetland and upland soils. *Curr. Opin. Environ. Sustain.* **3**: 379–388.

Bodelier, P.L. and Laanbroek, H.J. (2004) Nitrogen as a regulatory factor of methane oxidation in soils and sediments. *FEMS Microbiol. Ecol.* **47**: 265–277.

Bond-Lamberty, B. and Thomson, A. (2010) Temperature-associated increases in the global soil respiration record. *Nature.* **464**: 579–582.

Boot, C.M., Hall, E.K., Denef, K., and Baron, J.S. (2016) Long-term reactive nitrogen loading alters soil carbon and microbial community properties in a subalpine forest ecosystem. *Soil Biol. Biochem.* **92**: 211–220.

Borken, W. and Matzner, E. (2009) Reappraisal of drying and wetting effects on C and N mineralization and fluxes in soils. *Glob. Change Biol.* **15**: 808–824.

Born, M., Dorr, H., and Levin, I. (1990) Methane consumption in aerated soils of the temperate zone. *Tellus B.* **42**: 2–8.

Bowden, R.D., Davidson, E., Savage, K., Arabia, C., and Steudler, P. (2004) Chronic nitrogen additions reduce total soil respiration and microbial respiration in temperate forest soils at the Harvard Forest. *For. Ecol. Manag.* **196**: 43–56.

Bowden, R.D., Melillo, J.M., Steudler, P.A., and Aber, J.D. (1991) Effects of nitrogen additions on annual nitrous oxide fluxes from temperate forest soils in the northeastern United States. *J. Geophys. Res.* **96**: 9321.

Bowling, D.R., Pataki, D.E., and Randerson, J.T. (2008) Carbon isotopes in terrestrial ecosystem pools and CO_2 fluxes. *New Phytol.* **178**: 24–40.

Bradford, M.A., Keiser, A.D., Davies, C.A., Mersmann, C.A., and Strickland, M.S. (2013) Empirical evidence that soil carbon formation from plant inputs is positively related to microbial growth. *Biogeochemistry.* **113**: 271–281.

Bräuer, S.L., Cadillo-Quiroz, H., Ward, R.J., Yavitt, J.B., and Zinder, S.H. (2011) *Methanoregula boonei* gen. nov., sp. nov., an acidiphilic methanogen isolated from an acidic peat bog. *Int. J. Syst. Evol. Microbiol.* **61**: 45–52.

Bräuer, S.L., Cadillo-Quiroz, H., Yashiro, E., Yavitt, J.B., and Zinder, S.H. (2006) Isolation of a novel acidiphilic methanogen from an acidic peat bog. *Nature.* **442**: 192–194.

Bräuer, S.L., Yashiro, E., Ueno, N.G., Yavitt, J.B., and Zinder, S.H. (2006) Characterization of acid-tolerant H_2/CO_2-utilizing methanogenic enrichment cultures from an acidic peat bog in New York State. *FEMS Microbiol. Ecol.* **57**: 206–216.

Bronick, C.J. and Lal, R. (2005) Soil structure and management: A review. *Geoderma.* **124**: 3–22.

Brüggemann, N., Gessler, A., Kayler, Z., Keel, S.G., Badeck, F., Barthel, M. et al. (2011) Carbon allocation and carbon isotope fluxes in the plant-soil-atmosphere continuum: A review. *Biogeosciences.* **8**: 3457–3489.

Bühlmann, T., Hiltbrunner, E., Körner, C., Rihm, B., and Achermann, B. (2015) Induction of indirect N2O and NO emissions by atmospheric nitrogen deposition in (semi-)natural ecosystems in Switzerland. *Atmos. Environ.* **103**: 94–101.

Butterbach-Bahl, K. and Dannenmann, M. (2011) Denitrification and associated soil N_2O emissions due to agricultural activities in a changing climate. *Curr. Opin. Environ. Sustain.* **3**: 389–395.

Butterbach-Bahl, K. and Kiese, R. (2005) Significance of forests as sources for N_2O and NO. In *Tree Species Effects on Soils: Implications for Global Change: Proceedings of the NATO Advanced Research Workshop on Trees and Soil Interactions, Implications to Global Climate Change August 2004 Krasnoyarsk, Russia*, ed. B. Binkley, and O. Menyailo, 173–191. Dordrecht: Springer Netherlands.

Butterbach-Bahl, K., Baggs, E.M., Dannenmann, M., Kiese, R., and Zechmeister-Boltenstern, S. (2013) Nitrous oxide emissions from soils: How well do we understand the processes and their controls? *Philos. Trans. R. Soc. B Biol. Sci.* **368**: 20130122.

Butterbach-Bahl, K., Breuer, L., Gasche, R., Willibald, G., and Papen, H. (2002) Exchange of trace gases between soils and the atmosphere in Scots pine forest ecosystems of the northeastern German lowlands. *For. Ecol. Manag.* **167**: 123–134.

Butterbach-Bahl, K., Díaz-Pinés, E., and Dannenmann, M. (2012) Soil trace gas emissions and climate change. In *Global Environmental Change. Handbook of Global Environmental Pollution*, ed. B. Freedman, 325–334. Dordrecht: Springer Netherlands.

Butterbach-Bahl, K., Kock, M., Willibald, G., Hewett, B., Buhagiar, S., Papen, H., and Kiese, R. (2004) Temporal variations of fluxes of NO, NO_2, N_2O, CO_2 and CH_4 in a tropical rain forest ecosystem. *Glob. Biogeochem. Cycles.* **18**: GB3012.

Cadillo-Quiroz, H., Bräuer, S.L., Goodson, N., Yavitt, J.B., and Zinder, S.H. (2014) *Methanobacterium paludis* sp. nov. and a novel strain of *Methanobacterium lacus* isolated from northern peatlands. *Int. J. Syst. Evol. Microbiol.* **64**: 1473–1480.

Cadillo-Quiroz, H., Yavitt, J.B., and Zinder, S.H. (2009) *Methanosphaerula palustris* gen. nov., sp. nov., a hydrogenotrophic methanogen isolated from a minerotrophic fen peatland. *Int. J. Syst. Evol. Microbiol.* **59**: 928–935.

Cappenberg, T.E. (1974) Interrelations between sulfate-reducing and methane-producing bacteria in bottom deposits of a fresh-water lake. I. Field observations. *Antonie Van Leeuwenhoek.* **40**: 285–295.

Carey, J.C., Tang, J., Templer, P.H., Kroeger, K.D., Crowther, T.W., Burton, A.J. et al. (2016) Temperature response of soil respiration largely unaltered with experimental warming. *Proc. Natl. Acad. Sci.* **113**: 13797–13802.

Carney, K.M., Hungate, B.A., Drake, B.G., and Megonigal, J.P. (2007) Altered soil microbial community at elevated CO_2 leads to loss of soil carbon. *Proc. Natl. Acad. Sci.* **104**: 4990–4995.

Castaldi, S. and Tedesco, D. (2005) Methane production and consumption in an active volcanic environment of Southern Italy. *Chemosphere.* **58**: 131–139.

Castro, M.S., Melillo, J.M., Steudler, P.A., and Chapman, J.W. (1994) Soil moisture as a predictor of methane uptake by temperate forest soils. *Can. J. For. Res.* **24**: 1805–1810.

Castro, M.S., Steudler, P.A., Melillo, J.M., Aber, J.D., and Bowden, R.D. (1995) Factors controlling atmospheric methane consumption by temperate forest soils. *Glob. Biogeochem. Cycles.* **9**: 1–10.

Castro, M.S., Steudler, P.A., Melillo, J.M., Aber, J.D., and Millham, S. (1992) Exchange of N_2O and CH_4 between the atmosphere and soils in spruce-fir forests in the northeastern United States. *Biogeochemistry.* **18**: 119–135.

Chapuis-Lardy, L., Wrage, N., Metay, A., Chotte, J.-L., and Bernoux, M. (2007) Soils, a sink for N_2O? A review. *Glob. Change Biol.* **13**: 1–17.

Chen, W., Wolf, B., Zheng, X., Yao, Z., Butterbach-Bahl, K., Brüggemann, N. et al. (2011) Annual methane uptake by temperate semiarid steppes as regulated by stocking rates, aboveground plant biomass and topsoil air permeability. *Glob. Change Biol.* **17**: 2803–2816.

Cheng, H., and Wang, L. (2013) Lignocelluloses Feedstock biorefinery as petrorefinery substitutes. In *Biomass Now—Sustainable Growth and Use*, ed. M.D. Matovic, 347–388. Rijeka: InTech.

Christiansen, J.R., Levy-Booth, D., Prescott, C.E., and Grayston, S.J. (2016) Microbial and environmental controls of methane fluxes along a soil moisture gradient in a Pacific Coastal temperate rainforest. *Ecosystems.* **19**: 1255–1270.

Christiansen, J.R., Levy-Booth, D., Prescott, C.E., and Grayston, S. (2017) Different soil moisture control of net methane oxidation and production in organic upland and wet forest soils of the Pacific coastal rainforest in Canada. *Can. J. For. Res.* D.O.I. 1139/cjfr-2016-0390.

Ciais, P., Reichstein, M., Viovy, N., Granier, A., Ogée, J., Allard, V. et al. (2005) Europe-wide reduction in primary productivity caused by the heat and drought in 2003. *Nature.* **437**: 529–533.

Ciais, P., Sabine, C., Bala, G., Bopp, L., Brovkin, V., Canadell, J.G. et al. (2013) Carbon and other biogeochemical cycles. In *Climate Change 2013: The Physical Science Basis. Contribution of Working Group I to the Fifth Assessment Report of the Intergovernmental Panel on Climate Change*, ed. T. Stocker, D. Qin, G. Plattner, M. Tignor, S. Allen, J. Boschung et al., 465–570. Cambridge and New York: Cambridge University Press.

Cicerone, R.J. and Oremland, R.S. (1988) Biogeochemical aspects of atmospheric methane. *Glob. Biogeochem. Cycles.* **2**: 299–327.

Conrad, R. (1996) Soil microorganisms as controllers of atmospheric trace gases (H_2, CO, CH_4, OCS, N_2O, and NO). *Am. Soc. Microbiol.* **60**: 609–640.

Conrad, R. (2009) The global methane cycle: Recent advances in understanding the microbial processes involved. *Environ. Microbiol. Rep.* **1**: 285–292.

Cotrufo, M.F., Wallenstein, M.D., Boot, C.M., Denef, K., and Paul, E. (2013) The Microbial Efficiency-Matrix Stabilization (MEMS) framework integrates plant litter decomposition with soil organic matter stabilization: Do labile plant inputs form stable soil organic matter? *Glob. Change Biol.* **19**: 988–995.

Cowan, N.J., Famulari, D., Levy, P.E., Anderson, M., Reay, D.S., and Skiba, U.M. (2014) Investigating uptake of N_2O in agricultural soils using a high-precision dynamic chamber method. *Atmos Meas Tech.* **7**: 4455–4462.

Cowan, N.J., Levy, P.E., Famulari, D., Anderson, M., Drewer, J., Carozzi, M. et al. (2016) The influence of tillage on N_2O fluxes from an intensively managed grazed grassland in Scotland. *Biogeosciences.* **13**: 4811–4821.

Creamer, C.A., Jones, D.L., Baldock, J.A., Rui, Y., Murphy, D.V., Hoyle, F.C., and Farrell, M. (2016) Is the fate of glucose-derived carbon more strongly driven by nutrient availability, soil texture, or microbial biomass size? *Soil Biol. Biochem.* **103**: 201–212.

Crowther, T.W., Todd-Brown, K.E.O., Rowe, C.W., Wieder, W.R., Carey, J.C., Machmuller, M.B. et al. (2016) Quantifying global soil carbon losses in response to warming. *Nature.* **540**: 104–108.

Crutzen, P. and Lelieveld, J. (2001) Human impacts on atmospheric chemistry. *Annu. Rev. Earth Planet. Sci.* **29**: 17–45.

Daims, H., Lebedeva, E.V., Pjevac, P., Han, P., Herbold, C., Albertsen, M. et al. (2015) Complete nitrification by Nitrospira bacteria. *Nature.* **528**: 504–509.

Davidson, E.A. (1992) Sources of nitric oxide and nitrous oxide following wetting of dry soil. *Soil Sci. Soc. Am. J.* **56**: 95–102.

Davidson, E.A. and Janssens, I.A. (2006) Temperature sensitivity of soil carbon decomposition and feedbacks to climate change. *Nature.* **440**: 165–173.

Davidson, E.A. and Kanter, D. (2014) Inventories and scenarios of nitrous oxide emissions. *Environ. Res. Lett.* **9**: 105012.

Davidson, E.A., Belk, E., and Boone, R.D. (1998) Soil water content and temperature as independent or confounded factors controlling soil respiration in a temperate mixed hardwood forest. *Glob. Change Biol.* **4**: 217–227.

Davidson, E.A., Keller, M., Erickson, H.E., Verchot, L.V., and Veldkamp, E. (2000) Testing a conceptual model of soil emissions of nitrous and nitric oxides. *BioScience.* **50**: 667–680.

De Deyn, G.B., Cornelissen, J.H.C., and Bardgett, R.D. (2008) Plant functional traits and soil carbon sequestration in contrasting biomes. *Ecol. Lett.* **11**: 516–531.

Denman, K.L., Brasseur, A., Chidthaisong, A., Ciais, P., Cox, P.M., Dickinson, R.E. et al. (2007) Couplings between changes in the climate system and biogeochemistry. In *Climate Change 2007: The Physical Science Basis. Contribution of Working Group I to the Fourth Assessment Report of the Intergovernmental Panel on Climate Change*, ed. S. Solomon, D. Qin, M. Manning, Z. Chen, M. Marquis, K. Averyt et al. 499–587. Cambridge and New York: Cambridge University Press.

Díaz-Pinés, E., Schindlbacher, A., Godino, M., Kitzler, B., Jandl, R., Zechmeister-Boltenstern, S., and Rubio, A. (2014) Effects of tree species composition on the CO_2 and N_2O efflux of a Mediterranean mountain forest soil. *Plant Soil.* **384**: 243–257.

Dick, J., Skiba, U., Munro, R., and Deans, D. (2006) Effect of N-fixing and non N-fixing trees and crops on NO and N_2O emissions from Senegalese soils. *J. Biogeogr.* **33**: 416–423.

Dijkstra, F.A., Blumenthal, D., Morgan, J.A., Pendall, E., Carrillo, Y., and Follett, R.F. (2010) Contrasting effects of elevated CO_2 and warming on nitrogen cycling in a semiarid grassland. *New Phytol.* **187**: 426–437.

Dijkstra, F.A., Carrillo, Y., Pendall, E., and Morgan, J.A. (2013) Rhizosphere priming: A nutrient perspective. *Front. Microbiol.* **4**: 216.

Doetterl, S., Stevens, A., Six, J., Merckx, R., Van Oost, K., Casanova Pinto, M. et al. (2015) Soil carbon storage controlled by interactions between geochemistry and climate. *Nat. Geosci.* **8**: 780–783.

Donoso, L., Santana, R., and Sanhueza, E. (1993) Seasonal variation of N_2O fluxes at a tropical savannah site: Soil consumption of N_2O during the dry season. *Geophys. Res. Lett.* **20**: 1379–1382.

Dörr, H., Katruff, L., and Levin, I. (1993) Soil texture parameterization of the methane uptake in aerated soils. *Chemosphere* **26**: 697–713.

Drenovsky, R.E., Steenwerth, K.L., Jackson, L.E., and Scow, K.M. (2010) Land use and climatic factors structure regional patterns in soil microbial communities. *Glob. Ecol. Biogeogr.* **19**: 27–39.

Dunfield, P.F. and Knowles, R. (1995) Kinetics of inhibition of methane oxidation by nitrate, nitrite, and ammonium in a humisol. *Appl. Environ. Microbiol.* **61**: 3129–3135.

Dunfield, P.F., Yuryev, A., Senin, P., Smirnova, A.V., Stott, M.B., Hou, S. et al. (2007) Methane oxidation by an extremely acidophilic bacterium of the phylum Verrucomicrobia. *Nature.* **450**: 879–882.

Egerton-Warburton, L.M. and Allen, E.B. (2000) Shifts in arbuscular mycorrhizal communities along an anthropogenic nitrogen deposition gradient. *Ecol. Appl.* **10**: 484–496.

Ekschmitt, K., Kandeler, E., Poll, C., Brune, A., Buscot, F., Friedrich, M. et al. (2008) Soil-carbon preservation through habitat constraints and biological limitations on decomposer activity. *J. Plant Nutr. Soil Sci.* **171**: 27–35.

Erickson, H., Davidson, E.A., and Keller, M. (2002) Former land-use and tree species affect nitrogen oxide emissions from a tropical dry forest. *Oecologia.* **130**: 297–308.

Evans, S.E. and Wallenstein, M.D. (2014) Climate change alters ecological strategies of soil bacteria. *Ecol. Lett.* **17**: 155–164.

Farrer, E.C. and Suding, K.N. (2016) Teasing apart plant community responses to N enrichment: The roles of resource limitation, competition and soil microbes. *Ecol. Lett.* **19**: 1287–1296.

Firestone, M.K. and Davidson, E.A. (1989) Microbiological basis of NO and N_2O production and consumption in soil. In *Exchange of Trace Gases between Terrestrial Ecosystems and the Atmosphere*, ed. M.O. Andreae, D.S. Schimel, and G.P. Robertson, 7–21. New York: John Wiley and Sons Inc.

Fischer, E.M. and Knutti, R. (2014) Detection of spatially aggregated changes in temperature and precipitation extremes. *Geophys. Res. Lett.* **41**: 547–554.

Fischlin, A., Midgley, G.F., Price, J.T., Leemans, R., Gopal, B., Turley, C. et al (2007) Ecosystems, their properties, goods and services. In *Climate Change 2007: Impacts, Adaptation and Vulnerability. Contribution of Working Group II to the Fourth Assessment Report of the Intergovernmental Panel on Climate Change*, ed. M.L. Parry, O.F. Canziani, J.P. Palutikof, P.J. van der Linden, and C.E. Hanson, 211–272. Cambridge and New York: Cambridge University Press.

Flechard, C.R., Neftel, A., Jocher, M., Ammann, C., and Fuhrer, J. (2005) Bi-directional soil/atmosphere N_2O exchange over two mown grassland systems with contrasting management practices. *Glob. Change Biol.* **11**: 2114–2127.

Flechard, C.R., Nemitz, E., Smith, R.I., Fowler, D., Vermeulen, A.T., Bleeker, A. et al. (2011) Dry deposition of reactive nitrogen to European ecosystems: A comparison of inferential models across the NitroEurope network. *Atmospheric Chem. Phys.* **11**: 2703–2728.

Fontaine, S., Henault, C., Aamor, A., Bdioui, N., Bloor, J.M.G., Maire, V. et al. (2011) Fungi mediate long term sequestration of carbon and nitrogen in soil through their priming effect. *Soil Biol. Biochem.* **43**: 86–96.

Fontaine, S., Mariotti, A., and Abbadie, L. (2003) The priming effect of organic matter: A question of microbial competition? *Soil Biol. Biochem.* **35**: 837–843.

Forster, P., Ramaswamy, V., Artaxo, P., Berntsen, T., Betts, R., Fahey, D.W. et al. (2007) Changes in atmospheric constituents and in radiative forcing. In *Climate Change 2007. The Physical Science Basis,* ed. S. Solomon, D. Qin, M. Manning, Z. Chen, M. Marquis, K.B. Averyt et al. 129–234. Cambridge and New York: Cambridge University Press.

Fowler, D., Coyle, M., Skiba, U., Sutton, M.A., Cape, J.N., Reis, S. et al. (2013) The global nitrogen cycle in the twenty-first century. *Philos. Trans. R. Soc. B Biol. Sci.* **368**: 20130164.

Fowler, D., Pilegaard, K., Sutton, M.A., Ambus, P., Raivonen, M., Duyzer, J. et al. (2009) Atmospheric composition change: Ecosystems–atmosphere interactions. *Atmos. Environ.* **43**: 5193–5267.

Fowler, D., Steadman, C.E., Stevenson, D., Coyle, M., Rees, R.M., Skiba, U.M. et al. (2015) Effects of global change during the 21st century on the nitrogen cycle. *Atmospheric Chem. Phys.* **15**: 13849–13893.

Frey, S.D., Lee, J., Melillo, J.M., and Six, J. (2013) The temperature response of soil microbial efficiency and its feedback to climate. *Nat. Clim. Change.* **3**: 395–398.

Galloway, J.N. (2003) The Global Nitrogen Cycle. In *Treatise on Geochemistry*, ed. H. Holland and K. Turekian, 557–583. Amsterdam: Elsevier Ltd.

Galloway, J.N., Townsend, A.R., Erisman, J.W., Bekunda, M., Cai, Z., Freney, J.R. et al. (2008) Transformation of the nitrogen cycle: Recent trends, questions, and potential solutions. *Science.* **320**: 889–892.

Gambrell, R.P. and Patrick Jr, W.H. (1978) Chemical and microbiological properties of anaerobic soils and sediments. In *Plant Life in Anaerobic Environments*, ed. D.D Hook and R.M.M. Crawford, 375–423. Collingwood, Michigan: Ann Arbor Science Publishers, Inc.

Garcia, J.-L., Patel, B.K., and Ollivier, B. (2000) Taxonomic, phylogenetic, and ecological diversity of methanogenic Archaea. *Anaerobe* **6**: 205–226.

Geisseler, D., Lazicki, P.A., and Scow, K.M. (2016) Mineral nitrogen input decreases microbial biomass in soils under grasslands but not annual crops. *Appl. Soil Ecol.* **106**: 1–10.

Gelfand, I., Cui, M., Tang, J., and Robertson, G.P. (2015) Short-term drought response of N_2O and CO_2 emissions from mesic agricultural soils in the US Midwest. *Agric. Ecosyst. Environ.* **212**: 127–133.

Geraei, D.S., Hojati, S., Landi, A., and Cano, A.F. (2016) Total and labile forms of soil organic carbon as affected by land use change in southwestern Iran. *Geoderma Reg.* **7**: 29–37.

Geyer, K.M., Kyker-Snowman, E., Grandy, A.S., and Frey, S.D. (2016) Microbial carbon use efficiency: Accounting for population, community, and ecosystem-scale controls over the fate of metabolized organic matter. *Biogeochemistry.* **127**: 173–188.

Glatzel, S. and Stahr, K. (2001) Methane and nitrous oxide exchange in differently fertilised grassland in southern Germany. *Plant Soil.* **231**: 21–35.

Gödde, M., and Conrad, R. (1999) Immediate and adaptational temperature effects on nitric oxide production and nitrous oxide release from nitrification and denitrification in two soils. *Biol. Fertil. Soils.* **30**: 33–40.

Goebel, M.-O., Bachmann, J., Reichstein, M., Janssens, I.A., and Guggenberger, G. (2011) Soil water repellency and its implications for organic matter decomposition—Is there a link to extreme climatic events? *Glob. Change Biol.* **17**: 2640–2656.

Goldberg, S.D., and Gebauer, G. (2009) Drought turns a Central European Norway spruce forest soil from an N_2O source to a transient N_2O sink. *Glob. Change Biol.* **15**: 850–860.

Grandy, A.S., and Neff, J.C. (2008) Molecular C dynamics downstream: The biochemical decomposition sequence and its impact on soil organic matter structure and function. *Sci. Total Environ.* **404**: 297–307.

Gritsch, C., Egger, F., Zehetner, F., and Zechmeister-Boltenstern, S. (2016) The effect of temperature and moisture on trace gas emissions from deciduous and coniferous leaf litter: Climate effects on litter emissions. *J. Geophys. Res. Biogeosciences.* **121**: 1339–1351.

Groffman, P.M., Driscoll, C.T., Fahey, T.J., Hardy, J.P., Fitzhugh, R.D., and Tierney, G.L. (2001) Colder soils in a warmer world: A snow manipulation study in a northern hardwood forest ecosystem. *Biogeochemistry.* **56**: 135–150.

Hagerty, S.B., van Groenigen, K.J., Allison, S.D., Hungate, B.A., Schwartz, E., Koch, G.W. et al. (2014) Accelerated microbial turnover but constant growth efficiency with warming in soil. *Nat. Clim. Change.* **4**: 903–906.

Hanson, P.J., Edwards, N.T., Garten, C.T., and Andrews, J.A. (2000) Separating root and soil microbial contributions to soil respiration: A review of methods and observations. *Biogeochemistry.* **48**: 115–146.

Hanson, R.S., and Hanson, T.E. (1996) Methanotrophic bacteria. *Microbiol. Rev.* **60**: 439–471.

Harley, J.L., and Smith, S.E. (1983) *Mycorrhizal Symbiosis.* London: Academic Press.

Harter, J., Krause, H.-M., Schuettler, S., Ruser, R., Fromme, M., Scholten, T. et al. (2014) Linking N_2O emissions from biochar-amended soil to the structure and function of the N-cycling microbial community. *ISME J.* **8**: 660–674.

Hayatsu, M., Tago, K., and Saito, M. (2008) Various players in the nitrogen cycle: Diversity and functions of the microorganisms involved in nitrification and denitrification. *Soil Sci. Plant Nutr.* **54**: 33–45.

Hedderich, R. and Whitman, W. (2006) Physiology and biochemistry of the methane-producing Archaea. In *The Prokaryotes*, ed. M. Dworkin, S. Falkow, E. Rosenberg, K. Schleifer, and E. Stackebrandt, 1050–1079. New York: Springer.

Heimann, M., and Reichstein, M. (2008) Terrestrial ecosystem carbon dynamics and climate feedbacks. *Nature.* **451**: 289–292.

Henckel, T., Jäckel, U., Schnell, S., and Conrad, R. (2000) Molecular analyses of novel methanotrophic communities in forest soil that oxidize atmospheric methane. *Appl. Environ. Microbiol.* **66**: 1801–1808.

Hirsch, P.R. and Mauchline, T.H. (2015) The importance of the microbial N cycle in soil for crop plant nutrition. In *Advances in Applied Microbiology,* ed. S. Sariaslani and G.M. Gadd, 45–71. London: Academic Press.

Hofmann, K., Pauli, H., Praeg, N., Wagner, A.O., and Illmer, P. (2016) Methane-cycling microorganisms in soils of a high-alpine altitudinal gradient. *FEMS Microbiol. Ecol.* **92**: fiw009.

Hofmann, K., Reitschuler, C., and Illmer, P. (2013) Aerobic and anaerobic microbial activities in the foreland of a receding glacier. *Soil Biol. Biochem.* **57**: 418–426.

Högberg, P., Nordgren, A., Buchmann, N., Taylor, A.F., Ekblad, A., Högberg, M.N. et al. (2001) Large-scale forest girdling shows that current photosynthesis drives soil respiration. *Nature* **411**: 789–792.

Høj, L., Olsen, R.A., and Torsvik, V.L. (2005) Archaeal communities in High Arctic wetlands at Spitsbergen, Norway (78 N) as characterized by 16S rRNA gene fingerprinting. *FEMS Microbiol. Ecol.* **53**: 89–101.

Høj, L., Olsen, R.A., and Torsvik, V.L. (2008) Effects of temperature on the diversity and community structure of known methanogenic groups and other archaea in high Arctic peat. *ISME J.* **2**: 37–48.

Høj, L., Rusten, M., Haugen, L.E., Olsen, R.A., and Torsvik, V.L. (2006) Effects of water regime on archaeal community composition in Arctic soils. *Environ. Microbiol.* **8**: 984–996.

Holmes, A.J., Roslev, P., McDonald, I.R., Iversen, N., Henriksen, K., and Murrell, J.C. (1999) Characterization of methanotrophic bacterial populations in soils showing atmospheric methane uptake. *Appl. Environ. Microbiol.* **65**: 3312–3318.

Hörtnagl, L., and Wohlfahrt, G. (2014) Methane and nitrous oxide exchange over a managed hay meadow. *Biogeosciences.* **11**: 7219–7236.

Huang, T., Gao, B., Hu, X.-K., Lu, X., Well, R., Christie, P. et al. (2014) Ammonia-oxidation as an engine to generate nitrous oxide in an intensively managed calcareous Fluvo-aquic soil. *Sci. Rep.* **4**: 3950.

Hungate, B.A., Holland, E.A., Jackson, R.B., Stuart Chapin, F., Mooney, H.A., and Field, C.B. (1998) The fate of carbon in grasslands under carbon dioxide enrichment. *Nature.* **388**: 576–579.

Hurkuck, M., Brümmer, C., Mohr, K., Spott, O., Well, R., Flessa, H., and Kutsch, W.L. (2015) Effects of grass species and grass growth on atmospheric nitrogen deposition to a bog ecosystem surrounded by intensive agricultural land use. *Ecol. Evol.* **5**: 2556–2571.

Imachi, H., Sakai, S., Sekiguchi, Y., Hanada, S., Kamagata, Y., Ohashi, A., and Harada, H. (2008) Methanolinea tarda gen. nov., sp. nov., a methane-producing archaeon isolated from a methanogenic digester sludge. *Int. J. Syst. Evol. Microbiol.* **58**: 294–301.

Ingwersen, J., Butterbach-Bahl, K., Gasche, R., Papen, H., and Richter, O. (1999) Barometric process separation: New method for quantifying nitrification, denitrification, and nitrous oxide sources in soils. *Soil Sci. Soc. Am. J.* **63**: 117–128.

IPCC (2007) *Climate Change 2007. The Physical Science Basis. Contribution of Working Group I to the Fourth Assessment Report of the Intergovernmental Panel on Climate Change*, ed. S. Solomon, D. Qin, M. Manning, Z. Chen, M. Marquis, K.B. Averyt et al. Cambridge and New York: Cambridge University Press.

IPCC (2013) *Climate Change 2013: The Physical Science Basis. Contribution of Working Group I to the Fifth Assessment Report of the Intergovernmental Panel on Climate Change*, ed. TF Stocker, D. Qin, G.-K. Plattner, M. Tignor, S.K. Allen, J. Boschung et al. 3–29. Cambridge and New York: Cambridge University Press.

IPCC (2014) *Climate Change 2014. Impacts, Adaptation and Vulnerability: Part B: Regional Aspects. Contribution of Working Group II to the Fifth Assessment Report of the Intergovernmental Panel on Climate Change*, ed. V.R. Barros, C.B. Field, D.J. Dokken, M.D. Mastrandrea, K.J. Mach, T.E. Bilir et al. Cambridge and New York: Cambridge University Press.

Islam, K.R. and Weil, R.R. (2000) Land use effects on soil quality in a tropical forest ecosystem of Bangladesh. *Agric. Ecosyst. Environ.* **79**: 9–16.

Islam, T., Jensen, S., Reigstad, L.J., Larsen, Ø., and Birkeland, N.-K. (2008) Methane oxidation at 55 C and pH 2 by a thermoacidophilic bacterium belonging to the Verrucomicrobia phylum. *Proc. Natl. Acad. Sci.* **105**: 300–304.

Jamieson, N., Monaghan, R., and Barraclough, D. (1999) Seasonal trends of gross N mineralization in a natural calcareous grassland. *Glob. Change Biol.* **5**: 423–431.

Jangid, K., Williams, M.A., Franzluebbers, A.J., Schmidt, T.M., Coleman, D.C., and Whitman, W.B. (2011) Land-use history has a stronger impact on soil microbial community composition than aboveground vegetation and soil properties. *Soil Biol. Biochem.* **43**: 2184–2193.

Jobbágy, E.G. and Jackson, R.B. (2000) The vertical distribution of soil organic carbon and its relation to climate and vegetation. *Ecol. Appl.* **10**: 423–436.

Kaiser, C., Franklin, O., Dieckmann, U., and Richter, A. (2014) Microbial community dynamics alleviate stoichiometric constraints during litter decay. *Ecol. Lett.* **17**: 680–690.

Kaiser, C., Koranda, M., Kitzler, B., Fuchslueger, L., Schnecker, J., Schweiger, P. et al. (2010) Belowground carbon allocation by trees drives seasonal patterns of extracellular enzyme activities by altering microbial community composition in a beech forest soil. *New Phytol.* **187**: 843–58.

Kallenbach, C.M., Grandy, A., and Frey, S.D. (2016) Direct evidence for microbial-derived soil organic matter formation and its ecophysiological controls. *Nat. Commun.* **7**: 13630.

Kallenbach, C.M., Grandy, A.S., Frey, S.D., and Diefendorf, A.F. (2015) Microbial physiology and necromass regulate agricultural soil carbon accumulation. *Soil Biol. Biochem.* **91**: 279–290.

Karhu, K., Auffret, M.D., Dungait, J.A.J., Hopkins, D.W., Prosser, J.I., Singh, B.K. et al. (2014) Temperature sensitivity of soil respiration rates enhanced by microbial community response. *Nature.* **513**: 81–84.

Keiblinger, K.M., Hall, E.K., Wanek, W., Szukics, U., Hämmerle, I., Ellersdorfer, G. et al. (2010) The effect of resource quantity and resource stoichiometry on microbial carbon-use-efficiency. *FEMS Microbiol. Ecol.* **73**: 430–440.

Keiluweit, M., Nico, P.S., Kleber, M., and Fendorf, S. (2016) Are oxygen limitations under recognized regulators of organic carbon turnover in upland soils? *Biogeochemistry.* **127**: 157–171.

Keppler, F., Hamilton, J.T., Braß, M., and Röckmann, T. (2006) Methane emissions from terrestrial plants under aerobic conditions. *Nature* **439**: 187–191.

Keuschnig, C. (2016) Fungal and bacterial involvement in nitrogen cycling and N_2O production in soil. PhD diss, Ecole Centrale de Lyon, France.

Kiese, R. and Butterbach-Bahl, K. (2002) N_2O and CO_2 emissions from three different tropical forest sites in the wet tropics of Queensland, Australia. *Soil Biol. Biochem.* **34**: 975–987.

Kimmel, K. and Mander, U. (2010) Ecosystem services of peatlands: Implications for restoration. *Prog. Phys. Geogr.* **34**: 491–514.

King, J.S., Hanson, P.J., Bernhardt, E., DeAngelis, P., Norby, R.J., and Pregitzer, K.S. (2004) A multiyear synthesis of soil respiration responses to elevated atmospheric CO_2 from four forest FACE experiments. *Glob. Change Biol.* **10**: 1027–1042.

Kirschbaum, M. (2006) The temperature dependence of organic-matter decomposition—Still a topic of debate. *Soil Biol. Biochem.* **38**: 2510–2518.

Kitzler, B., Zechmeister-Boltenstern, S., Holtermann, C., Skiba, U., and Butterbach-Bahl, K. (2006) Nitrogen oxides emission from two beech forests subjected to different nitrogen loads. *Biogeosciences.* **3**: 293–310.

Klüber, H.D. and Conrad, R. (1998) Effects of nitrate, nitrite, NO and N_2O on methanogenesis and other redox processes in anoxic rice field soil. *FEMS Microbiol. Ecol.* **25**: 301–318.

Knief, C., Lipski, A., and Dunfield, P.F. (2003) Diversity and activity of methanotrophic bacteria in different upland soils. *Appl. Environ. Microbiol.* **69**: 6703–6714.

Kögel-Knabner, I., Guggenberger, G., Kleber, M., Kandeler, E., Kalbitz, K., Scheu, S. et al. (2008) Organo-mineral associations in temperate soils: Integrating biology, mineralogy, and organic matter chemistry. *J. Plant Nutr. Soil Sci.* **171**: 61–82.

Kolb, S. (2009) The quest for atmospheric methane oxidizers in forest soils. *Environ. Microbiol. Rep.* **1**: 336–346.

Könneke, M., Bernhard, A.E., José, R., Walker, C.B., Waterbury, J.B., and Stahl, D.A. (2005) Isolation of an autotrophic ammonia-oxidizing marine archaeon. *Nature* **437**: 543–546.

Kool, D.M., Wrage, N., Zechmeister-Boltenstern, S., Pfeffer, M., Brus, D., Oenema, O., and Van Groenigen, J.-W. (2010) Nitrifier denitrification can be a source of N2O from soil: A revised approach to the dual-isotope labelling method. *Eur. J. Soil Sci.* **61**: 759–772.

Koranda, M., Kaiser, C., Fuchslueger, L., Kitzler, B., Sessitsch, A., Zechmeister-Boltenstern, S., and Richter, A. (2013) Seasonal variation in functional properties of microbial communities in beech forest soil. *Soil Biol. Biochem.* **60**: 95–104.

Kotsyurbenko, O.R., Friedrich, M.W., Simankova, M.V., Nozhevnikova, A.N., Golyshin, P.N., Timmis, K.N., and Conrad, R. (2007) Shift from acetoclastic to H_2-dependent methanogenesis in a West Siberian peat bog at low pH values and isolation of an acidophilic Methanobacterium strain. *Appl. Environ. Microbiol.* **73**: 2344–2348.

Kramer, C. and Gleixner, G. (2008) Soil organic matter in soil depth profiles: Distinct carbon preferences of microbial groups during carbon transformation. *Soil Biol. Biochem.* **40**: 425–433.

Kuramae, E.E., Yergeau, E., Wong, L.C., Pijl, A.S., Veen, J.A., and Kowalchuk, G.A. (2012) Soil characteristics more strongly influence soil bacterial communities than land-use type. *FEMS Microbiol. Ecol.* **79**: 12–24.

Kuzyakov, Y. (2006) Sources of CO_2 efflux from soil and review of partitioning methods. *Soil Biol. Biochem.* **38**: 425–448.

Kuzyakov, Y. and Blagodatskaya, E. (2015) Microbial hotspots and hot moments in soil: Concept & review. *Soil Biol. Biochem.* **83**: 184–199.

Kuzyakov, Y., Ehrensberger, H., and Stahr, K. (2001) Carbon partitioning and below-ground translocation by Lolium perenne. *Soil Biol. Biochem.* **33**: 61–74.

Kuzyakov, Y. and Larionova, A.A. (2005) Root and rhizomicrobial respiration: A review of approaches to estimate respiration by autotrophic and heterotrophic organisms in soil. *J. Plant Nutr. Soil Sci.* **168**: 503–520.

Kuzyakov, Y., Friedel, J.K., and Stahr, K. (2000) Review of mechanisms and quantification of priming effects. *Soil Biol. Biochem.* **32**: 1485–1498.

Lacey, J. (1997) Actinomycetes in composts. *Ann. Agric. Environ. Med.* **4**: 113–121.

Lagomarsino, A., Moscatelli, M.C., De Angelis, P., and Grego, S. (2006) Labile substrates quality as the main driving force of microbial mineralization activity in a poplar plantation soil under elevated CO_2 and nitrogen fertilization. *Sci. Total Environ.* **372**: 256–265.

Lal, R. (2008) Carbon sequestration. *Philos. Trans. R Soc. B Biol. Sci.* **363**: 815–830.

LaRowe, D.E. and Van Cappellen, P. (2011) Degradation of natural organic matter: A thermodynamic analysis. *Geochim. Cosmochim. Acta.* **75**: 2030–2042.

Larsen, K.S., Andresen, L.C., Beier, C., Jonasson, S., Albert, K.R., Ambus, P. et al. (2011) Reduced N cycling in response to elevated CO_2, warming, and drought in a Danish heathland: Synthesizing results of the CLIMAITE project after two years of treatments. *Glob. Change Biol.* **17**: 1884–1899.

Lassaletta, L., Billen, G., Grizzetti, B., Garnier, J., Leach, A.M., and Galloway, J.N. (2014) Food and feed trade as a driver in the global nitrogen cycle: 50-year trends. *Biogeochemistry.* **118**: 225–241.

Latham, J., Cumani, R., Rosati, I., and Bloise, M. (2014) Global land cover share (GLC-SHARE) database beta-release version 1.0-2014. *FAO Rome Italy.*

Lauber, C.L., Ramirez, K.S., Aanderud, Z., Lennon, J., and Fierer, N. (2013) Temporal variability in soil microbial communities across land-use types. *ISME J.* **7**: 1641–1650.

Lauber, C.L., Strickland, M.S., Bradford, M.A., and Fierer, N. (2008) The influence of soil properties on the structure of bacterial and fungal communities across land-use types. *Soil Biol. Biochem.* **40**: 2407–2415.

Le Mer, J. and Roger, P. (2001) Production, oxidation, Emission and consumption of methane by soils: A review. *Eur J Soil Biol.* **37**: 25–50.

Le Quéré, C., Andrew, R.M., Canadell, J.G., Sitch, S., Korsbakken, J.I., Peters, G.P. et al. (2016) Global Carbon Budget 2016. *Earth Syst. Sci. Data.* **8**: 605–649.

Leifeld, J. and Fuhrer, J. (2010) Organic farming and soil carbon sequestration: What do we really know about the benefits? *AMBIO.* **39**: 585–599.

Lelieveld, J.O.S., Crutzen, P.J., and Dentener, F.J. (1998) Changing concentration, lifetime and climate forcing of atmospheric methane. *Tellus B.* **50**: 128–150.

Li, C., Aber, J., Stange, F., Butterbach-Bahl, K., and Papen, H. (2000) A process-oriented model of N_2O and NO emissions from forest soils: 1. Model development. *J. Geophys. Res. Atmospheres.* **105**: 4369–4384.

Lieberman, R.L. and Rosenzweig, A.C. (2004) Biological methane oxidation: Regulation, biochemistry, and active site structure of particulate methane monooxygenase. *Crit. Rev. Biochem. Mol. Biol.* **39**: 147–164.

Liu, C., Zheng, X., Zhou, Z., Han, S., Wang, Y., Wang, K. et al. (2010) Nitrous oxide and nitric oxide emissions from an irrigated cotton field in Northern China. *Plant Soil.* **332**: 123–134.

Liu, D.Y., Ding, W.X., Jia, Z.J., and Cai, Z.C. (2011) Relation between methanogenic archaea and methane production potential in selected natural wetland ecosystems across China. *Biogeosciences.* **8**: 329–338.

Liu, Y. and Whitman, W.B. (2008) Metabolic, phylogenetic, and ecological diversity of the methanogenic archaea. *Ann. N. Y. Acad. Sci.* **1125**: 171–189.

Liu, Y., Wang, C., He, N., Wen, X., Gao, Y., Li, S. et al. (2017) A global synthesis of the rate and temperature sensitivity of soil nitrogen mineralization: Latitudinal patterns and mechanisms. *Glob. Change Biol.* **23**: 455–464.

Lloyd, J. and Taylor, J.A. (1994) On the temperature dependence of soil respiration. *Funct. Ecol.* **8**: 315–323.

Lovett, G.M. and Rueth, H. (1999) Soil nitrogen transformations in beech and maple stands along a nitrogen deposition gradient. *Ecol. Appl.* **9**: 1330–1344.

Luo, G.J., Brüggemann, N., Wolf, B., Gasche, R., Grote, R., and Butterbach-Bahl, K. (2012) Decadal variability of soil CO_2, NO, N_2O, and CH_4 fluxes at the Höglwald Forest, Germany. *Biogeosciences.* **9**: 1741–1763.

Luo, Y. (2007) Terrestrial carbon–cycle feedback to climate warming. *Annu. Rev. Ecol. Evol. Syst.* **38**: 683–712.

von Lützow, M.V., Kögel-Knabner, I., Ekschmitt, K., Matzner, E., Guggenberger, G., Marschner, B., and Flessa, H. (2006) Stabilization of organic matter in temperate soils: Mechanisms and their relevance under different soil conditions—A review. *Eur. J. Soil Sci.* **57**: 426–445.

MacLeod, M., Moran, D., Eory, V., Rees, R.M., Barnes, A., Topp, C.F.E. et al. (2010) Developing greenhouse gas marginal abatement cost curves for agricultural emissions from crops and soils in the UK. *Agric. Syst.* **103**: 198–209.

Makoi, J.H. and Ndakidemi, P.A. (2008) Selected soil enzymes: Examples of their potential roles in the ecosystem. *Afr. J. Biotechnol.* **7**: 181–191.

Manzoni, S., Taylor, P., Richter, A., Porporato, A., and Ågren, G.I. (2012) Environmental and stoichiometric controls on microbial carbon-use efficiency in soils. *New Phytol.* **196**: 79–91.

Marr, A.G., Nilson, E.H., and Clark, D.J. (1963) The maintenance requirement of Eschericia coli. *Ann. N. Y. Acad. Sci.* **102**: 536–548.

Matson, P.A. and Vitousek, P.M. (1981) Nitrogen mineralization and nitrification potentials following clearcutting in the Hoosier National Forest, Indiana. *For. Sci.* **27**: 781–791.

McHale, P.J., Mitchell, M.J., and Bowles, F.P. (1998) Soil warming in a northern hardwood forest: Trace gas fluxes and leaf litter decomposition. *Can. J. For. Res.* **28**: 1365–1372.

McLeod, A.R., Fry, S.C., Loake, G.J., Messenger, D.J., Reay, D.S., Smith, K.A., and Yun, B.-W. (2008) Ultraviolet radiation drives methane emissions from terrestrial plant pectins. *New Phytol.* **180**: 124–132.

Medlyn, B.E., McMurtrie, R.E., Dewar, R.C., and Jeffreys, M.P. (2000) Soil processes dominate the long-term response of forest net primary productivity to increased temperature and atmospheric CO_2 concentration. *Can. J. For. Res.* **30**: 873–888.

Meijide, A., Gruening, C., Goded, I., Seufert, G., and Cescatti, A. (2017) Water management reduces greenhouse gas emissions in a Mediterranean rice paddy field. *Agric. Ecosyst. Environ.* **238**: 168–178.

Meixner, F.X., and Yang., W.X. (2006) Biogenic emissions of nitric oxide and nitrous oxide from arid and semi-arid land. In *Dryland Ecohydrology*, ed. P.D'Odorico and A.Porporato, 233–255. Dortrecht: Springer.

Merbold, L., Eugster, W., Stieger, J., Zahniser, M., Nelson, D., and Buchmann, N. (2014) Greenhouse gas budget (CO_2, CH_4 and N_2O) of intensively managed grassland following restoration. *Glob. Change Biol.* **20**: 1913–1928.

Metje, M. and Frenzel, P. (2005) Effect of temperature on anaerobic ethanol oxidation and methanogenesis in acidic peat from a northern wetland. *Appl. Environ. Microbiol.* **71**: 8191–8200.

Millennium Ecosystem Assessment (2005) *Ecosystems and Human Well-Being: Wetland and Water. Synthesis* Washington D.C.: World Resources Institute.

Miltner, A., Bombach, P., Schmidt-Brücken, B., and Kästner, M. (2012) SOM genesis: Microbial biomass as a significant source. *Biogeochemistry.* **111**: 41–55.

Mishurov, M. and Kiely, G. (2010) Nitrous oxide flux dynamics of grassland undergoing afforestation. *Agric. Ecosyst. Environ.* **139**: 59–65.

Mohan, J.E., Cowden, C.C., Baas, P., Dawadi, A., Frankson, P.T., Helmick, K. et al. (2014) Mycorrhizal fungi mediation of terrestrial ecosystem responses to global change: Mini-review. *Fungal Ecol.* **10**: 3–19.

Mohanty, S.R., Bodelier, P.L., and Conrad, R. (2007) Effect of temperature on composition of the methanotrophic community in rice field and forest soil. *FEMS Microbiol. Ecol.* **62**: 24–31.

Monks, P.S., Granier, C., Fuzzi, S., Stohl, A., Williams, M.L., Akimoto, H. et al. (2009) Atmospheric composition change—Global and regional air quality. *Atmos. Environ.* **43**: 5268–5350.

Mooshammer, M., Wanek, W., Zechmeister-Boltenstern, S., and Richter, A. (2014) Stoichiometric imbalances between terrestrial decomposer communities and their resources: Mechanisms and implications of microbial adaptations to their resources. *Front. Microbiol.* **5**: 1–10.

Moyano, F.E., Manzoni, S., and Chenu, C. (2013) Responses of soil heterotrophic respiration to moisture availability: An exploration of processes and models. *Soil Biol. Biochem.* **59**: 72–85.

Mulder, A., van de Graaf, A.A., Robertson, L.A., and Kuenen, J.G. (1995) Anaerobic ammonium oxidation discovered in a denitrifying fluidized bed reactor. *FEMS Microbiol. Ecol.* **16**: 177–183.

Myhre, G., Shindell, D., Bréon, F., Collins, W., Fuglestvedt, J., Huang, J. et al. (2013) Anthropogenic and natural radiative forcing. In *Climate Change 2013: The Physical Science Basis. Contribution of Working Group I to the Fifth Assessment Report of the Intergovernmental Panel on Climate Change*, ed. T. Stocker, D. Qin, G. Plattner, M. Tignor, S. Allen, J. Boschung et al., 659–740. Cambridge and New York: Cambridge University Press.

Nannipieri, P., Ascher, J., Ceccherini, M.T., Landi, L., Pietramellara, G., and Renella, G. (2003) Microbial diversity and soil functions. *Eur. J. Soil Sci.* **54**: 655–670.

Nazaries, L., Murrell, J.C., Millard, P., Baggs, L., and Singh, B.K. (2013) Methane, microbes and models: Fundamental understanding of the soil methane cycle for future predictions. *Environ. Microbiol.* **15**: 2395–2417.

Nisbet, R.E.R., Fisher, R., Nimmo, R.H., Bendall, D.S., Crill, P.M., Gallego-Sala, A.V. et al. (2009) Emission of methane from plants. *Philos. Trans. R Soc. B Biol. Sci.* **276**: 1347–1354.

Norby, R.J. and Zak, D.R. (2011) Ecological lessons from free-air CO_2 enrichment (FACE) experiments. *Annu. Rev. Ecol. Evol. Syst.* **42**: 181–203.

Oertel, C., Matschullat, J., Zurba, K., Zimmermann, F., and Erasmi, S. (2016) Greenhouse gas emissions from soils—A review. *Chem. Erde—Geochem.* **76**: 327–352.

Op den Camp, H.J., Islam, T., Stott, M.B., Harhangi, H.R., Hynes, A., Schouten, S. et al. (2009) Environmental, genomic and taxonomic perspectives on methanotrophic Verrucomicrobia. *Environ. Microbiol. Rep.* **1**: 293–306.

Papen, H., Von Berg, R., Hinkel, I., Thoene, B., and Rennenberg, H. (1989) Heterotrophic nitrification by Alcaligenes faecalis: NO_2^-, NO_3^-, N_2O, and NO production in exponentially growing cultures. *Appl. Environ. Microbiol.* **55**: 2068–2072.

Park, S., Croteau, P., Boering, K.A., Etheridge, D.M., Ferretti, D., Fraser, P.J. et al. (2012) Trends and seasonal cycles in the isotopic composition of nitrous oxide since 1940. *Nat. Geosci.* **5**: 261–265.

Paterson, E., Gebbing, T., Abel, C., Sim, A., and Telfer, G. (2007) Rhizodeposition shapes rhizosphere microbial community structure in organic soil. *New Phytol.* **173**: 600–610.

Paul, E.A. and Kucey, R.M.N. (1981) Carbon flow in plant microbial associations. *Science* **213**: 473–474.

Paustian, K., Lehmann, J., Ogle, S., Reay, D., Robertson, G.P., and Smith, P. (2016) Climate-smart soils. *Nature.* **532**: 49–57.

Peters, V. and Conrad, R. (1995) Methanogenic and other strictly anaerobic bacteria in desert soil and other oxic soils. *Appl. Environ. Microbiol.* **61**: 1673–1676.

Petersen, H. and Luxton, M. (1982) A comparative analysis of soil fauna populations and their role in decomposition processes. *Oikos.* **39**: 288–388.

Philippot, L., Hallin, S., and Schloter, M. (2007) Ecology of denitrifying prokaryotes in agricultural soil. *Adv. Agron.* **96**: 249–305.

Pilegaard, K. (2013) Processes regulating nitric oxide emissions from soils. *Philos. Trans. R. Soc. B Biol. Sci.* **368**: 20130126.

Pilegaard, K., Skiba, U., Ambus, P., Beier, C., Brüggemann, N., Butterbach-Bahl, K. et al. (2006) Factors controlling regional differences in forest soil emission of nitrogen oxides (NO and N_2O). *Biogeosciences.* **3**: 651–661.

Pinto, A.S., Bustamante, M.M., da Silva, M.R.S., Kisselle, K.W., Brossard, M., Kruger, R. et al. (2006) Effects of different treatments of pasture restoration on soil trace gas emissions in the Cerrados of Central Brazil. *Earth Interact.* **10**: 1–26.

Pittelkow, C.M., Liang, X., Linquist, B.A., van Groenigen, K.J., Lee, J., Lundy, M.E. et al. (2015) Productivity limits and potentials of the principles of conservation agriculture. *Nature.* **517**: 365–368.

Poeplau, C., Don, A., Vesterdal, L., Leifeld, J., Van Wesemael, B., Schumacher, J., and Gensior, A. (2011) Temporal dynamics of soil organic carbon after land-use change in the temperate zone—Carbon response functions as a model approach. *Glob. Change Biol.* **17**: 2415–2427.

Pol, A., Heijmans, K., Harhangi, H.R., Tedesco, D., Jetten, M.S., and Den Camp, H.J.O. (2007) Methanotrophy below pH 1 by a new Verrucomicrobia species. *Nature.* **450**: 874–878.

Praeg, N., Wagner, A.O., and Illmer, P. (2014) Effects of fertilisation, temperature and water content on microbial properties and methane production and methane oxidation in subalpine soils. *Eur. J. Soil Biol.* **65**: 96–106.

Prem, E.M., Reitschuler, C., and Illmer, P. (2014) Livestock grazing on alpine soils causes changes in abiotic and biotic soil properties and thus in abundance and activity of microorganisms engaged in the methane cycle. *Eur. J. Soil Biol.* **62**: 22–29.

Prendergast-Miller, M.T., Baggs, E.M., and Johnson, D. (2011) Nitrous oxide production by the ectomycorrhizal fungi Paxillus involutus and Tylospora fibrillosa: Nitrous oxide production by two ectomycorrhizal fungi. *FEMS Microbiol. Lett.* **316**: 31–35.

Priemé, A. and Christensen, S. (2001) Natural perturbations, drying-wetting and freezing-thawing cycles, and the emission of nitrous oxide, carbon dioxide and methane from farmed organic soils. *Soil Biol. Biochem.* **33**: 2083–2091.

Prior, S.D. and Dalton, H. (1985) The effect of copper ions on membrane content and methane monooxygenase activity in methanol-grown cells of Methylococcus capsulatus (Bath). *Microbiology.* **131**: 155–163.

Quin, P.R., Cowie, A.L., Flavel, R.J., Keen, B.P., Macdonald, L.M., Morris, S.G. et al. (2014) Oil mallee biochar improves soil structural properties—A study with x-ray micro-CT. *Agric. Ecosyst. Environ.* **191**: 142–149.

Raich, J.W. and Tufekciogul, A. (2000) Vegetation and soil respiration: Correlations and controls. *Biogeochemistry.* **48**: 79–90.

Ravishankara, A.R., Daniel, J.S., and Portmann, R.W. (2009) Nitrous oxide (N_2O): The dominant ozone-depleting substance emitted in the 21st century. *Science.* **326**: 123–125.

Reay, D.S., Radajewski, S., Murrell, J.C., McNamara, N., and Nedwell, D.B. (2001) Effects of land-use on the activity and diversity of methane oxidizing bacteria in forest soils. *Soil Biol. Biochem.* **33**: 1613–1623.

Reeburgh, W.S. (2006) Global methane biogeochemistry. In *The Atmosphere: Treatise on Geochemistry*, ed. R.K. Keeling, 65–89. Amsterdam: Elsevier.

Reinsch, S., Michelsen, A., Sárossy, Z., Egsgaard, H., Schmidt, I.K., Jakobsen, I., and Ambus, P. (2014) Short-term utilization of carbon by the soil microbial community under future climatic conditions in a temperate heathland. *Soil Biol. Biochem.* **68**: 9–19.

Repo, M.E., Susiluoto, S., Lind, S.E., Jokinen, S., Elsakov, V., Biasi, C. et al. (2009) Large N2O emissions from cryoturbated peat soil in tundra. *Nat. Geosci.* **2**: 189–192.

Robertson, G.P. and Groffman, P.M. (2015) Nitrogen Transformations. In *Soil Microbiology, Ecology and Biochemistry. Fourth Edition*, ed. E.A. Paul, 421–446. Boston: Academic Press.

Robinson, D.A., Jones, S.B., Lebron, I., Reinsch, S., Domínguez, M.T., Smith, A.R. et al. (2016) Experimental evidence for drought induced alternative stable states of soil moisture. *Sci. Rep.* **6**: 20018.

Rosenkranz, P., Brüggemann, N., Papen, H., Xu, Z., Seufert, G., and Butterbach-Bahl, K. (2006) N_2O, NO and CH_4 exchange, and microbial N turnover over a Mediterranean pine forest soil. *Biogeosciences.* **3**: 121–133.

Roy, R. and Conrad, R. (1999) Effect of methanogenic precursors (acetate, hydrogen, propionate) on the suppression of methane production by nitrate in anoxic rice field soil. *FEMS Microbiol. Ecol.* **28**: 49–61.

Rustad, L., Campbell, J., Marion, G., Norby, R., Mitchell, M., Hartley, A. et al. (2001) A meta-analysis of the response of soil respiration, net nitrogen mineralization, and aboveground plant growth to experimental ecosystem warming. *Oecologia.* **126**: 543–562.

Ryan, M.G. and Law, B.E. (2005) Interpreting, measuring, and modeling soil respiration. *Biogeochemistry.* **73**: 3–27.

Saad, O.A.L.O. and Conrad, R. (1993) Temperature dependence of nitrification, denitrification, and turnover of nitric oxide in different soils. *Biol. Fertil. Soils.* **15**: 21–27.

Saari, A., Martikainen, P.J., Ferm, A., Ruuskanen, J., De Boer, W., Troelstra, S.R., and Laanbroek, H.J. (1997) Methane oxidation in soil profiles of Dutch and Finnish coniferous forests with different soil texture and atmospheric nitrogen deposition. *Soil Biol. Biochem.* **29**: 1625–1632.

Saggar, S., Tate, K.R., Giltrap, D.L., and Singh, J. (2008) Soil-atmosphere exchange of nitrous oxide and methane in New Zealand terrestrial ecosystems and their mitigation options: A review. *Plant Soil.* **309**: 25–42.

Sakai, S., Ehara, M., Tseng, I.-C., Yamaguchi, T., Bräuer, S.L., Cadillo-Quiroz, H. et al. (2012) *Methanolinea mesophila* sp. nov., a hydrogenotrophic methanogen isolated from rice field soil, and proposal of the archaeal family *Methanoregulaceae* fam. nov. within the order *Methanomicrobiales*. *Int. J. Syst. Evol. Microbiol.* **62**: 1389–1395.

Sanz-Cobena, A., Lassaletta, L., Aguilera, E., Prado, A. del, Garnier, J., Billen, G. et al. (2017) Strategies for greenhouse gas emissions mitigation in Mediterranean agriculture: A review. *Agric. Ecosyst. Environ.* **238**: 5–24.

Saunois, M., Bousquet, P., Poulter, B., Peregon, A., Ciais, P., Canadell, J.G. et al. (2016) The global methane budget 2000–2012. *Earth Syst. Sci. Data.* **8**: 697–751.

Saunois, M., Jackson, R.B., Bousquet, P., Poulter, B., and Canadell, J.G. (2016) The growing role of methane in anthropogenic climate change. *Environ. Res. Lett.* **11**: 120207.

Savage, K., Phillips, R., and Davidson, E. (2014) High temporal frequency measurements of greenhouse gas emissions from soils. *Biogeosciences.* **11**: 2709–2720.

Schaufler, G., Kitzler, B., Schindlbacher, A., Skiba, U., Sutton, M.A., and Zechmeister-Boltenstern, S. (2010) Greenhouse gas emissions from European soils under different land use: Effects of soil moisture and temperature. *Eur. J. Soil Sci.* **61**: 683–696.

Schimel, D.S., Braswell, B.H., Holland, E.A., McKeown, R., Ojima, D.S., Painter, T.H. et al. (1994) Climatic, edaphic, and biotic controls over storage and turnover of carbon in soils. *Glob. Biogeochem. Cycles.* **8**: 279–293.

Schimel, J.P. and Bennett, J. (2004) Nitrogen mineralization: Challenges of a changing paradigm. *Ecology* **85**: 591–602.

Schindlbacher, A., Schnecker, J., Takriti, M., Borken, W., and Wanek, W. (2015) Microbial physiology and soil CO_2 efflux after 9 years of soil warming in a temperate forest—No indications for thermal adaptations. *Glob. Change Biol.* **21**: 4265–4277.

Schindlbacher, A., Zechmeister-Boltenstern, S., and Butterbach-Bahl, K. (2004) Effects of soil moisture and temperature on NO, NO_2 and N_2O emissions from European forest soils. *J. Geophys. Res.* **109**: D17302.

Schlamadinger, B. and Marland, G. (1996) The role of forest and bioenergy strategies in the global carbon cycle. *Biomass Bioenergy.* **10**: 275–300.

Schlesinger, W.H. and Andrews, J.A. (2000) Soil respiration and the global carbon cycle. *Biogeochemistry.* **48**: 7–20.

Schmidt, M.W.I., Torn, M.S., Abiven, S., Dittmar, T., Guggenberger, G., Janssens, I.A. et al. (2011) Persistence of soil organic matter as an ecosystem property. *Nature* **478**: 49–56.

Schönheit, P., Keweloh, H., and Thauer, R.K. (1981) Factor F420 degradation in Methanobacterium thermoautotrophicum during exposure to oxygen. *FEMS Microbiol. Lett.* **12**: 347–349.

Schulze, E.-D. (2000) The carbon and nitrogen cycle of forest ecosystems. In *Carbon and Nitrogen Cycling in European Forest Ecosystems*, ed. E.-D. Schulze, 3–13. Berlin, Heidelberg: Springer.

Selsted, M.B., Linden, L., Ibrom, A., Michelsen, A., Larsen, K.S., Pedersen, J.K. et al. (2012) Soil respiration is stimulated by elevated CO_2 and reduced by summer drought: Three years of measurements in a multifactor ecosystem manipulation experiment in a temperate heathland (CLIMAITE). *Glob. Change Biol.* **18**: 1216–1230.

Seneviratne, S.I., Nicholls, N., Easterling, D., Goodess, C.M., Kanae, S., Kossin, J. et al. (2012) Changes in climate extremes and their impacts on the natural physical environment. In *Managing the Risks of Extreme Events and Disasters to Advance Climate Change Adaption*, ed. C.B. Field, V. Barros, T.F. Stocker, D. Qin, D.J. Dokken, K.L. Ebi et al., 109–230. Cambridge and New York: Cambridge University Press.

Shaw, M.R. and Harte, J. (2001) Response of nitrogen cycling to simulated climate change: Differential responses along a subalpine ecotone. *Glob. Change Biol.* **7**: 193–210.

Shindell, D.T., Walter, B.P., and Faluvegi, G. (2004) Impacts of climate change on methane emissions from wetlands. *Geophys. Res. Lett.* **31**: L21202.

Shoun, H., Fushinobu, S., Jiang, L., Kim, S.-W., and Wakagi, T. (2012) Fungal denitrification and nitric oxide reductase cytochrome P450nor. *Philos. Trans. R. Soc. B Biol. Sci.* **367**: 1186–1194.

Silver, W.L., Herman, D.J., and Firestone, M.K. (2001) Dissimilatory nitrate reduction to ammonium in upland tropical forest soils. *Ecology*. **82**: 2410–2416.

Singh, B.K., Millard, P., Whiteley, A.S., and Murrell, J.C. (2004) Unravelling rhizosphere–microbial interactions: Opportunities and limitations. *Trends Microbiol.* **12**: 386–393.

Singh, J.S., Singh, S., Raghubanshi, A.S., Singh, S., Kashyap, A.K., and Reddy, V.S. (1997) Effect of soil nitrogen, carbon and moisture on methane uptake by dry tropical forest soils. *Plant Soil* **196**: 115–121.

Sinsabaugh, R.L., Manzoni, S., Moorhead, D.L., and Richter, A. (2013) Carbon use efficiency of microbial communities: Stoichiometry, methodology and modelling. *Ecol. Lett.* **16**: 930–939.

Six, J., Conant, R.T., Paul, E.A., and Paustian, K. (2002) Review: Stabilization mechanisms of soil organic matter: Implications for C-saturation of soils. *Plant Soil*. **241**: 155–176.

Six, J., Elliott, E.T., Paustian, K., and Doran, J.W. (1998) Aggregation and soil organic matter accumulation in cultivated and native grassland soils. *Soil Sci. Soc. Am. J.* **62**: 1367–1377.

Sizova, M.V., Panikov, N.S., Tourova, T.P., and Flanagan, P.W. (2003) Isolation and characterization of oligotrophic acido-tolerant methanogenic consortia from a Sphagnum peat bog. *FEMS Microbiol. Ecol.* **45**: 301–315.

Skiba, U.M., Sheppard, L.., Macdonald, J., and Fowler, D. (1998) Some key environmental variables controlling nitrous oxide emissions from agricultural and semi-natural soils in Scotland. *Atmos. Environ.* **32**: 3311–3320.

Slavíková, R., Püschel, D., Janoušková, M., Hujslová, M., Konvalinková, T., Gryndlerová, H. et al. (2017) Monitoring CO_2 emissions to gain a dynamic view of carbon allocation to arbuscular mycorrhizal fungi. *Mycorrhiza* **27**: 35–51.

Slemr, F. and Seiler, W. (1991) Field study of environmental variables controlling the NO emissions from soil and the NO compensation point. *J. Geophys. Res.* **96**: 13017.

Smith, K.A., ed. (2010) *Nitrous Oxide and Climate Change*. London and Washington, DC: Earthscan.

Smith, K.A., Ball, T., Conen, F., Dobbie, K.E., Massheder, J., and Rey, A. (2003) Exchange of greenhouse gases between soil and atmosphere: Interactions of soil physical factors and biological processes. *Eur. J. Soil Sci.* **54**: 779–791.

Smith, K.A., Dobbie, K.E., Ball, B.C., Bakken, L.R., Sitaula, B.K., Hansen, S. et al. (2000) Oxidation of atmospheric methane in Northern European soils, comparison with other ecosystems, and uncertainties in the global terrestrial sink. *Glob. Change Biol.* **6**: 791–803.

Snyder, C., Davidson, E., Smith, P., and Venterea, R. (2014) Agriculture: Sustainable crop and animal production to help mitigate nitrous oxide emissions. *Curr. Opin. Environ. Sustain.* **9–10**: 46–54.

Sørensen, L.I., Holmstrup, M., Maraldo, K., Christensen, S., and Christensen, B. (2006) Soil fauna communities and microbial respiration in high Arctic tundra soils at Zackenberg, Northeast Greenland. *Polar Biol.* **29**: 189–195.

Søvik, A.K., Augustin, J., Heikkinen, K., Huttunen, J.T., Necki, J.M., Karjalainen, S.M. et al. (2006) Emission of the greenhouse gases nitrous oxide and methane from constructed wetlands in Europe. *J. Environ. Qual.* **35**: 2360.

Spiro, S. (2012) Nitrous oxide production and consumption: Regulation of gene expression by gas-sensitive transcription factors. *Philos. Trans. R. Soc. B Biol. Sci.* **367**: 1213–1225.

Spohn, M., Pötsch, E.M., Eichorst, S.A., Woebken, D., Wanek, W., and Richter, A. (2016) Soil microbial carbon use efficiency and biomass turnover in a long-term fertilization experiment in a temperate grassland. *Soil Biol. Biochem.* **97**: 168–175.

Steenbergh, A.K., Meima, M.M., Kamst, M., and Bodelier, P.L. (2010) Biphasic kinetics of a methanotrophic community is a combination of growth and increased activity per cell. *FEMS Microbiol. Ecol.* **71**: 12–22.

Stevenson, F.J., Harrison, R.M., Wetselaar, R., and Leeper, R.A. (1970) Nitrosation of soil organic matter: III. Nature of gases produced by reaction of nitrite with lignins, humic substances, and phenolic constituents under neutral and slightly acidic conditions. *Soil Sci. Soc. Am. J.* **34**: 430–435.

Stoecker, K., Bendinger, B., Schöning, B., Nielsen, P.H., Nielsen, J.L., Baranyi, C. et al. (2006) Cohn's Crenothrix is a filamentous methane oxidizer with an unusual methane monooxygenase. *Proc. Natl. Acad. Sci. U. S. A.* **103**: 2363–2367.

Storz, G., Tartaglia, L.A., Farr, S.B., and Ames, B.N. (1990) Bacterial defenses against oxidative stress. *Trends Genet.* **6**: 363–368.

Strickland, M.S. and Rousk, J. (2010) Considering fungal: Bacterial dominance in soils—Methods, controls, and ecosystem implications. *Soil Biol. Biochem.* **42**: 1385–1395.

Subke, J.-A., Inglima, I., and Francesca Cotrufo, M. (2006) Trends and methodological impacts in soil CO_2 efflux partitioning: A metaanalytical review. *Glob. Change Biol.* **12**: 921–943.

Sutton, M.A., Howard, C.M., Nemitz, E., Arneth, A., Simpson, D., Mills, G. et al. (2015) *Effects of Climate Change on Air Pollution Impacts and Response Strategies for European Ecosystems.* Project Final Report, Seventh Framework Programme European Commission.

Sutton, M.A., Reis, S., Riddick, S.N., Dragosits, U., Nemitz, E., Theobald, M.R. et al. (2013) Towards a climate-dependent paradigm of ammonia emission and deposition. *Philos. Trans. R. Soc. B Biol. Sci.* **368**: 20130166–20130166.

Syakila, A. and Kroeze, C. (2011) The global nitrous oxide budget revisited. *Greenh. Gas Meas. Manag.* **1**: 17–26.

Talbot, J.M., Allison, S.D., and Treseder, K.K. (2008) Decomposers in disguise: Mycorrhizal fungi as regulators of soil C dynamics in ecosystems under global change. *Funct. Ecol.* **22**: 955–963.

Tang, X., Liu, S., Zhou, G., Zhang, D., and Zhou, C. (2006) Soil-atmospheric exchange of CO_2, CH_4, and N_2O in three subtropical forest ecosystems in southern China. *Glob. Change Biol.* **12**: 546–560.

Taylor, B.R., Parkinson, D., and Parsons, W.F.J. (1989) Nitrogen and lignin content as predictors of litter decay rates: A microcosm test. *Ecology.* **70**: 97–104.

Thauer, R.K. (1998) Biochemistry of methanogenesis: A tribute to Marjory Stephenson: 1998 Marjory Stephenson Prize Lecture. *Microbiology.* **144**: 2377–2406.

Thomson, A.J., Giannopoulos, G., Pretty, J., Baggs, E.M., and Richardson, D.J. (2012) Biological sources and sinks of nitrous oxide and strategies to mitigate emissions. *Philos. Trans. R. Soc. B Biol. Sci.* **367**: 1157–1168.

Throckmorton, H.M., Bird, J.A., Dane, L., Firestone, M.K., and Horwath, W.R. (2012) The source of microbial C has little impact on soil organic matter stabilisation in forest ecosystems. *Ecol. Lett.* **15**: 1257–1265.

Tiedje, J., Sexstone, A., Parkin, T., and Revsbech, N. (1984) Anaerobic processes in soil. *Plant Soil.* **76**: 197–212.

Treseder, K.K. (2008) Nitrogen additions and microbial biomass: A meta-analysis of ecosystem studies. *Ecol. Lett.* **11**: 1111–1120.

Trost, B., Prochnow, A., Drastig, K., Meyer-Aurich, A., Ellmer, F., and Baumecker, M. (2013) Irrigation, soil organic carbon and N_2O emissions. A review. *Agron. Sustain. Dev.* **33**: 733–749.

Trotsenko, Y.A. and Murrell, J.C. (2008) Metabolic aspects of aerobic obligate methanotrophy. *Adv. Appl. Microbiol.* **63**: 183–229.

Trumbore, S. (1996) Rapid exchange between soil carbon and atmospheric carbon dioxide driven by temperature change. *Science.* **272**: 393–396.

Tuomi, M., Vanhala, P., Karhu, K., Fritze, H., and Liski, J. (2008) Heterotrophic soil respiration—Comparison of different models describing its temperature dependence. *Ecol. Model.* **211**: 182–190.

Ussiri, D. and Lal, R. (2013) *Soil Emission of Nitrous Oxide and Its Mitigation.* Dordrecht and New York: Springer.

Velthof, G.L. and Oenema, O. (1995) Nitrous oxide fluxes from grassland in the Netherlands: I. Statistical analysis of flux-chamber measurements. *Eur. J. Soil Sci.* **46**: 533–540.

Vermeulen, S.J., Campbell, B.M., and Ingram, J.S.I. (2012) Climate change and food systems. *Annu. Rev. Environ. Resour.* **37**: 195–222.

Vigliotta, G., Nutricati, E., Carata, E., Tredici, S.M., De Stefano, M., Pontieri, P. et al. (2007) Clonothrix fusca Roze 1896, a filamentous, sheathed, methanotrophic γ-proteobacterium. *Appl. Environ. Microbiol.* **73**: 3556–3565.

Vitousek, P. and Howarth, R. (1991) Nitrogen limitation on land and in the sea: How can it occur? *Biogeochemistry.* **13**: 87–115.

Wallberg, A., Han, F., Wellhagen, B., Dahle, B., Kawata, M., Haddad, N. et al. (2014) A worldwide survey of genome sequence variation provides insight into the evolutionary history of the honeybee Apis mellifera. *Nat. Genet.* **46**: 1081–1088.

Wan, S., Norby, R.J., Ledford, J., and Weltzin, J.F. (2007) Responses of soil respiration to elevated CO_2 air warming, and changing soil water availability in a model old-field grassland. *Glob. Change Biol.* **13**: 2411–2424.

Wang, F.L. and Bettany, J.R. (1995) Methane emission from a usually well-drained prairie soil after snowmelt and precipitation. *Can. J. Soil Sci.* **75**: 239–241.

Wang, J.-T., Cao, P., Hu, H.-W., Li, J., Han, L.-L., Zhang, L.-M. et al. (2015) Altitudinal distribution patterns of soil bacterial and archaeal communities along Mt. Shegyla on the Tibetan Plateau. *Microb. Ecol.* **69**: 135–145.

Wang, Y., Hao, Y., Cui, X.Y., Zhao, H., Xu, C., Zhou, X., and Xu, Z. (2014) Responses of soil respiration and its components to drought stress. *J. Soils Sediments.* **14**: 99–109.

Wen, Y., Chen, Z., Dannenmann, M., Carminati, A., Willibald, G., Kiese, R. et al. (2016) Disentangling gross N_2O production and consumption in soil. *Sci. Rep.* **6**: 36517.

West, A.E. and Schmidt, S.K. (1999) Acetate stimulates atmospheric CH_4 oxidation by an alpine tundra soil. *Soil Biol. Biochem.* **31**: 1649–1655.

West, A.E. and Schmidt, S.K. (2002) Endogenous methanogenesis stimulates oxidation of atmospheric CH_4 in alpine tundra soil. *Microb. Ecol.* **43**: 408–415.

Whalen, S.C. and Reeburgh, W.S. (1996) Moisture and temperature sensitivity of CH_4 oxidation in boreal soils. *Soil Biol. Biochem.* **28**: 1271–1281.

Wiant, H.V. (1967) Has the contribution of litter decay to forest "soil respiration" been overestimated? *J. For.* **65**: 408–409.

Wieder, W.R., Bonan, G.B., and Allison, S.D. (2013) Global soil carbon projections are improved by modelling microbial processes. *Nat. Clim. Change.* **3**: 909–912.

Wieder, W.R., Grandy, A.S., Kallenbach, C.M., and Bonan, G.B. (2014) Integrating microbial physiology and physio-chemical principles in soils with the MIcrobial-MIneral Carbon Stabilization (MIMICS) model. *Biogeosciences.* **11**: 3899–3917.

Wolf, B., Zheng, X., Brüggemann, N., Chen, W., Dannenmann, M., Han, X. et al. (2010) Grazing-induced reduction of natural nitrous oxide release from continental steppe. *Nature.* **464**: 881–884.

Wrage, N., Velthof, G., van Beusichem, M., and Oenema, O. (2001) Role of nitrifier denitrification in the production of nitrous oxide. *Soil Biol. Biochem.* **33**: 1723–1732.

Xu, M. and Shang, H. (2016) Contribution of soil respiration to the global carbon equation. *J. Plant Physiol.* **203**: 16–28.

Xu, X., Schimel, J.P., Thornton, P.E., Song, X., Yuan, F., and Goswami, S. (2014) Substrate and environmental controls on microbial assimilation of soil organic carbon: A framework for Earth system models. *Ecol. Lett.* **17**: 547–555.

Yang, W.H., Teh, Y.A., and Silver, W.L. (2011) A test of a field-based ^{15}N-nitrous oxide pool dilution technique to measure gross N_2O production in soil. *Glob. Change Biol.* **17**: 3577–3588.

Yashiro, Y., Sakai, S., Ehara, M., Miyazaki, M., Yamaguchi, T., and Imachi, H. (2011) Methanoregula formicica sp. nov., a methane-producing archaeon isolated from methanogenic sludge. *Int. J. Syst. Evol. Microbiol.* **61**: 53–59.

Zechmeister-Boltenstern, S., Keiblinger, K.M., Mooshammer, M., Peñuelas, J., Richter, A., Sardans, J., and Wanek, W. (2015) The application of ecological stoichiometry to plant-microbial-soil organic matter transformations. *Ecol. Monogr.* **85**: 133–155.

Zeeman, M.J., Mauder, M., Steinbrecher, R., Heidbach, K., Eckart, E., and Schmid, H.P. (2017) Reduced snow cover affects productivity of upland temperate grasslands. *Agric. For. Meteorol.* **232**: 514–526.

Zhalnina, K., de Quadros, P.D., Gano, K.A., Davis-Richardson, A., Fagen, J.R., Brown, C.T. et al. (2013) Ca. Nitrososphaera and Bradyrhizobium are inversely correlated and related to agricultural practices in long-term field experiments. *Front. Microbiol.* **4**: 104.

Zhang, C., Niu, D., Hall, S.J., Wen, H., Li, X., Fu, H. et al. (2014) Effects of simulated nitrogen deposition on soil respiration components and their temperature sensitivities in a semiarid grassland. *Soil Biol. Biochem.* **75**: 113–123.

Zinder, S.H. (1993) Physiological ecology of methanogens. In *Methanogenesis,* ed. J.G. Ferry, 128–206. Boston: Springer.

Zona, D., Janssens, I.A., Gioli, B., Jungkunst, H.F., Serrano, M.C., and Ceulemans, R. (2013) N_2O fluxes of a bio-energy poplar plantation during a two years rotation period. *GCB Bioenergy.* **5**: 536–547.

Zou, J., Huang, Y., Zheng, X., and Wang, Y. (2007) Quantifying direct N_2O emissions in paddy fields during rice growing season in mainland China: Dependence on water regime. *Atmos. Environ.* **41**: 8030–8042.

Zumft, W.G. (1997) Cell biology and molecular basis of denitrification. *Microbiol. Mol. Biol. Rev.* **61**: 533–616.

3 Regionally Diverse Land-Use Driven Feedbacks from Soils to the Climate System

Hermann F. Jungkunst, Thomas Horvath, Stefan Erasmi,
Jan Paul Krüger, Katharina H.M. Meurer, Klaus Schützenmeister,
Thomas Guillaume, Thomas Scholten, Frank Baumann,
Per-Marten Schleuss, Jin-Sheng He, Peter Kühn,
Jessica Henkner, Jens Boy, Thomas Kätterer, and Julia Schneider

CONTENTS

3.1 INTRODUCTION

Hermann F. Jungkunst, Thomas Horvath, Stefan Erasmi, Jan Paul Krüger,
Katharina Meurer, and Klaus Schützenmeister

Terrestrial ecosystems contain large amounts of carbon (C) and nitrogen (N) in their soils (Trumper et al. 2009). Actually, C stocks in soils are larger than those that currently exist in the atmosphere and in plants combined (Ciais et al. 2013; Macías and Camps Arbestain 2010). If or when these stocks are released as greenhouse gases or GHGs (CO$_2$, CH$_4$ and N$_2$O—directly or indirectly), they would have uncontrollable consequences for the climate system. These forces from soils are being set in motion by land use and particularly land-use change (Achard et al. 2004; van Minnen et al. 2009; Houghton et al. 2012). Fortunately, humans are aware of this frightening scenario, and therefore scientists across the world work hard to keep the climatic consequences of any land use and land-use change within a manageable frame.

Organic matter (OM) is the main carrier in soils for reactive C and N that can be converted into GHGs, predominantly through microbial processes. Consequently, myriads of experiments, field observations and modelling of the turnover of soil organic matter (SOM) have been conducted (O'Brien and Stout 1978; Körschens et al. 1998; Berg 2000; Six et al. 2000; Schlesinger and Andrews 2000; Christensen 2001; Krull et al. 2003; Lutzow et al. 2006; Sollins et al. 2007; Schmidt et al. 2011; Allison 2012; Wieder et al. 2013; Lehmann and Kleber 2015; Rousk and Frey 2015; Fiedler et al. 2016; Keiluweit et al. 2016).

Despite the scientific effort, the question remains—why are the uncertainties of predictions of soil C and N releases still too high? First, soil C stocks vary greatly across the globe, i.e., regional differences exist (Figure 3.1). Therefore, their feedback potential to the climate system differs greatly in a spatial manner (Figure 3.2). This rather trivial fact is likely ignored when global budgets are estimated. What is often forgotten is that these different pools of SOM also differ in their processes by which SOM is converted to GHGs and exchanged with the atmosphere. Second, extrapolating plot-based observations for regional C balances ignores the differences of uptake and losses on the spatial and temporal scale (Koerner 2003). For example, depending on the grade and intensity of the disturbance, C losses can occur rapidly—within

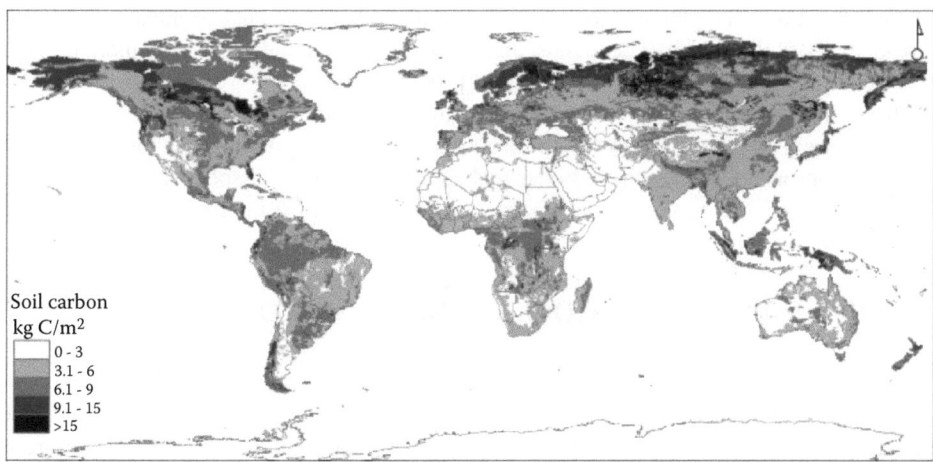

FIGURE 3.1 **(See color insert.)** Global soil organic carbon density in kg Carbon/m^2 to 1 m depth (©IGBP-DIS (1998) Soil Data (V.0) A program for creating global soil-property databases, IGBP Global Soils Data Task, France), 0.5 degree resolution.

FIGURE 3.2 **(See color insert.)** Hotspots of carbon release under changing climate conditions (based on rcp4.5 scenario for the period 2020 to 2039 and soil carbon stores). Red areas indicate increase of carbon release; green areas indicate decrease of carbon release.

hours, e.g., as a result of forest fires. In contrast, SOC stock equilibration after land-use changes have a rather long reaction time—up to more than 100 years (Poeplau et al. 2011). Natural disturbances, such as fires, drought, or wind throw, have a high potential to affect the mobilization of C (Jandl et al. 2007). However, findings differ among regions, and while Kramer et al. (2004) found decreased C contents in watershed areas in Alaska disturbed by wind throws, no or only a slight negative impact on SOC stocks has been observed in the High Tatra (Don et al. 2012). Thus, the uncertainties may be the result of a lack of recognition of the importance of regional (or smaller-scale) mechanisms.

Large-scale estimates may require refinement, but they have taught us that environmental conditions as well as the molecular structure of the SOM create a very complex system of feedbacks. Additionally, humans are constantly altering the physical structure of soils through land-use changes, which adds a layer of complexity to an already globally complex system. The objective of this chapter is to explore these environmental and structural factors contributing to global heterogeneity of active SOM processes. The focus is on the effects that (1) land-use change imparts on SOM processes, especially in the sensitive peatlands, and (2) the physical conditions of temperature and soil moisture are strong forcing mechanisms on SOM processes.

The discussion is designed to demonstrate that these key factors change in terms of their relevance at a relatively regional scale. Thus, the proposal or even the hypothesis is that a regional rather than a global approach is necessary to better understand the global perspective. Specific examples are presented from Sweden, Greenland, European Russia, Tibet, Sumatra, and Brazil to demonstrate the importance of a regional approach rather than just extrapolating up to a global scale.

3.2 GLOBAL SOIL CARBON STOCKS (MINERAL AND ORGANIC SOILS)

3.2.1 LAND USE CHANGES AND SOM

Several studies have been conducted to determine the effect of different 1 land uses and land-use changes and management practices on the losses and sequestration of SOC. For example, forest ecosystems beyond just the tropical rain forests are also facing threats. Assuming that some boreal and temperate forests exhibit high C sequestration rates (10–12 g C m^{-2} yr^{-1}; Schlesinger 1990), then changes in these regions could cause meaningful impacts on global C cycles. Indeed, the available information indicates that these regions are experiencing major land-use changes (Figure 3.3). With regard to Brazil, the pressure on the Amazonian rainforest has declined during the last decade, but the expansion of cropland in the neighboring Cerrado biome and accompanying conversion of forest areas resulted in an offset of 5–7% of the climate benefits (C emissions), which were avoided from reduced Amazon deforestation rates during 2011–2013 (Noojipady et al. 2017). In this context, Noojipady et al. (2017) mentioned the federal state Matopiba as the new frontier in agricultural production.

As forest soils are known to store more than 40% of the total SOC in terrestrial ecosystems (Intergovernmental Panel on Climate Change (IPCC) 2007), conversion to agricultural land goes along with decreasing C stocks. However, Wei et al. (2014) showed that this reaction is highly dependent on the region and forest type: largest decreases in SOC stocks could be found in temperate regions (52%), followed by tropical (41%) and boreal regions (31%). Generally, most of the SOC losses occur rapidly after conversion, and annual C losses have been found to decline after a few years of initial cultivation (e.g., Davidson and Ackerman 1993). In order to prevent further losses and

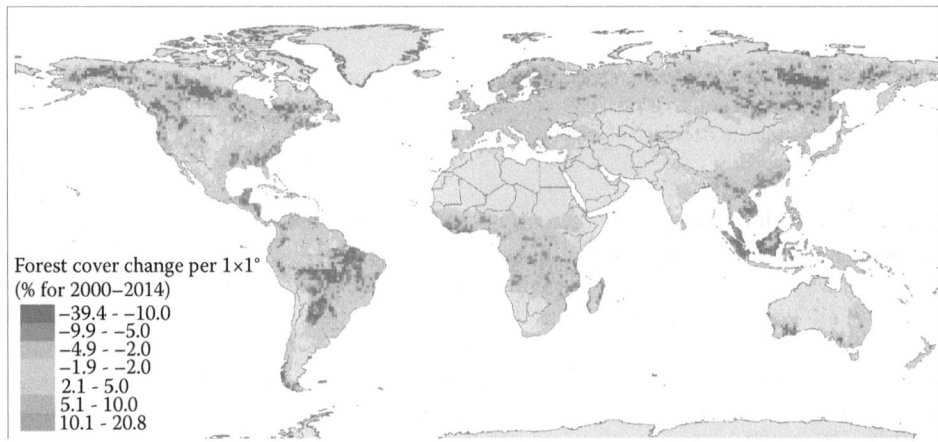

FIGURE 3.3 **(See color insert.)** Forest cover change per 1 × 1° grid cell for the period 2000 to 2014 (in percent with regard to the baseline 2000). Calculations are based on the Global Forest Change maps by Hansen et al. (2013). (From Hansen, M.C., Potapov, P.V., Moore, R., Hancher, M., Turubanova, S.A., Tyukavina, A., Thau, D., Stehman, S.V., Goetz, J., Loveland, T.R., Kommareddy, A., Egorov, A., Chini, L., Justice, C.O., & Townshend, J.R.G., *Science*, 342 (6160), 850–853, 2013.)

promote C sequestration, agricultural practices (e.g., no till or NT) can be adapted and integrated crop–livestock systems can be established (Selenobaldo et al. 2016). Conversion from conventional to conservational agriculture (CA) practices has the potential to sequester notable amounts of C in the upper top soil layers, as is shown by several regional (e.g., González-Sánchez et al. 2012; Aguilera et al. 2013; Land et al. 2017) and global meta-analyses (West and Post 2002; Alvarez 2006; Angers and Eriksen-Hamel 2008; Virto et al. 2012). However, this ability only takes effect on the long term and global annual sequestration potentials range from 23 (Virto et al. 2012) to 48 g C m^{-2} yr^{-1} (West and Post 2002) after changing from conventional to NT and the effect on total SOC stocks may be null or insignificant (Govaerts et al. 2009; Angers and Ericksen-Hamel 2008; VandenBygaart et al. 2003; Luo et al. 2010). Accumulation rates due to reforestation or grassland establishment, as soon as the agricultural is no longer used for cultivation, have been reported to be about 33 g C m^{-2} yr^{-1} on average (West and Kwon 2000).

Considering this, models have been used to give estimates the reaction of SOC to land-use changes and management practices. The CENTURY model (Parton et al. 1988) is one of the most commonly used SOM models and has been widely used to determine effects of management on SOC stock changes under different climatic conditions. Foereid and Hogh-Jensen (2004), for example, used the model to predict the effect of conversion to organic agriculture on SOM under Northern European agricultural conditions. The model predicted increasing SOM during the first 50 years (10 0 g C m^{-2} yr^{-1}) and stable level 100 years after conversion. However, this study compares conventional crop rotations with only annual crops with organic rotation including perennials. Not surprisingly, the rotation with perennial will result in higher SOC stocks. Most likely at a larger scale that is more relevant in the climate change context, a much larger area would be needed when switching to organic production for producing the same amount of feed/food. The resulting land-use change would cause a lot of C emissions. A similar modeling approach has been used by Farage et al. (2007) concerning the impact of different agricultural scenarios for drylands in Africa and Latin America (between 8 and 17 g C m^{-2} yr^{-1} over 50 years). Tornquist et al. (2009) and Bortolon et al. (2011) used the model to estimate the impact of crop rotations and management practices on SOC in Southern Brazil. Considerably larger soil disturbances in the form of land-use changes (conversion of rainforest to cattle pastures) have, for instance, been simulated for the Amazon region by Cerri et al. (2004, 2007).

3.2.2 Focus on Peatlands

Organic soils contain a large proportion of the global C stock. With the biomass production being higher than the decomposition of organic material, peatlands accumulate C in a natural state (Turunen et al. 2002). Most peat is found in the northern latitudes (50–70° north), especially in boreal and subarctic regions of Canada and Russia, where decomposition is slow. The humid and cold climatic conditions in these regions are suitable for a high rate of peat accumulation. The overall net radiative forcing of northern peatlands during the Holocene is a net cooling of about 0.2 to 0.5 W m^{-2} (Frolking and Roulet 2007).

Since the last glacial maximum, northern peatlands have accumulated more than 500 Pg C in their soils (Gorham 1991, Yu 2012, Jungkunst et al. 2012). In these water-saturated soils, anoxic conditions inhibit OM decomposition and favor peat accumulation (Clymo 1984). The rate of C accumulation for northern peatlands averaged 23 ± 2 g C m^{-2} yr^{-1} during the Holocene (Loisel et al., 2014). However, C accumulation rates have a high spatial and temporal variation. The highest rates of C accumulation (25–28 g C m^{-2} yr^{-1}) are recorded during the early Holocene when the climate was warmer than the present (Loisel et al. 2014), whereas lower C accumulation rates were observed in the northern peatlands during the Neoglacial period characterized by cooler and wetter conditions (Vitt et al. 2000). The onset of permafrost aggradation during this period, which happens mainly during the Little Ice Age, reduced the C-sink capacity of northern peatlands (Vitt et al. 2000; Krüger et al. 2017).

Beside the huge C stock of northern peatlands, more than 1000 Pg of C is stored in permafrost soils (Hugelius et al. 2014). However, this estimation has a high uncertainty as there are substantial regional gaps in the soil data of this region. Especially peatlands and cryoturbated soils in these regions are highly variable and difficult to assess. Permafrost thaw can release large amounts of SOC by decomposition, which is actually stored and protected against decomposition in frozen soil.

In a review by Yu (2012), the estimated C stocks of peatlands in different regions is about 178 Pg in North America, 70 Pg in Russia, and 2 Pg in Finland (Yu 2012 and references therein). Another estimation of total peat C storage in similar regions is 152 Pg and 119 Pg for Russia and Canada, respectively (Byrne et al. 2004). Yu (2012) desired a more precise and comprehensive peatland C stock estimation and argued with the best guess of about 500 ± 100 Pg C stored in northern peatlands. The total peat C storage for European peatlands has been estimated at 42 Pg C with the largest C stocks in Finland, Sweden, Norway, UK, and the European part of Russia (Byrne et al. 2004). However, it is notable that most of the natural peatland area in central Europe has decreased in the last decades/centuries. In Europe, more than 50% of the natural peatland area has been lost due to land-use change and management (Joosten and Clarke 2002).

But peat deposits are also found in the tropics between 20° south and 10° north (Aselmann and Crutzen 1989) with approximately 11% of the global peatland area. Moreover, the second largest concentration of C with about 85 Pg or 15–19% of the global peatland C stock is present in tropical peatlands. Most of the tropical peatlands occur in Southeast Asia (with about 5% of the global peatland area), and with the largest C stock in Indonesia (Page et al. 2011). However, recently a huge peatland was found in the central Congo Basin with an area of about 145,500 km^2 (Dargie et al. 2017). This peatland alone stores approximately 30 Pg of carbon and increases the tropical peatland carbon stocks by more than 35% (Dargie et al. 2017). This shows that even in the 21st century huge carbon stocks of peatlands are undiscovered and precise quantitative C stocks of soils and especially peatlands are still in progress. Tropical peatlands have the lowest C accumulation rates during the Holocene, but these peatlands started to form much earlier than those in the high northern latitudes (Yu et al. 2010). Frequently, these peatlands have built a more than 10m thick peat layer, storing a huge amount of C.

In the recent decades, the research focus has also been set on peatlands in the southern hemisphere. Peatlands in the southern hemisphere, mainly in Patagonia, South America, have accumulated approximately 15 Pg of C since the last glacial maximum (Yu et al. 2010). For the Patagonia region, the total C stock was estimated to be about 7.6 Pg C (Loisel and Yu 2013). These are the first estimations of C-stocks for the whole southern hemisphere. Mean C accumulation rates in Patagonian peatlands are 0.022 kg C m^{-2} yr^{-1} with low rates in the early Holocene and higher rates in the late Holocene (Loisel and Yu 2013).

The N stocks and N cycles are often in the background of the peatland C cycle and little is known about the N budget in peatlands. Using the total C stock of northern peatlands and assuming a mean C/N ratio of 45, the peat N stock is about 9.7 Pg—roughly corresponding to 10% of the global soil N stock (Loisel et al. 2014). This is similar to the range of 8–15 Pg N in peatlands from a previous estimation (Limpens et al. 2006). Holocene peatland N accumulation rate is about 0.5 g N m^{-2} yr^{-1} (Loisel et al. 2014). The fate of these large peat N stocks remains largely unknown under recent and projected warming. Indeed, the importance of permafrost (and potentially peatlands) as sources of nitrous oxide (N_2O) is just emerging (e.g., Marushchak et al. 2011; Repo et al. 2009), and studies have suggested that reduced surface moisture or increasing temperatures might significantly increase the N_2O emissions. In contrast, other studies assume that the potential increase in peatland N_2O emissions by a changing climate may not be significant relative to the global N_2O budget (Frolking et al. 2011; Martikainen et al. 1993). However, the application of fertilizer on managed peatlands and the N mineralization due to drainage, which is typically done in the temperate regions, and with this an increase in reactive N in the soil, has a significant effect on the N cycle. Overall, additional peat N cycling studies are needed to address these remaining questions

(Loisel et al. 2014). The peatland N cycle is less investigated in the past decades but contributes a significant amount of N to the global N pool, particularly with N_2O emissions. In general, for a better understanding of how key biogeochemical processes interact in peatlands, it is crucial to investigate the coupling between N and C cycling.

3.2.3 Land Use Change Alters the C Balance of Organic Soils

Due to a high C density and a large C reservoir in these ecosystems, peatlands are an important component of the global C cycle. With a view to the rising concentrations of GHGs, especially CO_2, the C storage function of peatlands is crucial: these ecosystems could release a huge amount of C to the atmosphere due to drainage and management activities. In 2008, for instance, drained peatlands emitted about 1.3 Pg CO_2 globally (Joosten 2010).

In the boreal region, a substantial proportion of peatlands in Fennoscandia and Russia was drained for forestry use with a total area of more than 100 000 km^2 (Minkkinen et al. 2008). In Finland for instance, more than half of the peatland area was drained during the 20th century, mainly for forestry use (Laine et al. 2006). Mean emission factors for ombrotrophic drained peatlands for forestry in the boreal region are 0.02 kg C m^{-2} yr^{-1}, whereas minerotrophic emit about 0.1 kg C m^{-2} yr^{-1} (IPCC 2013). However, the impact of draining boreal peatlands for forestry is a controversial debate (Krüger et al. 2016). Usually drained peatlands are C sources, but some studies have shown that peatlands are still C sinks after drainage for forestry usage mainly due to increased wood, root, and litter production (Minkkinen et al. 1999; Lohila et al. 2011; Ojanen et al. 2013). The sink/source function of drained boreal peatlands for forestry use depends mainly on the nutrient status and the drainage depth of the peatland (Minkkinen et al. 1999; Ojanen et al. 2014; Krüger et al. 2016). Nutrient rich peatlands, like fens, are usually a C source if they are drained for forestry usage, whereas nutrient poor peatlands, like bogs, are C neutral or even accumulate C after draining for forestry usage (Minkkinen et al. 1999; Krüger et al. 2016). Lohila et al. (2011) reported a C accumulation of 0.065 kg C m^{-2} yr^{-1} in a forestry drained peat soil in southern Finland. However, peatland drainage in the boreal region for grassland or cropland purpose usually changes a peatland from a C sink into a C source (C loss of 0.26 kg C m^{-2} yr^{-1} and 0.68 kg C m^{-2} yr^{-1}, respectively) (Couwenberg 2009; IPCC 2013).

Whereas peatlands in the boreal region are converted to forests, in the temperate region most of the peatland area has been drained and is managed as agriculture or grassland (Joosten and Clarke 2002). In the EU for example, more than 50% of the peatland area is used as agriculture (50%), forestry (30%) and peat extraction (10%) (Byrne et al. 2004). Mean GHG emissions from peatlands in the temperate region under grassland use are 0.6 kg C m^{-2} yr^{-1} for deeply drained areas and 0.4 kg C m^{-2} yr^{-1} for shallowly drained peatlands (IPCC 2013). Ranked by land-use intensity, intensively managed grasslands emit about 2.8 kg CO_{2eq} m^{-2} yr^{-1} and extensively managed grasslands emit between 0.2 and 2.0 kg CO_{2eq} m^{-2} yr^{-1} (depending on the water table) (Drösler et al. 2013). The GHG budgets of grasslands on peatlands were on average 2.9 kg CO_{2eq} m^{-2} yr^{-1} and were dominated by CO_2 fluxes followed by N_2O and CH_4 fluxes (Tiemeyer et al. 2016). N_2O emissions play an important role in peatlands managed as grassland or under agricultural usage. The application of fertilizer significantly increases the fluxes of N_2O by managed peatlands and contributed a critical amount to the GHG budget. On an integrated approach, drained peatlands managed as extensive and intensive grassland have emitted about 0.3–0.6 kg C m^{-2} yr^{-1} and 0.7–0.9 kg C m^{-2} yr^{-1}, respectively (Krüger et al. 2015a, b). Natural, unmanaged peatlands are almost climate-neutral (Drösler et al. 2013). However, dry bogs that are affected by past drainage activities emit up to 1.0 kg CO_{2eq} m^{-2} yr^{-1} (Drösler et al. 2013; Krüger et al. 2015a).

In the tropical region, mostly southeastern Asia, more than 13 M ha have been deforested and drained for large-scale oil palm and pulpwood (Acacia) plantations (Hooijer et al. 2010, 2012). Land-use change has a significant effect on GHG emissions with a net loss of soil C of about 1.1 kg C m^{-2} yr^{-1} and of about 1.7 kg C m^{-2} yr^{-1} including biomass when converting natural peat swamp forest into oil palm plantation (Hergoualch and Verchot 2011). Peatland drainage in Southeast Asia contributes about 1–3% of the current CO_2 emissions from global fossil fuel combustion (Hooijer et al. 2010). Furthermore, C losses from uncontrolled fires in the tropics ranged from 29 kg C m^{-2} for rice fields and 44 kg C m^{-2} for natural peat swamp forests (Hergoualch and Verchot 2011). Uncontrolled fires lead to large C losses from both vegetation and peat. Moreover, C losses from land-clearing fires lead to soil C losses of 11.3 and 5.7 kg C m^{-2} in oil palm and Acacia plantations, respectively (Hergoualch and Verchot 2011). However, a review on C losses in tropical peatlands by Hergoualch and Verchot (2011) demonstrated that many gaps remain in our understanding the C cycle in tropical peatlands.

Managed peatlands in the temperate and tropical region are hot spots of GHG emissions from organic soils. Drainage and intensive usage are the main factors of high GHG emissions from these soils. CO_2 is the main GHG released by managed organic soils, but in combination with fertilizer application these soils are a source of N_2O emissions. Rewetting has become a widespread tool to reduce GHGs from drained peatland soils in recent decades. However, flooding of formerly managed peatlands could release huge amounts of CH_4. A rewetting of previously drained peatlands has to be carefully evaluated and applied.

3.2.4 Climate Change and GHG Fluxes in Peatlands

Through the lens of climate change, peatlands can have positive or negative climate feedbacks. Warming without moisture stress is suggested to increase the net primary production (NPP) more than the peat decomposition and to enhance long-term soil C sequestration, leading to a negative climate feedback (Loisel et al. 2012; Loisel and Yu 2013). However, rising temperatures and changes in precipitation patterns could significantly influence the biogeochemical cycles of peatlands with an acceleration of C emissions from these soils and a potentially disturbance of these ecosystems, resulting in a positive climate feedback (Ise et al. 2008; Dise 2009; Dorrepaal et al. 2009).

In the northern discontinuous permafrost region, climate warming has accelerated emissions from these soils (Dorrepaal et al. 2009). Rising air temperatures, particularly in the high latitudes, leads to the thawing of permafrost and an increase of active layer thickness (Åkerman and Johansson 2008) and affects the hydrology of peatlands in this region. While it is certain that permafrost thaw will release C into the atmosphere and will accelerate climate change, the magnitude of this effect is highly uncertain to date (Schuur and Abbott 2011). For example, the alteration of palsa peatlands (peatlands within the discontinuous permafrost region) with a drastic change in vegetation patterns will change the C balance of this ecosystem, with a potential decrease in CO_2 emissions and an increase in CH_4 emissions (Bosiö et al. 2012).

The accelerated breakdown of soil organic matter by global warming, particularly in the high latitudes, can increase the release of GHGs from permafrost soils. But the magnitude and timing of the accelerated emissions of GHGs is uncertain (Schuur et al. 2015). Permafrost thaw leads to a shift in biogeochemical cycles of these soils towards faster decomposition and an increase in CH_4 emissions (Hodgkins et al. 2014). However, it is unknown which of the GHGs will contribute more to the GHG budget under thawing permafrost. The question is, will the C loss by CH_4 emissions of wetter peatlands overcompensate the release of CO_2 from drier peatland sites, and how will the proportion and distribution of wet and dry pattern look like. Nevertheless, the impact of permafrost thaw is able to release huge amounts of C to the atmosphere and intensify the feedback on climate change.

Drastic changes on peatlands and their C balance occur in the tropics, mainly southeastern Asia. The amount of GHGs emitted from degraded peatlands in the tropics varied widely according to the seasonal patterns of precipitation (Inubushi et al. 2003). It is projected that the seasonal variation in precipitation will change, which will have a significant impact on the GHG exchange. Peatland subsidence in the tropics is about 75 cm in the first year and remains constant at 5 cm yr^{-1} in advanced drainage status (Hooijer et al. 2010), whereas physical compaction not oxidation is usually triggering the first years. This brings the peat surface closer to the drainage base which results in decreased CO_2 emissions, but reduces the agricultural productivity (Hooijer et al. 2010). However, the effect of drainage ditches that draw down water levels has been present for decades. This has led to high GHG emissions and C losses from cultivated peatlands.

Also, large-scale weather patterns changes will alter the GHG exchange between peatland soils and the atmosphere. With increasing and more intensive El Niño events, the C losses from tropical peatland fires will increase. During the 1997 El Niño event, it was estimated that between 0.81 and 2.57 Pg C was released to the atmosphere as a result of burning peat and vegetation, which is equivalent to 13–40% of the mean annual global C emissions from fossil fuel (Page et al. 2002). Peatland fires are difficult to predict, but with an increase in the intensity of El Niño events and intensive drainage activities, it is likely that more peatland fires will occur. Thus, an increase of C loss from tropical peatlands can be expected.

Furthermore, small-scale changes on hydrological conditions in peatlands, including droughts and rewetting, accelerate C losses and destabilize the peatland C stocks (Fenner and Freemann 2011). For instance, organic soils in South Africa show a rapid reaction on rewetting with high CH_4 and N_2O emissions within a few days after the rewetting event (Krüger et al. 2014). These GHG emissions could contribute a significant amount to the local GHG budget.

Peatlands are vulnerable ecosystems in a changing climate. Climate change could increase the GHG emissions from organic soils resulting in a positive climate feedback and accelerate climate change. For instance, the magnitude and timing of permafrost thaw and the induced environmental changes in the high northern latitude remain uncertain. The vulnerable permafrost C will most likely be converted to GHGs and will be emitted to the atmosphere. However, the knowledge of soils in the permafrost region has advanced in recent decades, but still there are many open questions relating to the processes and the projection of GHG emissions from these soils in a changing climate.

3.3   TEMPERATURE AND SOIL MOISTURE

Global climate change models, despite variability in the intensity of changes, predict that temperature will continue to increase, with greater increases expected at higher latitudes. Given the vast amounts of terrestrial C stored in the soils, changes in climate (temperature and precipitation) could unleash enough GHGs (CO_2, CH_4 and N_2O) to produce uncontrollable positive feedback loops, as the sequestered C is exposed to decomposition processes. Such a scenario has been proposed for decades, but the mechanisms affecting releases are not quite understood, or at least are open to substantial criticism. To make matters worse, uncertainties even in the mechanisms that are somewhat understood through field and laboratory experiments and monitoring data, make attempts to model future scenarios difficult at best. Given the significant amount of attention that changing temperatures receives in climate change scenarios, it is worth looking at what is understood about how temperature could affect soil processes involved with mobilizing SOC.

Soils in poorly drained landscapes or permanently cold regions represent large C stocks. A simple mass balance approach suggests that if the input of C through primary production is greater than the output of C via respiration, then C accumulates. Accumulation scenarios have been the status quo because decomposition/respiration has been slowed by cold temperatures or extensively flooded landscapes. However, under climate change scenarios, northern biomes are predicted to warm. Thus, in soils where C loss was previously lower than C gains, the mass balance could reverse and soils in these biomes could exhibit a net loss of C. Thus, wetlands and peatlands, particularly in tundra and boreal biomes, have been the focus of temperature effects. However, temperature effects on mobilizing sequestered C in soils appear more complex than previously considered in models.

Whether or not the sequestered C becomes a significant contributor to atmospheric C, and thus a contributing factor to climate changes, depends on its availability. Microorganisms will be the primary biological component converting the OM to either CO_2 or CH_4. Thus, how available the organic C is to microbial processes will determine its fate. Davidson and Janssens (2006) argued that temperature sensitivity should be integrated into models to predict the mobility of the sequestered C. Temperature affects both the chemical properties of the reactions, which tend to be related to processes associated with kinetics of enzyme reactions, and the metabolic rates of microorganisms producing the extracellular enzymes. This seems rather straightforward as increases in temperature would lead to increased decomposition, and thus more CO_2 released into the atmosphere. The paradigm often cited is that a 10°C increase in temperature will result in a doubling of the SOM decomposition rate ($Q_{10} = 2$, with Q_{10} being the increase in reaction rate with a 10° increase in temperature). This is related to what Davidson and Janssens (2006) referred to as intrinsic temperature sensitivity, that is, the response of C decomposing related to the temperature-specific and kinematic properties of enzyme–substrate reactions.

Temperature alone, however, is not sufficient to explain the dynamics of SOM. Davidson and Janssens (2006) point out that the condition of the C source also is a significant factor. Labile C is decomposed more easily than recalcitrant forms. At the same time, the temperature sensitivity for recalcitrant OM seems to be greater than for labile OM (Lefévre et al. 2014). Thus, as temperatures increase, it will not only be the labile C that will be expected to contribute to lithosphere–atmosphere exchanges, but also the recalcitrant C stocks will likely play a major role.

To add another layer of complexity, soils are heterogeneous mixtures of OM "types" (conditions) and environmental conditions also affect the decomposition process. The C substrate has to be spatially available to the microorganisms or their extracellular enzymes. Because of the heterogeneous nature of soils, OM is not evenly available to optimal degradation. Here Davidson and Janssens (2006) list factors that can determine the availability of C to degradation. On a microscale, C molecules can be chemically and physically protected from enzyme activities. For example, OM can be adsorbed onto or imbedded within mineral particles, thus preventing enzymes from contacting the C substrate and slowing down reactions. Similarly, since enzyme mobility is dependent on the presence of water, environmental conditions affecting the hydrophobicity of soil particles can protect organic C from degradation. Conditions affecting presence or absence of oxygen availability also will determine the reaction pathways of enzymes—anaerobic pathways being slower than aerobic pathways. Thus, some C will be more stable and persistent than others (*sensu*; Schmidt et al. 2011). These conditions also change temporally, for example, with changing seasons and over longer time scales with climate.

Synergistic effects between decomposition of labile and recalcitrant SOM will also occur. When labile C is available, loss of recalcitrant C is also accelerated. Thus, as increasing temperature exposes large pools of labile C in previously frozen landscapes, the even larger pools of recalcitrant will become more "available" as well.

Taking the intrinsic temperature sensitivity together with the effects caused by other environmental conditions produces what Davidson and Janssens (2006) referred to as apparent temperature sensitivity. Determining the apparent temperature sensitivity of soil complexes is a daunting task. In order to improve modeling methods, however, it seems paramount to improve the scientific understanding of these complex and heterogeneous systems.

To complicate feedback from temperature, it is also predicted that increasing temperatures will increase the primary production in these thawing landscapes. It is also assumed that as these landscapes thaw, that flooded conditions may dominate (more thermokarstic water bodies), which may create large areas favoring anaerobic conditions. How exactly these will affect the mass balance of C exchange is unknown, but attempts to include them in modeling approaches are ongoing. In terms of positive feedbacks to GHG emissions, the switch from SOM primarily decomposed under anaerobic conditions, where CH_4 production occurs, to aerobic conditions decomposition, where CO_2 production is favored, may negate any acceleration of warming.

Until now, the focus has been on terrestrial systems, but similar processes should occur in aquatic ecosystems. Temperature in these systems also affects decomposition (Gudasz et al. 2015; Simcic et al. 2015) and primary production (Liboriussen and Jeppesen 2003), as do other limiting factors such as nutrient limitation (Ferreira et al. 2015). Thus, changes in temperature regimes in lakes and rivers will produce similarly complex scenarios for modeling approaches. Additionally, enzyme activities related to decomposition are up to 5 times greater in riverine sediments compared with terrestrial soils (Sinsabaugh et al. 2012).

3.4 SUMMARY

The massive stocks of OM in the terrestrial biomes have the potential to impart significant effects on the global climate system. Physical changes to the environment will significantly influence the rate of exchange between the two spheres. Global climate models predict changes in both temperature and precipitation, although the magnitude of change in both is uncertain.

Given that the relationships between soil C stocks and physical environment (here temperature and moisture) are fairly well understood, it is possible to position these relationships on top of a biome template. In Figure 3.4 one can visualize the dimensionality of these relationships using moisture as the driving physical property as an example. Soil C stocks will respond to increased soil moisture initially by decreasing as decomposition processes are favored. However, as soils become water inundated, the decomposition processes are impeded and soil C would be expected to accumulate, as long as rates of primary production continue to be high (this would be expected as one move along the biome gradient).

In addition, humans will play a role in affecting the balance of terrestrial and atmospheric C stocks. Conversion of land to agricultural uses or allowing formally used lands to return to a less managed state, indirectly affect the exposure of soil C. Anthropogenic actions are especially regionally relevant. It is pertinent to take a detailed look at six examples from different regions to point out why taking a regional approach makes sense.

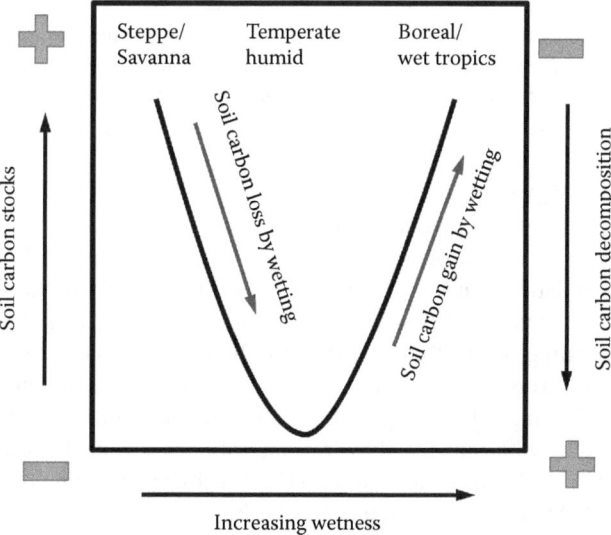

FIGURE 3.4   Conceptual response of soil carbon to increasing wetness: both sides of the story.

3.5 REGIONAL CASE STUDIES

3.5.1 SUMATRA: A HOTSPOT OF LAND-USE CHANGE SENSITIVE TO EXTREME CLIMATIC EVENTS

Thomas Guillaume

Specific Challenges That the Region Faces Due to Climate and Land-Use Change
Anyone stepping out of an airplane in Sumatra during the dry season gets at his first breath a sense of the global changes occurring in that region. The omnipresent burning smell leaves no doubt on the scale of human impacts resulting from the dynamism of the country. Each year during the dry season, land clearing by smallholder farmers and large multinational companies in Indonesia produces a thick haze that eventually affects neighboring countries, up north to Thailand. Forest, old plantations and peatlands fires are even exacerbated by El Niño Southern Oscillation (ENSO). During the 2015 ENSO event, CO_2 daily emission rates from Indonesian fires in September–October exceed fossil fuel CO_2 emission rate of the European Union (Huijnen et al. 2016). Indonesia is a hotspot of land-use change with deforestation rate in 2012 exceeding forest loss in Brazil (Margono et al. 2014). The following sections will highlight how soils and soil C stocks are affected by such drastic land-use changes and their impacts on climate change.

3.5.1.1 Environmental Setting

Sumatra is the sixth largest island in the world, located in the west of the Indonesian archipelago. Stretching out northwest to southeast across the equator, it experiences wet tropical climate with rainfall above 2000 mm. A drier season occurs usually between June and September during which potential evapotranspiration is eventually higher than rainfall (Laumonier et al. 2010; Merten et al. 2016).

Along the southwest coast, the narrow Barisan mountain range (50–100 km) extends on the whole length of the island (1700 km; up to 3800 m.a.s.l.). Moving eastward, elevation decreases rapidly below 300 m.a.s.l. and extends until the northeast coast, up to 300 km away from the mountain range. Sumatra's lowland represents about 80% of the island area. Tertiary sediments mainly cover this eastern peneplain except at the piedmont where younger quaternary sediments occur (Barber et al. 2005). The peneplain is dominated by highly weathered well-drained soils with low activity clay, mostly Acrisols and Ferralsols. Gleysols are frequent at lower landscape position and along the meandering rivers crossing the lowland from the mountain range to the NE coast. Sumatra alone encompasses 25% of the 27 Mha of lowland peatland present in Southeast Asia (Hooijer et al. 2010). Mostly located along the swampy northeast coast, few peatland areas are found in the Inland.

3.5.1.2 Land-Use Changes

Until the end of the 19th century, population density in Sumatra was low (<5 inhabitants/km²) and the island was nearly completely covered by primary forest (Feintrenie and Levang 2009). About a century later, the population density has increased by 20-fold and primary forest remains only on 28% of the island (Badan Pusat Statistik 2017; Margono et al. 2014). This rapid anthropization started at the beginning of the 20th century with the introduction of rubber cultivation in the form of agroforest (jungle rubber), followed by rubber (*Hevea Brasiliensis*) monocultures since the 1950s (Feintrenie and Levang 2009). Although oil palm was first introduced in 1911, the real oil palm boom started in the 1980s, enhanced by transmigration programs favoring the settlements of population from highly populated islands like Java (McCarthy and Cramb 2009; Corley and Tinker 2015; Euler et al. 2015). Area under oil palm (*Elaeis guineensis*) cultivation in Indonesia grew from 0.2 Mha in 1980 to 11.3 Mha in 2015 with 63% cultivated on Sumatra (Budan Pusat Statistiks 2015; FAO 2017). Assigning the responsibility of forest loss to specific land-use types is difficult because the final land-use type does not necessarily match the initial granted concession and concessions for a land-use type may be requested rather to benefit from land clearing, e.g., timber wood, than from

TABLE 3.1

Forest Cover and Deforestation Rates of Three Indonesian Islands

Island	Primary Forest Cover 2012[a]	Forest Loss Since 2000 (Mha)[a]	Lowland Primary Forest Cover 2012[a]	Ratio Wetland/ Lowland Forest Loss[a]
Sumatra	28%	2.9	17%	1.3
Kalimantan	52%	2.4	45%	0.7
Papua	86%	0.3	91%	0.3

[a] Modified from Margono, B.A., Potapov, P.V., Turubanova, S., Stolle, F., & Hansen, M.C., *Nature Climate Change*, 4, 730–735, 2014. doi.org/10.1038/nclimate2277. doi:10.1038/366051a0.

FIGURE 3.5   Oil palm plantations in Jambi province, Central Sumatra, on mineral soils (left) and organic soils (right). Note the decrease of visibility in the right picture due to the haze resulting from fires in October 2014.

the official land use. Nonetheless, half of forest losses in Sumatra between 2000 and 2010 occurred within oil palm, timber, fiber and mining industrial concessions (Abood et al. 2015). Nowadays, easily accessible forested lands are scarce in Sumatra with only 17% of lowland primary forest left in 2012. Accordingly, deforestation shifted from lowland mineral soil to wetland (Table 3.1). Unlike any other Indonesian island, deforestation rates in Sumatra between 2000 and 2012 were higher in wetland than in lowland. Pristine or degraded peat swamp forest remained only on 18% of the peatland area of Sumatra, while 66% were used equally by smallholders and industrial plantations (oil palm and pulp) in 2015 (Miettinen et al. 2016; Figure 3.5).

The last agricultural frontiers in Sumatra are mountainous area with up to 80% of the primary forest left. Nonetheless, these areas represent only a sixth of the highland and are not optimal for agriculture. The spatial extent of human footprint on Sumatra's ecosystems and soils reaches that observed in developing countries such as France or Germany, where forest covers about 30% of the land (Worldbank 2018). The vision of a luxuriant pristine tropical rainforest that could remain when evocating Sumatra is a relic from the past. Flying over the island shows the extent to which the landscape is structured. Agricultural expansion will not be any more achieved by encroaching new land but by using available agricultural land sustainably and efficiently. SOC will be a key resource to manage for achieving this goal by (1) keeping local soil fertility to reduce new forest loss and (2) decreasing soil CO_2 emission to mitigate global climate change. The following section will present the impacts on SOC of two main drivers of deforestation in Sumatra and other tropical regions: oil palm and rubber cultivation.

3.5.1.3   Effects of Land-Use Changes on SOC

3.5.1.3.1   Mineral Soils

The impact of rainforest conversion to oil palm and rubber plantations on SOC in mineral soils of Sumatra was studied within the EFForTS project. This project was launched in 2012 by a consortium of German and Indonesian universities with the aim of understanding the ecological and socioeconomic consequences of rainforest transformation into agricultural landscape in Sumatra (Drescher et al. 2016). Studies were conducted in two landscapes in the lowland of Jambi province, Central Sumatra, between the Barisan mountain range and the swampy northeast coast. The two landscapes, Harapan and Bukit Duabelas, experience a warm and humid (27°C, 2235 mm yr^{-1}) tropical climate and all research plots were established on similar heavily-weathered and well-drained Acrisols (WRB). In term of environmental settings, the two landscapes differed mainly by the soil texture, loamy in Harapan and clayish in Bukit Duabelas (Drescher et al. 2016). In each landscape, the effect on SOC of rainforest conversion to three common plantation types in Sumatra was studied by a space-for-time substitution approach. Plantation types represented a gradient of land-use intensity, including (1) jungle rubber, an extensive rubber agroforest, (2) rubber monocultures, and (3) oil palm monocultures (Figure 3.6). In contrast to monocultures, the establishment of jungle rubber plantations does not require full clearing of the land and other vegetation than rubber trees are present in the plantation (Feintrenie and Levang 2009).

Mineral soil under natural conditions stored 5.7–8.0 kg C m^{-2} in loam Acrisols and 8.4–9.1 kg C m^{-2} in clay Acrisols down to 50 cm depth (Guillaume et al. 2015). SOC stocks in the top 10 cm were similar in both landscapes but soils with higher clay content had higher SOC stocks in the subsoil. C content in the soil profile was closely associated with C content in organo-mineral horizon (Ah horizon) (Figure 3.7). SOC content decreased exponentially with depth, showing sharp decrease between the Ah horizon and the underlying eluviation horizons (E horizon). The decrease of SOC content with depth could be modeled with a single parameter and the average SOC content in the Ah horizon in all forest sites of the respective landscape. The model explained 91% of the variation of C content with depth in loam Acrisols. Less variation was explained in the clay Acrisols because the variability among sites was higher. When the model was applied separately to each soil, it explained between 81 and 99% of the C content with depth (Guillaume et al. 2015). Surprisingly, clay accumulation in Bt horizons, starting around 20–30 cm, did not lead to an accumulation of SOC in that layer. Nonetheless, the accumulation of SOC in the subsoil was higher in clay compared with loam Acrisols, resulting in much higher SOC stocks. Two main observations result from the distribution of SOC content with depth: (1) heavily weathered and well-drained mineral soils are poor in SOC except in a shallow surface layer and (2) despite being rich in SOC, the top layer contribution to total SOC stocks is small because of the huge subsoil volume. For instance, heavily weathered tropical Oxisols had at least twice as much SOC stocks from 0.3 to 4 m depth than in the top 0.3 m (Veldkamp et al. 2003).

FIGURE 3.6   Illustrations of plantations established on rainforest in Sumatra.

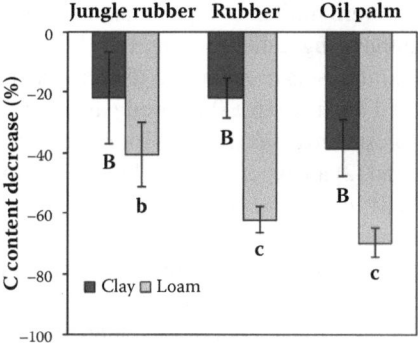

FIGURE 3.7 Relative decrease of C content in Ah horizons following rainforest conversion to plantations in two landscapes with contrasting soil texture. In clay Acrisols, forest conversion to plantations reduces significantly C content but differences are not significant between plantation types. In loam Acrisols, C content decreases are higher in intensive plantations (rubber and oil palm) as compared to extensive plantations (jungle rubber). Errors bars represent standard errors (n = 4). Differences tested by a priori ANOVA. (Modified from Guillaume, T., Damris, M., & Kuzyakov, Y., *Glob. Change Bio.*, 21, 3548–3560, 2015. doi.org/10.1111/gcb.12907.)

Oil palm and rubber cultivations had strong impact on the SOC rich surface layer (Guillaume et al. 2015). SOC content in the Ah horizon decrease by up to 70% in mature oil palm plantations (Figure 3.8). Oil palm and rubber monocultures had similar negative impact on SOC, whereas extensive rubber cultivation in jungle rubber limited C losses. SOC decrease and the effect of land-use intensity, however, were less strong in clay Acrisols, suggesting that clay mitigated the impact of land-use changes.

The SOC stored in the subsoil was not affected by land-use changes after less than two decades. Consequently, SOC stock losses down to 50 cm depth remain moderate and limited to the topsoil. In average, forest conversion lead to a loss of 1.3 kg C m^{-2} in the top 10 cm of loam Acrisols and 0.2 kg C m^{-2} in clay Acrisols. Strongest losses occurred under oil palm plantations with 1.5 kg C m^{-2} and 0.7 kg C m^{-2} in loam and clay Acrisols, respectively. Differences between plantations and differences with forest in clay Acrisols were not significant. Another study in Jambi province, but in other sites, found similar effects of forest conversion to oil palm and rubber monoculture on SOC

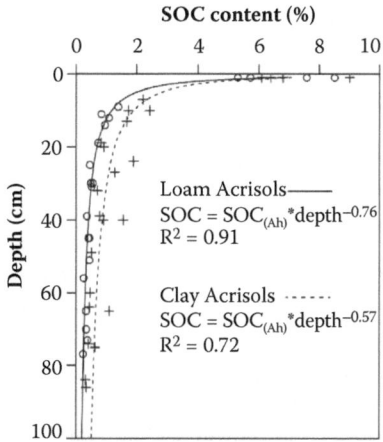

FIGURE 3.8 SOC depth distribution under natural rainforest. Circles: four studied sites in loam Acrisols. Crosses: four studied sites in clay Acrisols. SOC$_{(Ah)}$ correspond to the average SOC content in the Ah horizon of the four sites for each landscape. (Data from Guillaume, T., Damris, M., & Kuzyakov, Y., *Glob. Change Bio.*, 21, 3548–3560, 2015. doi.org/10.1111/gcb.12907.)

(van Straaten et al. 2015). Losses were limited to topsoil and reached 1.4 kg C m^{-2} and 0.7 kg C m^{-2} for oil palm and rubber. The authors suggested revising the IPCC tier 1 method that does not take into account SOC losses when rainforest is transformed to perennial crop. They proposed to modify the current SOC change factor of 1 and use 0.6 ± 0.1 for oil palm and 0.8 ± 0.3 for rubber plantations.

The decrease of SOC after forest conversion results from at least two main processes: (1) a strong decrease of C input to the soil, and (2) an export of SOC by soil erosion. The NPP remaining on site after harvesting decreases by half in both rubber and oil palm monoculture compared with forest (Kotowska et al. 2015). The C input to the soil decreases even more because the NPP allocated to the stem is not available for heterotrophic food chain since there is no wood mortality in healthy plantations. The decrease is even more pronounced in oil palm plantations because dead fronds are stacked on about 10–20% of the surface area of the plantations, leaving the rest with only root C input. The absence of litter layer in oil palm plantations or the thin litter layer in rubber planta-tions makes monoculture prone to soil erosion during the frequent heavy tropical rainfall event. Guillaume et al. (2015) estimated that up to 35 cm of soil has eroded since monoculture establish-ment, while erosion was absent form most jungle rubber sites. In Malaysia, erosion rates reached up to 9.9 cm yr^{-1} on field recently cleared and 1.6 cm yr^{-1} in mature monocultures (Gharibreza et al. 2013). It is difficult to determinate the proportion of SOC losses by soil erosion or mineralization because it depends on when SOC losses occurred during plantation maturation. Nonetheless, it is hypothesized that land clearing with heavy machinery and a long period of bare soils until canopy closure have strong detrimental effects on soil quality since early plantation stage.

The contribution of C losses from mineral soil remains relatively small compared with C losses from vegetation following forest conversion, which reached up to 17 kg C m^{-2} (Kotowska et al. 2015). Nonetheless, SOC losses had strong impact on soil quality, and especially microbial activ-ity (Guillaume et al. 2016a). Soil microbial activity, measured by microbial basal respiration and biomass, was resistant to SOC decrease when losses were small, i.e., SOC losses in percent of SOC under forest lead to a lower percentage of microbial activity loss (Figure 3.9).

However, at low SOC content, microbial activity was highly sensitive to SOC losses. The thresh-old at which microbial activity decreased stronger than SOC corresponded to 2.7% C. A survey of more than 200 jungle rubber, rubber and oil palm monocultures in Jambi province found that half

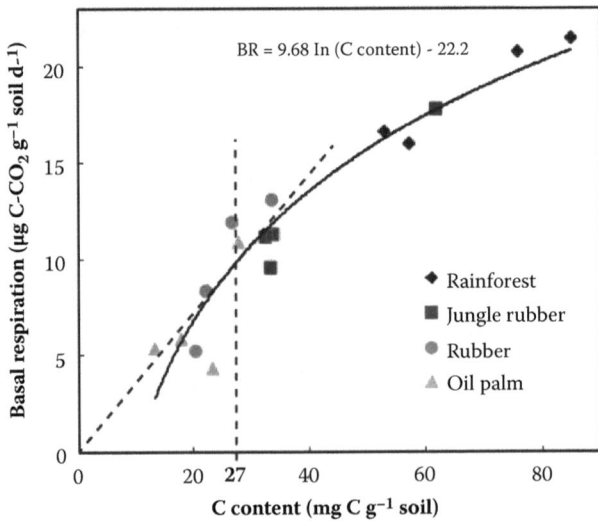

FIGURE 3.9 Microbial basal respiration in function of C content in Ah horizons under tropical plantations. Starting from 27 mg C g^{-1} soil, basal respiration decreases faster than C content. (Modified from Guillaume, T., Maranguita, D., Murtilaksono, K., & Kuzyakov, Y., *Ecological Indicators*, 67, 49–57, 2016a. doi.org/10.1016/j .ecolind.2016.02.039.)

of oil palm plantations had less than 2.2% C in the top 5 cm depth, with the highest frequency of plantations having between 1–2% C and a bulk density of 1.2–1.3 g cm^{-3}, close to the bulk density of a Bt horizon (Guillaume et al. 2016b). Additionally, soil degradation under oil palm plantation increased with plantations age. The potential loss of ecosystem services provided by the soils questions the ability of farmer to sustain the productivity of their plantations in the future. As a consequence, farmers might be forced to clear new forest area to compensate the decrease in productivity of old plantations. In conclusion, the biggest challenge to the climate associated with SOC losses in mineral soil might not result from direct GHG emission from soils but from the necessity to set new forest land under agriculture, driven by the loss of soil fertility.

3.5.1.3.2 Organic Soils

Tropical peatlands are one of the most C-rich ecosystems with C stocks reaching 300 kg C m^{-2} (Miettinen et al. 2017). The warm and humid conditions of tropical climate favor the NPP of tropical vegetation. Nonetheless, these conditions also favor soil microbial activity and OM mineralization. In well-drained soils, C accumulation in OM and its decomposition counterbalance each other, leading to fast OM turnover but to SOM stocks similar to the ones found in temperate soils. In waterlogged soils, OM mineralization by aerobic mircoorganisms is inhibited. In these conditions, the imbalance between high OM production and low mineralization results in huge accumulations of C in soils, peat thickness reaching up to 20 m in Indonesia (Page et al. 2011).

The use of peatland for agriculture, such as oil palm or acacia cultivation, triggers OM mineralization, turning the ecosystem from net C sinks to net C source. Peats are drained to maintain water table commonly below 0.7 m depth (Hooijer et al. 2010), releasing the inhibition of aerobic mineralization of SOC by microorganisms. Peat CO_2 emissions strongly depends on water table depth (Hooijer et al. 2010; Hirano et al. 2012; Jauhiainen et al. 2012). For instance, a decrease of 10 cm was associated with an increase of 0.91 kg CO_2 m^{-2} y^{-1} (Hooijer et al. 2010). C emission from Sumatra due to peat oxidation alone, i.e., without emission from fire, reached 79.5 Tg C y^{-1} in 2015 (Miettinen et al. 2017).

OM mineralization is further enhanced by other factors. SOM in peat has high carbon-to-nutrient ratios. Addition of nutrients by fertilizer application favors SOC mineralization. Fertilization effect on CO_2 emissions is even more marked as peat surface temperature increases; a consequence of reduced canopy cover in plantations compared to forests (Jauhiainen et al. 2014).

Loss of SOC by peat mineralization is a continuous process. Even though SOC is lost at a higher rate during the early stage after deforestation, SOC losses are not expected to stop as long as peats are drained (Hooijer et al. 2012). In contrast, peat fires result in short-term but huge peaks of CO_2 emissions. Peatland degradation for logging and/or cultivation increases the vulnerability of this ecosystem to fire (Field et al. 2009; Miettinen et al. 2012). In south Sumatra, almost all degraded peatland areas burnt within 3 years at the end of the 1990s (Miettinen et al. 2012). The origin of these fires is mostly anthropogenic as land is cleared by fire during the dry season. While fires happen each year, ENSO events in 1997, 2006, and 2015 resulted in abnormally dry conditions in Sumatra and large-scale uncontrolled fires (Page et al. 2002; Huijnen et al. 2016). Peat SOC losses from burning during these 3 years (1.2 Gt C) reached 60% of peat SOC losses from mineralization between 1997 and 2015 (Miettinen et al. 2017).

The long-term threat to degraded peatland is paradoxically water. The majority of Sumatra's peatland is located along the sea, a few meters above sea level. SOC mineralization and compaction results in a subsidence of peat surface, making plantations more and more susceptible to flooding (Miettinen et al. 2016). This will not increase SOC losses, but it will stop agricultural activities in frequently flooded area and shift them to new non-degraded peatland.

3.5.1.4 Conclusions

Climate change and land-use change interactions are strong in Sumatra and soils play a key role in their relationships. The most obvious positive feedback occurs between organic soils and climate. The projected increase of ENSO events frequency and the associated peat fires will increase GHG emissions if the authorities do not take strong measures against peat degradation and fire events. Nevertheless, SOC might also play a key role in climate – land-use change feedback in the long term not because of its function in sequestrating C but because of its maintenance of soil quality. A decrease of agricultural production consecutive to soil degradation would drive further deforestation and its associated GHG emissions.

3.5.2 TIBET: SOILS, CLIMATE, VEGETATION, AND LAND-USE FEEDBACKS ON THE TIBETAN PLATEAU

Thomas Scholten, Frank Baumann, Per-Marten Schleuss, and Jin-Sheng He

Specific Challenges That the Region Faces Due to Climate and Land-Use Change

The Tibetan Plateau is the third largest glaciated area of the world. This so-called third pole (Qui 2008) suffers dramatically from fast melting, degrading permafrost and alpine ecosystems, dust blow and recently also overgrazing. The effects are supposed to affect the Earth far beyond the plateau itself by altering the atmospheric circulation of the Northern hemisphere and the hydrological cycle of the adjacent regions of Central Asia inhabited by billions of people. Recent research has shown that soil is the key to understanding the process feedbacks that occur from climate and land-use change. The soil's pivotal role is due to its soil moisture and SOC regimes. Fast thawing of permafrost and increased temperature, about three times the global warming rate, result in a loss of soil moisture and an increase in soil respiration and consequently GHG emissions. Dust from the Taklamakan desert and from degraded grassland areas on the plateau itself absorb and reflect solar radiation, hamper soil fertility, and reduce SOC and N contents by mixing. This might affect the monsoon system over whole Asia and teleconnect to the Westerlies provoking severe land-use and ecosystem changes far beyond the Tibetan Plateau.

3.5.2.1   Environmental Settings

3.5.2.1.1   *General Settings, Climate, and Permafrost*

The Tibetan Plateau is the highest, largest, and youngest plateau on the globe. It covers an area of more than 2.4 million km^2 with an average altitude exceeding 4,000 m above sea level (Zheng et al. 2000). As a consequence of extreme environmental conditions due to strong and rapid tectonic uplift, highly adapted and sensitive ecosystems have developed. The Tibetan Plateau is considered a key area for the Earth's environmental evolution on regional and global scales and particularly sensitive to global climate warming (Liu and Zhang 1998; Liu and Chen 2000; Jin et al. 2007; Qiu 2008) and land-use changes (Yang et al. 2009b).

Climatically, the Tibetan Plateau acts as an anomalous mid-tropospheric heat source, thus being a major component of the Asian monsoonal system. The east-west stretching mountain ranges act as prominent barriers for the relatively moist tropical Indian Monsoon coming from the south. Kunlun Shan builds up the northern fringe, merging eastwards into the northwest–southeast ranging Bayan Har Shan. Tanggula Shan (approximately 33°N) is splitting the Tibetan Plateau roughly in half, likewise forming the southern border of continuous permafrost (Figure 3.10), which is restricted southwards to higher mountainous areas 4600 m above sea level. The Transhimalaya separates the cold and arid highlands from the comparably warm and moist southern Tibet with the Lhasa valley. At the eastern edge of the Tibetan Plateau, there is no such remarkable mountain range acting as a significant barrier. Instead, deep valleys of major rivers like Huanghe and Yangtze are distinctive landscape elements. The subtropical East Asian Monsoon transports warmer air masses with high water vapor content from the eastern lowlands to the Tibetan Plateau through the meridional flow furrows. The intensity of the East Asian Monsoon decreases westwards (Harris 2006), implying higher temperatures and precipitation in the south and east. During the cold and dry winters, extratropical westerlies occur together with the prevailing Mongolian-Siberian high pressure system (Domrös and Peng 1988). Due to the described large scale climatic patterns, precipitation decreases generally from SE to NW. However, this may vary considerably due to local mountain ranges, differences in altitude, and extreme relief positions. Overall, more than 80% of the mean annual

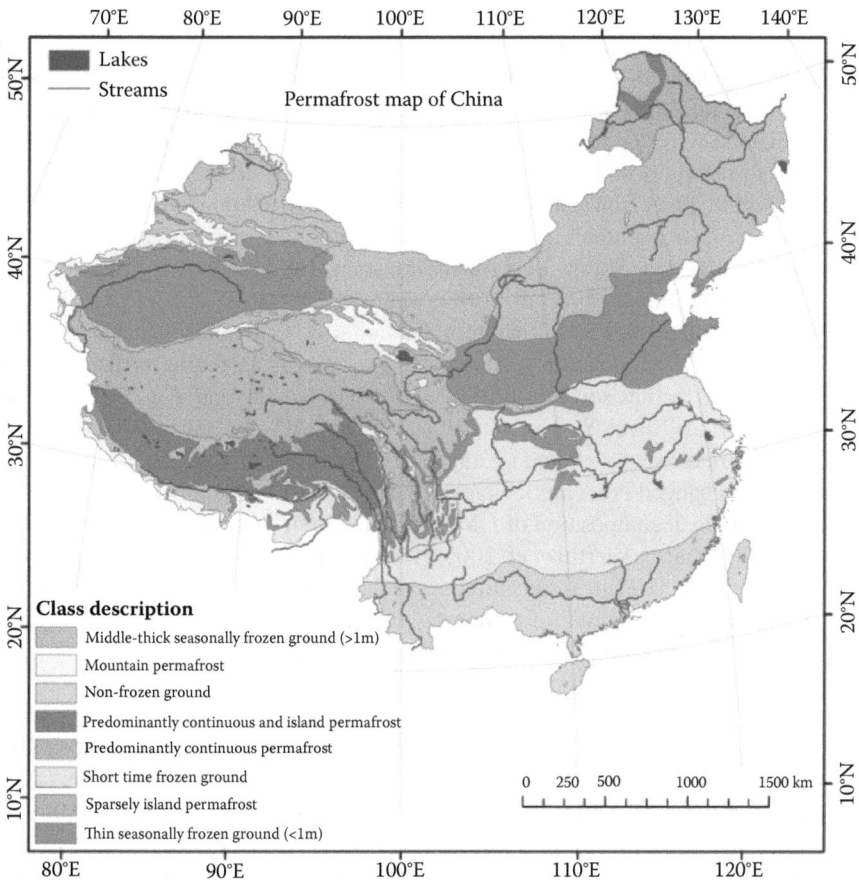

FIGURE 3.10 **(See color insert.)** Permafrost distribution of China. (Modified from Zhou, Y., Guo, D., Qiu, G., Cheng, G., & Li, S., *China Permafrost*, Science Press, Beijing, pp. 145–151, 2000.)

precipitation (MAP) occurs during summer months from July until September. Mean annual evaporation (MAE) ranges on average from 1400 to 1800 mm yr^{-1} for the whole Tibetan Plateau (Wang and French 1994; for details see Wang et al. 2001; Yao et al. 2000; Zhang et al. 2003). Despite variations of the provided values, it is essential to note that MAE largely exceeds MAP.

Due to the specific present-day climate conditions and climate evolution since the last ice age, two thirds of the Tibetan Plateau's total surface are affected by permafrost (Figure 3.10; Cheng 2005; Zhao et al. 2000). It represents the largest high-altitude and low-latitude permafrost region on earth, characterized by strong diurnal patterns, high insolation per surface area due to the comparably steep angle of solar radiation, and respective pronounced geothermal gradients (Wang and French 1994, 1995). Relatively high permafrost temperatures just below the freezing point are typical. Permafrost characteristics are closely linked to the mean annual soil temperature (MAST) gradient (Ping et al. 2004; Wang and French 1994). Accordingly, the active layer thickness increases from north to south and overall with lower altitude, averaging around 1–2 m in continuous permafrost (Cheng and Wu 2007; Wang et al. 2000). Zhao et al. (2000) report a negative relationship between biomass and active layer thickness in alpine meadow ecosystems. Patches of dense vegetation show isolating effects, hence protecting underlying permafrost leading to shallower active layer depths (Wang et al. 2009, 2010; Zhao et al. 2000). Generally, diurnal temperature fluctuations are extremely high compared to the high latitudes and can reach Δ25–40 °C leading to frequent daily freeze-thaw cycles (Ping et al. 2004).

Zhao et al. (2004) showed that climate warming is most distinct in the northeastern Tibetan Plateau, implying warming of air, surface temperatures as well as duration and depth of thawing. Rapid rises in mean annual air temperatures have been observed over the past decades (Wu et al. 2005; Yang et al. 2004a). Compared with high-latitude regions, permafrost degradation processes on the Tibetan Plateau were found to be even more intense (Wang et al. 2000; Yang et al. 2010a; Yang et al. 2004a) and proposed to be enhanced under future environmental change scenarios (Böhner and Lehmkuhl 2005; Nan et al. 2005). They led to the formation of numerous small depressions, where surface water accumulates or thermokarst lakes are formed (Niu et al. 2011). In the eastern part, under the influence of the East Asian Monsoon, discontinuous and unstable permafrost is evident. Taliks have developed (Jin et al. 2000), where soils freeze to a depth of 2–3 m, while the upper limit of the permafrost lies in 4-7 m depth, leading to a vertical disconnection of the permafrost. Many studies report degradation features, such as increasing permafrost temperatures and active layer thickness as well as soil moisture (SM) changes (Cheng and Wu 2007; Jin et al. 2000; Kang et al. 2010; Wu et al. 2010b; Yang et al. 2010a; Yi et al. 2011). Accordingly, soil hydrology and related SM regimes are distinctly altered (Wang et al. 2008a; Yang et al. 2011; Zhang et al. 2003), leading to essentially changed preconditions for soil development (Chapin III et al. 2000; Vitousek 1997). The latter include decomposition of OM (Stokstad 2004; Wang et al. 2009), nutrient supply (Anderson et al. 2006; Dharmakeerthi et al. 2005; Hook and Burke 2000) as well as weathering processes (Baumann et al. 2014). Hence, soils are the crucial connecting link between climate change, permafrost degradation and ecosystem functioning, having influence on vegetation, hydrology and consequently in turn also on land use, C sequestration, and GHG emissions.

3.5.2.1.2   *Vegetation*

The Tibetan Plateau hosts the largest and the highest alpine range in the world (Chang 1981; Lu et al. 2004). On the highlands two of the world's largest alpine ecosystems predominate: (a) the alpine steppe ecosystems of the arid southwestern Tibetan Plateau that covers about 800,000 km², and the *Kobresia pygmaea* pastures of the more humid eastern Tibetan Plateau, which extend across 450,000 km² (Figure 3.11; Miehe et al. 2011). The *Kobresia* ecosystem is also often termed as "alpine meadow" ecosystem and is used synonymously.

In the alpine steppe ecosystem, where stipa species (i.e., *Stipa purpurea*) are the predominant plants, the vegetation cover rarely exceeds 10 to 30% and the productivity is low (Suttie et al. 2005). In contrast the *Kobresia* pastures have often a closed vegetation cover and provide high quality forage (Long et al. 1999; Holzner and Kriechbaum 2001). However, both ecosystems face pressures from a set of abiotic factors limiting the NPP (Berdanier and Klein 2011). In particular, these include the low mean annual temperature and, the short vegetation period, high solar radiation, and nutrient deficiency (Callaway et al. 2002; Körner 2003; Hermans et al. 2006).

3.5.2.1.3   *Soils*

Soil formation on the Tibetan Plateau is largely controlled by geomorphological, cryogenic (solifluction), and erosive processes, frequently interrupted by fresh, mainly aeolian sedimentation, thus leading to a great variety of substrates for soil formation (Baumann et al. 2014). The parent material for soil formation are in most cases aeolian loess-like sediments being mainly of local origin (Feng et al. 2011) or parent rock reworked by periglacial processes mixed with aeolian derived material (Schlütz and Lehmkuhl 2009). The composition of the aeolian sediments can be related to the corresponding altitude: typical loess sediments occur up to 3,600 m a.s.l, whereas in higher regions sandy loess is more dominant (Fang et al. 2003). This is caused by stronger blowout processes, mainly due to sparse vegetation, lower lake levels and higher wind velocities that transport the silty components into the higher atmosphere (Klinge and Lehmkuhl 2005). Moreover, Pleistocene sand dunes occur frequently around lakes and rivers also in higher regions, often accompanied by lacustrine sediments of palaeo-lakes or fluvial sediments. These sediments can be again sources for blowout of aeolian material (Lehmkuhl 1997). In the periglacial environments of the Tibetan Plateau, physical weathering is predominant

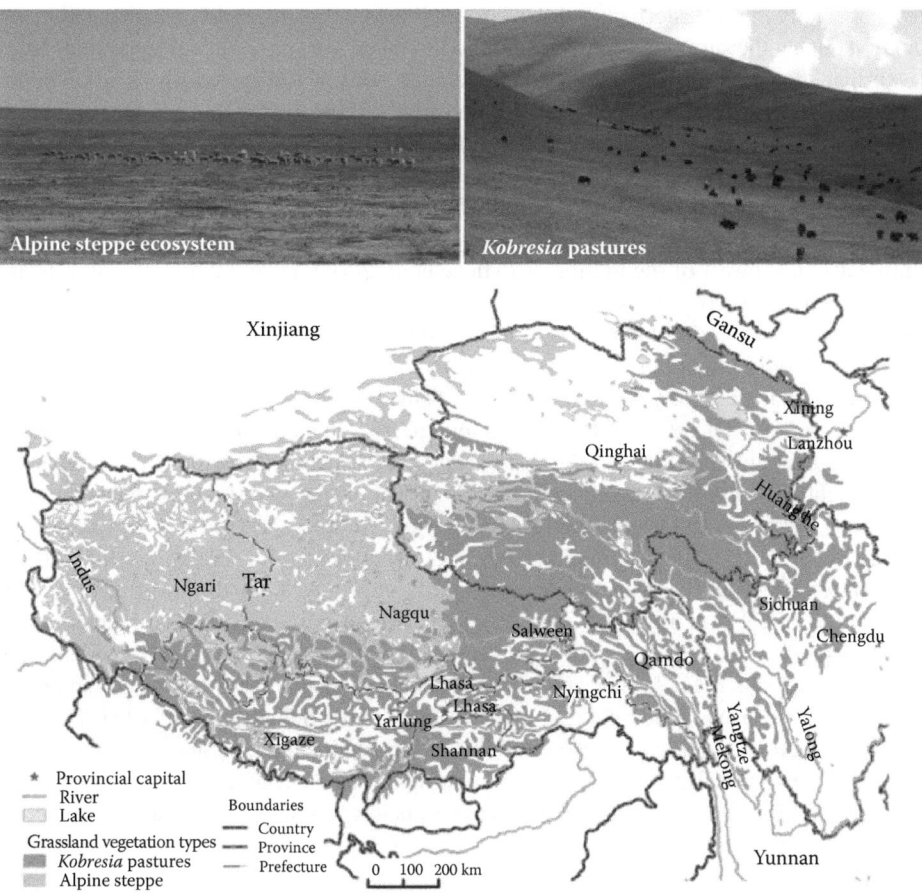

FIGURE 3.11   (See color insert.) The distributions of the two predominate ecosystems on the Tibetan Plateau, the alpine steppes in the northwest and the *Kobresia* pastures in the southeastern part of the highlands. (Modified from Miehe, G., Miehe, S., Kaiser, K., Liu, J., & Zhao, X., *Ambio*, 37, 272–279, 2008; cartography done by L. Lehnert and C. Enderle.)

(Baumann et al. 2014). Along with aeolian sedimentation, this results in sandy and silty substrates with low clay contents. Consequently, important soil functions usually associated with clay fractions, such as nutrient supply and water storage, have to be performed by the humic matter and various site parameters are directly dependent and influenced by SOM contents (Dörfer et al. 2013).

Overall, soil development is strongly influenced by water availability, which is in turn related to relief position and permafrost. On steep upper slopes and terraces, weakly developed soils, such as Leptosols, Leptic Cambisols, Haplic Regosols, and Mollic Cryosols are prevalent, whereas Gleysols and Gleyic Fluvisols commonly occur in morphological depressions and next to lakes or rivers (Kaiser 2004, Kaiser et al. 2008). At sites where permafrost is evident and under more stable conditions, Gelic Gleysols, Gelic Cambisols, Cambic Cryosols, and Permagelic/Gelic Histosols are developed. In regions influenced by discontinuous and sporadic permafrost, or generally speaking under warmer and moister conditions, also well-developed Cambisols are evident (IUSS Working Group WRB 2006). In cold alpine meadows, felty topsoils commonly occur (Kaiser 2004; Kaiser et al. 2008).

Since land-surface stability is closely related to permafrost distribution and degradation (cf. Section 3.5.1.1), aeolian sedimentation processes are typical for the Tibetan Plateau at large. Accordingly, soil formation is closely connected to these processes (Baumann et al. 2009). Soils on aeolian sediments are mostly young showing frequently polygenetic formation and partly strong degradation features. Especially in the eastern part of the Tibetan Plateau, this instability is enhanced by intense precipitation during the summer months leading to fluvial erosion and alluvial accumulation, partly also to laminar sheet floods along gentle slopes. Aeolian erosion and re-deposition is forced in such areas during winter by the dry winter monsoon and sparse vegetation (Dietze et al. 2012; Xue et al. 2009). Consequently, buried relict and mostly humic horizons can be observed.

3.5.2.2   Land-Use: Natural or Cultural Landscapes

The alpine steppes on the Tibetan Plateau are mainly controlled by water scarcity and thus prevail in the more arid northwest of the highlands (Miehe et al. 2008). Here, man-made impacts are historically low and these ecosystems can be widely considered as natural landscapes. In contrast, the *Kobresia* ecosystem is expected to have originated from a long moderate grazing history by Tibetan nomads. Paleo records and pollen analysis indicate the grazing-induced origin of this partly cultural landscape for more than 8000 years ago, and at least since the domestication of the yak, 4000 years ago (Miehe et al. 2014; Qiu et al. 2015). Over a long period, this has favored plants that are highly adapted to livestock grazing such has *Kobresia pygmaea*. Its specific growth form near ground level and small height (~2 cm) often prevents complete removal of the shoot biomass during grazing. At the same time, it does not hinder fruiting, since *K. pygmaea* is able to fruit in dwarf heights of 1 to 2 cm (Miehe et al. 2008). As germination rates of *K. pygmaea* are generally low (i.e., Miao et al. 2008; Huang et al. 2009), propagation occurs mainly vegetatively in the form of clonal growth (Deng et al. 2001; Seeber et al. 2016). The clones can extend over several square meters (probably more) and partial overgrazing does not ultimately lead to the death of a single individual. Besides, *Kobresia pygmaea* shows far higher below than aboveground investments (root to shoot ratio of 20) and thus forms dense and distinct root mats (*Kobresia* turf; Schleuss et al. 2015). These in turn strongly cushions the trampling of livestock and lowers mechanical degradation (Miehe et al. 2008). Moreover, the dense root network (a) ensures an efficient nutrient uptake, (b) outcompetes other plants and microbes for resource uptake (c), serves as belowground storage, and (d) provides a fast regrowth following grazing events (Schleuss et al. 2015; Ingrisch et al. 2015). Further indications for a cultural-induced ecosystem were derived from grazing enclosure studies (Li et al. 1996; Gao et al. 2011; Qiao et al. 2012). Assuming that the *Kobresia* pastures are a natural ecosystem would imply that *K. pygmaea* is the climax vegetation and consequently exclusion of livestock should have no effect. However, a decade of grazing exclusion triggered a change from a *Kobresia*-dominated to Poaceae-dominated grassland of 30 cm to 50 cm in height, with most of these changes occurring in the first year (Figure 3.12).

In summary, the adaptation to anthropogenic-induced stress mainly in form of grazing activities has contributed to the wide distribution of the *Kobresia* ecosystem since the past millennia. Yet the

FIGURE 3.12   Grazing exclosure (GE) experiments to test for change effects of vegetation and soil characteristics across the southeastern part of the Tibetan Plateau. GE were established in 1995, 1999, 2002, and 2010 for the Lhasa site, the Retting site, the Xinghai site, and the Naqu site, respectively.

ecosystem extends cross an elevational range of approximately 3,000 m, where this single sedge dominates and in some areas covers up to 98% of the total plant cover (Miehe et al. 2008).

3.5.2.3 Land-Use Change and Grassland Degradation

Based on the short growing season of the Tibetan Plateau, lasting for only about 3 months, and the associated lack of profitable crop farming, these areas are mainly used for livestock grazing. Estimations state that roughly 5 million pastoralists rely on livestock products from 12 million yaks and 30 million goats and sheep (Suttie et al. 2005). Stocking rates of the *Kobresia* pastures in the Qinghai region range from 28 to 70 animals per km^2, whereas for low productive alpine steppes stocking rates are considerably lower (8 to 9 animals per km^2, Schaller 1998). However, traditional land-use of transhumant herding (nomadism) has shifted since the 1960's toward sedentarization of herdsman and the privatization of local pastures, followed by increasing stocking rates since then (Suttie et al. 2005, Miehe et al. 2008). These management changes have caused serious levels of pasture degradation especially around cities and villages (Niu 1999; Wei et al. 2005; Zhou et al. 2005; Harris 2010; Babel et al. 2014; Qiu 2016). Accordingly, policy makers and local authorities initiated several programs (i.e., "retire livestock and restore pastures" approved in 2004, Yan and Lu 2015), to alleviate grassland degradation mainly by restricting livestock numbers (Figure 3.13), proceeding pasture privatization, establishing fences and reseeding with indigenous grasses; so far with variable success and unknown ecological feedbacks (Li et al. 1996; Wu and Du 2007; Wang et al. 2008).

Current estimates state that about 30–60% of these grasslands are moderately to severely degraded (Holzner and Kriechbaum 2001), whereby these numbers often rely on inconsistent definitions, use different indicators for surveys, and include different temporal and spatial scales (Wang and Wesche 2016). Nevertheless, degradation patterns are diverse and widely distributed over the entire Tibetan Plateau, yet they are seldom mechanistically explained or even classified. Overgrazing on the one hand, has caused direct trampling damages and on the other hand has led to a permanent removal of the photosynthetically active biomass. Together with the harsh environment, it places pressure on the ecosystem, reduces plant productivity and initiates a dying and decomposition of the *Kobresia* turf. Then repeated drying/rewetting and freezing/thawing cycles lead to soil contraction and expansion, gradually cracking the *Kobresia* turf, already weekend by overgrazing. This encourages soil erosion by wind and water and triggers a complete removal of most fertile and SOC enriched topsoil.

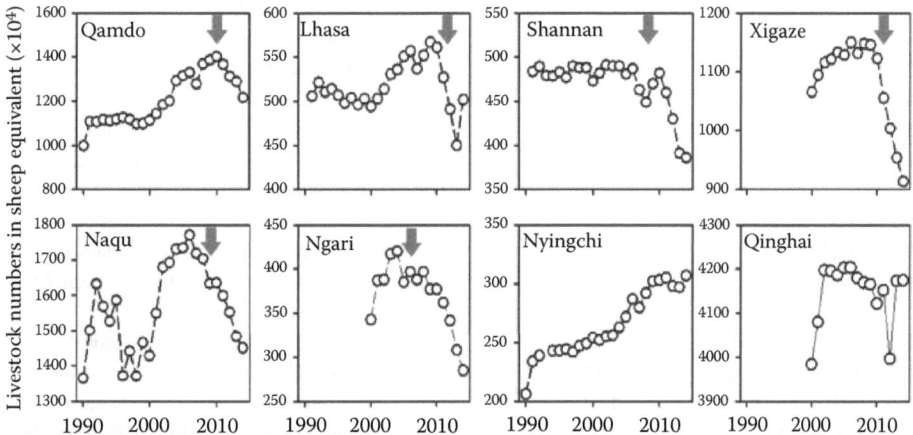

FIGURE 3.13 Livestock numbers for prefectures of the Tibetan Autonomous Region (TAR) and for Qinghai. Livestock numbers were converted into sheep equivalents according to Lehnert et al. (2016). Gray arrows indicate recent changes of livestock numbers due several political programs. Prefecture borders are illustrated in Figure 3.11.

3.5.2.4 Storage and Pools of Soil Organic and Inorganic Carbon

SOC contents of grassland soils exceed all other well-aerated soils with an estimated 10% of the worldwide C; some authors even calculate 30% (140-450 Pg) (Hall et al. 2000). According to Wang et al. (2002), about 33.52 Pg SOC (Tab.3.2) are stored in grassland soils of the Tibetan Plateau down to a depth of 70 cm. Combined, alpine meadow and alpine steppe soils have a share of about 23.2 Pg SOC, which represents 2.5% of the global soil C pool. Alpine meadows make up 38.2% of total grassland soil C in Chinese grassland soils (Ni 2002). Generally, SOC stocks on the Tibetan Plateau show high variation ranging from 2.6 to 13.7 kg m^{-2} for the top 30 cm of soil, 17.3 to 53.1 kg m^{-2} for the upper 75 cm of the soil and 9.1 to 20.0 kg m^{-2} for 0–100 cm of soil (Table 3.2). A comparison of such data is challenging, since the results are partly field data measured with various analytical methods (Chen et al. 2015), partly extrapolations and estimations by modeling with great differences in random quantities. Deviations can be explained by large, small-scale variability, especially in microclimate, vegetation type and soil properties, and a limited number of sampling sites. Highest values were found in tussock tundra and marsh meadow (Table 3.2). Lower values were reported for alpine meadow and steppe, except by Wang et al. (2002) who found very high values for alpine meadows in Qinghai, which is located at the wetter and warmer northeastern border of the Tibetan Plateau. Similar to arctic tundra, Tibetan alpine grasslands are characterized by large BGB proportions of total biomass (Liu et al. 2012) and related high SOC densities in soils (Geng et al. 2012a). Humus accumulation mainly occurs in the root area, which has a higher density than in other ecosystems (ca. 90% of the phytomass is allocated belowground). This is particularly evident for the alpine meadow ecosystems with higher SOC stocks compared with alpine steppe,

TABLE 3.2
Comparison of Soil Organic Carbon Stocks in High-Altitude and High-Latitude Permafrost-Affected Ecosystems

Study	Mean SOC Stocks [kg m^{-2}]	Depth [cm]	Ecosystem Type	Region
Uhlirova et al. (2007)	16.3	30	Tussock Tundra	Siberia, Russia
Wang et al. (2008)	9.3	30	Alpine steppe	Tibetan Plateau, China
Wang et al. (2008)	9.8	30	Alpine meadow	Tibetan Plateau, China
Wang et al. (2008)	10.7	30	Alpine marsh meadow	Tibetan Plateau, China
Yang et al. (2008)	6.2	30	Alpine meadow	Tibetan Plateau, China
Ohtsuka et al.(2008)	2.6 to 13.7	30	Alpine meadow	Tibetan Plateau, China
Yang et al. (2010c)	9.2	40	Alpine meadow	Tibetan Plateau, China
Dörfer et al. (2013)	7.6	30	Alpine meadow	Tibetan Plateau, China
Dörfer et al. (sub.)	9.7	30	Alpine meadow, Alpine steppe	Tibetan Plateau, China
Wang et al. (2002)	53.1	75	Alpine meadow	Qinghai, China
Wang et al. (2002)	29.0	75	Alpine meadow	Tibetan Plateau, China
Dörfer et al. (sub.)	17.3	75	Alpine meadow Alpine steppe	Tibetan Plateau, China
Post et al. (1982)	21.8	100	Tundra	Global
Gundelwein et al. (2007)	30.7	100	Tussock Tundra	Siberia, Russia
Jobbágy/Jackson (2000)	14.2	100	Tundra	Canada
Yang et al. (2008)	9.1	100	Alpine meadow	Tibetan Plateau, China
Yang et al. (2010c)	12.4	100	Alpine meadow	Tibetan Plateau, China
Dörfer et al. (sub.)	20.0	100	Alpine meadow Alpine steppe	Tibetan Plateau, China

frequently developing felty topsoils with extremely high belowground biomass (Kaiser et al., 2008). Accordingly, poor aeration under grassland vegetation is common, leading to sequestration of OM (Stevenson and Cole 1999).

The comparison of discontinuous and continuous permafrost (Dörfer et al. 2013) clearly shows higher SOC stocks for discontinuous (10.4 kg m^{-2}) than for continuous permafrost (3.4 kg m^{-2}). Highest values occur at water-saturated soils (19.3 kg m^{-2}), causing positive feedbacks to even higher SOC accumulation, if in turn denser vegetation isolates the soil. The colder and dryer climate in continuous permafrost areas leads to a lower productivity and an allocation of belowground biomass mainly in the top 10 cm of the soil. This is supported by studies conducted in the Shule River basin (Liu et al. 2012), also characterized by low mean annual temperature and precipitation, showing average SOC stocks of 7.7 kg m^{-2}. Moreover, different vegetation types can be distinguished clearly, ranging from 4.4 kg m^{-2} under desert vegetation to 19.8 kg m^{-2} under partly water-saturated alpine marsh meadow. Baumann et al. (2009) investigated 47 sites along a 1200 km transect across the high-altitude and low-latitude permafrost region of the central-eastern Tibetan Plateau. The transect crossed different climatic, hydrological and geomorphological regions containing continuous, discontinuous and sporadic permafrost as well as areas without permafrost or heavily degraded permafrost. It could be clearly shown that differences in SOC and soil organic N (SON) contents between sites are related to the presence of permafrost and specific relief positions with a clear trend to higher C and N stocks of topsoils with advanced soil development (Figure 3.14). These patterns are directly related to soil hydrological regimes, based on different soil drainage classes. Soil texture has a high influence on SOC and SON contents, showing strong correlations for sand and silt, whereas clay has only a weak impact (Baumann et al. 2009). This is due to the fact that the Tibetan Plateau is dominated by sandy and silty substrates, which exhibit an extremely low negative correlation among each other (r = −0.95, p < 0.01). Thus, distinct processes control the genesis of sand and silt: recent aeolian sedimentation contributes rather fine proximal generated sands, whereas mature soils are preferably developed on silty material originated from older, mainly glacial loess-like sediments.

Regarding SOC pools, about 18% of the total SOC contents under discontinuous and 14% under continuous permafrost were contributed by free particulate organic matter (FPOM), whereas the occluded organic matter portions (OPOM) were the same for both (8%) and the mineral associated organic matter fractions (MOM) contributed 74 and 78 per cent to the total SOC stocks (Dörfer et al. 2013).

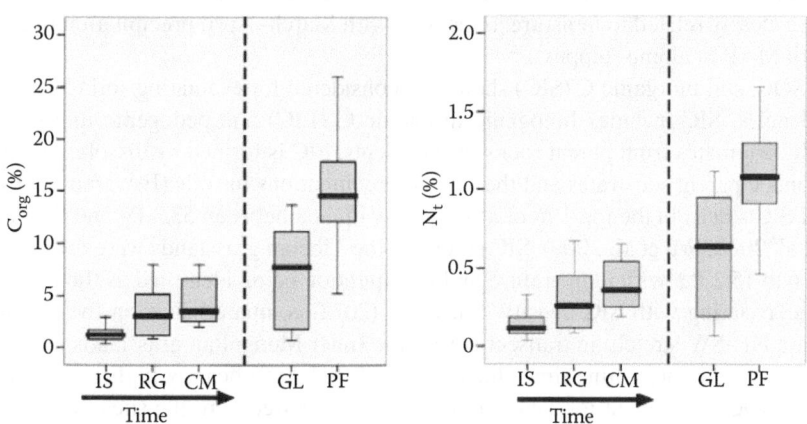

FIGURE 3.14 SOC (= Corg) and SON (= Nt) stocks related to the degree of soil development. IS, initially formed soils; RG, Regosols; CM, Cambisols; GL, groundwater influenced; PF, permafrost influenced. The boxplots show median, 25% and 75% quartiles with the error bar indicating 5–95% range of the observation. (Taken from Baumann, F., He, J.-S., Schmidt, K., Kühn, P., & Scholten, T., *Global Change Bio.*, 15, 3001–3017, 2009. doi.org/10.1111/j.1365-2486.2009.01953.x.)

Substantially higher portions of FPOM occur at water-saturated profiles (19.3 kg m^{-2}), which are especially vulnerable to climate change owing to shorter turnover rates. Higher SOC contents (320 g kg^{-1}) were found in OPOM, whereas MOM had the lowest SOC contents (29 g kg^{-1}). Due to their lower density, the easily decomposable fractions FPOM and OPOM contribute 27% (discontinuous permafrost) and 22% (continuous permafrost) to the total SOC stocks. Mean SOC stocks (0–30 cm depth) account for 10.4 kg m^{-2}, and between 39 and 53% of the SOC is stored in the upper 10 cm. Again, highest POM values of 36% occurred in profiles with high soil moisture content.

N pools are directly linked to SOC contents (Stevenson and Cole 1999) and semi-natural ecosystems, like the alpine grasslands on the Tibetan Plateau, are generally limited in plant available nutrients (Baumann et al. 2009). Hence, the productivity of alpine grasslands is determined by the available N pool, the amount of N input as well as by N fixation (Gerzabek et al. 2004; Körner 2003) and modified by water availability to plants over the year (Baumann et al. 2009). Mineralized plant available N can be found almost exclusively as NH_4^+–N. The highest contents occur at water saturated sites, frequently underlain by permafrost (Baumann et al. 2009). It is most relevant that the permafrost main soil group (PF) shows by far the highest portions of NH_4^+–N, thus differing clearly from soils solely influenced by groundwater. This supports the assumption that specific OM and moisture characteristics are responsible for the high ammonium-N availability (Nadelhoffer et al. 1991). NO_3^-–N contents are overall very low, but still significant relationships with temperature could be detected, emphasizing the strong temperature dependency of nitrification processes (Chapin et al. 2002; Robinson 2002).

Recent studies have reported inconsistent changes in SOC stocks in grasslands of the Tibetan Plateau over the past few decades (Chen et al. 2017). For example, Yang et al. (2009b) found slightly decreasing SOC stocks for alpine grasslands on the Tibetan Plateau at a rate of 0.6 g C m^{-2} yr^{-1} between the 1980s and 2004 due to grassland degradation. Contrarily, flux measurements demonstrated that the alpine meadows on the eastern edge of the Tibetan Plateau act as a strong C sink in the early 21st century (Yu et al. 2013). Repeated soil sampling to assess the changes in SOC stocks in the alpine grasslands across the Tibetan Plateau (Chen et al. 2017) exhibited a significant increase from 2002 to 2011 at an overall rate of 4.66 g C m^{-2} yr^{-1}. Whereas mesic and low-temperature-limited alpine meadows showed an average C gain, the relatively dry alpine steppes exhibited a slight C loss. These changes in SOC stocks were related to the original SOC stocks whereas soils with low C tended to gain C. Following the results of Baumann et al. (2009) on SOC contents, the changes were closely related to moisture associated with March–April precipitation in alpine meadows and with MAP in alpine steppes.

Besides SOC, soil inorganic C (SIC) should be considered for evaluating soil C stocks and their interdependencies. SIC includes lithogenic inorganic C (LIC) and pedogenic inorganic C (PIC). Whereas LIC originates from parent rocks or sediments, PIC is formed by dissolution and weathering of carbonatic parent substrates and the related precipitations in soils (Eswaran et al. 2000). For China, total SIC stocks in the top 1 m of a soil are estimated between 53.3 Pg and 77.9 Pg (Li et al. 2007; Mi et al. 2008; Wu et al. 2009). SIC stocks of the Tibetan grasslands were estimated by Yang et al. (2010b) to 15.2 Pg, with temperature and precipitation being identified as the most significant parameters correlating with SIC density. Shi et al. (2012) compared sites on the Tibetan Plateau with sites of a NE-SW stretching transect across the Inner Mongolian grasslands and showed that SOC exceeds approximately nine times the content of SIC. Significantly higher concentrations for both SIC and SOC are evident for Tibetan grasslands compared with the sites in Inner Mongolia, at site level as well as for depth increments. Since there are no significant differences of soil pH between Tibetan and Inner Mongolian grasslands, two other explanations are likely. First, soils on the Tibetan Plateau are relatively young, thus parent material and substrates trace their characteristics to related soils more strongly leading to higher LIC contents. In addition, carbonate migration is lower. Second, the much higher elevation of the Tibetan Plateau has a conceivable influence due to lower CO_2 partial pressure, which is enhanced by the lower soil respiration. The lower CO_2 partial

TABLE 3.3
SOC Stocks for Intact and Degraded Soil across the *Kobresia* Pastures on the Tibetan Plateau

Coordinates		Altitude (m)	SOC Stock (kg m⁻²)	SOC Stock (kg m⁻²)	SOC Loss (%)
			Intact Soil	**Degraded Soil**	
N35° 32′20.36″	E99° 50′56.89″	3430	7.1[a]	7.0[a]	−1.5
N35° 49′37.20″	E99° 39′05.34″	3760	5.7[a]	4.8[a]	−16.8
N33° 50′51.4″	E97° 10′49,704″	4431	7.4[a]	5.5[a]	−26.0
N33° 25′46.2″	E96° 08′59.60″	4461	8.8[b]	6.0[b]	−32.6
N32° 52′54,31″	E91° 55′ 16,06″	4515	4.5[a]	2.6[a]	−42.7
N32° 06′3.81″	E91° 41′36.60″	4735	4.7[a]	3.6[a]	−22.0
N31° 05′07,65″	E91° 41′ 19.61″	4735	4.6[a]	4.6[a]	+0.7

[a] SOC stock for 0–25 cm.
[b] SOC stock for 0–15 cm, (unpublished data).

pressure moves the equilibrium towards more precipitation of carbonates, benefitting the formation of PIC. Below pH 7, MAP and SM become more important, since with higher acidity, SIC tends to be in form of dissolved bicarbonate. Contrarily, above pH 7, thermal factors show stronger negative influence due to positive effects on biological activity and soil respiration (SR), implying increasing soil CO_2 partial pressure, hence inhibiting the formation of carbonates in topsoils (Shi et al. 2012).

The *Kobresia* pastures store huge amounts of SOC and it is estimated that a quarter of the organic carbon stored in China's grasslands is fixed within this alpine pastoral ecosystem (Ni 2002). A transect study across the *Kobresia* pastures revealed that on average 6.1 kg C were stored in the upper 25 cm of soil (Table 3.3), which is roughly 2 times higher than the global average (Batjes 1996, Wang et al. 2002). Nevertheless, the SOC stocks at the research sites near Naqu were up to 10 kg C m⁻² in the upper 30 cm presumably because of the dense and distinct root mat in this specific region (Kaiser et al. 2008; Schleuss et al. 2015). For this region, Ingrisch et al. (2015) showed that the *Kobresia* turf was the main compartment for the incorporation and the turnover of recently assimilated C using a ¹³C pulse labeling approach. In particular, the highly dynamic root system with short residence times of about 20 days was mainly driving the C cycling and thus contributing to C storage in this ecosystem.

3.5.2.5   Pattern and Control of Soil Organic and Inorganic Carbon
SOM in soils of the Tibetan Plateau reacts particularly sensitive to environmental changes (Baumann et al. 2009; Bosch et al. 2017). Zhang et al. (2007a) applied global C models to describe negative responses of SOC contents to increasing temperature, whereas Wang et al. (2007) used remote sensing data and Wang et al. (2008b) SOC fractionation studies to analyze the role of climate change for SOC and found the same negative feedback. Yang et al. (2009b) found SOC stocks to remain relatively stable for the timeframe from 1980 to 2004, resulting from increased biomass production through enhanced grassland productivity. The differing results lead to the assumption that response of SOC to changing climate parameters vary spatially to a high extend and on a small scale. This may be caused by numerous environmental factors showing high variations across the landscape at different temporal and spatial scales like permafrost distribution, soil texture, and related water logging (Hobbie et al. 2000; Schuur et al. 2008; Baumann et al. 2009; Yang et al. 2010c).

Generally, soil temperature (ST) is considered as a key factor of many terrestrial biochemical processes (Davidson and Janssens 2006), but land-use changes also have to be considered when evaluating SOC patterns (cf. Section 3.5.2.4). However, large portions of the Tibetan Plateau are moisture-limited with evaporation largely exceeding precipitation (cf. Section 3.5.4.1), thus SM has to be considered as another important parameter (Baumann et al. 2009). Further, permafrost reduces water infiltration, often leading to water saturation in topsoils. In certain relief positions, such as troughs, depressions, and valleys, this causes ponding surface water during the summer months. If permafrost degrades, the active layer thickness would increase leading to drainage processes in the topsoils. As a consequence, higher aeration can be expected associated with enhanced soil respiration rates, triggered by the improvement of oxygen supply to microbial decomposition processes (Wagner et al. 2009). If high SM is combined with low soil temperatures, the described processes are amplified by accumulation of SOM (Wang et al. 2006) giving again a positive feedback due to the isolating effect of dense vegetation and weakly decomposed SOM. The latter explains the high C contents, even though overall C uptake of the grasslands on the Tibetan Plateau is much lower than of grasslands at lower altitudes. Accordingly, it can be assumed that SM has a substantial influence on C and N cycling besides commonly considered temperature variables since it was found that SM has a strong effect on C and N contents. Climate parameters showed only moderate correlation to SM, giving evidence that permafrost and relief position are the major determinants for SM in high-altitude periglacial ecosystems. The strong link between hydrological conditions and permafrost, and thus between SM, C, N, and in particular mineralized plant available N indicates that nutrient supply is especially crucial for limiting plant growth even under higher temperatures. Geng et al. (2012b) showed that narrow-ranging species tended to occur at high altitude with lower temperature, but higher soil nutrient concentrations compared with wide-ranging species and patterns of leaf–soil nutrient relationships changed significantly across levels of range size. Narrow-ranging species seemed to be more sensitive to variation in soil nutrient availability than wide-ranging species, resulting in a shift from a positive leaf–soil nutrient relationship for narrow-ranging plants to no relationship for wide-ranging plants.

Degraded sites exhibit low C and N contents basically caused by two processes—higher mineralization rates under warmer and dryer conditions, and deposition of proximal airborne sediments. Rising air temperatures would trigger further permafrost decay, accompanied by lower SM contents and hence nutrient supply. For SR in Tibetan alpine grasslands, it could be shown that belowground biomass and SM, but not soil temperature, best explain the large-scale patterns (Geng et al. 2012a). This supports the interpretation that SM controlling C and N dynamics on a landscape and continental scale can override temperature effects. These may be more relevant in ecosystems, where SM or other factors are not limiting or altering the relationships between temperature and soil processes (Craine et al. 1999; Reichstein et al. 2003). Soil temperature is in turn more likely to account for seasonal and diurnal variations at site scale level (Kato et al. 2006). Notably, comparing plot and landscape scale is essential to analyze the relevance of water availability and temperature regimes at different levels.

Further, Baumann et al. (2014) have clearly identified pedogenesis and weathering processes as another important predictor for C and N contents of soils in the periglacial environment of the Tibetan Plateau. Additionally, $CaCO_3$ and sand contents are controlling parameters (Baumann et al. 2009). $CaCO_3$ contents mainly control soil acidity and, as an important parameter of pedogenesis, support the strong interrelationships between C and N contents and pedogenesis on the Tibetan Plateau.

The results for SIC storage and pools (cf. Shi et al. 2012) imply that SIC will decrease by 53% in the Tibetan grasslands, given the acidification rate for Chinese grassland soils in the future is comparable to acidification trends that have been measured in cropland soils adjacent to the investigated grassland during the past two decades. Nevertheless, the negative relationship between soil pH and SOC (Baumann et al. 2009) leads to the assumption that decomposition of SOC would in turn be slowed down. Hence, no significant impact by soil acidification on total soil C stocks can be expected, notably also because of the small average proportions of SIC.

Identifying the effects of grazing on SOC storage for the Tibetan Plateau is challenging as several studies indicate positive effects on the SOC stocks following grazing (Gao et al. 2007; Hafner et al. 2012), whereas others state negative effects (Sun et al. 2011, Dong et al. 2012). The main problem is that grazing activities are often not standardized and are based on inconsistent definition. For example, the livestock density (animal number per area), the temporal variations of grazing pressure (constant, decreasing or increasing grazing activities), the dominant grazing animal (yak, sheep or goat) and the grazing season (winter or summer pasture) are all important factors with feedbacks on plant productivity and finally on the C-cycle.

Grazing exclosure experiments provide a powerful tool for studying effects on SOC stocks, although they are based on the explicit assumptions that an absence of grazing follows inverse directions of increasing grazing intensities with respective effects on specific ecosystem properties (i.e., SOC storage). Further, they ignore that most grasslands are not grazing-pristine, but rather grazing from non-domestically animals has been a natural component for millennia, and so for the Tibetan Plateau. However, a recent meta-analysis demonstrated mainly positive effects on SOC storage for China's grasslands in the absence of livestock grazing (Hu et al. 2016). Using this data set for Tibetan sites only revealed similar patterns (Figure 3.15a). The SOC stocks increased on average by 6.2% per year comparing grazed with ungrazed sites. Moreover, the annual SOC changes were positively correlated with MAP (Figure 3.15b). This reflects strong interrelations between grazing-driven SOC dynamics (land-use-effect) and spatial precipitation variations (climatic effect), in the way that SOC storage relatively increases for ungrazed compared with grazed sites when MAP increases. The threshold between positive and negative effects were calculated for MAP of about 400 mm per year, consequently for higher precipitations grazing exclosure relatively increases the SOC stocks compared with free-grazed pastures. It indicates that in dry regions with limited water availabilities and shorter vegetation periods, the potential to recover the SOC stocks due to periods of overgrazing is lower. This is consistent with findings from Lehnert et al. (2016), who observed increasing grassland degradation, estimated by a decreasing vegetation cover, towards the dryer regions on the southwestern parts of the Tibetan Plateau.

Most studies state overgrazing as an important degradation driver because a constant removal of the photosynthetic shoot biomass imbalances the maintenance of the high belowground C costs (Schleuss et al. 2015). If this happens too frequently, it causes plants to die, then reducing the belowground C input and, in combination with proceeding SOC decomposition, decreases the SOC stocks. For the *Kobresia* pastures, the partly decomposed root mats are then claimed by crust forming lichens (Unteregelsbacher et al. 2012). These are characterized by a decreasing root biomass as well as declining amounts of organic C and N (Liu et al. 2016; Zhang et al. 2017). Identifying the SOC losses for these patches along a transect across the eastern part of the Tibetan Plateau revealed

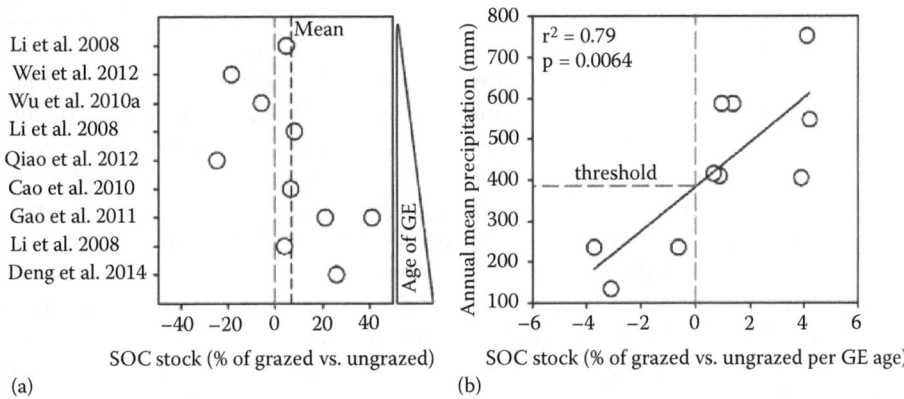

(a)   (b)

FIGURE 3.15   (a) SOC stock changes following grazing exclosure (GE) on the Tibetan Plateau and (b) relationship between SOC stock changes per GE age with annual mean precipitation (AMP). (Data from Hu, Z., Li, S., Guo, Q., Niu, S., He, N., Li, L., & Yu, G., *Glob. Change Biol.*, 22 (4), 1385–1393, 2016. doi.org/10.1111 /gcb.13133.). In Figure 3.13a the studies are arranged according to the age of the GE experiment (young to old from top to bottom).

average losses of 20% compared with non-degraded sites (Table 3.3). Near Naqu, which is located in the core area of the *Kobresia pygmaea* distribution, about 15% of the surface area was covered by these degraded root mats. Here, eddy-covariance observations combined with chamber measurements and modeling approaches predicted a strong decrease of net ecosystem exchange (NEE) for the degraded root mats even on landscape scale (Babel et al. 2014).

3.5.2.6 Implications of Regional and Land-Use Driven Feedbacks from Soils to the Climate System

The Tibetan Plateau, the largest geomorphological area in Central Asia, is the major region of permafrost outside the polar regions (Figure 3.10). Within the main types of grassland (alpine meadow, steppe meadow and alpine steppe), the Tibetan Plateau stores the highest amount of organic C in Chinese soils (Wang and Zhou 2001; Wang et al. 2002). Due to this fact, periglacial environments of the Tibetan Plateau play a major role in the global C cycle, especially due to the pronounced sensitivity of this region to climate changes (Jin et al. 2007). It is obvious that land-use changes and permafrost degradation may have a significant influence on the mechanism and processes of C balance between the pedosphere and the atmosphere. Around 18.6% of the permafrost on the Tibetan Plateau has degraded in the past 30 years and up to 46% permafrost is predicted to disappear in 100 years (Cheng and Wu 2007). The presence and condition of permafrost significantly regulates the hydrological environment of Tibetan soils. Degradation of permafrost has led to a lowering of groundwater levels, shrinking lakes and wetlands, and noticeable changes of grassland ecosystems from alpine meadows to steppes (Jin et al. 2009). Through the analysis of 30-year record of soil and climate data on the Tibetan Plateau, Zhao and colleagues (2004) could show that climate warming is most pronounced in the northeastern area of the Tibetan Plateau, which includes warming of air and surface temperatures as well as duration and depth of seasonally frozen soil.

More than 62% of the Tibetan Plateau is covered by alpine meadow (about 30%) and steppe meadow (about 32%) grassland vegetation, corresponding to an area of 1.6 10^8 ha and 40% of the Chinese grassland area in total (Hou 1982; Wu 1980). Alpine meadow ecosystems are especially crucial for C storage, currently acting as a C sink (Kato et al. 2004). Given the above-described pronounced sensitivity of the region to climate change and prevalent permafrost degradation, periglacial grassland environments of the Tibetan Plateau most likely play an important role for the present and future global C and N cycles, having been identified as one of the most sensitive areas of SOM loss in the last 20 years (Xie et al. 2007).

Wetlands on the Tibetan Plateau occur particularly under permafrost influence in specific relief positions, because evaporation exceeds more than three times precipitation (Wang and French 1994; Wang et al. 2006). On a worldwide scale, these wetlands contain related to their size by far the highest C stocks, and relatively quick turnover rates of OM for the turf-like upper layers of only some tens of years can be expected (Hirota et al. 2006). The presence and spatial distribution of permafrost (Nan et al. 2005) implies complex land-surface hydrological processes due to seasonal and diurnal variations of SM and ST regimes. Rising air temperatures lead to a greater active layer thickness, thus modifying the strongly interrelated preconditions. These processes will be amplified, if warming is accompanied by drier conditions (Yang et al. 2004a). Wang et al. (2006) reports pronounced land cover and biogeochemical changes, coupled with transformations of alpine meadow and marsh into steppe-like meadows, alpine steppes or even desertified land. Overall, substantial losses of SOC and N of degraded grassland ecosystems have been observed on the Tibetan Plateau over the past 20–40 years (Dai et al. 2011; Wang et al. 2001; Wang et al. 2008a; Wang et al. 2007; Yang et al. 2009b) and considerably modifies the C and N cycles (Cheng and Wu 2007; Lin et al. 2011).

Land-use changes amplify the described grassland degradation processes. Beside construction measures, such as road building, settlement or mining, livestock grazing is the most important factor of direct human impact in the region (Cheng 2005; Pei et al. 2006; Wu and Tiessen 2002; Zhang et al. 2006). Particularly overgrazing stimulates the negative feedback loops of climate warming and permafrost degradation (Zhou et al. 2005). This is mainly due to livestock trampling, nutrient and SOC loss as well as triggered soil erosion presses (Wu et al. 2010a; Zhang et al. 2006; Zhou et al. 2005). High vegetation cover implies higher SM contents, thus reducing the impact of heat cycling on permafrost by changed thermal conductivity, heat capacity and latent heat (Shur and Jorgenson 2007; Wang et al. 2010). Consequently, the direct influence of climate change can be buffered for a certain time frame, if vegetation cover remains intact.

Triggered by permafrost degradation processes, shifts in sedimentary and geomorphological systems are common, implying broad varieties and alterations of weathering intensities. Soil development on the Tibetan Plateau (cf. Section 3.5.2.1.3) is closely associated with specific weathering intensities under distinct environmental conditions (Baumann et al. 2014). Related shifts in the ecosystems can occur quickly, notably if strong proximal sediment input from degraded areas buries intact, well-developed soil profiles including vegetation cover. In other cases, gradually drying up of profiles could lead to lower vegetation cover and finally activated sediment outblow accompanied by deposition in certain relief positions. These processes are mainly linked to an increasing active layer thickness (Yang et al. 2011). Related desertification processes have been reported in various studies (e.g., Wang et al. 2011; Xue et al. 2009; Yan et al. 2009). Hence, tools for chemically describing and differentiating the sediment's weathering intensities and soil development processes was proposed by Baumann and colleagues (2014). They found that weathering trends along the climatic gradients could be outlined by weathering indices, whereas pedogenic oxides and pedogenic oxide ratios rather account for small scale variations, describing significant differences of pedogenesis between continuous and discontinuous permafrost conditions and related sedimentation processes.

The degradation processes are not only relevant on local and regional scales, but also in a global context. Higher decomposition rates of SOM and fundamental modifications of the C cycle can be expected with a shift to warmer and wetter climate on the Tibetan Plateau (Li et al. 2016). Additionally, desertification and changing sedimentary processes may alter SOC contents as well as weathering and soil forming processes (Qi et al. 2001; Baumann et al. 2009, 2014). In this respect, GHG emissions from soils are one important factor controlling climate change. SR defined as the CO_2-efflux to the atmosphere; fundamentally impacts the global C cycle (Chen et al. 2010). Most large-scale predictions of SR are based on MAT, MAP and belowground biomass (Luo and Zhou 2006). Compared to field measurements of SR at single spots in different vegetation zones on the Tibetan Plateau with a peak of 1876.63 g C m^{-2} y^{-1} (Zhang et al. 2009), Bosch et al. (2016) predicted a maximum of 1765.13 g C m^{-2} y^{-1}. An area-wide estimation of future potential CO_2 emissions for the permafrost region on the Tibetan Plateau was presented by Bosch et al. (2017). They calculated four potential SR scenarios for 2050 and 2070 each and provide an approximation of total potential soil CO_2 emissions on a regional scale ranging from 737.90 g CO_2 m^{-2} y^{-1} to 4224.77 g CO_2 m^{-2} y^{-1}. The calculations as first estimate of thawing-induced CO_2 emissions from permafrost soils with an SOC content ranging from 2.42 g C kg^{-1} to 425.23 g C kg^{-1} increase general SR by at least one third on average at a temperature of 5 °C. Thawing-induced CO_2 emissions generally decrease over time comparing 2015, 2050 and 2070.

CH_4 is a second major GHG originating as the end product of the anaerobic degradation of SOM by microorganisms. The CH_4 fluxes on the Tibetan wetlands range from 9.6 to 214 mg CH_4 m^{-2} d^{-1} (Jin et al. 1999; Hirota et al. 2004; Cao et al. 2008; Chen et al. 2013) and are generally comparable to Arctic permafrost regions (e.g., Sachs et al. 2010). Both, alpine marshes and wet meadows are thus main sources of CH_4 emission on the Tibetan Plateau. The CH_4 emission rate varies with wetland types, depending on environmental variables such as hydrological settings, permafrost extent, SM, ST (Wei et al. 2012, 2015; Yang et al. 2017), vegetation type (Jin et al. 1999) and root-rhizome architecture (Hirota et al. 2004). A recent measurement using the newly developed LI-COR LI-7700 open-path gas analyzer found that annual CH_4 emissions ranged from 26.4 to 33.8 g CH_4 m^{-2}, indicating that the traditional chamber-based measurement may over-estimate the CH_4 emission. In addition, the non-growing season CH_4 emissions accounted for 43.2–46.1% of the annual emissions, highlighting an indispensable contribution that was often overlooked by previous studies (Song et al. 2015). Rapid climate change and intensified human activities have resulted in water table lowering in some of the alpine wetlands on the east of the Tibetan Plateau. A mesocosm experiment controlling water table (decreasing 20 cm relative to control) showed that water table lowering reduces CH_4 emissions by 57.4% averaged over three growing seasons compared with control plots but had no significant effect on net CO_2 uptake or N_2O flux (Wang et al. 2016).

Besides release of GHG, changes in water balance as well as vegetation coverage and composition have to be considered. Degradation of alpine meadows as described above is associated with changes of soil water contents, thus having the potential to hydrological alterations on a regional scale (Wang et al. 2008a). Due to hydrological deterioration, degeneration from wetlands to meadows or from meadows to steppes have been observed at local scales (Jin et al. 2009; Brierley et al. 2016), which has subsequently impaired their roles in regulating the flow of rivers and C stores (Cheng and Jin, 2013). A latest estimate suggested that Tibetan wetland loss have led to approximately 20% reduction of the total amount of CH_4 emissions from natural wetlands (Li et al. 2016). Given the prominent role in the Asian monsoon system, not only does the global temperature increase but surface-heat-fluxes are also crucial for understanding the future development of the Tibetan Plateau. Moreover, some of the world's largest rivers, such as Yangtze and Yellow River originate on the Tibetan Plateau. Environmental changes in the headwater's regions will have consequences for most of the Asian continent, both in terms of water supply and impacts on riverine sediment flux.

3.5.3 Permafrost-Affected Soils, Climate, and Feedbacks in West Greenland

Peter Kühn and Jessica Henkner

Specific Challenges That the Region Faces Due to Climate and Land-Use Change

Climate change affects traditional ways of life, many parts of the society and the communities in Greenland, the cryosphere, all ecosystems and the related marine environment including rising sea levels. The effects of the melting of the Greenland ice sheet will have also effects on the climate system particularly of the Northern hemisphere (AMAP 2012). Generally, feedbacks from melting of permafrost-affected soils are closely related to increasing natural GHG emissions, since they store large amount of SOC. The SOC distribution in Greenland is mainly influenced by the mountainous topography and small scale periglacial landforms (resulting from processes like cryoturbation, palsa and peat formation, ice-wedges, thermokarst, aeolian activity). Scientific knowledge of the spatial distribution of SOC of the ca. 384,850 km² of the ice free terrestrial environment in Greenland is limited. In permanently frozen soils most of the C is withdrawn from interactions with the atmosphere and biogeochemical cycling (Wagner and Liebner 2009), but C in the active layer is available for microbial turnover and ecosystem processes and thereby likely to be emitted. Thawing of permafrost is one main source of releases of GHGs. Decreasing snow cover forces permafrost thawing and changes in vegetation cover and the loss of entire habitats in Greenland. The size of the C stock and the intensity of climate forcing at high latitudes may lead to a strong C feedback (Schuur et al. 2008). It is therefore important to improve global estimations by conducting regional to local research. To get a better regional assessment of SOC stocks distribution local SOC measurements combined with the distribution of landscape types will lead to valuable results (e.g., Henkner et al. 2016; Ping et al. 2008) in Greenland as well.

3.5.3.1 Environmental Settings

3.5.3.1.1 General Settings, Climate, Permafrost, and Soils

Greenland is stretching ca. 2.5 mio km² from 60°N to 84°N having less than 20% of ice free landscapes along the coast. The climate is diverse with a wet Low Arctic climate in the south with a mean annual precipitation >1000 mm and High Arctic climate conditions with <250 mm in the north (Jones 2010). West Greenland can be considered as being situated in between these two climate extremes in Greenland. Additionally, the ice sheet causes a distinct climate gradient from the inland ice to the coast, i.e., a general trend is noticeable from drier climate conditions next to the ice margin to moister climate conditions at the coastline.

The area around Kangerlussuaq in West Greenland is characteristic for an inland situation with an arctic continental climate (period 1976–1999, Kangerlussuaq meteorological station) having an annual mean air temperature of –5.7°C and 149 mm annual mean precipitation (Cappelen et al. 2001; Carstensen and Jørgensen 2009). Mean summer day temperatures are about 15 °C (June–August), highest precipitation occurs in August (monthly mean: 33 mm). About 150 snow-free and 80 consecutive frost-free days occur in summer (Bullard 2011). Closer to the ice margin climate becomes slightly colder but sunnier, drier and a higher wind speed. The bimodal wind regime is dominated by katabatic winds from the ice sheet, channeling in east–west oriented valleys particularly in winter, and westerly winds generated by Atlantic storms, which are less frequent and less strong (Dijkmans and Törnqvist 1991).

During the past century, warming of Arctic climate (>60°N) was approximately 0.09 K per decade, whereas on the entire northern hemisphere temperatures rose only by 0.06 K per decade (McBean et al. 2005). Recent climate data for the period 1979–2008 (Boas and Wang 2011) already show warming and higher precipitation (mean annual air temperature –4.8°C, mean annual precipitation 257 mm) with a particular increase of winter temperature and summer precipitation. The strong melting processes of the inland ice particularly recorded during the last decade are caused by changes in the large-scale atmospheric circulation (Tedesco et al. 2016).

Continuous permafrost is distributed in Greenland north of 70°N. In West Greenland continuous permafrost occurs in areas with more continental climate (closer to the ice margin), whereas discontinuous permafrost is characteristic closer to the coast. Sporadic permafrost represents the region south of the Arctic Circle (Jones et al. 2010). Where permafrost is formed in thick sediments, several meters of organic-rich sediments can be accumulated during several thousands of years and SOC is being protected from decomposition.

The soils in Greenland have generally developed during the Holocene. Related to the timing of the deglaciation the soils become gradually older from the coast to the current ice margin of the inland ice, i.e., in West Greenland from the west to the east. Recent studies show that the current ice margin in West Greenland was not overridden by glacier advances within the last 7000 years (Levy et al. 2012; Storms et al. 2012). This means that soils and permafrost formed within the last around 7000 years in the areas, which are currently ice free in the inland. Following the glaciation history, the soils and the permafrost closer to the coast have an age of around 9,000–10,000 years (Levy et al. 2012; Young et al. 2013). The SOC stored in the cryosphere will become available because of the warming of the Arctic. Soil temperatures at the permafrost table of artic soils already increased by 3 K since the 1980s (Lemke et al. 2007; Vaughan and Comiso 2013).

3.5.3.1.2 Vegetation

West Greenland can be counted to bioclimatic subzone E having a vegetative cover of 80–100% by vascular plants. Bioclimatic subzone D occurs in west south and east Greenland. Bioclimatic Subzone C is characteristic for east Greenland and B only for North Greenland (CAVM Team 2003). Main vegetation units of subzone E are tussock-sedge (dominated by Eriophorum vaginatum), dwarf-shrub (<40 cm), moss tundra, erected-dwarf shrub predominatly with Salix glauca and low-shrub tundra with Salix glauca, Vaccinium uliginosum, and Betula nana. (CAVM Team 2003). Fell-fields, oligotrophic mires (dominated by Eriophorum scheuchzeri) and grassland or steppe (dominated by Kobresia myosuroides and Artemisia spec) are also occurring (Jensen and Christensen 2003). A cryptogamic cover is widely distributed in all mentioned vegetation units (not only on barren sites) and may play an important role as a source for SOC in Arctic ecosystems (Elbert et al. 2012).

Vegetation next to the next to the ice margin can be grouped into three communities, according to species composition (Figure 3.16): (1) dwarf shrub heath contains *Betula nana*, *Salix glauca*,

FIGURE 3.16 Characteristic distribution pattern of vegetation communities in West Greenland close to the ice-margin. The light areas indicate dry steppe community dominatd by *Kobresia myosuroides*. The darker areas indicate dwarf shrub communities dominated by *Salix glauca* and *Betula nana*. The distribution mirrors the influence of wind and topography with dwarf shrub at lee postions, with Cyperaceae and *Eriophorum scheuchzeri* in moister areas, with dry steppe vegetation at windward positions and deflation areas and here with only minor influence of muskoxen on the general vegetation pattern.

Equisetum arvense, *Rododendron lapponicum*, Cyperaceae, mosses, and herb;. (2) moist grassland consists of Cyperaceae and wetland vegetation in general and *Betula nana*; and (3) dry steppe of *Salix glauca*, *Pyrola grandiflora*, *Betula nana*, *Kobresia myosuroides*, Poaceae, herbs (cf. Elberling et al. 2008b; van Tatenhove and Olesen 1994).

3.5.3.1.3   Soils (WRB)

Cryosols, Cambisols, Histosols, Leptosols, Regosols, and Umbrisols are the dominating Reference Soil Groups (RSG, IUSS Working Group WRB 2015) in West Greenland. Further north, beginning with the area of the Disko Bay, Leptic and Turbic Crysols dominate the landscape. This counts also for East Greenland at the same latitude. Podzols dominate in the very south, whereas Leptosols and Cambisols are the predominating RSGs in the southwest and west of Greenland (Jones et al. 2010).

Dystric Cambisols or Mollic Umbrisols are mapped for the Kangerlussuaq region close to the ice margin in West Greenland (Jones et al. 2010). These RSGs were not found in a valley beginning at the ice margin southeast of Kangerlussuaq, because the base saturation always exceeds 80% (Förth 2012), which is contradictory to the Dystric qualifier and the definition of Umbrisols (Henkner et al. 2016). The most common RSG found in the valleys close to the ice margin of West Greenland is the Haplic Regosol occurring on ridges, slopes, and at the valley bottom. The suffix qualifiers Humic, Siltic, Eutric, and Turbic occur in this order of frequency. Cryosols frequently occur on north-facing slopes, toe slope positions and the valley bottom. They sometimes have organic horizons and rarely exceed an active layer depth of 50 cm. Prefix qualifiers of Cryosols range from frequently occurring Turbic and Glacic to Mollic, Histic and Folic indicating cryoturbation, low temperature, and high SOC content. All Cryosols show cryogenic properties like cryoturbation and sorting of soil material. Most Turbic Regosols and Cryosols occur in the moist valley bottom. Dry crest positions are dominated by Haplic Regosols without turbic properties. Gleysols are generally found very close to widely distributed lakes (Henkner et al. 2016).

Arenosols are characteristic for sand sheets and dune fields frequently situated in the east-west oriented valleys close to the meltwater streams (e.g., Brookfield 2011; Müller et al. 2016; Willemse et al. 2003).

3.5.3.2   Land-Use: Natural or Cultural Landscapes

Generally land use in Greenland is low compared with other Arctic regions. Hotspots of land use are of course around the settlements and mining areas, but lower with increasing distance from them, because long roads generally do not exist between settlements. Construction activities, clearing for fields, sheep grazing and vehicle driving affected the landscape in SW Greenland during the last century (Jørgensen et al., 2013). Thus, wide unglaciated areas of Greenland generally can be considered as natural landscape with low human impact. However, grazing of caribous and muskoxen has an increasing impact on vegetation. Since 27 muskoxen were introduced again to West Greenland in the 1960ies the population is now >4000 individuals (Linell et al. 2000). Where muskoxen occur in larger quantities they control the distribution and growth of shrubs because shrubs are the main part of their diet. This means also that the response of shrub abundance to climate warming can be overprinted by herbivory even in remote regions like in Greenland (Post and Pedersen, 2008). Jørgensen et al. (2013) described an increasing shrub cover at E and SE slopes in areas with human disturbances, but without muskoxen grazing.

3.5.3.3   Carbon Stocks in West Greenland

Local studies show higher SOC stocks related to different topographic positions and vegetative cover than the modelled mean SOC stock (0–100 cm) with 6–15.9 kg m^{-2} predicts for West Greenland. So the overall mean of SOC stocks is 9.9 kg SOC m^{-2} already for 0–30 cm in a valley in West Greenland and extrapolated for 0–100 cm 30 kg SOCm^{-2}, which is higher than in the Canadian tundra with 14 kg SOC m^{-2} and at the lower boundary of the SOC stocks of the Alaskan tundra with 33–53 kg SOC m^{-2} for the same depth increment (Table 3.4, with references).

A comparison of predicted SOC stocks derived from global data with local studies shows the same picture in other areas. The Zackenberg area in East Greenland is estimated to store about 16.0–23.9 kg SOC m^{-2} in 0–100 cm, Elberling et al. (2008a) found 12–21 kg SOC m^{-2} in 0–50 cm, which is about twice as much as was predicted. In north eastern Greenland SOC stocks were highest under salix snowbed and fen vegetation, which has a higher coverage than the compared Dryas and Cassiope heaths (Elberling et al., 2004). Hobbie et al. (2000) found that growth form influences litter decomposition most, whereas temperature or other abiotic factors play a minor role.

Vegetation classes seem to predict SOC stocks sufficiently (Hugelius and Kuhry 2009). Moss litter for example is very slow in decomposition and therefore accumulates therefore in the soil leading to high SOC stocks. At sites with a high coverage of moss, high SOC stocks should be found (e.g., north-facing slopes in Umimmalissuaq Valley in West Greenland: 9.2–13.4 kg SOC m^{-2} in 0–30 cm). Even if SOM contents are high, SOC stocks can be low because of low bulk density. Graminoid dominated vegetation have higher SOC stocks than shrub dominated vegetation around Kangerlussuaq and with 22–29 kg SC m^{-2} in 0–60 cm much higher than predicted for 0–100 cm (Table 3.4).

Little local information is available about SOC conserved in permafrost >100 cm in West Greenland. Hugelius et al. (2014) predict 5–15 kg SOC m^{-2} for 100–200 cm and 0.1–5 kg SOC m^{-2} for 200–300 cm.

It is generally accepted that SOC stocks increase with moisture content, since litter or OM accumulates instead of being rapidly decomposed as in well-drained ecosystems (Hugelius et al. 2010; Hobbie et al. 2000). Soil moisture was found to be the most influencing parameter for the horizontal and vertical distribution of SOC stocks in high altitude grassland ecosystems (Baumann et al. 2009; Liu et al. 2012). SOC stock variations in dry periglacial ecosystems in Greenland could not be explained by soil moisture alone. Soil moisture, bulk density, and especially vegetation cover show positive effects on SOC stocks in the Umimmalissuaq valley near Kangerlussuaq. Landscape units representing slope gradient, relief position and aspect, correlate well with SOC stocks (Henkner et al. 2016). The use of landscape units as grouping variables is proposed, because they include landforms and vegetative cover.

TABLE 3.4

Comparison of Soil Organic Carbon Stocks of Greenland and Other Arctic Regions. This List Is Not Intended to Be Exhaustive

Region, Landscape Unit, Ecosystem Type	Mean SOC Stocks [kg m^{-2} 0–30 cm]	Mean SOC Stocks [kg m^{-2} 0–100 cm]	Study
West Greenland, Kangerlussuaq area		6.0–15.9	(Jones et al. 2010)
West Greenland	5–15	5–15	(Hugelius et al. 2014)
West Greenland, Umimmalissuaq valley (mean)	9.9	30.0*	(Henkner et al. 2016)
West Greenland, Umimmalissuaq valley, valley bottom soils	14.2	44.9*	(Henkner et al. 2016)
West Greenland, Umimmalissuaq valley, north facing soils	11.5	36.2*	(Henkner et al. 2016)
West Greenland, Umimmalissuaq valley, south facing soils	8.4	25.5*	(Henkner et al. 2016)
West Greenland, Umimmalissuaq valley, crest	6.0	14.5*	(Henkner et al., 2016)
West Greenland, Kangerlussuaq (graminoid dominated)		29.0 (0–60 cm)	(Petrenko et al. 2016)
West Greenland, Kangerlussuaq (shrub dominated)		22.5 (0–60 cm)	(Petrenko et al. 2016)
NE Greenland, Zackenberg, grassland (total carbon)	7.6	12.3 (0–50 cm)	(Elberling et al. 2008a)
NE Greenland, Zackenberg, Salix snow-bed (total carbon)	10.5	21.2 (0–50 cm)	(Elberling et al. 2008a)
Russia, Peat	20.9	81.3	(Stolbovoy 2002)
Canada, Tulemalu lake area	16.3	33.8	(Hugelius et al. 2010)
Siberia, Tussock Tundra	16.3	25.5	(Uhlířová et al. 2007)
Russia, Pre-Tundra/northern taiga	13.3	26.9	(Stolbovoy 2002)
Russia, humic accumulative soil	11.7	20.2	(Stolbovoy 2002)
Russia, Tundra	11.6	16.6	(Stolbovoy 2002)
Siberia, Usa Basin	10.7	25.5	(Kuhry et al. 2002)
Siberia, Lena delta	7.7	25.7	(Zubrzycki et al. 2014)
Russia, Cryozems	4.6		(Stolbovoy 2002)
Russia, shallow soils	3.1		(Stolbovoy 2002)
Russia, Polar desert	2.9	2.9	(Stolbovoy 2002)
Arctic frozen peatlands		86.0 (total)	(Tarnocai et al. 2007)
Arctic unfrozen peatlands		43.0 (total)	(Tarnocai et al. 2007)
Alaska, wet tundra		68.0	(Bockheim and Tarnocai 1998)
Alaska, Arctic tundra, upland		53.3	(Johnson et al. 2011)
Alaska, Arctic		47	(Michaelson et al. 2013)
Alaska, Arctic tundra, wetland		44.2	(Johnson et al. 2011)
Arctic, frozen mineral soils		42.0	(Tarnocai et al. 2007)
Alaska, Arctic tundra, lowland		32.8	(Johnson et al. 2011)
Siberia, Tussock tundra		30.7	(Gundelwein et al. 2007)
Tundra		21.8	(Post et al. 1982)
Canada, Tundra		14.2	(Jobbagy and Jackson, 2000)
Arctic unfrozen mineral soils		12.0	(Tarnocai et al. 2007)

(Continued)

TABLE 3.4 (CONTINUED)

Comparison of Soil Organic Carbon Stocks of Greenland and Other Arctic Regions. This List Is Not Intended to Be Exhaustive

Region, Landscape Unit, Ecosystem Type	Mean SOC Stocks [kg m^{-2} 0–30 cm]	Mean SOC Stocks [kg m^{-2} 0–100 cm]	Study
Alaska, dry tundra		12.0	(Bockheim and Tarnocai 1998)
N circumpolar permafrost region	1.0	2.6	(Tarnocai et al. 2009)
N circumpolar, Histosols	22.5 ± 0.5	69.1 ± 0.9	(Hugelius et al. 2014)
N circumpolar, Histosols		69.9	(Tarnocai et al. 2009)
N circumpolar, Turbels	14.7 ± 1.5	33 ± 3.5	(Hugelius et al. 2014)
N circumpolar, Turbels		32.3	(Tarnocai et al. 2009)
N circumpolar, Orthels	15.8 ± 2.6	25.3 ± 4.1	(Hugelius et al. 2014)
N circumpolar, Histels	18.1 ± 3	49.3 ± 8.4	(Hugelius et al. 2014)
N circumpolar, High Arctic	9.8 ± 7.4	17.8 ± 11	(Hugelius et al. 2014)

Note: * = Estimation of SOC, ( ) = Differing Depth.

3.5.3.4   Implications of Feedbacks from Permafrost-Affected Soils to the Climate System

So far it is to a great extent unknown how much carbon is stored in permafrost-affected soils and permafrost in Greenland. Different climate models predict an increase of the global mean surface temperature of up to 4.8 K until 2100 (IPCC 2007). Especially winter air temperatures will be significantly warmer (4–7 K) across the terrestrial Arctic (N60°N Arctic Climate Impact Assessment, 2004). This will convert many areas of regions with continuous permafrost to discontinuous or sporadic permafrost and some permafrost-affected regions to regions without permafrost the thawing of permafrost will release a huge amount of ancient organic matter stored for many centuries or many thousand years.

With rising air temperature the active-layer depth will increase and alter hydrological conditions; SOM sequestered in permafrost and permafrost-affected soils is likely to decompose to a larger extend and become a source for additional release of GHGs like CO_2, CH_4 and N_2O to the atmosphere (e.g., Schuur et al. 2008; Elberling et al. 2010; Nielsen et al. 2017; Zimov 2006). This release is then mainly driven by microbiological processes related to the kind of decomposition, modification and digestion of SOM by microorganisms in Arctic ecosystems (e.g., Elster et al. 2016; Wagner and Liebner 2009). Nevertheless, there are still many questions unanswered related to the precise processes, e.g., the availability of recently thawed SOC to microorganisms, the velocity of changes of microbial communities and their function with changing temperatures. This will be important for the prediction of future GHG emissions related to permafrost thawing particularly in Greenland.

3.5.4   Brazil: Soil Type and Subsoil Matters for Land-Use Related Feedbacks to the Climate System in Central Brazil

Jens Boy

Looking out of an airplane above Mato Grosso, a central Brazilian state, would easily give you a clear first impression on the specific challenges. The midwest of the US would appear a small-scale farmland in comparison. Even though up to 80% should remain native ecosystems by law—the remnants of these are either along rivers as riparian buffer zones or isolated patches randomly scattered over the intensively used agricultural matrix. Less apparent from above is the missing infrastructure for providing the possibility of a more diverse agricultural industry. Farmers of a greater region are forced to grow the same products (here corn, cotton, soy bean) leading to incredible large-scale monocultures. Mato Grosso was some decades ago one of the largest agricultural frontiers in the world, but changed to a largely precision-farmed, stock-market driven protein fabric to saturate the lust for meat elsewhere on the planet. In the meanwhile, the agriculture frontier moved further north deep into the state of Pará, yielding a space for time gradient of change, both in land-use history as well as sociological and political processes. Due to the vast area of Brazil, appearing endless to its inhabitants, sustainable use of agricultural land is often not the most economical option as there is still so much forest to slash down behind the horizon. Thus, the agricultural frontier is proceeding unhindered by environmental concerns further and further into the pristine rainforest, following the ever-same chain of timber extraction, followed by pastures which finally transforms into crop-land. Nevertheless, Brazil is a good example that intensification of agricultural production must not be the least favorable management style, as the land use of the forest invaders at the agricultural frontier is both highly destructive and little productive. In other words, a lot of GHGs are added to the atmosphere, the sink capacity of the tropical rainforest is erased and still not many mouths fed by this action. On the other side of the gradient, where the agro-industry took over long ago, crop production is huge and a far better ratio between food produced and GHGs emitted is achieved, but for the price of low resilience to climate-caused disasters increasingly hitting the region. Brazil is ranked as the third largest emitting country of GHG from agriculture for the year 2012 (FAO 2017), and trends are rising. It is the doubled impact of the forest conversion which ensures this leading position, and decisions weighing the good against the evil for any land-use type is a prerequisite to intensify the agriculture where it does the least harm and spare nature where it is the most rewarding from the perspective of global change mitigation. For this, a closer look on the actual change in soil carbon stocks and fluxes of N_2O (the agricultural GHG) as a function of, e.g., soil types or land uses is vital and presented in the following.

3.5.4.1 Environmental Setting

Soil C stocks and N_2O fluxes from typical land-uses were measured along the federal highway BR-163 which connects agricultural industries in Mato Grosso via the Amazon to the world market. It is also the main cause for a south to north migration of the agricultural frontier. It spans the ecological zones from cerrado (savannah type forest-grassland patches) to rainforest. Three regions were chosen for investigations.

Region 1 (R1) lies in southern Mato Grosso close to Primavera do Leste close to the large city of Cuiabá. The MAT is 24°C and the MAP is 1500 mm. The dry period lasts from April to September. The native vegetation of southern Mato Grosso is tropical savanna (Cerrado), which is not an open grassland but represents a moving contiuum from dense forest (Savana Florestada—Cerradão) over Savana Arborizada (wooded savanna) to open savannah (Savana Gramíneo-Lenhosa, Campo-Limpo-de-Cerrado) (IBGE—Instituto Brasileiro de Geografia e Estatistica 2012). Savana Arborizada is the most common Cerrado vegetation within R1 and was considered as reference native vegetation. Region 2 (R2) lies in central Mato Grosso and is close to the town of Sinop. The MAT is 25°C and the MAP is 1700 mm. The semi-deciduous pristine forest is classified as Floresta Estacional Sempre-Verde (IBGE—Instituto Brasileiro de Geografia e Estatistica 2012) reaching a canopy height of approximately 30 m. Pasture were substituted by virtually undiversified cropland of soybean/corn or cotton. Region 3 (R3) lies within the state of Pará and encompasses the town of Novo Progresso. The MAT is 25°C and MAP is calculated to be 2450 mm. There is a dryer period between June and September (Patry et al. 2013). The tropical rainforest mapped as Floresta Ombrófila Densa/Aberta Submontana (IBGE—Instituto Brasileiro de Geografia e Estatistica 2012) has an upper canopy layer at about 30 m while single trees may reach 50 m. Pastures usually follow selective logging and the burning of the leftover vegetation. Ferralsols are the dominate soils of southern (R1) and central (R2) Mato Grosso, and Acrisols are found more commonly in Novo Progresso, Pará (R1). Acrisols cover around 25% of Mato Grosso. Clay content in Ferralsols averaged at 58% in topsoil and 63% in subsoil. For Acrisols the average was 24% in topsoil and 45% in subsoil.

3.5.4.2 Land-Use (History; Pristine or Old Cultural Landscape, etc.) and Land-Use Change

The south of Mato Grosso (R1) has the longest history of converting native vegetation to arable land and it is characterized by the highest mechanized agriculture of the study transects. Presently it is dominated by intensified soybean/corn crop rotations increasingly modified with cotton as substitute for corn. Lacking infrastructure for large-scale storage, transportation and local consumption, there is virtually no other land use besides some pasture remnants. The arable lands are managed under NT, but also at R1 harrowing is necessary every 3–5 years to avoid soil compaction (information from cooperating farmers). Throughout all three regions, non-native grass varieties virtually limited to *Brachiaria* spec. or *Panicum maximum* are used to saw pastures

R2: The name of the city Sinop goes back to "Sociedade Imobiliária Noroeste do Paraná" (Northwestern Paraná Real Estate Society), which funded the town in 1974. It intentionally enhanced agriculture in northern Mato Grosso as a measure of development. Agricutural practice today does not differ much of that in the more southern parts (R1), besides time, since land-use change is shorter and forest remnants are typically more abundant and tend to be larger while not being limited to the buffer zones towards the rivers.

The youngest agricultural history, still in the phase of actual deforestation, applies for R1. The main land-use following deforestation is livestock farming and early adoption of mechanized agriculture. Pastures usually are not fertilized and the clearing of bushes and re-grown trees, as well as harrowing and liming is rarely utilized. However, the creation of initial cropfield is mostly initiated by the removal of the remaining trees and roots, followed by tillage (to a depth of 20–30 cm) followed by fertilization and liming.

3.5.4.3 Effect of Land-Use Changes on SOC and N_2O (Reiche Ich Nach)

3.5.4.3.1 Soil Types

As it was explained in other chapters (cf. Section 3.5.1), scale and regional characteristics are important factors in SOC dynamics, if not confounding factors in many studies. Across their whole study site in the Amazon, Strey et al. (2016) found no large differences in SOM among the land-use types following conversion (Figure 3.17). However, when the soil type is more closely analyzed, distinct differences appear between soil types, which are also regionally (i.e., spatially) distributed.

The more weathered Ferrasols respond almost in complete contrast to the less weathered Acrisols (Figure 3.17). For example, young pasture lands converted from natural vegetation exhibited strong reductions in OC stocks in Ferrasols, whereas they exhibited increased OC stocks in Acrisols. These differences in soil types may give insight into why such contrasting results are abundant in the literature. Thus, it is possible to find some studies reporting increased OC stocks (e.g., Braz et al. 2013; Koutika et al. 1997; Moraes et al. 1996), while others report decreases (da Silva et al. 2004; Marchão et al. 2009). When such studies are combined, the larger-scale lack of differences in OC stock following land-use conversions may be the result of these soil-type differences cancelling each other out (e.g., Batlle-Bayer et al. 2010). Fujisaki et al. (2015) identified a similar phenomenon, from reviewing the effects of conversion of forest to pasture on OC in soils at a global scale. While they found OC stock changed from −50% to +160% within different regions and in different studies, at a global scale the changes were not significant.

This underlines the main take-home message of this chapter that universal upscaling from small to regional or even larger scales is problematic. There is an inherent risk of over-generalization even at regional aspects but certainly at global perspectives.

The observed OC stocks and concentrations of the two most frequent soils in the study region, Acrisols and Ferralsols, matched literature reports (Batjes 2005; Bernoux et al. 2002; Carvalheiro and Nepstad 1996; Desjardins et al. 2004; Zinn et al. 2005). The two soil types showed significant differences in soil C stocks under native vegetation. Therefore, capacity to store C differs between Ferralsols and Acrisols, which fits to the knowledge that soil types matter for C accounting (Jungkunst et al. 2012).

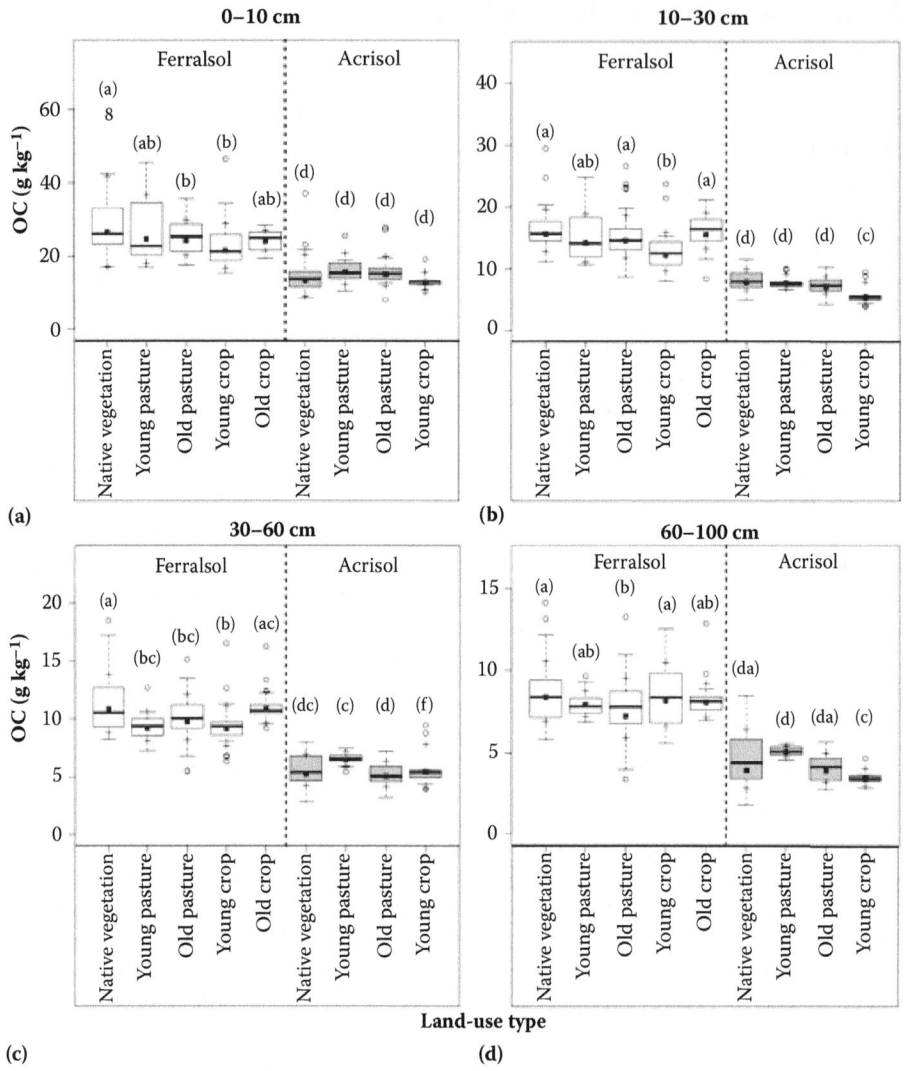

FIGURE 3.17 Land-use type and organic carbon content in soils of southern Amazonia.

3.5.4.4 Ferralsol

Ferrasols and Acrisols differ in many of their soil properties, some of which may explain the differences observed in OC stock dynamics. Reactive minerals play a key role here, as they may explain the initial differences of SOC stocks between different soil types. But what they can't be accused for are the different losses of forest-derived OC, nor the C newly incorporated into soil stemming from the conversion to pasture (Figure 3.18). The exposure of SOM to decomposition (oxidation) fundamentally differs between the two soil types. In Ferralsols, the soils are less compacted, thus allowing for better gas exchange and better aeration in subsurface layers (Buol and Eswaran 2000). In contrast, Acrisols tend to have higher bulk densities resulting in reduced aeration. Coupled with the fact that Acrisols also tend to be water-logged (Quesada et al. 2011), exacerbating the inhibition of aerobic decomposition, it should be expected that OC stocks from pasture-derived OC would accumulate in Acrisols (Davidson and Janssens 2006, Stockmann et al. 2013). Surface characteristics, such as soil compaction and crusting that are more common in Acrisols (West et al. 1998), also would tend to reduce OC loss. As in Section 3.5.3, the condition of the OC is also important

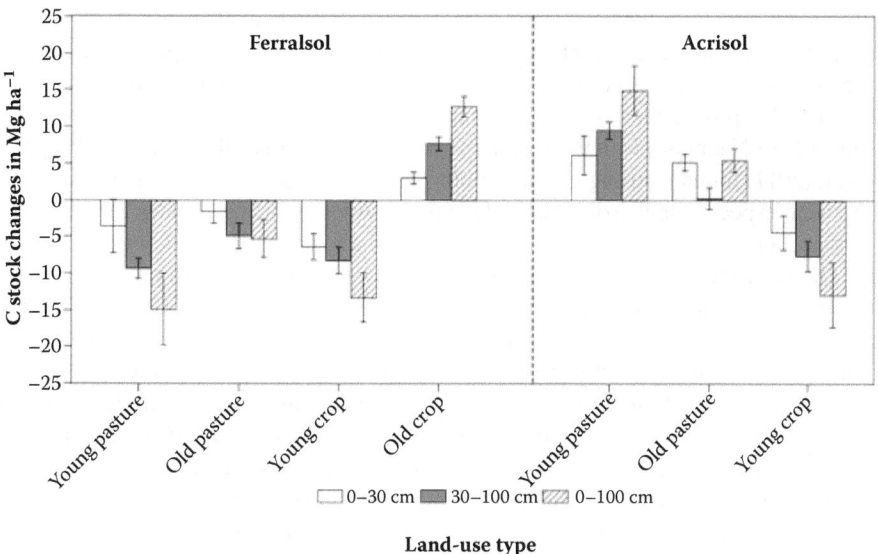

FIGURE 3.18 Organic carbon changes in soils under different land use in southern Amazonia.

for loss or gain rates in soils. The more labile OC in Ferrasols would be more readily decomposed than the more recalcitrant OC found in Acrisols. Thus, the loss rates are expected to be higher in Ferrasols from the beginning. It follows that the inherent soil characteristics, and their differences in this case, are key to understanding the OC stock dynamics among soil types. The argument that management practices of pasture or crop lands can be largely dismissed because these tended to be similar between soil types in the Amazon (Strey et al. 2016; Figure 3.19).

One should also be aware of the vertical distribution of OC and the OC dynamics. Most studies focus on the top 35 cm of soil as reported in Don et al. (2011). Strey et al. (2016) showed that most of the SOM is found at deeper soil depths, thus sampling to shallow depths can miss the most interesting part of the story if these depths are ignored. Changes to OC stocks at depths of 30–100 cm were even greater than those in the surface soils (Figure 3.19). Similar to more recent acknowledgement

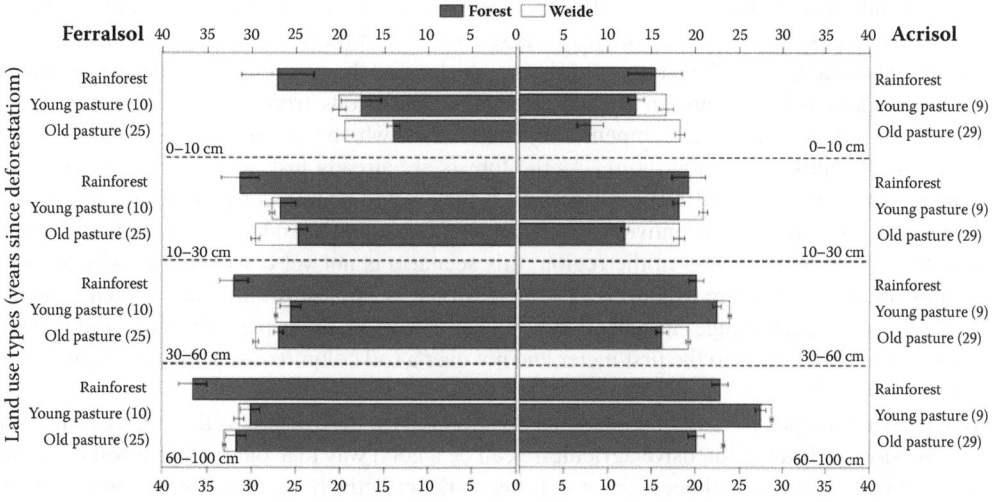

FIGURE 3.19 Soil organic carbon stocks in soil of southern Amazonia under different land uses.

that the deeper OC stocks in temperate and boreal soils will likely contribute to significant changes in terrestrial to atmospheric gas exchanges, rates of gas exchange in tropical soils also needs to be considered if it comes to OC dynamics in greater soil depths, although changes in SOC stocks seems to be not that pronounced on the first glance.

In terms of N_2O-N emissions in the regions of south Mato Grosso (R1) and Novo Progresso (R3), Meurer et al. (2017) found a strong effect of both land use and soil type. High emission rates from agriculture were expected as Brazil is ranked the third largest emitting country of GHGs from agriculture. Consequently, the Amazon virtually is used as a synonym for "everything is large here." However, a literature review revealed that measured N_2O-N emissions are low for Central Brazil, and natural rainforests tend to emit more than agricultural sites (Meurer et al. 2016). Therefore, a hint is given that for intact biogeochemical cycles, a certain loss of N_2O is to be expected as forests tend to emit similar amounts across the globe and disconntected cycles can also result in relative lower emissions compared to natural sites. At the landscape scale, higher emissions are most often associated with higher clay content. Sandy soils, such as Arenosols, are likely to emit less N_2O-N compared with fine-textured Ferralsols. Fluxes observed by Meurer et al. (2017) were the highest from a Ferralsol under rainforest (R3) (16 μg N m^{-2} h^{-1}) and actually negative on average from an Arenosol under Cerrado vegetation (R1) (−0.05 μg N m^{-2} h^{-1}). Cattle pastures were apparent in both regions (R1 and R3), and in Novo Progresso (Ferralsol and Acrisol) an age-related decrease in N_2O-N fluxes could be observed. Emissions from the cattle pasture in R1 (Arenosol) were close to zero (0.09 μg N m^{-2} h^{-1}). Overall, the amount of lost N in the form of N_2O-N was extremely low, and no emission peaks (i.e. short periods of high emissions) were detected following rainfall events and soil rewetting. However, these observations do not account for losses during actual land conversion, which can be expected to be magnitudes higher (see Schaldach et al. accepted). The observations by Meurer et al. (2017) and literature data (Meurer et al. 2016) orginate from dynamic steady states of forest or pasture, but not while land-use was changing. N_2O is known to react sensitively to external changes like frost/thaw and dry wet cycles. It therefore appears likely that most N_2O is emitted during the year or two when the forest is burned down or converted in any other way.

3.5.4.5 Conclusions

The observed results draw the attention to the difficulties arising in policies and management recommendation if only C-stock changes are taken into account. From the perspective of the soy farmers lobby groups, everything is perfect with their management style, as there are little changes in SOC stocks observed, thus no additional CO_2 seems to be emitted as compared to natural forests. But this is only true in a comparably short perspective (a few decades max), and nobody knows the fate of the forest-derived SOC which still is sequestered in the soils in deeper horizons, which won't be rejuvenated, as the forest now is missing. As long as the crop can refill these SOC stocks, this causes additional problems in REDD (**R**educing **E**missions from **D**eforestation and Forest **D**egradation) schemes, financial compensatory instruments which reward avoided GHG emissions, but not possible emissions in the future. As the forests are already lost, the farm lands could only gain increasing value under REDD perspectives, if they would be transformed into forests or plantations again, a strong false incentive to consume new forest and abandon yet unproductive fields. Due to the high productiveness in the region, this scenario is not yet common, but will certainly gain momentum in the future as judged by other regions' examples. Additionally, REDD schemes typically do not respect changes in subsoil SOC and are therefore favorable for agricultural systems, which concentrate biomass to the first meter and not over a soil column of several meters, as it is the case in tropical forests or Cerrados (Strey et al. 2017).

If seen from the perspective of forest conservation (forest destruction still is the real motor of GHG emissions in Brazil), intensive agriculture can be a good way to avoid further forest consumption by agriculture if political decisions and frameworks effectively regulate forest conversion in a land sparing approach (Boy et al. 2016). The latter means that larger bodies of natural forests would be strongly protected, while production on already "destroyed" lands is pushed to its upper limits.

This situation is already largely achieved in the Mato Grosso area (Region 1), but far away in Region 3, where vast areas of forest are destroyed for low productive meat production. Thus, especially in Brazil, the story of climate feedbacks is much more complex than CO_2 or C stocks in soil.

3.5.5 Long-Term Field Experiments in Sweden

Thomas Kätterer

Specific Challenges That the Region Faces Due to Climate and Land-Use Change

Large SOC losses have occurred during the historic expansion of agricultural land (Lal and Follett 2009). Soil C stocks most often lower in agricultural soil compared to grassland or forests (Poeplau et al. 2011). Agricultural soils depleted in SOC should therefore have the potential to sequester SOC and thereby mitigate climate change (Lal et al. 2011). It is likely that this can be achieved in most regions of the world when implementing region-specific best management practices (Minasny et al. 2017).

The response of SOC to changes in management is slow. Compared to the large amounts of C stored in soil and their spatial variability, annual changes in SOC are small and have to accumulate for decades for being measurable in the field (Kätterer et al. 2011). Therefore, long-term field experiments (LTEs) are indispensable source of information for quantifying the effect of agricultural management on SOC (Kätterer et al. 2012).

In 2003, the world-wide number of LTEs (>10 years) was estimated to 625 (Debreczeni and Körschens 2003). Since then this number has considerably increased. In a more recently published systematic map, 735 field studies with duration of at least ten years reporting on SOC changes have been identified solely in the boreo-temperate regions (Haddaway et al. 2015). Several meta-analyses are presently conducted using the comprehensive database presented in this map (Haddaway et al. 2016a,b) but the results are yet not published.

The Swedish University of Agricultural Sciences is hosting a considerable number of LTEs, focusing on management of arable land. Several of these experiments have been closed down during recent decades but still about 60 LTEs are on-going (Bergkvist and Öborn 2011). The management practices that are considered include cropping systems, rotations vs. monocultures, fertilization, liming, addition of organic amendments, tillage systems, subsidence of organic soils, water quality management including individually drained experimental plots. The majority of these LTEs belong to series of sites with identical or similar experimental layout at different locations throughout the country. In the following some highlights from recently published results from these experiments with focus on the effect of fertilization, crop residue handling, cover crops, rotations with different proportions of perennial vs. annual crops and recycling of organic wastes are presented.

Long-term application (>45 years) of mineral fertilizers resulted in opposite effects on long-term C storage in 10 meta-replicated LTEs. With N fertilization, SOC stocks followed yield increases. However, for all PK levels the response was the opposite (Poeplau et al. 2016a). The stimulating effect of P-fertilization on soil respiration was further confirmed in a follow-up study under controlled conditions (Poeplau et al. 2016b). The positive response of SOC to N fertilization was significantly and positively related to NPP. In average, SOC stocks increased by 1–2 kg C for each kg N applied (Kätterer et al. 2012). Similar responses to N fertilization were also reported from a compilation of 17 European, mostly German, LTEs (Körschens et al. 2013).

The positive effect of cereal straw incorporation on SOC balances increased with clay content (Poeplau et al. 2015a). Thus, on sandy soils, other management options than straw incorporation may be more efficient for increasing SOC. The use of straw as feedstock for bioenergy production and thereby substituting fossil energy sources could be an option where the conservation of soil quality is secured through other soil-improving management.

Data from 15 long-term sites showed that SOC stocks in ley-dominated rotations increased by about 0.5 Mg ha^{-1} yr^{-1} compared with those dominated by annual crops (Kätterer et al. 2013). The effect varied depending on plant traits, soil depth, management and the duration of the studies. The higher C allocation in perennial compared with annual plants to below-ground tissues was identified as the main driver for C sequestration in ley-dominated systems (Bolinder et al. 2007, 2012). Perennial forage crops are highly productive throughout the growing season. This leads to lower soil temperatures and soil water content and, therefore, to decreased decomposer activity (Kätterer and Andrén, 2009). Furthermore, and probably more importantly, roots have longer turnover times and contribute relatively more to the formation of stable SOC than above-ground crop residues (Hénin and Dupuis 1945; Kätterer et al. 2011). This is most likely also the main reason for the generally higher SOC stocks in grasslands compared to croplands (Kätterer et al. 2008).

Cover crops were introduced as erosion control and to reduce nutrient leaching, mainly in systems dominated of annual crops. In Sweden, ryegrass which is sown together with the main crop has been frequently used as cover crop. In this way, cropping systems dominated with annual plants get more similar to perennial systems because periods with bare soil are reduced and net primary production is increased. In Swedish experiments, incorporation of ryegrass resulted in C sequestration rates of 0.32 Mg C ha^{-1} yr^{-1} in average over the studies (Poeplau et al. 2015b). This is perfectly in line with the average sequestration rate found across all climatic zones and cover crop species (Poeplau and Don 2015). However, the net effect of cover cropping on the whole GHG balance of the system has still to be resolved. The surplus of easily decomposable biomass after incorporation of the cover crop may stimulate N_2O production and emissions from the soil (Petersen et al. 2011).

Application of organic amendments deriving from ex situ sources, such as farmyard manure, slurries and sludge, are frequently applied on agricultural fields. Their effect on SOC has been quantified in many LTEs and has been shown to vary depending on amendment properties and soil type. The Ultuna long-term SOM experiment is especially designed to quantify retention of C from different organic amendments because identical amounts of C have been added bi-annually since 1956 as straw, green manure, sawdust, peat, farmyard manure and sewage sludge. C retention after 53 years varied between these amendments from 15 to 59% (Kätterer et al. 2011). C retention from composted household-waste was as high as 85% after 13 years of repeated application to 40 cm depth according the analysis of another Swedish LTE (Kätterer et al. 2014).

Although application of organic amendments generally increases C stocks where they are applied, it has to be emphasized that this effect rarely can be scaled up to a C sequestration potential at regional or global scale. Increases in SOC through manure application, for example, are governed by the number of livestock in a given region. Whether this manure is uniformly distributed among fields or not will have minor impact on regional SOC balances. Only changes in livestock density in a given region will impact on regional SOC stocks. However, if amendments are deriving from sources that presently are not or only partly applied to soil, net SOC sequestration may occur. For example, slurries from biogas plants which use feedstocks such as household waste that previously have not been recycled have become a net source in several regions. The same applies to sewage sludge which presently is only partly recycled to agricultural fields. Nevertheless, the total C sequestration potential of sewage sludge is quite low and has been estimated to only about 10 kg C ha^{-1} year^{-1} under Swedish conditions (Kirchmann et al. 2017).

SOC sequestration is a time-dependent and reversible process (Andrén and Kätterer 2001). Management that results in C sequestration must be maintained for ever to avoid its reversal. Under Nordic conditions, it may take 100 years or more to reach a new steady state of SOC due to changes in land-use or agricultural management practices, whereas this can be reached within a few decades under warm humid climate conditions (Andrén et al. 2007). Thus, the duration of soil C responses differ between regions.

The management strategies that were identified to be most effective for SOC sequestration were all primarily determined by the total amount of CO_2 fixed through photosynthesis per unit area

(Kätterer et al. 2012). Minimizing the time with bare soil, is certainly an efficient management option (Kätterer et al. 2011). Sustainable intensification of crop production is therefor probably the most effective strategy for SOC sequestration and a prerequisite for preventing further areal expansion of agriculture at the global scale.

3.5.6 Russia: Regional Example Komi Republic, European Russia

Julia Schneider

The boreal zone of the Komi Republic is mainly covered by coniferous forest and peatlands. The human influence on the Russian forest ecosystems has been growing over the last decades and caused rapid forest cover changes. One of the main reasons for this is clear cutting or intense selective logging. These lead to changes in closely linked water and C cycles. The clear cutting might lead to paludification, which is a process of accumulation of poorly decomposed OM mostly originating from *Sphagnum* spp. which involves the formation of waterlogged conditions (Joosten and Clarke 2002). The thick moss organic layer on the forest floor and waterlogged soils favors production of the GHG CH_4. Peatlands are well known to be a long-term sink for atmospheric CO_2, but the CO_2 fluxes could be substantially altered in a changing climate, and peatlands may become a source of atmospheric C. Recently, some boreal peatlands were shown to be sinks of atmospheric CO_2; whereas, others were shown to be sources. Although it is unclear if the warmer projected climatic conditions for this part of Europe in the 21st century will be counterbalanced by an appropriate amount of increase in precipitation (Kirtman et al. 2013), the changing climate will also influence the methane fluxes of the peatlands.

3.5.6.1 Environmental Setting

The Ust-Pojeg investigation site is located in the Komi Republic, northwestern Russia. The climate is boreal continental and all-year humid with maximum precipitation in summer. Winter is characterized by freezing temperatures and continuous snow cover. The long-term (1986–2015) mean annual temperature in the region is 1.5°C, ranging from monthly averages of −14.2°C in January to 17.4°C in July, and mean annual precipitation is 630 mm, ranging from monthly averages of 33 mm in March to 77 in August (RIHMI-WDC 2016).

The peatland study site is a typical river valley peatland in a transitional state from fen to bog following paludification and so consists of minerogenous, ombrogenous, and transitional zones. It is not underlain by permafrost. Both the northern bog part (ombrogenous) and the southern fen part (minerogenous) of the study area at the Ust-Pojeg peatland are composed of a mosaic of different microforms defined by their situation within the microrelief (Figure 3.20). Hummocks which represent the driest conditions are elevated above the surrounding area and are covered by *Andromeda polifolia, Chamaedaphne calyculata, Betula nana,* and *Pinus sylvestris*. The hollows represent the wettest microforms and are occupied primarily by *Scheuchzeria palustris* and *Carex limosa*. The vegetation of lawns, intermediate microforms with respect to water level, consists of a mixture of species growing at hummocks and hollows. The differences in the nutrient conditions of the fen and bog part are indicated by the occurrence of *Menyanthes trifoliate* and *Utricularia intermedia* in the fen part of the investigation area. The transition zone between the fen and bog part is characterised by *Carex rostrata* lawns and ombrogenous lawns. A transect survey through the site revealed peat depths of up to 2 m.

According to WRB the soils of the peatland were classified as *Histosols*, which are soils built of organic material, especially soils of the boreal region that primarily consist of *Sphagnum* peat. The fen and bog part of the peatland are characterized by different decomposition rates of the peat. Though, the soil of the fen part is classified as *Fibric Histosol* and the bog part as *Fibric Lignic Histosol* (Langer 2012). *Lignic* describes the occurrence of intact wood fragments that compose at minimum a fourth of the soil volume in the upper 50 cm under the soil surface. *Fibric* describes that after rubbing, the organic material consists of at least two-thirds (volumic) visible plant residuals in

FIGURE 3.20   Investigation sites in Komi Republic. Upper left corner—map of Ust-Pojeg peatland and surrounding forest and clear cuts (cartography by I. Beil); upper right corner—hummock-hollow microrelief of the Ust-Pojeg peatland; bottom left corner—former forest site directly after clear cut; bottom right corner—5-year-old clear cut site.

the upper 100 cm under soil surface. The pH in the peat pore water is about 4 in the bog part and 5 in the fen part (Wolf 2009).

The Ust-Pojeg peatland is a pristine peatland. The presence of buried wood and forest humus at the peat-mineral subsoil transition and the absence of gyttja confirm that the investigated parts of the peatland were formed through paludification. This is the most widespread process for peatland initiation and consists of the lateral expansion of a peatland over an upland forest (Charman 2002). The radiocarbon dating of the basal peat suggests that the peat development was initiated about 9600 years ago (Pluchon et al. 2014).

The Ust-Pojeg peatland is surrounded by spruce forest (*Picea obovata*). At some patches trees were harvested for the paper mill situated in Syktyvkar, the capital of the Komi Republic. The most used harvesting method in Russia is the clear cut. For these studies, one investigation site was established 6 years after harvest. It is characterized by wooden remains of different size lying not only on the top of the soil, but also in the soil due to the heavy forest vehicles. Single trees of *Picea obovata* and *Betula pubescens* are growing in the tree layer, shrub and herb layer are well developed and *Betula pubescens* covers 70% of the area. The moss layer is not continuous, and consists of *Pleurozium schreberi*, *Polytrichum commune* and *Sphagnum* spp. (Figure 3.20).

According to WRB, the soils of the clear cut were classified as *Gleysol*. The term *Gleysols* describes soils that are affected by groundwater over a long period and show a characteristic greyish-bluish color pattern. The soils of the surrounding forests are classified as *Folic Gleysols* (Langer 2012). *Folic* described the occurrence of a well-aerated horizon, which was built of organic material and located at the soil surface. This horizon is missing at the clear cut site due to the use of heavy machinery during the forest logging.

3.5.6.2   Land Use

The middle taiga region of the Komi Republic is characterized by the boreal forests, which are greatly influenced by human activity. This includes different kinds of cuttings and other activities like firewood collecting and cattle grazing. These effects are most prevalent in and around densely populated areas and along the transport routes. The logging areas are usually left for natural regeneration. However, in remote regions primeval forest ecosystems still exist. Large areas of Komi Republic are covered by peatlands (up to 8% of the area), most of which are pristine with low intensity of use by berry pickers. Some peatlands or parts of peatlands have been drained for agriculture in the 1960s and 1970s. Recent peat excavation is not known to the author. Due to the severe climate and short growing season, agriculture does not play an important role in the Komi Republic's economy.

3.5.6.3   Carbon Stocks

Peatlands are known as C reservoirs. The C storage in Ust-Pojeg peatland sites varies from 13.5 kg C m^{-2} in the marginal parts to >100 kg C m^{-2} in the central parts of the peatland with the peat layer being the major reservoir of C (Pluchon et al. 2014). The aboveground phytomass C storage varies within the peatland and is on average 0.9 kg C m^{-2} in the marginal parts of the peatland area, with sparse and stunted trees, and 0.2 kg C C m^{-2} in the central, more open parts. Phytomass C storage in these different parts of the peatland represents 10% and 0.1% of the total ecosystem C storage, respectively (Pluchon et al. 2014).

In the forest sites, the C partitioning (storage in phytomass and in different soil layers) showed that aboveground phytomass represented about 25% of the total ecosystem C with a mean value of 4.0 kg C m^{-2} (Pluchon et al. 2014). These phytomass C storage values from Ust-Pojeg are comparable with the reported values of other boreal forest regions. The forest soil C storage for the reference 1 m depth ranged from 14.4 kg C m^{-2} to 40.3 kg C m^{-2} with a mean value of 21.5 kg C m^{-2}. This estimated forest soil C storage (down to 1 m depth) is higher in comparison with other reported values, but in a similar order of magnitude of values reported for moist (herbaceous) spruce forests in northern Komi (on average 20.3 kg C m^{-2}) by Hugelius and Kuhry (2009). These higher values could result from the proximity to the peatlands indicating early stages of paludification (Pluchon et al. 2014).

3.5.6.4 CO_2 and CH_4 Fluxes

Ust-Pojeg peatland was estimated to be a CO_2 source on an annual time scale (Schneider et al. 2012). However, great differences in NEE (sum of GPP and ecosystem respiration) were observed between the microform types. During the period of observations, hollows and minerogenous lawns were CO_2 sinks, and ombrogenous hummocks and lawns were CO_2 sources. *Carex* lawns and minerogenous hummocks could be either CO_2 sources or sinks depending on the degree of vegetation cover of each plot's dominant species. Plots with higher vegetation cover but similar water table depths usually show a stronger CO_2 uptake. CO_2 uptake is also influenced by vegetation composition. Another important factor in NEE is the water table level; it influences the vegetation composition and the behavior of plants regarding their gas exchange with the atmosphere, e.g., mosses in peatlands are adapted to high water table levels, but very high water content in *Sphagnum* mosses slows down the CO_2 and O_2 exchange to the photosynthetically active tissues (Wallen et al. 1988). Thus, the environmental conditions at minerogenous lawns with a lower water table level were more favorable for photosynthesis and respiration of mosses than at minerogenous hollows. Additionally, water table level determines the portion of the aerated peat profile. A higher portion leads to enhanced oxygen availability for microbial decomposition and root growth, and thus to higher respiration rates (Bubier et al. 2003). However, water table depth also influences the GPP. The effect of surface dryness, which occurred more often at hummocks, can lead to a reduction of photosynthesis of both vascular plants and *Sphagnum* mosses (Arneth et al. 2002).

Another well-known dominant controlling factor of NEE is air temperature. At Ust-Pojeg peatland seasonal patterns of NEE were well expressed at all microforms. Air temperature determines the timing of the snow melt. Several studies (Friborg et al. 1997; Aurela et al. 2004; Moore et al. 2006) showed that the spring period controls the annual CO_2 balance of peatlands. The snow melt timing is important as it determines the onset of the photosynthesis, which is associated with moss activity followed by the activity of the evergreen shrubs. Former studies (Alm et al. 1997; Moore et al. 2002; Sagerfors et al. 2008) reported that boreal fens and bogs were CO_2 sinks. In the investigation year, the spring was colder compared with other years. The snow melt also started later, which might have led to limited photosynthetic activity in spring and to a stronger CO_2 release. However, it is important to be aware that CO_2 exchange is sensitive to variations in environmental factors on the short-term scale, and a peatland can be a CO_2 sink in one year and a CO_2 source in another.

Boreal peatlands play also a crucial role in the global methane cycle. Similar to CO_2 flux dynamics, significant differences in CH_4 fluxes were observed at different microrelief positions in Ust-Pojeg peatland. A trend that a higher mean water table level leads to increased methane fluxes was detected. Moreover, it was observed that an increased vegetation cover of aerenchyimatous plants accounted for higher fluxes, and thus CH_4 fluxes were highest at minerogenous hollows followed by *Carex* lawns. The lowest mean values were measured at ombrogenous hummocks (Schneider et al. 2016). Another important factor affecting the CH_4 fluxes is the soil temperature. Overall, the winter fluxes were well within the range of what has been reported for the peatlands of other boreal regions before, but summer fluxes greatly exceeded the average range of 5–80 mg CH_4 $m^{-2}d^{-1}$ for the circumpolar boreal zone. Half of the measured fluxes ranged between 150 and 450 mg CH_4 $m^{-2}d^{-1}$. One possible reason for the higher fluxes compared with Scandinavia could be the continental climate (Schneider et al. 2016). It is most likely that temperature has an effect on substrate quality and quantity, e.g., on dissolved organic matter (DOM) in peatlands, and thus on the CH_4 fluxes. In Russia, due to very low temperatures, these substrates maybe preserved reducing microbial processes in winter. During higher temperatures in summer, this DOM is available for methanogenesis to a full extent. This climate type is also characterized by pronounced freeze–thaw cycles in the winter spring period, which increase the physical disruption of the organic soil, leading to higher fine root mortality. Both processes can result in higher dissolved organic carbon release, and thus to higher CH_4 fluxes. Further, well-described influencing factors are vegetation structure (e.g., Bubier 1995; Ström et al. 2003; Lai et al. 2014), ecosystem productivity (Ström et al. 2015) and recently fixed carbon (Chanton et al. 1995). Laine et al. (2015) showed that photosynthesis in Russian bogs was more intensive than in Ireland and Finland, which most likely leads to more recently fixed carbon in the rhizosphere, and thus to higher CH_4 fluxes. There are some ideas on which factors might lead to higher CH_4 in Russia, but more research is still needed.

In Russia, logging is usually implemented as clear cutting or intense selective logging. This kind of forest exploitation leads to dramatic changes in soil structure and microclimate, with consequent increases in mineralisation of organic matter and in heterotrophic respiration and decreases in evapotranspiration (Rannik et al. 2002). It was assumed that the removal of the forest cover led to a reduction in evapotranspiration and in the interception of rainfall also at the Ust-Pojeg clear cut investigation site. Researchers from Canada (Aust et al. 1997; Marcotte et al. 2008) described an increase of soil bulk density and flow path disturbances after harvesting, which led to water table elevations. Before the clear cut at the Ust-Pojeg site, the forest floor was characterised by an organic layer of minimum 10 cm thickness. After the logging timber was removed, there was still a lot of organic substrate in and on the soil. This poorly decomposed OM and waterlogged soils favor methanogenesis. That had been the hypothesis. The measurements at the Ust-Pojeg clear cut showed that CH_4 emission took place only in spring, when the water table level was high after the snow melt. Compared with Canadian sites, the water table dropped very fast in early summer due to coarser textured soils and less precipitation. The lower water table led to an increased thickness of aerated soil zone, which means that even if some methane was produced in the deeper soil layers, it was oxidised thereafter.

3.5.6.5 Conclusions Why Feedbacks Happen (Process Understanding)

A shift in the ecological function of peatlands from net C sink to net a C source could potentially enhance climatic change. An increased atmospheric loading of radiatively important gases such as

CO_2 and CH_4 might lead to higher air and soil temperatures. Increased soil temperature is expected to result in an increased C loss due to increased rates of microbial activity, and thus to higher microbial decomposition and to an extended growing season (Oechel et al. 1993; Carroll and Crill 1997). Higher soil temperatures might lead to lower water table levels and reduced soil moisture resulting in a shift from anaerobic to aerobic microbial decomposition, and a decrease in peat accumulation on the one hand. On the other hand, it would lead to an increase in C storage in living biomass of forested peatlands, which could develop on drier soils (Minkkinen and Laine 1998). It is important to take into account that drier surface peats are expected to increase fire frequency and intensity, which might significantly enhance gaseous C emissions. The distribution of ash and dissolved materials following fire increase nutrient availability and might also enhance microbial and plant productivity in peatlands that receive these materials (Hogg et al. 1992). Temperature also has an influence on the area covered by peatlands through control of surface evaporation and the soil water budget, and this effect can counterbalance the higher CH_4 fluxes due to higher soil temperatures (Ringeval et al. 2011). Not fully understood is also the interaction of increased CO_2 in the atmosphere and CH_4 fluxes, as it affects plant productivity and hence soil C availability for methanogenesis. Changes in plant structure (e.g., to higher coverage of *Carex* spp.) can lead to higher CH_4 fluxes due to the bypass of the oxidative zone in peatlands through the aerenchyma of plants.

Changes in vegetation cover after clear cut result in C cycle changes, including the changes of amount of C held in vegetation and soil. After clear cut, C stocks in above ground biomass are partly removed with the timber, and the wooden residues are exposed to microbial decomposition. This means, depending on the environmental conditions, CO_2 will be released sooner or later to the atmosphere. During the time required for the re-growth, which can be decades for trees, the above ground C storage will be smaller than their original value, especially in Russia where the man-made reforestation and cultivation of trees on clear cuts is uncommon. Due to missing trees the CO_2 remove from atmosphere through trees is lost after clear cut, but increase again as soon as new plant cover adapts. It is widely known that increased CO_2 concentration in the atmosphere leads to higher net primary production, which consistently exceeds the heterotrophic respiration and results in a stable CO_2 sink.

REFERENCES

Abood, S.a., Lee, J.S.H., Burivalova, Z., Garcia-Ulloa, J., & Koh, L.P. (2015). Relative contributions of the logging, fiber, oil palm, and mining industries to forest loss in Indonesia. *Conservation Letters*, 8, 58–67. doi.org/10.1111/conl.12103.

Achard, F., Eva, H.D., Mayaux, P., Stibig, H.-J., & Belward, A. (2004). Improved estimates of net carbon emissions from land cover change in the tropics for the 1990s. *Global Biogeochem. Cycles*, 18, GB2008. doi.org/10.1029/2003GB002142.

Aguilera, E., Lassaletta, L., Gattinger, A., & Gimeno, B.S. (2013). Managing soil carbon for climate change mitigation and adaptation in Mediterranean cropping systems: A meta-analysis. *Agriculture, Ecosystems & Environment*, 168, 25–36.

Åkerman, H.J., & Johansson, M. (2008). Thawing permafrost and thicker active layers in sub-arctic Sweden. *Permafrost and Periglacial Processes*, 19, 279–292, 10.1002/ppp.626. doi.org/10.1002/ppp.626.

Allison, S.D. (2012). A trait-based approach for modelling microbial litter decomposition. *Ecology Letters*, 15, 1058–1070.

Alm, J., Talanov, A., Saarnio, S., Silvola, J., Ikkonen, E., Aaltonen, H., Nykänen, H., & Martikainen, P.J. (1997). Reconstruction of the carbon balance for microsites in a boreal oligotrophic pine fen. *Finland Oecologia*, 110, 423–431. doi.org/10.1007/s004420050177.

Alvarez, R. (2006). A review of nitrogen fertilizer and conservation tillage effects on soil organic carbon storage. *Soil Use and Management*, 21, 38–52.

AMAP (2012). Arctic Climate Issues 2011: *Changes in Arctic Snow, Water, Ice and Permafrost*. SWIPA 2011 Overview Report. Arctic Monitoring and Assessment Programme (AMAP), Oslo. xi + 97pp.

Anderson, T.M., Dong, Y.A., & McNaughton, S.J. (2006). Nutrient acquisition and physiological responses of dominant Serengeti grasses to variation in soil texture and grazing. *J. Ecology*, 94 (6), 1164–1175. doi.org/10.1111/j.1365-2745.2006.01148.x.

Andrén, O., & Kätterer, T. (2001). Basic principles for soil carbon sequestration and calculating dynamic country-level balances including future scenarios. In: Lal, R., Kimble, J.M., Follett, R.F. and Stewart, B.A. *Assessment Methods for Soil Carbon*. Lewis Publishers, pp. 495–511.

Andrén, O., Kihara, J., Bationo, A., Vanlauwe, B., & Kätterer, T. (2007). Soil climate and decomposer activity in sub-Saharan Africa estimated from standard weather station data—A simple climate index for soil carbon balance calculations. *Ambio*, 36, 379–386.

Angers, D.A., & Eriksen-Hamel, N.S. (2008). Full-inversion tillage and organic carbon distribution in soil profiles: A meta-analysis. *Soil Sci. Soc. Am. J.*, 72, 1370–1374. doi.org/10.2136/sssaj2007.0342. doi.org/10.2136/sssaj2007.0342.

Arneth, A., Kurbatova, J., Kolle, O., Shibistova, O.B., Lloyd, J., Vygodskaya, N.N., & Schulze, E.-D. (2002). Comparative ecosystem-atmosphere exchange of energy and mass in a European Russian and central Siberian bog II. Interseasonal and interannual variability of CO_2 fluxes. *Tellus*, 54B, 514–530. doi.org/10.3402/tellusb.v54i5.16684.

Aselmann, I., & Crutzen, P.J. (1989). Global distribution of natural freshwater wetlands and rice paddies, their net primary productivity, seasonality and possible methane emissions. *J. Atmos. Chem.*, 8(4), 307–358. doi.org/10.1007/BF00052709.

Aurela, M., Laurila, T., & Tuovinen, J.-P. (2004). The timing of snow melt controls the annual CO_2 balance in a subarctic fen. *Geophys. Res. Letters*, 31. doi.org/10.1029/2004GL020315. doi:10.1029/2004GL020315.

Aust, W.M., Schoenholtz, S.H., Zaebst, T.W., & Szabo, B.A. (1997). Recovery status of a tupelo-cypress wetland 7 years after disturbance—Silvicultural implications. *Forest Ecology and Management*, 90, 161–169. doi.org/10.1016/S0378-1127(96)03899-6.

Babel, W., Biermann, T., Coners, H. et al. (2014). Pasture degradation modifies the water and carbon cycles of the Tibetan highlands. *Biogeosciences*, 11, 6633–6656. doi.org/10.5194/bg-11-6633-2014.

Badan Pusat Statistik (2017). https://www.bps.go.id/linkTableDinamis/view/id/842. Accessed June 4, 2018.

Barber, A.J., Crow, M.J., & Milsom, J.S. (2005). *Sumatra: Geology, Resources and Tectonic Evolution*. Geological Society, London, Memoirs, 31.

Batjes, N. (1996). Total carbon and nitrogen in the soils of the world. *Eur. J. Soil Sci.*, 47 (2), 151–163. doi.org/10.1111/ejss.12114_2.

Batjes, N.H. (2005). Organic carbon stocks in the soils of Brazil. *Soil Use Manag.*, 21, 22–24. doi.org/10.1079/SUM2005286. doi:10.1079/SUM2005286.

Batlle-Bayer, L., Batjes, N.H., & Bindraban, P.S. (2010). Changes in organic carbon stocks upon land-use conversion in the Brazilian Cerrado: A review. *Agric. Ecosyst. Environ.*, 137, 47–58. doi.org/10.1016/j.agee.2010.02.003. doi:10.1016/j.agee.2010.02.003.

Baumann, F., He, J.-S., Schmidt, K., Kühn, P., & Scholten, T. (2009). Pedogenesis, permafrost, and soil moisture as controlling factors for soil nitrogen and carbon contents across the Tibetan Plateau. *Global Change Bio.*, 15, 3001–3017. doi.org/10.1111/j.1365-2486.2009.01953.x.

Baumann, F., Schmidt, K., Dörfer, C., He, J.-S., Scholten T., & Kühn P. (2014). Pedogenesis, permafrost, substrate and topography: Plot and landscape scale interrelations of weathering processes on the central-eastern Tibetan Plateau. *Geoderma*, 226–227, 300–316. doi.org/10.1016/j.geoderma.2014.02.019.

Berdanier, A.B., & Klein, J.A. (2011). Growing season length and soil moisture interactively constrain high elevation aboveground net primary production. *Ecosystems*, 14, 963–974. doi.org/10.1007/s10021-011-9459-1.

Berg, B. (2000). Litter decomposition and organic matter turnover in northern forest soils, *Forest Ecol. and Manag.*, 133 (1), 13–22. doi.org/10.1016/S0378-1127(99)00294-7.

Bergkvist, G., & Öborn, I. (2011). Long-term field experiments in Sweden—What are they designed to study and what could they be used for? *Aspects Appl. Biol.*, 113, 75–85.

Bernoux, M., da Conceição Santana Carvalho, M., Volkoff, B., Cerri, C.C., & Carvalho da C.S., M. (2002). Brazil's soil carbon stocks. *Soil Sci. Soc. Am. J.*, 66, 888–896. doi.org/10.2136/sssaj2002.8880.

Boas, L., & Wang, P.R. (2011). Weather and climate data from Greenland 1958–2010: Observation data with description. Technical Report 11–15. Copenhagen, Danish Meteorological Institute, Ministry of Transport and Energy (http://www.dmi.dk/dmi/tr11-15.zip).

Bockheim, J.G., & Tarnocai, C. (1998). Recognition of cryoturbation for classifying permafrostaffected soils. *Geoderma*, 81, 281–293. doi.org/10.1016/S0016-7061(97)00115-8.

Böhner, J., & Lehmkuhl, F. (2005). Environmental change modelling for Central and High Asia: Pleistocene, present and future scenarios. *Boreas*, 34 (2), 220–231. doi.org/10.1111/j.1502-3885.2005.tb01017.x.

Bolinder, M.A., Janzen, H.H., Gregorich, E.G., Angers, D.A., & VandenBygaart, A.J. (2007). An approach for estimating net primary productivity and annual carbon inputs to soil for common agricultural crops in Canada. *Agric., Ecosys. & Envir.*, 118, 29–42. doi.org/10.1016/j.agee.2006.05.013.

Bolinder, M.A., Kätterer, T., André, O., & Parent, L.E. (2012). Estimating carbon inputs to soil in forage-based crop rotations and modelling the effects on soil carbon dynamics in a Swedish long-term field experiment. *Can. J. Soil Sci.*, 92, 821–833. doi.org/10.4141/cjss-2014-091.

Bortolon, E.S.O., Mielniczuk, J., Tornquist, C.G., Lopes, F., & Bergamaschi, H. (2011). Validation of the Century model to estimate the impact of agriculture on soil organic carbon in Southern Brazil. *Geoderma*, 167–168, 156–166. doi.org/10.1016/j.geoderma.2011.08.008.

Bosch, A., Dörfer, C., He, J.-S., Schmidt, K., & Scholten, T. (2016). Predicting soil respiration for the Qinghai-Tibet Plateau: An empirical comparison of regression models. *Pedobiologia—J. Soil Ecol.*, 59, 41–49. doi.org/10.1016/j.pedobi.2016.01.002.

Bosch, A., Dörfer, C., He, J.-S., Schmidt, K., & Scholten, T. (2017). Potential CO2 emissions from permafrost-affected soils of the Qinghai-Tibet Plateau under different scenarios of climate change in 2050 and 2070. *Catena*, 149, 221–231. doi.org/10.1016/j.catena.2016.08.035.

Bosiö, J., Johansson, M., Callaghan, T.V., Johansen, B., & Christensen, T.R. (2012). Future vegetation changes in thawing subarctic mires and implications for greenhouse gas exchange—A regional assessment. *Climatic Change*, 115, 379–398, 10.1007/s10584-012-0445-1. doi.org/10.1007/s10584-012-0445-1.

Boy, J., Strey, S., Schönenberg, R., Strey, R., Weber-Santos, O., Nendel, C., Klingler, M., Schumann, C., Hartberger, K., & Guggenberger, G. (2016). Seeing the forest not for the carbon: Why concentrating on land-use-induced carbon stock changes of soils in Brazil can be climate-unfriendly. *Regional Environmental Change*, 18, 63–75, doi.org/10.1007/s10113-016-1008-1.

Braz, S.P., Urquiaga, S., Alves, B.J.R., Jantalia, C.P., Guimarães, A.P., dos Santos, C.A., dos Santos, S.C., Machado Pinheiro, É.F., & Boddey, R.M. (2013). Soil carbon stocks under productive and degraded pastures in the Brazilian Cerrado. *Soil Sci. Soc. Am. J.*, 77, 914. doi.org/10.2136/sssaj2012.0269.

Brierley, G.J., Li, X., Cullum, C., & Gao, J. (2016). *Landscape and Ecosystem Diversity, Dynamics and Management in the Yellow River Source Zone*. Springer, Switzerland.

Brookfield, M.E. (2011). Aeolian processes and features in cool climates. In: Martin,i I.P., French, H.M., and Perez Alberti, A. (eds), *Ice-Marginal and Periglacial Processes and Sediments*. Geological Society of London (Special Publications 354), pp. 241–258.

Bubier, J.L. (1995). The relationship of vegetation to methane emission and hydrochemical gradients in northern peatlands. *J. Ecol.*, 83, 403–420.

Bubier, J.L., Crill, P.M., Mosedale, A., Frolking, S., & Linder, E. (2003). Peatland responses to varying inter-annual moisture conditions as measured by automatic CO2 chambers. *Glob. Biogeochem. Cycles*, 17, 1066. doi.org/10.1029/2002GB001946.

Budan Pusat Statistiks. (2015). Indonesian oil palm statistics 2015. Catalog: 5504003, ISSN Number: 1978-9947.

Bullard, J.E. (2011). Dust generation on a proglacial floodplain, West Greenland. *Aeolian Res.*, 43–54. doi.org/10.1016/j.aeolia.2011.01.002.

Buol, S.W., & Eswaran, H. (2000). Oxisols. *Adv. Agron.*, 68, 151–197.

Byrne, K.A., Chojnicki, B., Christensen, T.R., Drösler, M., Freibauer, A., Frolking, S., Lindroth, A., Mailhammer, J., Malmer, N., Selin, P., Turunen J., Valentini R., & Zetterberg, L. (2004). EU Peatlands: Current carbon stocks and trace gas fluxes. 1–58, Report 4/2004 to 'Concerted action: Synthesis of the European Greenhouse Gas Budget', Geosphere-Biosphere Centre, Univ.of Lund, Sweden.

Callaway, R.M., Brooker, R.W., Choler, P. et al. (2002). Positive interactions among alpine plants increase with stress. *Nature*, 417, 844–848. doi.org/10.1038/nature00812.

Cao, G., Xu, X., Long, R., Wang, Q., Wang, C., Du, Y., & Zhao, X. (2008). Methane emissions by alpine plant communities in the Qinghai-Tibet Plateau. *Biol. Letters*, 4, 681–684. doi.org/10.1098/rsbl.2008.0373.

Cappelen, J., Jørgensen, B.V., Laursen, E.V., Stannius, L.S., & Thomsen, R.S. (2001). *The observed climate of Greenland 1958–99—with climatological standard normals, 1961–90*. Technical Report 00–18. Danish Meteorological Institute, Ministry of Transport and Energy, Copenhagen (152 pp. http://www.vedrid.fo/fo/print/tr00-18.pdf).

Carroll, P., & Crill, P. (1997). Carbon balance of temperate poor fen. *Glob. Biogeochem. Cycles*, 11, 349–356. doi.org/10.1029/97GB01365.

Carstensen, L.S., & Jørgensen, B.V. (2009). *Weather and climate data from Greenland 1958–2008*. Dataset available for research and educational purposes. Descriptions and documentation of observations of temperature, precipitation, wind, could cover, air pressure, humidity and depth of snow. Technical Report 09–11. Danish Meteorological Institute, Ministry of Transport and Energy, Copenhagen (22 pp. http://www.dmi.dk/dmi/dcc07-02.pdf).

Carvalheiro, K., & Nepstad, D. (1996). Deep soil heterogeneity and fine root distribution in forests and pastures of eastern Amazonia. *Plant Soil*, 279–285. doi.org/10.1007/BF00029059.

CAVM Team. (2003). *Circumpolar Arctic Vegetation Map*. Scale 1:7,500,000. Conservation of Arctic Flora and Fauna (CAFF) Map No. 1. U.S. Fish and Wildlife Service, Anchorage, Alaska.

Cerri, C.E.P., Easter, M., Paustian, K., Killian, K., Coleman, K., Bernoux, M., Falloon, P., Pwlson, D.S., Batjes, N., Milne, E., & Cerri, C.C. (2007). Simulating SOC in 11 land-use change chronosequences from the Brazilian Amazon with RothC and Century models. *Agric., Ecosys. and Environ.*, 122, 46–57. Doi.org/10.1016/j.agee.2007.01.007.

Cerri, C.E.P., Paustian, K., Bernoux, M., Victoria, R.L., Melillo, J.M., & Cerri, C.C. (2004). Modeling changes in soil organic matter in Amazon forest to pasture conversion with the Century model. *Glob. Change Biol.*, 10, 815–832. doi.org/10.1111/j.1529-8817.2003.00759.x.

Chang, D.H.S. (1981). The vegetation zonation of the Tibetan Plateau. *Mountain Research and Development*, 1, 29–48. doi.org/10.2307/3672945.

Chanton, J.P., Bauer, J.E., Glaser, P.A., Siegel, D.I., Kelley, C.A., Tyler, S.C., Romanowicz, E.H., & Lazrus, A. (1995). Radiocarbon evidence for the substrates supporting methane formation within Northern Minnesota peatlands. *Geochimica et Cosmochimica Acta*, 59, 3663–3668. doi.org/10.1016/0016-7037(95)00240-Z.

Chapin III, F.S., Matson, P.A., & Mooney, H. (2002). *Principles of Terrestrial Ecosystem Ecology*. Springer, New York.

Chapin III, F.S., Zavaleta, E.S., Eviner, V.T., Naylor, R.L., Vitousek, P.M., Reynolds, H.L., Hooper, D.U., Lavorel, S., Sala, O.E., Hobbie, S.E., Mack, M.C., & Díaz, S. (2000). Consequences of changing biodiversity. *Nature*, 405 (6783), 234–242. doi.org/10.1038/35012241.

Charman, D.J. (2002). *Peatlands and Environmental Change*. John Wiley and Sons Ltd., Chichester.

Chen, H., Wu, N., Wang, Y. et al. (2013). Inter-annual variations of methane emission from an open fen on the Qinghai-Tibetan plateau: A three-year study. *PLOS ONE*, 8, e53878. doi.org/10.1371/journal .pone.0053878.

Chen, L.-T., Flynn, D.F.B., Jing, X., Kühn, P., Scholten, T., & He, J.-S. (2015). A comparison of two methods for soil organic carbon (SOC) of alpine grasslands on the Tibetan Plateau. *PLOS ONE*, 10 (5), e0126372. doi.org/10.1371/journal.pone.0126372.

Chen, L.-T., Jing, X., Flynn, D.F.B., Shi, Y., Kuehn, P., Scholten, T., & He, J.-S. (2017). Changes of carbon stocks in alpine grassland soils from 2002 to 2011 on the Tibetan Plateau and their climatic causes. *Geoderma*, 288, 166–274. doi.org/10.1016/j.geoderma.2016.11.016.

Chen, Q., Wang, Q., Han, X., Wan, S., & Li, L. (2010). Temporal and spatial variability and controls of soil respiration in a temperate steppe in northern China. *Glob. Biogeochem. Cycles*, 24 (2), n/a. doi.org /10.1029/2009GB003538.

Cheng, G. (2005). Permafrost studies in the Qinghai–Tibet Plateau for road construction. *Journal of Cold Regions Engineering* 19 (1), 19–29. doi.org/10.1061/(ASCE)0887-381X(2005)19:1(19).

Cheng, G., & Jin, H. (2013). Permafrost and groundwater on the Qinghai-Tibet Plateau and in northeast China. *Hydrogeol. J.*, 21, 5–23. doi.org/10.1007/s10040-012-0927-2.

Cheng, G., & Wu, T. (2007). Responses of permafrost to climate change and their environmental significance, Qinghai-Tibet Plateau. *J. Geophys. Res.*, 112 (F2). doi.org/10.1029/2006JF000631.

Christensen, B.T. (2001). Physical fractionation of soil and structural and functional complexity in organic matter turnover. *Eur. J. Soil Sci.*, 52, 345–353. doi.org/10.1046/j.1365-2389.2001.00417.x.

Ciais, P., Sabine, C., Bala, G., Bopp, L., Brovkin, V., Canadell, J., Chhabra, A., DeFries, R., Galloway, J., Heimann, M., Jones, C., Le Quéré, C., Myneni, R.B., Piao, S., & Thornton, P. (2013). Carbon and other biogeochemical cycles. In: Stocker, T.F., Qin, D., Plattner, G.-K., Tignor, M., Allen, S.K., Boschung, J., Nauels, A., Xia, Y., Bex, V., & Midgley, P.M. (eds.), *Climate Change 2013: The Physical Science Basis*. Contribution of Working Group I to the Fifth Assessment Report of the Intergovernmental Panel on Climate Change. Cambridge University Press, Cambridge, United Kingdom, and New York, NY, USA.

Clymo, R. (1984). The limits to peat bog growth, *Philosophical Transactions of the Royal Society of London. B. Biol. Sci.*, 303, 605–654. doi.org/10.1098/rstb.1984.0002.

Corley, R.H.V., & Tinker, P.B. (2015). *The Oil Palm*. John Wiley & Sons, Ltd, Chichester.

Couwenberg, J. (2009). *Emission Factors for Managed Peat Soils: An Analysis of IPCC Default Values*. Wetlands International, Ede.

Craine, J.M., Wedin, D.A., & Chapin III, F.S. (1999). Predominance of ecophysiological controls on soil CO2 flux in a Minnesota grassland. *Plant and Soil*, 207, 77–86. doi.org/10.1023/A:1004417419288.

Dai, F., Su, Z., Liu, S., & Liu, G. (2011). Temporal variation of soil organic matter content and potential determinants in Tibet, China. *Catena*, 85 (3), 288–294. doi.org/10.1016/j.catena.2011.01.015.

Dargie, G.C., Lewis, S.L., Lawson, I.T., Mitchard, E.T., Page, S.E., Bocko, Y.E., & Ifo, S.A. (2017). Age, extent and carbon storage of the central Congo Basin peatland complex. *Nature*, 542, 86–90 . doi.org/10.1038 /nature21048. doi.org/10.1038/nature21048.

da Silva, J., Resck, D.V., Corazza, E., & Vivaldi, L. (2004). Carbon storage in clayey Oxisol cultivated pastures in the "Cerrado" region. Brazil. *Agric. Ecosyst. Environ.*, 103, 357–363. doi.org/10.1016/j.agee .2003.12.007.

Davidson, E., & Ackerman, I. (1993). Changes in soil carbon inventories following cultivation of previously untilled soils. *Biogeochemistry*, 20(3), 161–193.

Davidson, E., & Janssens, I. (2006). Temperature sensitivity of soil carbon decomposition and feedbacks to climate change. *Nature*, 440, 165–73. doi.org/10.1038/nature04514.

Debreczeni, K., & Körschens, M. (2003). Long-term field experiments of the world. *Archives Agron. Soil Sci.*, 49, 465-483. doi.org/10.1080/03650340310001594754.

Deng, Z., Xie, X., Zhou, X., & Wang, Q. (2001). Study on reproductive ecology of Kobresia pygmaea population in alpine meadow. *Acta Botanica Boreali-Occidentalia Sinica*, 22, 344–349.

Desjardins, T., Barros, E., Sarrazin, M., Girardin, C., & Mariotti, A. (2004). Effects of forest conversion to pasture on soil carbon content and dynamics in Brazilian Amazonia. *Agric. Ecosyst. Environ.*, 103, 365–373. doi.org/10.1016/j.agee.2003.12.008.

Dharmakeerthi, R.S., Kay, B.D., & Beauchamp, E.G. (2005). Factors contributing to changes in plant available nitrogen across a variable landscape. *Soil Sci. Soc. Am. J.*, 69 (2), 453–462.

Dietze, E., Hartmann, K., Diekmann, B., IJmker, J., Lehmkuhl, F., Opitz, S., Stauch, G., Wünnemann, B., & Borchers, A. (2012). An end-member algorithm for deciphering modern detrital processes from lake sediments of Lake Donggi Cona, NE Tibetan Plateau, China. *Sedimentary Geology*, 243–244, 169–180. doi.org/10.1016/j.sedgeo.2011.09.014.

Dijkmans, J.W., & Törnqvist, T.E. (1991). *Modern periglacial Eolian Deposits and Landforms in the Sondre Stromfjord Area, West Greenland and Their Palaeoenvironmental Implications.* Meddelelser om Gronland, Geoscience 25. Museum Tusculanum Press, Copenhagen.

Dise, N.B. (2009). Peatland response to global change. *Science*, 326, 810. doi.org/10.1126/science.1174268.

Domrös, M., & Peng, G. (1988). *The Climate of China.* Springer, Berlin, Heidelberg, New York.

Don, A., Bärwolff, M., Kalbitz, K., Andruschkewitsch, R., Jungkunst, H.F., & Schulze E.D. (2012). No rapid soil carbon loss after a windthrow event in the High Tatra. *Forest Ecol. Manag.*, 276, 239–246. doi .org/10.1016/j.foreco.2012.04.010.

Don, A., Schumacher, J., & Freibauer A. (2011). Impact of tropical land-use change on soil organic carbon stocks—A meta-analysis. *Glob. Chang. Biol.*, 17, 1658–1670. doi.org/10.1111/j.1365-2486.2010.02336.x.

Dong, Q.M., Zhao, X.Q., Wu, G.L., Shi, J.J., Wang, Y.L., & Sheng, L. (2012). Response of soil properties to yak grazing intensity in a Kobresia parva-meadow on the Qinghai-Tibetan Plateau, China. *J. Soil Sci. Plant Nutrition*, 12, 535–546. doi.org/10.1126/science.1174268.

Dörfer, C., Kühn, P., Baumann, F., He, J.-S., & Scholten, T. (2013). Soil organic carbon pools and stocks in permafrost-affected soils on the Tibetan Plateau. *PLOS ONE*, 8 (2), e57024. doi.org/10.1126/science.1174268.

Dorrepaal, E., Toet, S., van Logtestijn, R.S.P., Swart, E., van de Weg, M.J., Callaghan, T.V., & Aerts, R. (2009). Carbon respiration from subsurface peat accelerated by climate warming in the subarctic. *Nature*, 460, 616–619. doi.org/10.1038/nature08216.

Drescher, J., Rembold, K., Allen, K. et al. (2016). Ecological and socio-economic functions across tropical land-use systems after rainforest conversion. *Phil. Trans. Royal Soc. B: Biolog. Sci.*, 371, 20150275. doi.org/10.1098 /rstb.2015.0275.

Drösler, M., Adelmann, W., Augustin, J., Bergman, L., Beyer, C., Chojnicki, B., Förster, C., Freibauer, A., Giebels, M., Görlitz, S., Höper, H., Kantelhardt, J., Liebersbach, H., Hahn-Schöfl, M., Minke, M., Petschow, U., Pfadenhauer, J., Schaller, L., Schägner, P., Sommer, M., Thuille A., & Werhan, M. (2013). *Klimaschutz durch Moorschutz, Schlussbericht des Vorhabens.* "Klimaschutz–Moornutzungsstrategien" 2006–2010, Freisingen.

Elberling, B., Christiansen, H.H., & Hansen, B. (2010). High nitrous oxide production from thawing permafrost. *Nature Geosci.*, 3 (5), 332–335. doi.org/10.1038/ngeo803.

Elberling, B., Jakobsen, B.H., Berg, P., Sondergaar, J., & Sigsgaard, C. (2004). Influence of vegetation, temperature, and water content on soil carbon distribution and mineralization in four high Arctic soils. *Arct. Antarct. Alp. Res.*, 36 (4), 528–538.

Elberling, B., Nordstroem, C., Groendahl, L., Soegaard, H., Friborg, T., Christensen, T.R., Stroem, L., Marchand, F., & Nijs, I. (2008a). High-arctic soil CO2 and CH4 production controlled by temperature, water, freezing and snow. In: Meltofte, H. (ed.), *High-Arctic Ecosystem Dynamics in a Changing Climate. Ten Years of Monitoring and Research at Zackenberg Research Station Northeast Greenland.* Advances in Ecological Research, Vol. 40. Elsevier, Amsterdam, Heidelberg, pp. 441–472.

Elberling, B., Tamstorf, M.P., Michelsen, A., Arndal, M.F., Sigsgaard, C., Illeris, L., Bay, C., Hansen, B.U., Christensen, T.R., Hansen, E.S., Jakobsen, B.H., & Beyens, L. (2008b). Soil and plant community-characteristics and dynamics at Zackenberg. In: Meltofte, H. (ed.), *High-Arctic Ecosystem Dynamics in a Changing Climate. Ten Years of Monitoring and Research at Zackenberg Research Station Northeast Greenland.* Advances in Ecological Research, Vol. 40. Elsevier, Amsterdam, Heidelberg, pp. 223–248.

Elbert, W., Weber, B., Burrows, S., Steinkamp, J., Büdel, B., Andreae, M.O., & Pöschl, U. (2012). Contribution of cryptogamic covers to the global cycles of carbon and nitrogen. *Nature Geosci.*, 5 (7), 459–462. doi .org/10.1038/ngeo1486.

Elster, J., Margesin, R., Wagner, D., & Häggblom, M. (2016). Editorial: Polar and Alpine Microbiology— Earth's cryobiosphere. *FEMS Microbiol. Ecol.*, 93 (1), fiw221–fiw221. doi.org/10.1093/femsec/fiw221. doi.org/10.1093/femsec/fiw221.

Eswaran, H., Reich, P.F., Kimble, J.M., Beinroth, F.H., Padmanabhan, E., & Moncharoen, P. (2000). Global carbon stocks. In: Lal, R., Kimble, J., Eswaran, H., and Stewart, B.A. (eds.), *Global Climate Change and Pedogenic Carbonates*, pp. 15–25. CRC Press, Boca Raton.

Euler, M., Schwarze, S., Siregar, H., & Qaim, M. (2015). Oil palm expansion among smallholder farmers in Sumatra, Indonesia. EFForTS Discussion Paper Series 8, Georg-August-Universität, Göttingen, Germany.

Fang, X., Lü, L., Mason, J.A., Yang, S., An, Z., Li, J., & Zhilong, G. (2003). Pedogenic response to millennial summer monsoon enhancements on the Tibetan Plateau. *Quaternary Intl.*, 106–107, 79–88. doi .org/10.1016/S1040-6182(02)00163-5.

FAO. 2017. Statistics Division, Food and Agricultural Organization, Rome.

Farage, P.K., Ardö, J., Olsson, L., Rienzi, E.A., Ball, A.S., & Pretty, J.N. (2007). The potential for soil carbon sequestration in three tropical dryland farming systems of Africa and Latin America: A modelling approach. *Soil Tillage Res.*, 94, 457–472. doi.org/10.1016/j.still.2006.09.006.

Feintrenie, L., & Levang, P. (2009). Sumatra's rubber agroforests; Advent, rise and fall of a sustainable cropping system. *Small-Scale Forestry*, 8, 323–335. doi.org/10.1007/s11842-009-9086-2.

Feng, J.-L., Hu, Z.-G., Ju, J.-T., & Zhu, L.-P. (2011). Variations in trace element (including rare earth element) concentrations with grain sizes in loess and their implications for tracing the provenance of eolian deposits. *Quaternary Inl.*, 236 (1–2), 116–126. doi.org/10.1016/j.quaint.2010.04.024.

Fenner, N., & Freeman, C. (2011). Drought-induced carbon loss in peatlands. *Nature Geosci.*, 4 (12), 895–900. doi.org/10.1038/ngeo1323.

Ferreira, V., Castagneyrol, B., Koricheva, J., Gulis, V., Chauvet, E., & Graca, M.A.S. (2015). A meta-analysis of the effects of nutrient enrichment on litter decomposition in streams. *Biol. Rev.*, 90 (3), 669–688. doi .org/10.1111/brv.12125.

Fiedler, S.R., Leinweber, P., Jurasinski, G., Eckhardt, K.-U., & Glatzel, S. (2016). Tillage-induced short-term soil organic matter turnover and respiration. *SOIL*, 2, 475–486. doi.org10.5194/soil-2-475-2016.

Field, R.D., van der Werf, G.R., & Shen, S.S.P. (2009). Human amplification of drought-induced biomass burning in Indonesia since 1960. *Nature Geosci.*, 2, 185–188. doi.org/10.1038/ngeo443.

Foereid, B., & Hogh-Jensen, H. (2004). Carbon sequestration potential of organic agriculture in northern Europe—A modelling approach. *Nutrient Cycling in Agroecosystems*, 68, 13–24. doi.org/10.1023 /B:FRES.0000012231.89516.80.

Förth, J. (2012). *Typische Vegetationsgesellschaften und Oberirdische Vertikalstruktur von Zwergsträuchern in Abhängigkeit Von Abiotischen Umweltfaktoren in West Grönland.* Diplomarbeit, Tübingen (119 pp.).

Friborg, T., Christensen, T.R., & Søgaard, H. (1997). Rapid response of greenhouse gas emission to early spring thaw in a subarctic mire as shown by micrometeorological techniques. *Geophys. Res. Lett.*, 24(23), 3061–3064. doi.org/10.1029/97GL03024.

Frolking, S., & Roulet, N.T. (2007). Holocene radiative forcing impact of northern peatland carbon accumulation and methane emissions. *Glob. Change Biol.*, 13, 1079–1088. doi.org/10.1111/j.1365-2486.2007.01339.x.

Frolking, S., Talbot, J., Jones, M.C., Treat, C.C., Kauffman, J.B., Tuittilaand, E.S., & Roulet, N. (2011). Peatlands in the Earth's 21st century climate system. *Environ. Rev.*, 19, 371–396.

Fujisaki, K., Perrin, A.-S., Desjardins, T., Bernoux, M., Balbino, L.C., & Brossard, M. (2015). From forest to cropland and pasture systems: A critical review of soil organic carbon stocks changes in Amazonia. *Glob. Change Biol*, 21, 2773–2786. doi.org/10.1111/gcb.12906.

Gao, Y.H., Luo, P., Wu, N., Chen, H., & Wang, G.X. (2007). Grazing intensity impacts on carbon sequestration in an alpine meadow on the eastern Tibetan Plateau. *Res. J. Agri. and Biol. Sci.*, 3, 642–647.

Geng, Y., Wang, Z., Liang, C., Fang, J., Baumann, F., Kühn, P., Scholten, T., & He, J.-S. (2012b). Effect of geographical range size on plant functional traits and the relationships between plant, soil and climate in Chinese grasslands. *Glob. Ecol. and Biogeograph.*, 21, 416–427. doi.org/ 10.1111/j.1466-8238.2011.00692.x.

Geng, Y., Wang, Y.-H., Yang, K., Wang, S.-P., Zeng, H., Baumann, F., Kuehn, P., Scholten, T., & He, J.-S. (2012a). Soil respiration in Tibetan alpine grasslands: Belowground biomass and soil moisture, but not soil temperature, best explain the large-scale patterns. *PLOS ONE*, 7 (4), e34968. doi.org/10.1371 /journal.pone.0034968.

Gerzabek, M.H., Haberhauer, G., Stemmer, M., Klepsch, S., & Haunold, E. (2004). Long-term behaviour of 15N in an alpine grassland ecosystem. *Biogeochemistry*, 70, 59–69. doi.org/10.1023/B:BIOG .0000049336.84556.62.

Gharibreza, M., Raj, J.K., Yusoff, I., Othman, Z., Tahir, W.Z.W.M., & Ashraf, M.A. (2013). Land-use changes and soil redistribution estimation using 137Cs in the tropical Bera Lake catchment, Malaysia. *Soil and Tillage Res.*, 131, 1–10. doi.org/10.1016/j.still.2013.02.010.

González-Sánchez, E.J., Ordónez-Fernández, R., Carbonell-Bojollo, R., Veroz-González, O., & Gil-Ribes, J.A. (2012). Meta-analysis on atmospheric carbon capture in Spain through the use of conservation agriculture. *Soil and Tillage Res.*, 122, 52–60. doi.org/10.1016/j.still.2012.03.001.

Gorham, E. (1991). Northern peatlands—Role in the carbon-cycle and probable responses to climate warming. *Ecol. Appl.*, 1, 182–195. doi.org/10.2307/1941811.

Govaerts, B., Verhulst, N., Castellanos-Navarrete, A., Sayre, K. D., Dixon J., & Dendooven L. (2009). Conservation agriculture and soil carbon sequestration: Between myth and farmer reality. *Crit. Rev. Plant Sci.*, 28:3, 97–122, doi.org/10.1080/07352680902776358

Gudasz, C., Sobek, S., Bastviken, D., Koehler, B., & Tranvik, L.J. (2015). Temperature sensitivity of organic carbon mineralization in contrasting lake sediments. *J. Geophys. Res.-Biogeosci.*, 120 (7), 1215–1225. doi.org/10.1002/2015JG002928.

Guillaume, T., Damris, M., & Kuzyakov, Y. (2015). Losses of soil carbon by converting tropical forest to plantations: Erosion and decomposition estimated by δ 13 C. *Glob. Change Bio.*, 21, 3548–3560. Doi .org/10.1111/gcb.12907.

Guillaume, T., Holtkamp, A.M., Damris, M., Brümmer, B., & Kuzyakov, Y. (2016b). Soil degradation in oil palm and rubber plantations under land resource scarcity. *Agric., Ecosys. & Environ.*, 232, 110–118. doi .org/10.1016/j.agee.2016.07.002.

Guillaume, T., Maranguita, D., Murtilaksono, K., & Kuzyakov, Y. (2016a). Sensitivity and resistance of soil fertility indicators to land-use changes: New concept and examples from conversion of Indonesian rainforest to plantations. *Ecological Indicators*, 67, 49–57. doi.org/10.1016/j.ecolind.2016.02.039.

Gundelwein, A., Müller-Lupp, T., Sommerkorn M., Haupt, E.T.K., Pfeiffer, E. et al. (2007). Carbon in tundra soils in the Lake Labaz region of arctic Siberia. *Europ. J. Soil Sci.*, 58 (5), 1164–1174. doi .org/10.1111/j.1365-2389.2007.00908.x.

Haddaway, N.R., Hedlund, K., Jackson, L.E., Kätterer, T., Lugato, E., Thomsen, I.K., Bracht Jørgensen, H., & Söderström, B. (2015). What are the effects of agricultural management on soil organic carbon in boreo-temperate systems? A systematic map. *Environ. Evidence,* 4, 23. doi.org/10.1186 /s13750-015-0049-0.

Haddaway, N.R., Hedlund, K., Jackson, L.E., Kätterer, T., Lugato, E., Thomsen, I.K., Bracht Jørgensen H., & Isberg, P.-E. (2016a). How does tillage intensity affect soil organic carbon (SOC) stocks? *Environ. Evidence*, 5, 1. doi.org/10.1186/s13750-016-0052-0.

Haddaway, N.R., Hedlund, K., Jackson, L.E., Kätterer, T., Lugato, E., Thomsen, I.K., Bracht Jørgensen, H., & Isberg, P.-E. (2016b). Which agricultural management interventions are most influential on soil organic carbon (using time series data)? *Environ. Evidence*, 5, 2. doi.org/10.1186/s13750-016-0053-z.

Hafner, S., Unteregelsbacher, S., Seeber, E., Becker, L., Xu, X., Li, X. et al. (2012). Effect of grazing on carbon stocks and assimilate partitioning in a Tibetan montane pasture revealed by 13CO2 pulse labeling. *Glob. Change Biol.*, 18 (2), 528–538. doi.org/10.1111/j.1365-2486.2011.02557.x.

Hall, D.O., Scurlock, J.M.O., Ojima, D.S., & Parton, W. (2000). Grasslands and the global carbon cycle: Modeling the effects of climate change. In: Wigley, T.M.L., and Schimel, D.S. (eds.), *The Carbon Cycle*, pp. 102–115. Cambridge University Press, Cambridge.

Hansen, M.C., Potapov, P.V., Moore, R., Hancher, M., Turubanova, S.A., Tyukavina, A., Thau, D., Stehman, S.V., Goetz, J., Loveland, T.R., Kommareddy, A., Egorov, A., Chini, L., Justice, C.O., & Townshend, J.R.G. (2013). High-resolution global maps of 21st-century forest cover change. *Science*, 342 (6160), 850–853.

Harris, N. (2006). The elevation history of the Tibetan Plateau and its implications for the Asian monsoon. *Palaeogeography, Palaeoclimatology, Palaeoecology*, 241 (1), 4–15. doi.org/10.1111/j.1365-2486.2011 .02557.x.

Harris, R.B. (2010). Rangeland degradation on the Qinghai-Tibetan plateau: A review of the evidence of its magnitude and causes. *J .Arid Environments*, 74, 1–12. doi.org/10.1016/j.jaridenv.2009.06.014.

Hénin, S., & Dupuis, M. (1945). Essai de bilan de la matière organique du sol. *Annales Agronomiques*, 15, 17–29.

Henkner, J., Scholten, T., & Kühn, P. (2016). Soil organic carbon stocks in permafrost-affected soils in West Greenland. *Geoderma*, 282, 147–159. doi.org/10.1016/j.geoderma.2016.06.021.

Hergoualch, K., & Verchot, L.V. (2011). Stocks and fluxes of carbon associated with land-use change in Southeast Asian tropical peatlands: A review. *Glob. Biogeochem. Cycles*, 25, GB2001. doi.org/10.1029/2009GB003718.

Hermans, C., Hammond, J.P., White, P.J., & Verbruggen, N. (2006). How do plants respond to nutrient shortage by biomass allocation? *Trends in Plant Sci.*, 11, 610–617. doi.org/10.1016/j.tplants.2006.10.007.

Hirano, T., Segah, H., Kusin, K., Limin, S., Takahashi, H., & Osaki, M. (2012). Effects of disturbances on the carbon balance of tropical peat swamp forests. *Glob. Change Biol.*, 18, 3410–3422. doi.org/10.1111/j.1365-2486.2012.02793.x.

Hirota, M., Tang, Y.H., & Hu, Q.W. (2004). Methane emissions from different vegetation zones in a Qinghai-Tibetan Plateau wetland. *Soil Bio. & Biochem.*, 36, 737–748. doi.org/10.1016/j.soilbio.2003.12.009.

Hirota, M., Tang, Y., Hu, Q., Hirata, S., Kato, T., Mo, W., Cao, G., & Mariko, S. (2006). Carbon dioxide dynamics and controls in a deep-water wetland on the Qinghai-Tibetan Plateau. *Ecosystems*, 9 (4), 673–688. doi.org/10.1007/s10021-006-0029-x.

Hobbie, S.E., Schimel, J.P., Trumbore, S.E., & Randerson, J.R. (2000). Controls over carbon storage and turnover in high-latitude soils. *Glob. Change Biol.*, 6 (S1), 196–210. doi.org/10.1046/j.1365-2486.2000.06021.x<.

Hodgkins, S.B., Tfaily, M.M., McCalley, C.K., Logan, T.A., Crill, P.M., Saleska, S.R., Rich, V.I., & Chanton, J.P. (2014). Changes in peat chemistry associated with permafrost thaw increase greenhouse gas production. *Proc. Nat. Acad. Sci.*, 111 (16), 5819–5824. doi.org/10.1073/pnas.131464111.

Hogg, E.H., Lieffers, V.J., & Wein, R.W. (1992). Potential carbon losses from peat profiles: Effects of temperature, drought cycles, and fire. *Ecologic. App.*, 2, 298–306. doi.org/10.2307/1941863.

Holzner, W., & Kriechbaum, M. (2001). Pastures in South and Central Tibet. II. Probable causes of pasture degradation. *Die Bodenkultur*, 52, 37–44.

Hooijer, A., Page, S., Canadell, J. G., Silvius, M., Kwadijk, J., Wösten, H., & Jauhiainen, J. (2010). Current and future CO2 emissions from drained peatlands in Southeast Asia. *Biogeosciences*, 7, 1505–1514. doi.org/10.5194/bg-7-1505-2010.

Hooijer, A., Page, S., Jauhiainen, J., Lee, W.A., Lu X.X., Idris, A., & Anshari, G. (2012). Subsidence and carbon loss in drained tropical peatlands. *Biogeosciences*, 9, 1053–1071. doi.org/10.5194/bg-9-1053-2012.

Hook, P.B., & Burke, I.C. (2000). Biogeochemistry in a shortgrass landscape: Control by topography, soil texture, and microclimate. *Ecology*, 81, 2686–2703. doi.org/10.2307/177334.

Hou, X.Y. (1982). Vegetation map of the People's Republic of China (1:4M). Chinese Map Publisher, Beijing.

Houghton, R.A., House, J.I., Pongratz, J., van der Werf, G.R., DeFries, R.S., Hansen, M.C., Le Quéré, C., & Ramankutty, N. (2012). Carbon emissions from land-use and land-cover change. *Biogeosciences*, 9, 5125–5142. doi.org/10.5194/bg-9-5125-2012.

Hu, Z., Li, S., Guo, Q., Niu, S., He, N., Li, L., & Yu, G. (2016). A synthesis of the effect of grazing exclusion on carbon dynamics in grasslands in China. *Glob. Change Biol.*, 22 (4), 1385–1393. doi.org/10.1111/gcb.13133.

Huang, J., Hu, T., & Zheng, H. (2009). The break dormancy and quantity of abscisic acid in Kobresia Willd. *Acta Agriculturae Boreali-Occidentalis Sinica*, 18, 152–155.

Hugelius, G., & Kuhry, P. (2009). Landscape partitioning and environmental gradient analyses of soil organic carbon in a permafrost environment. *Glob. Biogeochem. Cycles*, 23, GB3006. doi.org/10.1029/2008GB003419.

Hugelius, G., Kuhry, P., Tarnocai, C., & Virtanen, T. (2010). Soil organic carbon pools in a periglacial landscape: A case study from the central Canadian Arctic. *Permafrost Periglac. Process.*, 21, 16–29. doi.org/10.1002/ppp.677.

Hugelius, G., Strauss, J., Zubrzycki, S., Harden, J.W., Schuur, E.A.G., Ping, C.-L., Schirrmeister, L., Grosse, G., Michaelson, G.J., Koven, C.D., O'Donnell, J.A., Elberling, B., Mishra U., Camill, P., Yu, Z., Palmtag, J., & Kuhry, P. (2014). Estimated stocks of circumpolar permafrost carbon with quantified uncertainty ranges and identified data gaps. *Biogeosciences*, 11, 6573–6593. doi.org/10.5194/bg-11-6573-2014.

Huijnen, V., Wooster, M.J., Kaiser, J.W. et al. (2016). Fire carbon emissions over maritime southeast Asia in 2015 largest since 1997. *Scientific Reports*, 6, 26886. doi.org/ 10.1029/2008GB003419.

Ingrisch, J., Biermann, T., Seeber, E. et al. (2015). Carbon pools and fluxes in a Tibetan alpine Kobresia pygmaea pasture partitioned by coupled eddy-covariance measurements and $^{13}CO_2$ pulse labeling. *Science of the Total Env.*, 505, 1213–1224. doi.org/10.1029/2008GB003419.

Inubushi, K., Furukawa, Y., Hadi, A., Purnomo, E., & Tsuruta, H. (2003). Seasonal changes of CO2, CH4 and N2O fluxes in relation to land-use change in tropical peatlands located in coastal area of South Kalimantan. *Chemosphere*, 52 (3), 603–608.

IPCC (2013). *Supplement to the 2006 IPCC Guidelines for National Greenhouse Gas Inventories: Wetlands.* Hiraishi, T., Krug, T., Tanabe, K., Srivastava, N., Baasansuren, J., Fukuda, M. & Troxler, T.G. (eds). IPCC, Switzerland.

IPCC (2007). *Climate Change 2007: The Physical Science Basis. Contribution of Working Group I to the Fourth Assessment Report of the Intergovernmental Panel on Climate Change.* Cambridge University Press, Cambridge.

Ise, T., Dunn, A.L., Wofsy, S.C., & Moorcroft, P.R. (2008). High sensitivity of peat decomposition to climate change through water-table feedback. *Nature Geosci.,* 1, 763–766. doi.org/10.1038/ngeo331.

IUSS Working Group WRB. (2006). *World Reference Base for Soil Resources.* World Soil Resources Reports (103).

IUSS Working Group WRB. (2015). *World Reference Base for Soil Resources 2014 International Soil Classification System for Naming Soils and Creating Legends for SoilMaps.* Update 2015, Rome.

Jandl, R., Lindner, M., Vesterdal, L., Bauwens, B., Baritz, R., Hagedorn, F., Johnson, D.W., Minkkinen, K., & Byrne, K.A. (2007). How strongly can forest management influence soil carbon sequestration? *Geoderma,* 137, 253–268. doi.org/10.1016/j.geoderma.2006.09.003.

Jauhiainen, J., Hooijer, A., & Page, S.E. (2012). Carbon dioxide emissions from an Acacia plantation on peatland in Sumatra, Indonesia. *Biogeosciences,* 9, 617–630. doi.org/10.5194/bg-9-617-2012.

Jauhiainen, J., Kerojoki, O., Silvennoinen, H., Limin, S., & Vasander, H. (2014). Heterotrophic respiration in drained tropical peat is greatly affected by temperature—A passive ecosystem cooling experiment. *Environ. Res. Lett.,* 9, 105013. doi.org/10.1088/1748-9326/9/10/105013.

Jensen, D.B., & Christensen, K.D. (2003). *The Biodiversity of Greenland. A Country Study.* Pinngortitalerifik, Grønlands Naturinstitut. Technical Report No. 55, Nuuk. 165 p.

Jin, H., He, R., Cheng, G., Wu, Q., Wang, S., Lü, L., & Chang, X. (2009). Changes in frozen ground in the Source Area of the Yellow River on the Qinghai-Tibet Plateau, China, and their eco-environmental impacts. *Environ. Res. Lett.,* 4, 045206. doi.org/10.1088/1748-9326/4/4/045206.

Jin, H., Li, S., Cheng, G., Shaoling, W., & Li, X. (2000). Permafrost and climatic change in China. *Global and Planetary Change,* 26, 387–404. doi.org/10.1016/S0921-8181(00)00051-5.

Jin, H.J., Chang, X.L., & Wang, S.L. (2007). Evolution of permafrost on the Qinghai-Xizang (Tibet) Plateau since the end of the late Pleistocene. *J. Geophys. Res.,* 112, F02S09, doi.org/10.1016/S0921-8181(00)00051-5.

Jin, H.J., Wu, J., Cheng, G.D., Tomoko, N., & Sun, G.Y. (1999). Methane emissions from wetlands on the Qinghai-Tibet Plateau. *Chinese Sci. Bull.,* 44, 2282–2286. doi.org/10.1007/BF02885940.

Jobbagy, E., & Jackson, R.B. (2000). The vertical distribution of soil organic carbon and its relation to climate and vegetation. *Ecol. Appl.,* 10 (2), 423–436. doi.org/10.1890/1051-0761(2000)010[0423:TVDOSO]2.0.CO;2.

Jones, A., Stolbovoy, V., Tarnocai, C., Broll, G., Spaargaren, O., & Montanarella, L. (eds.) (2010). *Soil Atlas of the Northern Circumpolar Region.* Luxembourg (144 pp) Joosten, H. (2010). *The Global Peatland CO2 Picture: Peatland Status and Drainage Related Emissions in All Countries of the World.* Wetlands International, Ede.

Joosten, H., & Clarke, D. (2002). *Wise Use of Mires and Peatlands—background and Principles Including a Framework for Decision-Making.* International Mire Conservation Group and International Peat Society, Jyväskylä, Finland (303 pp.).

Jørgensen, R.H., Meilby, H., & Kollmann, J. (2013). Shrub expansion in SW Greenland under modest regional warming: Disentangling effects of human disturbance and grazing. *Arctic, Antarctic, and Alpine Res.,* 45 (4), 515–525. doi.org/10.1657/1938-4246-45.4.515.

Jungkunst, H.F., Krüger, J. P., Heitkamp, F., Erasmi, S., Fiedler, S., Glatzel, S., & Lal, R. (2012). *Accounting more precisely for peat and other soil carbon resources.* In: Lal, R., Lorenz, K., Hüttl, R.F.J., Schneider, B.U., abd von Braun, J. (eds.), *Recarbonization of the Biosphere—Ecosystems and the Global Carbon Cycle,* pp. 127–157. Springer, Amsterdam, Netherlands.

Kaiser, K. (2004). Pedogeomorphological transect studies in Tibet: Implications for landscape history and present-day dynamics. *Prace Geograficzne,* 200, 147–165.

Kaiser, K., Miehe, G., Barthelmes, A., Ehrmann, O., Scharf, A., Schult, M., Schlütz, F., Adamczyk, S., & Frenzel, B. (2008). Turf-bearing topsoils on the central Tibetan Plateau. China: Pedology, botany, geochronology. *Catena,* 73 (3), 300–311. doi.org/10.1016/j.catena.2007.12.001.

Kang, S., Xu, Y., You, Q., Flügel, W.-A., Pepin, N., & Yao, T. (2010). Review of climate and cryospheric change in the Tibetan Plateau. *Environ. Res. Lett.,* 5 (1), 15101. doi.org/10.1088/1748-9326/5/1/015101.

Kato, T., Tang, Y., Gu, S., Cui, X., Hirota, M., Du, M., Li, Y., Zhao, X., & Oikawa, T. (2004). Carbon dioxide exchange between the atmosphere and an alpine meadow ecosystem on the Qinghai–Tibetan Plateau, China. *Agricultural and Forest Meteorology,* 124 (1–2), 121–134. doi.org/10.1016/j.agrformet.2003.12.008.

Kato, T., Tang, Y., Gu, S., Hirota, M., Du, M., Li, Y., & Zhao, X. (2006). Temperature and biomass influences on interannual changes in CO2 exchange in an alpine meadow on the Qinghai-Tibetan Plateau. *Glob. Change Biol.*, 12 (7), 1285–1298. doi.org/10.1111/j.1365-2486.2006.01153.x.

Kätterer, T., & Andrén, O. (2009). Predicting daily soil temperature profiles in arable soils from air temperature and leaf area index. *Acta Agr. Scand. Section B—Soil & Plant Sci.*, 59, 77–86. doi.org/10.1080/09064710801920321.

Kätterer, T., Andersson, L., Andrén, O., & Persson, J. (2008). Long-term impact of chronosequential land-use change on soil carbon stocks on a Swedish farm. *Nutrient Cycling in Agroecosystems*, 81, 145–155. doi.org/10.1007/s10705-007-9156-9.

Kätterer, T., Börjesson, G., & Kirchmann, H. (2014). Changes in organic carbon in topsoil and subsoil and microbial community composition caused by repeated additions of organic amendments and N fertilisation in a long-term field experiment in Sweden. *Agriculture, Ecosystems and Environment*, 189, 110–118. doi.org/10.1016/j.agee.2014.03.025.

Kätterer, T., Bolinder, M.A., Andrén, O., Kirchmann, H., & Menichetti, L. (2011). Roots contribute more to refractory soil organic matter than aboveground crop residues, as revealed by a long-term field experiment. *Agriculture, Ecosystems and Environment*, 141, 184–192. doi.org/10.1016/j.agee.2011.02.029.

Kätterer, T., Bolinder, M.A., Berglund, K., & Kirchmann, H. (2012). Strategies for carbon sequestration in agricultural soils in northern Europe. *Acta Agr. Scand. Section A*, 62, 181–198. doi.org/10.1080/09064702.2013.779316.

Kätterer, T., Bolinder, M.A., Thorvaldsson, G. & Kirchmann, H. (2013). Influence of ley-arable systems on soil carbon stocks in Northern Europe and Eastern Canada. In: Helgadóttir, A., and Hopkins, A. (eds.), *The Role of Grasslands in a Green Future—Threats and Perspectives in Less Favoured Areas. Proceedings of the 17th Symposium of the European Grassland Federation, Akureyri, Iceland, 23–26 June 2013*, pp. 47–56. Grassland Science in Europe 18.

Keiluweit, M., Nico, P.S., Kleber, M., & Fendorf, S. (2016). Are oxygen limitations under recognized regulators of organic turnover in upland soils? *Biogeochemistry*, 127, 157. doi.org/10.1007/s10533-015-0180-6. doi:10.1007/s10533-015-0180-6.

Kirchmann, H., Börjesson, G., Kätterer, T., & Cohen, Y. (2017). From agricultural use of sewage sludge to nutrient extraction: A soil science outlook. *Ambio*, 46, 143–154. doi.org/10.1007/s13280-016-0816-3.

Kirtman, B. et al. (2013). Near-term climate change: Projections and predictability Climate Change 2013. In: Stocker, T.F. et al. (eds.), *The Physical Science Basis. Contribution of Working Group I to the V Assessment Report of the Intergovernmental Panel on Climate Change.* Cambridge University Press Cambridge

Klinge, M., & Lehmkuhl, F. (2005). Untersuchungen zur holozänen Bodenentwicklung und Geomorphodynamik in Tibet. In: Eidam, U. (ed.), *Hochgebirge und ihr Umland*, pp. 81–91. Berliner geographische Arbeiten 100. Geograph. Inst. der Humboldt-Univ., Berlin.

Koerner, C. (2003). Slow in, rapid out—Carbon flux Studies and Kyoto targets. *Science*, 300 (5623), 1242–1243. doi.org/10.1126/science.1084460.

Körner, C.H. (2003). *Alpine Plant Life: Functional Plant Ecology of High Mountain Ecosystems.* Springer, Berlin.

Körschens, M., Albert, E., Armbruster, M., Barkusky, D., Baumecker, M., Behle-Schalk, L., Bischoff, R., Čergan, Z., Ellmer, F., Herbst, F., Hoffmann, S., Hofmann, B., Kismanyoky, T., Kubat J., Kunzova, E., Lopez-Fando, C., Merbach, I., Merbach, W., Pardor, M.T., Rogasik, J., Rühlmann, J., Spiegel, H., Schulz, E., Tajnsek, A., Toth, Z., Wegener, H., & Zorn, W. (2013). Effect of mineral and organic fertilization on crop yield, nitrogen uptake, carbon and nitrogen balances, as well as soil organic carbon content and dynamics: Results from 20 European long-term field experiments of the twenty-first century. *Arch. Agron.and Soil Sci.*, 59, 1017–1040. doi.org/10.1080/03650340.2012.704548.

Körschens, M., Weigel, A., & Schulz, E. (1998). Turnover of soil organic matter (SOM) and long-term balances—Tools for evaluating sustainable productivity of soils. *Z. Pflanzenernaehr. Bodenk.*, 161, 409–424. doi.org/10.1002/jpln.1998.3581610409.

Kotowska, M.M., Leuschner, C., Triadiati, T., Meriem, S., & Hertel, D. (2015). Quantifying above- and belowground biomass carbon loss with forest conversion in tropical lowlands of Sumatra (Indonesia). *Glob. Change Bio.*, 21, 3620–3634. doi.org/10.1111/gcb.12979.

Koutika, L.-S., Bartoli, F., Andreux, F., Cerri, C.C., Burtin, G., Choné, T., & Philippy, R. (1997). Organic matter dynamics and aggregation in soils under rain forest and pastures of increasing age in the eastern Amazon Basin. *Geoderma*, 76, 87–112. doi.org/10.1016/S0016-7061(96)00105-X.

Koven, C., Friedlingstein, P., Ciais, P., Khvorostyanov, D., Krinner, G., & Tarnocai, C. (2009). On the formation of high-latitude soil carbon stocks: Effects of cryoturbation and insulation by organic matter in a land surface model. *Geophys. Res. Lett.*, 36, L21501, doi.org/10.1016/S0016-7061(96)00105-X.

Kramer, M.G., Sollins, P., & Sletten, R.S. (2004). Soil carbon dynamics across a Windthrow disturbance sequence in Southeast Alaska. *Ecology*, 85 (8), 2230–2244. doi.org/10.1890/02-4098.

Krüger, J.P., Alewell, C., Minkkinen, K., Szidat, S., & Leifeld, J. (2016). Calculating carbon changes in peat soils drained for forestry with four different profile-based methods. *Forest Ecology and Manag.*, 381, 29–36. doi.org/10.1890/02-4098.

Krüger, J.P., Beckedahl, H., Gerold, G., & Jungkunst, H.F. (2014). Greenhouse gas emission peaks following natural rewetting of two wetlands in the southern Ukhahlamba-Drakensberg Park, South Africa. *So. African Geograph. J.*, 96(2), 113–118. doi.org/10.1080/03736245.2013.847798.

Krüger, J.P., Conen, F., Leifeld, J., & Alewell, C. (2017). Palsa uplift identified by stable isotope depth profiles and relation of δ15N to C/N ratio. *Permafrost and Periglacial Processes*, 28, 485–492. doi.org/10.1002/ppp.1936.

Krüger, J.P., Leifeld, J., Glatzel, S., Szidat, S., & Alewell, C. (2015a). Biogeochemical indicators of peatland degradation – a case study of a temperate bog in northern Germany. *Biogeosciences*, 12, 2861–2871. Doi.org/10.5194/bg-12-2861-2015.

Krüger, J.P., Leifeld, J., Glatzel, S., & Alewell, C. (2015b). Soil carbon loss from managed peatlands along a land-use gradient–a comparison of three different methods. *Bull. BGS*, 36, 45–50.

Krull, E.S., Baldock, J.A., & Skjemstad, J.O. (2003). Importance of mechanisms and processes of the stabilisation of soil organic matter for modelling carbon turnover. *Funct. Plant Bio.*, 30, 207–222. doi.org/10.1071/FP02085.

Kuhry, P., Mazhitova, G., Forest, P.-A., Deneva, S.V., Virtanen, T., & Kulti, S. (2002). Upscaling soil organic carbon estimates for the Usa Basin (Northeast European Russia) using GIS-based landcover and soil classification schemes. *Geografisk Tidsskrift. Dan. J. Geogr.*, 102 (1), 11–25.

Lai, D.Y.F., Roulet, N.T., Tim, R., & Moore, T.R. (2014). The spatial and temporal relationships between CO2 and CH4 exchange in a temperate ombrotrophic bog. *Atmospheric Environment*, 89, 249–59.

Laine, A.M., Wilson, D., Alm, J., Schneider, J., & Tuittila, E.-S. (2015). Spatial variation in potential photosynthesis in Northern European bogs. *J. Vegetation Sci.*, 27, 365–376. doi.org/10.1111/jvs.12355.

Laine, J., Laiho, R., Minkkinen, K., & Vasander, H. (2006). Forestry and boreal peatlands. In: Wieder, R. K., and Vitt, D. H. (eds.), *Boreal Peatland Ecosystems*, pp. 331–357. Springer, Heidelberg.

Lal, R., & Follett, R.F. (2009). *Soil Carbon Sequestration and the Greenhouse Effect*. SSSA Special Publication 57, second edition. Soil Science Society of America, Madison, WI.

Lal, R., Delgado, J., Groffman, P., Millar, N., Dell, C., & Rotz, A. (2011). Management to mitigate and adapt to climate change. *J. Soil and Water Conserv.*, 66 (4), 276–85. doi.org/10.2489/jswc.66.4.276.

Land, M., Haddaway, N.R., Hedlund, K., Bracht Jørgensen, H., Kätterer, T., & Isberg, P.-E. (2017). How do selected crop rotations affect soil organic carbon in boreo-temperate systems? A systematic review protocol. *Environ. Evidence*, 6, 9. doi.org/10.1186/s13750-017-0086-y.

Langer, S.M. (2012). The Tracer Potential of Nutrient Elements for Lateral Matter Transport in a Boreal Forest-Peatland-Landscape. Diploma thesis, University of Hamburg.

Laumonier, Y., Uryu, Y., Stüwe, M., Budiman, A., Setiabudi, B., & Hadian, O. (2010). Eco-floristic sectors and deforestation threats in Sumatra: Identifying new conservation area network priorities for ecosystem-based land-use planning. *Biodiversity and Conserv.*, 19, 1153–1174. doi.org/10.1007/s10531-010-9784-2.

Lefèvre, R., Barré, P., Moyano, F.E., Christensen, B.T., Bardoux, G., Eglin, T., Girardin, C., Houot, S., Kätterer, T., van Oort, F., Chenu, C. (2014). Higher temperature sensitivity for stable than for labile soil organic carbon–Evidence from incubations of long-term bare fallow soils. *Glob. Change Bio.*, 20, 633–640. doi.org/10.1111/gcb.12402.

Lehmann, J., & Kleber, M. (2015). The contentious nature of soil organic matter. *Nature*, 528, 60–68. doi.org/10.1038/nature16069.

Lehmkuhl, F. (1997). The spatial distribution of loess and loess-like sediments in the mountain areas of Central and High Asia. *Zeitschrift für Geomorphologie*, 111, 97–116.

Lehnert, L. W., Wesche, K., Trachte, K., Reudenbach, C., & Bendix, J. (2016). Climate variability rather than overstocking causes recent large scale cover changes of Tibetan pastures. *Scientific Reports*, 6, 24367. doi.org/10.1038/srep24367.

Lemke, P., Ren, J., Alley, R., Allison, I., Carrasco, J., Flato, G., Fuji, Y., Kaser, G., Mote, P., Thomas, R., & Zhang, T. (2007). Observations: Changes in snow, ice and frozen ground. In: Solomon, S., Qin, D., Manning, M., Chen, Z., Marquis, M., Avereyt, K., Tignor, M., and Miller, H. (eds.), *Climate Change 2007: The Physical Science Basis. Contribution of Working Group I to the Fourth Assessment Report of the Inergovernmental Panel on Climate Change*, pp. 337–383. Cambridge University Press, Cambridge, New York.

Levy, L.B., Kelly, M.A., Howley, J.A., & Virginia, R.A. (2012). Age of the Ørkendalen moraines, Kangerlussuaq, Greenland: Constraints on the extent of the southwestern margin of the Greenland Ice Sheet during the Holocene. *Quaternary Sci. Rev.*, 52, 1–5. doi.org/10.1016/j.quascirev.2012.07.021.

Li, T., Zhang, Q., Cheng, Z., Ma, Z., Liu, J., Luo, Y., Xu, J., Wang, G., & Zhang, W. (2016). Modeling CH4 Emissions from natural wetlands on the Tibetan Plateau over the past 60 years: Influence of climate change and wetland loss. *Atmosphere*, 7, 90. doi.org/10.3390/atmos7070090.

Li, Z., Han, F., Su, Y., Zhang, T., Sun, B., Monts, D., & Plodinec, M. (2007). Assessment of soil organic and carbonate carbon storage in China. *Geoderma*, 138 (1–2), 119–126. doi.org/10.1016/j.geoderma.2006.11.007.

Liboriussen, L., & Jeppesen, E. (2003). Temporal dynamics in epipelic, pelagic and epiphytic algal production in a clear and a turbid shallow lake. *Freshwater Bio.*, 48 (3), 418–431. doi.org/10.1046/j.1365-2427.2003.01018.x.

Limpens, J., Heijmans, M.M., & Berendse, F. (2006). The nitrogen cycle in boreal peatlands. In: Wieder, R. K., and Vitt, D. H. (eds.), *Boreal Peatland Ecosystems*, pp. 195–230. Springer, Berlin, Heidelberg.

Lin, X., Zhang, Z., Wang, S., Hu, Y., Xu, G., Luo, C., Chang, X., Duan, J., Lin, Q., Xu, B., Wang, Y., Zhao, X., & Xie, Z. (2011). Response of ecosystem respiration to warming and grazing during the growing seasons in the alpine meadow on the Tibetan plateau. *Agricul. Forest Meteorol.*, 151 (7), 792–802. doi.org/10.1016/j.agrformet.2011.01.009.

Linell, J. D. C., Cuyler, C., Loison, A., Lund, P. M., Motzfeldt, K. G., Ingerslev, T., & Landa, A. (2000). The scientific basis for managing the sustainable harvest of caribou and muskoxen in Greenland for the 21st century: An evaluation and agenda. Technical Report 34. Greenland Institute of Natural Resources, Nuuk, Greenland (55 pp.).

Liu, S., Schleuss, P.-M., & Kuzyakov, Y. (2016). Carbon and nitrogen losses from soil depend on degradation of Tibetan Kobresia pastures. *Land Degradation and Develop.*, 28, 1253–1262, doi.org/10.1002/ldr.2522.

Liu, W.-J., Chen, S.-Y., Qi, X., Baumann, F., Scholten, T., Zhou, Z., Sun, W.-J., Zhang, T.-Z., Ren, J.-W., Qin, D. (2012). Storage, patterns, and controls of soil organic carbon and nitrogen in the northeastern margin of the Qinghai-Tibet Plateau. *Environ. Res. Lett.*, 7 (3), 035401. Doi.org/10.1088/1748-9326/7/3/035401.

Liu, X., & Chen, B. (2000). Climatic warming in the Tibetan Plateau during recent decades. *Int. J. Climatol.*, 20, 1729–1742. doi.org/10.1002/1097-0088(20001130)20:14<1729::AID-JOC556>3.0.CO;2-Y.

Liu, X., & Zhang, M. (1998). Contemporary climatic change over the Qinghai-Xizang Plateau and its response to the green-house effect. *Chinese Geograph. Sci.*, 8, 289–298. doi.org/10.1007/s11769-997-0034-9.

Lohila, A., Minkkinen, K., Aurel, a M., Tuovinen, J.P., Penttilä, T., Ojanen, P., & Laurila, T. (2011). Greenhouse gas flux measurements in a forestry-drained peatland indicate a large carbon sink. *Biogeosciences*, 8, 3203–3218. doi.org/10.5194/bg-8-3203-2011.

Loisel, J., & Yu, Z. (2013). Recent acceleration of carbon accumulation in a boreal peatland, south central Alaska. *J. Geophys. Res.: Biogeosci.*, 118, 41–53. doi.org/10.1029/2012JG001978.

Loisel, J., Gallego-Sala, A.V., & Yu, Z. (2012). Global-scale pattern of peatland Sphagnum growth driven by photosynthetically active radiation and growing season length. *Biogeosciences*, 9, 2737–2746. doi.org/10.5194/bg-9-2737-2012.

Loisel, J., Yu, Z., Beilman, D.W., Camill, P., Alm, J., Amesbury, M.J., Anderson, D., Andersson, S., Bochicchio, C., Barber, K., Belyea, L.R., Bunbury, J., Chambers, F.M., Charman, D.J., De Vleeschouwer, F., Fiałkiewicz-Kozieł, B., Finkelstein, S.A., Gałka, M., Garneau, M., Hammarlund, D., Hinchcliffe, W., Holmquist, J., Hughes, P., Jones, M.C., Klein, E.S., Kokfelt, U., Korhola, A., Kuhry, P., Lamarre, A., Lamentowicz, M., Large, D., Lavoie, M., MacDonald, G., Magnan, G., Mäkilä, M., Mallon, G., Mathijssen, P., Mauquoy, D., McCarroll, J., Moore, T.R., Nichols, J., O'Reilly, B., Oksanen, P., Packalen, M., Peteet, D., Richard, P.J.H., Robinson, S., Ronkainen, T., Rundgren, M., Sannel, A.B.K., Tarnocai, C., Thom, T., Tuittila, E.S., Turetsky, M., Väliranta, M., van der Linden, M., van Geel, B., van Bellen, S., Vitt, D., Zhao, Y., & Zhou, W. (2014). A database and synthesis of northern peatland soil properties and Holocene carbon and nitrogen accumulation. *Holocene*, 24, 1028–1042. doi.org/10.1177/0959683614538073.

Long, R. J., Apori, S.O., Castro, F.B., & Ørskov, E.R. (1999). Feed value of native forages of the Tibetan Plateau of China. *Animal Feed Sci. Technol.*, 80, 101–113.

Lu, H., Wu, N., Gu, Z., Guo, Z., Wang, L., Wu., H., Wang, G., Zhou, L., Han, J. & Liu, T. (2004). Distribution of carbon isotope composition of modern soils on the Qinghai-Tibetian Plateau. *Biogeochemistry*, 70, 275–299, doi.org/10.1023/B:BIOG.0000049343.48087.ac.

Luo, Y., & Zhou, X. (2006). *Soil Respiration and the Environment*. Elsevier, San Diego.

Luo, Z., Wang, E., & Sun, O.J. (2010). Can no-tillage stimulate carbon sequestration in agricultural soils? A meta-analysis of paired experiments. Agriculture, *Ecosystems & Environment*, 139, 224–231.

Lutzow, M.V., Kogel-Knabner, I., Ekschmitt, K., Matzner, E., Guggenberger, G., Marschner, B., & Flessa, H. (2006). Stabilization of organic matter in temperate soils: Mechanisms and their relevance under different soil conditions—A review. *Europ. J. Soil Sci.*, 57 (4), 426. doi.org/10.1023/B:BIOG.0000049343.48087.ac.

Macías, F., & Camps Arbestain, M. (2010). Soil carbon sequestration in a changing global environment. *Mitigation and Adaptation Strategies for Global Change*, 15, 511–529. doi.org/10.1007/s11027-010-9231-4.

Marchão, R.L., Becquer, T., Brunet, D., Balbino, L.C., Vilela, L., & Brossard, M. (2009). Carbon and nitrogen stocks in a Brazilian clayey Oxisol: 13-year effects of integrated crop–livestock management systems. *Soil Tillage Res.*, 103, 442–450. doi.org/10.1016/j.still.2008.11.002.

Marcotte, P., Roy, V., Plamondon, A.P., & Auger, I. (2008). Ten-year water table recovery after clearcutting and draining boreal forested wetlands in eastern Canada. *Hydrol.l Processes*, 22, 4163–4172. doi.org/10.1002/hyp.7020.

Margono, B.A., Potapov, P.V., Turubanova, S., Stolle, F., & Hansen, M.C. (2014). Primary forest cover loss in Indonesia over 2000–2012. *Nature Climate Change*, 4, 730–735. doi.org/10.1038/nclimate2277. doi:10.1038/366051a0.

Martikainen, P.J., Nykänen, H., Crill, P., & Silvola, J. (1993). Effect of a lowered water table on nitrous oxide fluxes from northern peatlands. *Nature*, 366 (6450), 51–53.

Marushchak, M.E., Pitkämäki, A., Koponen, H., Biasi, C., Seppälä, M., & Martikainen, P.J. (2011). Hot spots for nitrous oxide emissions found in different types of permafrost peatlands. *Glob. Change Bio.*, 17, 2601–2614. doi.org/10.1111/j.1365-2486.2011.02442.x.

McBean, G., Alekseev, G.V., Chen, D., Forland, E., Fyfe, J., Groisman, P.Y., King, R., Melling, H., Vose, R., & Whitfield, P.H. (2005). Arctic climate—Past and present In: Arctic climate impact assessment—Scientific report. Cambridge University Press, Cambridge, pp. 22–60.

McCarthy, JF., & Cramb, R.A. (2009). Policy narratives, landholder engagement, and oil palm expansion on the Malaysian and Indonesian frontiers. *Geograph. J.*, 175, 112–123. doi.org/10.1111/j.1475-4959.2009.00322.x.

Merten, J., Röll, A., Guillaume, T. et al. (2016). Water scarcity and oil palm expansion: Social views and environmental processes. *Ecol. and Soc.*, 21, 5. doi.org/10.5751/ES-08214-210205.

Meurer, K.H.E., Franko, U., Stange, C.F., Dalla Rosa, J., Madari, B., & Jungkunst, H.F. (2016). Direct nitrous oxide (N2O) fluxes from soils under different land use in Brazil—A critical review. *Environ. Res. Lett.*, 11, 023001, doi:10.1088/1748-9326/11/2/023001.

Meurer, K.H.E., Franko, U., Spott, O., Schuetzenmeister, K., Niehaus, E., Stange, C.F., & Jungkunst, H.F. (2017). Missing hot moments of greenhouse gases in Southern Amazonia. *Erdkunde*, 71(3), 195–211, doi: 10.3112/erdkunde.2017.03.03

Mi, N., Wang, S., Liu, J., Yu, G., Zhang, W., & Jobbágy, E. (2008). Soil inorganic carbon storage pattern in China. *Glob. Change Bio.*, 14 (10), 2380–2387. doi.org/10.1111/j.1365-2486.2008.01642.x.

Miao, Y., Xu, Y., Hu, T., Wang, Q., & Zang, J. (2008). Germplasm resources evaluation of Kobresia pygmaea in Tibet. *Pratacult, Anim. Husb.*, 11, 10–3.

Michaelson, G.J., Dai, X., & Ping, C.-L. (2004). Organicmatter and bioactivity in cryosols of Arctic Alaska. In: Kimble, J.M. (ed.), *Cryosols. Permafrost-Affected Soils*, pp. 463–477. Gardners Books, Eastbourne, UK.

Miehe, G., Miehe, S., Bach, K., Nölling, J., Hanspach, J., Reudenbach, C., Kaiser, K., Wesche, K., Mosbrugger, V., Yang, Y.P., & Ma, Y.M. (2011). Plant communities of central Tibetan pastures in the Alpine Steppe/ Kobresia pygmaea ecotone. *J. Arid Environ.*, 75, 711–723. doi.org/10.1016/j.jaridenv.2011.03.001.

Miehe, G., Miehe, S., Boehner, J., Kaiser, K., Hensen, I., Madsen, D., Liu, J., & Opgenoorth, L. (2014). How old is the human footprint in the world's largest alpine ecosystem? A review of multiproxy records from the Tibetan Plateau from the ecologists' viewpoint. *Quaternary Sci. Rev.*, 86, 190–209. doi.org/10.1016/j.quascirev.2013.12.004.

Miehe, G., Miehe, S., Kaiser, K., Liu, J., & Zhao, X. (2008). Status and dynamics of Kobresia pygmaea ecosystem on the Tibetan plateau. *Ambio*, 37, 272–279.

Miettinen, J., Hooijer, A., Vernimmen, R., Liew, S.C., & Page, S.E. (2017). From carbon sink to carbon source: Extensive peat oxidation in insular Southeast Asia since 1990. *Environ. Res. Lett.*, 12. doi.org/10.1088/1748-9326/aa5b6f.

Miettinen, J., Shi, C., & Liew, S.C. (2012). Two decades of destruction in Southeast Asia's peat swamp forests. *Frontiers in Eco. and Environ.*, 10, 124–128. doi.org/10.1890/100236.

Miettinen, J., Shi, C., & Liew, S.C. (2016). Land cover distribution in the peatlands of Peninsular Malaysia, Sumatra and Borneo in 2015 with changes since 1990. *Glob. Ecol. and Conserv.*, 6, 67–78. doi.org/10.1016/j.gecco.2016.02.004.

Minasny, B., Malone, B.P., McBratney, A.B., Angers, D.A., Arrouays, D., Chambers, A., Chaplot, V., Chen, Z.-S., Cheng, K., Das, B.S., Field, D.J., Gimona, A., Hedley, C.B., Hong, S.Y., Mandal, B., Marchant, B.P., Martin, M., McConkey, B.G., Leatitia Mulder, V., O'Rourke, S., Richer-de-Forges, A.C., Odeh, I., Padarian, J., Paustian, K., Pan, G., Poggio, L., Savin, I., Stolbovoy, V., Stockmann, U., Sulaeman, Y., Tsui, C.-C., Vågen, T.-G., van Wesemael, B., & Winowieck, L. (2017). Soil carbon 4 per mille. *Goderma*, 292, 59–86. doi.org/10.1016/j.geoderma.2017.01.002.

Minkkinen, K., & Laine, J. (1998). Long-term effect of forest drainage on the peat carbon stores of pine mires in Finland. *Can. J. Forest Res.*, 28 (9), 1267–1275. doi.org/10.1139/x98-104.

Minkkinen, K., Byrne, K.A., & Trettin, C. (2008). Climate impacts of peatland forestry. In: Strack, M. (ed.), *Peatland and Climate Change*, pp. 98–122. International Peat Society, Jyväskylä, Finland.

Minkkinen, K., Vasander, H., Jauhiainen, S., Karsisto, M., & Laine, J. (1999). Post-drainage changes in vegetation composition and carbon balance in Lakkasuo mire. *Central Finland, Plant and Soil*, 207, 107–120. doi.org/10.1023/a:1004466330076.

van Minnen, J.G., Klein Goldewijk, K., Stehfest, E., Eickhout, B., van Drecht, G., & Leemans, R. (2009). The importance of three centuries of land-use change for the global and regional terrestrial carbon cycle. *Clim. Change*, 97, 123–144. doi.org/10.1007/s10584-009-9596-0.

Moore, T.R, Bubier, J.L., Frolking, S.E., Lafleur, P.M., & Roulet, N.T. (2002). Plant biomass and production and CO2 exchange in an ombrotrophic bog. *J. Ecol.*, 90, 25–36. doi.org/10.1046/j.0022-0477.2001.00633.x.

Moore, T.R., Lafleur, P.M., Poon, D.M.I., Heumann, B.W., Seaquist, J.W., & Roulet, N.T. (2006). Spring photosynthesis in a cool temperate bog. *Glob. Change Biol.*, 12, 2323–2335. doi.org/10.1111/j.1365-2486.2006.01247.x.

Moraes, J.F.L., Volkoff, B., Cerri, C.C., & Bernoux, M. (1996). Soil properties under Amazon forest and changes due to pasture installation in Rondônia, Brazil. *Geoderma*, 70, 63–81. doi.org/10.1016/0016-7061(95)00072-0.

Müller, M., Thiel, C., & Kühn, P. (2016). Holocene palaeosols and aeolian activities in the Umimmalissuaq valley, West Greenland. *The Holocene*, 26 (7), 1149–1161. doi.org/10.1177/0959683616632885.

Nadelhoffer, K.J., Giblin, A.E., Shaver, G.R., & Laundre, J.A. (1991). Effects of temperature and substrate quality on element mineralization in six arctic soils. *Ecology*, 72, 242–253. doi.org/10.2307/1938918.

Nan, Z., Li, S., & Cheng, G. (2005). Prediction of permafrost distribution on the Qinghai-Tibet Plateau in the next 50 and 100 years. *Sci. in China Series D: Earth Sci.*, 48 (6), 797–804. doi.org/10.1360/03yd0258.

Ni, J. (2002). Carbon storage in grasslands of China. *J. Arid Environ.*, 50 (2), 205–218. doi.org/10.1006/jare.2001.0902.

Nielsen, C.S., Michelsen, A., Strobel, B.W., Wulff, K., Banyasz, I., & Elberling, B. (2017). Correlations between substrate availability, dissolved CH4, and CH4 emissions in an arctic wetland subject to warming and plant removal. *J. Geophys. Res. Biogeosci.*, 122, 645–660. doi:10.1002/2016JG003511.

Niu, Y. (1999). The study of environment in the Plateau of Qin-Tibet. *Progress in Geography*, 18, 163–171. doi.org/10.11820/dlkxjz.1999.02.010.

Niu, F., Lin, Z., Liu, H., & Lu, J. (2011). Characteristics of thermokarst lakes and their influence onpermafrost in Qinghai–Tibet Plateau. *Geomorphology*, 132 (3–4), 222–233. doi.org/10.1016/j.geomorph.2011.05.011.

Noojipady, P., Morton, D.C., Macedo, M.N., Victoria, D.C., Huang, C., Gibbs, H.K., & Bolfe, E.L. (2017). Forest carbon emissions from cropland expansion in the Brazilian Cerrado biome. *Envrion. Res. Lett.*, 12, 025004, doi.org/10.1088/1748-9326/aa5986. doi.org/10.1088/1748-9326/aa5986.

O'Brien, B.J., & Stout, J.D. (1978). Movement and turnover of soil organic matter as indicated by carbon isotope measurements. *Soil Bio. and Biochem.*, 10, 309–317. doi.org/10.1016/0038-0717(78)90028-7.

Oechel, W.C., Hastings, S.J., Vourtilis, G., Jenkins, M., Reichers, G., & Grulke, N. (1993). Recent change of Arctic tundra ecosystems from a net carbon sink to a source. *Nature*, 361, 520–523. doi org/10.1038/361520a0.

Ohtsuka, T., Hirota, M., Zhang, X., Shimono, A., Senga, Y., Du, M., Yonemura, S., Kawashima, S., & Tang, Y. (2008). Soil organic carbon pools in alpine to nival zones along an altitudinal gradient (4400–5300m) on the Tibetan Plateau. *Polar Sci.*, 2 (4), 277–285. doi.org/10.1016/j.polar.2008.08.003.

Ojanen, P., Lehtonen, A., Heikkinen, J., Penttilä, T., & Minkkinen, K. (2014). Soil CO2 balance and its uncertainty in forestry-drained peatlands in Finland. *Forest Ecol. Manag.*, 325, 60–73. doi.org/10.1016/j.foreco.2014.03.049.

Ojanen, P., Minkkinen, K., & Penttilä, T. (2013). The current greenhouse gas impact of forestry-drained boreal peatlands. *Forest Ecol. Manag.*, 289, 201–208. doi.org/10.1016/j.foreco.2012.10.008.

Page, S.E., Rieley, J.O., & Banks, C.J. (2011). Global and regional importance of the tropical peatland carbon pool. *Glob. Change Biol.*, 17, 798–818. Doi.org/10.1111/j.1365-2486.2010.02279.x.

Page, S.E., Siegert, F., Rieley, J.O., Boehm, H.D.V., Jaya, A., & Limin, S. (2002). The amount of carbon released from peat and forest fires in Indonesia during 1997. *Nature*, 420, 61–65. doi.org/10.1038/nature01131.

Parton, W.J., Stewart, J.W.B., & Cole, C.V. (1988). Dynamics of C, N, P and S in grassland soils: A model. *Biogeochemistry*, 5, 109-131. doi.org/10.1007/BF02180320.

Patry, C., Davidson, R., Lucotte, M., & Béliveau, A. (2013). Impact of forested fallows on fertility and mercury content in soils of the Tapajós River region, Brazilian Amazon. *Sci. Total Environ.*, 458–460, 228–37. doi.org/10.1016/j.scitotenv.2013.04.037.

Pei, S., Fu, H., Wan, C., Chen, Y., & Sosebee, R.E. (2006). Observations on changes in soil properties in grazed and nongrazed areas of Alxa Desert Steppe, Inner Mongolia. *Arid Land Res. Manag.*, 20 (2), 161–175. doi.org/10.1080/15324980600549257.

Petersen, S.O., Mutegi, J.K., Hansen, E.M., & Munkholm, L.J. (2011). Tillage effects on N2O emissions as influenced by a winter cover crop. *Soil Biol. Biochem.*, 43 (7), 1509–1517. doi.org/10.1016/j.soilbio.2011.03.028.

Petrenko, C.L., Bradley-Cook, J., Lacroix, E.M., Friedland, A.J., & Virginia, R.A. (2016). Comparison of carbon and nitrogen storage in mineral soils of graminoid and shrub tundra sites, western Greenland. *Arctic Sci.*, 2 (4), 165–182. doi.org/10.1139/as-2015-0023.

Ping, C.-L., Qiu, G., & Zhao, L. (2004). The periglacial environment of China. In: Kimble, J. (ed.), *Cryosols. Permafrost-Affected Soils*, pp. 275–291. Springer, Berlin.

Ping, C.L., Michaelson, G.J., Jorgenson, M.T., Kimble, J.M., Epstein, H., Romanovsky, V.E., & Walker, D.A. (2008). High stocks of soil organic carbon in the North American Arctic region, *Nat. Geosci.*, 1, 615–619.

Pluchon, N., Hugelius, G., Kuusinen, N., & Kuhry, P. (2014). Recent paludification rates and effects on total ecosystem carbon storage in two boreal peatlands of Northeast European. *Russia The Holocene*, 24, 1126–1136. doi.org/10.1177/0959683614523803.

Poeplau, C., & Don, A. (2015). Carbon sequestration in agricultural soils via cultivation of cover crops—A meta-analysis. *Agric. Ecosys. and Env.*, 200, 33–41. doi.org/10.1016/j.agee.2014.10.024.

Poeplau, C., Aronsson, H., Myrbeck, Å., & Kätterer, T. (2015b). Effect of perennial ryegrass cover crop on soil organic carbon stocks in southern Sweden. *Geoderma Regional*, 4, 126–133. doi.org/10.1016/j.geodrs.2015.01.004.

Poeplau, C., Bolinder, M.A., Kirchmann, H., & Kätterer, T. (2016a). Phosphorus fertilisation under nitrogen limitation can deplete soil carbon stocks—Evidence from Swedish meta-replicated long-term field experiments. *Biogeosciences*, 13, 1119–1127. doi.org/10.5194/bg-13-1119-2016.

Poeplau, C., Don, A., Vesterdal, L., Leifeld, J., van Wesemael, B., Schumacher, J., & Gensior A. (2011). Temporal dynamics of soil organic carbon after land-use change in the temperate zone—Carbon response functions as a model approach. *Glob. Change Bio.*, 17 (7), 2415–2427. doi.org/10.1111/j.1365-2486.2011.02408.x.

Poeplau, C., Herrmann, A.M., & Kätterer, T. (2016b). Opposing effects of nitrogen and phosphorus on soil microbial metabolism and the implications for soil carbon storage. *Soil Bio. & Biochem.*, 100, 83–9. doi.org/10.1016/j.soilbio.2016.05.021.

Poeplau, C., Kätterer, T., Bolinder, M.A., Börjesson, G., Berti, A., & Lugato, E. (2015a). Low stabilization of aboveground crop residue carbon in sandy soils of Swedish long-term experiments. *Geoderma*, 237–238, 246–255. doi.org/10.1016/j.geoderma.2014.09.010.

Post, E., & Pedersen, C. (2008). Opposing plant community responses to warming with and without herbivores. *Proc. Natl Acad. Sci.*, 105 (34), 12353–12358. doi.org/10.1073/pnas.0802421105.

Post, W.M., Emanuel, W.R., Zinke, P.J., & Stangenberger, A.G. (1982). Soil carbon pools and world life zones. *Nature*, 298 (5870), 156–159. doi.org/10.1038/298156a0.

Qi, F., Guoduong, C., & Masao, M. (2001). The carbon cycle of sandy lands in China and its global significance. *Climatic Change*, 48 (4), 535–549. doi.org/10.1023/A:1005664307625.

Qiu, J. (2008). China: The third pole. *Nature*, 454 (7203), 393–396. doi.org/10.1038/454393a.

Qiu, J. (2016). Trouble in Tibet. *Nature*, 529 (7585), 142.

Qiu, Q., Wang, L., Wang, K. et al. (2015). Yak whole-genome resequencing reveals domestication signatures and prehistoric population expansions. *Nature Commun.*, 6, 10283. doi.org/10.1038/ncomms10283.

Quesada, C.A., Lloyd, J., Anderson, L.O., Fyllas, N.M., Schwarz, M., & Czimczik, C.I. (2011). Soils of Amazonia with particular reference to the RAINFOR sites. *Biogeosciences*, 8, 1415–1440. doi.org/10.5194/bg-8-1415-2011.

Rannik, Ü., Altimir, N., Raittila, J., Suni, T., Gaman, A., Hussein, T., Hölttä, T., Lassila, H., Latokartano, M., Lauri, A., Natsheh, A., Petäjä, T., Sorjamaa, R., Ylä-Mella, H., Keronen, P., Berninger, F., Vesala, T., Hari, P., & Kulmala, M. (2002). Fluxes of carbon dioxide and water vapour over Scot pine forest and clearing agricultural and forest. *Meteorology*, 111, 187–202. doi.org/10.1016/S0168-1923(02)00022-9.

Reichstein, M., Rey, A., Freibauer, A., Tenhunen, J., Valentini, R., Banza, J., Casals, P., Cheng, Y., Grünzweig, J.M., Irvine, J., Joffre, R., Law, B.E., Loustau, D., Miglietta, F., Oechel, W., Ourcival, J.-M., Pereira, J.S., Peressotti, A., Ponti, F., Qi, Y., Rambal, S., Rayment, M., Romanya, J., Rossi, F., Tedeschi, V., Tirone, G., Xu, M., & Yakir, D. (2003). Modeling temporal and large-scale spatial variability of soil respiration from soil water availability, temperature and vegetation productivity indices. *Global Biogeochem. Cycles*, 17 (4), n/a. doi.org/10.1029/2003GB002035.

Repo, M.E., Susiluoto, S., Lind, S.E., Jokinen, S., Elsakov, V., Biasi, C., Virtanen, T., & Martikainen, P. J. (2009). Large N2O emissions from cryoturbated peat soil in tundra. *Nature Geosci.*, 2, 189–192. doi.org/10.1038/ngeo434.

RIHMI-WDC. (2016). Hydrometeorological data, Baseline Climatological Data Sets, Syktyvkar Site#23804 (http://meteo.ru/english/data/).

Ringeval, B., Friedlingstein, P., Koven, C., Cias, P., de Noblet-Ducoudré, N., Decharme, B., & Cadule, P. (2011). Climate-CH4 feedback from wetlands and its interaction with the climate-CO2 feedback. *Biogeosciences*, 8, 2137–2157. doi.org/10.5194/bg-8-2137-2011.

Robinson, C.H. (2002). Controls on decomposition and soil nitrogen availability at high latitudes. *Plant and Soil*, 242, 65–81. doi.org/10.1023/A:1019681606112.

Rousk, J., & Frey, S.D. (2015). Revisiting the hypothesis that fungal-to-bacterial dominance characterizes turnover of soil organic matter and nutrients. *Ecologic. Monographs*, 85, 457–472. doi.org/10.1890/14 -1796.1. doi.org/10.1023/A:1019681606112.

Sachs, T., Giebels, M., Boike, J., & Kutzbach, L. (2010). Environmental controls on CH4 emission from polyg- onal tundra on the microsite scale in the Lena river delta, Siberia. *Glob. Change Bio.*, 16, 3096–3110. doi.org/10.1111/j.1365-2486.2010.02232.x.

Sagerfors, J., Lindroth, A., Grelle, A., Klemendtsson, L., Weslien, P., & Nilsson, M. (2008). Annual CO2 exchange between nutrientpoor, minerotrophic, boreal mire and the atmosphere. *J. Geophys. Res.*, 113. doi.org/10.1029/2006JG000306.

Schaller, G.B. (1998). *Wildlife of the Tibetan Steppe*. University of Chicago Press, Chicago.

Schlesinger, W.H. (1990). Evidence from chronosequence studies for a low carbon-storage potential of soils. *Nature*, 348, 232–234. doi.org/10.1038/348232a0.

Schlesinger, W.H., & Andrews, J.A. (2000). Soil respiration and the global carbon cycle. *Biogeochemistry*, 48, 7–20. doi.org/10.1023/A:1006247623877.

Schleuss, P.-M., Heitkamp, F., Sun, Y., Miehe, G., Xu, X., & Kuzyakov, Y. (2015). Nitrogen uptake in an alpine Kobresia pasture on the Tibetan Plateau: Localization by 15N labeling and implications for a vulnerable ecosystem. *Ecosystems*, 18, 946–957. doi.org/10.1007/s10021-015-9874-9.

Schlütz, F., & Lehmkuhl, F. (2009). Holocene climatic change and the nomadic Anthropocene in Eastern Tibet: Palynological and geomorphological results from the Nianbaoyeze Mountains. *Quatern. Sci. Rev.*, 28 (15-16), 1449–1471. doi.org/10.1016/j.quascirev.2009.01.009.

Schmidt, M.W.I., Torn, M.S., Abiven, S., Dittmar, T., Guggenberger, G., Janssens, I.A., Kleber, M., Kögel-Knabner, I., Lehmann, J., Manning, D.A.C., Nannipieri, P., Rasse, D.P., Weiner, S., & Trumbore, S.E. (2011). Persistence of soil organic matter as an ecosystem property. *Nature*, 478 (7367), 49–56. doi .org/10.1038/nature10386.

Schneider, J., Jungkunst, H.F., Wolf, U., Schreiber, P., Gažovič, M., Miglovets, M., Mikhaylov O., Grunwald D., Erasmi S., Wilmking M., & Kutzbach, L. (2016). Russian boreal peatlands dominate the natural European methane budget. *Environ. Res. Lett.*, 11. doi.org/10.1088/1748-9326/11/1/014004.

Schneider, J., Kutzbach, L., & Wilmking, M. (2012). Carbon dioxide dynamics of a boreal peatland over a complete growing season, Komi Republic. NW Russia, Biogeochem., 111, 485–513. doi.org/ 10.1007/s10533-011-9684-x.

Schuur, E.A., & Abbott, B. (2011). High risk of permafrost thaw. *Nature*, 480, 32–33. doi.org/10.1038/480032a

Schuur, E.A.G., Bockheim, J.G., Canadell, J.G., Euskirchen, E., Field, C.B., Goryachkin, S.V., Hagemann, S., Kuhry, P., Lafleur, P.M., Lee, H., Mazhitova, G., Nelson, F.E., Rinke, A., Romanovsky, V.E., Shiklomanov, N.I., Tarnocai, C., Venevsky, S., Vogel, J.G., & Zimov, S.A. (2008). Vulnerability of per- mafrost carbon to climate change: Implications for the global carbon cycle. *Bioscience*, 58 (8), 701–714. doi.org/10.1641/B580807.

Schuur, E.A.G., McGuire, A.D., Schädel, C., Grosse, G., Harden, J.W., Hayes, D.J., Hugelius, G., Koven, C.D., Kuhry, P., Lawrence, D.M., Natali, S.M., Olefeldt, D., Romanovsky, V.E., Schaefer, K., Turetsky, M.R., Treat, T.T., & Vonk, J.E. (2015). Climate change and the permafrost carbon feedback. *Nature*, 520(7546), 171–179. doi.org/10.1038/nature14338.

Seeber, E., Miehe, G., Hensen, I., Yang, Y., & Wesche, K. (2016). Mixed reproduction strategy and poly- ploidy facilitate dominance of Kobresia pygmaea on the Tibetan Plateau. *J. Plant Eco.*, 9, 87–99. doi .org/10.1093/jpe/rtv035.

Selenobaldo, A.C. de Sant-Anna., Jantalia CPm Sá, J.M., Vilela, L., Marchao, R.L., Alves, J.R., Uequiaga, S., & Boddey, R.M. (2016). Changes in soil organic carbon during 22 years of pastures, cropping or integrated crop/livestock systems in the Brazilian Cerrado. *Nutr. Cycl. Agroecosyst.*, 108, 101–120, doi.org/10.1007 /s10705-016-9812-z.

Shi, Y., Baumann, F., Ma, Y.-L., Song, C., Kuehn, P., Scholten, T., & He, J.-S. (2012). Organic and inor- ganic carbon in the topsoil of the Mongolian and Tibetan grasslands: Pattern, control and implications. *Biogeosciences*, 9, 2287–2299. doi.org/10.5194/bg-9-2287-2012.

Shur, Y.L., & Jorgenson, M.T. (2007). Patterns of permafrost formation and degradation in relation to climate and ecosystems. *Permafrost Periglac. Process*, 18 (1), 7–19. doi.org/10.1002/ppp.582.

Simcic, T., Mori, N., Hossli, C., Robinson, C.T., & Doering, M. (2015). The response in floodplain respiration of an alpine river to experimental inundation under different temperature regimes. *Hydrological Processes*, 29 (26), 5438–5450. doi.org/10.1002/hyp.10584.

Sinsabaugh, R.L., Shah, J.J.F., Hill, B.H., & Elonen, C.M. (2012). Ecoenzymatic stoichiometry of stream sediments with comparison to terrestrial soils. *Biogeochemistry*, 111(1–3), 455–467. doi.org/10.1007/s10533-011-9676-x.

Six, J., Elliott, E.T., & Paustian, K. (2000). Soil macroaggregate turnover and microaggregate formation: A mechanism for C sequestration under no-tillage agriculture. *Soil Bio. Biochem.*, 32 (14), 2099–2103. doi.org/10.1016/S0038-0717(00)00179-6.

Sollins, P., Swanston, C., & Kramer, M. (2007). Stabilization and destabilization of soil organic matter—A new focus. *Biogeochemistry*, 85, 1–7. doi.org/10.1007/s10533-007-9099-x.

Song, W.M., Wang, H., Wang, G.S., Chen, L.T., Jin, Z.N., Zhuang. Q.L., & He, J.-S. (2015). Methane emissions from an alpine wetland on the Tibetan Plateau: Neglected but vital contribution of non-growing season. *J. Geophys. Res.: Biogeosci.*, 120, 1475–1490. doi.org/10.1002/2015JG003043.

Stevenson, F.J., & Cole, M.A. (1999). *Cycles of Soil: Carbon, Nitrogen, Phosphorus, Sulfur, Micronutrients*, 2nd ed. Wiley, New York, xviii, 427.

Stockmann, U., Adams, M.A., Crawford, J.W., Field, D.J., Henakaarchchi, N., Jenkins, M., Minasny, B., McBratney, A.B., De Courcelles, V.D.R., Singh, K., Wheeler, I., Abbott, L., Angers, D. A., Baldock, J., Bird, M., Brookes, P.C., Chenu, C., Jastrow, J.D., Lal, R., Lehmann, J., O'Donnell, A.G., Parton, W.J., Whitehead D., & Zimmermann, M. (2013). The knowns, known unknowns and unknowns of sequestration of soil organic carbon. *Agric. Ecosyst. Environ.*, 164, 80–99. doi.org/10.1016/j.agee.2012.10.001.

Stokstad, E. (2004). Defrosting the Carbon Freezer of the North. *Science*, 304 (5677), 1618–1620. doi.org/10.1126/science.304.5677.1618.

Stolbovoy, V. (2002). Carbon in Russian soils. *Clim. Chang.*, 55, 131–156. doi.org/10.1023/A:1020289403835.

Storms, J.E., Winter, I.L., Overeem, I., Drijkoningen, G.G., & Lykke-Andersen, H. (2012). The Holocene sedimentary history of the Kangerlussuaq Fjord-valley fill, West Greenland. *Quaternary Sci. Rev.*, 35, 29–50. doi.org/10.1016/j.quascirev.2011.12.014.

van Straaten, O., Corre, M.D., Wolf, K., Tchienkoua, M., Cuellar, E., Matthews, R.B., & Veldkamp, E. (2015). Conversion of lowland tropical forests to tree cash crop plantations loses up to one-half of stored soil organic carbon. *Proc. Natl. Acad. Sci.*, 112, 9956–9960. doi.org/10.1073/pnas.1504628112.

Strey, S., Boy, J., Strey, R., Weber O., & Guggenberger, G. (2016). Response of soil organic carbon to land-use change in central Brazil: A large-scale comparison of Ferralsols and Acrisols. *Plant and Soil.*, 408, 327–342, doi.org/10.1007/s11104-016-2901-6.

Ström, L., Ekberg, A., Mastepanov, M., & Christensen, T.R. (2003). The effect of vascular plants on carbon turnover and methane emissions from a tundra wetland. *Glob. Change Bio.*, 9, 1185–1192. doi.org/10.1046/j.1365-2486.2003.00655.x.

Sun, D.S., Wesche, K., Chen, D.D., Zhang, S.H., Wu, G.L., Du, G.Z., & Comerford, N.B. (2011). Grazing depresses soil carbon storage through changing plant biomass and composition in a Tibetan alpine meadow. *Plant Soil Environ.*, 57, 271–278.

Suttie, J.M., Reynolds, S.G., & Batello, C. (eds.) (2005). *Grasslands of the World*. Food and Agriculture Organization of the United Nations, Rome.

Tarnocai, C., Canadell, J.G., Schuur, E.A.G., Kuhry, P., Mazhitova, G., & Zimov, S.A. (2009). Soil organic carbon pools in the northern circumpolar permafrost region. *Global Biogeochem. Cycles,* 23, GB2023, doi.org/10.1029/2008GB003327.

Tarnocai, C., Ping, C.-L., & Kimball, J. (2007). Carbon cycles in the permafrost region of North America. In: King, A.W., Dilling, L., Zimmerman, G.P., Fairman, D.M., Houghton, R.A., Marland, G.H., Rose, A.Z., and Wilbanks, T.J. (eds.), *The First State of the Carbon Cycle Report (SOCCR). North American Carbon Budget and Implications for the Global Carbon Cycle*, Chapter 12. Synthesis and Assessment Product, Washington DC.

van Tatenhove, F.G.M., & Olesen, O.B. (1994). Ground temperature and related permafrost characteristics in West Greenland. *Permafr. Periglac. Process,* 5 (4), 199–215. doi.org/10.1002/ppp.3430050402.

Tedesco, M., Mote, T., Fettweis, X., Hanna, E., Jeyaratnam, J., Booth, J.F., Datta, R., & Briggs, K. (2016). Arctic cut-off high drives the poleward shift of a new Greenland melting record. *Nature Commun.*, 7, 11723. doi.org/10.1038/ncomms11723.

Tiemeyer, B., Albiac Borraz, E., Augustin, J., Bechtold, M., Beetz, S., Beyer, C., Drösler, M., Ebli, M., Eickenscheidt, T., Fiedler, S., Förster, C., Freibauer, A., Giebels, M., Glatzel, S., Heinichen, J., Hoffmann, M., Höper, H., Jurasinski, G., Leiber-Sauheitl, K., Peichl-Brak, M., Roßkopf, N., Sommer M., & Zeitz, J. (2016). High emissions of greenhouse gases from grasslands on peat and other organic soils. *Global Change Bio.*, 22 (12), 4134–4149. doi.org/10.1111/gcb.13303.

Tornquist C.G., Mielniczuk, J., & Cerri, C.E.P. (2009). Modeling soil organic carbon dynamics in Oxisols of Ibirubá (Brazil) with the Century Model. *Soil & Tillage Res.*, 105, 33–43. doi.org/10.1016/j .still.2009.05.005.

Trumper, K., Bertzky, M., Dickson, B., van der Heijden, G., Jenkins, M., & Manning, P. (2009). *The Natural Fix? The Role of Ecosystems in Climate Mitigation. A UNEP Rapid Response Assessment.* United Nations Environment Programme, UNEP-WCMC, Cambridge, UK.

Turunen, J., Tomppo, E., Tolonen, K., & Reinikainen, A. (2002). Estimating carbon accumulation rates of undrained mires in Finland–Application to boreal and subarctic regions. *Holocene*, 12, 69–80. doi.org/10.1191/0959683602hl522rp.

Unteregelsbacher, S., Hafner, S., Guggenberger, G., Miehe, G., Xu, X., Liu, J., & Kuzyakov, Y. (2012). Response of long-, medium- and short-term processes of the carbon budget to overgrazing-induced crusts in the Tibetan Plateau. *Biogeochemistry,* 111, 187–201, doi.org/10.1007/s10533-011-9632-9.

VandenBygaart, A.J., Gregorich, E.G. & Angers, D.A. (2003) Influence of agricultural management on soil organic carbon: A compendium and assessment of Canadian studies. *Can. J. Soil Sci.,* 83, 363–380, doi.org/10.4141/ S03-009.

Vaughan, D.G., & Comiso, J.C. (2013). Observations: Cryosphere. In: Intergovernmental Panel on Climate Change (IPCC) (ed.), *Climate Change 2013. The Physical Science Basis.* Working Group I Contribution to the Fifth Assessment Report of the Intergovernmental Panel on Climate Change. Cambridge University Press, Cambridge, New York, pp. 317–382.

Veldkamp, E., Becker, A., Schwendenmann, L., Clark, D.J., & Schulte-Bisping, H. (2003). Substantial labile carbon stocks and microbial activity in deeply weathered soils below a tropical wet forest. *Glob. Change Bio.*, 9, 1171–1184. doi.org/10.1046/j.1365-2486.2003.00656.x.

Virto, I., Barré, P., Burlot, A., & Chenu, C. (2012). Carbon input differences as the main factor explaining the variability in soil organic C storage in no-tilled compared to inversion tilled agrosystems. *Biogeochemistry*, 108, 17 – 26. doi.org/10.1007/s10533-011-9600-4.

Vitousek, P.M. (1997). Human domination of Earth's ecosystems. *Science*, 277 (5325), 494–499. doi.org/10 .1126/science.277.5325.494.

Vitt, D.H., Halsey, L.A., Bauer, I.E., & Campbell, C. (2000). Spatial and temporal trends in carbon storage of peatlands of continental western Canada through the Holocene. *Can. J. Earth Sci.*, 37 (5), 683–693. doi.org/10.1139/e99-097.

Wagner, D., & Liebner, S. (2009). Global warming and carbon dynamics in permafrost soils: Methane production and oxodation. In: Margesin, R. (ed.), *Permafrost Soils*, pp. 219–236. Springer, Berlin.

Wagner, D., Kobabe, S., & Liebner, S. (2009). Bacterial community structure and carbon turnover in permafrost-affected soils of the Lena Delta, northeastern Siberia. (This article is one of a selection of papers in the Special Issue on Polar and Alpine Microbiology.) *Can. J. Microbiol.*, 55 (1), 73–83. doi .org/10.1139/W08-121.

Wallen, B., Falkengren-Grerup, U., & Malmer, N. (1988). Biomass, productivity and relative rate of photosynthesis of Sphagnum at different water levels on a south Swedish peat bog. *Holarctic Ecol.*, 11, 70–76. doi.org/10.1111/j.1600-0587.1988.tb00782.x.

Wang, B., & French, H.M. (1994). Climate controls and high-altitude permafrost, Qinghai-Xizang (Tibet) Plateau, China. *Permafrost Periglac. Process.*, 5 (2), 87–100. Doi.org/10.1002/ppp.3430050203.

Wang, B., & French, H.M. (1995). Permafrost on the Tibet Plateau, China. *Quaternary Sci. Rev.*, 14, 255–274. doi.org/10.1016/0277-3791(95)00006-B.

Wang, C., Cao. G., Wang, Q., Jing, Z., Ding, L., & Long, R. (2008). Changes in plant biomass and species composition of alpine Kobresia meadows along altitudinal gradient on the Qinghai-Tibetan Plateau. *Sci. in China.*, 51, 86–94. doi.org10.1007/s11427-008-0011-2.

Wang, G., Bai, W., Li, N., & Hu, H. (2011). Climate changes and its impact on tundra ecosystem in Qinghai-Tibet Plateau, China. *Climatic Change*, 106 (3), 463–482. doi.org/10.1007/s10584-010-9952-0.

Wang, G., Ju, Q., Cheng, G., & Lai, Y. (2002). Soil organic carbon pool of grassland soils on the Qinghai-Tibetan Plateau and its global implication. *Sci. Total Environ.*, 291 (1–3), 207–217. doi.org/10.1016 /S0048-9697(01)01100-7.

Wang, G., Li, Q., Chang, G., & Shen, Y. (2001). Climate change and its impact on the eco-environment in the source regions of Yangtze and Yellow Rivers in recent 40 years. *J. Glaciol. Geocryol.*, 23, 346–352. doi.org/10.3390/ijerph121012057.

Wang, G., Li, Y., Hu, H., & Wang, Y. (2008a). Synergistic effect of vegetation and air temperature changes on soil water content in alpine frost meadow soil in the permafrost region of Qinghai-Tibet. *Hydrol. Process.*, 22 (17), 3310–3320. doi.org/10.1002/hyp.6913.

Wang, G., Li, Y., Wang, Y., & Wu, Q. (2008b). Effects of permafrost thawing on vegetation and soil carbon pool losses on the Qinghai–Tibet Plateau, China. *Geoderma*, 143 (1–2), 143–152. doi.org/10.1016/j.geoderma.2007.10.023.

Wang, G., Liu, L., Liu, G., Hu, H., & Li, T. (2010). Impacts of grassland vegetation cover on the active layer thermal regime, northeast Qinghai-Tibet Plateau, China. *Permafrost Periglac. Process,* 21 (4), 335–344. doi.org/10.1002/ppp.699.

Wang, G., Wang, Y., Li, Y., & Cheng, H. (2007). Influences of alpine ecosystem responses to climatic change on soil properties on the Qinghai–Tibet Plateau, China. *Catena*, 70 (3), 506–514. doi.org/10.1016/j.catena.2007.01.001.

Wang, G., Wang, Y., Qian, J., & Wu, Q. (2006). Land cover change and its impacts on soil C and N in two watersheds in the center of the Qinghai-Tibetan Plateau. *Mountain Res. Devel.*, 26 (2), 153–162. doi.org/10.1659/0276-4741(2006)26[153:LCCAII]2.0.CO;2.

Wang, G.X., Qian, J., Cheng, G.D., & Lai, Y.M. (2002). Soil organic carbon pool of grassland soils on the Qinghai-Tibetan Plateau and its global implication. *Sci. Total Env.*, 291, 207–217. doi.org/10.1016/S0048-9697(01)01100-7.

Wang, H., Yu, L.F., Zhang, Z.H., Liu, W., Chen, L.T., Cao, G.M., Yue, H.W., Zhou, J., Yang, Y.F., Tang, Y.H., & He, J.-S. (2016). Molecular mechanisms of water table lowering and nitrogen deposition in affecting greenhouse gas emissions from a Tibetan alpine wetland. *Glob. Change Biol.*, 23, 815–829, doi.org/10.1111/gcb.13467.

Wang, S., & Zhou, C. (2001). Estimation of soil organic carbon reservoir in China. *J. Georgr. Sci.*, 11, 3–13. doi.org/10.1007/BF02837371.

Wang, S., Jin, H., Li, S., & Zhao, L. (2000). Permafrost degradation on the Qinghai-Tibet Plateau and its environmental impacts. *Permafrost Periglac. Process*, 11, 43–53. doi.org/10.1002/(SICI)1099-1530(200001/03)11:1<43::AID-PPP332>3.0.CO;2-H.

Wang, W., Wang, Q., & Lu, Z. (2009). Soil organic carbon and nitrogen content of density fractionsand effect of meadow degradation to soil carbon and nitrogen of fractions in alpine Kobresia meadow. *Sci. China Ser. D—Earth Sci.*, 52 (5), 660–668. doi.org/10.1007/s11430-009-0056-5.

Wang, Y., & Wesche, K. (2016). Vegetation and soil responses to livestock grazing in Central Asian grasslands: A review of Chinese literature. *Biodivers. Conserv.*, 1–20. doi.org/10.1007/s10531-015-1034-1.

Wei, D., Xu-Ri, Tarchen, T., Dai, D.X., Wang, Y.S., & Wang, Y.H. (2015). Revisiting the role of CH4 emissions from alpine wetlands on the Tibetan Plateau: Evidence from two in situ measurements at 4758 and 4320m above sea level. *J. Geophys. Res.—Biogeosci.*, 120, 1741–1750. Doi.org/10.1002/2015JG002974.

Wie, X., Yang, P., Li, S., & Chen, H. (2005). Effects of over-grazing on vegetation degradation of the Kobresia pygmaea meadow and determination of degenerative index in the Naqu Prefecture of Tibet. *Acta Prataculturae Sinica*, 14, 41–49.

West, L.T., Beinroth, F.H., Summer, M.E., & Kang, B.T. (1998). Ultsiol; Characteristics and impacts on society. *Adv. Agron.*, 63, 163–224.

West, T.O., & Kwon, K.C. (2000). Soil Carbon Sequestration and Land-Use Change: Processes and Potential. *Glob. Change Biol.*, 6, 317–328. doi.org/10.1046/j.1365-2486.2000.00308.x.

West, T.O., & Post, W.M. (2002). Soil organic carbon sequestration rates by tillage and crop rotation: A global data analysis. *Soil Sci. Soc. Am. J.*, 66, 1930–1946. doi.org/10.3334/CDIAC/tcm.002.

Wieder, W.R., Bonan, G. B., & Allison, S.D. (2013). Global soil carbon predictions are improved by modelling microbial processes. *Nature Climate Change*, 3: 909–912. doi.org/10.1038/nclimate1951.

Willemse, N.W., Koster, E.A., Hoogakker, B. et al. (2003). A continuous record of Holocene eolian activity in West Greenland. *Quaternary Res.*, 59, 322–334. doi.org/10.1016/S0033-5894(03)00037-1.

Wolf, U. (2009). Above- and belowground methane dynamics of a boreal peatland ecosystem of varying vegetation composition during summer in the Republic of Komi, Russia. Master's thesis, University of Göttingen.

Worldbank (2018). http://databank.worldbank.org/data/reports.aspx?source=world-development-indicators, accessed 06.04.2018.

Wu, G., & Du, G. (2007). Germination is related to seed mass in grasses (Poaceae) of the eastern Qinghai-Tibetan Plateau, China. *Nordic J. Bot.*, 25, 361–365. doi.org/10.1111/j.0107-055X.2007.00179.x.

Wu, G.-L., Liu, Z.-H., Zhang, L., Chen, J.-M., & Hu, T.-M. (2010a). Long-term fencing improved soil properties and soil organic carbon storage in an alpine swamp meadow of western China. *Plant Soil*, 332 (1–2), 331–337. doi.org/10.1007/s11104-010-0299-0.

Wu, H., Guo, Z., Gao, Q., & Peng, C. (2009). Distribution of soil inorganic carbon storage and its changes due to agricultural land-use activity in China. *Agric., Ecosys. Env.*, 129 (4), 413–421. doi.org/10.1016/j.agee.2008.10.020.

Wu, J., Sheng, Y., Wu, Q., & Wen, Z. (2010b). Processes and modes of permafrost degradation on the Qinghai-Tibet Plateau. *Sci. China Ser. D–Earth Sci.*, 53 (1), 150–158. doi.org/10.1007/s11430-009-0198-5.

Wu, R., & Tiessen, H. (2002). Effect of land-use on soil degradation in Alpine grassland soil, China. *Soil Sci. Soc. Am. J.*, 66 (5), 1648. doi.org/10.2136/sssaj2002.1648.

Wu, S., Yin, Y., Zheng, D., & Yang, Q. (2005). Climate change in the Tibetan Plateau during the last three decades. *Acta Geographica Sinica*, 60, 3–11.

Wu, Z. (1980). *Vegetation of China*. Science Press, Beijing.

Xie, Z., Zhu, J., Liu, G., Cadisch, G., Hasegawa, T., Chen, C., Sun, H., Tang, H., & Zeng, Q. (2007). Soil organic carbon stocks in China and changes from 1980s to 2000s. *Glob. Change Biol.*, 13 (9), 1989–2007. doi.org/10.1111/j.1365-2486.2007.01409.x.

Xue, X., Guo, J., Han, B., Sun, Q., & Liu, L. (2009). The effect of climate warming and permafrost thaw on desertification in the Qinghai–Tibetan Plateau. *Geomorphology*, 108 (3–4), 182–190. doi.org/10.1016/j.geomorph.2009.01.004.

Yan, C., Song, X., Zhou, Y., Duan, H., & Li, S. (2009). Assessment of aeolian desertification trends from 1975's to 2005's in the watershed of the Longyangxia Reservoir in the upper reaches of China's Yellow River. *Geomorphology*, 112 (3–4), 205–211. doi.org/10.1016/j.geomorph.2009.06.003.

Yan, Y., & Lu, Xu. (2015). Is grazing exclusion effective in restoring vegetation in degraded alpine grasslands in Tibet, China? *PeerJ* 3, e1020. doi.org/10.7717/peerj.1020.

Yang, M., Nelson, F.E., Shiklomanov, N.I., Guo, D., & Wan, G. (2010a). Permafrost degradation and its environmental effects on the Tibetan Plateau: A review of recent research. *Earth-Sci. Rev.*, 103 (1–2), 31–44. doi.org/10.1016/j.earscirev.2010.07.002.

Yang, M., Wang, S., Yao, T., Gou, X., Lu, A., & Guo, X. (2004a). Desertification and its relationship with permafrost degradation in Qinghai-Xizang (Tibet) plateau. *Cold Regions Sci. Technol.*, 39 (1), 47–53. doi.org/10.1016/j.coldregions.2004.01.002.

Yang, S.Z., Liebner, S., Winkel, M., Alawi, M., Horn, F., Dörfer, C., Ollivier, J., He, J.S., Jin, H.J., Kuehn, P., Schloter, M., Scholten, T., & Wagner, D. (2017). In-depth analysis of core methanogenic communities from high elevation permafrost-affected wetlands. *Soil Biol Biochem*, 111, 66–77, doi.org/10.1016/j.soilbio.2017.03.007.

Yang, Y., Fang, J., Ji, C., Ma, W., Su, S., & Tang, Z. (2010b). Soil inorganic carbon stock in the Tibetan alpine grasslands. *Glob. Biogeochem. Cycles*, 24 (4), n/a. doi.org/10.1029/2010GB003804.

Yang, Y., Fang, J., Smith, P., Tang, Y., Chen, A., Ji, C., Hu, H., Rao, S., Tan, K., & He, J.-S. (2009b). Changes in topsoil carbon stock in the Tibetan grasslands between the 1980s and 2004. *Glob. Change Biol.*, 15 (11), 2723–2729. doi.org/10.1111/j.1365-2486.2009.01924.x.

Yang, Y., Fang, J., Tang, Y., Ji, C., Zheng, C., He, J.-S., & Zhu, B. (2008). Storage, patterns and controls of soil organic carbon in the Tibetan grasslands. *Glob. Change Biol.* 14 (7), 1592–1599. doi.org/10.1111/j.1365-2486.2008.01591.x

Yang, Y.H., Fang, J.Y., Guo, D.L., Ji, C.J., & Ma, W.H. (2010c). Vertical patterns of soil carbon, nitrogen and carbon: Nitrogen stoichiometry in Tibetan grasslands. *Biogeosci. Discuss.*, 7 (1), 1–24. doi.org/10.5194/bgd-7-1-2010.

Yang, Z., Ouyang, H., Zhang, X., Xu, X., Zhou, C., & Yang, W. (2011). Spatial variability of soil moisture at typical alpine meadow and steppe sites in the Qinghai-Tibetan Plateau permafrost region. *Environ. Earth Sci.*, 63 (3), 477–488. doi.org/10.1006/jare.2001.0828.

Yao, T., Liu, X., Wang, N., & Shi, Y. (2000). Amplitude of climatic changes in Qinghai-Tibetan Plateau. *Chin. Sci. Bull.*, 45 (13), 1236–1243. doi.org/10.1007/BF02886087.

Yi, S., Zhou, Z., Ren, S., Xu, M., Qin, Y., Chen, S., & Ye, B. (2011). Effects of permafrost degradation on alpine grassland in a semi-arid basin on the Qinghai–Tibetan Plateau. *Environ. Res. Lett.*, 6 (4), 45403. doi.org/10.1088/1748-9326/6/4/045403.

Young, N.E., Briner, J.P., Rood, D.H., Finkel, R.C., Corbett, L.B., & Bierman, P.R. (2013). Age of the Fjord Stade moraines in the Disko Bugt region, western Greenland, and the 9.3 and 8.2 ka cooling events. *Quaternary Sci. Rev.*, 60, 76–90. doi.org/10.1016/j.quascirev.2012.09.028.

Yu, G.R., Zhu, X.J., Fu, Y.L., He, H.L., Wang, Q.F., Wen, X.F., Li, X.R., Zhang, L.M., Zhang, L., Su, W., Li, S.G., Sun, X.M., Zhang, Y.P., Zhang, J.H., Yan, J.H., Wang, H.M., Zhou, G.S., Jia, B.R., Xiang, W.H., Li, Y.N., Zhao, L., Wang, Y.F., Shi, P.L., Chen, S.P., Xin, X.P., Zhao, F.H., Wang, Y.Y., & Tong, C.L. (2013). Spatial patterns and climate drivers of carbon fluxes in terrestrial ecosystems of China. *Glob. Change Biol.*, 19, 798–810. doi.org/10.1111/gcb.12079.

Yu, Z. (2012). Northern peatland carbon stocks and dynamics: A review. *Biogeosciences*, 9, 4071–4085. doi.org/10.5194/bg-9-4071-2012.

Yu, Z., Loisel, J., Brosseau, D.P., Beilman, D.W., & Hunt, S. J. (2010). Global peatland dynamics since the Last Glacial Maximum. *Geophys. Res. Lett.*, 37. doi.org/10.1029/2010GL043584.

Zhang, J.H., Liu, S.Z., & Zhong, X.H. (2006). Distribution of soil organic carbon and phosphorus on an eroded hillslope of the rangeland in the northern Tibet Plateau, China. *Eur. J. Soil Sci.*, 57 (3), 365–371. doi.org/10.1111/j.1365-2389.2005.00747.x.

Zhang, L., Unteregelsbacher, S., Hafner, S., Xu, X., Schleuss, P.-M., Miehe, G., & Kuzyakov, Y. (2017). Fate of organic and inorganic nitrogen in crusted and non-crusted kobresia grasslands. *Land Degradation and Development*, 28, 166–174. doi.org/10.1002/ldr.2582.

Zhang, P., Tang, Y., Hirota, M., Yamamoto, A., & Mariko, S. (2009). Use of a regression method to partition sources of ecosystem respiration in an alpine meadow. *Soil Biol. Biochem.*, 41, 663–670. doi.org/10.1016/j.soilbio.2008.12.026.

Zhang, Y., Ohata, T., & Kadota, T. (2003). Land-surface hydrological processes in the permafrost region of the eastern Tibetan Plateau. *J. Hydrol.*, 283 (1–4), 41–56. doi.org/10.1016/S0022-1694(03)00240-3.

Zhang, Y., Tang, Y., Jiang, J., & Yang, Y. (2007a). Characterizing the dynamics of soil organic carbon in grasslands on the Qinghai-Tibetan Plateau. *Sci China Ser D*, 50 (1), 113–120. doi.org/10.1007/s11430-007-2032-2.

Zhao, L., Cheng, G., Li, S., Zhao, X., & Wang, S. (2000). Thawing and freezing processes of active layer in Wudaoliang region of Tibetan Plateau. *Chin. Sci. Bull.*, 45, 2181–2187. doi.org/10.1007/BF02886326.

Zhao, L., Ping, C.-L., Yang, D., Cheng, G., Ding, Y., & Liu, S. (2004). Changes of climate and seasonally frozen ground over the past 30 years in Qinghai–Xizang (Tibetan) Plateau, China. *Global and Planetary Change*, 43 (1–2), 19–31. doi.org/10.1016/j.gloplacha.2004.02.003.

Zheng, D., Zhang, Q., & Wu, S. (2000). Mountain geoecology and sustainable development of the Tibetan Plateau. *The GeoJournal Library* 57. Kluwer Academic Publishers, Dordrecht, Boston, pp. xii, 393.

Zhou, H., Zhao, X., Tang, Y., Gu, S., & Zhou, L. (2005). Alpine grassland degradation and its control in the source region of the Yangtze and Yellow Rivers, China. *Grassland Sci.*, 51 (3), 191–203. doi.org/10.1111/j.1744-697X.2005.00028.x.

Zhou, Y., Guo, D., Qiu, G., Cheng, G., & Li, S. (2000). *China Permafrost*. Science Press, Beijing, pp. 145–151.

Zimov, S.A. (2006). Climate change: Permafrost and the global carbon budget. *Science*, 312, 1612–1613. doi.org/10.1126/science.1128908.

Zinn, Y.L., Lal, R., & Resck, D.V.S.S. (2005). Changes in soil organic carbon stocks under agriculture in Brazil. *Soil Tillage Res.*, 84, 28–40. doi.org/10.1016/j.geoderma.2005.02.010.

Zubrzycki, S., Kutzbach, L., & Pfeiffer, E.-M. (2014). Permafrost-affected soils and their carbon pools with a focus on the Russian Arctic. *Solid Earth*, 5 (2), 595–609. doi.org/10.5194/se-5-595-2014.

4 Conservation Agriculture
Maintaining Land Productivity and Health by Managing Carbon Flows

D. C. Reicosky and H. H. Janzen

CONTENTS

4.1 INTRODUCTION

4.1.1 AGRICULTURE'S FUTURE

As the world population increases and food demands rise, keeping our soil healthy and productive is of paramount importance for agriculture. The expanding global population, expected to reach 9.5 billion people by 2050 (United Nations 2014), is exerting mounting pressure on the finite land area and resources for growing food. In many places, traditional conventional agriculture with an emphasis on intensive tillage and monoculture practices has resulted in a slow environmental degradation that may ultimately jeopardize our food security. Meeting the needs of a growing population whilst minimizing impacts on the environment (Foley et al. 2011) will require the Sustainable

Intensification of agriculture (Tilman et al. 2011; Garnett et al. 2013). Conservation agriculture (CA) has been highlighted as a key avenue toward this urgent goal (Hobbs et al. 2008; Kassam et al. 2009; Erenstein et al. 2012; Kassam and Friedrich 2012; Corsi et al. 2012; Pretty and Bharucha 2014).

Meeting the twin challenges of growing more food and reducing environmental damage will require careful attention to the world's soils. Soils are a critical component of terrestrial ecosystems, and a fundamental constituent for sustaining all life on Earth. Food security rests on our living soils. We need climate-smart agriculture, making the transition to farming better suited to cope with impacts of climate extremes that jeopardize global food security and natural resources (Branca et al. 2011; Gattinger et al. 2011; Lal 2014).

4.1.2 CLIMATE EXTREMES CREATE NEW FARMING RISKS

Understanding weather and climate, because of their complexity and growing variability, is crucial for success in agriculture. Extreme rainfall events and flooding have increased during the last century, and these trends are expected to continue, causing erosion, declining water quality, and negative impacts on transportation, agriculture, human health, and infrastructure (Kunkel et al. 1999; Villarini et al. 2011a,b; Lal 2013). Evidence suggests that climate change will bring hotter temperatures, changing rainfall patterns, climate extremes, and more frequent natural disasters (Hatfield et al. 2011). Severe droughts, floods and heat waves at key times in the growing season are forcing farmers to innovate and invest in new technology to cope with dramatic year-to-year yield variations (Rosenzweig et al. 2002; Hatfield et al. 2011; Smith 2011; Andresen et al. 2012; Lal et al. 2012).

Ongoing loss and degradation of agricultural soil and water assets due to increasing extremes in precipitation will continue to challenge both rain-fed and irrigated agriculture unless innovative conservation methods are implemented (Larson et al. 1997; Smith 2011; Lal 2013). Conservation agriculture has increasingly been endorsed as Climate Smart Agriculture, contributing to both climate change adaptation and mitigation (Harvey et al. 2013; Pretty and Bharucha 2014). This will require a global orientation and perspective (Hobbs and Govaerts 2010; Kassam et al. 2009; Kassam and Friedrich 2012; Friedrich et al. 2012; Kirkegaard et al. 2014), a concerted, strategic action to protect our soil, water, and air quality through new more efficient conservation practices. Because of climate and other global changes, many of the practices once deemed appropriate may have been rendered obsolete.

4.1.3 10,000 YEARS OF ARABLE AGRICULTURE AND CONTINUING EROSION

Farmers around the world have relied on intensive tillage for the last 10,000 years (Lal et al. 2007; Reusser et al. 2015). Although this approach has had many benefits, it also has serious problems, notably the resulting susceptibility of soil to erosion. According to Pimentel et al. (1995), about 430 million ha—almost one-third of the global arable land area—has been lost to soil erosion. Efforts to control human-induced land degradation and soil erosion have been building on the ruins of the past tillage and monoculture concepts (Lal et al. 2007; Montgomery 2007a,b). In many places, tillage for planting and cultivation loosens the soil, leaving it untethered and vulnerable to transport by wind or water. In sloping lands, the tillage process itself can transport soil downslope (Lindstrom et al. 1990). Consequently, erosion rates from conventionally plowed agricultural fields can be 1–2 orders of magnitude higher than rates of soil production (Montgomery 2007b). We are losing soil faster than nature can make it.

Diamond (2005, p. 486) contends that the most severe environmental problems confronting human societies, now and in the past, fall into 12 groups. Of these, most are related, one way or another, to the functioning of soils in ecosystems. Both Diamond (2005) and Montgomery (2007a) agree: agricultural sustainability—particularly the conservation of soils—is critical to our long-term survival.

A fundamental facet of soil conservation is maintaining adequate reserves of carbon. Observations over thousands of years, as well as recent scientific studies, have shown that the productivity and functioning of a soil is directly related to its content of plant-derived organic matter (Allison 1973; Lal 2014). The plant depends on the soil and the soil depends on the plant. Carbon provides a critical linkage in this symbiotic relationship (Nielsen et al. 2011; Tardy et al. 2015). Moreover, soil carbon has also been promoted recently as a repository of excess CO_2, by various 'sequestration' strategies (Lal 2009). For these reasons, conserving land is inextricably linked to conserving carbon. Carbon is the "C" that starts "C"onservation.

Soil is alive and is as vital to human survival as air, water, and the sun; its protection and enrichment with energy and organic carbon (C) are needed for the future sustainability of our planet. Our objectives, therefore, in this overview are: to describe Conservation Agriculture (CA) and associated C management with its potential role in meeting food demands while minimizing environmental damage; and to examine how improvements in carbon cycling and energy flow might advance the effectiveness of CA systems in meeting these challenges.

4.2   CONSERVATION AGRICULTURE (CA)—AN OVERVIEW

4.2.1   Evolution of Conservation Agriculture Systems

Conventional tillage refers to the sequence of operations "most commonly or historically used in a given field to prepare a seedbed and produce a given crop" (Reeder 2000; Reicosky and Allmaras 2003; Mitchell et al. 2016). Strictly speaking, therefore, what is deemed "conventional" tillage varies from place to place, depending on local conditions and cultural influences. "Conventional tillage" in an arid extensive farming system, for example, is likely very different from "conventional tillage" in a humid intensive farming operation. As well, the nature of tillage deemed "conventional" has evolved over time, as farmers adapted to changing conditions, new technologies, and advancing equipment. Typically, however, under conditions like those in Midwest U.S. in the last 30 years, conventional tillage involved soil inversion using a sequence of deep moldboard plow, disk harrow, and field cultivator operations prior to planting (Figure 4.1). Other forms of conventional

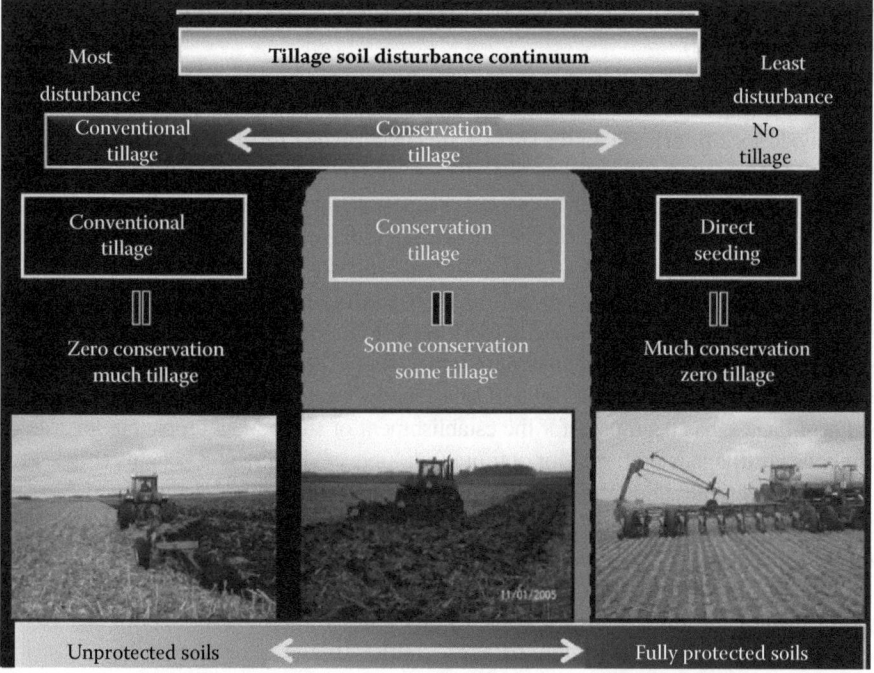

FIGURE 4.1   Three different tillage systems representing a continuum from maximum soil disturbance to minimum soil disturbance. These systems leave a parallel gradient from bare, unprotected soils with degradation to residue covered, well protected soils with regeneration and renewal.

tillage include "deep ripping," "disk plowing," and "chisel plowing" as the initial tillage pass followed by some form of secondary tillage in other geographical areas.

Tillage practices affect soil C, water pollution, and farmers' energy and pesticide use, and therefore data on tillage can be valuable for understanding the practice's role in reaching climate and other environmental goals. New stresses imposed by climate and other global changes now demand a re-assessment and re-configuration of the traditional soil conservation practices in many regions of the world (Reicosky et al. 2011; Reicosky 2015).

Conservation tillage is a generic term denoting any tillage system that reduces loss of soil and water compared with conventional tillage (Mannering and Fenster 1983; Hall 1998; Reeder 2000). Baker et al. (2002) defined conservation tillage as "the collective umbrella term commonly given to no-tillage, direct-drilling, minimum-tillage and/or ridge-tillage, to denote that the specific practice has a conservation goal of some nature." Usually, the retention of 30% surface cover by residues characterizes the lower limit of classification for conservation-tillage, but other conservation objectives for the practice include conservation of time, fuel, earthworms, soil water, soil structure and nutrients. Residue levels alone do not adequately describe all conservation tillage practices (Baker et al. 2002).

Conservation tillage has many variants. Baker et al. (2002) list and describe 14 different names including "minimum-tillage," "reduced-tillage," "ridge-tillage," "mulch-tillage," "strip tillage," "zero-tillage," and "no-tillage" without defining critical attributes of soil and residue mixing. Minimum tillage is based on the minimum soil manipulation necessary for crop production or meeting tillage requirements under the existing soil and climatic conditions; the tillage reduction can be in intensity of tillage, depth of tillage or time involved (number of machinery passes for all tillage operations). "Conservation tillage," in the broad sense, is sometimes oversold for its conservation benefits; in some instances, unacceptable soil degradation continues under practices deemed "conservation tillage" (Crovetto 1996; Derpsch et al. 2010; Derpsch et al. 2014; Reicosky 2015).

For the purpose of this review, conservation tillage is a broad umbrella term used to define *any* non-inversion tillage system with a primary objective of reducing soil and water loss that leaves 30% residue cover after planting, but without definition or limits of soil disturbance and residue incorporation (Figure 4.2).

The international literature on CA systems suffers from some semantic confusion. Difference of interpretation about the importance of "minimum soil disturbance" as a pillar in CA is an example. Some definitions provide specific criteria for the surface area that can be disturbed with shallow tillage and still be considered CA (Corsi et al. 2012). "Low disturbance" systems, including no-tillage and direct seeding, involve uninterrupted minimum mechanical soil disturbance, with maximum disturbed area of 15 cm wide or 25% of the cropped area (whichever is less). Strip tillage is considered CA if the disturbed area is less than these prescribed limits. Corsi et al. (2012) also allowed three levels of crop residue cover that qualify as CA. Given the significance of no-till/direct seeding in minimizing environmental impacts, this chapter treats no-tillage as a separate activity with minimum soil disturbance for seed placement and uses the term conservation tillage more narrowly to denote any other type of reduced tillage practice other than no-tillage, after Eagle et al. (2012). Occasionally, rotational tillage—alternating conservation tillage and no-tillage—has been used for the establishment of subsequent crops, but for true conservation, continuous no-tillage is preferred for optimum multiple environmental benefits (Eagle et al. 2012). These distinctions merit attention if CA systems are to be properly defined and adopted.

No-tillage (NT)/direct seeding (DS) and conservation tillage are often used interchangeably and as a result conservation tillage practices with some small amount of soil disturbance and residue mixing are equated to no-tillage. The definition of no-tillage has been used very loosely and inconsistently over the last 40 years. To some, no-till simply means no moldboard plowing. To others, the terms "conservation tillage," "minimum tillage," "mulch tillage," "reduced tillage," "strip tillage," "rotational tillage," "vertical tillage," and endless variations are lumped into the broad category of no-tillage, even though they may not meet the criteria of "minimum soil disturbance" in CA. No-till minimally disturbs soil for seed placement and crop residues remain in place, providing protection from erosion and minimizing the loss of soil organic carbon (SOC) to the atmosphere. For this

FIGURE 4.2 Photos illustrating the wide range of soil disturbance and residue incorporation categorized under the broad umbrella of "conservation tillage" tools and planters generally understood to have 30% residue cover after planting (Reicosky 2015).

review, no-till/direct seeding is defined as a system involving only the minimum disturbance of soil (Eagle et al. 2012). In no-till systems, crop residues cannot be burned and must be uniformly distributed over the field (USDA-NRCS 2013).

4.2.2 What Is a Conservation Agriculture System?

Conservation agriculture is more than avoidance of tillage—it is an ecosystem approach that involves progressive, system-wide change in the farmer's cultural practices, along with a change in mindset, to bypass the use of the plow. Rooted historically in the conservation tillage perception (Gebhardt et al. 1985; Dumanski et al. 2006; Hobbs 2007; Erenstein et al. 2012; Farooq and Siddique 2015; Lal 2015a), this new CA terminology enfolds continuously-improving varieties, technologies, and incorporating soil health principles—the outcome of continuing learning from over the last 100 years of crop production.

The dust bowl era in the U.S. led to the development of the Soil Conservation Service (SCS). During this evolutionary process, conservation terminology shifted from referring to "soil conservation" (reducing erosion) into "no-tillage farming" (avoiding erosion), then into "conservation tillage" which now, in the last 18 years, has evolved into what we call CA (Kassam et al. 2009; Kassam and Friedrich 2012). Conservation agriculture aims to conserve, improve and make more efficient use of natural resources through integrated management of available soil, water and biological resources combined with lower external inputs, thereby easing environmental damage as well as enhancing and sustaining agricultural production. A synonym for CA might be "resource-efficient" or "resource effective" agriculture (FAO 2011) with regenerative characteristics. As we will argue later, CA is C-based and C-focused; it is an ever-evolving, integrative approach which manages C flows to sustain manifold ecosystem functions, including the production of food over the long-term.

Conservation agriculture was introduced by the FAO (2008) as an innovative concept for resource-efficient agricultural crop production based on an integrated management of soil, water and biological resources combined with external inputs (Hobbs et al. 2008). Conservation agriculture concepts have been widely promoted with the expectation that they can help reconcile various competing objectives and contribute to sustainable intensification (Hobbs 2007; Hobbs et al. 2008; FAO 2011). As evident in recent reviews, CA has spread into other regions of the world, and now has become a global agricultural movement (Hobbs 2007; Hobbs et al. 2008; Dumanski et al. 2006; Baker et al. 2006; Goddard et al. 2008; Govaerts et al. 2009; Kassam et al. 2009; Erenstein et al. 2012; Kassam and Friedrich 2012; Friedrich et al. 2012; Jat et al. 2014; Kassam et al. 2014; Farooq and Siddique 2015; Lal 2015a,b; Kuhn et al. 2016). According to recent estimates (Table 4.1), CA has been adopted on almost 160 million ha, and this area is increasing by about 10 million ha yr^{-1} (Kassam et al. 2014).

The concept of CA, however, has not been without its controversies (Sumberg and Thompson 2012), mainly in sub-Saharan Africa. The positive yield and environmental benefits of CA have not been universally agreed upon (Giller et al. 2009; Giller et al. 2011; Sumberg et al. 2012; Andersson and Giller 2012; Erenstein 2012; Erenstein et al. 2012; Pittelkow et al. 2015; Giller et al. 2015).

TABLE 4.1
Area of Conservation Agriculture Worldwide as of 2013

Country	Area ('000 ha)	Country	Area ('000 ha)
Argentina	29,181	Namibia	0.34
Australia	17,695	Netherlands	0.5
Azerbaijan	1.3	New Zealand	162
Belgium	0.27	Paraguay	3,000
Bolivia	706	Portugal	32
Brazil	31,811	Moldova	40
Canada	18,313	Russia	4,500
Chile	180	Slovakia	35
China	6,670	South Africa	368
Colombia	127	Spain	792
Korea	23	Switzerland	17
Finland	200	Syria	30
France	200	Tunisia	8
Germany	200	Turkey	45
Ghana	30	Ukraine	700
Greece	24	United Kingdom	150
Hungary	5	Tanzania	25
India	1,500	United States	35,613
Iraq	15	Uruguay	1,072
Ireland	0.2	Uzbekistan	2.45
Italy	360	Venezuela	300
Kazakhstan	2000	Zambia	200
Kenya	33.1	Zimbabwe	332
Kyrgyzstan	0.7		
Lebanon	1.2		
Lesotho	2.0	Total	156,991
Madagascar	6		
Malawi	65		
Mexico	41		
Morocco	4		
Mozambique	152		

Source: FAO http://www.fao.org/ag/ca/6c.html; accessed January 28, 2017.

If the CA approach is to be applied globally, more agronomic and social research will be required to develop a more widely-adapted understanding and thereby develop global solidarity on the definition and utilization of CA for food security.

4.2.3 SYSTEMS EMPHASIS

The principles of CA are universal, but the solutions are local and revolve around carbon cycling using a systems approach (Lal 2015b). Conservation agriculture is a broad term to describe continuous no-tillage, continuous crop residue cover, and diverse agronomic and cover cropping systems that incorporate soil health concepts. While each of the principles may be considered a separate entity, it is the fluid integration of all three principles and their supportive soil health practices that are keys to effective CA. The natural interactions and diversity within the cropping system contribute to numerous economic and environmental benefits (Lal 2015b).

A first facet of CA systems is to preserve near continuous soil cover, composed of either live cover crops or crop residues accumulating as mulches on the soil surface (Figure 4.3). Retaining mulch between crops provides better protection against erosion and can also maintain higher soil moisture in dry regions, enrich the soil with organic matter, and, if the mulch is sufficiently dense, prevent the regrowth of weeds.

A second element of CA systems is a diversified cropping pattern (Figure 4.3), ideally consisting of at least three plant species including one legume (Chatterjee et al. 2016). To incorporate more diversity, some farmers are using up to 14 species in cover crop mixes.

The third facet of CA systems is minimal soil disturbance (Figure 4.3), typically achieved using no-tillage management strategies. Another approach is the planting of perennial crops such as forages which, by definition, involve no cultivation for the duration of their growth.

The principles of CA should ideally be integrated and applied continuously for improved carbon management and long-term sustainability. This means avoiding tillage whenever possible. While the principles of CA seem well-established, some soils may have "natural resilience" that allows an occasional tillage event with minimum impact on no-tillage soil properties. A spate of recent studies has examined the effect of "occasional tillage," or "strategic tillage" for controlling "herbicide resistant weeds" on soil carbon, microbial communities, and other indices of soil quality, but the findings are not always conclusive nor consistent (Garcia et al. 2007; Quincke et al. 2007; Rincon-Florez et al. 2016a,b; Liu et al. 2016a,b,c; Crawford et al. 2015; Dang et al. 2015; Wortmann et al. 2008, 2010; López-Garrido et al. 2011). For example, Wortmann et al. (2010) noted slight decreases in microbial biomass but concluded that "One-time tillage of NT can be done without measureable effects on yield or soil properties." In contrast, Stockfisch et al. (1999) observed that organic matter

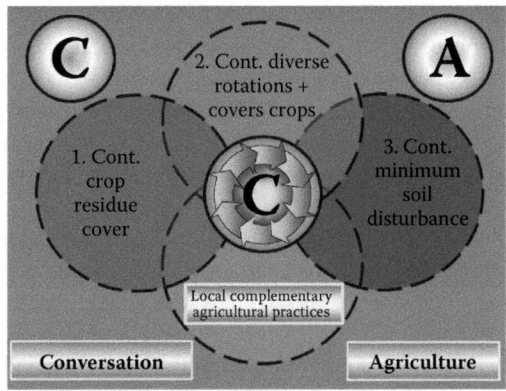

FIGURE 4.3 Schematic of a CA system with three integrated principles and complimentary practices focused on carbon cycling.

stratification and accumulation as a result of long-term minimum tillage (20 years) were completely lost by a single moldboard plow tillage. Grandy et al. (2006) similarly concluded that cultivating no-till systems can decrease soil aggregation and accelerate C and N losses so rapidly that years of soil C restoration can be undone within weeks to months supporting the minimum soil disturbance criteria of Corsi et al. (2012). This inconsistency among findings may reflect, in part, the uniqueness of the ecosystems studied; the effects of sporadic tillage may vary depending on properties of the site in question, subtle differences in research methods, and the complement of other agronomic practices applied there. While the tillage tradition has been around for 10,000 years (Lal et al. 2007), we still argue that no-till remains an important feature of CA, and the imposition of even occasional tillage may often leave the soil vulnerable to loss and depletion and environmental degradation.

Soil is a diverse biological habitat and genetic reserve for countless plants, animals, and other organisms that is very fragile and is greatly affected by tillage. Tillage disrupts the "soil microenvironment" by destroying the soil structure, breaking up the aggregates exposing available carbon releasing more carbon to the atmosphere through enhanced microbial activity. The larger soil fauna, for example earthworms and white grubs, are drastically reduced with intensive tillage destroying macro pores impacting infiltration and root penetration (Kladivko 2001; Kemper et al. 2011). Intensive tillage fragments fungal hyphae networks and upsets the balance between the fungi and bacteria impacting soil carbon loss, as reported by Bailey et al. (2002). They found those ecosystems that were tilled had lower fungal activities and lower stored C than those that were maintained under native or no-till systems. Six et al. (2006), indicate that most agricultural soils are dominated by bacterial activity. They found a quantitative and qualitative increase in soil organic matter (SOM) is generally observed in agricultural systems favoring a fungal dominated community suggesting the need for minimum soil disturbance to optimize fungal activity. de Vries et al. (2006) suggested that high soil fungal to bacterial (F/B) ratios are indicative of a more sustainable agricultural system. They found fungal and bacterial biomass and the F/B ratio were higher in grass than in grass-clover. The F/B ratio decreased with increasing N application rate that reflected decreasing fungal activity and increasing bacterial activity. The way we manage soil disturbance and fertilizer N can have a great impact on the soil biology and ecosystem that affects the soil carbon and energy flow.

While these three main principles—continuous soil cover, diversified cropping, and minimum disturbance—are general in their application, specific differences in each principle need to be defined clearly so that there is no confusion through the use of ambiguous jargon. Conservation agriculture derives many of its multiple benefits from synergistic simplicity of no-till (minimizes C and soil loss) and the use of diverse rotations and cover crop mixes (maximizes soil coverage and C input) for soil diversity protection and regeneration.

Conservation agriculture is a sustainable production system that can also enhance C management compared to typical tillage-based approaches. A unifying message needs to take a broad definition of CA that encompasses essential elements of food security and environmental protection. Transforming agricultural production from a threat to global biodiversity and ecosystem services to a means of promoting ecosystem integrity is unquestionably a key challenge of the twenty-first century. Often, eco-agriculture landscapes could help achieve the critical goals of agricultural sustainability, resilience of food systems and adaptation to climate change. Achieving these potentials of CA will require careful re-evaluation and coordination of priorities and strategies within the agricultural and conservation research and policy communities.

The goals of CA are to improve long-term productivity, profits, and food security, particularly under the threat of climate change. Because CA avoids tillage, it is less time-consuming and can be more cost-effective than conventional farming methods. Even without significant C sequestration, forms of CA are an important new technology that improves soil health processes, controls soil erosion and degradation, reduces production cost, maintains environmental quality, and provides food security in a sustainable manner (Govaerts et al. 2009; Reicosky and Saxton 2007a,b; Reicosky 2008; Reicosky et al. 2011). Conservation agriculture is a production system that can "weather the climate extremes"!

4.2.4   Biodiversity in Carbon Inputs Enhance Biotic Communities

Most natural systems derive strength and resilience from diversity. "Agricultural biodiversity is the first link in the food chain, developed and safeguarded by indigenous people throughout the world, and it makes an essential contribution to feeding the world," stated Nakhauka (2009). The living soil system includes a jungle of creatures, seen and unseen, working underfoot performing countless critical functions (Wall and Nielsen 2012; Bardgett and van der Putten 2014). Biodiversity is an element of sustainability and is necessary for harmony and stability in nature (Nielsen et al. 2011; Balvanera et al. 2014); understanding and nurturing this biodiversity is needed to ensure that management practices sustain ecosystem services (Pereira et al. 2012; Costello et al. 2013; Pascual et al. 2015).

One way of enhancing diversity of organisms in the soil is to grow a diversity of crops, thereby extending crop rotations (Poeplau and Don 2015; Chatterjee et al. 2016). Introducing a variety of cover crops including cool season and warm season species, legumes and grasses, shallow rooted and deep rooted crops (Anderson 2008; Burney et al. 2010; Kemper et al. 2011; Kätterer et al. 2011; Paustian et al. 2016) can also promote diversity (Figure 4.4a,b). Crop rotation also influences soil C dynamics by affecting the proportions of the decomposable organic compounds returned to the soil, leading to potential gains in soil carbon (Hutchinson et al. 2007; Ogle et al. 2005).

Increasing inter-specific diversity via CA also reduces vulnerability to damage from pests and extreme weather compared to monoculture systems. This resilience may become even more important in the future, with greater potential extremes in weather and the incursion of new pests under climate change. Using multiple species also has synergistic benefits for nutrient uptake and cycling and habitat for wildlife and pollinators (Figure 4.4b). The robust and resilient nature of CA systems may also help to protect stored C from climatic disturbances, increasing the permanence and benefits of C.

4.2.5   Site-Specific Adoption of CA, Locally Applied,
Complementary Supporting Practices

Every farmer, farm, soil type, and climate will lend itself to different cropping combinations. Conservation agriculture integrates management of available sun, soil, water, air and biological resources combined with site specific external inputs to create a system of biological agriculture, with a complementary suite of practices, uniquely conformed to each individual farm (Corsi et al. 2012). Conservation agriculture seeks to configure the farm—an ecosystem—specifically adapted and tailored to local conditions, taking into account the trade-offs associated with technology choice in the short and long term. Under such a far-sighted, systems-based perspective, the landscape will evolve into a dynamic, site-specific system for each farm and will reflect individual producers' personal decisions about managing farms and their resources (Corsi et al. 2012). Every farm and farmer reflects a unique combination of multiple management decisions. The cascading effects of these decisions on the agricultural ecosystem can result in synergies whereby one beneficial activity enhances or is additive to another, or in avoiding trade-offs whereby one activity reduces or eliminates the benefits of another.

Conservation agriculture systems need to be flexible and tailored to individual farmer's specific needs and local lands. No two farms are the same and no two farmers are the same; different management strategies, different goals, different personalities demand management flexibility in adapting practices of fertility, herbicides, pesticides, diversification and local farmer knowledge (Anderson 2008). While the three principal facets (Figure 4.3) are the foundation of CA systems, there is need for another "complementary/support component" to enable integration and incorporation of comprehensive agronomic and management practices to fine-tune the functional system (Corsi et al. 2012). The complementary practices include incorporation of animals and poultry into

(a)

(b)

FIGURE 4.4 (a) A variety of cover crop monocultures providing biodiversity and multiple functions with specific plant characteristics that contribute to conservation agriculture systems. (Photo credits: Dirceau Gassen, David Brandt, and NRCS.) (b) An example of a cover crop "cocktail," a mixture of seven species. (Photo credit: David Brandt.)

the production system with additional site-specific diversity supporting services including genetic biodiversity for use in breeding crops, soil formation and structure, soil fertility, nutrient cycling and the provision of irrigation water.

4.2.6 CARBON AS A UNIFYING PERSPECTIVE FOR CA SYSTEMS

The sun powers all life through the C cycle using plants as the main source of energy generation. The living soil is the silent engine fueled by C that recycles nutrients helping plants keep us and the planet alive. Given the numbers and diversity of microbes/fauna living in the soil, the diversity of agricultural lands, and the multiplicity of demands placed upon them, we need a unifying framework to provide a cohesive approach to developing and managing CA systems. We propose here that carbon offers such a unifying perspective. As described in the sections that follow, the flows and storage of carbon are tightly interwoven with most or all of the ecosystem services expected of agricultural ecosystems. Carbon is the universal carrier of solar energy, and its dynamics therefore are closely correlated with the functioning and health of ecosystems (Blaustein 2012; Pelletier et al. 2011; Volk 2003). On that premise, CA systems can be broadly defined as following management approaches that seek to maximize the amount of carbon captured in photosynthesis, and then using the energy within that biomass as wisely and efficiently as possible to provide the services expected of the specific land in question.

4.3 CARBON STORAGE AND DYNAMICS ON AGRICULTURAL LANDS

4.3.1 CARBON CYCLE IN AGRO-ECOSYSTEMS

Farming, in essence, is about managing carbon—exploiting photosynthesis to trap solar energy in carbon-rich plant materials that can be used as food, feed, fuel, or fiber in human societies. The plant communities used for this purpose vary widely among ecosystems. In many places, the crop plants are predominantly annual species, such as wheat, corn, rice, or soybean, which have been deliberately selected to maximize the proportion of photosynthate in the harvestable portion. Thus, for example, the grain may hold almost half of the carbon of the entire above-ground plant in many cereal crops (Bolinder et al. 2007). Perennial crops, by comparison, such as tree fruits and forage grasses and legumes, have to conserve carbon and energy in reserve for repeated cycles of regrowth, so that they allocate more carbon to roots and other vegetative components that persist from year to year (Rees et al. 2005). In most farming systems, however, the overriding aim is to maximize the long-term harvest (yield) of carbon for use by humans and their animals and machines.

The carbon cycle in agricultural ecosystems has been widely studied because of its fundamental agronomic and ecological role (Figure 4.5). Briefly, carbon diffuses from the atmosphere into chloroplasts of leaves, where it is converted into simples sugars by photosynthesis. These sugars are translocated throughout the plant and used as building blocks for various other organic forms: proteins, complex carbohydrates, lipids, and so on. A portion of this carbon—sometimes as much as half or more—is removed from the ecosystem, as harvested grain, forage, or other commodity, some of which may eventually be returned to the system as organic amendments, in the form of manure, bio-solids, and biochar. The unharvested carbon ("residues"), including root exudates,

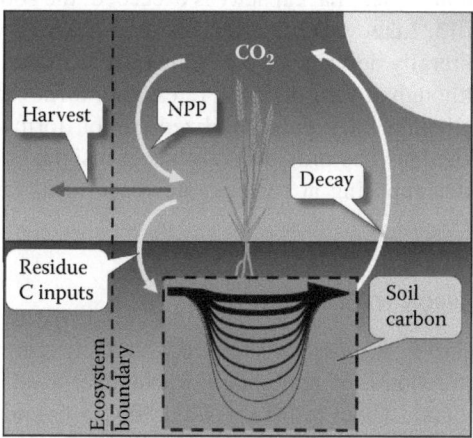

FIGURE 4.5 A schematic showing flows of carbon through the soil. Atmospheric CO_2 is fixed by photosynthesis into energy-enriched organic molecules. Some of these are removed in plant harvest; the rest enter the soil as plant litter, roots, and exudates. In soil, the organic carbon is decayed, at various rates, back to CO_2, completing the cycle.

along with any amendments, remains on or in the soil, and becomes part of the soil organic matter (SOM). Through the microbial process of decay, the carbon in residues, amendments, and SOM is oxidized back to CO_2 and returned to the air, thereby completing the cycle.

At any given time, an agricultural ecosystem holds carbon in one of three pools: (1) the atmosphere, above and within the soil, mostly as CO_2; (2) the vegetation; and (3) the soil organic matter. The air in the column above a hectare of land now contains about 17 Mg C as CO_2 (assuming CO_2 concentration = 405 ppmv), appreciably more than the C content of vegetation in most agricultural lands, which typically amounts to several Mg C (Goudriaan, Groot and Uithol 2001). By comparison, the soil contains amounts of carbon about an order of magnitude higher. Although highly variable, mineral agricultural soils may contain as much as 200 Mg C ha^{-1} or more to a depth of 1 m (Batjes 1996). Concentrations of C are usually highest at the surface, diminishing with depth, reflecting the origin of most added C on or near the surface.

The carbon content of soil is vital not only because of its sheer mass as a reservoir, but also because of its function (Murphy 2014; Lal 2015c; Milne et al. 2015; Lehman et al. 2015; Oldfield et al. 2015). This importance arises not from carbon itself, but from the broader assemblage—the soil organic matter—in which it is embedded. The carbon content of SOM, which also includes hydrogen, oxygen, and numerous nutrients, is remarkably constant (typically about 0.58 g C g^{-1} SOM), so carbon can be used conveniently as a measure and indicator of SOM (Allison 1973). Although sometimes referred to as one pool, the SOM contains a wide range of forms and is usually defined as all materials, living and dead, of biotic origin (Semenov et al. 2013; Franzluebbers 2010; Stockmann et al. 2013). Thus it includes living microbes, senescing vegetative materials including roots and exudates, as well as carbonaceous compounds derived from plant materials applied centuries or millennia ago. To reflect this diversity, SOM is often sub-divided into multiple fractions (Waksman 1925; Wander 2004), using a wide range of chemical and physical techniques, including acid hydrolysis, flotation, sieving, bioassay, and innumerable combinations thereof. The intent is to subdivide the SOM into various categories or pools, based on susceptibility to decay, ranging from highly-decomposable to the very stable or recalcitrant. This stability in SOM was once believed to be related to the chemical composition of the SOM molecules themselves, but growing evidence suggests that most organic materials are inherently decomposable, and their recalcitrance is conferred by association with mineral components of the soil through adsorptive or protective mechanisms (Dungait et al. 2012). For example, association of SOM with adsorptive sites on clay minerals may slow its decay, suggesting that soils have a finite 'capacity' for affording protection to added carbon. When that capacity is exceeded, the soils are said to be 'saturated' with C (Castellano et al. 2015; Lützow et al. 2006; Six et al. 2002; Stewart et al. 2007), though that metaphor may not be literally descriptive. Regardless of form and mechanisms involved, all SOM is subject to decay, although at very different rates, with turnover times ranging from days, for soluble exudates, to millennia for highly-stabilized carbon forms (Janzen 2005). Although dissecting it into various fractions has provided invaluable insights, organic matter is best seen as a continuum of forms, decomposing at various rates back to its simple mineral constituents (Lehmann and Kleber 2015).

4.3.2 IMPORTANCE OF CARBON TO SOIL FUNCTIONS

Organic matter, and hence carbon, is closely associated with many critical functions of soil. Firstly, the SOM contains a vast storehouse of nutrients. For example, a soil with 175 tonnes SOM/ha (100 tonnes C/ha), might contain about 9,100 kg N/ha, 800 kg organic P/ha, and 1,300 kg S/ha in organic form (based on Powlson et al. 2015). As the SOM decays, these nutrients are slowly released in mineral forms available for plant uptake. Because these nutrients are derived from added plant materials, they occur in SOM in proportions roughly proportional to plant demands. And because rates of decay are favored by moisture and temperature conditions similar to those for plant growth, the rate of nutrient release is often reasonably synchronized with temporal patterns of plant demands. Aside from its being a direct repository of nutrients, SOM can also help retain nutrient ions by enhancement of soil cation exchange capacity.

Secondly, SOM has benefits for physical properties of soil, notably by promoting aggregation which improves soil tilth and aeration while also suppressing erosion (Tisdall and Odes 1982; Carter and Stewart 1996; Carter 2002). Recently-added plant litter may be especially important in promoting microbial processes contributing to aggregate formation (Angers and Giroux 1996; Cotrufo et al. 2013). Soil water retention is also favored by SOM (Hudson 1994; Rawls et al. 2003), an important benefit in some drought-stressed soils (Allison 1973).

Thirdly, SOM is a source of energy for the microbial communities which mediate so many essential processes in soil (Sakrabani et al. 2012; Girla 2014). "Bacteria, fungi, and invertebrates live out their lives reproducing by the power of sun-sponsored photons captured in the green molecular traps set above" (Jackson 2011, 141). Or, as King (1907, p. 95) wrote long ago, "We should think of humus as the food of microscopic life in the soil". Through decay, solar energy captured by photosynthesis in plants, is delivered to soil biota, fueling countless essential processes in soil and ensuring the preservation of soil biodiversity, which is critical to ecosystem health (Wall et al. 2010). "Organic matter may well be considered as fuel for bacterial fires in the soil, ...," said Albrecht (1938). This perspective, however, creates a carbon dilemma (Janzen 2006): for any cohort of incoming carbon, you cannot both "sequester" it, and also use the energy released by its decay.

For all these reasons and more, higher SOM has often been linked to greater crop yield, an observation made millennia ago (Allison 1973) and also by more recent scientific investigations (Lal 2006). Beyond the direct benefits for productivity, SOM can also promote other ecosystem services, such as maintaining soil biodiversity and mitigating climate change. Consequently, the SOM content of soils is often considered the preeminent indicator of soil health, and the impetus for its enhancement has recently been further intensified (Lehman et al. 2015; Milne et al. 2015; Lal 2015c).

4.3.3 Effect of Management Practices on Soil Carbon

The trajectory of SOC stocks—whether they are increasing or declining—depends on the amounts of carbon entering the soil from plant litter relative to the amounts of carbon lost via decomposition of SOM. Over time, soil carbon content tends to approach a steady state content, where inputs are comparable to outputs (Odum 1969). This tendency toward an equilibrium occurs because decay follows first-order kinetics; that is, the amount of decay per unit time is roughly proportional to the amount of carbon present. Thus, for example, increasing the rate of litter C input will enhance SOC, but since CO_2 emission is proportional to amount of C present, its rate will increase, eventually again approaching the rate of C input, thereby reaching a new higher steady state. Most soils, maintained under consistent conditions or management regimes for more than a few decades tend to have minimal change in soil carbon over time because the inputs and losses are roughly equivalent. For example, soils historically under well-managed or 'natural' grasslands are likely to exhibit little discernible change in soil carbon over time (Smith 2014).

Once a steady state has been established, SOC will increase or decrease significantly only with the advent of a management change. Among the most influential such instigators is land use change to arable agriculture. For example, co-opting "natural" lands, such as grasslands, for cultivated agriculture almost invariably results in SOC loss (Schlesinger 1985; Houghton et al. 1999; Smith 2008). This loss of C occurs for several reasons (Odell et al. 1984; Darmody and Peck 1997; Buyanovsky and Wagner 1997). Firstly, the physical disruption and mixing soil may enhance decomposition by making the SOM more accessible to microbes (Six et al. 1998). Tillage itself results in the immediate release of gaseous CO_2 (Ellert and Janzen 1999; Reicosky and Lindstrom 1993; Rochette and Angers 1999), although much of this is previously-decayed C, entrapped in soil as CO_2. But accelerated emissions of CO_2 can continue, indicating a prolonged effect on subsequent decomposition of soil organic matter (Shahidi et al. 2014). Secondly, perhaps more importantly, agriculture is designed to *remove* as much harvestable biomass as possible from the ecosystem, so carbon inputs are usually much lower in agricultural systems than in the preceding natural condition (Janzen 2004; Lal et al. 2011). In some instances, SOC loss from a given site is exacerbated by water or wind erosion, but until this carbon is oxidized, this movement represents physical translocation of carbon, not loss from the landscape as a whole. Carbon removed from one site in the landscape may be buried elsewhere, under conditions that suppress rate of decomposition (VandenBygaart et al. 2012; VandenBygaart et al. 2015). Consequently, some debate remains about whether local C losses via erosion add to or subtract from the atmospheric CO_2 burden (Lal 2005; Van Oost et al. 2007; Kuhn et al. 2009; Sanderman and Chappell 2013; Galy 2015).

Cumulative losses of soil carbon from land-use change have been estimated to be about 50 Pg C or more, globally (Smith et al. 2016; Sanderman et al. 2017). This historic loss of SOC from agricultural lands around the world presents an opportunity for future gains in C (Paustian et al. 2016). Collectively, such gains are often referred to as "carbon sequestration," although the term remains somewhat ambiguous and not always tightly defined, as described later.

On arable lands subject to earlier losses, gains of carbon can be elicited by any change in practice that enhances carbon inputs or by suppressing decomposition (or both). A wide range of such practices have been shown to increase soil carbon on croplands, including: reducing tillage intensity; enhancing yield and residue inputs through fertilization or irrigation; restoring degraded lands; eliminating summer fallow; planting more perennial forages; and re-establishing permanent grassland (Janzen et al. 1998; Lal 2014; Srivastava et al. 2012). The rates of carbon gain in response to these practices, however, varies widely among sites; and not all practices are effective everywhere (Smith et al. 2008).

4.3.4 Defining Carbon Sequestration

Defining precisely what is meant by carbon "sequestration" is necessary to avoid ambiguity and potential confusion. Carbon sequestration has been defined as the "process of transfer and secure storage of atmospheric CO_2 into other long-lived C pools that would otherwise be emitted or remain in the atmosphere" (Lal 2008). Sedjo and Sohngen (2012), similarly, defined carbon sequestration as "the process of capture (through photosynthesis) and long-term storage of atmospheric carbon dioxide." The term, therefore, denotes two critical conditions: it involves net removal of CO_2 from the atmosphere by photosynthesis; and the removal must be "secure" or "long-term."

Both of these conditions merit consideration in any evaluation of soil carbon sequestration. The first condition means that local increases in soil carbon achieved by importing carbon from elsewhere do not necessarily qualify as carbon sequestration (Olson et al. 2014). For example, amending a soil with manure adds carbon, but this direct increase does not denote atmospheric CO_2 removal; it merely involves the re-distribution of already captured carbon (Schlesinger 2000; Powlson et al. 2011a,b).

More difficult is the second facet of sequestration—the stipulation that removal of carbon from the atmosphere is secure or long-term; that it exhibits "persistence" (Hutchinson et al. 2007). Organic carbon, by nature, is dynamic; a carbon atom trapped by photosynthesis and entering the soil as plant residue is likely to re-appear in the atmosphere again within a few months or years. Only about one third of crop residue-C remains in the soil after one year (Angers and Chenu 1997; Jenkinson and Rayner 1977; Voroney et al. 1989; Gregorich et al. 2017) and only 5 to 20% remain after two years (Buyanovsky and Wagner 1997; Broder and Wagner 1988). Just a small fraction of the C added to soil enters non-ephemeral pools (Paul et al. 1997; Lal, Follett, and Kimble 2003; Franzluebbers et al. 1998; Robert 2001), and even that "stabilized" carbon is not locked away permanently but remains vulnerable to gradual release over periods of decades to centuries. As a result, soil organic matter cannot be seen as a "storehouse" of carbon permanently locked away; instead, it may be better visualized as a "stream" of carbon flowing, at various rates, back to atmospheric CO_2 (Figure 4.6). The latter perspective, based on the endless cycling of carbon through soil,

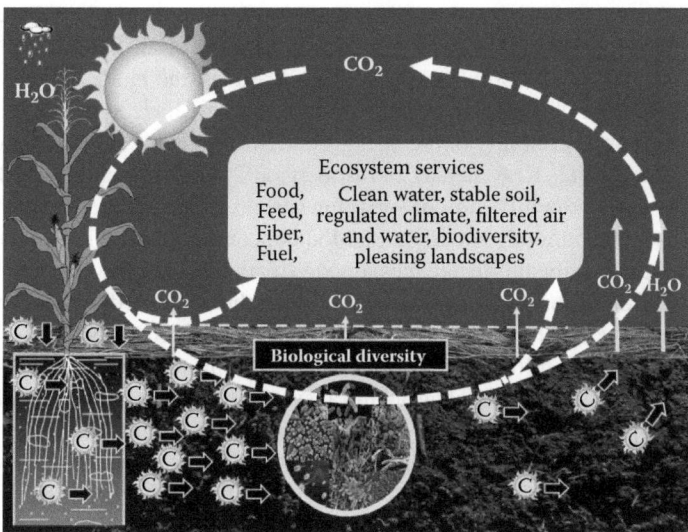

FIGURE 4.6 The carbon cycle in cropland, envisioned as a conduit of energy flow. Carbon in CO_2 from the atmosphere is invested with solar energy by plant photosynthesis. Much of this carbon, upon plant senescence, passes through the soil delivering solar energy to soil biota, which generate many ecosystem services. Once the carbon has been stripped of energy, it returns to the atmosphere, completing the cycle.

reflects the transient, thermodynamically-unstable nature of soil carbon, continually decayed by relentless microbial activity in soil (Kleber 2010; Kleber and Johnson 2010; Janzen 2015; Lehmann and Kleber 2015).

By virtue of its biological flux, little of the carbon present in soil can be considered truly "sequestered" or permanently withheld from the atmosphere. Consequently, instead of focusing on sequestering carbon, an alternative aim might be to "manage" carbon in such a way that its flow sustains the critical biological functions in the soil and that the instantaneous mass of carbon in that flow is maintained or increased over intervals of time.

To use an analogy: The soil carbon reservoir is a little like a mountain lake, continually fed by incoming streams and drained by an outgoing river. Individual water molecules are gradually displaced, but the lake itself remains. Ideally, the amount of water in the lake (stock) and the volume of the stream (flow) are both maintained.

4.3.5 POTENTIAL SOIL CARBON GAINS IN RESPONSE TO IMPROVED MANAGEMENT

The potential gains in soil carbon with adoption of improved management practices vary widely from site to site, depending on soil, climatic, and management factors (Smith et al. 2008). Of particular interest, in this review, is the potential increase in stored carbon under CA, especially in response to no-tillage, which is an important facet of CA systems. Many studies, world-wide, have documented higher carbon storage under no-till systems, compared to those under conventional tillage (Lal and Kimble 1997; Smith et al. 2008; West and Marland 2002; West and Post 2002). For this reason, CA, along with other low-disturbance systems, has been widely advocated as a way of capturing more soil carbon (e.g., Chivenge et al. 2007; Hobbs et al. 2008; Corsi et al. 2012; Fuentes et al. 2012; Palm et al. 2014; Lal 2015a,b).

The response of soil carbon to adoption of no-till, however, is not universal (DeLuca and Zabinski 2011; Blanco-Canqui and Lal 2008). Almost always, no-till results in accumulation of carbon near the soil surface, whereas plowing mixes the carbon inputs to the depth of tillage (Kuhn et al. 2016). Thus, sampling only the surface soil, in a comparison between tilled and un-tilled systems, will bias the results in favor of no-till, sometimes leading to over-estimates of the benefits of no-till for carbon sequestration (Baker et al. 2007; Manley et al. 2005; Angers and Eriksen-Hamel 2008; Christopher et al. 2009; Luo et al. 2010; Srinivasarao et al. 2015). Deep sampling is therefore required to properly compare carbon storage among tillage systems, but such sampling also increases the size and variability of the background SOC stock, reducing the precision of measurements (Ellert et al. 2002; Kravchenko and Robertson 2011; Baldock et al. 2012). In part, the ongoing debate may reflect the difficulty of verifying carbon change in the full profile; as the depth of sampling increases, the amount of carbon and the concomitant variability increases, making it more difficult to detect significant differences among treatments. So the question arises: Does the absence of significant difference in the entire profile mean there is no difference, or merely that any difference is obscured by the sheer mass and variability of measured carbon?

The response of soil carbon to no-till alone may also be variably influenced by agronomic, climatic, and soil factors specific to a local agricultural system (Corbeels et al. 2016). In some regions, for example, the adoption of no-till involves a yield penalty; a comprehensive meta-analysis of 5463 paired yield observations from 610 studies suggested an average yield penalty of around 10% (Pittelkow et al. 2015). Where yields are reduced under no-till, soil carbon replenishment may be constrained by reduced amounts of residue entering the soil.

Local climate and soil conditions may also variously affect response to no-till through effects on decomposition. For example, no-till systems have been shown to increase soil carbon in semi-arid western Canada, but not in more humid eastern Canada (VandenBygaart et al. 2003). This regional effect has been attributed to differences in climate: in western Canada, under semi-arid climate, no-till slows decay of incoming residues because they remain desiccated on the surface, resulting in accrual of soil carbon. In eastern Canada, conversely, conditions are sufficiently humid that residue decay on the soil surface is not constrained by desiccation, so there is little difference in decay rate between buried and surface residue (Helgason et al. 2014).

Because of such site-dependent factors, no-till or other elements of a CA system cannot be universally assumed to elicit soil carbon gains (Yang et al. 2013; Palm et al. 2014; Powlson et al. 2014, 2016). A meta-analysis of the effects of CA on SOC in Africa, for example, found little evidence of increases in SOC under CA (Govaerts et al. 2009). Even in the dry climates of Australia, inconsistent results have been obtained with various types of analysis (Luo et al. 2010; Lam et al. 2013; Kirkegaard et al. 2014). These studies, taken collectively, do not necessarily suggest that CA does not promote soil carbon storage; but they caution against assuming in advance that CA will always promote soil carbon and will effectively mitigate atmospheric CO_2. The effectiveness of CA in achieving this aim will depend on configuring the system, including the careful adoption of no-tillage (Eagle et al. 2012), to the lands upon which it is imposed (Govaerts et al. 2009). As already indicated, the system that best advances the aims CA must be malleable, tuned to meet the demands and conditions of local lands.

Typically, estimated rates of carbon gain in response to improved practices on arable lands amount to about 0 to 0.5 Mg C ha^{-1} (IPCC 2000; Smith et al. 2008). Higher rates are achievable in lands converted back to grasslands and other native vegetation. The latter approach, however, curtails agricultural productivity and hence cannot be widely applied without effects on food security.

Although scientists have been measuring soil carbon content for more than a century, we still have difficulty in precisely quantifying short-term changes in carbon stocks, despite the availability of accurate, efficient analytical instruments. A fundamental challenge is that the change in soil C stock is small, relative to a large and spatially-variable background (Conant and Paustian 2002; Baldock et al. 2012; Loveland et al. 2014). In response to a new management practice, for example, soil carbon may increase at a rate of 0.2 Mg C ha^{-1} yr^{-1}; but the amount of carbon already stored in the surface 1 m may conceivably be 200 +/– 40 Mg C ha^{-1}. Reliably measuring this change, therefore, requires either a very intensive sampling strategy, or waiting long enough for the cumulative annual increments from the new management to reach measurable amounts. An important facet of such strategies is to compare similar amounts of soil mass, not merely similar soil volumes. If a new management regime increases bulk density relative to the conventional management, then sampling to a fixed depth will consider more soil mass in the former case, leading to an over-estimate of carbon storage. Several approaches have been developed to correct for this effect, but all still involve some approximations (Ellert and Bettany 1995; Ellert et al. 2000; Wendt and Hauser 2013). The difficulties associated with accurately quantifying changes in soil carbon stocks do not necessarily diminish opportunities to bolster soil carbon reserves. They do, however, emphasize the importance of following well-conceived sampling and analytical strategies, and the need for carefully evaluating earlier measurements to ensure they are not inadvertently biased by methodological oversights (Dick et al. 1996; Robertson et al. 1999; Milne et al. 2010; Philibert et al. 2012; Derpsch et al. 2014; Brouder and Gomez-Macpherson 2014; Kladivko et al. 2014; Olson et al. 2014; White et al. 2013).

Globally, the technical potential for soil carbon sequestration has been estimated to be as high as several Pg C yr^{-1} (Lal 2011). Economically-viable potential, however, is somewhat smaller, perhaps in the range of 0.4 to 0.7 Pg C yr^{-1} (Smith 2008; 2016). Additional amounts of C may be stored from biochar amendments, which are relatively resistant to decay (Lehmann 2007). Paustian et al. (2016) recently suggested that mitigation potential from soil management, including that from biochar and enfolding also other greenhouse gases, might be as high as ~8 Pg CO_2(eq) yr^{-1}, equating to about 2 Pg C yr^{-1}, but they imply that additional incentives would be required to approach that potential. If a substantial portion of this potential is achievable, then soil carbon sequestration can offset a significant but minor fraction of the global emissions of fossil-fuel derived CO_2, now about 10 Pg C yr^{-1}, or a larger fraction of current emissions from land use change, now about 1.3 Pg C yr^{-1} (Le Quéré et al. 2016).

Such increases, however, can be sustained only for a limited time period (perhaps one or several decades) and can therefore only be considered an interim measure to the more important aim of curtailing fossil C emissions (Post et al. 2012). Soil C storage rates tend to decrease over time. For example, Mortenson et al. (2004) measured gains of 1.2 Mg CO_2 ha^{-1} yr^{-1}, 2.4 Mg CO_2 ha^{-1} yr^{-1}, and 5.7 Mg CO_2 ha^{-1} yr^{-1} in soil carbon 36, 14, and 3 years, respectively, after alfalfa inter-seeding on a northern mixed-grass rangeland in South Dakota. A related factor, often overlooked, is the management history; a soil which, in the past, has been subjected to carbon-losing practices may have a much higher potential for future gain than a soil which has always been managed under C-conserving conditions (IPCC 2000). Gains of soil carbon from adopting no-till have been widely reported (Boddey et al. 2010; Sommer and Bossio 2014; Grace et al. 2012; Ogle et al. 2012; Varvel and Wilhelm 2011; Virto et al. 2012; Peterson et al. 1998; Sisti et al. 2004). In an extensive review, for example, Eagle et al. (2012) separated true no-till from various forms of "conservation tillage" and showed substantial benefits for soil C with minimum soil disturbance for seed placement, emphasizing the importance of defining carefully the detailed attributes of CA systems.

Despite intensive international efforts to promote soil carbon sequestration, initiated two decades ago (e.g., Paustian et al. 1997; Bruce et al. 1999), the global contribution of soil carbon sequestration to reduced atmospheric carbon load is not easily confirmed. This uncertainty likely reflects several ongoing realities: the notorious difficulty of verifying soil carbon gains on landscapes (they may be occurring but we cannot easily quantify them); social and economic barriers to adoption of carbon-conserving practices; and possible overestimation of the effectiveness of some carbon-conserving practices in storing more carbon (Powlson et al. 2011).

Although storing soil carbon to mitigate climate remains a worthy, laudable goal, aiming for carbon management rather than focusing specifically on carbon sequestration may offer wider benefits. A growing body of science indicates switching to no-till farming practices can offer enormous environmental benefits, but net C sequestration from adoption of such practices depends on a variety of factors such as existing soil C, soil type, climate and management practices (Smith et al. 2007). If soils are managed as C sinks, then SOC will increase only over a limited time, up to the point when a new SOC equilibrium is reached (Sommer and Bossio 2014). The larger concentration near the surface in no-till is generally beneficial for soil properties that often, though not always, translate into improved crop growth.

4.3.6 CARBON DYNAMICS AND STORAGE FOR MANIFOLD FUNCTIONS—A BROADER PERSPECTIVE

Much of the research on soil carbon in recent decades has been motivated by prospects of building soil organic matter to mitigate climate change by withdrawing CO_2 from the atmosphere (Lal 2004). This research thrust has powerfully enhanced our understanding of soil carbon processes and provided new insights into carbon contents and dynamics in soils worldwide. But climate mitigation is only one of many ecosystem services driven by carbon dynamics in soils (Powlson et al. 2016). As indicated earlier, the flows of carbon through soils have wide-ranging benefits for many dimensions of ecosystem performance, including productivity, biodiversity, economics, human health, and land health. The management of carbon in agricultural ecosystems, therefore, should reflect a holistic perspective, including manifold soil functions arising from carbon dynamics and storage (Murphy 2014). In some cases, this may mean that the "best" management practices for carbon may not always be the practice that stores the most carbon.

To devise optimal ways of managing soil carbon it is important to distinguish between carbon stocks and carbon flows. The former indicates the amount of carbon stored in soil at any one time; the latter refers to the passage of carbon through the soil. One is a quantity; the other is a rate. Some of the ecosystem benefits of soil carbon arise from the presence of carbon in soil (from the stock); for example, cation exchange is directly related to the amount of organic carbon present in the soil. Most of the benefits, however, arise from the flows (the dynamics) of carbon through soil. For example, microbial activity and release of nutrients from organic matter are both driven by the decay of carbon, and its oxidation to CO_2. Clearly, for a given biomass input into soil, maximizing preservation of carbon and maximizing decay of carbon are mutually exclusive aims. Managing carbon for multiple ecosystem functions (Murphy 2014; Noellemeyer and Six 2015) therefore involves carefully weighing and sometimes compromising between potential benefits of maximum carbon stocks and maximum carbon flows.

4.4 CONSERVATION AGRICULTURE: HARVESTING AND WISELY USING SOLAR ENERGY

Researchers have long been preoccupied, justifiably, with conserving carbon in soils. In many respects, however, carbon is an indicator of a more fundamental facet of ecosystem performance: solar energy. The primary driving force of all ecosystems on Earth is photosynthesis, which invests solar energy in CO_2 from the atmosphere, forming energy-rich compounds with C–C and C–H bonds which then shuttle the stored solar energy in a cascade of energy to other organisms in the ecosystems (Volk 2003; Pelletier et al. 2011; Blaustein 2012) (Figure 4.6). In CA ecosystems the goal is to maximize energy capture in photosynthesis: some of this stored energy is harvested for food to fuel human metabolism, some is harvested for feed as energy for domestic animals, and some is harvested as biofuel for industry and automobiles. A large share of the stored solar energy, however, is returned to the soil in the form of crop residues, roots, and root exudates (Gessner et al. 2010). These materials are quickly decayed by soil biota, who extract from them the solar energy needed to drive their own metabolism, with the release of CO_2 back to the atmosphere. A small and ever-shrinking proportion of the energy from added biomass remains behind in soil organic matter.

Nature runs on sunlight. Land conservation, therefore, depends fundamentally on maintaining the flows of solar energy into and through the soil. In many ways, it is solar energy that matters; carbon is merely the measurable indicator of stored solar energy. Perhaps the most important (and often overlooked) aim of land conservation is to direct sufficient solar energy to the soil biotic communities that catalyze countless essential and still somewhat mysterious beneficial processes in ecosystems (Pereira et al. 2012). Increasingly, we are aware of the importance of active and diverse microbial and faunal communities in soil (Wall et al. 2010); these depend on the continual flux of solar energy into soil in the form of carbonaceous residues invested with solar energy by photosynthesis in plants. By definition, if decomposable organic matter is diminished in a soil, the activity of beneficial soil microbial communities must be suppressed.

The delivery of solar energy into soil, aside from direct enhancement of microbial activities, also furnishes numerous corollary benefits. For example, the release of energy by mineralizing of carbon is accompanied also by release of nutrients also stored in the biomass entering soil (Powlson et al. 2015). Thus, the decay of biomass, with release of energy, is also an essential means of recycling nutrients for re-use by successive crops. Further, the microbial processes of decay, driven by the demand for solar energy, also reconfigures the chemical structure of remaining organic substrates and, notably, its affiliation with mineral components of the soil, creating physical arrangements (aggregates) that improve soil physical structure, enhance aeration and infiltration, and reduce erodibility. Through these and other mechanisms, the continual cyclical flux of carbon (solar energy) into and through the soil maintains soil health and multiple ecosystem functions.

Although it is the flow of carbon (solar energy) that drives ecosystem function, it is the stock of carbon (stored solar energy) that instills resilience. A large pool of organic matter in the soil serves as a buffer of energy and nutrients to sustain the ecosystem and its organisms through short-term stresses, deficiencies, and shortfalls. Thus, the soil organic matter serves as a battery, storing solar energy (Schramski et al. 2015). If maintained more-or-less fully charged, the ecosystem can be carried through periods of short-term shortages and upheavals. If allowed to discharge over time, the ecosystem has little energy and nutrient reserve (little resilience) to endure prolonged periods of stress. Clearly, there is a conflict between maximizing flows of energy and maximizing storage of energy; you cannot both keep the battery fully charged and also expect maximum use of the energy. Optimum management of land, therefore, involves aiming for the best compromise between maintaining flows to drive the ecosystem, and holding back sufficient solar energy in organic matter as a buffer to ensure resilience.

We propose that CA be framed on the premise of promoting carbon management for optimum solar energy flows through lands while, at the same time, maintaining sufficient reserves of energy to achieve long-term resilience. This revised perspective shifts the emphasis beyond merely conserving as much carbon as possible, to ensuring the continuing capacity of lands to maintain or increase future capture of solar energy via photosynthesis. Thus 'conservation' now applies not to carbon itself, but to the capacity of lands to acquire more solar energy stored in carbon-rich compounds.

4.5 CONSERVATION AGRICULTURE: PRACTICAL IMPLICATIONS OF THE ENERGY PERSPECTIVE

4.5.1 SYNERGISTIC BENEFITS OF SUSTAINING FLOWS OF CARBON

If CA is defined from an energy perspective, that has several practical ramifications. Firstly, it implies a strong emphasis on managing lands in such a way as to maximize solar energy capture by plants (Elton 1958). Thus, CA is closely allied with the aim of maintaining high yields, which are a measure of solar energy capture. Innovative and effective management of plant nutrients, water, and pests is therefore an essential aspect of CA. Further, practices like appropriate crop rotations and/or intercropping, using perennial crops or diverse cover crop mixes, may also be critical elements of CA because they can extend the photosynthesizing period and capture more solar energy.

Secondly, this energy perspective emphasizes the importance of returning as much as possible of the energy stored in plant biomass to the soil and its biological community. Although some harvest and removal of energy is inevitable in agriculture (that is the point of farming, after all), a fundamental aim, always, is to maximize the return within economic limits. This return of solar energy may be direct and immediate, in the form of residues, or it may be indirect and delayed, as in the form of animal manures, biochar, compost or other organic materials which originate on the land but are returned by circuitous routes.

Thirdly, this perspective emphasizes the importance of maintaining a sizable buffer of stored energy in soil organic matter. Such a buffer may be especially important in coming decades with a multitude of growing stresses and pressures of climate extremes on agricultural lands. Maintaining a robust buffer in the form of soil organic matter has the added benefit of also withholding carbon from the atmosphere. Thus carbon sequestration becomes a secondary (but worthy) benefit of CA.

Fourthly, the proposed framework emphasizes the application of local wisdom to the development of CA systems. Energy capture and its wise dispersal are dependent on innumerable conditions of the local land, including properties of climate, soil, landscape, economics, and social factors. Given this diversity of factors, practices leading to CA cannot be dictated as universal decrees from afar, but need to evolve with scientific, economic and social insights gleaned and applied locally by innovative farmers.

4.6   SUMMARY

As food production demands rise, we must keep our soil healthy and productive to maintain food security that rests on our living soils requiring a continuous flow of energy. We need a new climate-smart agriculture to protect our soils from erosion, making the transition to farming better suited to cope with impacts of soil loss and degradation associated with climate extremes.

Soil organisms play a key role in C dynamics. Conventional plow tillage can disrupt the "living soil biology" by destroying the soil structure breaking up the aggregates and earthworms, exposing available carbon and releasing more carbon to the atmosphere through enhanced microbial activity. Further, tillage fragments fungal hyphae networks and upsets the balance between the fungi and bacteria, impacting soil carbon loss.

Conservation agriculture has been proposed as a resilient type of agriculture to protect our natural resources that points to the need for better carbon (C) management. Conservation agriculture, C-based and C-focused, makes better use of agricultural resources through the integrated management of available sun, soil, water, air and biological resources, combined with limited external inputs. It integrates system concepts based on three key principles: (1) continuous crop residue cover on the soil surface; (2) continuous minimum soil disturbance (no-tillage); and (3) diverse crop rotations and cover crop mixes with location-specific complementary practices, all important elements of CA. Enhanced C management enables interactive synergies between the biological, physical, and chemical properties and processes with multiple economic and environmental benefits.

The sun is the primary energy driving force of all ecosystems on Earth through photosynthesis, which invests solar energy and CO_2 from the atmosphere, forming energy-rich compounds which then shuttle the stored solar energy in a cascade of energy to other organisms in the ecosystem. In agricultural ecosystems, this stored energy is harvested for human food, some is harvested for feed domestic animals, some is harvested as fiber, and some is harvested as biofuel. A large share of the captured solar energy, however, is returned to the soil in the form of crop residues, roots, and root exudates. These materials are quickly decayed by soil biota extracting the solar energy needed to recycle nutrients with the release of CO_2 back to the atmosphere.

Biodiversity with C cycling and flow is an element of sustainability and is necessary for harmony and stability in nature. This biodiversity provided by crop rotations and cover crop mixes must be understood and nurtured to enhance soil biodiversity and resilient soil ecosystem services. Soil organic matter is also closely associated with many critical functions that include a vast storehouse of nutrients, promoting aggregation which improves soil structure and aeration while also suppressing erosion, and a source of energy for the microbiomes.

Carbon cycling and sequestration related problems require a common understanding or definition of our production system and clear communication to avoid confusion and misunderstanding. Carbon sequestration describes long-term storage of C usually to mitigate and avoid climate extremes. Occasional debate over the concept of soil C storage/sequestration and associated "jargon" is often inconsistent and needs to be balanced with a view of soil C dynamics and cycling in soils. Understanding the types and sizes of soil C stocks and energy fluxes is critical for improved C management in CA systems. Understanding the complexities and benefits of carbon flows may be more important than only maximizing carbon stocks.

Net C storage can depend on a variety of factors such as existing soil C, soil type, climate and management practices that provide a clear understanding of the benefits when C is considered a proxy for energy and viewed from the perspective of energy flow through the soil versus energy stored in the soil. Following the carbon flows through the system helps clarify the C cycle and provides a new perspective on improved C management for soil functions. Increasing the efficiency of C flow through agricultural production systems starts with maximizing C capture per unit land area through photosynthesis. The management of C in agricultural ecosystems, therefore, should reflect a holistic perspective, including manifold functions arising from C dynamics, flow and storage.

We propose that CA be framed on the premise of promoting solar energy flows through lands while, at the same time, maintaining sufficient reserves of energy to achieve long-term resilience. The goals of CA are to improve long-term productivity, profits and food security, particularly under the threat of climate change. Because CA avoids tillage, it is less time-consuming and can be more cost-effective than conventional farming methods. Forms of CA are an important innovative technology that improves soil processes, controls soil erosion and degradation, reduces production cost, maintains environmental quality, and provides food security in a sustainable manner. The synergistic simplicity of minimum soil disturbance (minimizes C and soil loss) and the use of diverse rotations and cover crop mixes (maximizes soil coverage and C input) for soil diversity protection and regeneration benefits CA.

Conservation agriculture is a major player in our survival through carbon cycling and soil carbon storage. Global food security requires efficient carbon cycling and global environmental preservation may require soil carbon storage as the main goal for improved C flow management of sustainable farming systems. These two major functions are agriculture's "carbon conundrum" and will require a practical compromise that may depend on society's decision to support climate mitigation or to support food security or both. Future research and innovative farmers using CA principles and techniques for improved carbon management must find a workable solution.

4.7 ACKNOWLEDGMENT

The authors gratefully acknowledge the assistance of Trina Burgess and Yuka Hopkins for editorial assistance.

REFERENCES

Albrecht, W. A. 1938. Loss of soil organic matter and its restoration. In *Soils and Men: Yearbook of Agriculture 1938*, 347–60. Washington, DC: United States Department of Agriculture.

Allison, F. E. 1973. *Soil Organic Matter and Its Role in Crop Production*. Amsterdam: Elsevier Scientific Publishing Company.

Anderson, R. L. 2008. Diversity and no-till: Keys for pest management in the U.S. great plains. *Weed Science* 56, no. 1: 141–5. doi.org/10.1614/WS-07-007.1.

Andersson, J. A. and K. E. Giller. 2012. On heretics and God's blanket salesman: Contested claims for conservation agriculture and the politics of its promotion in African smallholder farming. In *Contested Agronomy: Agricultural Research in a Changing World*, ed. J. Sumberg and J. Thompson. London: Routledge.

Andresen, J., S. Hilberg, and K. Kunkel. 2012. Historical climate and climate trends in the midwestern USA. In *U.S. National Climate Assessment Midwest Technical Input Report*. J. Winkler, J. Andresen, J. Hatfield, D. Bidwell, and D. Brown, coordinators. Available from the Great Lakes Integrated Sciences and Assessments (GLISA) Center. http://glisa.umich.edu/function/publications#pubs-nca.

Angers, D. A. and C. Chenu. 1997. Dynamics of soil aggregation and C sequestration. In *Soil Processes and the Carbon Cycle*, ed. R. Lal, 199–206. Boca Raton, FL: CRC Press.

Angers, D. A. and M. Giroux. 1996. Recently deposited organic matter in soil water-stable aggregates. *Soil Science Society of America* 60, no. 5: 1547–51.

Angers, D. A. and N. S. Eriksen-Hamel. 2008. Full-inversion tillage and organic carbon distribution in soil profiles: A meta-analysis. *Soil Science Society of America* 72, no. 5: 1370–4.

Bailey, V. L., J. L. Smith, and H. J. Bolton. 2002. Fungal-to-bacterial ratios in soils investigated for enhanced carbon sequestration. *Soil Biology and Biochemistry*. 34, no. 7: 1385–9.

Baker, C. J., K. E. Saxton, and W. R. Ritchie. 2002. *No-Tillage Seeding: Science and Practice. 2nd ed*. Oxford, UK: CAB International.

Baker, C. J., K. E. Saxton, W. R. Ritchie et al. 2006. *No-Tillage Seeding in Conservation Agriculture, 2nd ed.* Oxford, UK: CAB International.

Baker, J. M., T. E. Ochsner, R. T. Venterea, and T. J. Griffis. 2007. Tillage and soil carbon sequestration—What do we really know? *Agriculture, Ecosystems and Environment* 118, no. 1: 1–5.

Baldock, J. A., I. Wheeler, N. McKenzie, and A. McBrateny. 2012. Soils and climate change: Potential impacts on carbon stocks and greenhouse gas emissions, and future research for Australian agriculture. *Crop and Pasture Science* 63, no. 3: 269–83.

Balvanera, P., I. Siddique, L. Dee et al. 2014. Linking biodiversity and ecosystem services: Current uncertainties and the necessary next steps. *BioScience* 64, no. 1: 49–57. doi.org/10.1093/biosci/bit003.

Bardgett, R. D. and W. H. Van der Putten. 2014. Belowground biodiversity and ecosystem functioning. *Nature* 515, no. 7528: 505–11.

Batjes, N. H. 1996. Total carbon and nitrogen in the soils of the world. *European Journal of Soil Science* 47, no. 2: 151–63.

Blanco-Canqui, H. and R. Lal. 2008. No-tillage and soil-profile carbon sequestration: An on-farm assessment. *Soil Science Society of America* 72, no. 3: 693–701.

Blaustein, R. 2012. Can biology transform our energy future? *Bioscience* 62, no. 2: 115–9.

Boddey, R. M., C. P. Jantalia, P. C. Conceicao et al. 2010. Carbon accumulation at depth in Ferralsols under zero-till subtropical agriculture. *Global Change Biology* 16, no. 2: 784–95.

Bolinder, M. A., H. H. Janzen, E. G. Gregorich, D. A. Angers, and A. J. VandenBygaart. 2007. An approach for estimating net primary productivity and annual carbon inputs to soil for common agricultural crops in Canada. *Agriculture, Ecosystems and Environment* 118, no. 1: 29–42.

Branca, G. N. McCarthy, L. Lipper, and M. C. Jolejole. 2011. Climate-smart agriculture: A synthesis of empirical evidence of food security and mitigation benefits from improved cropland management. *Mitigation of Climate Change in Agriculture series 3*. Food and Agriculture Organization of the United Nations (FAO).

Broder, M. D. and G. H. Wagner. 1988. Microbial colonization and decomposition of corn, wheat and soybean residue. *Soil Science Society of America* 52, no. 1: 112–7.

Brouder, S. M. and H. Gomez-Macpherson. 2014. The impact of conservation agriculture on smallholder agricultural yields: A scoping review of the evidence. *Agriculture, Ecosystems and Environment* 187: 11–32.

Bruce, J. P., M. Frome, E. Haites, H. Janzen, R. Lal, and K. Paustian. 1999. Carbon sequestration in soils. *Journal of Soil and Water Conservation* 54, no. 1: 382–9.

Burney, J. A., S. J. Davis, D. B. Lobell, and G. P. Robertson. 2010. Greenhouse gas mitigation by agricultural intensification. *Proceedings of the National Academy of Sciences* 107, no. 26: 12052–7.

Buyanovsky, G. A. and G. H. Wagner. 1997. Crop residue input to soil organic matter on Sanborn field. In *Soil Organic Matter in Temperate Agroecosystems: Long-Term Experiments in North America,* ed. E. A. Paul, K. H. Paustian, E. T. Elliot, and C. V. Cole, 73–83. Boca Raton, FL: CRC Press.

Carter, M. R. 2002. Soil quality for sustainable land management: Organic matter and aggregation interactions that maintain soil functions. *Agronomy Journal* 94, no. 1: 38–47.

Carter, M. R. and B. A. Stewart. 1996. *Structure and Organic Matter Storage in Agricultural Soils.* Boca Raton, FL: CRC Press, Lewis Publishers.

Castellano, M. J., K. E. Mueller, D. C. Olk, J. E. Sawyer, and J. Six. 2015. Integrating plant litter quality, soil organic matter stabilization, and the carbon saturation concept. *Global Change Biology* 21, no. 9: 3200–9.

Chatterjee, A., K. Cooper, A. Klaustermeier, R. Awale, and L. J. Cihacek. 2016. Does crop species diversity influence soil carbon and nitrogen pools? *Agronomy Journal* 108, no. 1: 427–32. doi.org/10.2134/agronj2015.0316.

Chivenge, P. P., H. K. Murwira, K. E. Giller, P. Mapfumo, and J. Six. 2007. Long-term impact of reduced tillage and residue management on soil carbon stabilization: Implications for conservation agriculture on contrasting soils. *Soil and Tillage Research* 94, no. 2: 328–37.

Christopher, S. F., R. Lal, and U. Mishra. 2009. Regional study on no-till effects on carbon sequestration in the Midwestern United States. *Soil Science Society of America* 73, no. 1: 207–16.

Conant, R. T. and K. Paustian. 2002. Spatial variability of soil organic carbon in grasslands: Implications for detecting change at different scales. *Environmental Pollution* 116: S127–S135.

Corbeels, M., R. L. Marchão, M. S. Neto et al. 2016. Evidence of limited carbon sequestration in soils under no-tillage systems in the Cerrado of Brazil. *Scientific Reports* 6: 21450. doi.org/10.1038/srep21450.

Corsi, S., T. Friedrich, A. Kassam, M. Pisante, and J. de M. Sà. 2012. *Soil Organic Carbon Accumulation and Greenhouse Gas Emission Reductions from Conservation Agriculture: A Literature Review.* Rome, Italy: Food and Agriculture Organization of the United Nations (FAO).

Costello, M. J., R. M. May, and N. E. Stork. 2013. Can we name Earth's species before they go extinct? *Science* 339, no. 6118: 413–6.

Cotrufo, M. F., M. D. Wallenstein, C. M. Boot, K. Denef, and E. Paul. 2013. The microbial efficiency-matrix stabilization (MEMS) framework integrates plant litter decomposition with soil organic matter stabilization: Do labile plant inputs form stable soil organic matter? *Global Change Biology* 19, no. 4: 988–95.

Crawford, M. H., V. Rincon-Florez, A. Balzer et al. 2015. Changes in the soil quality attributes of continuous no-till farming systems following a strategic tillage. *Soil Research* 53, no. 3: 263–73.

Crovetto, C. 1996. *Stubble over the Soil: The Vital Role of Plant Residue in Soil Management to Improve Soil Quality.* Madison, WI: American Society of Agronomy.

Dang, Y. P., P. W. Moody, M. J. Bell et al. 2015. Strategic tillage in no-till farming systems in Australia's northern grains-growing regions: II. Implications for agronomy, soil and environment. *Soil and Tillage Research* 152: 115–23.

Darmody, R. G. and T. R. Peck. 1997. Soil organic carbon changes through time at the University of Illinois Morrow Plots. In *Soil Organic Matter in Temperate Agroecosystems: Long-Term Experiments in North America*, ed. E. A. Paul, K. H. Paustian, E. T. Elliot, and C. V. Cole. Boca Raton, FL: CRC Press.

De Luca, T. H. and C. A. Zabinski. 2011. Prairie ecosystems and the carbon problem. *Frontiers in Ecology and the Environment* 9, no. 7: 407–13.

Derpsch, R., T. Friedrich, A. Kassam, and L. Hongwen. 2010. Current status of adoption of no-till farming in the world and some of its main benefits. *International Journal of Agricultural and Biological Engineering* 3, no. 1: 1–25.

Derpsch, R., A. J. Franzluebbers, S. W. Duiker et al. 2014. Why do we need to standardize no-tillage research? *Soil and Tillage Research* 137: 16–22.

de Vries, F. T., E. Hoffland, N. van Eekeren, L. Brussaard, J. Bloem et al. 2006. Fungal/bacterial ratios in grasslands with contrasting nitrogen management. *Soil Biology and Biochemistry* 38, no. 8: 2092–103.

Diamond, J. M. 2005. *Collapse: How Societies Choose to Fail or Succeed. Revised Edition.* New York: Penguin Books.

Dick, R. P., D. R. Thomas, and J. J. Halvorson. 1996. Standardized methods, sampling, and sample retreatment. In *Methods for Assessing Soil Quality*, ed. J. W. Doran and A. J. Jones, 107–21. Madison, WI: Soil Science Society of America.

Dumanski, J., R. Peiretti, J. R. Benites, D. McGarry, and C. Pieri. 2006. The paradigm of conservation agriculture. *Proceedings of World Association of Soil and Water Tillage*, no. P1–P7: 58–64.

Dungait, J. A. J., D. W. Hopkins, A. S. Gregory and A. P. Whitmore. 2012. Soil organic matter turnover is governed by accessibility not recalcitrance. *Global Change Biology* 18, no. 6: 1781–96.

Eagle, A., L. Olander, L. R. Henry, K. Haugen-Kozyra, N. Millar, and G. P. Robertson. 2012. *Greenhouse Gas Mitigation Potential of Agricultural Land Management in the United States: A Synthesis of the Literature, 3rd ed.* Durham, NC: Nicholas Institute for Environmental Policy Solutions, Duke University.

Ellert, B. H. and J. R. Bettany. 1995. Calculation of organic matter and nutrients stored in soils under contrasting management regimes. *Canadian Journal of Soil Science* 75, no. 4: 529–38.

Ellert, B. H. and H. H. Janzen. 1999. Short-term influence of tillage on CO_2 fluxes from a semi-arid soil on the Canadian Prairies. *Soil and Tillage Research* 50, no. 1: 21–32.

Ellert, B. H., H. H. Janzen and B. G. McConkey. 2000. Measuring and comparing soil carbon storage. In *Assessment Methods for Soil Carbon,* ed. R. Lal, J. M. Kimble, R. F. Follett, and B. A. Stewart, 131–146. Boca Raton, FL: CRC Press, Lewis Publishers.

Ellert, B. H., H. H. Janzen, and T. Entz. 2002. Assessment of a method to measure temporal change in soil carbon storage. *Soil Science Society of America Journal* 66, no. 5: 1687–95.

Elton, C. S. 1958. *The Ecology of Invasion by Animals and Plants.* Chicago: University of Chicago Press.

Erenstein, O. 2012. Conservation agriculture-based technologies and the political economy: Lessons from South Asia. In *Contested Agronomy: Agricultural Research in a Changing World*, ed. J. Sumberg and J. Thompson. London: Routledge.

Erenstein, O., K. Sayre, P. Wall, J. Hellin, and J. Dixon. 2012. Conservation agriculture in maize- and wheat-based systems in the (sub)tropics: Lessons from adaptation initiatives in South Asia, Mexico, and Southern Africa. *Journal of Sustainable Agriculture* 36, no. 2: 180–206.

FAO (Food and Agriculture Organization of the United Nations). FAO, 2008. Conservation Agriculture. 2008-07-08 http://www.fao.org/ag/ca/index.html.

—. 2011. Save and grow: A policymaker's guide to the sustainable intensification of smallholder crop production. Rome: Food and Agriculture Organization of the United Nations. http://www.fao.org/docrep/014/i2215e/i2215e00.htm.

Farooq, M. and K. Siddique. 2015. Conservation agriculture: concepts, brief history, and impacts on agricultural systems. In *Conservation Agriculture*, ed. M. Farooq and K. Siddique, 3–17 Switzerland: Springer International Publishing.

Foley, J. A., N. Ramankutty, K. A. Brauman et al. 2011. Solutions for a cultivated planet. *Nature* 478, no. 7369: 337–342. doi.org/10.1038/nature10452.

Franzluebbers, A. J. 2010. Will we allow soil carbon to feed our needs? *Carbon Management* 1, no. 2: 237–51.

Franzluebbers, A. J., F. M. Hons and D. A. Zuberer. 1998. In situ and potential CO_2 evolution from a Fluventic Ustochrept in southcentral Texas as affected by tillage and cropping intensity. *Soil and Tillage Research* 47, no. 3: 303–8.

Friedrich, T., R. Derpsch and A. H. Kassam. 2012. Overview of the global spread of Conservation Agriculture. *Field Actions Science Reports* no. 6. http://factsreports.revues.org/1941#text.

Fuentes, M., C. Hidalgo, J. Etchevers et al. 2012. Conservation agriculture, increased organic carbon in the top-soil macro-aggregates and reduced soil CO_2 emissions. *Plant and Soil* 355, no. 1–2: 183–97.

Galy, V., B. Peucker-Ehrenbrink, and T. Eglinton. 2015. Global carbon export from the terrestrial biosphere controlled by erosion. *Nature* 521, no. 7551: 204–7.

Garcia, J. P., C. S. Wortmann, M. Mamo, R. Drijber, and D. Tarkalson. 2007. One-time tillage of no-till: Effects on nutrients, mycorrhizae, and phosphorus uptake. *Agronomy Journal* 99, no. 4: 1093–103.

Garnett, T., M. C. Appleby, A. Balmford et al. 2013. Sustainable intensification in agriculture: Premises and policies. *Science* 341, no. 6141: 33–4. doi.org/10.1126/science.1234485.

Gattinger, A., J. Jawtusch, A. Muller, and P. Mäder. 2011. No-till agriculture—A climate smart solution? Climate Change and Agriculture Report No. 2, MISEREOR e.V. Aachen, Germany.

Gebhardt, M. R., T. C. Daniel, E. E. Schweizer, and R. R. Allmaras. 1985. Conservation tillage. *Science* 230, no. 4726: 625–30.

Gessner, M. O., C. M. Swan, C. K. Dang, B. G. McKie, R. D. Bardgett, D. H. Wall, and S. Hättenschwiler. 2010. Diversity meets decomposition. *Trends in Ecology and Evolution* 25, no. 6: 372–80.

Giller, K. E., J. A. Andersson, M. Corbeels et al. 2015. Beyond conservation agriculture. *Frontiers in Plant Science* 6, no. October: 1–14. doi.org/10.3389/fpls.2015.00870.

Giller, K. E., P. Tittonell, M. C. Rufino et al. 2011. Communicating complexity: Integrated assessment of trade-offs concerning soil fertility management within African farming systems to support innovation and development. *Agricultural Systems* 104, no. 2: 191–203. doi.org/10.1016/j.agsy.2010.07.002.

Giller, K. E., E. Witter, M. Corbeels, and P. Tittonell. 2009. Conservation agriculture and smallholder farming in Africa: The heretics' view. *Field Crops Research* 114, no. 1: 23–34. doi.org/10.1016/j.fcr.2009.06.017.

Girla, D. 2014. Energy Status of Soil Agro-ecosystems. In *Soil as World Heritage*, ed. D. Dent, 57–9. Dordrecht: Springer.

Goddard, T., M. Zoebisch, Y. Gan, W. Ellis, A. Watson, and S. Sombatpanit, eds. 2008. *No-Till Farming Systems*. Special Publication No. 3, World Association of Soil and Water Conservation (WASWC), Bangkok.

Goudriaan, J., J. J. R. Groot, and P. W. J. Uithol. 2001. Productivity of agro-ecosystems. In *Terrestrial Global Productivity*, ed. J. Roy, B. Saugier, and H. A. Mooney, 301–13. San Diego: Academic Press.

Govaerts, B., N. Verhulst, A. Castellanos-Navarrete, K. D. Sayre, J. Dixon, and L. Dendooven. 2009. Conservation agriculture and soil carbon sequestration: Between myth and farmer reality. *Critical Reviews in Plant Sciences* 28, no. 3: 97–122.

Grace, P. R., J. Antle, P. K. Aggarwal, S. Ogle, K. Paustian, and B. Basso. 2012. Soil carbon sequestration rates and associated economic costs for farming systems of the Indo-Gangetic Plain. *Agriculture, Ecosystems & Environment* 146, no. 1: 137–46.

Grandy, A. S., G. P. Robertson, and K. D. Thelen. 2006. Do productivity and environmental trade-offs justify periodically cultivating no-till cropping systems? *Agronomy Journal* 98, no. 6: 1377–83.

Gregorich, E. G., H. Janzen, B. H. Ellert et al. 2017. Litter decay controlled by temperature, not soil properties, affecting future soil carbon. *Global Change Biology* 23: 1725–34.

Hall, A. 1998. Sustainable agriculture and conservation tillage: Managing the contradictions. *Canadian Review of Sociology/Revue Canadienne de Sociologie* 35, no. 2: 221–51.

Harvey, C. A., M. Chacón, C. I. Donatti et al. 2013. Climate-smart landscapes: Opportunities and challenges for integrating adaptation and mitigation in tropical agriculture. *Conservation Letters* 7, no. 2: 77–90. doi.org/10.1111/conl.12066.

Hatfield, J. L., K. J. Boote, B. A. Kimball et al. 2011. Climate impacts on agriculture: Implications for crop production. *Agronomy Journal* 103, no. 2: 351–70. doi.org/10.2134/agronj2010.0303.

Helgason, B. L., E. G. Gregorich, H. H. Janzen, B. H. Ellert, N. Lorenz, and R. P. Dick. 2014. Long-term microbial retention of residue C is site-specific and depends on residue placement. *Soil Biology and Biochemistry* 68: 231–40.

Hobbs, P. R. 2007. Conservation agriculture: What is it and why is it important for future sustainable food production? *The Journal of Agricultural Science* 145, no. 2: 127–37.

Hobbs, P. and B. Govaerts. 2010. How conservation agriculture can contribute to buffering climate change. In *Climate Change and Crop Production,* ed. M. Reynolds, 177–99. Wallingford, Oxfordshire, UK: CABI.

Hobbs, P. R., K. Sayre, and R. Gupta. 2008. The role of conservation agriculture in sustainable agriculture. *Philosophical Transactions of the Royal Society B* 363, no. 1491, 543–55. doi.org/10.1098/rstb.2007.2169.

Houghton, R. A., J. L. Hackler, and K. T. Lawrence. 1999. The U.S. carbon budget: Contributions from land-use change. *Science* 285, no. 5427: 574–7.

Hudson, B. D. 1994. Soil organic matter and available water capacity. *Journal of Soil and Water Conservation.* 49, no. 2: 189–94.

Hutchinson, J. J., C. A. Campbell, and R. L. Desjardins. 2007. Some perspectives on carbon sequestration in agriculture. *Agricultural and Forest Meteorology* 142, no. 2: 288–302. doi.org/10.1016/j.agrformet.2006.03.030.

IPCC. 2000. *Land Use, Land-Use Change, and Forestry*, ed. R. T. Watson, I. R. Noble, B. Bolin, N. H. Ravindranath, D. J. Verardo, and D. J. Dokken A Special Report of the Intergovernmental Panel on Climate Change. Cambridge University Press.

Jackson, W. 2011. *Nature as Measure: The Selected Essays of Wes Jackson.* Berkeley, CA: Counterpoint.

Janzen, H. H. 2004. Carbon cycling in earth systems—A soil science perspective. *Agriculture, Ecosystems & Environment* 104, no. 3: 399–417.

—. 2005. Soil carbon: A measure of ecosystem response in a changing world? *Canadian Journal of Soil Science* 85, Special Issue: 467–80. doi: 10.4141/S04-081.

—. 2006. The soil carbon dilemma: Shall we hoard it or use it? *Soil Biology and Biochemistry* 38, no. 3: 419–24.

—. 2015. Beyond carbon sequestration: Soil as conduit of solar energy. *European Journal of Soil Science* 66, no. 1: 19–32. http://onlinelibrary.wiley.com/doi/10.1111/ejss.12194/full.

Janzen, H. H., C. A. Campbell, R. C. Izaurralde et al. 1998. Management effects on soil C storage on the Canadian prairies. *Soil and Tillage Research* 47, no. 3–4: 181–95.

Jat, R. A., K. L. Sahrawat, A. H. Kassam et al. 2014. Conservation agriculture for sustainable and resilient agriculture: Global status, prospects and challenges. In *Conservation Agriculture: Global Prospects and Challenges,* ed R. A. Jat, K. L. Sahrawat, and A. H. Kassam, 1–25. Wallingford, UK: CAB International.

Jenkinson, D. S. and J. H. Rayner. 1977. The turnover of soil organic matter in some of the Rothamsted classical experiments. *Soil Science* 123, no. 5: 298–305.

Kassam, A., R. Derpsch and T. Friedrich. 2014. Global achievements in soil and water conservation: The case of conservation agriculture. *International Soil and Water Conservation Research* 2, no. 1: 5–13.

Kassam, A. and T. Friedrich. 2012. An ecologically sustainable approach to agricultural production intensification: Global perspectives and developments. *Field Actions Science Reports* no. 6. http://factsreports.revues.org/1382.

Kassam, A., T. Friedrich, F. Shaxson, and J. Pretty. 2009. The spread of conservation agriculture: Justification, sustainability and uptake. *International Journal of Agricultural Sustainability* 7, no. 4: 292–320.

Kätterer, T., M. A. Bolinder, O. Andrén, H. Kirchmann, and L. Menichetti. 2011. Roots contribute more to refractory soil organic matter than above-ground crop residues as revealed by a long-term field experiment. *Agriculture, Ecosystems and Environment* 141, no. 1: 184–92.

Kemper, W. D., N. N. Schneider, and T. R. Sinclair. 2011. No-till can increase earthworm populations and rooting depths. *Journal of Soil and Water Conservation.* 66, no. 1: 13A–17A.

King, F. H. 1907. *The Soil: Its Nature, Relations, and Fundamental Principles of Management.* London: The MacMillan Company.

Kirkegaard, J. A., M. K. Conyers, J. R. Hunt, C. A. Kirkby, M. Watt, and G. J Rebetzke. 2014. Sense and nonsense in conservation agriculture: Principles, pragmatism and productivity in Australian mixed farming systems. *Agriculture, Ecosystems and Environment* 187: 133–45. doi.org/10.1016/j.agee.2013.08.011.

Kladivko, E. J. 2001. Tillage systems and soil ecology. *Soil and Tillage Research* 61, no. 1: 61–76.

Kladivko, E. J., M. J. Helmers, L. J. Abendroth et al. 2014. Standardized research protocols enable trans-disciplinary research of climate variation impacts in corn production systems. *Journal of Soil and Water Conservation* 69, no. 6: 532–42. doi.org/10.2489/jswc.69.6.532.

Kleber, M. 2010. What is recalcitrant soil organic matter? *Environmental Chemistry* 7, no. 4: 320–32.

Kleber, M. and M. G. Johnson. 2010. Advances in understanding the molecular structure of soil organic matter: Implications for interactions in the environment. *Advances in Agronomy* 106: 77–142.

Kravchenko, A. N. and G. P. Robertson. 2011. Whole-profile soil carbon stocks: The danger of assuming too much from analyses of too little. *Soil Science Society of America Journal* 75, no. 1: 235–40.

Kuhn, N. J., T. Hoffmann, W. Schwanghart, and M. Dotterweich. 2009. Agricultural soil erosion and global carbon cycle: Controversy over? *Earth Surface Processes and Landforms* 34, no. 7: 1033–8.

Kuhn, N. J., Y. Hu, L. Bloemertz, J. He, H. Li, and P. Greenwood. 2016. Conservation tillage and sustainable intensification of agriculture: Regional vs. global benefit analysis. *Agriculture, Ecosystems and Environment* 216: 155–65.

Kunkel, K. E., K. Andsager, and D. R. Easterling. 1999. Long-term trends in extreme precipitation events over the conterminous United States and Canada. *Journal of Climate* 12, no. 8: 2515–27.

Lal, R. 2004. Soil carbon sequestration to mitigate climate change. *Geoderma* 123: 1–22.

—. 2005. Soil erosion and carbon dynamics. *Soil Tillage and Research* 81, no. 2: 137–42.

—. 2006. Managing soils for feeding a global population of 10 billion. *Journal of the Science of Food and Agriculture* 86, no. 14: 2273–84.

—. 2008. Carbon sequestration. *Philosophical Transactions of the Royal Society B: Biological Sciences* 363, no. 1492: 815–30.

—. 2009. Sequestering atmospheric carbon dioxide. *Critical Reviews in Plant Sciences* 28, no. 3: 90–6.

—. 2011. Sequestering carbon in soils of agro-ecosystems. *Food Policy* 36: S33–9.

—. 2013. Intensive agriculture and the soil carbon pool. In *Combating Climate Change: An Agricultural Perspective,* ed M.S Kang and S.S. Banga, 59–72. Boca Raton, FL: CRC Press, Taylor and Francis Group.

—. 2014. Climate strategic soil management. *Challenges* 5, no. 1: 43–74. doi.org/10.3390/challe5010043.

—. 2015a. Sequestering carbon and increasing productivity by conservation agriculture. *Journal of Soil and Water Conservation* 70, no. 3: 55A–62A. doi.org/10.2489/jswc.70.3.55A.

—. 2015b. A system approach to conservation agriculture. *Journal of Soil and water Conservation* 70, no. 4: 82A–88A. doi.org/10.2489/jswc.70.4.82A.

—. 2015c. Restoring soil quality to mitigate soil degradation. Sustainability 7, no. 5: 5875–95.

Lal, R. and J. M. Kimble. 1997. Conservation tillage for carbon sequestration. *Nutrient Cycling in Agroecosystems* 49, no. 1: 243–53.

Lal, R., J. A. Delgado, P. M. Groffman, N. Millar, C. Dell, and A. Rotz. 2011. Management to mitigate and adapt to climate change. *Journal of Soil and Water Conservation* 66, no. 4: 276–85.

Lal, R., J. A. Delgado, J. Gulliford, D. Nielsen, C. W. Rice, and R. S. Van Pelt. 2012. Adapting agriculture to drought and extreme events. *Journal of Soil and Water Conservation* 67, no. 6: 162A–6A.

Lal, R., R. F. Follett, and J. M. Kimble. 2003. Achieving soil carbon sequestration in the US: A challenge to policy makers. *Soil Science* 168, no. 12: 827–45.

Lal, R., D. C. Reicosky, and J. D. Hanson. 2007. Evolution of the plow over 10,000 years and the rationale for no-till farming. *Soil Tillage Re*search 93, no. 1: 1–12.

Lam, S. K., D. Chen, A. R. Mosier and R. Roush. 2013. The potential for carbon sequestration in Australian agricultural soils is technically and economically limited. *Scientific Reports* 3. doi.org/10.1038/srep02179.

Larson, W. E., M. J. Lindstrom, and T. E. Schumacher. 1997. The role of severe storms in soil erosion: A problem needing consideration. *Journal of Soil Water Conservation* 52, no. 2: 90–5.

Lehmann, J. 2007. A handful of carbon. *Nature* 447, no. 7141: 143–4.

Lehman, R. M., C. A. Cambardella, D. E. Stott et al. 2015. Understanding and enhancing soil biological health: The solution for reversing soil degradation. *Sustainability* 7, no. 1: 988–1027. doi.org/10.3390/su7010988.

Lehmann, J. and M. Kleber. 2015. The contentious nature of soil organic matter. *Nature* 528, no. 7580: 60–8. doi.org/10.1038/nature16069.

Le Quéré, R. M. Andrew, J. G. Canadell et al. 2016. Global carbon budget 2016. *Earth System Science Data* 8: 605–49.

Lindstrom, M. J., W. W. Nelson, T. E. Schumacher, and G. D. Lemme. 1990. Soil movement by tillage as affected by slope. *Soil and Tillage Research* 17, no. 3: 255–64.

Liu, H., L. C. Carvalhais, M. Crawford, Y. P. Dang, P. G. Dennis, and P. M. Schenk. 2016a. Strategic tillage increased the relative abundance of acidobacteria but did not impact overall soil microbial properties of a 19-year no-till Solonetz. *Soil and Tillage Research* 52, no. 7: 1021–35.

Liu, H., L. C. Carvalhais, V. Rincon-Florez et al. 2016b. One-time strategic tillage does not cause major impacts on soil microbial properties in a no-till Calcisol. *Soil and Tillage Research* 158: 91–99.

Liu, H., M. Crawford, L. C. Carvalhais, Y. P. Dang, P. G. Dennis, and P. M. Schenk l. 2016c. Strategic tillage on a Grey Vertosol after fifteen years of no-till management had no short-term impact on soil properties and agronomic productivity. *Geoderma* 267: 146–153.

Loveland, P. J., F. Conen, and B. Van Wesemael. 2014. Commentary on the impact of Batjes (1996). *European Journal of Soil Science* 65, no. 1: 4–9.

López-Garrido, R., E. Madejón, J. M. Murillo, and F. Moreno. 2011. Soil quality alteration by mouldboard ploughing in a commercial farm devoted to no-tillage under Mediterranean conditions. *Agriculture, Ecosystem and Environment* 140, no. 1: 182–90.

Luo, Z., E. Wang, and O. J. Sun. 2010. Can no-tillage stimulate carbon sequestration in agricultural soils? A meta-analysis of paired experiments. *Agriculture Ecosystems and Environment* 139, no. 1: 224–31.

Lützow, M. V, I. Kogel-Knabner, K. Ekschmitt, E. Matzner, G. Guggenberger, B. Marschner and H. Flessa. 2006. Stabilization of organic matter in temperate soils: Mechanisms and their relevance under different soil conditions—A review. *European Journal of Soil Science* 57, no. 4: 426–45.

Manley, J., G. C. van Kooten, K. Moeltner, and D. W. Johnson. 2005. Creating carbon offsets in agriculture through no-till cultivation: A meta-analysis of costs and carbon benefits. *Climatic Change* 68, no. 1–2: 41–65.

Mannering, J. V. and C. R. Fenster. 1983. What is conservation tillage? *Journal of Soil and Water Conservation* 38, no. 3: 140–3.

Milne, E. M. Sessay, K. Paustian et al. 2010. Towards a standardized system for the reporting of carbon benefits in sustainable land management projects. Grassland carbon sequestration: Management, policy and economics (Proceedings of the Workshop on the role of grassland carbon sequestration in the mitigation of climate change, Rome, April 2009). *Integrated Crop Management* 11: 105–17.

Milne, E., S. A. Banwart, E. Noellemeyer et al. 2015. Soil carbon, multiple benefits. *Environmental Development* 13: 33–8.

Mitchell, J. P., L. M. Carter, D. C. Reicosky et al. 2016. A history of tillage in California's Central Valley. *Soil and Tillage Research* 157: 52–64.

Montgomery, D. R. 2007a. *Dirt: The Erosion of Civilizations*. Berkeley: University of California Press.

Montgomery, D. R. 2007b. Soil erosion and agricultural sustainability. *Proceedings of the National Academy of Sciences of the United States of America* 104, no. 33: 13,268–72.

Mortenson, M. C., G. E. Schuman, and L. J. Ingram. 2004. Carbon sequestration in rangelands interseeded with yellow-flowering alfalfa (Medicago sativa ssp. falcata). *Environmental Management* 33, no. 1: S475–81.

Murphy, B. W. 2014. *Soil Organic Matter and Soil Function—Review of the Literature and Underlying Data*. Australian Government, Department of the Environment, Canberra, Australia.

Nakhauka, E. B. 2009. Agricultural biodiversity for food and nutrient security: The Kenyan perspective. *International Journal of Biodiversity and Conservation* 1, no. 7: 208–14.

Nielsen, U. N., E. Ayres, D. H. Wall, and R. D. Bardgett. 2011. Soil biodiversity and carbon cycling: A review and synthesis of studies examining diversity-function relationships. *Global Change Biology* 62, no. 1: 105–16.

Noellemeyer, E. and J. Six. 2015. Basic principles of soil carbon management for multiple ecosystem benefits. In *Soil Carbon: Science, Management and Policy for Multiple Benefits*, ed S.A. Banwart, E. Noellemeyer, and E. Milne, 265–76. Wallingford, UK: CAB International.

Odell, R. T., S. W. Melsted, and W. M. Walker. 1984. Changes in organic carbon and nitrogen of Morrow Plot soils under different treatments. *Soil Science* 137, no. 3: 160–71.

Odum, E. P. 1969. The strategy of ecosystem development. *Science* 164, no. 3877: 262–70.

Ogle, S. M., F. J. Breidt, and K. Paustian. 2005. Agricultural management impacts on soil organic carbon storage under moist and dry climatic conditions of temperate and tropical regions. *Biogeochemistry* 72, no. 1: 87–121.

Ogle, S. M., A. Swan, and K. Paustian. 2012. No-till management impacts on crop productivity, carbon input and soil carbon sequestration. *Agriculture, Ecosystems and Environment* 149: 37–49.

Oldfield, E. E., S. A. Wood, C. A. Palm, and M. A. Bradford. 2015. How much SOM is needed for sustainable agriculture? *Frontiers in Ecology and the Environment* 13, no. 10: 527.

Olson, K. R., M. M. Al-Kaisi, R. Lal, and B. Lowery. 2014. Experimental consideration, treatments, and methods in determining soil organic carbon sequestration rates. *Soil Science Society of America Journal* 78, no. 2: 348–60. doi.org/10.2136/sssaj2013.09.0412.

Palm, C., H. Blanco-Canqui, F. De Clerck, L. Gatere, and P. Grace. 2014. Conservation agriculture and eco-system services: An overview. *Agriculture, Ecosystems and Environment* 187: 87–105. doi.org/10.1016/j.agee.2013.10.010.

Pascual, U., M. Termansen, K. Hedlund et al. 2015. On the value of soil biodiversity and ecosystem services. *Ecosystem Services* 15: 11–8.

Paul, E. A., K. Paustian, E. T. Elliott, and C. V. Cole. 1997. *Soil Organic Matter in Temperate Agroecosystems: Long-Term Experiments in North America.* Boca Raton, FL: CRC Press.

Paustian, K., O. Andrén, H. H. Janzen et al. 1997. Agricultural soil as a sink to mitigate CO_2 emissions. *Soil Use and Management* 13, 230–244. doi.org/10.1111/j.1475-2743.1997.tb00594.x.

Paustian, K., J. Lehmann, S. Ogle, D. Reay, G. P. Robertson, and P. Smith. 2016. Climate-smart soils. *Nature* 532, no. 7597: 49–57. doi.org/10.1038/nature17174.

Pelletier, N., E. Audsley, S. Brodt et al. 2011. Energy intensity of agriculture and food systems. *Annual Review of Environment and Resources* 36, no. 1: 223–46.

Pereira, H. M., L. M. Navarro and I. S. Martins. 2012. Global biodiversity change: The bad, the good, and the unknown. *Annual Review of Environment and Resources* 37, no. 1: 25–50.

Peterson, G. A., A. D. Halvorson, J. L. Havlin, O. Jones, D. J. Lyon, and D. L. Tanaka. 1998. Reduced tillage and increasing cropping intensity in the Great Plains conserves soil C. *Soil and Tillage Research* 47, no. 3: 207–18.

Philibert, A., C. Loyce, and D. Makowski. 2012. Assessment of the quality of meta-analysis in agronomy. *Agriculture, Ecosystems and Environment* 148: 72–82.

Pimentel, D., C. Harvey, P. Resosudarmo et al. 1995. Environmental and economic cost of soil erosion and conservation benefits. *Science* 267, no. 5201: 1117–23.

Pittelkow, C. M., X. Liang, B. A. Linquist et al. 2015. Productivity limits and potentials of the principles of conservation agriculture. *Nature* 517, no. 7534: 365–68. doi.org/10.1038/nature13809.

Poeplau, C. and A. Don. 2015. Carbon sequestration in agricultural soils via cultivation of cover crops—A meta-analysis. *Agriculture, Ecosystems and Environment* 200: 33–41.

Post, W. M., R. C. Izaurralde, T. O. West, M. A. Liebig, and A. W. King. 2012. Management opportunities for enhancing terrestrial carbon dioxide sinks. *Frontiers in Ecology and the Environment* 10, no. 10: 554–61.

Powlson, D. S., P. C. Brookes, A. P. Whitmore, K. W. T. Goulding, and D. W. Hopkins. 2011a. Soil organic matters. *European Journal of Soil Science* 62, no. 1: 1–4.

Powlson, D. S., Z. Cai, and P. Lemanceau. 2015. Soil carbon dynamics and nutrient cycling. In *Soil Carbon: Science, Management and Policy for Multiple Benefits,* ed S. A. Banwart, E. Noellemeyer, and E. Milne, 98–107. Wallingford, UK: CAB International.

Powlson, D. S., C. M. Stirling, M. L. Jat et al. 2014. Limited potential of no-till agriculture for climate change mitigation. *Nature Climate Change* 4, no. 8: 678–83. doi.org/10.1038/nclimate2292.

Powlson, D. S., C. M. Stirling, C. Thierfelder, R. P. White, and M. L. Jat. 2016. Does conservation agriculture deliver climate change mitigation through soil carbon sequestration in tropical agro-ecosystems? *Agriculture, Ecosystems and Environment* 220: 164–74.

Powlson, D. S., A. P. Whitmore, and K. W. T. Goulding. 2011b. Soil carbon sequestration to mitigate climate change: A critical re-examination to identify the true and the false. *European Journal of Soil Science* 62, no. 1: 42–55.

Pretty, J. and Z. P. Bharucha. 2014. Sustainable intensification in agricultural systems. *Annals of Botany* 114, no. 8: 1571–1596. doi.org/10.1093/aob/mcu205.

Quincke, J. A., C. S. Wortmann, M. Mamo, T. Franti, and R. A. Drijber. 2007. Occasional tillage of no-till systems: Carbon dioxide flux and changes in total and labile soil organic carbon. *Agronomy Journal* 99, no. 4: 1158–68.

Rawls, W. J., Y. A. Pachepsky, J. C. Ritchie, T. M. Sobecki, and H. Bloodworth. 2003. Effect of soil organic carbon on soil water retention. *Geoderma* 116, no. 1: 61–76.

Reeder, R. 2000. *Conservation Tillage Systems and Management: Crop Residue Management with No-Till, Ridge-Till, Mulch-Till And Strip-Till.* Ames, IA: MidWest Plan Service, Iowa State University.

Rees, R. M., I. J. Bingham, J. A. Baddeley, and C. A. Watson. 2005. The role of plants and land management in sequestering soil carbon in temperate arable and grassland ecosystems. *Geoderma* 128, no. 1: 130–54.

Reicosky, D. C. 2008. Carbon sequestration and environmental benefits from no-till systems. In *No-Till Farming Systems. Special Publication No. 3,* ed. T. Goddard, M. A. Zoebisch, Y. T. Gan, W. Ellis, A. Watson, and S. Sombatpanit, 43–58 Bangkok, Thailand: World Association of Soil and Water Conservation.

—. D. C. 2015. Conservation tillage is not conservation agriculture. *Journal of Soil and Water Conservation* 70, no. 5: 103A–8A.

Reicosky, D. C. and R. R. Allmaras. 2003. Advances in tillage research in North American cropping systems. *Journal of Crop Production* 8, no. 1–2: 75–125.

Reicosky, D. C. and M. J. Lindstrom. 1993. Fall tillage method: Effect on short-term carbon dioxide flux from soil. *Agronomy Journal* 85, no. 6: 1237–43.

Reicosky, D. C., T. J. Sauer, and J. L. Hatfield. 2011. Challenging balance between productivity and environmental quality: Tillage impacts. In *Soil Management: Building a Stable Base for Agriculture*, ed. J. L. Hatfield and T. J. Sauer, 13–38. Portland, OR: American Society of Agronomy and Soil Science Society of America.

Reicosky, D. C. and K. E. Saxton. 2007a. The benefits of no-tillage. In *No-tillage Seeding in Conservation Agriculture, 2nd ed.*, ed. C. J. Baker, K. E. Saxton, W. R. Ritchie et al., 11–20. Rome, Italy: FAO and CAB International.

—. 2007b. Reduced environmental emissions and carbon sequestration. In *No-tillage Seeding in Conservation Agriculture, 2nd ed.*, ed. C. J. Baker, K. E. Saxton, W. R. Ritchie et al., 257–67. Rome, Italy: FAO and CAB International.

Reusser, P., P. Bierman, and D. Rood. 2015. Quantifying human impacts on rates of erosion and sediment transport at a landscape scale. *Geology* 43, no. 2: 171–4. doi.org/10.1130/G36272.1.

Rincon-Florez, V. A., Y. P. Dang, M. H. Crawford, P. M. Schenk, and L. C. Carvalhais. 2016a. Occasional tillage has no effect on soil microbial biomass, activity and composition in Vertisols under long-term no-till. *Biology and Fertility of Soils* 52, no. 2: 191–202.

Rincon-Florez, V. A., C. Ng, Y. P. Dang, P. M. Schenk, and L. C. Carvalhais. 2016b. Short-term impact of an occasional tillage on microbial communities in a Vertosol after 43 years of no-tillage or conventional tillage. *European Journal of Soil Biology* 74: 32–8.

Robert, M. 2001. *Soil Carbon Sequestration for Improved Land Management*. World Soil Resources Reports 96. Rome: Food and Agriculture Organization of the United Nations.

Robertson, G. P., D. C. Coleman, C. S. Bledsoe et al. (ed.). 1999. *Standard Soil Methods for Long-Term Ecological Research*. New York: Oxford University Press.

Rochette, P. and D. A. Angers. 1999. Soil surface carbon dioxide fluxes induced by spring, summer, and fall moldboard plowing in a sandy loam. *Soil Science Society of America Journal* 63, no. 3: 621–8.

Rosenzweig, C., F. N. Tubiello, R. Goldberg, E. Mills, and J. Bloomfield. 2002. Increased crop damage in the US from excess precipitation under climate change. *Global Environmental Change* 12, no. 3: 197–202.

Sakrabani, R., L. K. Deeks, M. G. Kibblewhite, and K. Ritz. 2012. Impacts of agriculture upon soil quality. In *Environmental Impacts of Modern Agriculture,* ed. R. E Hester and R. M. Harrison, 35–56. London, UK: Royal Society of Chemistry.

Sanderman, J. and A. Chappell. 2013. Uncertainty in soil carbon accounting due to unrecognized soil erosion. *Global Change Biology* 19, no. 1: 264–72.

Sanderman, J., T. Hengl, and G. J. Fiske. 2017. Soil carbon debt of 12,000 years of human land use. *Proceedings of the National Academy of Sciences of the United States of America* 114, no. 36: 9575–80.

Schlesinger, W. H. 1985. Changes in soil carbon storage and associated properties with disturbance and recovery. In *The Changing Carbon Cycle: A Global Analysis,* ed. J. R. Trabalha and D. E. Reichle, 194–220. New York: Springer-Verlag.

Schlesinger, W. H. 2000. Carbon sequestration in soils: Some cautions amidst optimism. *Agriculture, Ecosystems and Environment* 82, no. 1–3: 121–7.

Schramski, J. R., D. K. Gattie, and J. H. Brown. 2015. Human domination of the biosphere: Rapid discharge of the earth-space battery foretells the future of humankind. *Proceedings of the National Academy of Sciences* 112, no. 31: 9511–7.

Sedjo, R. and B. Sohngen. 2012. Carbon sequestration in forests and soils. *Annual Review of Resource Economics* 4, no. 1: 127–44.

Semenov, V. M., A. S. Tulina, N. A. Semenova, and L. A. Ivannikova. 2013. Humification and nonhumification pathways of the organic matter stabilization in soil: A review. *Eurasian Soil Science* 46, no. 4: 355–68.

Shahidi B. M. R., M. Dyck, and S. S. Malhi. 2014. Carbon dioxide emissions from tillage of two long-term no-till Canadian prairie soils. *Soil and Tillage Research* 144: 72–82.

Sisti, C. P. J., H. P. dos Santos, R. Kohhann, B. J. Alves, S. Urquiaga, and R. M. Boddey. 2004. Change in carbon and nitrogen stocks in soil under 13 years of conventional or zero tillage in southern Brazil. *Soil and Tillage Research* 76, no. 1: 39–58.

Six, J., R. T. Conant, E. A. Paul, and K. Paustian. 2002. Stabilization mechanisms of soil organic matter: Implications for C-saturation of soils. *Plant and Soil* 241, no. 2: 155–176.

Six, J., E. T. Elliott, K. Paustian, and J. W. Doran. 1998. Aggregate and soil organic matter accumulation in cultivated and native grassland soils. *Soil Science Society of America Journal* 62: 1367–1377.

Six, J., S. D. Frey, R. K. Thiet, and K. M. Batten. 2006. Bacterial and fungal contributions to carbon sequestration in agro-ecosystems. *Soil Science Society of America Journal* 70, no. 2: 555–69.

Smith, M. D. 2011. The ecological role of climate extremes: Current understanding and future prospects. *Journal of Ecology* 99, no. 3: 651–5.

Smith, P. 2008. Land use change and soil organic carbon dynamics. *Nutrient Cycling in Agroecosystems* 81, no. 2: 169–78.

—. 2014. Do grasslands act as a perpetual sink for carbon? *Global Change Biology* 20, no. 9: 2708–11.

—. 2016. Soil carbon sequestration and biochar as negative emission technologies. *Global Change Biology* 22: 1315–1324.

Smith, P., D. Martino, Z. Cai et al. 2007. Agriculture. In *Climate Change 2007: Mitigation. Contribution of Working Group III to the Fourth Assessment Report of the Intergovernmental Panel on Climate Change*, ed. B. Metz, O.R. Davidson, P. R. Bosch, R. Dave, and L.A. Meyer, 499–540. Cambridge, UK and New York: Cambridge University Press.

Smith, P., D. Martino, Z. Cai et al. 2008. Greenhouse gas mitigation in agriculture. *Philosophical Transactions of the Royal Society B: Biological Sciences* 363, no. 1492: 789–813.

Smith, P., J. I. House, M. Bustamante et al. 2016. Global change pressures on soils from land use and management. *Global Change Biology* 22, no. 3: 1008–28.

Sommer, R. and D. Bossio. 2014. Dynamics and climate change mitigation potential of soil organic carbon sequestration. *Journal of Environmental Management* 144: 83–7.

Srinivasarao, C., R. Lal, S. Kundu, and P. B. Thakur. 2015. Conservation agriculture and soil carbon sequestration. In *Conservation Agriculture,* ed. M. Farook and K.H.M. Siddique, 479–523. Switzerland: Springer International Publishing.

Srivastava, P., A. Kumar, S. K. Behera, Y. K. Sharma, and N. Singh. 2012. Soil carbon sequestration: An innovative strategy for reducing atmospheric carbon dioxide concentration. *Biodiversity and Conservation* 21, no. 5: 1343–58.

Stewart, C. E., K. Paustian, R. T. Conant, A. F. Plante, and J. Six. 2007. Soil carbon saturation: Concept, evidence and evaluation. *Biogeochemistry* 86, no. 1: 19–31.

Stockfisch, N., T. Forstreuter, and W. Ehlers. 1999. Ploughing effects on soil organic matter after twenty years of conservation tillage in Lower Saxony, Germany. *Soil and Tillage Research* 52, no. 1: 91–101.

Stockmann, U., M. A. Adams, J. W. Crawford et al. 2013. The knowns, known unknowns and unknowns of sequestration of soil organic carbon. *Agriculture, Ecosystems and Environment* 164: 80–99.

Sumberg, J. E. and J. Thompson, ed. 2012. *Contested Agronomy: Agricultural Research in a Changing World.* New York: Earthscan.

Sumberg, J., J. Thompson, and P. Woodhouse. 2012. Contested agronomy: Agricultural research in a changing world. In *Contested Agronomy: Agricultural Research in a Changing World*, ed. J. Sumberg and J. Thompson. London: Routledge.

Tardy, V., A. Spor, O. Mathieu et al. 2015. Shifts in microbial diversity through land use intensity as drivers of carbon mineralization in soil. *Soil Biology and Biochemistry* 90: 204–13.

Tilman, D., C. Balzer, J. Hill, and B. L. Befort. 2011. Global food demand and the sustainable intensification of agriculture. *Proceedings of the National Academy of Sciences* 108, no. 50: 20260–4. doi.org/10.1073/pnas.1116437108.

Tisdall, J. M. and J. M. Oades. 1982. Organic matter and water-stable aggregates in soils. *Journal of Soil Science* 33, no. 2: 141–63.

United Nations. 2014. Concise Report on the World Population Situation 2014. United Nations Department of Economic and Social Affairs, Washington D.C.

USDA-NRCS. 2013. Conservation practice standard: Residue and tillage management, reduced till. (Ac.) Code 345.

VandenBygaart, A. J., E. G. Gregorich, and D. A. Angers. 2003. Influence of agricultural management on soil organic carbon: A compendium and assessment of Canadian studies. *Canadian Journal of Soil Science* 83, no. 4: 363–80.

VandenBygaart, A. J., E. G. Gregorich, and B. L. Helgason. 2015. Cropland C erosion and burial: Is buried soil organic matter biodegradable? *Geoderma* 239–240: 240–249.

VandenBygaart, A. J., D. Kroetsch, E. G. Gregorich, and D. Lobb. 2012. Soil C erosion and burial in cropland. *Global Change Biology* 18, no. 4: 1441–52.

Van Oost, K., T. A. Quine, G. Govers et al. 2007. The impact of agricultural soil erosion on the global carbon cycle. *Science* 318, no. 5850: 626–9.

Varvel, G. E. and W. W. Wilhelm. 2011. No-tillage increases soil profile carbon and nitrogen under long-term rainfed cropping systems. *Soil and Tillage Research* 114, no. 1: 28–36.

Villarini, G., J. A. Smith, M. L. Baeck, and W. F. Krajewski. 2011a. Examining flood frequency distributions in the midwest U.S. *Journal of the American Water Resources Association* 47, no. 3: 447–63.

Villarini, G., J. A. Smith, M. L. Baeck, R. Vitolo, D. B. Stephenson, and W. F. Krajewski. 2011b. On the frequency of heavy rainfall for the midwest of the United States. *Journal of Hydrology* 400, no. 1: 103–20.

Virto I., P. Barré, A. Burlot, and C. Chenu. 2012. Carbon input differences as the main factor explaining the variability in soil organic C storage in no-tilled compared to inversion tilled agrosystems. *Biogeochemistry* 108: 17–26.

Volk, T. 2003. *Gaia's Body—Toward a Physiology of Earth.* Cambridge MA.: MIT Press.

Voroney, R. P., E. A. Paul, and D. W. Anderson. 1989. Decomposition of wheat straw and stabilization of microbial producers. *Canadian Journal of Soil Science* 69, no. 1: 63–77.

Waksman, S. A. 1925. What is humus? *Proceedings of the National Academy of Sciences* 11, no. 8: 463–8.

Wall, D. H., R. D. Bardgett, and E. F. Kelly. 2010. Biodiversity in the dark. *Nature Geoscience* 3, no. 5: 297–8.

Wall, D. H. and U. N. Nielsen. 2012. Biodiversity and ecosystem services: Is it the same below ground? *Nature Education Knowledge* 3, no. 12: 8. http://www.nature.com/scitable/knowledge/library/biodiversity-and-ecosystem-services-is-it-the-96677163.

Wander, M. 2004. Soil organic matter fractions and their relevance to soil function. In *Soil Organic Matter in Sustainable Agriculture,* ed. F. Magdoff and R. R. Weil, 67–102. Boca Raton, FL: CRC Press.

Wendt, J. W. and S. Hauser. 2013. An equivalent soil mass procedure for monitoring soil organic carbon in multiple soil layers. *European Journal of Soil Science* 64, no. 1: 58–65.

West, T. O. and W. M. Post. 2002. Soil organic carbon sequestration rates by tillage and crop rotation: A global data analysis. *Soil Science Society of America Journal* 66, no. 6: 1930–46.

West, T. O. and G. Marland. 2002. A synthesis of carbon sequestration, carbon emissions, and net carbon flux in agriculture: Comparing tillage practices in the United States. *Agriculture, Ecosystems and Environment* 91, no. 1: 217–32.

White, J. W., L. A. Hunt, K. J. Boote et al. 2013. Integrated description of agricultural field experiments and production: The ICASA Version 2.0 data standards. *Computers and Electronics in Agriculture* 96: 1–12.

Wortmann, C. S., J. A. Quincke, R. A. Drijber, M. Mamo, and T. Franti. 2008. Soil microbial community change and recovery after one-time tillage of continuous no-till. *Agronomy Journal* 100, no. 6: 1681–6.

Wortmann, C. S., R. A. Drijber, and T. G. Franti. 2010. One-time tillage of no-till crop land five years post-tillage. *Agronomy Journal* 102, no. 4: 1302–7.

Yang, X., C. F. Drury, and M. M. Wander. 2013. A wide view of no-tillage practices and soil organic carbon sequestration. *Acta Agriculturae Scandinavica. Section B—Soil and Plant Science* 63, no. 6: 523–30.

5 Nutrient Requirements for Soil Carbon Sequestration

B. A. Stewart

CONTENTS

5.1 INTRODUCTION

Between 1950 and 2016, the world population increased from about 2.5 billion to 7.4 billion people, and is expected to reach 9.7 billion by 2050. Not only is the world population increasing rapidly, many developing countries are becoming more prosperous so inhabitants are shifting to diets containing more animal based protein. Therefore, food production, particularly cereals, must increase at a significantly higher rate than population. This is a huge challenge, particularly when many soils of the world have become seriously degraded. Lal (2016) stated that conversion of natural ecosystems to managed agroecosystems leads to a reduction of the antecedent soil organic carbon (SOC) pool by 30% to 50% over 50 years in temperate climates, and as much as 75% in 10 to 25 years in the tropics. The loss of SOC generally results in declining soil health and lower productivity. Sanchez and Swaminathan (2005), in devising a plan to cut world hunger in half, said that restoring soil health is often the first entry point for increasing agricultural productivity. They were focusing on farmers with small holdings and stated that soil restoration could double or triple yields of cereals but would require applying appropriate combinations of mineral and organic fertilizers, using leguminous green manures and agroforestry fertilizer trees, returning crop residues to the soil, and using improved methods of soil conservation. Lal (2016) also states that the potential is great for restoring SOC. He stated that, theoretically, there could be as much as 62 t ha^{-1} over the next 50 to 75 years with a total C sink capacity of about 88 Gt on the 1400 Mha of cropland in the world. In addition, there is potential for sequestering SOC in grazing lands, forest lands, and degraded and desertified lands. The USDA Natural Resources Conservation Service (NRCS), based on work with farmers from 2005 to 2014, concluded that with proper management, conservation practices could sequester from 1.4 to 19.2 Mg C ha^{-1} in 20 years (Chambers et al. 2016). There have been countless other publications over the past few decades advocating the importance of C sequestration to restore soil health and to reduce CO_2 emissions to the atmosphere. About 30% to 35% of global greenhouse gases (GHG) emissions are contributed to agriculture (IPCC 2014). Although agriculture is more responsible for methane and nitrous oxide emissions than CO_2, sequestering C in soil is looked on as a beneficial way of reducing agriculture's effect on global warming while at the same time restoring soil health. In spite of the support and optimistic estimates of amounts of C that can be sequestered in soils, results have been mixed. The goal of this chapter is to look at the nutrient requirements for sequestering C in soil.

5.2   WHAT IS CARBON SEQUESTRATION?

Carbon sequestration is a term that is widely used but seldom specifically defined. Therefore, it is sometimes difficult to clearly interpret results and statements in the literature. The author's view is that C is not sequestered in soil but that carbonaceous materials present in soil are biologically decomposed to form organic compounds that become increasingly resistant with further decomposition. These organic compounds consist of C, H, O, N, P, S, K, and other chemical elements. Some of the compounds become bound with clay particles and extremely resistant to further decomposition while others are subject to further biological breakdown under certain conditions. The important point is that there is no specific point that SOC is not subject to biological activity. In many of the reports projecting the amounts of C that can be sequestered in soils over time, it is implied that the projected amounts of C sequestered in a year remains long term. Chambers et al. (2016) list a number of conservation practices used in the U.S. on cropland that on average will sequester varying amounts of SOC in soil. For example, residue and tillage management practices will sequester from 0.15 to 0.27 Mg C ha^{-1} yr^{-1} and then this number is used to estimate how much is sequestered in 20 years. The "4 per Thousand" worldwide initiative is based on the hypothesis that SOC can be increased at the compounded rate of 0.4% until stocks plateau at a new equilibrium (Chambers et al. 2016). Lal (2016) states that the world's cropland soils have the theoretical potential of sequestering as much as 0.8 to 1.2 t ha^{-1} yr^{-1} for the next 50 to 75 years. In reality, some of the SOC sequestered in a given year is partially decomposed the next year. Historically, the amount of SOC in ecosystem soils that were converted to cropland was essentially a constant value because it had reached equilibrium where the carbon inputs were equal to the carbon outputs. The amount of SOC sequestered for a given ecosystem depended largely on the soil and climate. When the ecosystem was converted to cropland, the amount of SOC generally decreased rapidly during the first several years because tillage and other management practices drastically altered the soil and the biological activity. Tillage breaks aggregates and exposes SOC to microorganisms that had largely been protected. The loss of SOC affects the soil in many ways. Soil physical properties are generally degraded, but decomposition of SOC provides nutrients for crop growth and this is the reason that tillage was and still is considered desirable by many farmers. As the amount of SOC decreases, the soil becomes less productive because soil fertility decreases, soil physical properties become less favorable, and biological activity declines.

Carbon sequestration is in many ways a nebulous term, particularly when applied to soil. The FAO (2017) states that atmospheric CO_2 concentrations can be lowered by reducing emissions or by taking CO_2 out of the atmosphere and storing it in terrestrial, oceanic, or freshwater aquatic ecosystems. They define a sink as a process or an activity that removes greenhouse gas from the atmosphere. The FAO further states that the long-term conversion of grassland and forestland to cropland (and grazing lands) has resulted in historic losses of soil carbon worldwide but there is a major potential for increasing soil carbon through restoration of degraded soils and widespread adoption of soil conservation practices. However, there is no indication given as to how long this sequestered C will remain in storage. The Ecological Society of America (ESA 2017) states that C is the major building block for life on Earth. It exists predominately as plant biomass, SOC, and as CO_2 in the atmosphere and dissolved in seawater, and C sequestration is the long-term storage of C in oceans, soils, vegetation, and geologic formations. Although oceans store most of the Earth's carbon, soils (to 1-m depth) contain approximately 75% of the carbon pool on land–three times more than the amount stored in living plants and animals (ESA 2017). Lal (2009) stated that the process of CO_2 sequestration into terrestrial ecosystems is based on the natural process of photosynthesis. It involves adoption of land use and soil/vegetation management systems which enhance photosynthesis and transfer some of the biomass into SOC as stable humic substances with long residence time.

The difficulty with most definitions of carbon sequestration is that they imply that the C remains in the soil indefinitely and there no mention that other elements are also sequestered. Carbon alone is not sequestered by photosynthesis. Varying amounts of all other elements required for plant growth are also sequestered, and a portion of these elements will remain sequestered as long as some of the C remains in storage. Thus it is critical to understand that, in reality, it is soil organic matter (SOM) that is being sequestered rather than SOC, and C is only one of the many elements contained in SOM. It is also well understood that SOM is a very complex mixture of compounds that consists of labile (easily decomposed) materials and other materials estimated to persist in soil for as long as 500 to 5000 years (Brady and Weil 2008). They state that SOC consists of living organisms (biomass), identifiable dead tissue (detrius), and nonliving, nontissue substances (humus). The humus portion of SOM is considered stable because radioactive studies have shown that some SOC incorporated into humus thousands of years ago is still present in soils. However, without annual additions of sufficient plant residues, microbial oxidation of humus that make up 60% to 80% of SOM will result in reduced SOM levels (Brady and Weil 2008). Soil management, particularly tillage, can also reduce SOM levels.

The SOM concentration of natural ecosystem soils is generally considered to be the equilibrium concentration and represents the maximum amount for the soil texture and mineral makeup, and the climate of the region. This implies that the decomposition of existing SOC is equal to the formation of SOC from newly formed humus. Therefore, although humus is considered stable, it is evident that some humus decomposition is always occurring and is an important process for recycling nutrients.

5.3 SOIL WARMING EFFECT ON SOIL ORGANIC CARBON

While C sequestration has gained prominence as a means of reducing CO_2 in the atmosphere to reduce climate change, the effect of climate change on accelerating the decomposition of SOC is seldom mentioned. However, average annual temperature of an area has a profound effect on the SOM content. Haas et al. (1957) determined N and C contents in virgin and cropped soils in the U.S. Great Plains. A fine sandy loam virgin soil in Mandan, North Dakota, with a mean temperature of 5°C contained 2.11 SOM compared to 0.72 for a similar soil in Dalhart, Texas, with a mean temperature of 12.4°C. The average annual temperature for North Dakota has increased more than 2°C since the 1980s (NOAA 2017) and has almost certainly resulted in a higher rate of soil humus decomposition. This would apply to all land and could conceivably negate a portion or even exceed the amount of humus formed from improved management practices on cropland.

On a global scale, Crowther et al. (2016) attempted to quantify SOC losses in response to warming. They stated, "We find that the effects of warming are contingent on the size of the initial soil carbon stock, with considerable losses occurring in high-latitude areas. By extrapolating this empirical relationship to the global scale, we provide estimates of soil carbon sensitivity to warming that may help to constrain Earth system model projections. Our empirical relationship suggests that global soil carbon stocks in the upper soil horizons will fall by 30 ± 30 petagrams of carbon to 203 ± 161 petagrams of carbon under one degree of warming, dpending on the rate at which the effects of warming are realized. Under the conservative asumption that the response of soil carbon to warming occurs within a year, a business-as-usual climate scenario would drive the loss of 55 ± 50 petagrams of carbon from the upper soil horizons by 2050. The value is around 12%–17% of the expected anthropogenic emissions over ths period. Despite the considerable uncertainty in our estimates, the direction of the global soil carbon response is consistent across all scenarios. This provides strong emperical support for the idea that rising temperatures will stimulate the net loss of soil carbon to the atmospere, driving a positive land carbon-climate feedback that could accelerate climate change."

Pries et al. (2017) reported that there are about 3000 Pg C stored as SOC on the planet. While most SOC studies have focused on the 0–20 cm topsoil layers, they stated that about 50% of the total SOC is in the subsoil layers below 20 cm. In their deep warming experiment in mineral soil, CO_2 production from all soil depths to 100 cm increased from 34–37% with 4°C warming. They concluded that this potentially large subsoil response to warming should not be ignored because it would be roughly 3% of current ecosystem respiration and about 30% of current anthropogenic emissions.

5.4   HUMUS FORMATION ANNUALLY IN U.S. CROPLAND SOILS

Humus formation and decomposition occur simultaneously in all soils, but generally at much faster rates in cropland soils. When a natural ecosystem is converted to cropland, and particularly when extensive tillage is used, the decomposition rate usually greatly exceeds the formation of new humus. Thus, the SOC content of the soil decreases rapidly the first several years before the SOC concentration in the soil tends to equilibrate. As already stated, Lal (2016) reported that conversion of natural ecosystems to managed agroecosystems leads to a reduction of the antecedent soil organic carbon (SOC) pool by 30% to 50% over 50 years in temperate climates, and as much as 75% in 10 to 25 years in the tropics. It is also known that improved soil management practices can result in humus formation exceeding humus decomposition to increase the SOC concentration. Chambers et al. (2016) estimated the amounts of C sequestered annually by a variety of practices used in the U.S. on croplands and grazing lands but made no estimates of decomposition rates. Thus, this implies that total SOC is increased by the amount newly formed humus. In reality, the net increase is almost certainly less.

Corn (*Zea mays*) and wheat (*Triticum aestivum*) are two of the most widely grown crops in the U.S. Together, they are grown on about 59 M ha, which is approximately 35% of all U.S. cropland. For this chapter, estimated amounts of humus formed annually from the crop residues remaining in the field for each Mg of grain removed from the field are calculated. Certain assumptions are made based on the literature as documented in Table 5.1. Although the author believes these estimates are reasonable for somewhat average conditions, actual values will vary considerably depending on yield level. The estimated amount of C sequestered in humus formed from the decomposition of corn stover is 193 kg for each Mg of grain produced. In addition to the C sequestered, 16.1 kg N, 3.9 kg P, and 2.8 kg S will be sequestered. For wheat straw, estimated amounts of C, N, P, and S sequestered for each Mg of grain produced are 236, 9.9, 2.3, and 2.7 kg, respectively. The average corn grain yield in the U.S. is approximately 10.5 Mg ha^{-1} and the average wheat yield is about 3 Mg ha^{-1}. Therefore, the estimated amounts of C, N, P, and S sequestered by the formation of humus from the crop residues are 2027, 169, 41, and 29 kg ha^{-1}, respectively, for an average U.S. corn field. For an average U.S. wheat field, the estimated sequestered amounts are 708, 30, 7, and 8 kg ha^{-1}. The estimated amounts of C sequestered are in line with estimates by Lal (2016) that worldwide SOC sequestration could be from 800 to 1200 kg ha^{-1} yr^{-1}, and Lal et al. (1998) who estimated the C sequestration potential for U.S. cropland ranged from an average of 450 to 1270 kg ha^{-1} yr^{-1}.

Himes (1997) stated that the challenge of sequestering C is to increase the concentration of humus while producing good crop yields. While the estimates presented above for the amounts of C sequestered by decomposition of corn stover and wheat straw are somewhat similar to earlier estimates, it is not realistic to assume that the C sequestered during a given year remains in storage long-term. Based on the values in Table 5.1, approximately 28.2 kg N, 5.0 kg P, and 3.0 kg S are required to produce the grain and stover for each 1000 kg ha^{-1} of grain, so the amounts of N, P, and S in the grain and stover for an average U.S. yield of 10.5 Mg ha^{-1} would be 296 kg N, 52.5 kg P, and 31.5 kg S. This far exceeds the amounts of fertilizer applied so it is apparent that much of the N, P, and S in succeeding corn crops is recycled from the decomposition of soil humus that was previously sequestered. The approximate average amount of fertilizer N applied to U.S. corn is 160 kg ha^{-1} yr^{-1} and for P, the amount is 43 kg ha^{-1} yr^{-1} (USDA-ERS 2017).

TABLE 5.1

Estimated Amounts of C Sequestered in Humus from Corn Stover and Wheat Straw Produced for Each Mg of Grain

Crop	Estimated Sequestration (kg C/ha for Each Mg of Grain)	Description	Reference
Corn			
	1222	Corn stover based on assumed average harvest index of 0.45.	Prihar and Stewart (1990)
	550	Based on assumed 45% C content.	Himes (1997)
	193	In humus produced from the 550 kg C in stover based on 35% of C in stover becoming incorporated into humus.	Jenkinson (1981)
	333	Based on C content of humus being 58% C.	Himes (1997)
	–	16.1 kg N, 3.9 kg P, and 2.8 kg S will be sequestered in the 333 kg of humus based on C:N, C:P, and C:S ratios of 12:1, 50:1, and 70:1, respectively.	Eakin (1972)
	–	The N, P, and S concentrations of the corn stover were assumed to be 1% N, 0.18% P, and 0.15% S, so the 1222 kg contained 12.2 kg N, 2.2 kg P, and 1.8 kg S. These amounts are less than those needed for maximum humus formation so additional amounts are required.	Eakin (1972)
	–	The N, P, and S concentrations of corn grain are assumed to be 1.6%, 0.28%, and 0.12%, respectively, which is equal to 16 kg, 2.8 kg, and 1.2 kg for each 1000 kg grain	Eakin (1972)
Wheat			
	1500	Wheat straw based on assumed harvest index of 0.4.	Prihar and Stewart (1990)
	675	Based on assumed 45% C content.	Himes (1997)
	236	Contained in the 625 kg C in straw based on 35% of C in straw becoming incorporated into humus.	Jenkinson (1981)
	407	Based on C content of humus being 58% C.	Himes (1997)
	33.9	19.7 kg N, 4.7 kg P, and 3.4 kg S will be sequestered in the 407 kg humus based on C:N, C:P, and C:S ratios of 12:1, 50:1, and 70:1, respectively.	Himes (1997)
	–	The N, P, and S concentrations of the wheat straw were assumed as 0.66% N, 0.15% P, and 0.18% S, so the 1500 kg wheat straw contained 9.9 kg N, 2.3 kg P, and 2.7 kg S. These amounts are less than those needed for maximum humus formation so additional amounts are required.	Eakin (1972)
	–	The N, P, and S concentrations of wheat grain are assumed to be 2.1%, 0.46%, and 0.13%, respectively, which is equal to 21 kg, 4.6 kg, and 1.3 kg for each 1000 kg grain.	Eakin (1972)

5.5 WHAT FACTORS DETERMINE LONG-TERM STORAGE OF C SEQUESTERED IN HUMUS?

Estimates like the ones for corn and wheat discussed in Section 5.3 can be made for any crop based on values obtained from the literature. It can also be reasonably assumed that the estimated amounts of C, N, P, and S sequestered are independent of where the crops are grown because the primary factor is the amount of biomass C available for microorganisms. As already mentioned, however, the C sequestered in the humus is composed of many different and complex compounds that vary widely in their resistance to decomposition. Once C is sequestered into humus, the time that it remains sequestered will be mostly controlled by climate and soil manipulation. Even though the decomposition of plant residues on croplands is continuously adding newly formed humus to the soil, it is difficult to show in many cases any increase in SOC. In fact, many soils, particularly in warm areas, continue to show a decline in SOC even though grain yields have increased as a result of improved management practices and plant genetics.

In the past few decades, the Food and Agriculture Organization (FAO) and other national and international agencies have widely promoted conservation agriculture as a technology to significantly increase C sequestration and crop yields. Conservation agriculture is based on the simultaneous use of three principles: (1) minimum tillage and soil disturbance; (2) permanent soil cover with crop residues and live mulches; and (3) crop rotation and intercropping (FAO 2014). The program has been highly successful in some areas but disappointedly low in other areas. When tillage is eliminated or greatly reduced, newly formed humus is more stable and there is less mineralization of N and P and crop yields can be reduced by low fertility. It is often reported that additional fertilizer is required to maintain yields the first few years following the initiation of a conservation agriculture program. After a few years, however, there is often enough build-up of humus so that significant amounts of nutrients are released for crop growth and fertilizer rates can be reduced. These observations are not surprising in view of the discussion in Section 5.3 that suggests that the formation of humus from crop residues to sequester C theoretically sequesters all N, P, and S contained in the residue and most likely also some additional nutrients from the soil when available. This again emphasizes the importance of understanding that sequestration of C contained in plant residues also sequesters N, P, S and other nutrients that remain sequestered as long as the C remains stored. While the benefits resulting from increased amounts of humus in soils are well documented and significant, large amounts of N, P, and other nutrients are required. Chambers et al. (2016) estimated that a 10-year phase-in of the "4 per Thousand" program in the U.S. could increase soil C stocks nationally by 68 Tg C and by 2050 could store 75 Tg C yr^{-1}. Based on the C:N and C:P ratios in humus (Himes 1997), 6.25 Tg yr^{-1} of N and 1.5 Tg yr^1 would also be sequestered. These amounts are equivalent to about 50% of the N and 80% of the P applied as fertilizer in the U.S. in 2014 (FAOSTAT 2017). This is not to imply that every opportunity feasible should not be used to maximize the formation and retention of humus in soils. It does, however, suggest that there are many challenges associated with increasing soil humus. This seems particularly true in warm climates. An example is the work of Schwartz et al. (2015) at Bushland, TX on a Pullman silty clay loam with an annual precipitation of about 500 mm and an average temperature of 13.1°C. An adjacent grassland with no history of cultivation was compared to field following 86 years of cultivation. The SOC in the surface 0.30 m declined 41% with half of the change occurring during the first 20 years. In the historical stubble-mulch under a wheat-fallow rotation, SOC and total soil N levels significantly increased with decreasing tillage intensities in 1977 for treatments imposed in 1941. On graded terrace plots under a wheat-sorghum (*Sorghum bicolor* (L.) Moench)-fallow rotation, SOC and total soil N stored under no-tillage were not significantly different from stubble-mulch tillage 30 years after treatments were imposed. This illustrates the difficulty, particularly in warm climates with limited precipitation, of increasing SOC even though grain yields may have increased significantly because of improved water management technologies and plant genetics. In contrast, there are many examples (FAO 2014) showing large increases in SOC when the principles of conservation agriculture are practiced, particularly in favorable climates and when legumes are incorporated into the rotations.

Han et al. (2016) conducted a global meta-analysis of published data on the responses of SOC to fertilizer management in 1741 paired field experiments. They concluded that SOC in the topsoil increased 0.9 (0.7–1.0) g kg^{-1} (10% relative change), 1.7 (1.2–2.3) g kg^{-1} (15.4%), 2.0 (1.9–2.2) g kg^{-1} (19.5%), and 3.5 (3.2–3.8) g kg^{-1} (36.2%) with use of unbalanced chemical fertilizers, balanced chemical fertilizers, balanced chemical fertilizers and straw, and chemical fertilizers combined with manure applications, respectively. However, they stated that at least 2.0 Mg ha^{-1} yr^{-1} C input was needed to maintain the SOC level in approximately 85% of the cases. They also showed that high variability across climatic regions with SOC being more resilient in cooler than warmer regions. The estimated 2.0 Mg C yr^{-1} threshold amount needed to just maintain the SOC level is greater than available in many parts of the world, and to make matters worse, crop residues are used for fuel or animal feed in many developing countries. The world average yield for maize in 2014 was 5620 kg ha^{-1} and 3080 kg ha^{-1} for wheat. Using the values listed in Table 5.1, the C contained in the maize stover would be 3.1 Mg ha^{-1} and 2.2 Mg ha^{-1} for the wheat straw. Therefore, based on the analysis of Han et al. (2016), increasing the SOC content on much of the world's cropland will be challenging although they stated a great C sequestration potential existed for either maintaining or improving current SOC stocks across all agro-ecosystems with the use of fertilizers and particularly so when fertilizers are applied in combination with crop residues and manure.

5.6 CONCLUSION

Carbon sequestration in soils is widely heralded as having a high potential for reducing CO_2 emissions to the atmosphere and thus reduce the greenhouse gas effect commonly believe to be a major cause of climate change. As much as one-third of the GHG is contributed either directly or indirectly by agriculture, and C sequestration in soils is considered favorably as a sink for C that will not only reduce CO_2 in the atmosphere but simultaneously improve soil health and increase crop yields. While it is an indisputable fact that C can be sequestered in soil, the amount and the length of time it remains sequestered are highly variable and poorly understood. This chapter stresses that in reality it is organic matter rather than C that is being sequestered, and C is just one component. A common definition of C sequestration is that it is the C contained in humus which is the substance remaining in soil after crop residues or other organic C materials are decomposed by biological activity to the point that there is no longer active biological activity. On average, the humus contains about 58% C, but many other elements are present in the compounds. Since humus contains the decomposed remains of the countless numbers and kinds of microorganisms that require many elements to grow, most of the compounds containing C also contain N, P, S, and other elements required for biological activity to flourish. For every 100 units of C, there are generally about 12 units of N, 2 units of P, and 1.4 units of S. Therefore, C sequestration in soil only occurs when there is simultaneous sequestration of N, P, S and other elements required by microorganisms. Another commonly accepted guideline is that about 35% of the C added to the soil in organic substances such as plant residues will remain in the soil as humus and this is commonly assumed to be the amount of C sequestered. Therefore, it is relatively simple to estimate the amounts of C, N, P, and S that will be sequestered in the soil and this is largely independent of the soil or climate. The length of time that these sequestered elements remain in storage, however, is highly dependent on soil texture, soil water conditions, temperature, humidity, and a host of other variables. This makes estimating the cumulative amounts of C sequestration extremely difficult, but should not deter efforts to develop, promote, and apply technologies that maximize to the extent feasible the formation and retention of soil humus. It can only be speculated as to what impact increasing soil humus will have on climate change but it can only be positive, and it can also only have a positive effect on soil health. It is conceivable, however, that if the global land and ocean average temperature continues to increase, the SOC content of all land areas could gradually decrease because of increasing biological activity and emit more CO_2 to the atmosphere.

REFERENCES

Brady, N.C. and R.R. Weil. 2008. *The Nature and Properties of Soils*. Pearson Prentice Hall.

Chambers, A., R. Lal, and K. Paustian. 2016. Soil carbon sequestration potential of US croplands and grass-lands: Implementing the 4 per Thousand Initiative. *Journal Soil and Water Conservation* 71: 68A–74A. doi.org/10.2489/jswc.71.3.68A.

Crowther, T.W., K.E.O. Todd-Brown, C.W. Rowe, W.R, Wieder, J.C. Carey, N,B, Machmuller, B.L. Snoek, S. Fang, G. Zhou, S.D. Allison, J.M. Blair, S.D. Bridgham, A.J. Burton, Y. Carrillo, P.B. Reich, J.S. Clark, A.T. Classen, F.A. Dijkstra, B. Elberling, B.A. Emmert, m. Estiarte, S.D. Frey, J. Guo, J. Harte, L. Hang, B.R. Johnson, G. Kröel-Dulay, K.S. Larsen, H. Laudon, J.M. Lavallee, Y. Luo, M. Lupascu, L.N. Ma, S. Marhan, A. Michelsen, J. Mohan, S. Niu, E. Pendall, J. Peñuelas, L. Pfeifer-Meister, C. Poll, S. Reinsch, L.L. Reynolds, L.K. Schmidt, S. Sistla, N.W. Sokol, P.h. Templer, K.K. Treseder, J.M. Welker and M.A. Bradford. 2016. Quantifying global soil carbon losses in response to warming. *Nature* 540: 104–108. doi.org/10.1038/nature20150.

Eakin, J.H. 1972. Food and fertilizers. In: The Fertilizer Handbook, 1–21. The Fertilizer Institute, Washington, DC.

ESA. 2017. What is carbon sequestration? Ecological Society of America. Washington, DC. Available at: https://www.esa.org/esa/wp-ontent/uploads/2012/12/carbonsequestrationinsoils.pdf.

FAO. 2014. Conservation Agriculture. Food and Agriculture Organization of the United Nations, Rome. Available at: http://www.fao.org/ag/ca/.

FAO. 2017. What is soil carbon sequestration? Food and Agriculture Organization of the United Nations, Rome. Available at: http://www.fao.org/soils-portal/soil-management/soil-carbon-sequestration/en/.

FAOSTAT. 2017. Food and agriculture data. Food and Agriculture Organization of the United Nations, Rome. Available at: http://www.fao.org/faostat/en/#home.

Haas, H.J., C.E. Evans, and E.F. Miles. 1957. Nitrogen and carbon changes in great plains soils as influenced by cropping and soil treatments. Technical Bulletin 1164. U.S. Department of Agriculture, Washington, DC.

Han, W.Z., G. Wang, W. Sun, and Y. Huang. 2016. Changes in soil organic carbon in croplands subjected to fertilizer management: A global meta-analysis. Scientific Reports 6, Article number 27199. doi.org/10.1038/srep27199.

Himes, F.L. 1997. Nitrogen, sulfur and phosphorus and the sequestration of carbon. In: R. Lal, J.M. Kimble, R.F. Follett, and B.A. Stewart (eds.), *Soil Processes and the Carbon Cycle*, 315–319. CRC Press, Boca Raton, FL.

IPCC. 2014. Climate change 2014: Synthesis Report. The Fifth Assessment Report of Intergovermental Panel on Climate Change. Intergovernmental Panel on Climate Change (IPCC), Geneva, Switzerland.

Jenkinson, D.S. 1981. The fate of plant and animal residues in soil. In: D.J. Greenland and M.H. B. Hayes (eds.), *The Chemistry of Soil Processes*, 505–561. John Wiley & Sons, New York.

Lal, R. 2009. Sequestering atmospheric carbon dioxide. *Journal Critical Reviews in Plant Science* 28: 90–96. doi.org/10.1080/07352680902782711/.

Lal, R. 2016. Beyond COP21: Potential and challenges of the "4 per Thousand" initiative. *Journal Soil and Water Conservation* 71: 20A–25A. doi.org/10.2489/jswc.71.1.20A.

Lal, R., J.M. Kimble, R.F. Follett, and C.V. Cole. 1998. *The Potential of U.S. Cropland to Sequester Carbon and Mitigate the Greenhouse Effect*. Ann Arbor Press, Chelsea, MI.

NOAA. 2017. Climate at a Glance. National Centers for Environmental Information, National Oceanic and Atmospheric Administration. https://www.ncdc.noaa.gov/cag/time-series/global.

Pries, C.E.H., C. Castahna, R.C. Porras, and M.S. Torn. 2017. The whole-soil carbon flux in response to warming. *Science* 31 Vol. 355 Issue 6332 pp. 1420–1423. DOI: 10.1126/science.aa11319.

Prihar, S.S. and B.A. Stewart. 1990. Using upper-bound slope through origin to estimate genetic harvest index. *Agronomy Journal* 82: 1160–1165.

Sanchez, P.A. and M.S. Swaminathan. 2005. Cutting world hunger in half. *Science* 307: 357–359. doi.org/10.1126/Science.1109067.

Schwartz, R.C., R.L. Baumhardt, B.R. Scanlon, J.M. Bell, R.G. Davis, N. Ibragimov, O.R. Jones, and R.C. Reedy. 2015. Long-term changes in soil organic carbon and nitrogen under semiarid tillage and crop-ping practices. *Soil Science Society of America Journal* 79: 1771–1781.

USDA-ERS. 2017. Fertilizer use and price. Economic Research Service, U.S.D.A., Washington, DC. Available at: https://www.ers.usda.gov/data-products/fertilizer-use-and-price/.

6 Physical Protection and Mean Residence Time of Soil Carbon

Jennifer A.J. Dungait, Asmeret Asefaw Berhe,
Andrew S. Gregory, and David W. Hopkins

CONTENTS

6.1 INTRODUCTION

Soils comprise the planet's largest terrestrial carbon (C) pool, containing both organic and inorganic C. In recent times, the significance of soil inorganic carbon (SIC) has been relatively overlooked compared to soil organic C (SOC). One of the objectives of this chapter is to raise the awareness of the SIC pool in soil. In order to avoid confusion, the term "soil C" will be used to mean SIC or SOC or both.

The mean residence time of soil C is modulated by a complex and interacting set of biotic and abiotic processes all of which are likely to be affected by climate change to some extent. For a long time, SOC has been regarded as a reservoir trapped in soil, but Janzen (2015) has recently provided us with an alternative perspective in which it may be better regarded as "a stream of C atoms flowing through soil, propelled relentlessly by thermodynamics to the higher entropy of CO_2" temporarily arrested by a variety of biological, chemical, and physical mechanisms. The same rationale can be applied equally to SIC, although the mechanisms and the rates will differ. The dynamic three-dimensional matrix of soil provides habitats for microorganisms (Figure 6.1), the formation of which they influence in part, and in which they may be protected from desiccation and predation for some of the time, while still having access to nutrients, oxygen, and water. Crawford et al. (2012) refer to this matrix and the role of the soil organisms in contributing to its structure as "soil self-organization." A consequence of this dynamic three-dimensional structure is that soil organisms may also be periodically isolated from the flow of energy represented by SOC (Dungait et al. 2012), and access to SOC by the organisms operates at different scales, which correspond to the size and motility of the organisms, the ability of the organisms to disrupt the mechanisms of protection,

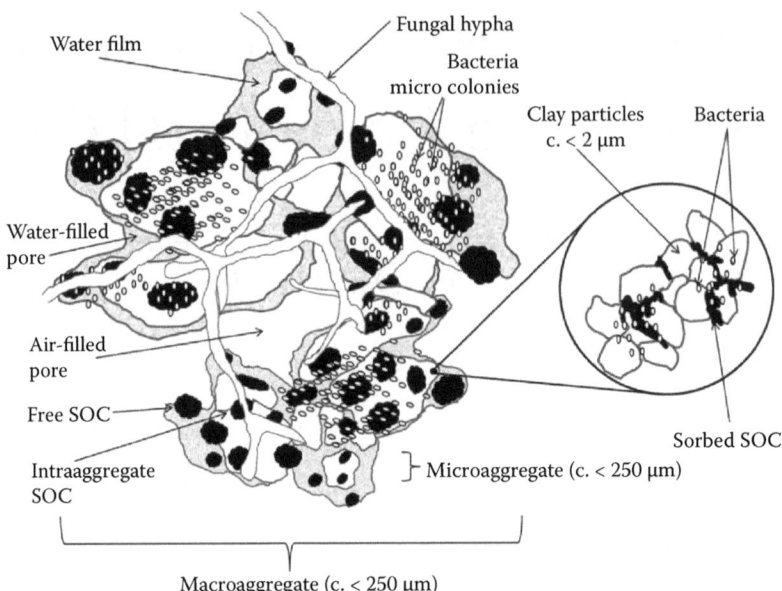

FIGURE 6.1 Spatial organization at the macro- and microaggregate levels of soil biotic and abiotic components.

and the longevity of the structures. Physical protection of SOC is recognized as a dominant control on longer-term residence times of SOC at the decadal-to-millennial scale.

In the context of this chapter, physical protection refers to any set of conditions, other than "recalcitrance" (i.e., short-term stability conferred by molecular structure), which limits the rate at which SOC that was originally fixed by photoautotrophs from either atmospheric CO_2 or dissolved bicarbonates is released back to the inorganic pool. We review the major mechanisms of physical protection of soil C and consider how climate change will affect those processes through its impact on meteorological processes, including anthropogenic land use change.

6.1.1 Soil Organic Carbon (SOC)

Most SOC is the product of photosynthesis, driven by solar energy, in the chloroplasts of higher plants, algae and some bacteria including the cyanobacteria and lichens. The proportions of organic C fixed by different groups of photoautotrophs varies markedly between ecosystems with, for example higher plants being the overwhelming source in most terrestrial ecosystems, but cyano-bacteria and algae, including the contributions from the symbionts in lichens, predominating in some environmentally harsh ecosystems such as deserts and mudflats. Heterotrophic soil organisms can utilise SOC directly as a C source for biosynthesis and oxidize it to harness the energy, a set of processes variously referred to as C mineralization, heterotrophic respiration or soil respira-tion, and decomposition. However, it is important to recognize that SOC that is assimilated by soil heterotrophs is not immediately decomposed or mineralized, but rather released as CO_2 when the decomposer organisms die and are themselves decomposed or when their waste materials are mineralized. Carbon mineralization gives rise to a variety of intermediate products with different mean residence times in the soil, and, under aerobic conditions, ultimately CO_2 which returns to the atmospheric pool (Dungait et al. 2012).

The tendency of SOC to become protected from mineralization is affected by the interactions between the SOC and the soil mineral fraction, and the spatial arrangements of the soil mineral components around the SOC (Baldock and Skjemstad 2000), as well as environmental factors that regulate decomposition, such as temperature and moisture availability. Through the interaction of soil physical, chemical and biological processes, SOC may become sorbed to mineral surfaces and/or incorporated within aggregates. Soils with high concentrations of clay-sized particles and reactive colloids have substantial surface area on which the organic molecules can be absorbed, and therefore tend to have a larger SOC content relative to those with less clay. Clay content and mineralogy are also related to the formation of stable aggregates that facilitate C storage and protection compared to soils with less stable structure. Aggregate stability is itself partially derived from SOC because of the cohesive effects of the organic molecules and because SOC sustains soil organisms which are agents of soil aggregation. Thus, perhaps counterintuitively, C mineralization can promote C sequestration if the C being lost by mineralization is replaced by new incoming organic C from, for example, plant and animal remains.

6.1.2 Soil Inorganic Carbon (SIC)

There are three main types of SIC: first, lithogenic carbonates derived directly from parent calcareous geology, e.g., limestone (Kraimer et al. 2005); second, biogenic carbonates formed by organisms, e.g., calcite granules (biospheroliths) produced by earthworms (Lambkin et al. 2011); and third, pedogenic carbonates formed by dissolution and re-precipitation of dissolved SIC due to abiotic processes, e.g,. carbonate nodules, or biotic processes, e.g., rhizoliths (Klappa 1980). The capture of respired CO_2 by pedogenic carbonate formation provides a potential route for the physical protection of C fixed by photoautotrophs in the geological C pool, with very long mean residence times ranging up to nearly 100 thousand years (Schlesinger 1985), compared with estimates of 1000 to 2000 years for the mean residence times of SOC (Jenkinson et al. 2008). Re-emergence of interest in the contribution of biological processes to SIC cycling has highlighted its importance as a terrestrial C pool with a millennial mean residence time, especially in semi-arid and arid climates, where SIC may comprise the largest soil C pool (Zamanian et al. 2016).

6.2 CURRENT PERSPECTIVES ON CONTROLS IN SOC STABILIZATION AND DECOMPOSITION

A paradigm shift in the understanding of SOC cycling has been facilitated by the advances in analytical technologies using stable and radio-isotopes over the last few decades that have allowed the quantification of the turnover rates of specific SOC pools and the fluxes between them (Dungait et al. 2012). It seems that if soil microorganisms can access SOC, they can degrade it relatively rapidly almost regardless of its molecular structure (Kleber 2010; Schmidt et al. 2011), marginalizing the concept of recalcitrance as a control in all but the most extreme cases (von Lützow and Kögel-Knabner 2010), such as, hydrophobicity caused by coatings of lipids around soil particles (Ellerbrock and Gerke 2004). This has led to calls for the evolution of SOC turnover models from those that rely on distinct pools with computed mean residence times corresponding to kinetically-defined SOC fractions, to those that conceptualise SOC as a continuum of turnover times with turnover moderated by physical constraints on accessibility by organisms and extracellular enzymes (Dungait et al. 2012; Lehmann and Kleber 2015). In particular, it has called into question the existence of a specific "humic" or inert pool, the presumed existence of which in mineral soils was necessary for models such as Roth-C and Century to run, but which has no reliable foundation in modern analytical chemistry (Lehmann and Kleber 2015).

The major C stabilization mechanisms in soils are now recognized to be "biologically non-preferred soil spaces" (Ekschmitt et al. 2008), where SOC is physically-protected from microbial activity regardless of its molecular structure (Kleber et al. 2011). Van Veen and Kuikman (1990) and von Lützow et al. (2006) listed the mechanisms involved in physical preservation of SOC. First, occlusion within aggregates; second, adsorption onto minerals, and third, substrate-driven "biological rate limitation" (e.g., Ekschmitt et al. 2005). Biological rate limitation refers to circumstances where the conditions limit exploitation of SOC by decomposer organisms by factors other than C supply. For example, decomposition may be constrained by lack of oxygen, water or available nutrients, or by extremes of temperature or acidity. This new understanding of the controls on SOC turnover is typified by the new perspectives on dissolved SOC (DOC) cycling described by Kaiser and Kalbitz (2012). The simple, polar compounds constituting most of the DOC are assumed to be the most biologically-available pool in soils (Boddy et al. 2007), yet they appear deep in the soil profile and have relatively old radiocarbon ages (Tipping et al. 2012). The persistence of DOC in the subsoil despite its inherent "labile" chemical properties suggests it has been protected from decomposition by association with the mineral phase followed by desorption, and/or the incorporation and release from the associated microbial pool during its transit through the profile (Kaiser and Kalbitz 2012). For example, Saggar et al. (1999) reported rapid incorporation of C from ^{14}C-glucose into other sugars, peptides, and amino acids sorbed to clay particles within hours confirming the role of microbial metabolism in the rapid transformation of labile glucose-C into stabilized clay-bound organic compounds.

6.3 MECHANISMS OF PHYSICAL PROTECTION

The predominant mechanisms providing physical preservation of soil C are adsorption of SOC onto minerals and occlusion within aggregates (Six et al. 2001, 2002), but there must be a threshold at which such C storage sites are completely filled, referred to as the SOC saturation threshold, defined by the physicochemical properties of the soil (Hassink and Whitmore 1997; Six et al. 2002), or occupation by competing inorganic moieties (Lima et al. 2010). Once the SOC saturation threshold is reached, further increments of SOC are not physically protected by association with the soil mineral fraction but may be protected from decomposition because of other environment constraints, such as lack of oxygen as occurs in peat.

6.3.1 ADSORPTION ONTO MINERALS

Organo–mineral complexes are likely to be a primary mechanism for long term SOC storage, because the adsorption affinity between SOC and the large, charged surface areas of clay minerals and amorphous colloids exceeds that of the capacity of organisms to capture SOC before it becomes associated with the mineral fraction. The fact that clay, silt, and sand fractions further combine to form soil aggregates enhances protection because a combination of adsorption and occlusion operate together (Martens et al. 2003). The protection of SOC is proportional to the decreasing size and increasing density of the soil fractions (Zimmermann et al. 2007), because a substantial amount of SOC with old radiocarbon ages is associated with the silt and especially clay-sized fractions (Marschner et al. 2008).

The behavior of both SOC and soil inorganic colloids is dominated by the physical and chemical properties distributed over their relatively large surface areas as described by the classical DLVO model (named after Derjaguin, Landau, Verwey, and Overbeek) for interactions at the surfaces of colloids. Most mineral and organic colloids in soil carry a net negative surface charge and thus, at close proximity, are repelled from each other (Figure 6.2). Opposing this repulsion, van der Waal's forces of attraction operate over slightly longer ranges (Figure 6.2). Therefore, the proximity of SOC to a mineral colloid surface is determined by the balance between the two opposing forces, with a primary position of minimum interaction close to the surface and secondary minimum more remote. Bacteria also behave as organic colloids, and this influences their ability to access SOC at this most intimate scale. The classical DLVO theory assumes soil colloids have flat surfaces, but recently Bradford et al. (2017) have proposed that for rough and irregular colloid surfaces, which are likely to predominate for soil colloids, the position of the secondary minimum is much reduced. At the nanoscale, the interlamellar spaces of certain clays have hygroscopic properties which draw

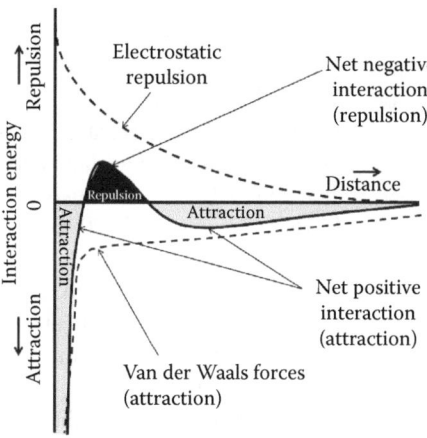

FIGURE 6.2 Interaction of electrostatic repulsion and van der Waals forces of attraction at colloid surfaces.

in and trap dissolved SOC. The SOC is physically protected from decomposition if the dimensions of these spaces are too small for soil microorganisms or extracellular enzymes to access the SOC. At the micro- and sub-millimeter scale, adsorption of SOC onto mineral surfaces of clay and silt particles because of a range of properties including adhesion, cation bridging and hydrogen bonding leads to persistent complexes between SOC and mineral surfaces (Dungait and Hopkins 2016).

6.3.2 OCCLUSION WITHIN AGGREGATES

SOC occluded within aggregates has longer mean residence times than so-called "free" SOC because microbial decomposition becomes restricted due to environmental constraints, such as the rate of diffusion of gases and nutrients in water, largely related to soil pore neck size (Kravchenko and Guber 2017) and physical access to the SOC by the decomposer organisms (Young and Ritz 2000). The practice of considering SOC based on physically-defined fractions is well established (Six et al. 2001). Separation using procedures that discriminate on the basis of size, solubility, or density, and sometimes several in combination, follows the rationale that the fragment size of SOC decreases with time and the density of the fraction in which SOC is recovered increases with time because of increasing association with relatively inorganic soil components (Dungait and Hopkins 2016). Microaggregate-associated SOC is the most stable because of the inverse relationship between aggregate size and energy needed to disrupt them (Stewart et al. 2009) and can be further enhanced by incorporation into macroaggregates. The adhesive properties of SOC, microbial extracellular polymeric substances, root exudates and both secretions and excretions from soil invertebrates, adsorb strongly to negatively charged soil particles through cation bridging and, along with the enmeshing action of fungi and fine roots, bind together individual soil particles and microaggregates into macroaggregates in a continuum of size ranging from microaggregates to large peds. The size of macroaggregates is increasingly controlled by physical mechanisms that enforce a maximum size limit, e.g., water infiltration, erosion, and management (such as tillage and compaction). Stable aggregates contribute to the stabilization of SOC (Watts et al. 2001, 2005) but the maintenance of stable aggregates depends at least in part on the turnover of SOC in aggregates to sustain soil microorganisms. Therefore, if the supply of SOC is restricted, aggregate stability declines and the contribution to SOC protection by occlusion diminishes (Hirsch et al. 2009). Thus, physical protection of SOC by occlusion mechanism relies on SOC turnover and is a prime example of the "use it or hoard it" principle (Janzen 2006).

6.3.3 LANDSCAPE PROCESSES

The Earth's land surface is dominated by sloping landscapes, and every year soil erosion distributes about 75 Gt of topsoil (Sanderman and Berhe 2017). The process of soil erosion induces terrestrial sequestration of atmospheric CO_2 when C eroded from up-slope is replaced by new input of organic C to soil from plant production and/or stored in more stable pools in lowland depositional positions (Sanderman and Berhe 2017; Harden et al. 1999; Stallard 1998). Over the last two decades, it has been proposed that erosion leads to loss of SOC (for example, Bajracharya et al. 2000; Jacinthe and Lal 2001; Lal 2003; Lal 2005), while others suggest that erosion typically leads to net loss of C from soil only when part of the landscape (i.e., eroding slopes) are considered (Sanderman and Berhe 2017; Doetterl et al. 2016; Van Oost et al. 2007). Erosional redistribution of topsoil the changes amount and nature of physically protected C in soil. In eroding landform positions raindrop impact and wear of aggregates during runoff can lead to breakdown of aggregates and exposure of previously physically protected C to loss through decomposition and leaching. By contrast, in depositional landform positions, the opposite occurs, with deposition of eroded topsoil C being accompanied by deposition of reactive minerals (including iron and aluminium oxyhydroxides) eroded from upslope positions. Deposition of reactive minerals and C in typically poorly-drained depositional landform positions leads to aggregate formation and stabilization, which along with burial, protects deposited SOC from loss (Berhe and Kleber 2013).

6.3.4 BIOGENIC AND PEDOGENIC CARBONATE FORMATION

Biological mechanisms in secondary carbonate formation can lead to the capture of C originating from autotrophs and incorporation into carbonate minerals, usually calcite or aragonite. Considering the potentially millennial mean residence times of SIC, this accumulation of pedogenic SIC can be considered as a stable sink of atmospheric CO_2 (Bughio et al. 2016). Biogenic carbonates are those synthesised in the bodies of organisms. For example, many earthworms excrete by calcium carbonate ($CaCO_3$) granules up to 2 mm in diameter (biospheroids) produced in calciferous glands (Barta 2011). The reason for their production is assumed to be linked to excretion of excess Ca, neutralisation of gut pH and regulation of CO_2 (Lambkin et al. 2011). Briones et al. (2008) showed that CO_2 capture is the primary function of $CaCO_3$ granule production in *Lumbricus terrestris* using $^{13}CO_2$-labeling in a microcosm experiment. They estimated that under average earthworm population densities the $CaCO_3$ granules could yield up to 25 kg C ha^{-1} a^{-1} and therefore play a small, but important role in the physical protection of soil C because they can be stored in soils under appropriate conditions for millennia (Versteegh et al. 2013).

Pedogenic carbonates are formed by dissolution and re-precipitation of dissolved SIC due to biotic processes, e.g., rhizoliths (Klappa 1980). Carbon mineralization increases the partial pressure of soil CO_2 and the decomposition of SOC itself is a source of Ca^{2+} and Mg^{2+} (Lal and Shukla 2004). The biologically-derived CO_2 reacts with soil water creating carbonic acid (H_2CO_3) and bicarbonate (HCO_3^{3-}), which react with Ca^{2+} present in the soil and precipitate as $CaCO_3$ (Lal and Kimble 2000).

- $CO_2 + H_2O \Leftrightarrow H^+ + HCO_3 \Leftrightarrow H_2CO_3$
- $Ca^{2+} + 2HCO_3 \Leftrightarrow CaCO_3 + H_2CO_3$
- $Ca^{2+} + HCO_3 \Leftrightarrow CaCO_3 + H^+$

The stable C isotopic signature of pedogenic carbonates indicates that they are precipitated in isotopic equilibrium with soil CO_2 released predominantly from root and microbial respiration and reflect the photosynthetic pathway of the vegetation present during their formation (Gocke et al. 2012). The roots of some plants (mostly grass species) undergo calcification and form calcified root cell structures and/or hypocoatings and rhizosheaths on viable plant roots due to the formation of zones of supersaturated CO_2 in the rhizosphere caused by the suction effect of the roots formed during the root's lifespan or shortly thereafter (Gocke et al. 2014).

6.4 ENVIRONMENTAL CONSTRAINTS ON SOC DECOMPOSITION

Even under circumstances where both SOC and heterotrophic soil organisms are simultaneously co-located, physical conditions including temperature, water availability and oxygen availability may still restrict C mineralization, because biological processes can only operate within a range of environmental tolerances. Where conditions in the soil exceed these thresholds heterotrophic activity may be repressed, but if inputs from photosynthesis proceed SOC may accumulate.

6.4.1 TEMPERATURE

Biological processes, including C mineralization in soils, respond to changes in temperature. Assuming that there is no water limitation, the optimum temperature range for metabolic activity is around typically 25–30°C. The magnitude of change to the rate of SOC decomposition as temperature increases by 10°C is described by the Q_{10} parameter and usually has a value of around 2, except at the limits of biological activity. The upper limit for biological activity in soils is around 50°C, because most proteins that control metabolic function (including the enzymes that catalyse SOC decomposition) denature above this limit and Q_{10} falls rapidly as the upper temperature limit is approached. At low temperatures, biological activity increases rapidly as the temperature increases away from 0°C. Although some biological processes can be detected even at sub-zero temperatures, their contribution to net C mineralization is negligible. Moreover, under natural circumstances, it is difficult to separate the effect of extreme temperatures from water availability in soils which is limiting at both high and low temperatures. Where photosynthesis proceeds but C mineralization is retarded by temperature, SOC may accumulate. Thus, the large stocks of SOC in the cryosols are particularly vulnerable to warming, and in the boreal regions which contain large SOC stocks protected from C mineralization by low temperatures, the effect of warming will be particularly marked (Karhu et al. 2014).

6.4.2 WATER AND OXYGEN

At temperatures above 0°C and below 100°C, the availability of water and the availability of oxygen are closely linked. In general, the amount of water in the soil is inversely related to the amount of oxygen in the soil. Most soil organisms are obligate aerobes, requiring oxygen to act as the terminal electron acceptor during respiration. When soils are waterlogged, the rate of diffusion of oxygen through water to biological organisms is too slow to allow aerobic respiration to proceed. However, facultative and obligate anaerobes become active under these conditions allowing reduced rates of C mineralization to proceed by fermentation, which ultimately leads to the production of gaseous methane, and other incompletely oxidised terminal products in solution. At the scale of the individual aggregate, water and oxygen are not uniformly distributed. Water tends to occupy the pores with the smallest neck diameters under non-saturated conditions in soils, drawn and held strongly in place by capillarity, restricting the movement of oxygen. Consequently, locations at the interior of aggregates may be persistently anoxic, regardless of the moisture content of the wider soil environment. Anoxic conditions are likely to arise rapidly in newly-formed aggregates where fresh organic C is newly-encapsulated. These include the casts of earthworms which are important macroinvertebrate "soil engineers" in many soils (Jouquet et al. 2006).

6.5 ANTHROPOGENIC EFFECTS ON PHYSICAL PROTECTION OF SOIL CARBON

Human activity over the last couple of millennia, but most significantly in the last 200 years (Certini and Scalenghe 2011; Richter et al. 2015) has affected the physical protection of soil C both directly and indirectly (Figure 6.3). Physical protection is not a single process, but rather the result of multiple

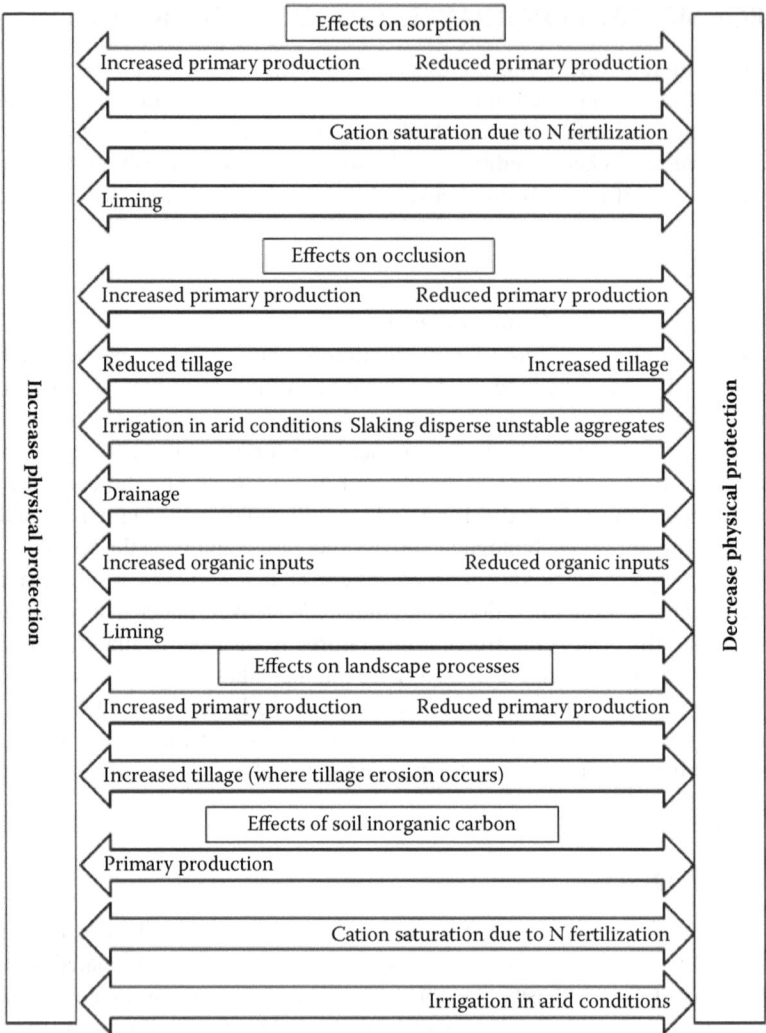

FIGURE 6.3 Generalized effects of different environmental and management effects on physical protection of soil organic and inorganic carbon.

interacting processes operating over a range of spatial and temporal scales. Making predictions about the effects of climate change on the amount of SOC and SIC protected by physical mechanisms and the detail of the specific mechanisms is not straightforward, not least because some of the changes to the climate are not shifts in the average conditions (i.e., consistently warmer or drier or wetter), but changes in the frequency and intensity of the extreme conditions leading to flooding or droughts and in turn significant erosional redistribution and land slips. The summary in Figure 6.3 represents generalizations and should be interpreted as though all other factors are equal and that the details will differ subtly between soils. It is nonetheless clear, that the key components of physical protection are encapsulation of the SOC at sites where it is isolated from microorganisms or where the environmental conditions retard decomposition, or stabilization on colloid surfaces retarding microbial attack, or because of the overall restrictions on decomposition activity imposed by the wider environmental conditions such as temperature or oxygen availability. Thus, in addition to being of fundamental interest to advance understanding of SOC dynamics, appreciation of the sensitivity of SOC to changes in the physical protection mechanisms has important applied and soil management implication.

6.6 ACKNOWLEDGMENTS

This work was supported as part of Rothamsted Research's Institute Strategic Programme – Soil to Nutrition (BB/PO1268X/1), funded by the UK Biotechnology and Biological Sciences Research Council which is acknowledged, as is the support from The Royal Agricultural University.

REFERENCES

Baldock, J., Skjemstad, J. 2000. Role of the soil matrix and minerals in protecting natural organic materials against biological attack. *Organic Geochemistry* 31, 697710.

Bajracharya, R.M., Lal, R., Kimble, J.M. 2000. Diurnal and seasonal CO_2–C flux from soil as related to erosion phases in central Ohio. *Soil Science Society of America Journal* 64, no 1: 286–293.

Barta, G. 2011. Secondary carbonates in loess-paleosoil sequences: A general review. *Open Geosciences* 3, 129–146.

Berhe, A.A., Kleber, M. 2013. Erosion, deposition, and the persistence of soil organic matter: mechanistic considerations and problems with terminology. *Earth Surface Processes and Landforms* 38, 908–912.

Boddy, E., Hill, P.W., Farrar, J., Jones, D.L. 2007. Fast turnover of low molecular weight components of the dissolved organic carbon pool of temperate grassland field soils. *Soil Biology & Biochemistry* 39, 827–835.

Bradford, S., Kim, H., Shen, C., Sasidharan, S., Shang, J. 2017. Contributions of nanoscale roughness to anomalous colloid retention and stability behavior. *Langmuir* 33, no 38: 10094–10105.

Briones, M.J.I., Ostle, N.J., Piearce, T.G. 2008. Stable isotopes reveal that the calciferous gland of earthworms is a CO_2-fixing organ. *Soil Biology & Biochemistry* 40, 554–557.

Bughio, M.A., Wang, P., Meng, F., Qing, C., Kuzyakov, Y., Wang, X., Junejo, S.A. 2016. Neoformation of pedogenic carbonates by irrigation and fertilization and their contribution to carbon sequestration in soil. *Geoderma* 262, 12–19.

Certini, G., Scalenghe, R. 2011. Anthropogenic soils are the golden spikes for the Anthropocene. *The Holocene* 21, 1269–1274.

Crawford, J., Deacon, L., Grinev, D., Harris, J., Ritz, K., Singh, B., Young, I. 2012a. Microbial diversity affects self-organization of the soil-microbe system with consequences for function. *Journal of the Royal Society Interface* 9, no 71: 1302–1310.

Doetterl, S., Berhe, A.A., Nadeu, E., Wang, Z., Sommer, M., Fiener, P. 2016. Erosion, deposition and soil carbon: A review of process-level controls, experimental tools and models to address C cycling in dynamic landscapes. *Earth-Science Reviews* 154, 102–122.

Dungait, J.A.J., Hopkins, D.W. 2016. Carbon: Physical protection. Lal, R. (Ed.), *Encyclopedia of Soil Science*, Third Edition. CRC Press, pp. 327 Lal, R. (Ed.), 330.

Dungait, J.A.J., Hopkins, D.W., Gregory, A.S., Whitmore, A.P. 2012. Soil organic matter turnover is governed by accessibility not recalcitrance. *Global Change Biology* 18, 1781–1796.

Ekschmitt, K., Kandeler, E., Poll, C., Brune, A., Buscot, F., Friedrich, M., Gleixner, G., Hartmann, A., Kästner, M., Marhan, S. 2008. Soil carbon preservation through habitat constraints and biological limitations on decomposer activity. *Journal of Plant Nutrition and Soil Science* 171, 27–35.

Ekschmitt, K., Liu, M., Vetter, S., Fox, O., Wolters, V. 2005. Strategies used by soil biota to overcome soil organic matter stability—Why is dead organic matter left over in the soil? *Geoderma* 128, 167–176.

Ellerbrock, R., Gerke, H. 2004. Characterizing organic matter of soil aggregate coatings and biopores by fourier transform infrared spectroscopy. *European Journal of Soil Science* 55, no 2: 219–228.

Gocke, M., Gulyás, S., Hambach, U., Jovanović, M., Kovács, G., Marković, S.B., Wiesenberg, G.L.B. 2014. Biopores and root features as new tools for improving paleoecological understanding of terrestrial sediment-paleosol sequences. *Palaeogeography, Palaeoclimatology, Palaeoecology* 394, 42–58.

Gocke, M., Pustovoytov, K., Kuzyakov, Y. 2012. Pedogenic carbonate formation: Recrystallization versus migration—Process rates and periods assessed by ^{14}C labeling. *Global Biogeochemical Cycles* 26, GB1018.

Harden, J., Sharpe, J., Parton, W., Ojima, D., Fries, T., Huntington, T., Dabney, S. 1999. Dynamic replacement and loss of soil carbon on eroding cropland. *Global Biogeochemical Cycles* 13, no 4: 885–901.

Hassink, J., Whitmore, A.P. 1997. A model of the physical protection of organic matter in soils. *Soil Science Society of America Journal* 61, 131–139.

Hirsch, P.R., Gilliam, L.M., Sohi, S.P., Williams, J.K., Clark, I.M., Murray, P.J. 2009. Starving the soil of plant inputs for 50 years reduces abundance but not diversity of soil bacterial communities. *Soil Biology and Biochemistry* 41, 2021–2024.

Jacinthe, P.A., Lal, R. 2001. A mass balance approach to assess carbon dioxide evolution during erosional events. *Land Degradation & Development* 12, 329–339.

Janzen, H. 2006. The soil carbon dilemma: Shall we hoard it or use it? *Soil Biology & Biochemistry* 38, 419–424.

Janzen, H. 2015. Beyond carbon sequestration: Soil as conduit of solar energy. *European Journal of Soil Science* 66, 19–32.

Jenkinson, D.S., Poulton, P.R., Bryant, C. 2008. The turnover of organic carbon in subsoils. Part 1. Natural and bomb radiocarbon in soil profiles from the Rothamsted long-term field experiments. *European Journal of Soil Science* 59, 391–399.

Jouquet, P., Dauber, J., Lagerlöf, J., Lavelle, P., Lepage, M. 2006. Soil invertebrates as ecosystem engineers: Intended and accidental effects on soil and feedback loops. *Applied Soil Ecology* 32, 153–164.

Kaiser, K., Kalbitz, K. 2012. Cycling downwards—Dissolved organic matter in soils. *Soil Biology and Biochemistry* 52, 29–32.

Karhu, K., Auffret, M.D., Dungait, J.A.J., Hopkins, D.W., Prosser, J.I., Singh, B.K., Subke, J.-A., Wookey, P.A., Agren, G.I., Sebastia, M.-T., Gouriveau, F., Bergkvist, G., Meir, P., Nottingham, A.T., Salinas, N., Hartley, I.P. 2014. Temperature sensitivity of soil respiration rates enhanced by microbial community response. *Nature* 513, 81–84.

Klappa, C.F. 1980. Rhizoliths in terrestrial carbonates: Classification, recognition, genesis and significance. *Sedimentology* 27, 613–629.

Kleber, M. 2010. What is recalcitrant soil organic matter? *Environmental Chemistry* 7, 320–332.

Kleber, M., Nico, P.S., Plante, A.F., Filley, T., Kramer, M., Swanston, C., Sollins, P. 2011. Old and stable soil organic matter is not necessarily chemically recalcitrant: Implications for modeling concepts and temperature sensitivity. *Global Change Biology* 17, 1097–1107.

Kraimer, R.A., Monger, H.C., Steiner, R.L. 2005. Mineralogical distinctions of carbonates in desert soils. *Soil Science Society of America Journal* 69, 1773–1781.

Kravchenko, A.N., Guber, A.K. 2017. Soil pores and their contributions to soil carbon processes. *Geoderma* 287, 31–39.

Lal, R. 2003. Soil erosion and the global carbon budget. *Environment International* 29, 437–450.

Lal, R. 2005. Soil erosion and carbon dynamics. *Soil and Tillage Research* 81, 137–142.

Lal, R., Shukla, M.K. 2004. *Principles of Soil Physics*. CRC Press. p. 73.

Lal, R., Kimble, J.M., 2000. Pedogenic carbonates and the global carbon cycle. In *Global Climate Change and Pedogenic Carbonates*, pp. 1–14.

Lambkin, D.C., Gwilliam, K.H., Layton, C., Canti, M.G., Piearce, T.G., Hodson, M.E. 2011. Production and dissolution rates of earthworm-secreted calcium carbonate. *Pedobiologia* 54, S119–S129.

Lehmann, J., Kleber, M. 2015. The contentious nature of soil organic matter. *Nature* 528, 60–68.

Lima, I., Ro, K., Boateng, A., Klasson, K. 2010. Biochars from agricultural residuals as adsorbents for environmental remediation. *Abstracts of Papers of the American Chemical Society* 239.

Marschner, B., Brodowski, S., Dreves, A., Gleixner, G., Gude, A., Grootes, P.M., Hamer, U., Heim, A., Jandl, G., Ji, R., Kaiser, K. 2008. How relevant is recalcitrance for the stabilization of organic matter in soils? *Journal of Plant Nutrition and Soil Science* 171, 91–110.

Martens, D.A., Reedy, T.E., Lewis, D.T. 2003. Soil organic C content and composition of 130-year crop, pasture and forest land-use managements. *Global Change Biology* 10, 65–78.

Richter, D., de B., Bacon, A.R., Brecheisen, Z., Mobley, M.L. 2015. Soil in the Anthropocene. IOP Conference Series: *Earth and Environmental Science* 25, 012010.

Saggar, S, Parshotam, A., Hedley, C., Salt, G. 1999. ^{14}C-labelled glucose turnover in New Zealand soils. *Soil Biology and Biochemistry* 31, 2025–2037.

Sanderman, J., Berhe, A.A. 2017. Biogeochemistry: The soil carbon erosion paradox. *Nature Climate Change* 7, 317.

Schlesinger, W. 1985. The formation of caliche in soils of the Mojave Desert, California. *Geochimica Et Cosmochimica Acta* 49, no 1: 57–66.

Schmidt, M.W.I., Torn, M.S., Abiven, S., Dittmar, T., Guggenberger, G., Janssens, I.A., Kleber, M., Kogel-Knabner, I., Lehmann, J., Manning, D.A.C., Nannipieri, P., Rasse, D.P., Weiner, S., Trumbore, S.E. 2011. Persistence of soil organic matter as an ecosystem property. *Nature* 478, 49–56.

Six, J., Conant, R., Paul, E., Paustian, K. 2002. Stabilization mechanisms of soil organic matter: Implications for C-saturation of soils. *Plant and Soil* 241, 155–176.

Six, J., Guggenberger, G., Paustian, K., Haumaier, L., Elliott, E.T., Zech, W. 2001. Sources and composition of soil organic matter fractions between and within soil aggregates. *European Journal of Soil Science* 52, 607–618.

Stallard, R. 1998. Terrestrial sedimentation and the carbon cycle: Coupling weathering and erosion to carbon burial. *Global Biogeochemical Cycles* 12, no 2: 231–257.

Stewart, C.E., Paustian, K., Conant, R.T., Plante, A.F., Six, J. 2009. Soil carbon saturation: Implications for measurable carbon pool dynamics in long-term incubations. *Soil Biology and Biochemistry* 41, 357–366.

Tipping, E., Chamberlain, P.M., Fröberg, M., Hanson, P.J., Jardine, P.M. 2012. Simulation of carbon cycling, including dissolved organic carbon transport, in forest soil locally enriched with ^{14}C. *Biogeochemistry* 108, 91–107.

Van Oost, K., Quine, T., Govers, G., De Gryze, S., Six, J., Harden, J., Ritchie, J., Mccarty, G., Heckrath, G., Kosmas, C., Giraldez, J., Da Silva, J., Merckx, R. 2007. The impact of agricultural soil erosion on the global carbon cycle. *Science* 318, no 5850: 626–29.

Van Veen, J., Kuikman, P. 1990. Soil structural aspects of decomposition of organic matter by micro-organisms. *Biogeochemistry* 11, 213–233.

Versteegh, E.A.A., Black, S., Canti, M.G., Hodson, M.E. 2013. Earthworm-produced calcite granules: A new terrestrial palaeothermometer? *Geochimica et Cosmochimica Acta* 123, 351–357.

Von Lutzow, M., Kogel-Knabner, I. 2010. Response to the concept paper: "What is recalcitrant soil organic matter?" by Markus Kleber. *Environmental Chemistry* 7, no 4: 333–335.

Von Lützow, M., Kögel Knabner, I., Ekschmitt, K., Matzner, E., Guggenberger, G., Marschner, B., Flessa, H. 2006. Stabilization of organic matter in temperate soils: Mechanisms and their relevance under different soil conditions—A review. *European Journal of Soil Science* 57, 426–445.

Watts, C., Whalley, W., Longstaff, D., White, R., Brook, P., Whitmore, A. 2001. Aggregation of a soil with different cropping histories following the addition of organic materials. *Soil Use and Management* 17, 263–268.

Watts, C.W., Whalley, W.R., Brookes, P.C., Devonshire, B.J., Whitmore, A.P. 2005. Biological and physical processes that mediate micro-aggregation of clays. *Soil Science* 170, 573–583.

Young, I., Ritz, K. 2000. Tillage, habitat space and function of soil microbes. *Soil and Tillage Research* 53, 201–213.

Zamanian, K., Pustovoytov, K., Kuzyakov, Y. 2016. Pedogenic carbonates: Forms and formation processes. *Earth-Science Reviews* 157, 1–17.

Zimmermann, M., Leifeld, J., Schmidt, M., Smith, P., Fuhrer, J. 2007. Measured soil organic matter fractions can be related to pools in the RothC model. *European Journal of Soil Science* 58, 658–667.

7 Nitrogen Cycling and Dynamics in Terrestrial Ecosystems

David A.N. Ussiri and Rattan Lal

CONTENTS

7.1 INTRODUCTION

Nitrogen (N), the fifth most abundant element in the earth, is the most abundant element in the atmosphere, comprising 78% by volume as N_2 gas. However, it is unavailable to most organisms because of the strength of the triple bond that binds the two N atoms together. Nitrogen is an important element in biogeochemical cycles and plays a pivotal role in regulating species composition, diversity, dynamics, and function (Vitousek et al. 1997a). All organisms require N in order to live, and it ranks fourth behind oxygen (O), carbon (C), and hydrogen (H) as the most common element in living tissues, and depending on the life form, for every 100 atoms of C incorporated into cells, 2 to 20 atoms of N are needed (Sterner and Elser 2002). In living organisms, N plays two major roles: assimilation (i.e., acquisition of matter for incorporation into biomass) and dissimilation (i.e., extraction of energy from the environment). N is unique among the major elements required for life in that its cycling includes a vast atmospheric reservoir of N_2 gas that must be fixed before it can be used by most organisms. Therefore, the productivity of many terrestrial ecosystems, both unmanaged and managed, including agricultural and pastures are generally limited by the supply of biologically available N. The net primary production (NPP)—the productive capacity of the ecosystem—is generally regulated by the plant useable forms of N in soils. In addition, the supply of the fixed N, in part, controls the C storage and species composition of many ecosystems.

Over the evolutionary history, only a limited number of species of bacteria and archaea have evolved the ability to convert N_2 to reactive N (Nr, which includes NH_3, NH_4^+, NO, NO_2, HNO_3, N_2O, NO_3^-, and organic compounds—urea, amines, proteins, nucleic acids; Table 7.1) which, in addition to their role in biological and ecosystem functions in terrestrial and marine ecosystems,

TABLE 7.1

Nitrogenous Compounds and Their Role in Biosphere N Cycling

Compound	Chemical Formula	Natural Sources	N Cycling Processes Involved
Nitrogen gas	N_2	Nitrification Denitrification	• Biological N fixation • Denitrification
Nitrate/Nitrite Nitric acid/nitrous acid	NO_3^-/NO_2^- HNO_3/HNO_2	Nitrification Denitrification Lightning	• Dry deposition • Wet deposition • Leaching to groundwater, rivers, oceans • Atmospheric reactions • Plant uptake
Nitric oxide	NO	Lightning Nitrification Denitrification	• Dry deposition • Atmospheric reactions
Nitrogen dioxide	NO_2		• Atmospheric reactions • Nitrite reduction in acid soils
Nitrous oxide	N_2O	Nitrification Denitrification	• Stratospheric reactions • Greenhouse effects
Nitrogen pentoxide	N_2O_5		
Organic nitrogen	Variable	Biomass production Organic matter	• Dry deposition • Volatilization • Organic matter decomposition
Ammonia/Ammonium	NH_3/NH_4^+	Biological N fixation	• Plant uptake • Nitrification • Fertilizer production • Dry deposition • Wet deposition • Leaching to groundwater, rivers and oceans • Biomass combustion • Volatilization • Stratospheric reactions • Organic matter decomposition

Source: Adapted from Ussiri, D.A.N. and Lal, R., *Soil Emmisions of Nitrous Oxide and Its Mitigation*, Springer, Dordrecht, Netherlands and New York, USA, 378 p., 2013.

also becomes widely distributed in the atmosphere and cryosphere and are available for the synthesis of amino acids and other metabolic products to support growth. The global N cycle is central to the biogeochemistry of the Earth with large natural and anthropogenic flows of N from the atmosphere into terrestrial and marine ecosystems through N fixation, where largely unreactive molecular N_2 is reduced to ammonium (NH_4^+) compounds. The fixed N is subsequently transformed into a range of organic and inorganic compounds including amino acids, proteins, and oxidized compounds by microorganisms and finally returned to the atmosphere as molecular N through microbial denitrification in soils, freshwaters, and sediments (Galloway et al. 2004). Many terrestrial ecosystems are adapted to conditions of low N availability, a condition that often leads to plant communities with high diversity (Bobbink et al. 1998). Even with adaptations to use N efficiently, primary production in many ecosystems of the world is limited by N (Falkowski 1997, 2008; Vitousek et al. 1997a), mainly because: (i) available N is almost completely tied to biological N (as organic N which is immobile and unavailable for plant uptake) rather than inorganic forms that are available for chemical processes, and (ii) unlike other nutrients, N almost never accumulates in soils in inorganic form for any length of time. The limited availability of Nr enabled the natural processes to be effective and efficient at using N as an important nutrient resource and created interactions among plants, animals, and soil life rich in biodiversity (De Vries et al. 2006).

In the absence of human influence, biological N fixation (BNF) and the production of NO_x by lightning were the only sources of new Nr in the environment, and the quantity of new Nr added to the existing pool of Nr that cycles among the living and nonliving components of Earth's ecosystems represented only a small fraction. One of the important features of pre-industrial N cycle is that N fixation—largely by BNF and lightning—was largely balanced by deep sedimentation, denitrification, anammox, and other processes which converted Nr back to N_2, with little accumulation of Nr in environmental reservoirs (Canfield et al. 2010; Galloway et al. 2004, 2013, 2014). Therefore, Nr was mainly used by organisms in terrestrial and aquatic ecosystems to sustain life and efficiently recycled through the food web system. The N availability was the main limiting factor for productivity in terrestrial, aquatic, and marine ecosystems, and Nr concentrations remained low in most environmental systems. The only significant anthropogenic process used to create new Nr to grow crops was the cultivation of legumes, and this was on a small scale. For example, in 1850, only ~20% of Nr creation on land was due to cultivation-induced BNF (Galloway et al. 2013). As the human population grew, the demand for food increased. In addition, the increased demand for energy to power the Industrial Revolution was met by increased fossil fuel combustion, which also contributed to anthropogenic Nr creation.

Human activities began to substantially contribute to the global N cycle at the beginning of the 20th century, both intentionally through the synthesis of industrial N fertilizers by the Haber–Bosch process (Erisman et al. 2008), implementing new agricultural practices that boost crop yields, and the cultivation of crops that support BNF, and also unintentionally through fossil fuel combustion (Galloway et al. 2008; Canfield et al. 2010). Therefore, humans, who played an insignificant role in the creation of Nr, became the dominant source of Nr. The industrial fixation of Nr through the Haber–Bosch process increased from <10 Tg yr^{-1} in 1950s to current levels at 115 Tg N yr^{-1} (Erisman et al. 2008) and is expected to increase to nearly 135 Tg N yr^{-1} by 2030 (Galloway and Cowling 2002). The global N cycle has been perturbed by anthropogenic activity over the past 100 years, with approximately two thirds of annual flux of Nr entering the biosphere at the beginning of the 21st century being of anthropogenic origin (Galloway et al. 2004; Fowler et al. 2013, 2015). Global Nr creation continues to rise with the increasing global population (Erisman et al. 2011). Of the total Nr created by the Haber–Bosch process, approximately 80% is used in the production of agricultural fertilizers (Galloway et al. 2008). Global N fertilizer use increased steadily from early 1960s through the mid-1980s and then declined through the early 1990s before starting to rise again (Figure 7.1a). Global cereal production and fertilizer N use are closely correlated (Figure 7.1b), and about 30%–50% of cereal yield increase has been attributed to N application through mineral fertilizer (Stewart et al. 2005; Stewart and Roberts 2012). About 50% of industrial Nr produced by the Haber–Bosch process is applied to three major cereals: maize (*Zea mays*) 16%, rice (*Oryza sativa*) 16%, and wheat (*Triticum vulgaris*) 18% (Ladha et al. 2016). These provide the bulk of human food calories and protein consumed either directly as grain or indirectly through livestock products (Ladha et al. 2016). Increased N availability increases productivity and biomass accumulation substantially at least on a short-term basis. The increase in Nr has been exploited in agriculture to increase crops yield and provide food for the growing global population. Almost half of population at the beginning of the 21st century depends on fertilizer N for their food production (Erisman et al. 2008). As a result, billions of people have been fed by increased productivity in most regions facilitated by the production of fertilizers through the industrial Haber–Bosch process, even though the main motivation for the industrial synthesis of ammonia from its elements was to provide raw materials for the explosives to be used in weapons (Enrisman et al. 2008).

The biogeochemistry of N is almost entirely dependent on microbially mediated reduction–oxidation (redox) reactions and to a lesser extent on long-term recycling through the geosphere. In the ecosystem, N exists in more chemical forms than most other elements (Table 7.1) and undergoes a series of unique chemical transformations, most of them mediated by microorganisms as part of

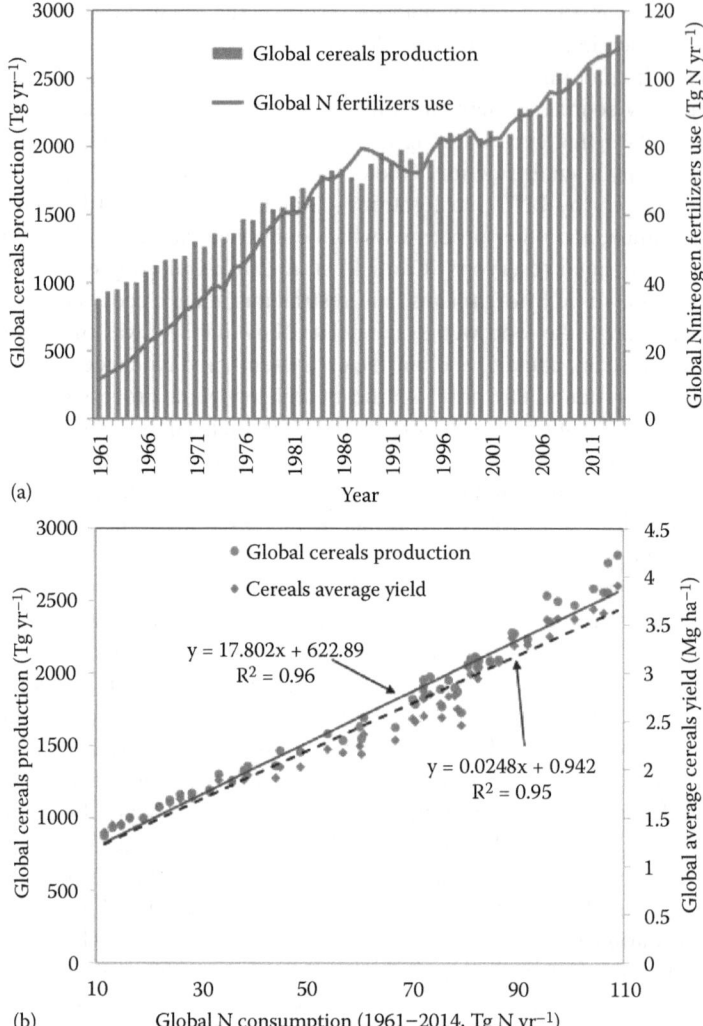

FIGURE 7.1 **(See color insert.)** World cereal production and fertilizer N use, 1961–2014. (a) Annual cereal production and its relation to annual fertilizer N use. (b) Relationship between fertilizer N use, cereal production (blue dots) and cereal yield in Mg ha^{-1} (red diamonds). (Data from FAOSTAT, World Food and Agriculture Statistics. Food and Agriculture Organization of the United Nations, Statistical Division, Rome, Italy. Accessed April 2017, 2017.)

their metabolism either to harness energy for their growth or as an alternative electron acceptor. This allows the same molecule of Nr to take part in a series of effects, both positive and negative, before being transformed back to molecular N$_2$ and returned to the atmospheric reservoir. This transformation has been termed the N cascade (Galloway et al. 2003). The large number of N compounds in the biosphere and their different transformations complicate an assessment of the pathways and effects of Nr in the environment. The extensive conversion of Nr in the environment mediated by biological and chemical processes is sensitive to environmental conditions and thus is likely to respond to climate change over the coming decades since the processes that regulate transfers between the atmosphere, terrestrial and marine reservoirs of Nr are sensitive to aspects of climate that are expected to change (Fowler et al. 2015).

In addition to an increase in food production, human alteration of the global N cycle has led to widespread negative consequences by directly contributing to the reduction in biodiversity at the regional scale in terrestrial ecosystems, damaging human health through aerosols and ozone production, and directly contributing to radiative forcing of climate (Erisman et al. 2013). A large fraction of N fertilizer applied to cropland—generally greater than 50%—is not used by the crops and is lost into the air, surface and groundwater, and downstream and downwind habitats, polluting landscapes and water sources. Consequently, the understanding of Nr and the global N cycle has shifted from how to promote food production to the realization that agricultural intensification damages the environmental systems (Vitousek et al. 1997a). The damage by Nr to ecosystems, human health, and climate change results from the leakage of N compounds from its use in agriculture, industry, and transport (Erisman et al. 2013). The magnitude of anthropogenic Nr creation estimated at 210 Tg N yr^{-1} is so large such that it has doubled the global cycling of N compared to natural sources estimated to contribute 222 ± 50 Tg N yr^{-1}, which existed before the Industrial Revolution (Fowler et al. 2013). This review describes the current global N cycle and quantifies major terms of the global N budget with the emphasis on the role of soils and anthropogenic activities impacting the alteration of the natural global N cycle. Some aspects of marine and geological processes are also covered briefly for the purpose of presenting more balanced budget.

7.2 GLOBAL NITROGEN INVENTORY

Approximately 60% of the N resides in the Earth's mantle and another 11–16% is stored in the continental and oceanic crust combined (Johnson and Goldblatt 2015; Bebout et al. 2016). Therefore, about 73% of N in the Earth's system is in its core and the mantle and is out of circulation (Table 7.2). The release of N from the mantle is negligible and almost equivalent to the reincorporation of surface volatiles at subduction zones (Walker 1977). The atmospheric reservoir accounts for 27–30% of the total N budget of the Earth (Palya et al. 2011). Overall, the surface reservoirs of N receiving the most attention (i.e., ocean, soils, and biota) contain far less than 1% N by mass (Table 7.2; Bebout et al. 2013; Galloway 2014). Although N is relatively inert and has a volatile behavior, it can occur in certain minerals in the crust and mantle, especially when it is structurally bound as NH_4^+ (Busigny and Bebout 2013). The NH_4^+-bound N in micas and feldspars is one of the efficient carriers of initially organic N from surface reservoirs into deep Earth subduction zones during the formation of the continental crust. Similar to C (Ussiri and Lal, 2017), N is cycled in the earth system as a fast biologic N cycle involving ocean, soil, biota, and atmosphere with fast turnover, and a slow geologic N cycle involving sediment, continental and ocean crust, upper and lower mantle, biosphere, and atmosphere (Figure 7.2), with N turnover times of many millions of years (Berner 2006).

7.2.1 GEOLOGICAL N CYCLE

The discriminating feature of the geologic N cycle from the biological N cycle is that the geologic N cycle involves the participation of N stocks stored in rocks. The principal processes involved in geologic N cycle are (i) burial of organic N and traces of NH_4^+ substituting for K^+ in clay minerals into sedimentary rocks and the ultimate transformation to silicate minerals (Boyd 2001), (ii) weathering of sedimentary organic matter and the liberation of N to surface biological systems (Holloway and Dahlgren 2002), and (iii) emission of N_2 and other nitrogenous gases from volcanic and metamorphic degassing (Giggenbach and Matsuo 1991). The N in the mantle, ocean, and continental crust forms a slow geologic N cycle that exchanges N in geological time periods, commonly through volcanic eruptions. The transfer of N between rocks and the surface reservoirs may involve conversion between several N oxidation states. Geologic N is frequently ignored in N budgets. However, evidence indicates that weathering of certain bedrocks can contribute significant inputs of N to marine and terrestrial ecosystems, which affects the long-term accumulation of soil N (Holloway et al. 1998, 2001).

N is present as a number of species in the solid Earth, and its speciation is primarily controlled by redox, with the temperature and pH playing roles in stability and solubility (Johnson and Goldblatt

TABLE 7.2
Estimates of Earth Nitrogen Budget Partitioned into Different Reservoirs

Source	Reservoir Size (Pg N)	Reference
Surface and Near-Surface		
Atmosphere	3.95×10^6	Galloway 2014
Sedimentary rocks	1×10^6	
Ocean N_2	2×10^4	
Ocean NO_3^-	5.7×10^2	
Soil organics	1.9×10^2	
Land biota	1.0×10	
Marine biota	5×10^{-1}	
Total	4.97×10^6	
Earth N Budget		
Atmosphere	3.9×10^6	Palya et al. 2011
Continental crust	1.1×10^6	
Ocean crust	3.2×10^5	
Upper mantle	1.7×10^5	
Lower mantle	7.8×10^6	
Deep ocean	6×10^2	
Surface ocean	6×10	
Soils	1×10^2	
Biomass	4.3	
Marine biota	3×10^{-1}	
Terrestrial vegetation	4	
Total	13.29×10^6	
Major Earth Reservoirs of N		
Atmosphere	4×10^6	Goldblatt et al. 2009
Continental crust	2.1×10^6	
Oceanic crust	3.2×10^5	
Mantle	8.4×10^6	
Total	14.82×10^6	

2015); the chemical form of the N depends on the reservoir (Galloway 2014; Mackenzie 2010). The N_2 is the most common form of volcanic gas emitted at a rate of ~2 Tg N yr^{-1} (Jaffee 1992). Once emitted, it either remains in the atmosphere or is deposited to the Earth's surface and continues the process of bio-geochemical cycling. The important N species in solid Earth are (i) N_2 (fluid incursions and degassing magma) (Marty 1995), (ii) NH_3 (in reduced fluids) (Li and Keppler 2014), (iii) NH_4^+ (stable and bound in mineral lattices (Itihara and Honma 1979), and (iv) nitrides, e.g., FeN (Adler and Williams 2005).

The N in the rocks and the NH_4^+ in the Earth's crust are derived from the atmosphere through biotic and abiotic N fixation (Boyd 2001) and include N in igneous rocks—where NH_4^+ is frequently present in the lattice of silicate minerals. The important path for N incorporation into minerals is substitution of trace amounts of NH_4^+. The NH_4^+ has the same charge and similar ionic radius as K$^+$ (i.e., <0.2 A° compared to the ionic radius of K$^+$, 1.61–1.69 A°) and it can readily substitute into mineral lattice sites that are normally occupied by K$^+$ bearing minerals (Whittaker and Muntus 1970; Khan and Baur 1972). Clay minerals, micas, and K-feldspars are important mineral hosts of N. In some instances, N is introduced into the mineral lattice by ammonium-rich hot spring waters (e.g., buddingtonite). The average for igneous rocks is 25 mg N kg^{-1} (Wedepohl 2012).

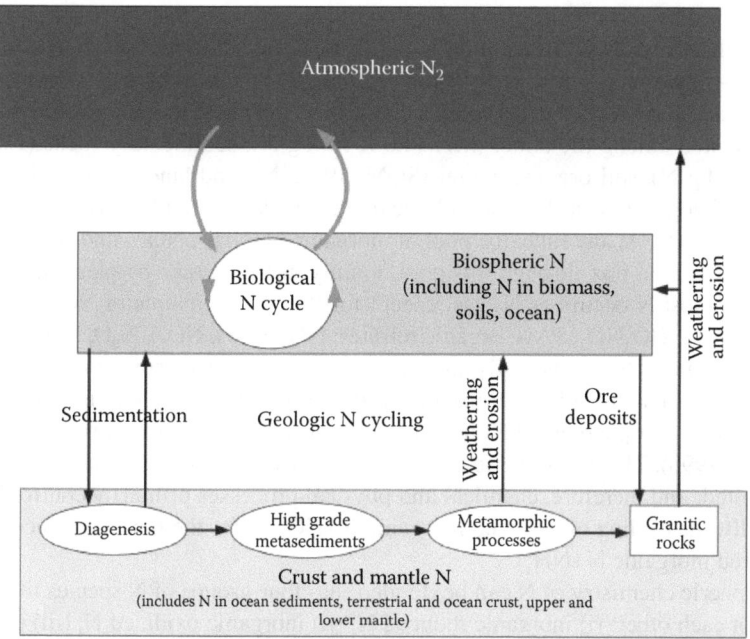

FIGURE 7.2 A simplified conceptual diagram illustrating the interaction between fast biologic N cycling and slow geologic N cycling. (Based on Johnson, B., Goldblatt, C., *Earth-Science Reviews* 148, 150–173, 2015; Bebout, G.E., Lazzeri, K.E., Geiger, C.A., *American Mineralogist* 101, 7–24, 2016.)

The N concentration in rocks varies based on rock formation conditions. In general, quartz contains the least NH_4^+ and biotite the most, while feldspar falls in between the two. In sedimentary and metamorphic rocks, NH_4^+ is the dominant form of N. In sedimentary rocks, N is derived from organic sediments, which lost the amino groups during conversion of sediments to rock or are retained by adsorption. The average N content of sedimentary rocks is 490 mg kg^{-1}, but the range of concentration is broad—70 mg kg^{-1} limestones, 150 mg kg^{-1} sandstones and greywackes, and 600 mg kg^{-1} in shales (Wedepohl 2012). Some C-rich sedimentary rocks contain large quantities of undestroyed organic N, which can constitute 75% to 85% of total N (Holloway et al. 2001). Metamorphism generally destroys organic materials and drives off NH_4^+ along with other volatiles. The N content of metamorphic rocks is generally comparable with that of igneous rocks. N entrained in sediments, rocks, and crust is generally carried into subduction zones where it is either volatilized and removed from the down-going plate or carried into the mantle past the subduction barrier depending on temperature (Johnson and Goldblatt 2015; Elkins et al. 2006; Mitchell et al. 2010). Volatilized N either oxidizes to N_2 and escapes via volcanism or is incorporated into intrusive igneous rocks. The N that is not returned to the surface becomes entrained in mantle circulation. Therefore, it has the potential to be the tracer of the processes linking surface Earth to different reservoirs in the solid planet (Johnson and Goldblatt 2015).

7.2.2 BIOLOGICAL N CYCLE

The circulation of N between atmosphere, soils, ocean, terrestrial, and marine biota forms a fast biological N cycle (Figure 7.2), commonly referred to as global N cycle in most scientific literature involving the interaction of <1% of N in the biosphere and 27% of N in the atmosphere with minor contribution from sedimentary rocks. The biological N cycle involves N fixation naturally by bacteria and archaea, and also by lightning or anthropogenically. The fixed N is then released to the environment primarily as NH_4^+ and NO_3^-. NH_4^+ is quickly oxidized to NO_3^- in a bacterially mediated nitrification. The primary return of N to the atmosphere is through denitrification, where NO_3^- is used by certain bacteria as the electron acceptors in the electron transport chain and converted to either N_2 or N_2O. In addition to denitrification, additional reaction anaerobic ammonium oxidation—anammox— also returns fluxes of N_2 to the atmosphere (Thamdrup 2012).

The surface and near-surface reservoirs account for 27–30% of the total N budget of the Earth, and N is distributed among the atmosphere (4.0×10^6 Pg N), sedimentary rocks (1.0×10^6 Pg N), oceans (2×10^4 Pg N), soil organic matter (SOM, 19 Pg N), and land biota (10 Pg N) (Galloway 2014). The geobiology of N in the fast biological cycle is dominated by large inert reservoirs and small biological fluxes. At any time, the pool of inorganic N (NH_4^+, NO_3^-, and NO_2^-) is rather small despite the large annual flux through this pool, mainly due to uptake by plants and organisms. In the atmosphere, most N occurs as N_2 gas, except for the trace amounts of NO_y (i.e., NO_x, HNO_3, HONO, N_2O_5, NO_3^-, HO_2NO_2, PAN, organic nitrates; $NO_x = NO$, NO_2), N_2O, NH_x (i.e., NH_3, NH_4), and particulate organic N. In the soils and oceans, N occurs primarily as organic N, NO_3^-, and NH_4^+. Atmospheric and hydrologic transports provide the primary paths for the exchange of N between reservoirs. A triple bonded N_2 must be transformed into reactive form to become bioavailable (Galloway, 1998). The atmospheric cycling of N is a simple cycle because the direct influence of biota is limited, and therefore, chemical and physical processes primarily control N transformations. Generally, the cycling of oxidized inorganic N (NO_y) is for the most part decoupled from the cycle of reduced inorganic N (NH_x).

The atmospheric chemistry of N can be divided into four groups of N species that are generally independent of each other: (i) inorganic reduced N, (ii) inorganic oxidized N, (iii) organic reduced N, and (iv) organic oxidized N. The atmospheric reactions generally occur within the grouping. The inorganic reduced N consists of two species: NH_3 and NH_4^+, with the primary species emitted to the atmosphere being NH_3 produced during OM decomposition or volatilization of NH_4^+ fertilizers, and emitted when partial pressure in the soil, water, or plant is greater than that in the atmosphere. NH_3 is the most common atmospheric gaseous base, and once in the atmosphere it can be converted to an aerosol in an acid–base reaction with a gas such as HNO_3 or aerosol such as H_2SO_4.

$$NH_3 + HNO_3 \rightarrow NH_4NO_3 \tag{7.1}$$

$$NH_3 + H_2SO_4 \rightarrow NH_4HSO_4 \tag{7.2}$$

$$NH_3 + NH_4HSO_4 \rightarrow (NH_4)_2SO_4 \tag{7.3}$$

These species are readily removed from the atmosphere by atmospheric deposition. NH_3 is primarily removed by dry deposition—often close to the source—while aerosol NH_4^+ is primarily removed by wet deposition through hydroscopic aerosol as a cloud-condensation nuclei. When NH_x is lifted above the planetary boundary layer, it can be transported large distances—1000 km or more—with emissions impacting receptors far downwind.

The inorganic oxidized N has many species and valence states ranging from 2 to 5 (Table 7.3), but the most of oxidized N in the atmosphere is part of NO_y—which include NO_x, HNO_3, particulate ammonium nitrate and other trace oxidized species. All the species are relatively reactive, with the atmospheric lifetimes of minutes to days. NO is commonly emitted from several sources including high temperature combustion, conversion of N_2 to NO by fossil fuels, lightning, and conversion of Nr species to NO through fire or microbial activity. Approximately 40% of the global emissions of NO results from the combustion of fossil fuels, which almost exclusively leads to emissions directly into the planetary boundary layer (Fowler et al. 1997). In the atmosphere, NO is rapidly oxidized to NO_2 and then oxidized to HNO_3, which react with NH_3 to form aerosol (Figure 7.3). Nitrogen oxides play a key role in atmospheric radical chemistry that leads to the oxidation of reactive trace gases and to the photochemical formation

TABLE 7.3

Examples of Main Forms of Nitrogen in Soil and Their Oxidation States

Name	Chemical Formula	Oxidation State
Nitrate	NO_3^-	+5
Nitrogen dioxide (gas)	NO_2	+4
Nitrite	NO_2^-	+3
Nitric oxide (gas)	NO	+2
Nitrous oxide (gas)	N_2O	+1
Nitrogen (gas)	N_2	0
Ammonia (gas)	NH_3	−3
Ammonium	NH_4^+	−3
Organic nitrogen	R_{NH}	−3

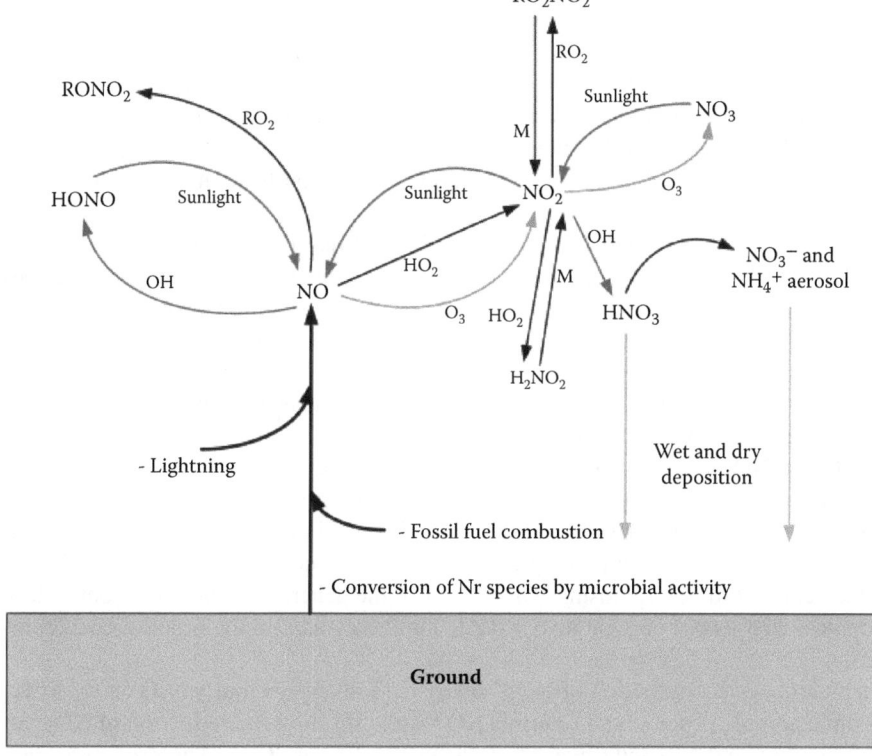

FIGURE 7.3 Schematic flowchart showing simplified atmospheric reactions of nitrogen compounds. (Modified from Fowler, D. et al., Ozone in the United Kingdom—Fourth Report of the Photochemical Oxidants Review Group. Annual Report. Department of the Environment, Transport and the Regions. London, U.K. 234 p., 1997; Monks, P.S. et al., *Atmospheric Environment* 43, 5268–5350, 2009.)

of ozone (Monks et al. 2009). In addition, deposition of gas phase N compounds plays an important role as an input of nutrients to the vegetation (Fowler et al. 2009). The oxidized inorganic N which is not part of NO_y is N_2O, produced during nitrification and denitrification processes in the terrestrial and marine ecosystems, and is globally distributed. It has relatively high stability and an atmospheric residence time of ~100 yrs. In the stratosphere, it is converted to NO by ultraviolet radiation, where the NO produced destroys stratospheric ozone (O_3). Increasing concentrations of atmospheric N_2O contribute to two current environmental issues: (i) greenhouse gas warming potential, and (ii) O_3 destruction.

Reduced atmospheric organic N occurs as bacteria, particulate matter, and soluble species such as amines which are emitted to the atmosphere through low temperature processes (e.g., turbulence and high temperature processes such as biomass burning). Many of these species can react quickly with HO and organic nitrates and are not transported far from their emission points (Neff et al. 2002). The atmospheric organic oxidized N species are generally formed in the atmosphere as the end products of reactions of hydrocarbons with NO_x (Neff et al. 2002). Hydrocarbons can also form organic radicals—RO, RO_2—through reaction with light, OH, or O_3, where the resulting species can react with NO_2 to form $RONO_2$.

N in the atmosphere, land, and water (0.14% of the earth N budget) exists predominantly as Nr. The atmospheric N_2O concentration is currently at 327.1 ± 0.1 ppb, with a mass of 1.57×10^3 Tg N_2O-N (Trenberth and Guillemot 1994; Montzka and Reimann 2010; Ciais et al. 2013).

The N cycling is driven by a combination of assimilatory (i.e., acquisition of matter for incorporation into biomass) and dissimilatory (i.e., processes that are associated with the extraction of energy from the environment) biological transformations. The dissimilatory transformations are largely carried out by prokaryotes—bacteria and archaea—but sometimes with the contribution of eukaryotes. Biotic transformations such as atmospheric N fixation associated with lightning and chemo-denitrification make only small contributions relative to those of the microbial transformations (Gruber and Galloway 2008). The microbial processes have a strong and, in many cases, controlling influence on the biogeochemistry of Nr in biosphere systems. The individual microbial pathways and processes in the global N cycle include: (i) N_2 fixation, (ii) NH_3 assimilation, (iii) nitrification, (iv) assimilatory NO_3 reduction, (v) ammonification, (vi) denitrification, and (vii) anammox.

N fixation is the process by which atmospheric N_2 is converted to any N and has a nonzero oxidation state. Historically, the most common process has been the biologically driven reduction of N_2 to NH_3 or NH_4^+. However, anthropogenically enhanced N fixation dominates on continents. Ammonia assimilation is the uptake of NH_3 or NH_4^+ by an organism into its biomass in the form of an organic N compound. For the organisms that can assimilate reduced inorganic N, this becomes an efficient process to incorporate N into the biomass. Nitrification is the aerobic process by which microorganisms oxidize NH_3 to NO_2^- (e.g., *Nitrosomonas*) after which a different group (e.g., *Nitrobacter*) oxidizes NO_2^- to NO_3^-. Assimilatory nitrate reduction is the uptake of NO_3^- by an organism and incorporation as biomass through NO_3^- reduction. It is an important process because it allows the mobile NO_3^- ion to be NH_3 for subsequent uptake. It is an important input of N for many plants and organisms. Ammonification is the primary process that converts reduced organic N (R-NH_2) through the action of microorganisms. This is part of the general process of decomposition where heterotrophic microbes use organic matter (OM) for energy and, in the process, convert organic N to NH_4^+.

Denitrification is the reduction of NO_3^- to any gaseous N, normally N_2O or N_2. This process requires an anaerobic process and requires NO_3^- and OM. It is the reduction of NO_3^- and NO_2^- to gases NO, N_2O, and N_2 (Seitzinger et al. 2006; Groffman 2012). The process is carried out primarily (but not exclusively) by facultative anaerobic bacteria that normally respires O_2 but in its absence respire the N oxides. Most denitrifying bacteria are heterotrophs, requiring organic compounds as an energy source. This process is of great interest because it can significantly reduce pools of N_2 as well as productivity in the ecosystems, and because NO_3^-, NO, and N_2O cause diverse water and air pollution problems (Davidson et al. 2012). Global mass balance analyses (Seitzinger et al. 2006) suggest that the biggest global sink for anthropogenic Nr is terrestrial denitrification. Modeling estimates that global N_2 production from denitrification may increase from 96 Tg N yr^{-1} in 2000 to 142 Tg N yr^{-1} in 2050 due to increased Nr inputs in the global agricultural systems (Bouwmann et al. 2013a). Microorganisms use NO_3^- as an oxidant to obtain energy from OM. It is prevalent in waterlogged soils and is the primary process that converts Nr back to N_2.

Anammox (i.e., anaerobic ammonium oxidation) is a reaction that oxidizes ammonium to dinitrogen gas using nitrite as the electron acceptor under anoxic conditions. Anammox organisms have the ability to combine NH_4^+ and NO_2^- back to N_2. It is an anaerobic process and its discovery led to the realization that a substantial part of the enormous N losses that are observed in the marine environment, as high as 50% of the total N turnover, can be due to the activity of these bacteria (Kuenen 2008). Anammox has been predicted to be a more thermodynamically favorable process than aerobic ammonium oxidation. Although anammox bacteria have been detected in various natural habitats, such as anoxic marine sediments and water columns, freshwater sediments and water columns, terrestrial ecosystems, and some special ecosystems (Hu et al. 2011), it is considered to be critically important in the marine N cycle. The relative contribution of the anammox process to the total production of N_2 gas has been estimated to be 50% in the ocean (Jetten et al. 2009).

7.3 GLOBAL SOURCES OF REACTIVE N CREATION

7.3.1 LIGHTNING

Atmospheric N_2 is converted to Nr by four basic processes: (i) lightning, (ii) high temperature combustion, (iii) BNF, and (iv) Haber–Bosch industrial process. Lightning is the natural process which creates Nr in the form of NO_x and introduces Nr to relatively remote regions of the troposphere. Lightning produces high electrical energy in the atmosphere which overcomes the energy barrier to break the triple bond of molecular N_2 ($N\equiv N$) and double bonds of molecular O_2 ($O=O$), whereby the two molecules combine to form NO in the atmosphere. Lightning produces momentary conditions of high pressure and temperature in the lightning channel, which causes thermal dissociation of O_2 and N_2 and allows them to combine and form NO (Equation 7.4). The NO is photochemically converted to NO_y.

$$N_2 + O_2 + \text{ electrical energy} \rightarrow 2NO \qquad (7.4)$$

In the atmosphere, NO is transformed into various compounds of N that have an average atmospheric residence time of <10 days before they are deposited on the Earth's surface either as wet deposition in precipitation or dry deposition in aerosol or gas associated with particulate matter. Lightning, therefore, introduces Nr into ecosystems, especially over the tropical regions. At an early point of Earth's history, lightning was an important process in creating Nr from N_2, but it has become less important globally under currently anthropogenic alteration of N cycling, although it still converts an appreciable quantity of NO. The lightning-induced formation of NO still remains important in areas of deep convective activities such as those occurring in tropical continental regions, however.

The lightning activity has been investigated using direct measurements and supported by satellite remote sensing, and global production estimated using available data and models. However, substantial uncertainties of the estimate remain, in part due to difficulties in up-scaling. Possible effects of climate change on the rates of NO_x production from lightning also have been considered and estimated increase in the range of 3% to 12% in NO_x per °C increase is predicted (Brasseur et al. 2006). Estimates of the global source strength of lightning Nr range from 2 to 10 Tg N yr^{-1} (Tie et al. 2002). More recent values estimate lightning NO_x closer to 5 Tg N yr^{-1} (Figure 7.4; Ehhalt et al. 2001).

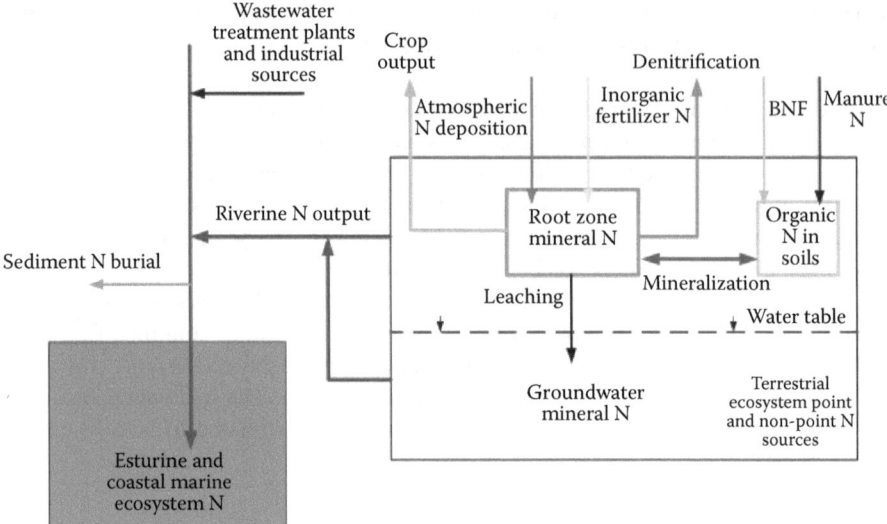

FIGURE 7.4 Conceptual schematic diagram showing fluxes and stores of reactive N (Nr) under human impacted N dynamics for both point and non-point N sources.

7.3.2 HIGH TEMPERATURE COMBUSTION

High temperature pyrolysis (>1800°C) and pressure formed during fossil fuel combustion, primarily via the internal combustion engines industrial power plants, especially for electricity generation, provide the energy to convert N_2 to NO through reaction with O_2:

$$N_2 + O_2 + \text{fossil energy} \rightarrow 2NO \tag{7.5}$$

High temperature combustion is currently contributing 40 Tg N yr^{-1} (Figure 7.4; van Vuuren et al. 2011). NO_x also can be formed from the oxidation of fossil organic N, which is primarily found in fossil fuels. However, this is not the creation of new Nr but rather the mobilization of Nr that had been sequestered for millions of years (Socolow 1999). Biomass burning also represents an important global contribution of Nr to the atmosphere, but the Nr created is primarily from N contained in the biomass and does not represent new Nr fixation. Emissions of nitric oxide from soils also contribute to atmospheric NO_x emissions, but this also is not an additional source of Nr since it is the result of microbial transformation of existing Nr in soil through nitrification and denitrification (Skiba et al. 1993). Emissions from biomass combustion and soil NO are estimated at 5 Tg N yr^{-1} each. It is projected that control measures on NO_x emissions will reduce total NO_x emissions from internal combustion engines by the middle of the century to ~30 Tg N yr^{-1}.

7.3.3 BIOLOGICAL N FIXATION

BNF is a process restricted to the prokaryotes of the domains Archaea and Bacteria. It is a microbially-mediated reductive process that occurs in several types of archaea, bacteria and blue-green algae that transforms N_2 to NH_3 (Bottomley and Myrold 2015). This process uses the enzyme nitrogenase and occurs in an anaerobic environment. The microbes can be free-living or in a symbiotic association with higher plants. Legumes are the best-known example of symbiotic association (Schlesinger and Bernhardt 2013). The following equation represents an overall N_2 conversion to NH_3:

$$2N_2 + 6H_2O \rightarrow 4NH_3 + 3O_2 \tag{7.6}$$

Given that over half of fixed N input that sustains the Earth's population is supplied biologically, there has been intense interest in understanding how the nitrogenase enzyme accomplishes the task of N_2 fixation at the ambient temperature and pressure (Burgess and Lowe 1996; Eady 1996; Varley et al. 2015; Hoffman et al. 2014). An understanding of BNF may serve as the formulation of genetically endowing higher plants with the capacity to fix their own N (Beatty and Good 2011; Godfray et al. 2010; Rubio and Ludden 2008) and also developing improved synthetic catalysts based on biological mechanisms (Jia and Quadrelli 2014; MacLeod and Holland 2013; Tanabe and Nishibayashi 2013).

Being a microbially-mediated process that occurs in several types of bacteria, archaea, and blue-green algae (cyanobacteria and cyanophyta), BNF occurs only when other forms of inorganic N are unavailable in solution. The lack of inorganic N provokes physiological and biochemical changes in organisms which gives them the ability to fix atmospheric N_2. In the case of cyanobacteria, the lack of inorganic N leads to differentiation of specialized cells called "heterocysts" while in legumes, it provokes formation of root nodules (Boyd 2001). This biologically irreversible reaction is catalyzed by an extremely conserved heterodimeric enzyme complex nitrogenase, which is inhibited by O_2; therefore, BNF is strictly an anaerobic process (Postgate 1998). Nitrogenase enzyme complex consists of two separate enzymes—dinitrogenase reductase and dinitrogenase—which are involved in the conversion of N_2 to NH_4^+. This heterodimeric enzyme complex serves as a catalyst by hydrolyzing 16 adenosine triphosphate (ATP) molecules for each molecule of N_2 fixed (Canfield et al. 2010). The electrons required for this process come from respiration of OC (Howard and Rees 2006). Because N fixation is rather expensive in terms of energy, this capability is only an advantageous to microorganisms when inorganic N is scarce, and the N fixing capability lies dormant during times of plentiful N supply. The microbes can be free-living (non-symbiotic) bacteria and algae which are widespread in freshwater, oceans, and uncultivated soils and also form mutualistic association with a range of plant species, or symbiotic bacteria, mostly belonging to the genus *Rhizobium*, which forms a symbiotic association with roots of higher plants in the terrestrial ecosystems—mostly belonging to the family Leguminosae.

The majority of symbiotic BNF occurs via a root nodule symbiosis. In this relationship, the plant maintains N fixer by acting as a C source and the microbes excrete the fixed N for the benefit of the plant. Legumes and *Rhizobium* spp. are the best-known examples of such a symbiotic relationship and play a large role in BNF in modern agricultural soils. Non-symbiotic N fixation includes fixation by the true free-living N fixers as well as by autotrophic and heterotrophic organisms not in direct symbioses with vascular plants, including cyanobacteria symbiotic in lichens, bryophytes, or associative N_2 fixing systems. In addition, actinomycetes (*Frankia*) can fix N as free-living microorganisms or in symbiosis with a number of non-leguminous vascular species. The free living N fixing microorganisms are also known as rhizosphere diazotrophs—these are most active near plant roots where decomposable C supply is high—but are not found within living tissues. These microorganisms are ubiquitous in soils and are highly diverse. The process uses the enzyme nitrogenase to convert N_2 to NH_3:

$$N_2 + 8H^+ + 8e^- + 16ATP \rightarrow 2NH_3 + H_2 + 16ADP + 16P_i \tag{7.7}$$

$$N_2 + 6H^+ + 6e^- \rightarrow 2NH_3 \tag{7.8}$$

The reduction of N_2 is an exergonic reaction, but it requires formidable activation energy to break the triple bond and a catalyst to overcome the energy barrier. The energy required to break the triple bond of an N_2 molecule is 226 kcal mol^{-1}. Although nitrogenase enzyme is widely distributed among prokaryotic lineages, most prokaryotic organisms cannot fix N but rather obtain their N directly as NH_4^+, organic N from the environment or reduction of NO_3^- to NH_4^+ through assimilatory

NO_3^- reduction. BNF can occur in both managed and unmanaged ecosystems. In managed ecosystems, cultivation of legume crops enhances BNF and is the main source of human-induced BNF. In the pristine biosphere, N fixation by terrestrial BNF is the dominant source of newly fixed N in the terrestrial ecosystems (Cleveland et al. 1999).

There remain important limitations in understanding BNF; especially since the capability to fix N biologically is widespread in ecosystems, it is not clear why organisms do not fix more N, since this would provide competitive advantages over the competitor organisms that lack N-fixing capability. Current knowledge of processes and controls of BNF has not provided unambiguous answers, although widespread application of N fertilizers on farmland and atmospheric deposition has decreased non-agricultural BNF.

BNF provides an important reference when quantifying the importance of anthropogenic inputs to the global N cycle since it is the primary non-anthropogenic input of Nr (Vitousek et al. 2002). The recently estimated pre-industrial BNF in terrestrial ecosystems ranges from 40 to 100 Tg N yr^{-1} (Vitousek et al. 2013). These estimates are smaller than earlier estimates that suggested pre-industrial BNF in the range of 100–290 Tg N yr^{-1} (Cleveland et al. 1999), although this was later revised down to 128 Tg N yr^{-1} (Galloway et al. 2004). The recent estimates are based on the hydrological losses of N from terrestrial ecosystems and the fraction of N denitrified in streams and rivers may have been overestimated (Fowler et al. 2013). Cultivation-induced BNF (C-BNF) occurs in several agricultural ecosystems including crops, pastures, and fodder legumes (Smil 1999). Recent global estimates from agricultural crops and grazed savannahs range from 50 to 70 Tg N yr^{-1} (Table 7.4; Herridge et al. 2008). This value is further disaggregated into grain legumes, forage legumes, and other croplands, with grain legume crops (peas, *Pisum sativum*, beans, *Phaseolus vulgaris* etc.) and forage legumes (alfalfa, *Medicago sativa*) contributing 21 and 19 Tg N yr^{-1}, respectively (Table 7.4; Herridge et al. 2008). Other minor inputs of N by BNF in agriculture include symbiotic N fixation from tropical savannahs used for grazing estimated at 14 Tg N yr^{-1} and free-living microorganisms associated with rice (*Oryza sativa*) paddies (5 Tg N yr^{-1}) and sugar cane (*Saccharum officinarum*) (0.5 Tg N yr^{-1}) (Fowler et al. 2015). Increased soybean (*Glycine max*) and meat production since 2000 may have also increased global C-BNF (Galloway et al. 2004). Future growth of legume crops will be constrained by land area available to agriculture, and increases may only occur when legumes replace other crops. There exists substantial uncertainty in BNF from agriculture, however, and more precise data are needed. The N_2O emissions resulting from the legume crops are generally lower compared to those from other crops, and the Intergovernmental Panel on Climate Change (IPCC) guidelines on GHG reporting assumes that N input resulting from legume production is not associated with N_2O (IPCC 2006). Therefore, legume cultivation generally has been promoted as way of reducing N_2O emissions from agriculture (Luescher et al. 2014). The current terrestrial ecosystem estimated BNF contribution of Nr range from 77 to 387 Tg N yr^{-1} with an average value

TABLE 7.4

Terrestrial Contribution to Biological N Fixation in the 21st Century

Ecosystem	Annual N Fixation (Tg N yr^{-1})	Reference
Grain legumes (*Rhizobium* spp)	21 (10–21)	(Smil 1999; Herridge et al. 2008)
Forage (*Rhizobium* spp)	18.5 (12–25)	(Herridge et al. 2008;
Rice (*Azolla* spp)	5 (4–6)	Smil 1999; Herridge et al. 2008)
Other croplands (free-living)	3.5	(Herridge et al. 2008)
Tropical savannah (free-living)	12 (5–42)	(Cleveland et al. 1999)
Non-agricultural (rhizobium and free living bacteria and algae)	128 (44–290)	(Cleveland et al. 1999; Galloway et al. 2004; Vitousek et al. 2013)
Total	188 (77–387)	

of 188 Tg N yr^{-1} (Table 7.4; Cleveland et al. 1999; Galloway et al. 2004; Vitousek et al. 2013; Fowler et al. 2015) and is the largest single source of Nr, although there are significant uncertainties about the magnitude and distribution of BNF-Nr fluxes. Estimates of the global terrestrial BNF are highly uncertain due to higher uncertainty in available data of BNF rates at plot scale, methodological differences, and non-uniform spatial coverage of important N fixing species and locational biases in different studies (Cleveland et al. 1999). In addition, in the terrestrial tropical regions (i.e., Africa, Asia, and South America), where BNF is likely the important source of Nr, there are the least measurements of terrestrial BNF rates. Estimates of the global BNF rates are also highly uncertain, mainly because of scant relevant data for natural vegetated ecosystems. In terrestrial environments, a wide diversity of symbiotic and free-living N fixers contribute to BNF in non-agricultural soils, but lack of reliable measurements results in large uncertainties in reported values (Cleveland et al. 1999; Galloway et al. 2004). Most studies present BNF estimates as global values based on few but broad components such as forests, grasslands, and others (Paul 2015). Such coarse divisions average large land areas that contain significant variation in datasets and in biome types, which diminish the usefulness and credibility of the data.

Recent measurements of BNF by methanotrophs in pristine peatland at high latitude suggest appreciable fixation in these environments which have not been included in global estimates (Vile et al. 2014). These peatlands contain approximately 3% of global land surface but contain about 25% of global soil C, suggesting an additional source of 4.8 to 62.3 kg N ha^{-1}, with an annual mean value of 25.8 kg N ha^{-1} yr^{-1} (Vile et al. 2014). Based on net C uptake, lichens and bryophytes have also been suggested as significant contributors of global N fixation (Elbert et al. 2012; Porada et al. 2014).

BNF associated with non-agricultural ecosystems is susceptible to changes in environmental conditions. In addition, there is coupling between N, C, and phosphorus (P) cycling in environments (Vitousek et al. 2002a,b). Free-living and symbiotic organisms with the potential to fix N tend to be at a selective advantage in environments with low P availability. However, the high energy costs of BNF require an adequate supply of available fixed C. In addition, the nitrogenase enzyme responsible for the N fixation process is sensitive to temperature, with the optimum temperature of ~25.2°C, and the steep decline in rates of fixed N below 5°C and above 40°C (Houlton et al. 2008; Sheffer et al. 2015). Increases in global temperatures resulting from climate change are, therefore, likely to be associated with increases in BNF, provided water remains sufficient to maintain NPP. However, other environmental changes could counteract increases in fixed N resulting from climate change.

7.3.4 INDUSTRIAL N FIXATION BY HABER–BOSCH PROCESS

In 1908, Fritz Haber discovered the right process conditions that made it feasible to eventually scale up the process of reactions of atmospheric N_2 with H_2 to industrial scale in the presence of Fe at high pressure and temperature (Smil 2001). Carl Bosch subsequently developed this process on an industrial scale, for which he was awarded the Nobel Prize in 1931. This industrial process, known as Haber–Bosch industrial N fixation, uses natural gas methane (CH_4) to produce H_2, which is then combined with N_2 to form gaseous NH_3 under high temperature (500 to 600°C) and pressure with metallic Fe catalyst:

$$CH_4 + 6H_2O \rightarrow 3CO_2 + 12H_2 \tag{7.9}$$

$$4N_2 + 12H_2 \rightarrow 8NH_3 \tag{7.10}$$

The resultant NH_3 is refined and compounded to make the nitrogenous fertilizers that modern agriculture depends on. Although the Haber–Bosch process was developed for military use and was used extensively during World War I to produce munitions, since the early 1950s it has become the world's largest source of N fertilizers (Smil 2001). The NH_3 synthesis is important because it is the primary ingredient in N fertilizers, without which modern agriculture would be impossible. The Haber–Bosch process produces virtually all of N fixed industrially (IFA 2000). Sometimes called the most important technological advancement of the 20th century, the Haber–Bosch process has significantly lowered the energy requirements and is substantially cheaper, and therefore forms the basis of an alternative expanding supply of Nr. It has also boosted the production of many expensive or rare compounds such as dyes and artificial fibers and has the greatest impact on the production of explosives and fertilizers (Smil 2001).

Chemical N fertilizers are one of the major facets of the Green Revolution, which has resulted in many-fold increases in crop yields which feed more than 50% of the world population (Mann 1999; Erisman et al. 2008). Synthetic inorganic fertilizer N production supplies >120 Tg N yr^{-1} to agricultural soils (Schlesinger and Hartley 1992; Matthews 1994; Potter et al. 2010; FAOSTAT 2017). About 40% of the large annual increase in crop production during the post–Green Revolution period is attributed to the increase in use of synthetic fertilizer N (Brown 1999). In addition, fertilizer N is required for bioenergy and biofuel production, which contributes about 10% of global energy requirement and 1.5% of fuel (Erisman et al. 2008). The production of synthetic fertilizers for food production began rising rapidly by the 1950s, and by 1970 the creation of Nr by the Haber–Bosch process became more important than BNF in unmanaged ecosystems, where humans surpassed nature in introducing Nr in the environment.

About 80% of Nr manufactured by the Haber–Bosch process is used in agricultural fertilizes (Galloway et al. 2008). The total Nr production through the Haber–Bosch process that is used for agriculture is estimated at 102 ± 12 Tg N yr^{-1} as NH_3 (Fowler et al. 2015; FAOSTAT 2017) and it represents the largest single contribution to Nr formation through anthropogenic activity. The use of N-fixing crops contributes additional 60 ± 10 Tg N yr^{-1} which enters crop and soil N cycling of Nr (Herridge et al. 2008). A large portion of this is lost to the environment, where it cascades through atmospheric, terrestrial, aquatic, and marine pools before eventually being denitrified to N_2 or stored as fossil and Nr. Volatilization of NH_3 or leaching of NO_3^- to adjacent natural ecosystems is environmentally important. Its deposition leads to unintentional fertilization and loss of terrestrial biodiversity (Hesterberg et al. 1996). Transfer of Nr from terrestrial to coastal systems has led to algal blooms and decline in the quality of surface and groundwaters. In the atmosphere, Nr alters the balance of GHG, enhances tropospheric O_3, decreases stratospheric O_3, increases soil acidification, and stimulates formation of secondary particulate matter in the atmosphere. All these have negative effects on people and the environment (Erisman et al. 2008).

7.4 NITROGEN DYNAMICS IN THE TERRESTRIAL ECOSYSTEMS

Atmospheric N_2 is transformed by N fixation and imported into terrestrial and aquatic ecosystems. Although generally low, N availability can fluctuate greatly in both space and time due to several factors including precipitation, temperature, wind, soil type, and pH. Soil N is available for plant uptake by roots as NO_3^-, NH_4^+, amino acids, or peptides, and both NO_3^- and NH_4^+ are highly mobile in soil. The preferred form in which N is taken up depends on plant adaptation to soil conditions (Miller et al. 2009). In non-aerobic soil conditions NH_4^+ is often the prevalent form of N. Generally, plants adapted to low pH and reducing soils tend to take up NH_4^+ or amino acids, whereas plants adapted to higher pH and more aerobic soils prefer NO_3^- (Maathuis 2009). The use of N by plants involves several steps, including uptake, assimilation, translocation, and when the plant is aging, recycling, and remobilizing. N assimilation requires the reduction of NO_3^- to NH_4^+ followed by NH_4^+ assimilation into amino acids. Two important enzymes—nitrate reductase and nitrite reductase—ensure that the prevalent form of N is taken up and NO_3^- is converted to NH_4^+. The NO_3^- reduction can take place in both roots and shoots. The main function of N is to provide amino groups in amino acids. N is also prolific in nucleic acids, energy homeostasis, and biochemistry of many non-protein compounds such as coenzymes, photosynthetic pigments, and secondary metabolites (Maathuis 2009). Animals, humans, and other organisms obtain N through conversion and transformation from plants. The Nr is returned to soils through biomass, and its emission pathways mainly include denitrification in soils, anammox, and gaseous N emission processes.

As early as the late 1980s, Aber et al. (1989) indicated two major pathways that Nr enters ecosystems as pulse fertilizer N addition and the chronic atmospheric deposition. As a result of human modification of the global N cycle, the amount of Nr entering an ecosystem over several years exceeds the capacity of an ecosystem for plant uptake and retaining it within soils, plants, and microbes, resulting in N saturation and potentially leaching into surface and groundwater (Aber et al. 1989, 1998; Kopacek et al. 2013a). There is a growing evidence that planetary boundaries for maintaining human and ecosystem health have been exceeded (Carpenter et al. 2011; Rockstrom et al. 2009; Steffen et al. 2015). The N level and dynamic phase in any ecosystem passes through four stages: (i) N limitation, (ii) N balance, (iii) N saturation, and (iv) excessive N (Kopacek et al. 2013a). In N limitation, all available N forms in soils tend to be immobilized by plant and microbial biomass, and this was likely the state in the pre-industrial period. During this period, the C:N ratio decreases only slowly due to the net ecosystem N gain from N_2 fixation and low Nr deposition, which was to a large extent balanced by terrestrial exports of recalcitrant OM, while NO_3^- leaching was probably small or non-existent. A similar situation was also common in current N-limited terrestrial ecosystems (Perakis and Hedin 2002). Over time, however, soil biogeochemistry and fertility of ecosystems adapted to the steady state conditions began to change due to rapidly increasing NO_x and NH_3 emissions from food production and Nr deposition, mostly due to increased anthropogenic Nr creation, and a new steady-state during the industrial period has not yet been established because Nr emission and deposition rates have not stabilized, since the ecosystem response tends to be slow (Kopacek and Posch 2011). In response to increased Nr deposition the ecosystems reduce the rate of more energy demanding N_2 fixation, accompanied with alleviation of natural Nr limitation for the unmanaged ecosystems. This triggers a cascade of biologically mediated changes in both terrestrial and aquatic ecosystems, which include changes in ecosystem biodiversity (Bobbink et al. 2010), microbial composition (Hogberg et al. 2007a), internal N cycling (Aber et al. 1998), and NO_3^- leaching. In current agricultural and pasture systems, N balance can be defined by input from atmospheric Nr deposition, fertilizer and manure addition, and N fixation minus output from yields of crops and pastures.

In addition to its role as a nutrient, changes in Nr in agricultural ecosystems is also associated with the role of NO_3^- as a strong acid anion in soil solution and as an electron acceptor in microbial energy metabolism. Therefore, changes in Nr inputs affect C cycling and pools of bioavailable dissolved organic C (DOC) for soil microorganisms (Alewell et al. 2008). The terrestrial N and C cycles are tightly coupled, as evidenced by the constrained flexibility of the ecosystem C: N stoichiometry (Zaehle 2013). Heterotrophic microorganisms relying on bioavailable DOC are responsible for most of N transformations in soils, implying that environmental changes affecting OC solubility and bioavailability in soils could also affect N cycling.

The creation of large hypoxic zones and the resulting loss of habitat and species diversity in estuarine and coastal marine ecosystems has been one of the most significant impacts of increased Nr flows (Howarth et al. 2011). While the need to manage N flows and their associated ecological impacts has been recognized, implementation of conservation measures to reduce stream N concentrations has had only limited success (Kopacek et al. 2013b). Increasing evidence suggests that the lack of success could be attributed to diffuse legacy sources of N that can lead to time lags between management changes and measurable improvement in water quality, as the diffuse sources continue to impair water quality even after agricultural inputs have ceased. Both regional and continental scale studies of N dynamics suggest that an inefficient use of N is common in heavily fertilized agricultural watersheds, leading to a large N surplus (i.e., N inputs minus usable output) (Leip et al. 2011). A portion of surplus N exists in terrestrial ecosystems within the watershed and may exit the watershed as riverine output (Figure 7.4), although its quantity remains largely uncertain, and also its fate and long-term implication remain unknown. However, denitrification and subsurface Nr storage constitute a well-known pathway by which Nr may exit watersheds or be retained for long period, respectively, and is frequently referred to as N retention. Direct, large-scale evidence of Nr accumulation in the root zone of agricultural soils has been observed based on long-term soil data (1957 to 2010) analysis throughout the Mississippi River Basin which revealed the accumulation Nr in croplands ranging from 25 to 70 kg N ha^{-1} yr^{-1} (Van Meter et al. 2016), demonstrating that agriculture soils can act as net N sinks as a result of anthropogenic perturbation of soil N cycling.

Denitrification, a microbial-mediated process that occurs in soils, stream sediments, and marine ecosystems, is a process by which Nr is removed from the ecosystems through N_2O and N_2 gas (Seitzinger et al. 2006; Canfield et al. 2010). Denitrification is inherently difficult to measure directly, and considerable uncertainty exists regarding the denitrification rates in terrestrial systems (Seitzinger et al. 2006). As a result, denitrification is often estimated as a balancing/error term in mass balance studies, and this assumption has shown to be no longer applicable with current high inputs of Nr in intensively managed landscapes (Galloway et al. 2004, 2008; Canfield et al. 2010; Gruber and Galloway 2008). A conceptualized subsurface environment is composed of three major N pools: (i) dissolved NO_3^- in the root zone, (ii) dissolved NO_3^- in groundwater aquifers, and (iii) organic N in the soil profile (Figure 7.4). A large store of root zone inorganic N has been demonstrated in deserts and semi-arid regions, and its magnitude varies as a function of rainfall, tillage, and irrigation history (Walvoord et al. 2003; Scanlon et al. 2008), while the existence of significant groundwater reservoir is based on observations of increasing groundwater N in the US and Europe (Puckett et al. 2011; Worrall et al. 2015). The soil organic N (SON) in the root zone remains the largest pool of N in most terrestrial ecosystems, however, and at the current N inputs, it has been suggested that terrestrial N sequestration may be occuring at the global scale on the order of 20 to 100 Tg N yr^{-1} (Fowler et al. 2013, 2015; Galloway et al. 2004).

The major factor that drives the change in the global N cycle is the increased Nr creation rate due to increased human demands for food and energy. For example, since 1970, Nr creation has increased by 120% and continues to increase each year, mainly as a result of intensive agriculture and livestock development, but fossil energy use also plays an important role (Galloway et al. 2008). The anthropogenic Nr creation in 1860 was 15 Tg N yr^{-1}; it increased to 156 Tg N yr^{-1} by 1995 and increased further to 187 Tg N yr^{-1} in 2005 (Galloway et al. 2008), and 222 Tg N yr^{-1} by 2012 (Fowler et al. 2013, 2015). The global cereal and meat production increased by 20.5% and 26%, respectively, from 2000 to 2013 (FAOSTAT 2017), mostly resulting from further increase in Nr creation. This increase is a result of population growth accompanied by increase in per capita food consumption.

Anthropogenic Nr can be emitted to the atmosphere as NO_x, NH_3, and organic N (Dentener et al. 2006; Galloway et al. 2004). The major sources of gases are combustion of fossil fuels and biomass for NO_x, fertilizers and manure for NH_3, and both natural and anthropogenic for organic N. Dry and wet deposition of Nr have become a significant component of the global N cycling and has led to severe N imbalance in ecosystems. Furthermore, anthropogenic Nr input from fertilizers, manure, and fossil fuel burning is the dominant contributor to dry and wet depositions at local, regional, and global scales (Galloway et al. 2008; Paulot et al. 2013). The annual emission from terrestrial ecosystems in the form of NO_x, NH_3, and N_2O is estimated at 40, 60, and 13 Tg N yr^{-1}, respectively (Table 7.5). After undergoing chemical conversion and physical transport, NO_x and NH_3 are removed from the atmosphere by wet and dry depositions to terrestrial and marine ecosystems. A fraction of Nr that is returned to the land and ocean accounts for 60% to 80% of the total Nr emitted to the atmosphere (Zhang et al. 2012; Fowler et al. 2013, 2015).

As the input of Nr in the ecosystems increases, ecosystem requirements of additional N are reduced. An early ecosystem response is the reduction in the rates of BNF (DeLuca et al. 2008). Subsequent symptoms of sustained elevated Nr inputs include the alleviation of the natural N limitation in the unmanaged ecosystems, which can trigger a cascade of changes in both terrestrial and aquatic ecosystems (Galloway et al. 2004). Consequences of long-term N enrichment include changes in ecosystem biodiversity (Bobbink et al. 2010), microbial composition (Hogberg et al. 2007a,b), internal N cycling variation (Aber et al. 1998), and NO_3^- leaching (Stoddard 1994).

TABLE 7.5
The Global Nr Fluxes

Source of Nr	Flux (Tg N yr⁻¹)	Reference
	N fixation	
Industrial fertilizer	122	(Galloway et al. 2008; Fowler et al. 2015)
Chemical industry	20	(Bouwman et al. 2013b)
Natural ecosystems BNF	128	(Fowler et al. 2015)
Agricultural crops BNF	60	(Fowler et al. 2015)
Ocean BNF	140	(Voss et al. 2013)
Lightning Nr fixing	5	(Fowler et al. 2015)
Combustion NO_x emissions	40	(Fowler et al. 2015)
Soil emissions of NO	5	(Pilegaard 2013)
Emissions of N_2O from soils	18.8	(Fowler et al. 2015)
Denitrification on terrestrial	100	(Fowler et al. 2013)
Wet and dry Nr deposition on land	70	(Fowler et al. 2013)
Wet and dry Nr deposition on sea	30	(Duce et al. 2008)
Terrestial NH_3 emissions	60	(Bouwman et al. 2011)
Ocean NH_3 emissions	9	(Voss et al. 2013)
Nr burial	20	(Voss et al. 2013)
Denitrification to N_2 in oceans	100–280	(Duce et al. 2008; Voss et al. 2013)

Generally, these biologically mediated responses are associated with the general role of N as a limiting nutrient in the terrestrial ecosystems as well as the inability of terrestrial ecosystems to retain excessive N due to the mechanisms associated with NO_3^- as a strong acid anion in soil solution.

A significant fraction of anthropogenic Nr mobilized from watersheds enters groundwater and surface water and is transported to coastal and marine systems, resulting in negative environmental impacts. In the land, atmospheric release of Nr affects human health and water pollution, and loss of habitat and biodiversity. In marine ecosystems, Nr leads to fresh water and coastal eutrophication, and increased frequency and severity of harmful algal blooms, hypoxia, and fish kills due to oxygen depletion (Diaz and Rosenberg 2008). In extreme cases, it leads to the development of "dead zones"—areas depleted of oxygen such that they cannot support any marine life (Diaz and Rosenberg 2008; Perrings et al. 2014). Marine dead zone have now been reported in over 400 systems affecting an estimated total area of 245,000 km² (Diaz and Rosenberg 2008; Perrings et al. 2014).

7.5 NITROGEN CYCLING IN TERRESTRIAL ECOSYSTEMS

At the beginning of the 20th century humans began to have significant impact on the global N cycle through fossil fuel combustion which creates fixed N as NO_x, the Haber–Bosch industrial process of creating Nr, and the implementation of new agricultural practices to increase the crop yields (Vitousek et al. 1997b). Anthropogenic perturbations of the global N cycle disturbed the pre-existed equilibrium between natural Nr creation and Nr removal by denitrification and burial of N in sediments and rocks that existed, and the amount of anthropogenic Nr converted back to non-reactive N_2 by denitrification became smaller compared to Nr produced each year. Approximately two-thirds of the annual fluxes of Nr entering the atmosphere at the beginning of the 21st century are of anthropogenic origin (Galloway et al. 2004; Fowler et al. 2013). Additionally, 30% to 75% of the total Nr each year is not denitrified back to N_2 (Bouwman et al. 2013a; Canfield et al. 2010). Many cereal crops that use high N fertilizers have low N use efficiency (NUE), typically below 40%, and most of the applied fertilizer N is either stored in the root zone as missing N in N budgets

(Boyer et al. 2002; Leip et al. 2011; Van Breemen et al. 2002; Van Meter et al. 2016), leaches out of the root zone (Howarth et al. 2011), or is lost to the atmosphere by denitrification or volatilization of NH_3 before it is assimilated into plant biomass (Sutton et al. 2013b). Therefore, excessive use of N in agriculture initiates a cascade of large-scale environmental impacts (Galloway et al. 2008). The endpoint of cascading is ultimately the emission of N_2 or N_2O to the atmosphere. During the cascade, Nr can influence GHG exchanges with the atmosphere, aerosol production, tropospheric O_3, or increase biological productivity, which require C. All these processes have impact and significant implication to global climate.

Globally, the Haber–Bosch industrial process created 122 Tg N yr^{-1} of new Nr as synthetic NH_3 in 2012 (Figure 7.5; FAOSTAT 2017), of which, 102 Tg was used to sustain crop and grass production and 20 Tg N as feedstock for many industrial processes (Table 7.5; Galloway et al. 2008; (Fowler et al. 2013, 2015), including nylon and explosives. In addition, 50–70 Tg N yr^{-1} is fixed biologically by agricultural systems (Table 7.5; Herridge et al. 2008). Only 20–30% of N introduced in agricultural soils ends up in food for human consumption such as cereals, vegetables, and fruits, while the rest is used to sustain livestock production (Billen et al. 2013). Similarly, only a small fraction of N input to livestock is consumed by humans as meat, whereas the larger fraction is lost or recycled by agricultural soils (Billen et al. 2013). In the analysis of the fate of Nr introduced in the global ecosystems, Smil (1999) estimated that 50% is removed by harvested crops, 23% is leached N, 6% is volatilized as NH_3, and 6% is released into the atmosphere as NO_x and N_2O. Only 10% of Nr is denitrified and converted into inert N_2 within the ecosystem. The remaining ~90% is recycled within the ecosystems. Between 1961 and 2008, fertilizer N application to agricultural lands increased at about 1.72 Tg N yr^{-1} to match the food demand for the world population, which grew by about 72 million yr^{-1}. Until the 1970s, most of the industrial N fertilizers were applied in developed countries. At present, however, fertilizer N use in developed countries has stabilized, but application is increasing dramatically in developing countries (Zhu et al. 2005). The momentum of growth in the human population and the increase in urbanization as well as changing dietary needs ensure that industrial N fixation will continue to grow at even higher rates for decades to come because of the increase in demand for food.

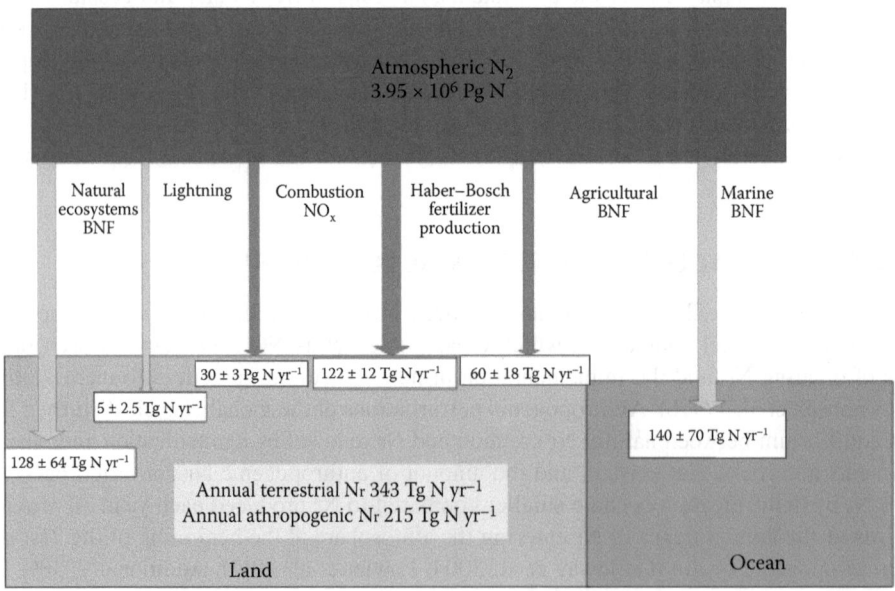

FIGURE 7.5 The global reactive N (Nr) fixation showing both anthropogenic (red) and natural in oxidized and reduced forms.

Losses from agricultural soils are mainly NH_3 emissions from N fertilizer applications and live-stock production. Increased consumption of meat and dairy products requires more livestock and fertilizers, resulting in NH_4 deposition in some areas by greater than 40%. The global NH_3 emissions are estimated at 37 Tg N yr^{-1} (Sutton et al. 2013a). Soil denitrification is estimated at 25 Tg N yr^{-1} (Billen et al. 2013) and N leaching and runoff at 95 Tg N yr^{-1} (Billen et al. 2013). Humans also contribute to the loss of N in the environment through food waste representing a third of the food produced globally (Gustavsson 2011) and partially treated sewage discharge, which produces unintended N export to fresh waters and emissions to the atmosphere. Fresh waters receive 39–95 Tg N yr^{-1} from agricultural soils (Bouwman et al. 2013b; Billen et al. 2013), with part of it remaining in superficial aquifers, a part that is denitrified contributing to N_2O emissions to the atmosphere, and a part exported to coastal waters estimated at 40–66 Tg N yr^{-1} (Seitzinger et al. 2005) where it enhances water eutrophication and hypoxia. The burning of fossil fuels for energy production has also increased the amount of Nr in the environment, since NO_x is formed as a by-product of combustion. Through these activities, humans are currently dominating the introduction of Nr to terrestrial environment.

Nearly 90% of the N fertilizer used globally is in the form of NH_4^+, and when applied to the soil, nitrifying bacteria can convert it into highly mobile NO_3^- which in turn can leach into rivers, lakes, and aquifers, resulting in N loss, while also leading to the eutrophication of coastal waters and creating large hypoxic zones around the world (Diaz and Rosenberg 2008). The Nr is highly mobile in the ecosystems. Most of it dissipates into the environment and cascades through air, waters, and terrestrial ecosystems where it contributes to multiple effects, including adverse impacts on human health, ecosystem services, biodiversity, and climate change (Fowler et al. 2009; Delon et al. 2008). During the cascade, Nr can influence GHG exchanges with the atmosphere, aerosol production, tropospheric O_3, or increased biological productivity, which require C. All these processes have an impact on global climate. The endpoint of the cascade is its conversion back to N_2. Since Nr is currently being produced more rapidly than it can be converted back to N_2, Nr tends to accumulate in the environment in many regions.

Other negative consequences of increased Nr are its contributions to radiative forcing of climate, reductions in biodiversity at regional scales in terrestrial ecosystems, and damage to human health through aerosols and ozone (O_3) production (Erisman et al. 2013). In wet soils (i.e., anoxic conditions), in addition to N_2, denitrification produces N_2O, a GHG which has nearly 300 times the warming potential of CO_2 per molecule and also destroys ozone (O_3) through its stratospheric reactions (Butler et al. 2016; Ravishankara et al. 2009). A fraction of the formed N_2O is lost to the atmosphere and contributes to increases in atmospheric N_2O concentrations. N_2O is formed as an intermediate product of nitrification, and excessive use of N fertilizers in agricultural systems represent a large source of N_2O emissions to the atmosphere. Agricultural systems account for about one-fourth of the global N_2O emissions (Ciais et al. 2013). Agricultural sources of Nr also produce atmospheric emissions of NH_3 and NO_x. When released to lower atmosphere, NO_x can increase tropospheric O_3 formation, smog, particulate matter, and aerosols. Inhalation of the NO_x by-products causes health disorders including asthma, respiratory disorders, inflammation of airways, reduced lung functions, bronchitis and cancers (Erisman et al. 2013). Anthropogenic modification of the global N cycle also has some substantial benefits. Increased use of N fertilizers has allowed for the production of food that is necessary to support a rapidly increasing global population and for increasing per-capita consumption of meat and milk (Galloway and Cowling 2002). Increased Nr in marine and terrestrial ecosystems also stimulates global CO_2 sequestration biologically in these systems (Zaehle 2013). For example, Nr addition to the terrestrial biosphere has increased global productivity by ~2.6 Pg C yr^{-1}, an approximately 12% increase since pre-industrial times. About 0.2 Pg C yr^{-1} of this increase in production is sequestered in the terrestrial biosphere (Zaehle 2013). The ultimate fate of NO_x and NH_x is their removal by the wet and dry depositions on terrestrial and aquatic ecosystems. The deposition of Nr induces a cascade of effects (Galloway et al. 2004). The vast majority of the land area of the northern hemisphere is receiving increased N deposition due to anthropogenic activity. For example, deposition to forests ranges from 5 kg N ha^{-1} yr^{-1} in northern Europe to >60 kg N ha^{-1} yr^{-1} in central Europe, and the range for short vegetation is about half (Dise et al. 2009). Ammonium is the dominant form of atmospheric N input in Europe. Ammonium originates from NH_3 emitted primarily from animal husbandry, while NO_3^- originates from N oxides emitted by fossil fuel combustion and automotive exhaust. At remote sites in the southern hemisphere, wet deposition is <1 kg N ha^{-1} yr^{-1} (Vitousek et al. 1997a). Total pre-industrial inputs to forests were therefore <2 kg N ha^{-1} yr^{-1}. The NPP of most terrestrial ecosystems is limited by N availability and Nr deposition may enhance ecosystem productivity (Vitousek et al. 1997a) with possible consequences to the global C cycle (Prentice et al. 2001).

Based on a survey of the existing literature from natural terrestrial ecosystems, Galloway et al. (2004) estimated that prior to large anthropogenic alteration of global N cycle during industrial era; the BNF of natural terrestrial ecosystems contributed 100 to 290 Tg N yr^{-1}. Of the 11.5 Mha of natural vegetated land (Mackenzie 2010), it is estimated that 0.76 Mha had been altered by anthropogenic action by the beginning of the industrial era in 1860—including land clearing for cultivation and conversion of forests to pastures (Houghton 1999). Therefore, the natural BNF was estimated at 120.0 Tg N yr^{-1} (Figure 7.5). The BNF also created 40.0 to 140.0 Tg N yr^{-1} in marine ecosystems (Galloway 1998). However, N fixed in marine ecosystems is not transported to terrestrial ecosystems, except in small quantities volatilized as NH_3 and subsequently deposited on the land, which is difficult to quantify.

Anthropogenic activities influence the N budgets in two ways, by (i) increasing the mobilization of existing Nr and (ii) creating new Nr. As hunter-gatherers, anthropogenic impact on the N cycle was limited to the mobilization of existing forms of Nr (such as biomass burning), and the extent of the impact was limited relative to the natural processes. Anthropogenic activities that created Nr during the pre-industrial era were mainly cultivation of legumes and fossil fuel combustion for energy, especially coal. It is estimated that fossil fuel combustion of coal generated 0.6 Tg N yr^{-1} in the form of NO_x in 1890 (van Aardenne et al. 2001). Scaling back these estimates by population and other factors such as energy demand (Galloway et al. 2004) indicates that fossil fuel combustion in 1860 created 0.3 Tg N yr^{-1} in the form of N_x, (mainly through coal combustion at the beginning of industrial era). The advent of legume cultivation ~5000 yrs. ago initiated crop-BNF on Earth. At the beginning of the industrial era, crop BNF was ~15.0 Tg N yr^{-1} (Galloway et al. 2004; Figure 7.5).

Emissions of Nr (N_y, NH_x) to the atmosphere are a key driver to the atmospheric chemistry and composition (Monks et al. 2009; Fowler et al. 2009), as well as the productivity of freshwater, terrestrial, and marine ecosystems. Atmospheric chemistry and climate change are coupled. Atmospheric composition influences climate by regulating the radiation budget. Industrial development and agricultural production emit large quantities of NO_x (i.e., $NO + NO_2$) and NH_3 to the atmosphere (Vet et al. 2014). Different Nr compounds and their reaction products have widely varying chemical and physical properties, thus the processes that chemically convert species of oxidized as well as reduced Nr compounds are important for their transport, influence on atmospheric chemistry, and deposition to the surface. The main sources of NO_x are soils, natural fires, transport from stratosphere, and combustion of fossil fuels. The sinks are microbial uptakes in soils as well as reactions with OH in the atmosphere (Miyazaki et al. 2012). Emissions of NO_x (oxidized N) for 2000 are estimated at 40 Tg N yr^{-1} (van Vuuren et al. 2011; Figure 7.6). Fossil fuel combustion is the dominant source of NO_x, contributing 30 Tg N yr^{-1} distributed among road transport (10 Tg N yr^{-1}), the energy sector (7.5 Tg N yr^{-1}), the shipping and aviation (6 Tg N yr^{-1}) industry (4.5 Tg N yr^{-1}), and buildings (3 Tg N yr^{-1}). In addition, the biomass burning contribution is estimated to release about 6 Tg N yr^{-1}. Soil NO emissions are the highest in agricultural fields receiving fertilizer, followed by grasslands, while NO emissions from forests and other natural ecosystems are generally lower (Ludwig et al. 2001). Estimates of global NO_x emissions from soils based on SOC content, soil pH, land cover type, climate, N input, soil temperature, soil moisture regime, and vegetation fire were estimated to be 7.43 Tg N yr^{-1}, and decreased to 4.94Tg N yr^{-1} after the canopy reduction (Yan et al. 2005).

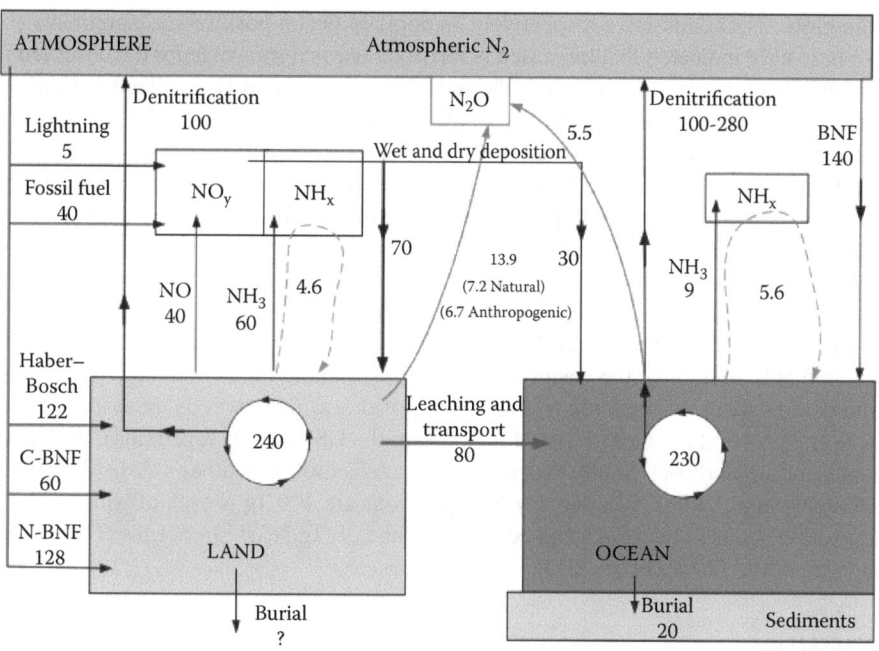

FIGURE 7.6 A simplified global N cycling showing the processing of reactive N (Nr) in terrestrial, atmosphere, and marine systems. Numbers indicate the estimated fluxes in Tg N yr^{-1} and arrows indicate the direction of fluxes flow. (Adapted with modification from Ussirri, D.A.N., Lal, R., *Soil Emissions of Nitrous Oxide and its Mitigation.* Springer, Dordrecht; Netherlands, and New York; USA, 378 p., 2013.)

Ammonia is emitted from agricultural sources, mainly the breakdown of animal excreta and release from synthetic N fertilizers (Bouwman et al. 1997), and has a typical atmospheric residence time of about 24 hrs. NH_3 react with H_2SO_4 and HNO_3 and other acidic compounds in the atmosphere to form particulate NH_4^+. Both NH_3 and gaseous NO_x form aerosol phase compounds (NH_4^+ and NO_3^- respectively) which have atmospheric residence times of several days. The global emissions of NH_3 (reduced N) from soils in 2008 are estimated at 56.8 Tg N yr^{-1} and range from 46 to 85 Tg N yr^{-1} (Sutton et al. 2013b), of which, anthropogenic emissions are estimated at 40 Tg yr^{-1} (Fowler et al. 2013). Agricultural soils and crops—including grazing and land application of animal manure—are estimated at 28 Tg N yr^{-1}, while excreta from domestic animals excluding (emissions from land application of manure) is estimated at 9.0 Tg N yr^{-1}. The Nr fixed industrially through N fertilizer synthesis is currently at 120 Tg N yr^{-1}; thus, the emissions to the atmosphere from agriculture (crops and livestock) represent about a quarter of the annual fertilizer production, which is the unintentional leakage of Nr from farming systems due to NH_3 volatilization, effectively fertilizing the atmosphere. An increase in global meat per-capita consumption has magnified the fertilizer requirements and NH_3 emissions (Westhoek et al. 2014; Erisman et al. 2008). Other major sources of NH_3 emissions (Tg N yr^{-1}) include 5.5 from biomass burning (including savannah, agricultural waste, forest, grassland and peat fires), 4.4 from waste, 5 from soil under natural vegetation and excreta from wild animals, 3.3 from human waste and pets, and 1.6 from fossil fuel burning and industrial activities. In addition, global ocean and volcanic activities contribute 8.6 Tg N yr^{-1} (Sutton et al. 2013a).

Microbial denitrification and nitrification processes are responsible for 87% of the global annual N_2O budget of 18.8 Tg N yr^{-1} in 2006 (Syakila and Kroeze 2011). Non-biological sources including fossil fuel combustion, biofuel and biomass burning, and industrial processes contribute the remaining 13%. The average annual increase in N_2O emissions is estimated at 3.6 ± 0.15 Tg N (N_2O) yr^{-1} (Ciais et al. 2013). Agriculture contributes 27%, while natural soils and oceans contribute 35 and

25% of the global N_2O emissions, respectively. Isotopic N_2O composition measurements ($^{14/15}N$) in the atmosphere have indicated that increased N fertilizer use is responsible for the observed increase in atmospheric N_2O concentration during the industrial period, and N fertilized agricultural soils are responsible for ~16% of the global annual N_2O emissions (Park et al. 2012).

After a series of chemical transformations and physical transport processes in the atmosphere, NH_3 and NO_x are removed through dry deposition and wet scavenging and deposited on the Earth's surface (Dentener et al. 2006). High quality measurements of wet and dry Nr deposition remain a major weakness of modeling of the atmospheric N cycle, since deposition depends on the hydrological cycle and wind, which are highly uncertain. Wet N deposition has been researched intensively due to the ease of observation compared to dry N deposition (Du et al. 2014). Using a multi-model analysis, Dentener et al. (2006) estimated that 36% to 51% of all NO_y and NH_x is deposited over the ocean, and 50% to 80% of the fraction of deposition on land falls on natural vegetation. About 11% of the world's natural vegetation receives N deposition which exceeds the critical threshold of 1000 mg N m^{-2} yr^{-1}, and the most affected regions are the US (20% of vegetation), Western Europe (30%), Eastern Europe (80%), South Asia (60%), East Asia (40%), Southeast Asia (30%), and Japan (50%) (Dentener et al. 2006). Wet and dry N depositions are 100 Tg N yr^{-1}, of which, 70 Tg N yr^{-1} of the deposition occur in the terrestrial ecosystems and 30 Tg N yr^{-1} in oceans (Duce et al. 2008; Dentener et al. 2006; Fowler et al. 2013).

7.6 SYNTHESIS

The fate of anthropogenic Nr input into the terrestrial ecosystems is relatively well understood and has low uncertainty. However, the fate of Nr into other ecosystems, including aquatic, marine, and atmospheric systems, is not well known. About 18% of Nr added into the terrestrial ecosystems is exported to rivers, estuaries, and coastal waters and is denitrified; 13% is deposited in oceans from marine atmosphere, and 4% is emitted as N_2O (Galloway et al. 2004). Therefore, 65% of Nr added into the terrestrial ecosystem either accumulates in soils, vegetation, and groundwater or is denitrified into N_2. The denitrification rates in soils are not well understood, and as a result the quantity of Nr stored in the terrestrial ecosystems is uncertain. The Nr input into terrestrial ecosystems continues to rise, and the impacts of increase in Nr input to terrestrial ecosystems are not fully understood. There is a possibility that as the terrestrial ecosystems become saturated with Nr, more of it will be exported into coastal waters, or more may be denitrified, increasing the risks of global warming. The amount of Nr used to produce food is on average about 10-fold higher than its consumption, due to inefficient food production–processing–consumption chain (Sutton et al. 2011; Galloway et al. 2008). The most direct effects of increased Nr to the environment are through the formation of N_2O, which is responsible for anthropogenic radiative forcing. Other consequences include NH_3 and NO_x emissions from agriculture to the air, and NO_3^- to ground streams and water (Galloway et al. 2008). Pathways and impacts of Nr are also influenced by its chemical form, a factor that highlights the need for a better understanding of the fate of oxidized versus reduced forms of N. Nr is highly mobile, and most of it dissipates into the environment and cascades through air and ecosystems where it contributes to multiple effects including human and human health, ecosystem services biodiversity, and climate change (Galloway et al. 2008).

N is affected by and affects the climate, and the net contributions of anthropogenic Nr to climate change are vigorously debated (Sutton et al. 2007). The N_2O is formed during industrial fertilizer production, incomplete combustion, or microbial denitrification and nitrification, especially after fertilizer and manure application to soils. The Nr increases radiative forcing in the troposphere, principally by the production of N_2O and tropospheric O_3. In contrast, it can also have a cooling effect, largely through tropospheric aerosols and stratospheric O_3 decline (Denman et al. 2007). In addition, Nr can interact with the C cycle, leading to global effects on atmospheric CO_2 and CH_4 (Hungate et al. 2003; Gruber and Galloway 2008). Elevated levels of Nr deposition may stimulate plant growth in N limited ecosystems and increase CO_2 uptake. The terrestrial sink of CO_2 is influenced by increased Nr availability. The C and N cycles are linked through the ecosystem C:N ratio. Uptake of atmospheric CO_2 is therefore limited by N availability. Effects of increase in N on C storage need to be assessed, however. The acceleration of N cycling because of global warming is one of the mechanisms to further increase N availability to the ecosystem. This mechanism works through faster decomposition of organic matter and associated N mineralization rates. In ruminants, an increased Nr supply has also been associated with the reduction of CH_4 from these animals through the increased digestibility of their diets (Erisman et al. 2013).

An increase in Nr in the atmosphere influences oxidation capacity, radiation balance, and acidity. Many of these effects are specific to individual N compounds (e.g., oxidative effects of NO_x, radiative balance of N_2O, acidification of some NO_x, and acid neutralizing capacity of NH_x). However, the biogeochemistry of Nr is intriguing because under appropriate conditions, almost any species of Nr can be converted into any other form of Nr. Therefore, one atom of N can have cascading impacts. For example, NO emitted from fossil fuel combustion can change the oxidative capacity of the atmosphere, increase the acidity of aerosols, change the radiative properties of the atmosphere, increase the acidity of precipitation and ecosystems, and enhance the NPP of the ecosystems, etc. The impacts of Nr on a species or ecosystem depends on several factors, including the duration of exposure, total amount and forms of N, the sensitivity of species, and intrinsic ecosystem properties such as fertility and acid neutralizing capacity (Bobbink et al. 2010). These cascading effects can continue as the same atoms are converted from one species into another until they are either stored in a long-term reservoir or are denitrified into N_2. Such cascading effects of Nr make its anthropogenic increase of great interest to biogeochemists, environmentalists, agriculturalists, and atmospheric scientists as well as policy makers. Commonly, Nr acts directly on organisms through factors such as nutrient enrichment, O_2, depletion—especially in aquatic systems, soil, or water acidification, altering nutrient ratios, or intensifying the impact of other stressors such as pathogens or climate change.

The effects of Nr input into tropical regions in terrestrial and aquatic ecosystems must be quantified. Due to BNF, P, and cations deficiency, many tropical ecosystems tend to be relatively N-rich, suggesting that increase in Nr may lead to a much different response compared to that in the temperate regions, probably leading to more N_2O emissions (Matson et al. 1999). Unlike the situation in temperate ecosystems, where the anthropogenic N stimulates plant growth while also creating a significant sink for excessive CO_2 (Townsend et al. 1996), the addition of anthropogenic N is likely to have little direct effect on C uptake in most tropical forests. Studies of N cycling and trace gas emissions across a range of tropical ecosystems suggest that N gas fluxes as a proportion of N mineralization are greater in the tropical than in the temperate systems, and that, quantitatively, N trace gas fluxes are on average much greater in tropical than in the temperate ecosystems (Hall et al. 1996). Soils of old-growth tropical forests are typically rich in available N but poor in available P, since tropical forests could accumulate N through leaching and immobilization during soil forming processes (Vitousek et al. 2010), thereby aggravating P limitation over time (Matson et al. 1999; Li et al. 2016).

Rapidly growing biofuel development may create yet new and rapidly changing dimensions of the anthropogenic influence on the global N cycle. In addition to increased emissions of NO_x and NH_x, the increase in use of biofuel to reduce the dependence of fossil fuel as an energy source may increase the use of fertilizer N to meet the new demand for fuel crops for use in energy feedstock. The production of biofuel may increase the release of Nr into an ecosystem, accentuating the contribution to change in the N cycle. It may also increase N_2O emissions from fertilizers and combusted biomasses, which may negate the savings in CO_2 emissions due to changes from fossil fuel use into biofuel (Crutzen et al. 2008). It can also lead to tropospheric O_3 production. Therefore, before recommending changes in the energy policy in favor of agro-biofuel, there is a need for better analysis to quantify the true emissions saving based on global warming potential of CO_2 and N_2O.

Overall, the anthropogenic alteration of the global N cycle has multiple consequences in the atmosphere, terrestrial, and marine ecosystems as well as human health. These include (i) increasing the concentration of the N_2O globally; (ii) substantial contribution to acid rain and photochemical smog that afflicts global urban and agricultural production areas (Vitousek et al. 1997a); (iii) increases in fluxes of Nr (i.e., more than two-thirds of global NO and NH_3 emissions are human-caused); (iv) acidification; (v) eutrophication; (vi) foliar damage, particularly to lower plants; (vii) susceptibility to stress; (viii) reduced allocation of OC from the vegetation to mycorrhizal fungi; (ix) reduced plant richness in a broad ranges of ecosystems; and (x) ozone exposure to natural ecosystems (Erisman et al. 2013). Emissions of NO and CH_3 to the atmosphere have increased by about fivefold since pre-industrial times (Galloway et al. 2008). Approximately 40% of fertilizer N lost to the environment is denitrified back to unreactive atmospheric N_2 (Galloway et al. 2004), while the rest of the excess N escapes into environmental terrestrial, aquatic, and marine pools before eventually being denitrified or stored as fossil Nr. Atmospheric deposition of reactive N and S into the aquatic ecosystems with low acid neutralizing capacity—primarily freshwater—results in acidification. A sharp decline in sulfur emissions beginning in the early 1980s makes Nr the major component of acidic deposition in many areas of Europe and North America, and it is becoming a recurring problem in many developing countries. Coastal eutrophication has recently emerged as a global issue of major concern due to steady growth in the extent and persistence of eutrophic, hypoxic, and anoxic coastal waters (Rabalais et al. 2002) and incidence of toxic algal blooms. About 415 eutrophic and hypoxic coastal systems have been identified worldwide and only 13 systems are in recovery (Selman et al. 2008).

Circulation of anthropogenic Nr in Earth's atmosphere, hydrosphere, and biosphere causes a variety of consequences, and these are magnified with time as Nr moves along biogeochemical pathway. For example, the same atom of Nr can cause multiple effects in the atmosphere, terrestrial ecosystems, freshwater, and marine systems as well as human health through a sequence generally termed as "nitrogen cascade" (Galloway et al. 2003). The Nr that is emitted to the atmosphere is deposited downwind, where it can influence the dynamics of the recipient ecosystems. In the N deficient regions, added N generally leads to unintentional fertilization, leading to an increase in ecosystem productivity and C storage within the ecosystems, and ultimately increasing losses of N and cations from soils through "N saturation" (Vitousek et al. 1997b). Increased productivity of the ecosystems as a result of added N also causes loss of terrestrial diversity. Anthropogenically created Nr also moves from agriculture, sewage systems, and N-saturated terrestrial ecosystems to streams, rivers, groundwater, lakes, and ultimately the oceans. Fluxes of N through streams and rivers have increased markedly as human alteration of the global N cycle has accelerated (Gruber and Galloway 2008). Generally, river NO_3^- is highly correlated with the human population of the river basins (Howarth et al. 2011). Increases in river Nr drive the eutrophication of estuaries, causing algal blooms that threaten the sustainability of marine fisheries (Vitousek et al. 1997b). Many of the coastal ecosystems receiving increased Nr loading are generally N-limited, leading to harmful algal blooms and a decline in the quality of surface and groundwaters. In addition to the ecosystem-level disturbances, Nr alters the balance of GHGs, enhances tropospheric ozone, decreases stratospheric ozone, increases soil acidification, and stimulates the formation of secondary particulate matter in the atmosphere. All these have negative effects on people and the environment.

The increasing population and increase in per-capita food and energy consumption will lead to a continued increase in Nr generation and N mobilization in the future because of the use of Nr for agricultural production (Figure 7.6) and increase in energy demand to match the population growth. It is possible to slow down the increase in fertilizer N use by increasing N fertilizer use efficiency, since crops do not utilize about 50% of fertilizer N applied (Smil 1999). There are several interventions to reduce the creation of Nr, improve N use efficiency with which it is used or converting the excess Nr to non-reactive atmospheric N_2, and minimize the cascading effects of Nr (Smil 1999). These include: (i) choice of appropriate fertilizing compounds based on soil testing, (ii) maintenance of proper nutrient ratios, (iii) attention to timing and N placement of fertilizer, (iv) increasing NUE in food production, (v) altering human diets, and (vi) improving the treatment of human and animal waste (Galloway et al. 2008). In addition, other indirect approaches that can reduce the need for synthetic fertilizers or increase the use efficiency of N are (i) planting of leguminous crops and optimizing of conditions to favor BNF organisms, (ii) maximizing the recycling of organic wastes, (iii) adopting strategies of integrated use of organic and synthetic fertilizers, (iv) improving soil management by reducing soil erosion, (v) maintaining adequate soil moisture, and (vi) controlling crop pests (Smil 1999).

7.7 CONCLUSIONS

Nitrogen is the important element that controls the productivity and functioning of terrestrial and marine ecosystems. The nitrogen (N) cycle and its balance have been altered primarily by anthropogenic activities and environmental changes at scales ranging from local landscape and watershed to regional and global scales. These alterations have drastically increased both the availability and mobility of N over large regions of Earth and affected the structure and functions of natural and agro-ecosystems and the associated aquatic systems. The mobility of N means that although most of the deliberate applications of N occur locally, the influences extend beyond local environments and spread regionally and even globally. The importance of N as a nutrient for all biota, the ever-increasing rates of its anthropogenic inputs in terrestrial agro-ecosystems to increase crop yields, its resultant losses to the environment, and the complexity of the biological, physical, and chemical factors that regulate N cycling processes all contribute to the necessity to further understand the alteration of N cycling. The global N cycle has been significantly modified by human activity and is the most perturbed among the biogeochemical cycles on the planet. Transformation of the global N cycle is also predicted to continue at a record pace, reflecting the growing demand for reactive N in agriculture, increased combustion of fossil fuels, and the increased demand for Nr in industry. Global N fixation contributes 413 Tg of Nr to terrestrial and marine ecosystems annually, of which anthropogenic activities are responsible for half. The major inputs of reactive N (Nr) from human activities to ecosystems are fertilizer N input to agriculture, biological N fixation (BNF) by increased legume cultivation, and combustion-related NO_x estimated to add 222 Tg N yr^{-1} in 2012, which is approximately equal to the natural unmanaged terrestrial and marine ecosystems. The major driving factor for this growth is the increased agricultural demand for food to match with the population growth. Leakage from agriculture use of fertilizer N and manure contributes to nitrate (NO_3^-) in drainage waters from farmland and emissions of trace Nr compounds to the atmosphere. Emissions of ammonia (NH_3) from land and NO_x from combustion related emissions contribute 100 Tg N yr^{-1} to the atmosphere which is transported and processed within the atmosphere and generates secondary pollutants which include ozone, photochemical oxidants, and aerosols. The Nr generation trend is still accelerating and is unlikely to change in the foreseeable future, because of the growing demand for food for the growing global population. N is influenced by and influences climate. The majority of transformations of natural and anthropogenic Nr occur on land, within soils and vegetation, and in the atmosphere. Reactive N can directly increase radiative forcing in the troposphere through the production of N_2O and tropospheric O_3, but Nr can also have a cooling effect through tropospheric aerosol formation and decreasing of stratospheric O_3. In addition, N has a strong interaction with C, and elevated Nr deposition may stimulate plant growth in N-limited ecosystems, thereby increasing CO_2 sequestration. The anthropogenic influences on the N cycling, its availability, and importance in biogeochemistry and chemical interactions with climate system vary across regions of the globe and remain uncertain. To date, the largest changes in N cycling have occurred in developed countries in temperate zones. However, this is changing as N fertilizer use and fossil fuel combustion increase in South Asia, East Asia, and tropical South America. The significant consequences of anthropogenic acceleration of the N cycle is the eutrophication of estuaries and coastal waters, leading to hypoxic zones in many areas, as well as increased global inventories of N_2O. Further research is needed to evaluate the overall consequences of increased Nr to the ecosystem. Other aspects of N cycling alteration, which need further research, include (i) the ultimate fate of Nr in both terrestrial and marine ecosystem need to be refined, and (ii) effects increasing Nr on human health must be quantified.

REFERENCES

Aber, J., McDowell, W., Nadelhoffer, K., Magill, A., Berntson, G., Kamakea, M., McNulty, S., Currie, W., Rustad, L., Fernandez, I., 1998. Nitrogen saturation in temperate forest ecosystems—Hypotheses revisited. *Bioscience* 48 (11), 921–934.

Aber, J.D., Nadelhoffer, K.J., Steudler, P., Melillo, J.M., 1989. Nitrogen saturation in northern forest ecosystems. *Bioscience* 39, 378–286.

Adler, J.F., Williams, Q., 2005. A high-pressure x-ray diffraction study of iron nitrides: Implications for Earth's core. *Journal of Geophysical Research—Solid Earth* 110, B01203.

Alewell, C., Paul, S., Lischeid, G., Storck, F.R., 2008. Co-regulation of redox processes in freshwater wetlands as a function of organic matter availability? *Science of the Total Environment* 404 (2–3), 335–342.

Beatty, P.H., Good, A.G., 2011. Future prospects for cereals that fix nitrogen. *Science* 333 (6041), 416–417.

Bebout, G.E., Fogel, M.L., Cartigny, P., 2013. Nitrogen: Highly volatile yet surprisingly compatible. *Elements* 9 (5), 333–338.

Bebout, G.E., Lazzeri, K.E., Geiger, C.A., 2016. Pathways for nitrogen cycling in Earth's crust and upper mantle: A review and new results for microporous beryl and cordierite. *American Mineralogist* 101, 7–24.

Berner, R.A., 2006. Geological nitrogen cycle and atmospheric N_2 over Phanerozoic time. *Geology* 34, 413–415.

Billen, G., Garnier, J., Lassaletta, L., 2013. The nitrogen cascade from agricultural soils to the sea: Modelling nitrogen transfers at regional watershed and global scales. *Philosophical Transactions of the Royal Society B—Biological Sciences* 368, 1–13.

Bobbink, R., Hicks, K., Galloway, J., Spranger, T., Alkemade, R., Ashmore, M., Bustamante, M., Cinderby, S., Davidson, E., Dentener, F., Emmett, B., Erisman, J.W., Fenn, M., Gilliam, F., Nordin, A., Pardo, L., De Vries, W., 2010. Global assessment of nitrogen deposition effects on terrestrial plant diversity: A synthesis. *Ecological Applications* 20, 30–59.

Bobbink, R., Hornung, M., Roelofs, J.G.M., 1998. The effects of air-borne nitrogen pollutants on species diversity in natural and semi-natural European vegetation. *Journal of Ecology* 86, 717–738.

Bottomley P.J., Myrold D.D., 2015. Biological N inputs. In: Paul, E.A. (Ed), *Soil Microbiology, Ecology and Biochemistry*. 4th Ed. Elsevier Academic Press, London, U.K. and San Diego, USA, pp. 447–471.

Bouwman, A.F., Beusen, A.H.W., Griffioen, J., Van Groenigen, J.W., Hefting, M.M., Oenema, O., Van Puijenbroek, P., Seitzinger, S., Slomp, C.P., Stehfest, E., 2013a. Global trends and uncertainties in terrestrial denitrification and N_2O emissions. *Philosophical Transactions of the Royal Society B—Biological Sciences* 368, 1–12.

Bouwman, L., Goldewijk, K.K., Van Der Hoek, K.W., Beusen, A.H.W., Van Vuuren, D.P., Willems, J., Rufino, M.C., Stehfest, E., 2013b. Exploring global changes in nitrogen and phosphorus cycles in agriculture induced by livestock production over the 1900–2050 period. *Proceedings of the National Academy of Sciences of the United States of America* 110, 20882–20887.

Bouwman, A.F., Lee, D.S., Asman, W.A.H., Dentener, F.J., VanderHoek, K.W., Olivier, J.G.J., 1997. A global high-resolution emission inventory for ammonia. *Global Biogeochemical Cycles* 11, 561–587.

Boyd, S.R., 2001. Nitrogen in future biosphere studies. *Chemical Geology* 176, 1–30.

Boyer, E.W., Goodale, C.L., Jaworsk, N.A., Howarth, R.W., 2002. Anthropogenic nitrogen sources and relationships to riverine nitrogen export in the northeastern USA. *Biogeochemistry* 57, 137–169.

Brasseur, G.P., Schultz, M., Granier, C., Saunois, M., Diehl, T., Botzet, M., Roeckner, E., Walters, S., 2006. Impact of climate change on the future chemical composition of the global troposphere. *Journal of Climate* 19, 3932–3951.

Brown, L.R., 1999. Feeding Nine Billion. In: Brown, L.R., Flavin, C., French, H.F. (Eds.), *State of the World 1999*: A Worldwatch Institute Report on Progress Toward a Sustainable Society, New York, USA, London, U.K.

Burgess, B.K., Lowe, D.J., 1996. Mechanism of molybdenum nitrogenase. *Chemical Reviews* 96, 2983–3011.

Busigny, V., Bebout, G.E., 2013. Nitrogen in the silicate Earth: Speciation and isotopic behavior during mineral-fluid interactions. *Elements* 9, 353–358.

Butler, A.H., Daniel, J.S., Portmann, R.W., Ravishankara, A.R., Young, P.J., Fahey, D.W., Rosenlof, K.H., 2016. Diverse policy implications for future ozone and surface UV in a changing climate. *Environmental Research Letters* 11, 064017.

Canfield, D.E., Glazer, A.N., Falkowski, P.G., 2010. The evolution and future of Earth's nitrogen cycle. *Science* 330, 192–196.

Carpenter, S.R., Stanley, E.H., Vander Zanden, M.J., 2011. State of the world's freshwater ecosystems: Physical, chemical, and biological changes. *Annual Review of Environment and Resources* 36, 75–99.

Ciais, P., Sabine, C., Bala, G., Bopp, L., Brovkin, V., Canadell, J., Chhabra, A., DeFries, R., Galloway, J., Heimann, M., Jones, C., Quéré, C.L., Myneni, R.B., Piao, S., Thornton, P., 2013. Carbon and Other Biogeochemical Cycles. In: Stocker, T.F., Qin, D., Plattner, G.-K., Tignor, M., Allen, S.K., Boschung, J., Nauels, A., Xia, Y., Bex, V., Midgley, P.M. (Eds.), *Climate Change 2013: The Physical Science Basis. Contribution of Working Group I to the Fifth Assessment Report of the Intergovernmental Panel on Climate Change.* Cambridge University Press, Cambridge, United Kingdom, and New York, NY, USA., pp. 465–570.

Cleveland, C.C., Townsend, A.R., Schimel, D.S., Fisher, H., Howarth, R.W., Hedin, L.O., Perakis, S.S., Latty, E.F., Von Fischer, J.C., Elseroad, A., Wasson, M.F., 1999. Global patterns of terrestrial biological nitrogen (N-2) fixation in natural ecosystems. *Global Biogeochemical Cycles* 13, 623–645.

Crutzen, P.J., Mosier, A.R., Smith, K.A., Winiwarter, W., 2008. N_2O release from agro-biofuel production negates global warming reduction by replacing fossil fuels. *Atmospheric Chemistry and Physics* 8, 389–395.

Davidson, E.A., David, M.B., Galloway, J.N., Goodale, C.L., Haeuber, R., Harrison, J.A., Howarth, R.W., Jaynes, D.B., Lowrance, R.R., Nolan, B.T., Peel, J.L., Pinder, R.W., Porter, E., Snyder, C.S., Townsend, A.R., and Ward, M.H., 2012. Excess nitrogen in the U.S. environment: Trends, risks, and solutions. *Issues in Ecology* 15, 1–16.

De Vries, W., Reinds, G.J., Gundersen, P., Sterba, H., 2006. The impact of nitrogen deposition on carbon sequestration in European forests and forest soils. *Global Change Biology* 12, 1151–1173.

Delon, C., Reeves, C.E., Stewart, D.J., Serca, D., Dupont, R., Mari, C., Chaboureau, J.P., Tulet, P., 2008. Biogenic nitrogen oxide emissions from soils—Impact on NOx and ozone over West Africa during AMMA (African Monsoon Multidisciplinary Experiment): Modelling study. *Atmospheric Chemistry and Physics* 8, 2351–2363.

DeLuca, T.H., Zackrisson, O., Gundale, M.J., Nilsson, M.C., 2008. Ecosystem feedbacks and nitrogen fixation in boreal forests. *Science* 320, 1181. doi: 10.1126/science.1154836.

Denman, K.L., Brasseur, G., Chidthaisong, A., Ciais, P., Cox, P.M., Dickinson, R.E., Hauglustaine, D., Heinze, C., Holland, E., Jacob, D., Lohmann, U., Ramachandran, S., Dias, P.L.d.S., Wofsy, S.C., Zhang, X., 2007. Couplings Between Changes in the Climate System and Biogeochemistry. In: Solomon, S., Qin, D., Manning, M., Chen, Z., Marquis, M., Averyt, K.B., M.Tignor, Miller, H.L. (Eds.), *Climate Change 2007: The Physical Science Basis Contribution of Working Group I to the Fourth Assessment Report of the Intergovernmental Panel on Climate Change.* Cambridge University Press, Cambridge, United Kingdom, and New York, NY, USA, pp. 501–587.

Dentener, F., Drevet, J., Lamarque, J.F., Bey, I., Eickhout, B., Fiore, A.M., Hauglustaine, D., Horowitz, L.W., Krol, M., Kulshrestha, U.C., Lawrence, M., Galy-Lacaux, C., Rast, S., Shindell, D., Stevenson, D., Van Noije, T., Atherton, C., Bell, N., Bergman, D., Butler, T., Cofala, J., Collins, B., Doherty, R., Ellingsen, K., Galloway, J., Gauss, M., Montanaro, V., Mueller, J.F., Pitari, G., Rodriguez, J., Sanderson, M., Solmon, F., Strahan, S., Schultz, M., Sudo, K., Szopa, S., Wild, O., 2006. Nitrogen and sulfur deposition on regional and global scales: A multimodel evaluation. *Global Biogeochemical Cycles* 20, GB4003.

Diaz, R.J., Rosenberg, R., 2008. Spreading dead zones and consequences for marine ecosystems. *Science* 321, 926–929.

Dise, N.B., Rothwell, J.J., Gauci, V., van der Salm, C., de Vries, W., 2009. Predicting dissolved inorganic nitrogen leaching in European forests using two independent databases. *Science of the Total Environment* 407, 1798–1808.

Du, E.Z., de Vries, W., Galloway, J.N., Hu, X.Y., Fang, J.Y., 2014. Changes in wet nitrogen deposition in the United States between 1985 and 2012. *Environmental Research Letters* 9, 1–8.

Duce, R.A., LaRoche, J., Altieri, K., Arrigo, K.R., Baker, A.R., Capone, D.G., Cornell, S., Dentener, F., Galloway, J., Ganeshram, R.S., Geider, R.J., Jickells, T., Kuypers, M.M., Langlois, R., Liss, P.S., Liu, S.M., Middelburg, J.J., Moore, C.M., Nickovic, S., Oschlies, A., Pedersen, T., Prospero, J., Schlitzer, R., Seitzinger, S., Sorensen, L.L., Uematsu, M., Ulloa, O., Voss, M., Ward, B., Zamora, L., 2008. Impacts of atmospheric anthropogenic nitrogen on the open ocean. *Science* 320, 893–897.

Eady, R.R., 1996. Structure-function relationships of alternative nitrogenases. *Chemical Reviews* 96, 3013–3030.

Ehhalt, D., Prather, M., Dentener, F.J., Derwent, R., Dlugokencky, E.J., Holland, E.A., Isaksen, I., Katima, J., Kirchhoff, V., Matson, P.A. 2001. Atmospheric chemistry and greenhouse gases. In: J. T. Houghton, Y. Ding, D. J. Griggs, M. Noguer, P. J. vander Linden, X. Dai, K. Maskell,C. A. Johnson (Eds.) *Climate Change 2001: The Scientific Basis. Contribution of Working Group I to the Third Assessment Report of the Intergovernmental Panel on Climate Change* Cambridge University Press, Cambridge, United Kingdom, and New York, NY, USA., pp. 239–287.

Elbert, W., Weber, B., Burrows, S., Steinkamp, J., Buedel, B., Andreae, M.O., Poeschl, U., 2012. Contribution of cryptogamic covers to the global cycles of carbon and nitrogen. *Nature Geoscience* 5, 459–462.

Elkins, L.J., Fischer, T.P., Hilton, D.R., Sharp, Z.D., McKnight, S., and Walker, J., 2006. Tracing nitrogen in volcanic and geothermal volatiles from the Nicaraguan volcanic front. *Geochimica et Cosmochimica Acta* 70, 5215–5235.

Erisman, J.W., Galloway, J.N., Seitzinger, S., Bleeker, A., Dise, N.B., Petrescu, A.M.R., Leach, A.M., de Vries, W., 2013. Consequences of human modification of the global nitrogen cycle. *Philosophical Transactions of the Royal Society B—Biological Sciences* 368, 20130116.

Erisman, J.W., Sutton, M.A., Galloway, J., Klimont, Z., Winiwarter, W., 2008. How a century of ammonia synthesis changed the world. *Nature Geoscience* 1, 636–639.

Falkowski, P.G., 1997. Evolution of the nitrogen cycle and its influence on the biological sequestration of CO_2 in the ocean. *Nature* 387, 272–275.

Falkowski, P.G., Fenchel, T., Delong, E.F., 2008. The microbial engines that drive Earth's biogeochemical cycles. *Science* 320, 1034–1039.

FAOSTAT, 2017. World Food and Agriculture Statistics. Food and Agriculture Organization of the United Nations, Statistical Division, Rome, Italy.Accessed April 2017.

Fowler, D., Coyle, M., Anderson, R., Ashmore, M.R., Bower, J.S., Burgess, R.A., Cape, J.N., Cox, R.A., Derwent, R.G., Dollard, G.J., Grennfelt, P., Harrison, R.M., Hewitt, C.N., Hov, O., Jenkin, M.E., Lee, D.S., Maynard. R.L., Penkett, S.A., Smith, R.I., Stedman, J.R., Weston, K.J., Williams, M.L., Woods, P.J., 1997. Ozone in the United Kingdom—Fourth Report of the Photochemical Oxidants Review Group. Annual Report. Department of the Environment, Transport and the Regions. London, U.K. 234 p.

Fowler, D., Coyle, M., Skiba, U., Sutton, M.A., Cape, J.N., Reis, S., Sheppard, L.J., Jenkins, A., Grizzetti, B., Galloway, J.N., Vitousek, P., Leach, A., Bouwman, A.F., Butterbach-Bahl, K., Dentener, F., Stevenson, D., Amann, M., Voss, M., 2013. The global nitrogen cycle in the twenty-first century. *Philosophical Transactions of the Royal Society B—Biological Sciences* 368, 20130164.

Fowler, D., Pilegaard, K., Sutton, M.A., Ambus, P., Raivonen, M., Duyzer, J., Simpson, D., Fagerli, H., Fuzzi, S., Schjoerring, J.K., Granier, C., Neftel, A., Isaksen, I.S.A., Laj, P., Maione, M., Monks, P.S., Burkhardt, J., Daemmgen, U., Neirynck, J., Personne, E., Wichink-Kruit, R., Butterbach-Bahl, K., Flechard, C., Tuovinen, J.P., Coyle, M., Gerosa, G., Loubet, B., Altimir, N., Gruenhage, L., Ammann, C., Cieslik, S., Paoletti, E., Mikkelsen, T.N., Ro-Poulsen, H., Cellier, P., Cape, J.N., Horvath, L., Loreto, F., Niinemets, U., Palmer, P.I., Rinne, J., Misztal, P., Nemitz, E., Nilsson, D., Pryor, S., Gallagher, M.W., Vesala, T., Skiba, U., Brueggemann, N., Zechmeister-Boltenstern, S., Williams, J., O'Dowd, C., Facchini, M.C., de Leeuw, G., Flossman, A., Chaumerliac, N., Erisman, J.W., 2009. Atmospheric composition change: Ecosystems-atmosphere interactions. *Atmospheric Environment* 43, 5193–5267.

Fowler, D., Steadman, C.E., Stevenson, D., Coyle, M., Rees, R.M., Skiba, U.M., Sutton, M.A., Cape, J.N., Dore, A.J., Vieno, M., Simpson, D., Zaehle, S., Stocker, B.D., Rinaldi, M., Facchini, M.C., Flechard, C.R., Nemitz, E., Twigg, M., Erisman, J.W., Butterbach-Bahl, K., Galloway, J.N., 2015. Effects of global change during the 21st century on the nitrogen cycle. *Atmospheric Chemistry and Physics* 15, 13849–13893.

Galloway, J.N., 1998. The global nitrogen cycle: Changes and consequences. *Environmental Pollution* 102, 15–24.

Galloway, J.N., 2014. The global nitrogen cycle. In: Holland, H.D., Turekian, K.K. (Eds.), *Treatise on Geochemistry.* 2nd Ed. Elsevier, Oxford, pp. 475–498.

Galloway, J.N., Cowling, E.B., 2002. Reactive nitrogen and the world: 200 years of change. *Ambio* 31, 64–71.

Galloway, J.N., Aber, J.D., Erisman, J.W., Seitzinger, S.P., Howarth, R.W., Cowling, E.B., Cosby, B.J., 2003. The nitrogen cascade. *Bioscience* 53, 341–356.

Galloway, J.N., Dentener, F.J., Capone, D.G., Boyer, E.W., Howarth, R.W., Seitzinger, S.P., Asner, G.P., Cleveland, C.C., Green, P.A., Holland, E.A., Karl, D.M., Michaels, A.F., Porter, J.H., Townsend, A.R., Vorosmarty, C.J., 2004. Nitrogen cycles: Past, present, and future. *Biogeochemistry* 70, 153–226.

Galloway, J.N., Leach, A.M., Bleeker, A., Erisman, J.W., 2013. A chronology of human understanding of the nitrogen cycle. *Philosophical Transactions of the Royal Society B—Biological Sciences* 368, 20130120. doi.org/10.1098/rstb.2013.0120.

Galloway, J.N., Townsend, A.R., Erisman, J.W., Bekunda, M., Cai, Z., Freney, J.R., Martinelli, L.A., Seitzinger, S.P., Sutton, M.A., 2008. Transformation of the nitrogen cycle: Recent trends, questions, and potential solutions. *Science* 320, 889–892.

Galloway, J.N., Winiwarter, W., Leip, A., Leach, A.M., Bleeker, A., Erisman, J.W., 2014. Nitrogen footprints: Past, present and future. *Environmental Research Letters* 9.

Giggenbach, W.F., Matsuo, S., 1991. Evaluation of results from 2nd and 3rd IAVCEI Field Workshops on Volcanic Gases, Mt. Usu, Japan, and White Island, New-Zealand. *Applied Geochemistry* 6, 125–141.

Godfray, H.C.J., Crute, I.R., Haddad, L., Lawrence, D., Muir, J.F., Nisbett, N., Pretty, J., Robinson, S., Toulmin, C., Whiteley, R., 2010. The future of the global food system. *Philosophical Transactions of the Royal Society B—Biological Sciences* 365, 2769–2777.

Goldblatt, C., Claire, M.W., Lenton, T.M., Matthews, A.J., Watson, A.J., Zahnle, K.J., 2009. Nitrogen-enhanced greenhouse warming on early Earth. *Nature Geoscience* 2, 891–896.

Groffman, P.M., 2012. Terrestrial denitrification: Challenges and opportunities. *Ecological Processes* 1, 1–11.

Gruber, N., Galloway, J.N., 2008. An Earth-system perspective of the global nitrogen cycle. *Nature* 451, 293–296.

Gustavsson, J., 2011. Global food losses and food waste: Extent, causes and prevention. Food and Agriculture Organization (FAO), Rome, Italy, p. 29.

Hall, S.J., Matson, P.A., Roth, P.M., 1996. NO_x emissions from soil: Implications for air quality modeling in agricultural regions. *Annual Review of Energy and the Environment* 21, 311–346.

Herridge, D.F., Peoples, M.B., Boddey, R.M., 2008. Global inputs of biological nitrogen fixation in agricultural systems. *Plant and Soil* 311, 1–18.

Hesterberg, R., Blatter, A., Fahrni, M., Rosset, M., Neftel, A., Eugster, W., Wanner, H., 1996. Deposition of nitrogen-containing compounds to an extensively managed grassland in central Switzerland. *Environmental Pollution* 91, 21–34.

Hoffman, B.M., Lukoyanov, D., Yang, Z.-Y., Dean, D.R., Seefeldt, L.C., 2014. Mechanism of nitrogen fixation by nitrogenase: The next stage. *Chemical Reviews* 114, 4041–4062.

Hogberg, M.N., Chen, Y., Hogberg, P., 2007a. Gross nitrogen mineralisation and fungi-to-bacteria ratios are negatively correlated in boreal forests. *Biology and Fertility of Soils* 44, 363–366.

Hogberg, M.N., Hogberg, P., Myrold, D.D., 2007b. Is microbial community composition in boreal forest soils determined by pH, C-to-N ratio, the trees, or all three? *Oecologia* 150, 590–601.

Holloway, J.M., Dahlgren, R.A., 2002. Nitrogen in rock: Occurrences and biogeochemical implications. *Global Biogeochemical Cycles* 16, 1118. doi.org/10.1029/2002GB001862.

Holloway, J.M., Dahlgren, R.A., Casey, W.H., 2001. Nitrogen release from rock and soil under simulated field conditions. *Chemical Geology* 174, 403–414.

Holloway, J.M., Dahlgren, R.A., Hansen, B., Casey, W.H., 1998. Contribution of bedrock nitrogen to high nitrate concentrations in stream water. *Nature* 395, 785–788.

Houghton, R.A., 1999. The annual net flux of carbon to the atmosphere from changes in land use 1850-1990. *Tellus Series B—Chemical and Physical Meteorology* 51, 298–313.

Houlton, B.Z., Wang, Y.-P., Vitousek, P.M., Field, C.B., 2008. A unifying framework for dinitrogen fixation in the terrestrial biosphere. *Nature* 454, 327–334.

Howard, J.B., Rees, D.C., 2006. How many metals does it take to fix N-2? A mechanistic overview of biological nitrogen fixation. *Proceedings of the National Academy of Sciences of the United States of America* 103, 17088–17093.

Howarth, R., Chan, F., Conley, D.J., Garnier, J., Doney, S.C., Marino, R., Billen, G., 2011. Coupled biogeochemical cycles: Eutrophication and hypoxia in temperate estuaries and coastal marine ecosystems. *Frontiers in Ecology and the Environment* 9, 18–26.

Hu, B.-l., Shen, L.-d., Xu, X.-y., and Zheng, P., 2011. Anaerobic ammonium oxidation (anammox) in different natural ecosystems. *Biochemical Society Transactions* 39, 1811–1816.

Hungate, B.A., Dukes, J.S., Shaw, M.R., Luo, Y.Q., Field, C.B., 2003. Nitrogen and climate change. *Science* 302, 1512–1513.

IFA, 2000. Mineral fertilizer production and the environment. International Fertilizer Association (IFA) of the United Nations Environment Programme, Paris, France, p. 53.

IPCC. 2006. 2006 IPCC guidelines for national greenhouse gas inventories. Prepared by the National Greenhouse Gas Inventories Programme. (Eds) H.S. Eggleston, L. Buendia, K. Miwa, T. Ngara, K. Tanabe. Hayama, Japan: Institute for Global Environmental Strategies (IGES).

Itihara, Y., Honma, H., 1979. Ammonium in biotite from metamorphic and granitic rocks of Japan. *Geochimica et Cosmochimica Acta* 43 (4), 503–509.

Jaffee, D.A., 1992. The global nitrogen cycle. In: S.S. Butcher, G.H. Orians, R.J. Charlston, G.V. Wolfe, (Eds.), *Global Biogeochemical Cycles*. Academic Press, London, U.K., pp. 263–284.

Jetten, M.S.M., Niftrik, L.v., Strous, M., Kartal, B., Keltjens, J.T., and Op den Camp, H.J.M., 2009. Biochemistry and molecular biology of anammox bacteria. *Critical Reviews in Biochemistry and Molecular Biology* 44, 65–84.

Jia, H.-P., Quadrelli, E.A., 2014. Mechanistic aspects of dinitrogen cleavage and hydrogenation to produce ammonia in catalysis and organometallic chemistry: Relevance of metal hydride bonds and dihydrogen. *Chemical Society Reviews* 43, 547–564.

Johnson, B., Goldblatt, C., 2015. The nitrogen budget of Earth. *Earth-Science Reviews* 148, 150–173.

Khan, A.A., Baur, W.H., 1972. Salt hydrates. VII. The crystal structures of sodium ammonium orthochromate dihydrate and magnesium diammonium bis(hydrogen orthophosphate) tetrahydrate and a discussion of the ammonium ion. *Acta Crystallographica Section B* 28, 683–693.

Kopacek, J., Posch, M., 2011. Anthropogenic nitrogen emissions during the Holocene and their possible effects on remote ecosystems. *Global Biogeochemical Cycles* 25, GB2017. Doi.org/10.1029/2010GB003779.

Kopacek, J., Cosby, B.J., Evans, C.D., Hruska, J., Moldan, F., Oulehle, F., Santruckova, H., Tahovska, K., Wright, R.F., 2013a. Nitrogen, organic carbon and sulphur cycling in terrestrial ecosystems: Linking nitrogen saturation to carbon limitation of soil microbial processes. *Biogeochemistry* 115, 33–51.

Kopacek, J., Hejzlar, J., Posch, M., 2013b. Factors Controlling the export of nitrogen from agricultural land in a large Central European Catchment during 1900–2010. *Environmental Science & Technology* 47, 6400–6407.

Kuenen, J.G., 2008. Anammox bacteria: From discovery to application. *Nature Reviews Microbiology* 6, 320–326.

Ladha, J.K., Tirol-Padre, A., Reddy, C.K., Cassman, K.G., Verma, S., Powlson, D.S., van Kessel, C., Richter, D. de B., Chakraborty, D., Pathak, H., 2016. Global nitrogen budgets in cereals: A 50-year assessment for maize, rice, and wheat production systems. *Scientific Reports* 6, 19355. doi.org/10.1038/srep19355.

Leip, A., Britz, W., Weiss, F., de Vries, W., 2011. Farm, land, and soil nitrogen budgets for agriculture in Europe calculated with CAPRI. *Environmental Pollution* 159, 3243–3253.

Li, Y., Keppler, H., 2014. Nitrogen speciation in mantle and crustal fluids. *Geochimica Et Cosmochimica Acta* 129, 13–32.

Li, Y., Niu, S., and Yu, G., 2016. Aggravated phosphorus limitation on biomass production under increasing nitrogen loading: A meta-analysis. *Global Change Biology* 22, 934–943.

Ludwig, J., Meixner, F.X., Vogel, B., Förstner, J., 2001. Soil-air exchange of nitric oxide: An overview of processes, dnvironmental vactors, and modeling studies. *Biogeochemistry* 52, 225–257.

Luescher, A., Mueller-Harvey, I., Soussana, J.F., Rees, R.M., Peyraud, J.L., 2014. Potential of legume-based grassland-livestock systems in Europe: A review. *Grass and Forage Science* 69, 206–228.

Maathuis, F.J., 2009. Physiological functions of mineral macronutrients. *Current Opinion in Plant Biology* 12, 250–258.

Mackenzie, F.T., 2010. *Our Changing Planet: An Introduction to Earth System Science and Global Environmental Change.* 4th ed. Prentice Hall, Upper Saddle River, New Jersey, USA.

MacLeod, K.C., Holland, P.L., 2013. Recent developments in the homogeneous reduction of dinitrogen by molybdenum and iron. *Nature Chemistry* 5, 559–565.

Mann, C.C., 1999. Crop scientists seek a new revolution. *Science* 283, 310–314.

Marty, B., 1995. Nitrogen content of the mantle inferred from N_2-Ar correlation in oceanic basalts. *Nature* 377, 326–329.

Matson, P.A., McDowell, W.H., Townsend, A.R., Vitousek, P.M., 1999. The globalization of N deposition: Ecosystem consequences in tropical environments. *Biogeochemistry* 46, 67–83.

Matthews, E., 1994. Nitrogenous fertilizers: Global distribution of consumption and associated emissions of nitrous oxide and ammonia. *Global Biogeochemical Cycles* 8, 411–439.

Miller, A.J., Shen, Q., Xu, G., 2009. Freeways in the plant: Transporters for N, P and S and their regulation. *Current Opinion in Plant Biology* 12, 284–290.

Mitchell, E.C., Fischer, T.P., Hilton, D.R., Hauri, E.H., Shaw, A.M., de Moor, J.M., Sharp, Z.D., Kazahaya, K., 2010. Nitrogen sources and recycling at subduction zones: Insights from the Izu-Bonin-Mariana arc. *Geochemistry, Geophysics, Geosystems* 11, 1–24.

Miyazaki, K., Eskes, H.J., Sudo, K., 2012. Global NOx emission estimates derived from an assimilation of OMI tropospheric NO_2 columns. *Atmospheric Chemistry and Physics* 12, 2263–2288.

Monks, P.S., Granier, C., Fuzzi, S., Stohl, A., Williams, M.L., Akimoto, H., Amann, M., Baklanov, A., Baltensperger, U., Bey, I., Blake, N., Blake, R.S., Carslaw, K., Cooper, O.R., Dentener, F., Fowler, D., Fragkou, E., Frost, G.J., Generoso, S., Ginoux, P., Grewe, V., Guenther, A., Hansson, H.C., Henne, S., Hjorth, J., Hofzumahaus, A., Huntrieser, H., Isaksen, I.S.A., Jenkin, M.E., Kaiser, J., Kanakidou, M., Klimont, Z., Kulmala, M., Laj, P., Lawrence, M.G., Lee, J.D., Liousse, C., Maione, M., McFiggans, G., Metzger, A., Mieville, A., Moussiopoulos, N., Orlando, J.J., O'Dowd, C.D., Palmer, P.I., Parrish, D.D., Petzold, A., Platt, U., Poschl, U., Prevot, A.S.H., Reeves, C.E., Reimann, S., Rudich, Y., Sellegri, K., Steinbrecher, R., Simpson, D., ten Brink, H., Theloke, J., van der Werf, G.R., Vautard, R., Vestreng, V., Vlachokostas, C., von Glasow, R., 2009. Atmospheric composition change—Global and regional air quality. *Atmospheric Environment* 43, 5268–5350.

Montzka, S.A., Reimann, S., 2010. Ozone-depliting substances (ODSs) and related chemicals. In: Ajavon, A.L.N., Newman, P.A., Pyle, J.A., Ravishankara, A.R. (Eds.), *Scientific Assessment of Ozone Depletion: 2010.* World Meteorological Organization, Global Ozone Research Monitoring, Geneva, Switzerland, pp. 1–108.

Neff, J.C., Holland, E.A., Dentener, F.J., McDowell, W.H., Russell, K.M., 2002. The origin, composition and rates of organic nitrogen deposition: A missing piece of the nitrogen cycle? *Biogeochemistry* 57, 99–136.

Palya, A.P., Buick, I.S., Bebout, G.E., 2011. Storage and mobility of nitrogen in the continental crust: Evidence from partially melted metasedimentary rocks, Mt. Stafford, Australia. *Chemical Geology* 281, 211–226.

Park, S., Croteau, P., Boering, K.A., Etheridge, D.M., Ferretti, D., Fraser, P.J., Kim, K.R., Krummel, P.B., Langenfelds, R.L., van Ommen, T.D., Steele, L.P., Trudinger, C.M., 2012. Trends and seasonal cycles in the isotopic composition of nitrous oxide since 1940. *Nature Geoscience* 5, 261–265.

Paul, E.A., 2015. *Soil Microbiology, Ecology and Boiochemistry.* Elsevier Academic Press, New York, USA.

Paulot, F., Jacob, D.J., Henze, D.K., 2013. Sources and processes contributing to nitrogen deposition: An adjoint model analysis applied to biodiversity hotspots worldwide. *Environmental Science & Technology* 47, 3226–3233.

Perakis, S.S., Hedin, L.O., 2002. Nitrogen loss from unpolluted South American forests mainly via dissolved organic compounds. *Nature* 415, 416–419.

Perrings, C., Kinzig, A., Halkos, G., 2014. Sustainable development in an N-rich/N-poor world. *Ambio* 43, 891–905.

Pilegaard, K., 2013. Processes regulating nitric oxide emissions from soils. *Philosophical Transactions of the Royal Society B—Biological Sciences* 368, 20130126. doi.org/10.1098/rstb.2013.0126.

Porada, P., Weber, B., Elbert, W., Poeschl, U., Kleidon, A., 2014. Estimating impacts of lichens and bryophytes on global biogeochemical cycles. *Global Biogeochemical Cycles* 28, 71–85.

Postgate, R., 1998. *Nitrogen Fixation.* 3rd ed. Cambridge University Press, Cambridge, United Kingdom.

Potter, P., Ramankutty, N., Bennett, E.M., Donner, S.D., 2010. Characterizing the spatial patterns of global fertilizer application and manure production. *Earth Interactions* 14, 1–22.

Prentice, I.C., Farquhar, G.D., Fasham, M.J.R., Goulden, M.L., Heimann, M., Jaramillo, V.J., Kheshgi, H.S., Quéré, C.L., Scholes, R.J., Wallace, D.W.R., 2001. The carbon cycle and atmospheric carbon dioxide. In: J.T. Houghton, Y. Ding, D.J. Griggs, M. Noguer, P.J. van der Linden, X. Dai, K. Maskell, C.A. Johnson (Eds.) *Climate Change 2001: The Scientific Basis. Contribution of Working Group I to the Third Assessment Report of the Intergovernmental Panel on Climate Change.* Cambridge University Press, Cambridge, United Kingdom and New York, NY, USA, pp. 183–237.

Puckett, L.J., Tesoriero, A.J., Dubrovsky, N.M., 2011. nitrogen contamination of surficial aquifers—A growing legacy. *Environmental Science & Technology* 45, 839–844.

Rabalais, N.N. 2002. Nitrogen in aquatic ecosystems. *Ambio* 31, 102–112.

Ravishankara, A.R., Daniel, J.S., Portmann, R.W., 2009. Nitrous oxide (N2O): The dominant ozone-depleting substance emitted in the 21st century. *Science* 326, 123–125.

Rockström, J., Steffen, W., Noone, K., Persson, Å., Chapin III, F.S., Lambin, E.F., Lenton, T.M., Scheffer, M., Folke, C., Schellnhuber, H.J., Nykvist, B., de Wit, C.A., Hughes, T., van der Leeuw, S., Rodhe, H., Sörlin, S., Snyder, P.K., Costanza, R., Svedin, U., Falkenmark, M., Karlberg, L., Corell, R.W., Fabry, V.J., Hansen, J., Walker, B., Liverman, D., Richardson, K., Crutzen, P., Foley, J.A., 2009. A safe operating space for humanity. *Nature* 461, 472–475. doi:10.1038/461472a.

Rubio, L.M., Ludden, P.W., 2008. Biosynthesis of the iron-molybdenum cofactor of nitrogenase. *Annual Review of Microbiology* 62, 93–111.

Scanlon, B.R., Reedy, R.C., Bronson, K.F., 2008. Impacts of land use change on nitrogen cycling archived in semiarid unsaturated zone nitrate profiles, Southern High Plains, Texas. *Environmental Science & Technology* 42, 7566–7572.

Schlesinger, W.H., Bernhardt, E.S., 2013. *Biogeochemistry: An Analysis of Global Change.* Elsevier Academic Press, Waltham, MA, USA.

Schlesinger, W.H., Hartley, AE., 1992. A global budget for atmospheric NH3. *Biogeochemistry* 15, 191–211.

Seitzinger, S., Harrison, J.A., Bohlke, J.K., Bouwman, A.F., Lowrance, R., Peterson, B., Tobias, C., Van Drecht, G., 2006. Denitrification across landscapes and waterscapes: A synthesis. *Ecological Applications* 16, 2064–2090.

Seitzinger, S.P., Harrison, J.A., Dumont, E., Beusen, A.H.W., Bouwman, A.F., 2005. Sources and delivery of carbon, nitrogen, and phosphorus to the coastal zone: An overview of Global Nutrient Export from Watersheds (NEWS) models and their application. *Global Biogeochemical Cycles* 19, GB4S01. doi.org/10.1029/2005GB002606.

Selman, M., Sugg, Z., Greenhlgh, S., Diaz, R.J., 2008. Eutrophication and hypoxia in coastal areas: A global assessment of the state of knowledge. World Resources Institute Report, Washington, D.C.

Sheffer, E., Batterman, S.A., Levin, S.A., Hedin, L.O., 2015. Biome-scale nitrogen fixation strategies selected by climatic constraints on nitrogen cycle. *Nature Plants* 1, 15182. doi: 10.1038/nplants.2015.182.

Skiba, U., Smith, K.A., Fowler, D., 1993. Nitrification and denitrification as sources of nitric-oxide and nitrous-oxide in a sandy loam soil. *Soil Biology & Biochemistry* 25, 1527–1536.

Smil, V., 1999. Nitrogen in crop production: An account of global flows. *Global Biogeochemical Cycles* 13, 647–662.

Smil, V., 2001. *Enriching the Earth: Fritz Haber, Carl Bosch, and the Transformation of World Food Production.* MIT Press, Cambridge, Massachusetts, USA.

Socolow, R.H., 1999. Nitrogen management and the future of food: Lessons from the management of energy and carbon. *Proceedings of the National Academy of Sciences of the United States of America* 96, 6001–6008.

Steffen, W., Richardson, K., Rockström, J., Cornell, S.E., Fetzer, I., Bennett, E.M., Biggs, R., Carpenter, S,R,, de Vries W., de Wit, C.,A., Folke, C., Gerten. D., Heinke, J., Mace, G.M., Persson. L.M., Ramanathan, V., Reyers, B., Sörlin, S., 2015. Planetary boundaries: Guiding human development on a changing planet. *Science* 347, 1259855. doi.org/10.1126/science.1259855.

Sterner, R.W., Elser, J.J., 2002. *Ecological Stoichiometry: The Biology of Elements from Molecules to the Biosphere.* Princeton University Press, Princeton, NJ, USA.

Stewart, W.M., Roberts, T.L., 2012. Food security and the role of fertilizer in supporting it. *Procedia Engineering* 46, 76–82.

Stewart, W.M., Dibb, D.W., Johnston, A.E., Smyth, T.J., 2005. The contribution of commercial fertilizer nutrients to food production. *Agronomy Journal* 97, 1–6. doi.org/10.2134/agronj2005.0001.

Stoddard, J.L., 1994. Long-term changes in watershed retention of nitrogen. In: Baker, L.A. (Ed.), *Environmental Chemistry of Lakes and Reservoirs.* Advances in Chemistry Series. American Chemical Society, Washington, D.C., USA, pp. 223–284.

Sutton, M.A., Bleeker, A., Howard, C.M., Bekunda, M., Grizzetti, B., de Vries, W., van Grinsven, H.J.M., Abrol, Y.P., Adhya, T.K., Billen, G., Davidson, E.A., Datta, A., Diaz, R., Erisman, J.W., Liu, X.J., Oenema, O., Palm, C., Raghuram, N., Reis, S., Scholz, R.W., Sims, T., Westhoek, H., Zhang, F.S., 2013a. Our nutrient world: The challenge to produce more food and energy with less pollution. Global overview of nutrient management Centre for Ecology and Hydrologyon behalf of the Global Partnership on Nutrient Management and the International Nitrogen Initiative, Edinburgh, UK. 128 p. Available at: http://www.unep.org/gpa/documents/publications/ONW.pdf.

Sutton, M.A., Nemitz, E., Erisman, J.W., Beier, C., Bahl, K.B., Cellier, P., de Vries, W., Cotrufo, F., Skiba, U., Di Marco, C., Jones, S., Laville, P., Soussana, J.F., Loubet, B., Twigg, M., Famulari, D., Whitehead, J., Gallagher, M.W., Neftel, A., Flechard, C.R., Herrmann, B., Calanca, P.L., Schjoerring, J.K., Daemmgen, U., Horvath, L., Tang, Y.S., Emmett, B.A., Tietema, A., Penuelas, J., Kesik, M., Brueggemann, N., Pilegaard, K., Vesala, T., Campbell, C.L., Olesen, J.E., Dragosits, U., Theobald, M.R., Levy, P., Mobbs, D.C., Milne, R., Viovy, N., Vuichard, N., Smith, J.U., Smith, P., Bergamaschi, P., Fowler, D., Reis, S., 2007. Challenges in quantifying biosphere-atmosphere exchange of nitrogen species. *Environmental Pollution* 150, 125–139.

Sutton, M.A., Oenema, O., Erisman, J.W., Leip, A., van Grinsven, H., Winiwarter, W., 2011. Too much of a good thing. *Nature* 472, 159–161.

Sutton, M.A., Reis, S., Riddick, S.N., Dragosits, U., Nemitz, E., Theobald, M.R., Tang, Y.S., Braban, C.F., Vieno, M., Dore, A.J., Mitchell, R.F., Wanless, S., Daunt, F., Fowler, D., Blackall, T.D., Milford, C., Flechard, C.R., Loubet, B., Massad, R., Cellier, P., Personne, E., Coheur, P.F., Clarisse, L., Van Damme, M., Ngadi, Y., Clerbaux, C., Skjoth, C.A., Geels, C., Hertel, O., Kruit, R.J.W., Pinder, R.W., Bash, J.O., Walker, J.T., Simpson, D., Horvath, L., Misselbrook, T.H., Bleeker, A., Dentener, F., de Vries, W., 2013b. Towards a climate-dependent paradigm of ammonia emission and deposition. *Philosophical Transactions of the Royal Society B—Biological Sciences* 368, 20130166. doi.org/10.1098/rstb.2013.0166.

Syakila, A., Kroeze, C., 2011. The global nitrous oxide budget revisited. *Greenhouse Gas Measurement and Management* 1, 17–26.

Tanabe, Y., Nishibayashi, Y., 2013. Developing more sustainable processes for ammonia synthesis. *Coordination Chemistry Reviews* 257, 2551–2564.

Thamdrup, B., 2012. New pathways and processes in the global nitrogen cycle. *Annual Review of Ecology, Evolution, and Systematics* 43, 407–428.

Tie, X.X., Zhang, R.Y., Brasseur, G., Lei, W.F., 2002. Global NOx production by lightning. *Journal of Atmospheric Chemistry* 43, 61–74.

Trenberth, K.E., Guillemot, C.J., 1994. The total mass of the atmosphere. *Journal of Geophysical Research-Atmospheres* 99, 23079–23088.

Townsend, A.R., Braswell, B.H., Holland, E.A., and Penner, J.E., 1996. Spatial and temporal patterns in terrestrial carbon storage due to deposition of fossil fuel nitrogen. *Ecological Applications* 6, 806–814.

Ussirri, D.A.N., Lal, R., 2013. *Soil Emissions of Nitrous Oxide and its Mitigation.* Springer, Dordrecht; Netherlands, and New York; USA, 378 p.

Ussiri, D.A.N., Lal, R., 2017. *Carbon Sequestration for Climate Change Mitigation and Adaptation.* Springer, Dordrecht; Netherlands, and New York; USA, 549 p.

van Aardenne, J.A., Dentener, F.J., Olivier, J.G.J., Goldewijk, C., Lelieveld, J., 2001. A 1 degrees x 1 degrees resolution data set of historical anthropogenic trace gas emissions for the period 1890–1990. *Global Biogeochemical Cycles* 15, 909–928.

Van Breemen, N., Boyer, E.W., Goodale, C.L., Jaworski, N.A., Paustian, K., Seitzinger, S.P., Lajtha, K., Mayer, B., Van Dam, D., Howarth, R.W., Nadelhoffer, K.J., Eve, M., Billen, G., 2002. Where did all the nitrogen go? Fate of nitrogen inputs to large watersheds in the northeastern USA. *Biogeochemistry* 57, 267–293.

Van Meter, K.J., Basu, N.B., Veenstra, J.J., Burras, C.L., 2016. The nitrogen legacy: Emerging evidence of nitrogen accumulation in anthropogenic landscapes. *Environmental Research Letters* 11, 035014. doi. org/10.1088/1748-9326/11/3/035014.

van Vuuren, D.P., Bouwman, L.F., Smith, S.J., Dentener, F., 2011. Global projections for anthropogenic reactive nitrogen emissions to the atmosphere: An assessment of scenarios in the scientific literature. *Current Opinion in Environmental Sustainability* 3, 359–369.

Varley, J.B., Wang, Y., Chan, K., Studt, F., Norskov, J.K., 2015. Mechanistic insights into nitrogen fixation by nitrogenase enzymes. *Physical Chemistry Chemical Physics* 17, 29541–29547.

Vet, R., Artz, R.S., Carou, S., Shaw, M., Ro, C.U., Aas, W., Baker, A., Bowersox, V.C., Dentener, F., Galy-Lacaux, C., Hou, A., Pienaar, J.J., Gillett, R., Forti, M.C., Gromov, S., Hara, H., Khodzher, T., Mahowald, N.M., Nickovic, S., Rao, P.S.P., Reid, N.W., 2014. A global assessment of precipitation chemistry and deposition of sulfur, nitrogen, sea salt, base cations, organic acids, acidity and pH, and phosphorus. *Atmospheric Environment* 93, 3–100.

Vile, M.A., Wieder, R.K., Zivkovic, T., Scott, K.D., Vitt, D.H., Hartsock, J.A., Iosue, C.L., Quinn, J.C., Petix, M., Fillingim, H.M., Popma, J.M.A., Dynarski, K.A., Jackman, T.R., Albright, C.M., Wykoff, D.D., 2014. N_2 fixation by methanotrophs sustains carbon and nitrogen accumulation in pristine peatlands. *Biogeochemistry* 121, 317–328.

Vitousek, P.M., Aber, J.D., Howarth, R.W., Likens, G.E., Matson, P.A., Schindler, D.W., Schlesinger, W.H., Tilman, D., 1997a. Human alteration of the global nitrogen cycle: Sources and consequences. *Ecological Applications* 7, 737–750.

Vitousek, P.M., Porder, S., Houlton, B.Z., and Chadwick, O.A., 2010. Terrestrial phosphorus limitation: Mechanisms, implications, and nitrogen–phosphorus interactions. *Ecological Applications* 20, 5–15.

Vitousek, P.M., Cassman, K., Cleveland, C., Crews, T., Field, C.B., Grimm, N.B., Howarth, R.W., Marino, R., Martinelli, L., Rastetter, E.B., Sprent, J.I., 2002a. Towards an ecological understanding of biological nitrogen fixation. *Biogeochemistry* 57, 1–45.

Vitousek, P.M., Hattenschwiler, S., Olander, L., Allison, S., 2002b. Nitrogen and nature. *Ambio* 31, 97–101.

Vitousek, P.M., Menge, D.N.L., Reed, S.C., Cleveland, C.C., 2013. Biological nitrogen fixation: Rates, patterns and ecological controls in terrestrial ecosystems. *Philosophical Transactions of the Royal Society B—Biological Sciences* 368, 20130119. doi.org/10.1098/rstb.2013.0119.

Vitousek, P.M., Mooney, H.A., Lubchenco, J., Melillo, J.M., 1997b. Human domination of Earth's ecosystems. *Science* 277, 494–499.

Voss, M., Bange, H.W., Dippner, J.W., Middelburg, J.J., Montoya, J.P., Ward, B. 2013. The marine nitrogen cycle: Recent discoveries, uncertainties and the potential relevance of climate change. *Philosophical transactions of the Royal Society of London B—Biological Sciences* 368, 20130121.

Walker, J.C.G., 1977. *Evolution of the Atmosphere.* Macmillan Publishing Co., New York, NY, USA.

Walvoord, M.A., Phillips, F.M., Stonestrom, D.A., Evans, R.D., Hartsough, P.C., Newman, B.D., Striegl, R.G., 2003. A reservoir of nitrate beneath desert soils. *Science* 302, 1021–1024.

Wedepohl, K.H., 2012. *Handbook of Geochemistry.* Springer, Berlin Heidelberg.

Westhoek, H., Lesschen, J.P., Rood, T., Wagner, S., De Marco, A., Murphy-Bokern, D., Leip, A., van Grinsven, H., Sutton, M.A., Oenema, O., 2014. Food choices, health and environment: Effects of cutting Europe's meat and dairy intake. *Global Environmental Change—Human and Policy Dimensions* 26, 196–205.

Whittaker, E.J.W., Muntus, R., 1970. Ionic radii for use in geochemistry. *Geochimica et Cosmochimica Acta* 34, 945–956.

Worrall, F., Howden, N.J.K., Burt, T.P., 2015. Evidence for nitrogen accumulation: The total nitrogen budget of the terrestrial biosphere of a lowland agricultural catchment. *Biogeochemistry* 123, 411–428.

Yan, X.Y., Ohara, T., Akimoto, I., 2005. Statistical modeling of global soil NOx emissions. *Global Biogeochemical Cycles* 19, GB3019. doi.org/10.1029/2004GB002276.

Zaehle, S., 2013. Terrestrial nitrogen—Carbon cycle interactions at the global scale. *Philosophical Transactions of the Royal Society B—Biological Sciences* 368, 20130125. doi.org/10.1098/rstb.2013.0125.

Zhang, L., Jacob, D.J., Knipping, E.M., Kumar, N., Munger, J.W., Carouge, C.C., van Donkelaar, A., Wang, Y.X., Chen, D., 2012. Nitrogen deposition to the United States: Distribution, sources, and processes. *Atmospheric Chemistry and Physics* 12, 4539–4554.

Zhu, Z., Xiong, Z., Xing, G., 2005. Impacts of population growth and economic development on the nitrogen cycle in Asia. *Science in China Series C: Life Sciences* 48, 729–737.

8 Biochar for Climate Change Mitigation

Navigating from Science to Evidence-Based Policy

Dominic Woolf, Johannes Lehmann, Annette Cowie, Maria Luz Cayuela, Thea Whitman, and Saran Sohi

CONTENTS

8.1 INTRODUCTION

8.1.1 OVERVIEW OF BIOCHAR SYSTEMS AS A CLIMATE MITIGATION STRATEGY

Biochar is the carbon-rich solid formed by heating biomass in an anaerobic environment (a process called pyrolysis). This pyrogenic carbonized material is typically known as biochar when it is intended as a soil amendment or to provide related environmental benefits. The concept of using carbonized biomass on a large scale as a climate-change mitigation approach can be traced back to two independent fields of study: (1) Seifritz (1993) discussed the climate-mitigation potential of industrial production and the burial of charcoal in landfills but did not consider its use as a soil amendment. (2) Charcoal was found to contribute a major component of the stable carbon in Amazonian dark earth soils known also as Terra Preta de Indio (Glaser et al. 2001, 2004). Sombroek et al. (1993) described the historical use of organic matter additions by indigenous peoples in the Amazon region to improve soil fertility, the resulting large accumulation of carbon (C), and its consequent potential as a climate-change mitigation strategy. The conjunction of these strands—that large scale charcoal production can be used to sequester C, and that charcoal can be used as a soil fertility amendment—has driven a rapid expansion and interest in biochar research over the past two decades, with only one journal article in the period 1990–1999 using the term "biochar," rising to 78 in 2000–2009 and soaring to over 4000 since 2010 (Web of Science citations). It has been estimated that biochar systems can mitigate up to 1.8 Pg CO_2C equivalent yr^{-1} (12% of anthropogenic CO_2, CH_4, and N_2O emissions), without endangering food security, habitat or soil conservation—a larger climate-change mitigation potential than using the same biomass for bioenergy (Woolf et al. 2010).

8.1.2 MAIN IMPACTS OF BIOCHAR SYSTEMS ON CLIMATE CHANGE

Biochar's climate-change mitigation potential stems primarily from its slower decomposition than the raw biomass from which it is generated, thus lowering the rate at which photosynthetically-fixed C is returned to the atmosphere (Wang et al. 2016) (Figure 8.1). It is this *difference* in decomposition rates that is critical in determining how net carbon stocks evolve over time (Figure 8.2 and Whitman et al. 2010). Although approximately half of the carbon in a biomass feedstock is emitted as CO_2 during biochar production; by comparison, more readily-decomposed un-pyrolysed biomass will rapidly return most of its carbon to the atmosphere if allowed to decompose. Therefore, the carbon stocks remaining over time are larger for biochar than for raw biomass, leading to a net increase in soil carbon stocks. Thus, although embedding carbon in biochar is, in one sense, a redistribution of biomass carbon rather than newly fixed carbon, nonetheless the greater persistence of the biochar drives a net sequestration of carbon. Most studies have concluded that this persistence-derived carbon sequestration is the largest influence of biochar on net greenhouse gas balances, while other mechanisms serve to mediate this primary influence (Fowles 2007; Gaunt & Lehmann 2008; Roberts et al. 2010; Whitman et al. 2010; Woolf et al. 2010; Cowie et al. 2015).

In addition, biochar's overall impact on climate change mitigation depends also on a range of other secondary mechanisms. It can reduce nitrous oxide (N_2O) emissions from soil (Cayuela et al. 2013). It can alter methane emission or oxidation rates in soil (Jeffery et al. 2016). Conversion of biomass to biochar can avoid emissions of N_2O and/or methane (CH_4) that would have arisen from the decomposition or combustion of that biomass (Woolf et al. 2010). Biochar can enhance plant growth (Crane-Droesch et al. 2013; Jeffery et al. 2016), this enhanced productivity providing a positive feedback that further enhances the amount of CO_2 removed from the atmosphere, particularly if that increased biomass increases the feedstock available for further biochar production (Woolf et al. 2010). Biochar can alter the turnover rate of native soil organic matter (SOM), thus potentially either increasing or decreasing stocks of non-pyrogenic soil C (Zimmerman et al. 2011).

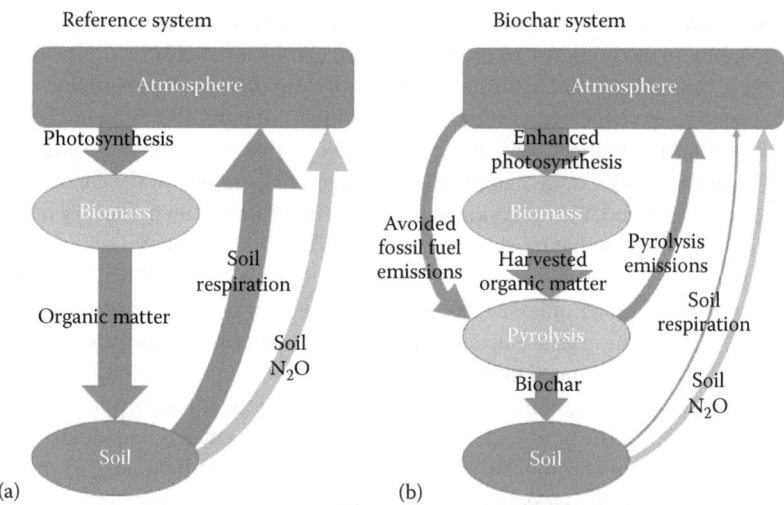

Reference system | Biochar system

(a) | (b)

FIGURE 8.1 Main impacts of biochar on greenhouse gas (GHG) fluxes. In the reference system without biochar (a), plants remove CO_2 from the atmosphere by photosynthesis with fixed carbon being returned to the atmosphere as plants decompose in the soil. A fraction of the reactive nitrogen in the soil is also released to the atmosphere in the form of nitrous oxide, a potent GHG. In the biochar system (b), approximately half of the biomass carbon is returned to the atmosphere during pyrolysis, but the remaining carbon in the biochar decomposes more slowly than raw biomass, leading to an overall reduction in the rate of CO_2 emission to the atmosphere. Simultaneously, nitrous oxide emissions from soil can be reduced by up to 80%. Draw down of CO_2 by photosynthesis may be increased by improved soil fertility leading to an increase in net primary production. Finally, co-production of bioenergy in the pyrolysis process can displace fossil fuels that would otherwise have provided that energy (indicated as a net reduction in atmospheric CO_2 on the diagram).

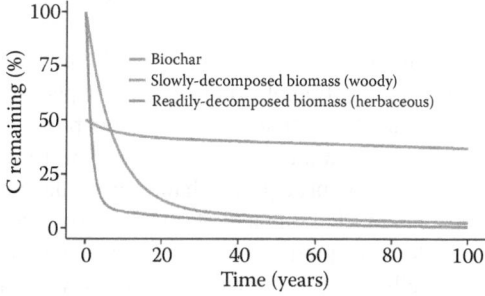

FIGURE 8.2 (See color insert.) Conceptual comparison of un-mineralized biomass carbon (C) remaining from different grades of organic matter, as a function of time. The lines are modeled using a two pool exponential decay model, comparing slow-turnover woody biomass (green line), fast-turnover herbaceous biomass (blue line), and biochar (red line). Assumed representative fast and slow fraction half lives, respectively, were 4 and 25 years for woody biomass, 1 and 25 years for herbaceous biomass, and 5 and 500 years for biochar. It was also assumed that half of the initial biomass carbon is lost during biochar production, hence the carbon remaining in biochar starts at 50% at time equal to zero.

As with *any* mechanism that increases the soil's organic C (SOC) stocks, biochar reduces soil albedo, thus increasing radiative forcing on bare soils (Verheijen et al. 2013). Slow decay of biochar in soils, together with tillage and transport activities, return a small amount of CO_2 to the atmosphere (Roberts et al. 2010). Reduced fertilizer requirements from improved nutrient-use efficiency (NUE) can reduce associated greenhouse gases (GHGs) from fertilizer production, transport and

N$_2$O emissions. Coproduction of bioenergy can offset fossil-fuel emissions, while returning about half of the C fixed by photosynthesis to the atmosphere (Woolf et al. 2014).

Although these secondary mechanisms are typically smaller than persistence-derived carbon sequestration, they are, nonetheless, pivotal in determining whether the climate-change abatement from biochar is greater or smaller than can be achieved by other ways (such as bioenergy) the same biomass could be used to offset or displace GHG emissions. Depending on the carbon intensity of the energy supply that is displaced, the reduction in fossil fuel emissions from bioenergy may be comparable to the carbon sequestration that biochar provides (Fowles 2007; Gaunt & Lehmann 2008; Roberts et al. 2010; Woolf et al. 2010; Cowie et al. 2015; Woolf, Lehmann & Lee 2016). Thus, the question of whether biochar provides greater or lesser abatement than bioenergy hinges largely on the size of these secondary climate-change mitigation impacts and the carbon intensity of any displaced fossil fuels.

The net impact of all these mechanisms is highly variable, depending on the characteristics of the entire biochar production system (from feedstock provision, through pyrolysis conditions, to soil biogeochemistry) and also on the characteristics of the assumed reference system to which it is compared (i.e., how the land, biomass and energy systems would be managed in the absence of biochar production). Life cycle assessments (LCAs) have indicated that the net mitigation impact of biochar systems commonly range from -0.6 to $+1.75$ Mg CO$_2$e Mg^{-1} feedstock (-0.3 to $+1.3$ Mg C Mg^{-1} feedstock-C) (Cowie et al. 2015). This wide variability is only partly attributable to uncertainties that remain in the size and longevity of some effects (especially, on plant growth, N$_2$O emissions from soil, and stabilization of non-pyrogenic SOC). The variability in impacts is mainly dependent on differences in the scenarios that are modeled (type of feedstock, pyrolysis conditions, and the reference energy system). Although some options are estimated to provide minimal or even negative climate-change abatement, one thing emerges clearly from these LCA studies: provided care is taken to avoid controllable detrimental impacts (such as indirect land use change), then the net GHG abatement is typically greater than the C sequestration alone. That is, the positive feedbacks on GHG abatement from factors such as reduced CH$_4$ and N$_2$O emissions, increased net primary production (NPP), reduced fossil fuel consumption, and increased soil C stabilization tend to outweigh the negative feedbacks from factors such as energy required for transporting biochar and incorporating it into the soil.

Thus, the overall picture that emerges is nuanced with both knowns and unknowns. While much remains to be learned about how long different types of biochar persist in different environments, nonetheless enough is known to be confident that woody biochars produced at high temperatures are sufficiently persistent to justify their use in managing the C cycle over the centennial timescales that are most relevant to navigating humanity through the challenges of the coming century. While much uncertainty remains in the impact of biochar on yield of different crops in different soils and agroecological zones, and it is expected to be as predictably variable as any other soil management such as fertilizer or lime applications, nonetheless, it is known that sandy, low cation exchange capacity (CEC) and highly acidic soils typically benefit the most from biochar applications; whereas, already-fertile soils will see little or no improvement, and adding alkaline biochar to alkaline soils can depress yields.

In the remainder of this chapter we investigate what is currently known and what remains to be researched about the major impacts of biochar on climate change. Finally, the policy implications of the current state of knowledge about biochar and climate are also discussed to draw recommendations about what types of biochar systems can already be confidently deployed, under what conditions biochar systems may be suboptimal or other forms of organic matter (OM) management would be preferred, and what are the most pressing research directions that need to be addressed to improve the decision making tools for biochar systems.

8.2   PERSISTENCE

A greater persistence of charred than uncharred OM is the foundation of biochar systems' capacity to reduce GHG emissions, and thereby mitigate climate change (Lehmann et al. 2006;

Lehmann et al. 2015). Such greater persistence means that the C in the charred biomass is returning to the atmosphere much slower than that in charred biomass, leading to more C in soil (or wherever the biochar is applied to) than in the atmosphere (Lehmann 2007). In addition to this most obvious impact, increased persistence has three other main effects: (1) the C in biochar does not generate CH_4, a much more potent GHGs than CO_2, if landfilled or added to soil (Houghton et al. 1997); (2) biochar added to soil (or compost) may reduce CH_4, N_2O, or CO_2 production from C or nitrogen (N) otherwise originating from soil (Van Zwieten et al. 2015; Whitman et al. 2015a,b)—the longevity of these impacts being contingent on the persistence of the charred material; and (3) biochar added to soil may increase plant growth through a variety of mechanisms (Jeffery et al. 2014), whose longevity also depends on biochar persistence. Except for the remediation of CH_4 emission from biomass (point 1 above) which is instantaneous and quasi permanent, greater persistence is the basis for greater emission reductions and therefore the basis for the role of biochar systems in climate change mitigation. However, it is important to recognize that, while greater persistence inferred by charring is a necessary precondition for climate change mitigation by biochar systems, it is not sufficient, because life-cycle emission reductions will hinge on additional factors such as the fate of biomass under business as usual, transportation, and energy use among other factors discussed in the sections below.

The importance of persistence with respect to C trading and dangerous climate change over the coming decades and century hinges not so much on the absolute persistence than on the persistence relative to the scenario were the biomass not charred. When considering centennial assessment horizons, the extent of emission reductions is remarkably insensitive to the extent of biochar persistence, both for emissions of N_2O (Gaunt and Lehmann 2008), as well as the C in the biochar (Roberts et al. 2010), provided that the mean residence time (MRT) remains above 100 years. This importance over the coming century has to be distinguished from the millennial C balance (Lehmann et al. 2010), which requires MRT in the range of millennia for biochar systems to safeguard against return of C to the atmosphere on the long term. Since biochar may have a role to play in an energy and C management transition (Woolf, Lehmann, and Lee 2016), MRT of above 100 years appear to be sufficient in a first approximation for contribution to climate stabilization over the coming century.

Evidence of biochar persistence has been forthcoming from naturally occurring pyrogenic organic C generated by vegetation fires, from anthropogenic forest clearing and by-product management, from field trials with deliberate additions of biochar, and from incubation studies of biochars. Each of these approaches to the quantification of its persistence has advantages and disadvantages (Lehmann et al. 2015). As with uncharred OM, both material properties as well as environmental factors play a role in the decomposition rate of biochar. Material properties play a greater role in the short term of months and decades, while interaction with the soil matrix emerges as an important control over the long term, with factors such as temperature and moisture being the most important control during any period. Over the period of decades and a century that is most relevant to the current discussion and need for climate change mitigation, the most important determinants of biochar persistence are material properties—specifically the extent of thermal transformation to a higher degree of what is called aromatic condensation, i.e., to what extent and in which configuration C atoms in the biochar are fused together during charring (Lehmann et al. 2015). Different biochars therefore have vastly different mineralization rates, with a MRT of about 556 years of 97% of its mass (Wang et al. 2016) for production conditions that typically generate biochars with greater aromatic condensation. The degree of aromatic condensation can be studied using spectroscopic techniques (Wiedemeier et al. 2015) but is often estimated simply by their elemental oxygen-to-organic carbon (Spokas 2010) or hydrogen-to-organic carbon ratios (Lehmann et al. 2015) that are to a certain degree interchangeable (Enders et al. 2012). An important nuance to recognizing and analytically establishing the material properties is that biochars (as any plant residue) are not homogeneous but typically contain some portion that is extremely easily metabolized to CO_2, while others require enzymatic expenditures by microorganisms that are energetically costly and are therefore typically mineralized only very slowly.

Two points deserve special attention: first, no matter what MRT is measured for a certain biochar, more important than the absolute mineralization is the difference in mineralization between the uncharred OM and the charred biochar. While biochar mineralization varies depending on all the factors mentioned above including environmental factors, under identical soil and climatic conditions charring has always been shown to confer greater persistence to organic residues, typically by more than one order of magnitude. Second, the charring process itself releases CO_2 (or gases and particulates with even greater climate forcing than CO_2) to the atmosphere (Whitman et al. 2013) that has to be compensated by slower mineralization of the biochar compared to the uncharred feedstock. In addition, the released gases and volatiles can be utilized as an energy or bio-product and offset other bioenergy or even fossil energy use, which would create a very favorable emission balance depending on the fuel that is offset (Woolf et al. 2010). Third, both biochar persistence in absolute terms (such as through calculation of its MRT) as well as the effect that charring has on organic residues in relative terms will always be highly variable in time and space, simply because of the multitude of controls that are at play (as discussed above). This variability must not be confused with uncertainty about biochar persistence (Lehmann and Rillig 2014). While the variability in mineralization rates between different biochars or between sites can be very large, such variability is predictable given the knowledge of mineralization processes.

Therefore, it is evident that knowing the production conditions (mainly pyrolysis temperature and duration) and the feedstock type will be sufficient to be certain of the degree of aromatic condensation that will reduce mineralization of a given plant residue under otherwise identical soil and climatic conditions. This aromatic condensation can be verified using said elemental ratios (oxygen-to-organic C or hydrogen-to-organic C). For example, woody biomass residues pyrolysed above 500°C for at least 10 min will have molar hydrogen-to-organic C ratios of below 0.4 (Enders et al. 2012) and therefore under climate and soil conditions where biochar is applied in an agricultural context show MRT of more than 1000 years (Lehmann et al. 2015). The myriad of possible different feedstocks, conversion technologies and conditions, soil properties and management, as well as climate and weather conditions and their combinations will generate a wide range of persistence values. Information about all possibilities will emerge as science progresses, but sufficient information is already available for certain combinations of factors that have been studied in greater detail and for which natural analogs exist (e.g., from vegetation fires or historic anthropogenic additions).

From a scientific perspective, even though it is a foundational and necessary property of biochar, persistence can be managed and verified, and will not constitute a constraint for designing biochar systems for climate change mitigation. From an industry, policy and public perspective, this is different. Trust in the extent of biochar persistence remains a constraint that will likely only be mitigated by improved communication and additional experimentation, with a focus on long-term field experiments.

The persistence of emission changes of other greenhouse gases from soils (CO_2, CH_4, N_2O) and growth responses by plants that could all increase or decrease the life-cycle emission balance, also benefit from further and long-term (decadal) field experimentation. However, sufficient information is available for many different combinations of factors that appropriate estimates can be made to sufficiently constrain projections over the coming century (Woolf et al. 2010). Therefore, it is appropriate that the first publically available C trading methodology has been established (Placer 2015). For a truly long-term C sequestration (aka millennial or semi-permanent from a geological perspective), more ambitious criteria to biochar persistence have to be applied than for arresting dangerous climate change over the coming century (Lehmann et al. 2010). The scientific desire and justification to establish a global network of biochar field experiments to generate long-term (more than 20 years) data on full-field emission balances should not be seen as an argument against implementing biochar systems for climate change mitigation now, in order to start gathering relevant data beyond but including biochar persistence at scale of implementation.

8.3   SOIL NITROUS OXIDE EMISSIONS

Nitrous oxide (N_2O) is now considered the third most important long-lived GHG after CO_2 and CH_4 (Davidson and Kanter, 2014). The emission projections considering business-as-usual scenarios estimate that anthropogenic N_2O emissions could double in 2050 (from 5.3 to 9.7 Tg N_2O-N y^{-1}). Importantly these emissions are mostly associated with food production, because agriculture currently accounts for 56%–81% of gross anthropogenic N_2O emissions (Davidson and Kanter, 2014). Since arable soils are by far the dominant N_2O source, reducing N_2O fluxes from fertilized soils would represent a substantial mitigation opportunity in the agricultural sector (Paustian et al., 2016).

N_2O is produced in soils by microbiological processes and it is generally believed that the principal factor responsible for agricultural N_2O emissions is a lack of synchronization between crop N demand and soil N supply (Venterea et al. 2012). Microorganisms utilize N compounds, not only to incorporate N into their cell structures, but also in many catabolic reactions that imply N redox transformations. It is within these transformations where several trace N gases (including N_2O and NO) are formed and emitted to the atmosphere (Conrad 1996). Thus, whenever N enters the soil matrix a competition starts between plant roots and soil microorganisms to use that N (Hodge et al. 2000; Kuzyakov & Xu 2013). This implies that some N_2O emissions associated with fertilized croplands is inevitable, and that even when engineered high N-use-efficiency cultivars are used and N is applied in synchronization with plant N demand, there will always be a vigorous competition for N in the rhizosphere. Depending on the circumstances, this can lead to substantial N_2O emissions. Ultimately, mitigating N_2O emissions without limiting N supplied to plants can be achieved only by methods that either (1) increase the proportion of added N that is assimilated by crops, so that less N-fertilizer inputs are needed (Venterea et al. 2012), or (2) promote the consumption of N_2O in soil, i.e., its reduction to N_2 (Richardson et al. 2009)—both routes where biochar has been found to have a substantial impact (Cayuela et al. 2013; Zheng et al. 2013; Quin et al. 2015).

Soils emit more N_2O to the atmosphere than any other source (Paustian et al. 2016), which is largely the result of the extensive use of N-based fertilizers in agricultural lands (Stein & Yung 2003; Smith et al. 2008; Hu et al. 2015; van Groenigen et al. 2015). In spite of the numerous research efforts over the past decades, N_2O mitigation remains a serious challenge, since few practices have been found to be both effective and consistent across agroecosystems. Recently, the use of biochar is being investigated as a means to mitigate N_2O emissions from arable soils (Spokas & Reikosky 2009; Van Zwieten et al. 2009), and there is strong evidence of the potential of biochar to decrease N_2O emissions from fertilized soils. A recent meta-analysis summarizing results from 56 studies (published between 2009 and 2014) found that biochar reduced soil N_2O emissions by an average of 49% (Cayuela et al. 2015). However, the N_2O mitigation was highly contextual, with some soils undergoing 90% reductions (Spokas & Reikosky 2009; Cayuela et al. 2013; Nelissen et al. 2014) and others none or even an increase in N_2O emission (Sánchez-García et al. 2014; Wells & Baggs, 2014). This fact is clearly linked to different N_2O formation pathways in different soils, since N_2O is both an intermediary and a by-product in several N chemical reactions. To date, the individual N_2O formation pathways on which biochar operates have not been studied to a sufficient extent. Aligned to a paucity of data on biochar mechanisms are gaps in the scientific understanding on the processes leading to N_2O production (and consumption) themselves irrespective of biochar additions, with new studies challenging previous assumptions being published every year (Sanford et al. 2012; Yang et al. 2012; Zhu et al. 2013; Phillips et al. 2016).

In spite of the complexity of the subject, there is some evidence about which types of biochar might work best for N_2O mitigation. So far, the highest N_2O reductions have been found with biochars made of lignocellulosic or woody feedstocks by slow pyrolysis and at relatively high temperatures (500–700°C). These biochars generally have low ash content (Cayuela et al. 2013), medium-to-high organic C to N ratios (C_{org}:N >30) (Cayuela et al. 2014), and low atomic H:C_{org} (<0.5) (Cayuela et al. 2015). The dose of application is also relevant and needs to be at least 1% in dry weight (equivalent to approximately 10 Mg ha^{-1}) in order to reach significant reductions (Cayuela et al. 2014), beyond which threshold N_2O response increases linearly with dose at 0.17% to 0.91% of initial emissions per Mg ha^{-1} biochar applied (Woolf et al. 2016; Cayuela et al. 2014). As regards environmental settings, the best results have been reported with conditions promoting denitrification, i.e., high soil NO_3^- concentration and moisture content (Nelissen et al. 2014; Thomazini et al. 2015). Interestingly a correlation has been found between N_2O total cumulative emissions and biochar mitigation: the higher the N_2O emissions, the higher the proportion of reductions with biochar (Cayuela et al. 2013; Thomazini et al. 2015).

Several hypotheses have been suggested about the potential mechanisms underpinning N_2O mitigation with biochar. However, the number of studies focusing on unveiling the involvement of a particular mechanism remains conspicuously low. A brief summary of the most studied mechanisms to date is given below. It is important to note that different mechanisms could be more or less relevant depending on the specific soil and prevailing environmental conditions.

8.3.1   INCREASING SOIL pH

Lower soil pH is generally associated to higher N_2O emissions from both nitrification and denitrification processes (Mørkved et al. 2007; Liu et al. 2010). The role of biochar alkalinity on N_2O mitigation was studied by Cayuela et al. (2013), who demonstrated that the alkalizing effect of a series of biochars could explain between 6-65% of the N_2O decrease in an organic acid soil. Subsequently, Obia et al. (2015) quantified NO, N_2O and N_2 production at high temporal resolution under full denitrification conditions and found a clear link between biochar alkalizing effect and the reduction in N_2O/N_2 ratio in acidic soils. However, they acknowledged that pH alone could not explain the observed reductions and that other factors may have contributed to the suppression of N_2O in addition to the pH effect. In a field experiment Hüppi et al. (2015) compared biochar with lime amendment (adjusting to the same pH) to a slightly acid soil (pH 6.3) and found 52% reductions with biochar whereas no reductions with lime. They concluded that there is no evidence that reduced N_2O emissions with biochar is merely caused by a higher soil pH.

8.3.2   ALTERING MICROBIAL COMPOSITION AND/OR FUNCTIONING OF COMMUNITIES

Biochar changes microbial abundance and community composition, with known impacts on soil N transformations (Lehmann et al. 2011). It was initially speculated that biochar's toxicity (due to its polyaromatic hydrocarbon (PAH) or dioxin content) could decrease the total activity of denitrifiers in soil (Wang et al. 2013). Nonetheless, this hypothesis was rebutted by Alburquerque et al. (2015), who demonstrated that the presence of PAHs at typical biochar concentrations did stimulate, rather than inhibit, N_2O emissions and could not explain N_2O reductions with biochar. Harter et al. (2014) were the first to report an increase in the relative gene and transcript copy numbers of the $nosZ$-encoded bacterial N_2O reductase in a biochar amended soil, a fact that was accompanied by a decrease in the N_2O/N_2 ratio. In a more recent study, Harter et al. (2016) found that biochar not only altered the 16S rRNA gene-based community composition and structure, but it also led to the development of distinct functional traits capable of N_2O reduction containing typical and atypical $nosZ$ genes.

8.3.3   FAVORING BIOTIC/ABIOTIC ELECTRON TRANSPORT

Several recent articles underline the importance of biochar electron transport properties, which might have a bigger impact on soil biogeochemical processes than previously thought (Chen et al. 2014; Kappler et al. 2014; Prévoteau et al. 2016; Saquing et al. 2016; Sun et al. 2017). Abiotic interactions would imply a purely chemical redox interaction of biochar with N compounds. For instance, Quin et al. (2015) measured N2O reduction by injecting 15N-N_2O in sterilized soil columns and demonstrated that biochar took part in abiotic redox reactions reducing N_2O to dinitrogen

(N_2), in addition to adsorption of N_2O. Biotic interactions would imply the ability of biochar to directly accept or donate electrons from/to soil microorganisms. This ability has been demonstrated with microorganisms carrying out N redox transformations, like dissimilatory nitrate reduction to ammonium (Saquing et al. 2016), but to date they have not been explored in N_2O studies.

8.3.4 SORPTION OF C AND/OR N COMPOUNDS

Since denitrification requires organic C as an electron donor, any modification of the concentration of easily mineralizable C_{org} in soil will have an impact on denitrification rates. Biochar (especially when produced at high temperature) provides little easily mineralizable C_{org}. On the contrary, it is known to interact with native or added C_{org}, reducing its availability for soil microorganisms (Kasozi et al. 2010). Although this mechanism has been suggested (Borchard et al. 2014), its importance for N_2O emission reductions has not been systematically evaluated. On the other hand, the $C_{org}:NO_3^-$ ratio is crucial in determining the stoichiometry of denitrification products, with higher ratios favoring the last step of denitrification and therefore decreasing N_2O emissions. The adsorption of NO_3^- by biochar could potentially increase this ratio and this mechanism has also been postulated to explain N_2O mitigation. This hypothesis is supported by Hagemann et al. (2017b) who reported that biochar is able to capture nitrate (NO_3^-), although no isotope labeling was employed in this study leaving the question of the source of sorbed N unresolved. Other studies have indicated that NO_3^- is typically not retained by biochar (Hollister et al. 2013; Wang et al. 2016).

Few studies have looked at the long-term N_2O mitigation capacity of biochar and the results to date are contradictory. The first study investigating how biochar aging might modify its ability to decrease N_2O emissions was carried out by Spokas (2013). This study concluded that 3 years of weathering negated the suppression of N_2O production that was originally observed from the fresh biochar. Subsequently Felber et al. (2014) found a decreasing capacity of biochar to reduce N_2O emissions after one year in grassland. Conversely, Hagemann et al. (2017a) reported that N_2O emissions were still effectively reduced by biochar in the third year after application in a field experiment with corn. In summary, the knowledge on long-term N_2O mitigation with biochar is clearly insufficient, as the number of long-term studies is still low.

8.4 PRIMING

In order to fully characterize the net climate impact of biochar in a given system, its effects on non-biochar soil organic carbon (SOC) stocks must also be accounted for. Although biochar itself usually decomposes very slowly in soils, it can affect existing SOC stocks, changing their rate of mineralization. These changes in mineralization rate are often referred to as "priming," where "positive priming" indicates that SOC is mineralized faster with biochar additions than without, and "negative priming" indicates that SOC is mineralized more slowly than it would have been without biochar additions (Bingeman et al. 1953; Woolf & Lehmann 2012). This section provides an overview of observations of priming in biochar systems, possible mechanisms that may drive priming, interactions with other C sources in soils, including plant roots and added OM, and implications for the expected magnitude and duration of priming.

A number of papers have recently summarized biochar effects on SOC stocks (Wang et al. 2016; Whitman et al. 2015b; Maestrini et al. 2015; Sagrilo et al. 2015). A meta-analysis of 21 studies reported a mean decrease of 3.8% in SOC mineralization with biochar additions (Wang et al. 2016), although the 95% confidence interval included zero. This study builds on a meta-analysis conducted two years earlier, which included 16 studies, and found a mean increase of 15% in SOC mineralization 1 year after biochar additions to soil (Maestrini et al. 2015). The apparent contrast between these two studies reflects both the larger number of observations and the increase in the number of longer-term studies available to the more recent (Wang et al. 2016) meta-analysis. The inclusion of more long-term data is particularly relevant here, as both positive and negative priming effects can coexist, with initial positive priming often transitioning to later negative priming. (Zimmerman et al. 2011; Woolf & Lehmann 2012; Singh and Cowie 2014; Weng et al. 2015; Maestrini et al. 2015; Wang et al. 2016). These metastudies also highlight the wide range of potential SOC responses to biochar, and the challenge of succinctly summarizing effects from a relatively small number of studies, given the wide diversity of potential soil–biochar systems and timescales. Factors such as soil texture or mineralogy, study duration, and biochar feedstock and temperature, all influence the net effect of biochar on SOC mineralization (Maestrini et al. 2015; Wang et al. 2016; Whitman et al. 2015b). To thoroughly characterize and understand these effects, further studies will be necessary, covering a wider range of systems, including more field studies and long-term studies, and focusing on understanding the mechanisms behind these effects.

Part of the challenge of predicting the long-term effects of biochar on SOC stocks is that there are diverse mechanisms by which biochar may affect SOC mineralization. The brief summary given below is in the context of theories of SOC stabilization (Six & Paustian 2014; Cotrufo et al. 2013). Determinants of SOC persistence in soils include environmental controls (e.g., optimal moisture, temperature, oxygen), physicochemical stabilization (e.g., SOC stabilization on mineral surfaces and occlusion by soil aggregation), chemical recalcitrance of SOM (e.g., lignin vs. simple sugars), and biological factors (e.g., microbial community composition and functional potential). A brief discussion follows on how biochar could affect each of these factors. Any impact of biochar on soil moisture or temperature could be predicted to have an impact on soil microbial activity, and SOC mineralization rates. For example, if biochar increases water retention in soils (Abel et al. 2013) or shifts soil pH to be more favorable for microbial activity, priming could occur. Physicochemical stabilization of SOC could be affected in at least two ways by biochar additions. First, a commonly cited explanation for longer-term negative priming is that biochar surfaces sorb SOM, making it less easily available to microbes (Weng et al. 2015; Woolf & Lehmann 2012; Zimmerman et al. 2011; Kerré et al. 2017). Second, in theory, if biochar additions to soil affected aggregation dynamics (e.g., by promoting or inhibiting fungal growth or other aggregate-stabilizing factors), then SOC protection by aggregates could be affected. While biochar wouldn't likely directly affect the chemical composition of SOM, it could change the relative "appeal" of a given substrate (Whitman et al. 2014b). For example, Whitman et al. (2014a) observed very short-term (less than one week) negative priming, and suggested that substrate switching may have occurred, where the small but relatively easily-mineralizable fraction of biochar was preferentially used as a C source by soil microbes. Conversely, that same fraction of biochar has been postulated to be responsible for short-term positive priming, where the input of easily-mineralizable C temporarily increases total soil microbial activity, increasing SOC mineralization as well (Zimmerman et al. 2011; Whitman et al. 2014b). Finally, biochar additions could shift the microbial community composition (Whitman et al. 2016; Xu et al. 2014), which could have positive or negative implications for microbial activity.

Modeling can play a key role in helping us predict the long-term effects of biochar on SOC mineralization. While short-term experimental studies may give us a rate of change in SOC mineralization, predicting the long-term net effects of biochar additions to soil is not simply a matter of extrapolating from the short-term rate. Depending on the mechanism by which biochar affects SOC mineralization, it is possible to predict whether the effects of biochar on SOC will saturate or will persist over time. For example, if biochar itself is responsible for directly sorbing and stabilizing SOC, thereby decreasing its mineralization rate (negative priming), the net potential effect on SOC will depend on the number of potential sorption sites on the total added biochar but would be expected to persist while the biochar remains in the soil. In contrast, if biochar additions increase SOC mineralization (positive priming) due to the stimulation of the microbial community by the relatively small easily-mineralizable fraction of biochar, this effect might be expected to persist only as long as the easily-mineralizable fraction of biochar exists. Woolf & Lehmann (2012) modeled these assumptions, modifying the RothC model (Coleman & Jenkinson 2008) to represent positive priming by changing mineralization rates, and to represent negative priming by changing the fraction of C entering the stable carbon pool. The model predicted that negative priming would have a greater net impact on SOC stocks than positive priming, over 100 years. Archontoulis et al. (2015) used a similar modeling approach and identified the predicted duration of priming effects as a key knowledge gap for predicting net effects of biochar on SOC stocks.

In addition to direct biochar–SOC interactions, researchers have begun to investigate interactions between biochar, SOC, and other OM inputs, including litter and plant root C inputs. While there are still relatively few studies, this research is essential, as complex three- or more-way interactions likely occur in these more realistic systems. For example, Cui et al. (2017) found that biochar caused positive priming of SOM, but that priming was less than was caused by fresh OM additions, and when the two were added together, SOM decomposition rates were lower than with fresh OM alone. Ventura et al. (2015) found that the presence of roots increased biochar decomposition rates. Weng et al. (2015) found that biochar induced negative priming of SOC when plants were present but had no effect on SOC mineralization in soils without plants. Similarly, Keith et al. (2015) found that biochar reduced positive priming effects on SOC by roots. However, this effect was limited to only one of two soils studied. In contrast, Whitman and Lehmann (2015) found evidence that biochar may enhance positive priming of SOC by plants, although this effect was just observed at a single timepoint.

The challenge for researchers in the coming years will be to continue to disentangle the myriad potential mechanisms that could affect biochar–SOC interactions in real-world systems. Isolating and testing for specific mechanisms, controlling for or quantifying all potential drivers of priming in an experiment, and determining when each factor is most important will be key to developing a predictive understanding of biochar's effects on SOC stocks. Understanding the underlying mechanisms will improve the ability to predict long-term priming effects of biochar on SOC.

8.5 ALBEDO

The GHG emissions are not the only means by which human activity can influence climate change. Another forcing mechanism arises from anthropogenic alteration of the Earth's albedo (the fraction of incident solar radiation that is reflected). Soil darkening from increased soil C stocks can reduce the local surface albedo, particularly during bare fallows and while leaf area index is low. While this is true of any mechanism that increases soil carbon stocks (for example Meyer et al. 2012 found that compost reduced soil albedo by the same amount as biochar, per unit C), it is nonetheless worthy of discussion in this chapter due to the greater attention that albedo has received in the biochar literature than in relation to other methods of SOC sequestration.

The earliest published discussion of the impacts of biochar on albedo probably appears in Woolf (2008), who noted the potential for biochar amendments to lower local surface albedo, and also for airborne dust from biochar storage, transport, application, and wind erosion to contribute to tropospheric aerosols and black carbon (BC) deposition on snow and ice. Woolf (2008) also suggested that measures should be taken to reduce airborne dust (such as pelletization of biochar) and to mitigate potential impacts on surface albedo by limiting applications in regions with light soils and extended periods of bare soils.

Woolf (2011) provided a first order estimate of the impact of biochar application on albedo at the global scale, by assuming that the albedo of a soil-biochar mixture is approximately linearly interpolated between the albedos of biochar and soil, on a volumetric mixing ratio basis. In practice, the combined albedo of a soil–biochar mixture will also depend on their particle size distributions, physical interactions such as the coating of biochar particles with mineral deposits or the coating of soil particles with organic compounds from the biochar, biological effects such as the effect of biochar on soil microfauna and also on the extent of vertical transport of biochar within the soil column (Serbin et al. 2009). Nonetheless the linear mixing model provides at least an indication of the expected order of magnitude and also provides a useful baseline to compare measured albedo impacts to.

The global mean albedo of cropland soils is 0.144, derived by combining spatial datasets of soil albedo (Wilson & Henderson-Sellers 1985) and cropland density (Erb 2007). The albedo of charcoal is 0.04 (Serbin et al. 2009). Woolf (2011) thus estimated that a 5% v/v biochar-to-soil mixture (equivalent to 25 g kg^{-1}, or 50 Mg BC ha^{-1} to a depth of 0.15 m) would reduce the mean soil albedo by approximately 5%. The associated radiative forcing (RF) arising from this change in soil albedo can then be calculated according to (Lenton and Vaughan 2009):

$$RF = -0.579 S_0 f_e f_s \ \alpha_s, \qquad (8.1)$$

where S_0 is the annual mean flux of solar radiation at the top of the atmosphere (342 W m^{-2}); $\Delta\alpha_s$ is the change in soil albedo; f_e is the fraction of the Earth's surface over which the change in albedo occurs; and f_s is the fraction of surface albedo attributable to soil albedo (which depends on the vegetation canopy density), estimated to be 0.2 according to Serbin et al. (2009). A 5% reduction in mean soil albedo thus corresponds to a net radiative forcing of 6×10^{-3} W m^{-2}, if biochar were applied to all 1.5 Gha (giga hectare) of global cropland (Woolf 2011). This is equivalent to 1.1% of the -0.4 W m^{-2} radiative forcing from avoided CO_2 emissions attributable to the same biochar applications (Lenton & Vaughan 2009).

Relatively few studies have quantified the change in soil albedo from biochar additions experimentally, with somewhat variable results. Genesio et al. (2012) measured surface albedo in a winter durum wheat crop (*Triticum durum*) following application of biochar (0, 30 and 60 Mg ha^{-1}), and incorporation in the top 0.1 m with a rotary hoe. They found that, in the first year, soil albedo in both the 30 and 60 Mg ha^{-1} treatments was 0.062 ± 0.001 compared to the contrsol albedo of 0.208 ± 0.004. This marked reduction in albedo ($\Delta\alpha_s$) on the biochar plots is approximately one order of magnitude greater than would be predicted by the linear mixing model and corresponds to a soil albedo that is almost as low as that of pure charcoal (0.04; Serbin et al. 2009). However, Genesio et al. (2012) also found that $\Delta\alpha_s$ declined to 0.05 after 18 months, and completely disappeared following a second tillage operation in the second year. The observation that soil albedo was almost as low as pure charcoal in the first year but became indistinguishable from soil albedo after tillage in the second year suggests that the profound reduction in albedo in the first year may have been caused by incomplete mixing, with a higher concentration of biochar remaining at the soil surface. This hypothesis is compatible with the use of a rotary hoe for biochar incorporation, which is designed for weeding rather than pedoturbation.

Other authors have reported significantly lower values of $\Delta\alpha_s$ than observed in the first year by Genesio et al. (2012). Oguntunde et al. (2008) found that surface deposits of charcoal remaining at charcoal production sites 2 to 14 months after cessation of operation produced a $\Delta\alpha_s$ of $0.03 - 0.06$. Meyer et al. (2012) measured $\Delta\alpha_s$ to be 0.01 from 31.5 Mg ha^{-1} biochar tilled to 0.1 m depth in a wheat field trial in Donndorf, Germany. Notwithstanding the much lower value observed for $\Delta\alpha_s$ in their field trial, Meyer et al. (2012) ("conservatively") estimated the long term radiative forcing of albedo by assuming that $\Delta\alpha_s$ in the first year was 65% of the value measured by Genesio et al. (2012) in the first year, declining to 22% by year 3 and thereafter declining only slowly with a half-life of 500 years. Based on this assumption, Meyer et al. (2012) estimated that the overall climate change mitigation of biochar could be reduced by 13%–22% by the albedo impact.

Verheijen et al. (2013) found a 0.2% reduction in soil albedo per Mg ha^{-1} biochar application mixed to 0.15 m depth in wet soils, and a 0.35% albedo reduction per Mg ha^{-1} biochar in dry soils. Based on these measurements, they estimated a 5%–11% reduction in the mitigation potential of biochar due to the albedo impact. Verheijen et al. (2013) also found that surface application of biochar without incorporation caused a much more profound in decrease in soil albedo that reduced its mitigation impact by 11%–23%. Similar to Verheijen et al. (2013), Bozzi et al. (2015) measured a 0.25% reduction in soil albedo per Mg ha^{-1} biochar application. Combining this with time-averaged satellite measurements of surface albedo over agriculture areas with different soil albedos, they estimated the biochar application rates that would result in the same soil albedo difference as that present between the different soil types. Based on these inferred equivalent biochar application rates Bozzi et al. (2015)

calculated that the measured time-averaged radiative forcing for dark soils relative to bright soils of 0.1–2.0 W m^{-2} would imply a 1.1%–29.7% reduction in the mitigation potential of biochar.

Zhang et al. (2015) found that 45 Mg ha^{-1} biochar reduced surface albedo by 23% at seeding of maize (0.5% per Mg ha^{-1}), declining to 20% (0.4% per Mg ha^{-1}) in the jointing stage, and no significant difference from the heading stage to mature stage. Zhang et al. (2013) observed that 4.5 Mg ha^{-1} biochar caused a 2%–7% *increase* in soil reflectance in the 350–495 nm (ultraviolet to blue) range and a 1%–6% decrease in the 496–2474 nm range.

Usowicz et al. (2016) found that surface application (without incorporation into the soil profile) of 30 Mg ha^{-1} biochar on grassland reduced surface albedo by 30%, from 0.17 to 0.10 (1% per Mg ha^{-1} biochar) when the grass was short (0.06 m). When the grass had grown to 0.10–0.15 m height there was no significant difference in albedo between biochar and control plots, and when the grass reached 0.15–0.50 m, the effect was reversed with surface albedo increased 15% from 0.20 to 0.23 (0.5% per Mg ha^{-1} biochar). On bare fallow plots, in which 30 Mg ha^{-1} biochar was incorporated to 0.115 m by rotary tillage, Usowicz et al. (2016) observed a reduction in soil albedo of 0.02–0.03 (0.5% per Mg ha^{-1} biochar).

The large variability in these results indicates that the effects on albedo are mediated by a number of both controllable and uncontrollable factors. These include:

- Initial soil albedo, with the albedo of lighter colored soils being reduced more than those of darker ones.
- Hydrology, with dry soil albedo being reduced more than wet soils.
- Vegetation, with increasing leaf area reducing the impact of soil albedo on surface albedo. The interaction between vegetation cover, time, and atmospheric diffusivity is also important, with bare soil being less important to dark sky albedo at lower solar declination.
- Time, with some evidence that initial reductions in albedo may become negligible after as little as one year. This may be caused either by vertical transport of biochar into the soil column away from the surface, and/or coating of biochar surfaces with minerals or OM.
- Depth of incorporation, with mixing of biochar into deeper soil reducing its albedo impact. As noted by Verheijen et al. (2013), below surface application of biochar would prevent any albedo reduction but is more expensive and future mechanical soil operations could yet expose biochar to the surface.

Given that under unfavorable conditions the albedo impact might be a significant part (possibly up to 30%) of the mitigation potential of biochar, whereas under favorable conditions the albedo impact is negligible (Figure 8.3), it would be prudent while more long-term data are acquired to take

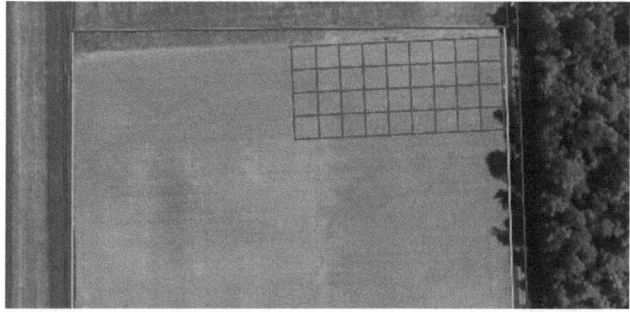

FIGURE 8.3 Satellite image (April 14, 2017) of bare soil on a biochar field trial in a continuous maize cropping system in New York State (42.7305 N, 76.655 W). Biochar was applied in 2006 at rates of 0 to 50 Mg ha^{-1} in a randomized block design in the blocks outlined in purple. The field was plowed annually in spring. No treatment differences in albedo are discernable by the naked eye 10 years after application.

active measures to limit its importance when planning and implementing biochar systems. Steps that will mitigate albedo impacts include avoiding surface application without incorporation (particularly where soil is likely to be bare for an extended period), ensuring thorough mixing of biochar into the soil, and mixing biochar deeper into the soil. These steps are consistent with optimal use of biochar to improve soil fertility. Although albedo impacts can also be mitigated by preferentially using biochar in soils with already lower albedo and/or high vegetation cover, this must be balanced against the fact that the more degraded soils that can most benefit from biochar typically have lower OM content that can give them a lighter color. In addition to the potential impacts of biochar on surface albedo, Genesio et al. (2016) remind us that airborne BC aerosols from wind erosion of fine particulate BC may not be entirely negligible and note that quantification of this effect has not received sufficient attention. Pelletization of biochar prior to field application (Woolf et al. 2010) or charring of pelleted feedstock has been suggested as one way to limit the production of airborne dust.

It is clear that the albedo impacts of biochar warrant further research, particularly into finding optimal strategies to mitigate the impact, and into the how changes in soil albedo develop over time as biochar is translocated vertically or horizontally, and the surface properties of the biochar change due to coating with minerals and organic matter and occlusion within soil aggregates.

As a final note, it is important to reiterate that albedo reduction is an issue for all types of soil C sequestration, not only biochar. Although there is little research to quantify the impact of non-pyrogenic SOM on albedo, Meyer et al. (2012) found no difference between the reduction of soil albedo by compost or biochar, per unit mass of C added.

8.6 PLANT RESPONSE TO BIOCHAR

Many studies have shown positive responses of crop or biomass yield to biochar amendment, although negative responses are also sometimes seen (Figure 8.4 and Jeffery et al. 2011; Macdonald et al. 2014; Mukherjee & Lal 2014). Even though the focus of this chapter is on the climate-change impacts of biochar rather than on its agronomic use, nonetheless it is pertinent to include a discussion of biochar's impact on plant growth, because altered rates of plant growth are not only of

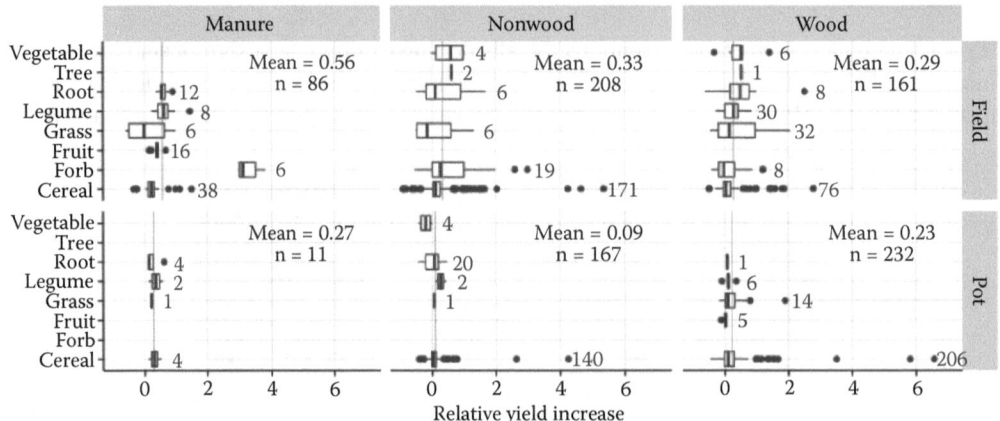

FIGURE 8.4 (See color insert.) Yield response of crops to biochar additions (difference between biochar amended yield and control, expressed as a fraction of control yield). Data from 865 treatments from 74 published articles are broken down by field trial versus pot trial, by feedstock type (manure, wood or non-wood), and by crop type. Vertical red line on each panel shows the mean crop response. Mean response and number treatments for each panel are also given in red text. Numbers shown in blue adjacent to each box indicate the number of treatments in the sub-category.

interest from a crop production standpoint, but also affect climate. There are five primary ways in which the soil fertility and crop productivity impacts of biochar relate to climate-change mitigation:

1. Sustainable intensification of cropping systems from increased yields may reduce pressure on land use and have a positive impact on indirect land use change (Wicke et al. 2012) (see also Section 7.4 below).
2. Biochar can improve fertilizer use efficiency, thereby reducing emissions associated with fertilizer manufacture and transport (Sohi et al. 2010). The manufacture of N fertilizer emits approximately 3.2 Mg CO_2 Mg^{-1} N (West & Marland 2002), which is approximately equal to the 3.0 Mg CO_2e Mg^{-1} N direct N_2O emissions from fertilizer application (Del Grosso et al. 2006).
3. Increased NPP can increase biomass available for mitigation through increased inputs available for building SOC stocks (Roberts et al. 2010), producing more biochar (Woolf et al. 2010), generating bioenergy, or other biomass-based mitigation approaches (Woolf et al. 2016).
4. In perennial (particularly woody) systems, increased NPP can increase biomass carbon stocks (Scharenbroch et al. 2013; Ghosh et al. 2015; Thomas & Gale 2015).
5. Last, but not least, the value of increased yield and reduced fertilizer requirement is critical in making biochar systems economically viable (McCarl et al. 2009; Field et al. 2013). Unless biochar provides a long-term improvement in soil fertility it is not economically competitive with other uses of the same biomass to provide climate-change mitigation, such as bioenergy or bioenergy with carbon capture and storage (Woolf et al. 2016).

There are several mechanisms by which biochar can affect crop yields. These include direct provision of nutrients; altering soil pH; increasing CEC, which in turn can improve fertilizer use efficiency and thus nutrient uptake for a given fertilizer application rate; and by increasing the water holding capacity (WHC) of sandy soils or the drainage of clayey soils (Jeffery et al. 2011; Sohi et al. 2010; Atkinson et al. 2010). The impacts of biochar's nutrient content and liming potential are likely to be short-lived effects, whereas CEC and WHC impacts are associated with the C matrix provided by biochar and are likely to persist while the biochar remains. Indeed, the CEC of biochar increases over time as its surface develops oxygenated functional groups through oxidation from exposure to oxygen and water (Liang et al. 2006). Effect on WHC varies with soil, biochar, and application rate. Biochar has been shown to increase WHC by up to 84%, with the greatest increases being on sandy soils, or to reduce WHC by up to 45%, with the greatest decreases being on clayey soils (Masiello et al. 2015).

The complex interaction of the various mechanisms by which crop yields are affected means that biochars have variable impacts on plant growth, depending on the biochar physical and chemical properties (which in turn depend on both feedstock and production conditions; Enders et al. 2012), properties of the soil, requirements of the target crop, application rate, depth of incorporation into soil, and time since application.

Although many studies have measured the impact of biochar on crop yields, the number of studies that directly investigate the underlying mechanisms driving these responses remains low. To fill this gap, some meta-analyses have attempted to discern general relationships from the corpus of published data. Jeffrey et al. (2011) found that biochar altered yields by −28% to +39%, with a mean increase of 10%. In a similar updated meta-analysis with more recent data, Jeffrey et al. (2015) found the mean yield increase was 18% relative to controls without biochar. Negative yield responses, where they have been observed, can generally be attributed to use of biochar with chemical properties that are inappropriate for the soil, crop or production system. Most often this involves the application of high pH biochar to soils that already have a neutral or alkaline pH, or the use of biochar whose high C:N ratio and high easily mineralizable -C content can give rise to N immobilization. These types of negative impacts can be avoided through the provision of appropriate decision-making tools or guidelines to agronomists and farmers. The greatest positive effects on crop yield were seen in acidic to neutral pH soils, and coarse to medium textured soils. Jeffrey et al. (2011) suggested, on this basis, that the main mechanisms for yield increase may be liming, improved WHC, and improved nutrient availability. WHC has been shown to The biochar feed-stocks that showed the greatest yield increase was poultry litter, whereas biosolids were the only feedstock showing a statistically significant negative effect (Jeffery et al. 2011). Crane-Droesch et al. (2013) found soil CEC and organic C were the strongest predictors of yield response, with low cation exchange and low C associated with positive response, and that yield response increased over time since initial application. Biochar characteristics, on the other hand, were not significant predictors of yield impact (Crane-Droesch et al. 2013). Woolf et al. (2016) conducted a meta-analysis on a sub-set of published data that excluded biochars made from manures, thus excluding nutrient provision as a mechanism for yield impact, with the intention to better understand the extent to which longer lasting mechanisms affect yield. They found that soil pH and CEC were overwhelmingly the most important predictors of yield response, with soil texture, biochar carbon fraction, and fertilizer and biochar application rates of secondary importance.

While these meta-analyses are helpful to shed some light on the mechanisms by which biochar affects yield, they are of necessity limited by the extent of available published data, with most stud-ies being only short term (relating to the first or second cropping season following biochar appli-cation), and the full range of soil and crop types and environmental and management conditions remaining relatively unexplored.

Notwithstanding these shortcomings in available data and lack of studies specifically designed to test for mechanisms, some clear patterns are already apparent. It is important in this regard to take care to distinguish between uncertainty and manageable variability (Lehmann & Rillig 2014), with much of the observed variation in crop response being attributable to predictable biochar–soil–crop interactions. Yield benefits are generally greatest in poor soils, especially light-textured acidic or degraded soils, particularly those with low soil CEC. Although a correlation between higher crop response and low soil CEC suggests that biochar may offer long term improvement to the fertility of low CEC soils (Woolf et al. 2016), experimental data on long term effects remain sparse. Furthermore, in some studies, unfavorable changes in soil chemical, physical and biological properties, and reductions in crop yields have been reported (Mukherjee & Lal 2014). An improved understanding of the mechanisms underlying crop response to biochar will help to direct biochar applications into systems they can most benefit, and also avoid the application of unsuitable types of biochar in cropping systems where they may do harm.

In conclusion, although there is a substantial body of evidence indicating that biochar can improve yields in infertile or degraded soils, a high uncertainty remains in the expected long-term (>3 years) response of specific soil-crop systems to biochar amendments. Further research is needed to provide long term data on crop response in a variety of soils, cropping systems and agroecologi-cal zones. A greater focus on research that aims to understand the mechanistic causes of observed impacts should also be a priority. This will support the development of decision-support models that can ensure use of different biochar types in cropping systems they can most benefit.

8.7 GHG ACCOUNTING FOR BIOCHAR

Methods for quantifying GHG emissions fluxes from biochar systems are required for GHG inven-tory, for GHG accounting against performance targets, for C footprint quantification and for cal-culating abatement at project level. Applicable methods depend on the purpose and the specific sources and sinks that are included.

8.7.1 GHG Inventory and Reporting

GHG inventory may include national level inventory undertaken for reporting under the United Nations Framework Convention on Climate Change (UNFCCC), for which the Intergovernmental

Panel on Climate Change (IPCC) prepares guidance, and may also refer to organizational inventory, undertaken for sub-national reporting or for regulatory agencies and corporations to establish compliance or to track progress towards emission reduction goals.

A GHG inventory is usually undertaken on annual basis, estimating all GHG emissions and removals that occur within the national or organizational boundaries within a specified period. To avoid double-counting, the inventory is usually confined to those facilities for which the country or organization has control, that is, to emissions that occur within their boundaries.

National GHG inventory reporting is sector-based, and GHG fluxes associated with biochar systems will be counted in different sectors: fuel use in the energy sector; C stock changes in biomass and soil, in the land use, land use change and forestry (LULUCF) sector; and emissions of N_2O and CH_4 from soils and manure management in agriculture (Cowie et al. 2012). Methods for inventory and reporting for biochar systems would need to integrate with existing quantification methods for each of these sectors. Quantification of some sources and sinks is straight forward, while others require adaptation to recognize the mitigation benefits of biochar. The IPCC publishes Tier 1 default methods (equations and emissions factors), and many countries have developed Tier 2 (country-specific emissions factors) or Tier 3 (measured or comprehensively modelled) methods for their key source categories.

It is important that the national inventory is accurate at the national scale. High spatial precision is not necessary for national scale assessment of net emissions and progress towards mitigation targets, although monitoring and verification of spatially heterogeneous C stocks and GHG fluxes may be facilitated by spatially-explicit reporting. Whether or not high spatial resolution is provided, comprehensive coverage of emissions sources and sinks is desirable to assist policy-makers in gauging the success of our collective efforts, globally, to curb emissions.

8.7.2 GHG Accounting

GHG accounting refers to assessing performance with respect to a GHG emissions reduction target, such as the commitments made by Annex I countries under the Kyoto Protocol. GHG accounting may be less comprehensive than GHG inventory reporting, as it may focus on certain sources or sinks targeted by the policy. As with GHG inventory reporting, GHG accounting is usually conducted on an annual basis, and the same quantification methods can be applied. For accounting, the objective is to ensure that results reflect effort in pursuing agreed mitigation activities, and high precision is less important.

8.7.3 Project and Product Accounting

Project-level accounting is undertaken to assess the mitigation benefits of an activity, such as within emissions trading schemes where abatement activities earn offset credits. The C footprinting quantifies the net GHG emissions associated with a specific product and may be used for product labelling to inform consumers, or for business-to-business communication to inform downstream customers. In both cases the intention is to encourage a change in behavior to reduce net GHG emissions. For each of these applications it is necessary to take a full life cycle approach, considering emissions and removals across the supply chain and also indirect effects, so that the full climate change impacts of the biochar system can be quantified. Ideally, all significant emissions sources and sinks directly or indirectly affected by the activity should be included.

For project- and product-level GHG accounting the most important criteria are that the method is cost-effective whilst being sufficiently accurate to ensure credibility that abatement has occurred, to encourage maximum participation and therefore maximum abatement. Methods for project- and product-level accounting should therefore be as simple as possible, conservative (i.e., tending to underestimate rather than to overestimate GHG abatement), and readily audited under a verification process. Methods based on emissions factors linked to specific practices are much more readily applied and verified than methods based on physical measurements at the site where biochar is applied. Project methods must also be widely applicable, or at least their applicability should be readily specified and assessed. For example, a GHG source may be excluded if the project activity is restricted to ensure that the source is not a high risk. For example, methane emitted from biomass decomposition during storage could be minimized by requiring that the biomass is dried before storage and stored in dry conditions under cover, thereby avoiding the need to quantify CH_4 emissions.

To quantify the climate change impacts of a biochar system for carbon footprinting and project-level accounting it is necessary to quantify the life cycle emissions of the biochar system and the processes and products that it displaces. For projects, this involves comparing the biochar system with the "no-biochar" reference scenario (Figure 8.5). Note that Figure 8.1 shows a reference and biochar system from the perspective of the fate of the biomass. Figure 8.5, in contrast, illustrates the services supplied by biochar system, and the corresponding provision of the same services in the reference system. The reference, also known as baseline, could be the situation at start of project, a BAU (business as usual) projection, or a forward counterfactual that envisages the no-biochar future. In research, or to inform policy development, it can be enlightening to compare several alternative scenarios.

Carbon footprinting of products is based on LCA methodology, which is standardized through ISO (ISO 2006a,b, 2013). LCA is commonly undertaken using proprietary or open source LCA software that facilitates the construction of models, access to databases of inventory data, and agreed models for analysis.

LCA approaches can be distinguished as consequential (CLCA) or attributional (ALCA). CLCA considers the direct and indirect effects of producing an additional unit of the product, while ALCA quantifies the impacts of the average unit of production, focusing on the direct supply chain emissions. CLCA applies system expansion and substitution to handle co-products, giving the studied product a credit for emissions avoided due to products displaced by the co-product (ISO 2015). ALCA handles co-products through allocation, dividing the supply chain emissions between the different products according to economic value or physical features such as energy content or mass. CLCA gives more accurate results, though often with greater uncertainty, and is recommended for

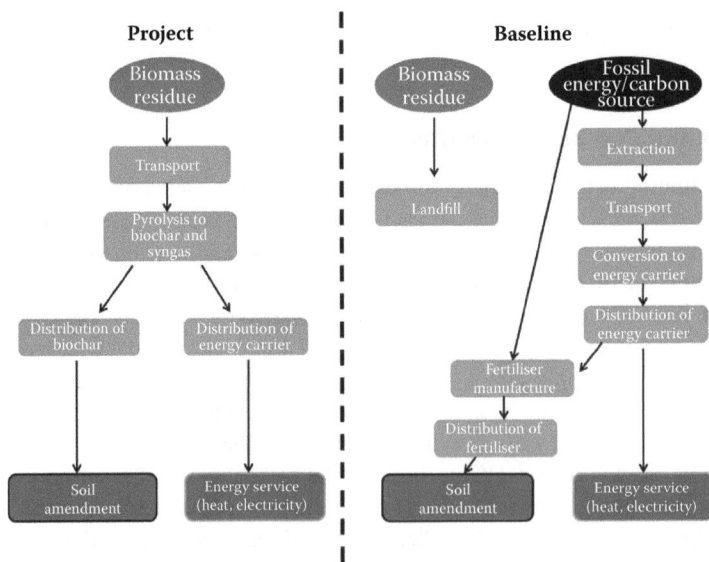

FIGURE 8.5 Project-level GHG accounting compares the life cycle emissions from the biochar system with those of the reference system providing the same services.

policy development (Brandão et al. 2014). ALCA is often employed for policy implementation, as it is considered easier, and more appropriate because it reflects the emissions and removals over which the operator has control.

8.7.4 METHODS FOR QUANTIFYING GHG FLUXES FROM BIOCHAR SYSTEMS

Biochar systems can affect sources and sinks, and can contribute to abatement through carbon sequestration, direct reduction in GHG emissions, and indirect reduction in GHG emissions. Below, each of these processes is considered with respect to its readiness for inclusion in GHG accounting, and, for abatement processes considered ready, a method for quantification is briefly described.

8.7.4.1 Carbon Sequestration

Increased persistence: Biochar systems that include the production of biomass and production of biochar remove CO_2 from the atmosphere and sequester C in pyrolysed biomass, which is more persistent than the raw biomass, thus delaying emissions for tens to thousands of years, as described in Section 8.2. Stabilization of biomass C is the most important contributor to abatement in most LCA studies of biochar systems (Cowie et al. 2015). Budai et al. (2013) reviewed alternative methods to assess the persistence of biochar, and concluded that $H:C_{org}$ ratio is a suitable measure for application in project-level accounting. They proposed the metric BC_{+100}, the quantity of biochar C remaining after 100 years and defined cut-offs at 0.4 and 0.7 in order to identify stability classes: for an $H:C_{org}$ value lower than or equal to 0.4, which is considered "highly stable", at least 70% of the biochar C is predicted to remain in soil for ≥100 years ($BC_{+100} = 70\%$), whereas for an $H:C_{org}$ value greater than 0.4 and lower than or equal to 0.7, considered "stable," a BC_{+100} of 50% can be conservatively expected.

    This method for estimating abatement has several advantages. First, the method can be applied to any biochar technology and production conditions (kiln temperature, heating rate and residence time); these factors will affect biochar persistence, and these impacts will be reflected in the measured persistence. Second, it can be applied at the point of production of the biochar, and so does not require a mechanism to track the fate of the biochar. However, it is necessary to ensure that the biochar is not combusted, such as for cooking fuel. This can be assured when the biochar is produced from biomass mixtures that include manure, for example. With clean feedstocks, measures may be required to confirm application to soil.

Negative priming: As discussed in Section 8.4, biochar addition to soil can delay the decomposition of native SOM and newly added add OM, such as plant litter and root exudates. While negative priming has been demonstrated in several soil types, agricultural systems and environments, knowledge is currently too immature to allow this benefit to be predicted in GHG accounting. A change in SOC stock could be measured directly, through soil sampling and analysis such as applied to other soil C enhancement projects (for example, the Australian Emissions Reduction Fund methodology, Australian Government 2014). Care should be taken to avoid double-counting: if C persistence is estimated as in 1a, then the amount of C added in biochar should be deducted from the measured C stock. If an offsets credit is granted on the basis of measured changes, then the permanence of that sequestration needs to be assured through a program that requires maintenance of the sequestered C. Permanence measures can include a make-good requirement in the event of loss; a buffer of unsold abatement; or measures that allow the maintenance obligation to be transferred to other sequestration activities.

Increased plant growth: Biochars have variable impacts on plant growth, depending on the feedstock, properties of the receiving soil and requirements of the target crop, and negative responses are sometimes seen (Section 8.4 above and Macdonald et al. 2014).

Yield benefits are generally greatest in poor soils, especially light-textured acidic or degraded soils. Because the result varies with soil type, biochar type, crop, and rate of application (see Section 6), it is not possible to provide a generically-applicable method to estimate response. Measured increase in plant growth could be included in the calculation of project abatement, expressed as the increase in average C stock.

8.7.4.2   Emissions Reduction

Reduced nitrous oxide emissions from soil: Application of biochar can reduce emissions of N_2O from soil, through a range of mechanisms discussed above (section 3), with meta-analysis showing an average reduction of 54% (Cayuela et al. 2014). Cayuela et al. (2015) have identified a relationship between decrease in N_2O emissions and the molar H/C_{org} and N/C_{org} ratios of the biochar, which hold promise as a basis for a quantification method similar to that used to estimate carbon stability. However, given that the long-term persistence of the N_2O abatement has not been well established, conservative accounting of N_2O impacts would assume that it is a short-lived effect (Section 8.3).

Reduced fuel use: Biochar can enhance water holding capacity, thereby reducing the need for pumping irrigation water, and may reduce soil strength, decreasing fuel use. These effects cannot readily be generalized. Other factors could increase fuel use in the biochar system compared to the reference system: collection and processing of biomass, construction of pyrolysis facilities, transport of biochar. To the extent that reduction in use of electricity and/or liquid fuels relative to the reference system can be documented, a reduction in GHG emissions could be included in project accounting.

Reduced emissions during composting: Adding biochar to compost can reduce emissions of CH_4 and N_2O during composting (e.g., Agyarko-Mintah et al. 2016). However, there is currently insufficient knowledge to develop a model to estimate this reduction, and direct measurement is impractical on a routine basis.

8.7.4.3   Indirect Avoided Emissions

Avoided fossil fuel use: Pyrolysis gases can be used to displace fossil fuels for heat and electricity. IPCC Tier 1 or national Tier 2 emission factors can be used with activity data to quantify avoided emissions.

Avoided biomass decomposition: Pyrolysis of biomass may avoid emissions of CH_4 from the decomposition of biomass such as manure that would otherwise be stockpiled, crop residues that would be burned or processing residues that would be landfilled. IPCC methods (equations and emission factors) used to quantify these emissions sources can be used to estimate emissions avoided.

Reduced fertilizer manufacture: Biochars can reduce the requirement for chemical fertilizer by enhancing NUE. If a reduction in chemical fertilizer use per unit crop produced can be demonstrated, avoided emissions from fertilizer production could be calculated using life cycle inventory data for the specific fertilizer products avoided.

There are also factors that reduce the abatement that must be included:

Emission from pyrolysis: CH_4 and N_2O emissions are produced during pyrolysis at variable rates depending on the design of the facility, feedstock properties and skill of the operator. For small scale facilities such as cookstoves and drum ovens, a conservative estimate of emissions based on published literature (MacCarty et al. 2008; Whitman et al. 2011; Sparrevik et al. 2015) should be included. For larger scale facilities, gases should be flared or combusted to produce heat or electricity. Furthermore, emissions should be measured to establish the average emissions for each feedstock processed at the facility, and these emissions should be deducted from the calculated abatement.

Positive priming: Positive priming, that is, the stimulation of turnover of native SOC has been suggested to reduce the climate-change mitigation benefits of biochar systems (Wardle et al. 2008; Luo et al. 2011). However, more recent research (reviewed by Whitman et al. 2015b and discussed above in section 4), suggests that positive priming is not likely to cause a significant loss of abatement in mineral soils. However, acknowledging the small risk of positive priming it is conservatively suggested that the quantity of biochar C stabilized should be discounted by 5%, as proposed by Koper et al. (2013).

Carbon stock loss in biomass and soil: If additional biomass is harvested for biochar, this may lead to a reduction in biomass C or soil C at the harvest site. Any reduction in average C stock should be included the calculation of abatement.

Indirect land use change (iLUC): If biomass crops are grown to produce biochar the biomass crops may displace other crops producing food, feed or fiber. The continued demand for these crops may lead to land use change elsewhere to provide land for these crops. iLUC is challenging to prove or quantify, so it is difficult to include in the quantification of abatement at project level. Because of this difficulty in quantification, current international standards for LCA and climate finance do not require the inclusion of iLUC (Finkbeiner 2014). Nevertheless, some biochar studies have estimated iLUC (e.g., Roberts et al. 2010), and methods for inclusion of iLUC in LCA have been proposed (Schmidt et al. 2015). iLUC risk can be avoided by using biomass residues for biochar production. Risk of iLUC should be assessed and iLUC should be included in a sensitivity analysis in studies conducted to inform policy (Muñoz et al. 2015). Ideally, iLUC associated with biochar should be investigated in comprehensive global modelling using partial or general-equilibrium economic models that take into consideration market-mediated interactions between land use, energy sector, food prices and other macroeconomic indicators of production, consumption and trade. Equilibrium models such as FAPRI-CARD (Food and Agricultural Policy Research Institute - Center for Agricultural and Rural Development), GCAM (Global Change Assessment Model), and GTAP (Global Trade Analysis Project) that account for factors such as land, food, fiber and energy prices, maps of land suitability, proximity to transport infrastructure and existing crop production have been applied previously to the question of quantifying iLUC due to biofuels (Searchinger et al. 2008; Hertel et al. 2008; Schmidt et al. 2015; Flugge et al. 2017), and could be applied also to biochar systems. Despite the inherent, and possibly intractable, difficulties in rigorously quantifying iLUC impacts of specific projects, the general philosophy outlined above that GHG accounting should be conservative indicates that either (a) iLUC should be accounted for with a sensitivity analysis that includes conservative estimates in its range, or (b) projects should demonstrate that iLUC is not a risk, because non-competitive biomass such as unused crop or forestry residues is being sourced as feedstock.

8.7.5 SUMMARY OF GHG ACCOUNTING FOR BIOCHAR

Quantification of GHG fluxes differs depending on the purpose. National inventory applies IPCC methods to quantify annual emissions and removals. In contrast, GHG accounting for biochar projects compares the life cycle emissions and removals for a biochar system with those for a reference system providing the same services. Ideally, methods for project-level accounting are based on cost-effective methods utilizing scientifically-based models. Biochar knowledge has improved markedly over the last decade, providing the basis for estimation methods for the key processes that contribute to abatement. Thus, the molar $H:C_{org}$ ratio has been proposed as a simple indicator of C persistence, with the BC_{+100}, the quantity of biochar C remaining after 100 years, proposed as the metric to distinguish "highly stable" and "stable" biochars. The molar $H:C_{org}$ and $N:C_{org}$ ratios also appear to provide a suitable basis for estimating the reduction in N_2O emissions from soil, although the longevity of such impacts remains uncertain. Other abatement processes including reduction in emissions from manure handling, fuel use in irrigation and cultivation, and fertilizer manufacture can be estimated using emissions factors, life cycle inventory data and activity data. Inclusion of negative priming and enhanced biomass production will require on-site sampling and measurement, which will substantially increase transaction costs and effort required for GHG estimation and verification.

8.8 DISCUSSION AND CONCLUSIONS

8.8.1 Evidence-Based Policy

The discussion presented in this chapter shows that, although many uncertainties remain, nonetheless a large body of evidence has been developed in the scientific literature over the last decade which allows us to move forwards with some clear policy guidelines. Although some data are still lacking to predict the climate-change mitigation performance of biochar in all environmental conditions, soils types and management systems, enough is known now to design biochar systems that adopt best practice recommendations and are applied in soils and cropping systems where a positive outcome is expected. The data presented in this chapter provides clear guidance on what types of best management practices (BMPs) will ensure that biochar projects avoid the potential pitfalls of negative impacts. These recommendations include:

1. Use of pyrolysis temperatures and/or reaction times that are high enough to create biochar that is sufficiently persistent to guarantee carbon sequestration over centennial timescales or longer. The H:C$_{org}$ ratio of the biochar providing a simple and reliable proxy for its persistence, with values less than 0.4 indicating high persistence.
2. The same pyrolysis conditions that offer improved biochar persistence also tend to offer the best mitigation of soil N$_2$O emissions, although the longevity of this impact remains to be proven.
3. Surface application of biochar without incorporation should be avoided, as this may lead to reductions in albedo that have a warming effect on the climate and may give rise to airborne dust. Additionally, incorporation of biochar will increase its agronomic effectiveness.
4. Soil fertility and crop productivity benefits of biochar will be optimized by applying biochar in soils that have some combination of low CEC, low SOM, low pH and low WHC, as these are the primary constraints that can be alleviated by biochar. The corollary is that adding high rates of biochar to soils with high pH can suppress yields and should be avoided, and adding biochar to soils that already have high CEC is unlikely to show any benefits in cycling or availability of nutrients.
5. The same pyrolysis conditions that optimize persistence and N$_2$O abatement will also eliminate the risk of lowering crop yield through nitrogen immobilization arising from easily mineralizable OM with a high C:N ratio.
6. Use of biomass feedstocks that are non-competitive with other demands will avoid emissions associated with direct or indirect land use change.

Existing GHG accounting life cycle assessment methodologies provide a sound basis to develop accounting methodologies for biochar, and there is sufficient evidence to parameterize the methodologies for biochar systems using conservative estimates. Thus, despite remaining uncertainties, use of conservative values means that we can confidently move forwards with biochar assessment protocols that estimate the lower range of potential mitigation impacts. Using this approach is prudent in that it provides assurance that biochar projects should perform better rather than worse than predicted, thus reducing or eliminating the risk that anticipated climate abatement goals will not be achieved.

It is also clear from the evidence provided in this chapter that policy development must be informed by comprehensive analyses that include biophysical, economic and social factors, so that market-mediated effects are also included. Policy development should also model a range of alternative scenarios to fully understand the range of possible outcomes for a given policy and also to compare predicted outcomes from a range of alternative policy options.

With regard to informing climate change mitigation policy at the regional or global scales, there is a clear need to include biochar in Integrated Assessment Models (IAMs) (Smith 2016; Woolf et al. 2016). Use of IAMs is the best way to understand where biochar fits into an overall portfolio of mitigation options in terms of both its technical and economic performance. IAMs and/or general equilibrium models will also provide the best means to assess potential impacts of biochar policy frameworks on broader outcomes such as land use and land use change, and on food and energy prices.

8.8.2 Conclusion

In summary, it is concluded that biochar offers high potential as a climate change mitigation technology, but that careful design and monitoring of projects, policy frameworks, and agricultural extension advice will be required to optimize results and to avoid negative outcomes from poor implementation practices. Robust and conservative GHG accounting methodologies can already be applied based on the existing evidence base and LCA. These accounting methodologies can be improved over time by ongoing research, particularly with respect to improving predictability of long-term impacts on priming, soil N_2O emissions, and crop yield responses. However, the large body of scientific evidence that has been accumulated over the last decade means that well-designed biochar projects can already be deployed at low risk while we continue to learn more about the mechanisms involved.

ACRONYMS

C	Carbon
C_{org}	Organic carbon
N	Nitrogen
H	Hydrogen
O	Oxygen
N_2O	Nitrous oxide
CH_4	Methane
CO_2	Carbon dioxide
CO_2e	Carbon dioxide equivalent
SOM	Soil organic matter
SOC	Soil organic carbon
NUE	Nutrient-use efficiency
GHG	Greenhouse gas
LCA	Life cycle assessment
NPP	Net primary production
CEC	Cation exchange capacity
OM	Organic matter
MRT	Mean residence time
PAH	Polyaromatic hydrocarbon
NO_3	Nitrate
BC	Black carbon
RF	Radiative forcing
WHC	Water holding capacity

UNFCCC   United Nations Framework Convention on Climate Change
LULUCF   Land use, land use change, and forestry
IPCC   Intergovernmental Panel on Climate Change
BAU   Business as usual
ISO   International Standards Organization
CLCA   Consequential LCA
ALCA   Attributional LCA
iLUC   Indirect land use change
LUC   Land use change
FAPRI   Food and Agricultural Policy Research Institute
CARD   Center for Agricultural and Rural Development
GCAM   Global Change Assessment Model
GTAP   Global Trade Analysis Project
IAM   Integrated Assessment Model

REFERENCES

Abel, S., Peters, A., Trinks, S., Schonsky, H., Facklam, M., and Wessolek, G. 2013. Impact of Biochar and Hydrochar Addition on Water Retention and Water Repellency of Sandy Soil. *Geoderma* 202–203 (July): 183–91. doi.org/10.1016/j.geoderma.2013.03.003.

Agyarko-Mintah, E., Cowie, A., Singh, B. P., Joseph, S., Van Zwieten, L., Cowie, A., Harden, S., and Smillie, R. 2016. Biochar Increases Nitrogen Retention and Lowers Greenhouse Gas Emissions When Added to Composting Poultry Litter. *Waste Management* 61: 138–149.

Alburquerque, J. A., Sánchez-Monedero, M. A., Roig, A., Cayuela, M. L., 2015. High Concentrations of Polycyclic Aromatic Hydrocarbons (Naphthalene, Phenanthrene And Pyrene) Failed to Explain Biochar's Capacity to Reduce Soil Nitrous Oxide Emissions. *Environmental Pollution* 196: 72–77.

Archontoulis, Sotirios V., Huber, I., Miguez, F. E., Thorburn, P. J., Rogovska, N., and Laird, D. 2015. A Model for Mechanistic and System Assessments of Biochar Effects on Soils and Crops and Trade-Offs. *GCB Bioenergy.* http://onlinelibrary.wiley.com/doi/10.1111/gcbb.12314/pdf.

Atkinson, Christopher J., Fitzgerald, Jean D., and Hipps, Neil A. 2010. Potential Mechanisms for Achieving Agricultural Benefits from Biochar Application to Temperate Soils: A Review. *Plant and Soil* 337 (1–2): 1–18.

Australian Government. 2014. Carbon farming initiative. Soil Sampling Design—Method and Guidelines. Australian Government Department of the Environment. http://www.environment.gov.au/system/files /pages/b341ae7a-5ddf-4725-a3fe-1b17ead2fa8a/files/cfi-soil-sampling-design-method-and-guidelines.pdf

Bingeman, C. W, Varner, J. E., and Martin, W. P. 1953. The Effect of the Addition of Organic Materials on the Decomposition of an Organic Soil. *Soil Science Society of America Journal* 17 (1): 34–38.

Borchard, N., Spokas, K., Prost, K., Siemens, J., 2014. Greenhouse Gas Production in Mixtures of Soil with Composted and Noncomposted Biochars Is Governed by Char-Associated Organic Compounds. *Journal of Environmental Quality* 43: 971–979.

Bozzi, E., Genesio, L., Toscano, P., Pieri, M., and Miglietta, F.. 2015. Mimicking Biochar-Albedo Feedback in Complex Mediterranean Agricultural Landscapes. *Environmental Research Letters* 10 (8): 084014.

Brandão, M., Clift, R., Cowie, A., and Greenhalgh, S. 2014. The Use of Life Cycle Assessment in the Support of Robust (Climate) Policy Making: Comment on "Using Attributional Life Cycle Assessment to Estimate Climate-Change Mitigation." *Journal of Industrial Ecology* 18: 461–463.

Budai, A., Zimmerman, A., Cowie, A., Webber, J., Singh, B., Glaser, B., Masiello, C., Andersson, D., Shields, F., and Lehmann, J. 2013. Biochar Carbon Stability Test Method: An Assessment of Methods to Determine Biochar Carbon Stability. *Carbon Methodology, IBI.* http://www.biochar-international.org /sites/default/files/IBI_Report_Biochar_Stability_Test_Method_Final.pdf.

Cayuela, M. L., van Zwieten, L., Singh, B. P., Jeffery, S., Roig, A., and Sánchez-Monedero, M. A. 2014. Biochar's Role in Mitigating Soil Nitrous Oxide Emissions: A Review and Meta-Analysis. *Agriculture, Ecosystems & Environment* 191: 5–16.

Cayuela, M., Jeffery, S., and Van Zwieten, L. 2015. The Molar H: Corg Ratio of Biochar Is a Key Factor in Mitigating N_2O Emissions from Soil. *Agriculture, Ecosystems & Environment* 202: 135–138.

Cayuela, M. L., Sánchez-Monedero, M. A., Roig, A., Hanley, K., Enders, A., and Lehmann, J. 2013. Biochar and Denitrification in Soils: When, How Much and Why Does Biochar Reduce N$_2$O Emissions? *Scientific Reports* 3: 1732.

Chen, S., Rotaru, A.-E., Shrestha, P. M., Malvankar, N. S., Liu, F., Fan, W., Nevin, K. P., and Lovley, D. R. 2014. Promoting Interspecies Electron Transfer with Biochar. *Scientific Reports* 4: 5019.

Coleman, K., and Jenkinson, D. S. 2008. *ROTHC-26.3*. Rothamsted Research, Harpenden, Herts, UK.

Conrad, R., 1996. Soil Microorganisms as Controllers of Atmospheric Trace Gases (H2, CO, CH4, OCS, N2O, and NO). *Microbiological Reviews* 60: 609–640.

Cotrufo, M. Francesca, Wallenstein, Matthew D., Boot, Claudia M., Denef, Karolien, and Paul, Eldor. 2013. The Microbial Efficiency-Matrix Stabilization (MEMS) Framework Integrates Plant Litter Decomposition with Soil Organic Matter Stabilization: Do Labile Plant Inputs Form Stable Soil Organic Matter? *Global Change Biology* 19 (4): 988–995.

Cowie, A., Woolf, D., Gaunt, J., Brandão, M., Anaya de la Rosa, R., Cowie, A., Lehmann, J., and Joseph, S. 2015. Biochar, Carbon Accounting and Climate Change. In *Biochar for Environmental Management: Science, Technology and Implementation*, 763–794. Lehmann and Joseph (Eds), Routledge (Oxon, UK): 763–794.

Cowie, A., Eckard, R., and Eady, S. 2012. Greenhouse Gas Accounting for Inventory, Emissions Trading and Life Cycle Assessment in the Land-Based Sector: A Review. *Crop and Pasture Science* 63: 284–296.

Crane-Droesch, Andrew, Abiven, Samuel, Jeffery, Simon, and Torn, Margaret S. 2013. Heterogeneous Global Crop Yield Response to Biochar: A Meta-Regression Analysis. *Environmental Research Letters* 8 (4): 044049.

Cui, Jun, Ge, Tida, Kuzyakov, Yakov, Nie, Ming, Fang, Changming, Tang, Boping, and Zhou, Chunlin. 2017. Interactions between Biochar and Litter Priming: A Three-Source 14 C and δ 13 C Partitioning Study. *Soil Biology and Biochemistry* 104: 49–58.

Davidson, E. A., and Kanter, D. 2014. Inventories and Scenarios of Nitrous Oxide Emissions. *Environmental Research Letters* 9: 105012.

Del Grosso, S. J., Parton, W. J., Mosier, A. R., Walsh, M. K., Ojima, D. S., and Thornton, P. E. 2006. DAYCENT National-Scale Simulations of Nitrous Oxide Emissions from Cropped Soils in the United States. *Journal of Environmental Quality* 35 (4): 1451–1460. doi.org/10.2134/jeq2005.0160.

Enders, Akio, Hanley, Kelly, Whitman, Thea, Joseph, Stephen, and Lehmann, Johannes. 2012. Characterization of Biochars to Evaluate Recalcitrance and Agronomic Performance. *Bioresource Technology* 114: 644–653.

Erb, K. H. 2007. A Comprehensive Global 5 Min Resolution Land-Use Data Set for the Year 2000 Consistent with National Census Data. *Journal of Land Use Science* 2 (3): 191–224.

Felber, R., Leifeld, J., Horák, J., and Neftel, A. 2014. Nitrous Oxide Emission Reduction with Greenwaste Biochar: Comparison of Laboratory and Field Experiments. *European Journal of Soil Science* 65: 128–138.

Field, John L., Keske, Catherine M. H., Birch, Greta L., DeFoort, Morgan W., and Cotrufo, M. Francesca. 2013. Distributed Biochar and Bioenergy Coproduction: A Regionally Specific Case Study of Environmental Benefits and Economic Impacts. *GCB Bioenergy* 5 (2): 177–191.

Finkbeiner, Matthias. 2014. Indirect Land Use Change—Help beyond the Hype? *Biomass and Bioenergy* 62 (March): 218–221.

Flugge, M., Lewandrowski, J., Rosenfeld, J., Boland, C., Hendrickson, T., Jaglo, K., Kolansky, S., Moffroid, K., Riley-Gilbert, M., and Pape, D. 2017. A Life-Cycle Analysis of the Greenhouse Gas Emissions of Corn-Based Ethanol. Report prepared by ICF under USDA Contract No. AG-3142-D-16-0243. January 30, 2017.

Fowles, M. (2007). Black Carbon Sequestration as an Alternative to Bioenergy. *Biomass and Bioenergy*, 31 (6): 426–432.

Gaunt, J. L., and Lehmann, J. 2008. Energy Balance and Emissions Associated with Biochar Sequestration and Pyrolysis Bioenergy Production. *Environmental Science & Technology* 42 (11): 4152–4158.

Genesio, L., Miglietta, F., Lugato, E., Baronti, S., Pieri, M., and Vaccari, F. P. 2012. Surface Albedo Following Biochar Application in Durum Wheat. *Environmental Research Letters* 7 (1): 014025.

Genesio, Lorenzo, Vaccari, Francesco Primo, and Miglietta, Franco. 2016. Black Carbon Aerosol from Biochar Threats Its Negative Emission Potential. *Global Change Biology* 22 (7): 2313–2314.

Ghosh, S., Fern Ow, L., and Wilson, B. 2015. Influence of Biochar and Compost on Soil Properties and Tree Growth in a Tropical Urban Environment. *International Journal of Environmental Science and Technology* 12 (4): 1303–1310.

Glaser, B., Haumaier, L., Guggenberger, G., and Zech, W. 2001. The 'Terra Preta' Phenomenon: A Model for Sustainable Agriculture in the Humid Tropics. *Naturwissenschaften* 88 (1): 37–41.

Glaser, B., Zech, W., and Woods, W. I. 2004. History, Current Knowledge and Future Perspectives of Geoecological Research Concerning the Origin Of Amazonian Anthropogenic Dark Earths (Terra Preta). In *Amazonian Dark Earths: Explorations in Space and Time* (pp. 9–17). Springer, Berlin, Heidelberg.

Hagemann, N., Harter, J., Kaldamukova, R., Guzman-Bustamante, I., Ruser, R., Graeff, S., Kappler, A., and Behrens, S. 2017a. Does Soil Aging Affect the N2O Mitigation Potential of Biochar? A Combined Microcosm and Field Study. *GCB Bioenergy* 9 (5): 953–964.

Hagemann, N., Kammann, C. I., Schmidt, H.-P., Kappler, A., Behrens, S. 2017b. Nitrate Capture and Slow Release in Biochar Amended Compost and Soil. *PLOS ONE* 12: e0171214.

Harter, J., Krause, H.-M., Schuettler, S., Ruser, R., Fromme, M., Scholten, T., Kappler, A., and Behrens, S., 2014. Linking N2O Emissions from Biochar-Amended Soil to the Structure and Function of the N-Cycling Microbial Community. *ISME Journal* 8: 660–674.

Harter, J., Weigold, P., El-Hadidi, M., Huson, D.H., Kappler, A., and S., 2016. Soil Biochar Amendment Shapes the Composition of N2O-Reducing Microbial Communities. *Science of The Total Environment* 562: 379–390.

Hertel, T., Rose, S., and Tol, R. 2008. Land Use in Computable General Equilibrium Models: An Overview. GTAP Technical Paper. https://www.gtap.agecon.purdue.edu/resources/download/3659.pdf.

Hodge, A., Robinson, D., and Fitter, A, 2000. Are Microorganisms More Effective Than Plants at Competing for Nitrogen? *Trends in Plant Science* 5–308.

Hollister, C. C., Bisogni, J. J., and Lehmann, J. 2013. Ammonium, Nitrate, and Phosphate Sorption to and Solute Leaching from Biochars Prepared from Corn Stover (L.) and Oak Wood (spp.). *Journal of Environmental Quality* 42 (1): 137–144.

Houghton, J. T., Meira Filho, L. G., Lim, B., Treanton, K., Mamaty, I. et al. 1997. Revised 1996 IPCC Guidelines for National Greenhouse Gas Inventories. v. 1: Greenhouse Gas Inventory Reporting Instructions.—v. 2: Greenhouse Gas Inventory Workbook.—v. 3: Greenhouse Gas Inventory Reference Manual. http://agris.fao.org/agris-search/search.do?recordID=XF2015041434.

Hu, H.-W., Chen, D., and He, J.-Z. 2015. Microbial Regulation of Terrestrial Nitrous Oxide Formation: Understanding the Biological Pathways for Prediction of Emission Rates. *FEMS Microbiology Reviews* 39: 729–749.

Hüppi, R., Felber, R., Neftel, A., Six, J., and Leifeld, J. 2015. Effect of Biochar and Liming on Soil Nitrous Oxide Emissions from a Temperate Maize Cropping System. *SOIL* 1: 707–717.

ISO (2006a). ISO 14040:2006 Environmental Management-Life Cycle Assessment-Principles and Framework. International Organisation for Standardization, Geneva, Switzerland.

ISO (2006b). ISO 14044:2006 Environmental Management—Life Cycle Assessment—Principles and Framework. Vol. ISO 14040:2006(E), pp. 28. International Organization for Standardization, Geneva, Switzerland, Switzerland.

ISO (2013). ISO/TS 14067:2013 Greenhouse Gases—Carbon Footprint of Products—Requirements and Guidelines for Quantification and Communication. International Organization for Standardization, Geneva, Switzerland.

ISO (2015). ISO 13065:2015 Sustainability Criteria for Bioenergy. International Organization for Standardization, Geneva, Switzerland.

Jeffery, S., Abalos, D., Spokas, K. A., and Verheijen, F. G. 2015. Biochar Effects on Crop Yield. In *Biochar for Environmental Management: Science, Technology and Implementation*, Vol. 2. Lehmann and Joseph (Eds), Routledge (Oxon, UK).

Jeffery, S., Verheijen, F. G., Van Der Velde, M., and Bastos, A. C. 2011. A Quantitative Review of the Effects of Biochar Application to Soils on Crop Productivity Using Meta-Analysis. *Agriculture, Ecosystems & Environment* 144: 175–187.

Jeffery, S., Verheijen, F. G. A., Kammann, C., and Abalos, D. 2016. Biochar Effects on Methane Emissions from Soils: A Meta-Analysis. *Soil Biology and Biochemistry* 101: 251–258.

Jeffery, S., Verheijen, F. G. A., Bastos, A. C., and Van Der Velde, M. 2014. A Comment on "Biochar and Its Effects on Plant Productivity and Nutrient Cycling: A Meta-Analysis": On the Importance of Accurate Reporting in Supporting a Fast-Moving Research Field with Policy Implications. *GCB Bioenergy* 6 (3): 176–179.

Kappler, A., Wuestner, M. L., Ruecker, A., Harter, J., Halama, M., and Behrens, S. 2014. Biochar as an Electron Shuttle between Bacteria and Fe(III) Minerals. *Environmental Science & Technology* Letters 1: 339–344.

Kasozi, G. N., Zimmerman, A. R., Nkedi-Kizza, P., and Gao, B. 2010. Catechol and Humic Acid Sorption onto a Range of Laboratory-Produced Black Carbons (Biochars). *Environmental Science & Technology* 44: 6189–6195.

Keith, A., Singh, B., and Dijkstra, F. A. 2015. Biochar Reduces the Rhizosphere Priming Effect on Soil Organic Carbon. *Soil Biology and Biochemistry* 88: 372–379.

Kerré, B., Willaert, B., and Smolders, E. 2017. Lower Residue Decomposition in Historically Charcoal-Enriched Soils Is Related to Increased Adsorption of Organic Matter. *Soil Biology and Biochemistry* 104: 1–7.

Koper, T., Weisberg, P., Lennie, A., Driver, K., Simons, H., Rodriguez, M., Reed, D., Jirka, S., and Gaunt, J. 2013. Methodology for Biochar Projects, version 1.0. Public Comment Draft submitted to the American Carbon Registry (ACR). http://americancarbonregistry.org/carbon-accounting/methodology -for-biochar-projects

Kuzyakov, Y., and Xu, X. 2013. Competition between Roots and Microorganisms for Nitrogen: Mechanisms and Ecological Relevance. *New Phytologist* 198: 656–669.

Lehmann, J., Rillig, M. C., Thies, J., Masiello, C. A., Hockaday, W. C., and Crowley, D. 2011. Biochar Effects on Soil Biota—A Review. *Soil Biology and Biochemistry* 43: 1812–1836.

Lehmann, J., and Rillig, M. 2014. Distinguishing Variability from Uncertainty. *Nature Climate Change* 4 (3): 153–153.

Lehmann, J., Amonette, J. E., and Roberts, K. 2010. Role of Biochar in Mitigation of Climate Change." *Handbook of Climate Change and Agroecosystems: Impacts, Adaptation, and Mitigation* (pp. 343–363). Imperial College Press, London.

Lehmann, J., Gaunt, J., and Rondon, M. 2006. Bio-Char Sequestration in Terrestrial Ecosystems – A Review. *Mitigation and Adaptation Strategies for Global Change* 11 (2): 395–419.

Lehmann, J., Abiven, S., Kleber, M., Pan, G., Singh, B. P., Sohi, S. P., Zimmerman, A. R., Lehmann, J., and Joseph, S. 2015. Persistence of Biochar in Soil. *Biochar for Environmental Management: Science, Technology and Implementation*, 233–280.

Lehmann, J. 2007. A Handful of Carbon. *Nature* 447 (7141): 143–144.

Lenton, T. M., and Vaughan, N. E. 2009. The Radiative Forcing Potential of Different Climate Geoengineering Options. *Atmospheric Chemistry and Physics Discussions* 9: 2559–2608.

Liang, B., Lehmann, J., Solomon, D., Kinyangi, J., Grossman, J., O'Neill, B., Skjemstad, J. O. et al. 2006. Black Carbon Increases Cation Exchange Capacity in Soils. *Soil Science Society of America Journal* 70 (5): 1719–1730.

Liu, B., Mørkved, P. T., Frostegård, Å., and Bakken, L. R. 2010. Denitrification Gene Pools, Transcription and Kinetics of NO, N_2O and N_2 Production as Affected by Soil pH. *FEMS Microbiological Ecology* 72: 407–417.

Luo, Y., Durenkamp, M., De Nobili, M., Lin, Q., and Brookes P. C. 2011. Short Term Soil Priming Effects and the Mineralisation of Biochar Following Its Incorporation to Soils of Different pH. *Soil Biology and Biochemistry* 43: 2304–2314.

MacCarty, N., Ogle, D., Still, D., Bond, T., and Roden, C. 2008. A Laboratory Comparison of the Global Warming Impact of Five Major Types of Biomass Cooking Stoves. *Energy for Sustainable Development* 12: 56–65.

Macdonald, L. M., Farrell, M., Van Zwieten, L., and Krull, E. S. 2014. Plant Growth Responses to Biochar Addition: An Australian Soils Perspective. *Biology and Fertility of Soils* 50: 1035–1045.

Maestrini, B., Nannipieri, P., and Abiven, S. 2015. A Meta-Analysis on Pyrogenic Organic Matter Induced Priming Effect. *GCB Bioenergy* 7 (4): 577–590.

Masiello, C., Dugan, B., Brewer, C., Spokas, K., Novak, J. M., Liu, Z., and Sorrenti, G. 2015. Biochar Effects on Soil Hydrology. *Biochar for Environmental Management*, 2nd ed. Earthscan, Routledge.

McCarl, B., Peacocke, C., Chrisman, R., Chih-Chun, K., and Sands, R. 2009. Economics of Biochar Production, Utilisation and Emissions. In *Biochar for Environmental Management: Science and Technology*, Lehmann, J. & Joseph, S. (Eds). Earthscan.

Meyer, S., Bright, R. M., Fischer, D., Schulz, H., and Glaser, B. 2012. Albedo Impact on the Suitability of Biochar Systems To Mitigate Global Warming. *Environmental Science & Technology* 46 (22): 12726–12734.

Mørkved, P.T., Dörsch, P., and Bakken, L.R. 2007. The N2O Product Ratio of Nitrification and Its Dependence on Long-Term Changes in Soil pH. *Soil Biology and Biochemistry* 39: 2048–2057.

Mukherjee, A., and Lal, R. 2014. The Biochar Dilemma. *Soil Research* 52 (3): 217–230.

Muñoz, I., Schmidt, J. H., Brandão, M., and Weidema, B. P. 2015. Rebuttal to "Indirect Land Use Change (iLUC) within Life Cycle Assessment (LCA)—Scientific Robustness and Consistency with International Standards." *GCB Bioenergy* 7 (4): 565–566.

Nelissen, V., Saha, B. K., Ruysschaert, G., and Boeckx, P. 2014. Effect of Different Biochar and Fertilizer Types on N2O and NO Emissions. *Soil Biology Biochemistry* 70: 244–255.

Obia, A., Cornelissen, G., Mulder, J., and Dörsch, P. 2015. Effect of Soil pH Increase by Biochar on NO, N_2O and N_2 Production during Denitrification in Acid Soils. *PLoS ONE* 10, e0138781.

Oguntunde, P. G., Abiodun, B. J., Ajayi, A. E., and van de Giesen, N. 2008. Effects of Charcoal Production on Soil Physical Properties in Ghana. *Journal of Plant Nutrition and Soil Science* 171 (4): 591–596. doi.org/10.1002/jpln.200625185.

Paustian, K., Lehmann, J., Ogle, S., Reay, D., Robertson, G.P., and Smith, P. 2016. Climate-Smart Soils. *Nature* 532: 49–57.

Phillips, R. L., Song, B., McMillan, A. M. S., Grelet, G., Weir, B. S., Palmada, T., and Tobias, C., 2016. Chemical Formation of Hybrid Di-Nitrogen Calls Fungal Codenitrification into Question. *Scientific Reports* 6: 39077.

Placer County. 2015. Biochar Production Project Reporting Protocol: GHG Emission Reduction Accounting. Version 3.4. September 10. https://www.placer.ca.gov/~/media/apc/documents/apcd%.

Prévoteau, A., Ronsse, F., Cid, I., Boeckx, P., and Rabaey, K., 2016. The Electron Donating Capacity of Biochar Is Dramatically Underestimated. *Scientific Reports* 6: 32870.

Quin, P., Joseph, S., Husson, O., Donne, S., Mitchell, D., Munroe, P., Phelan, D., Cowie, A., and Van Zwieten, L., 2015. Lowering N2O Emissions from Soils Using Eucalypt Biochar: The Importance of Redox Reactions. *Scientific Reports* 5: 16773.

Richardson, D., Felgate, H., Watmough, N., Thomson, A., and Baggs, E., 2009. Mitigating Release of the Potent Greenhouse Gas N2O from the Nitrogen Cycle—Could Enzymic Regulation Hold The Key? *Trends in Biotechnology* 27: 388–397.

Roberts, K. G., Gloy, B. A., Joseph, S., Scott, N. R., and Lehmann, J. 2010. Life Cycle Assessment of Biochar Systems: Estimating the Energetic, Economic, and Climate Change Potential. *Environmental Science & Technology* 44 (2): 827–833. doi:10.1021/es902266r.

Sagrilo, E., Jeffery, S., Hoffland, E., and Kuyper, T. W. 2015. Emission of CO2 from Biochar-Amended Soils and Implications for Soil Organic Carbon. *GCB Bioenergy* 7 (6): 1294–1304.

Sánchez-García, M., Roig, A., Sanchez-Monedero, M. A., Cayuela, M. L., 2014. Biochar Increases Soil N2O Emissions Produced by Nitrification-Mediated Pathways. *Frontiers in Environmental Science* 2: 25.

Sanford, R. A., Wagner, D. D., Wu, Q., Chee-Sanford, J. C., Thomas, S. H., Cruz-García, C., Rodríguez, G., Massol-Deyá, A., Krishnani, K. K., Ritalahti, K. M., Nissen, S., Konstantinidis, K. T., and Löffler, F.E. 2012. Unexpected Nondenitrifier Nitrous Oxide Reductase Gene Diversity and Abundance sn Soils. *Proceedings of the National Academy of Science* 109: 19709–19714.

Saquing, J. M., Yu, Y.-H., and Chiu, P. C., 2016. Wood-Derived Black Carbon (Biochar) as a Microbial Electron Donor and Acceptor. *Environmental Science & Technology Letters* 3: 62–66.

Scharenbroch, B. C., Meza, E. N., Catania, M., and Fite, K. 2013. Biochar and Biosolids Increase Tree Growth and Improve Soil Quality for Urban Landscapes." *Journal of Environmental Quality* 42 (5): 1372–1385. doi.org10.2134/jeq2013.04.0124.

Schmidt, J. H., Weidema, B. P., and Brandão, M. (2015). A Framework for Modelling Indirect Land Use Changes in Life Cycle Assessment. *Journal of Cleaner Production* 99: 230–238.

Searchinger, T., Heimlich, R., Houghton, R. A., Dong, F., Elobeid, A., Fabiosa, J., Tokgoz, S., Hayes, D., and Yu, T-H. 2008. Use of US Croplands for Biofuels Increases Greenhouse Gases through Emissions from Land Use Change. *Science* 319 (5867): 1238–1240.

Seifritz, W. 1993. Should We Store Carbon in Charcoal? *International Journal of Hydrogen Energy* 18 (5): 405–407.

Serbin, G., Daughtry, C. S. T., Hunt, E. R., Reeves, J. B., and. Brown, D. J. 2009. Effects of Soil Composition and Mineralogy on Remote Sensing of Crop Residue Cover. *Remote Sensing of Environment* 113 (1): 224–238.

Singh, B. P., and Cowie, A. L. 2014. Long-Term Influence of Biochar on Native Organic Carbon Mineralisation in a Low-Carbon Clayey Soil. *Scientific Reports*, 4. https://doi.org/10.1038/srep03687.

Six, J., and Paustian, K. 2014. Aggregate-Associated Soil Organic Matter as an Ecosystem Property and a Measurement Tool. *Soil Biology and Biochemistry* 68: A4–A9.

Smith, P., Martino, D., Cai, Z., Gwary, D., Janzen, H., Kumar, P., McCarl, B., Ogle, S., O'Mara, F., Rice, C., Scholes, B., Sirotenko, O., Howden, M., McAllister, T., Pan, G., Romanenkov, V., Schneider, U., Towprayoon, S., Wattenbach, M., and Smith, J. 2008. Greenhouse Gas Mitigation in Agriculture. *Philosophical Transactions of the Royal Society of London B: Biological Sciences* 36: 789–813.

Smith, P. 2016. Soil Carbon Sequestration and Biochar as Negative Emission Technologies. *Global Change Biology*, January, 22(3): 1315–1324. doi.org/10.1111/gcb.13178.

Sohi, S. P., Krull, E., Lopez-Capel, E., and Bol, R. 2010. A Review of Biochar and Its Use and Function in Soil." In *Advances in Agronomy*, Vol. 105. D. L. Sparks (Ed.) Academic Press (Cambridge, MA). (pp. 47–82).

Sombroek, W. G., Nachtergaele, F. O., and Hebel, A. 1993. Amounts, Dynamics and Sequestering of Carbon in Tropical and Subtropical Soils. *Ambio*. 22 (7): 417–426.

Sparrevik, M., Adam, C., Martinsen, V., and Cornelissen, G. 2015. Emissions of Gases and Particles from Charcoal/Biochar Production in Rural Areas Using Medium-Sized Traditional and Improved "Retort" Kilns. *Biomass and Bioenergy* 72: 65–73.

Spokas, K. 2010. Review of the Stability of Biochar in Soils: Predictability of O:C Molar Ratios. *Carbon Management* 1 (2): 289–303. doi.org/10.1016/j.carbpol.2010.10.007.

Spokas, K. A., 2013. Impact of Biochar Field Aging on Laboratory Greenhouse Gas Production Potentials. *GCB Bioenergy* 5: 165–176.

Spokas, K. A., and Reikosky, D. C., 2009. Impacts of Sixteen Different Biochars on Soil Greenhouse Gas Production. *Annals of Environmental Science* 3: 179–193.

Stein, L. Y., and Yung, Y. L., 2003. Production, Isotopic Composition and Atmospheric Fate of Biologically Produced Nitrous Oxide. *Annual Review of Earth and Planetary Sciences* 31: 329–356.

Sun, T., Levin, B. D. A., Guzman, J. J. L., Enders, A., Muller, D. A., Angenent, L. T., and Lehmann, J. 2017. Rapid Electron Transfer by the Carbon Matrix in Natural Pyrogenic Carbon. *Nature Communications* 8: 14873.

Thomas, S. C., and Gale, N. 2015. Biochar and Forest Restoration: A Review and Meta-Analysis of Tree Growth Responses. *New Forests* 46 (5–6): 931–946.

Thomazini, A., Spokas, K., Hall, K., Ippolito, J., Lentz, R., and Novak, J. 2015. GHG Impacts of Biochar: Predictability for the Same Biochar. *Agriculture, Ecosystems & Environment* 207: 183–191.

Usowicz, B., Lipiec, J., Łukowski, M., Marczewski, W., and Usowicz, J. 2016. The Effect of Biochar Application on Thermal Properties and Albedo of Loess Soil under Grassland and Fallow. *Soil and Tillage Research* 164: 45–51.

van Groenigen, J. W., Huygens, D., Boeckx, P., Kuyper, T. W., Lubbers, I. M., Rütting, T., and Groffman, P. M., 2015. The Soil N Cycle: New Insights and Key Challenges. *SOIL* 1: 235–256.

Van Zwieten, L., Kammann, C., Cayuela, M. L., Singh, B. P., Joseph, S., Kimber, S., Clough, T., and Spokas, K. 2015. Biochar Effects Emissions of Non-CO 2 GHGs from Soil. In *Biochar for Environmental Management: Science and Technology II*, Lehmann and Joseph (Eds), Routledge (Oxon, UK), (Chapter 17, pp. 489–520).

Van Zwieten, L., Singh, B., Joseph, S., Kimber, S., Cowie, A., and Chan, K., 2009. Biochar and Emissions of Non-CO2 Greenhouse Gases from Soil. In *Biochar for Environmental Management: Science and Technology*, Lehmann and Joseph (Eds), Routledge (Oxon, UK): 227–249.

Venterea, R. T., Halvorson, A. D., Kitchen, N., Liebig, M. A., Cavigelli, M. A., Grosso, S. J. D., Motavalli, P. P., Nelson, K. A., Spokas, K. A., Singh, B. P., Stewart, C. E., Ranaivoson, A., Strock, J., and Collins, H., 2012. Challenges and Opportunities for Mitigating Nitrous Oxide Emissions from Fertilized Cropping Systems. *Frontiers in Ecology and the Environment* 10: 562–570.

Ventura, Maurizio, Alberti, G., Viger, M., Jenkins, J. R., Girardin, C., Baronti, S., Zaldei, A. et al. 2015. Biochar Mineralization and Priming Effect on SOM Decomposition in Two European Short Rotation Coppices. *GCB Bioenergy* 7 (5): 1150–1160.

Verheijen, F. G. A., Jeffery, S., van der Velde, M., Penížek, V., Beland, M., Catarina Bastos, A., and Keizer, J. J. 2013. Reductions in Soil Surface Albedo as a Function of Biochar Application Rate: Implications for Global Radiative Forcing. *Environmental Research Letters* 8 (4): 044008.

Wang, Jinyang, Xiong, Z., and Kuzyakov, Y. 2016. Biochar Stability in Soil: Meta-Analysis of Decomposition and Priming Effects. *GCB Bioenergy* 8 (3): 512–523.

Wang, Z., Zheng, H., Luo, Y., Deng, X., Herbert, S., and Xing, B. 2013. Characterization and Influence of Biochars on Nitrous Oxide Emission from Agricultural Soil. *Environmental Pollution* 174: 289–296.

Wardle, D. A., Nilsson, M.-C., & Zackrisson, O. 2008. Fire-Derived Charcoal Causes Loss of Forest Humus. *Science* 320: 629.

Wells, N. S., and Baggs, E. M. 2014. Char Amendments Impact Soil Nitrous Oxide Production During Ammonia Oxidation. *Soil Science Society of America Journal* 78: 1656–1660.

Weng, Z. H., Van Zwieten, L., Pal Singh, B., Kimber, S., Morris, S., Cowie, A., and Macdonald, L. M. 2015. Plant-Biochar Interactions Drive the Negative Priming of Soil Organic Carbon in an Annual Ryegrass Field System. *Soil Biology and Biochemistry* 90: 111–121.

West, T. O., and Marland, G. 2002. A Synthesis of Carbon Sequestration, Carbon Emissions, and Net Carbon Flux in Agriculture: Comparing Tillage Practices in the United States. *Agriculture, Ecosystems & Environment* 91 (1–3): 217–232.

Whitman, T., Nicholson, C. F., Torres, D., and Lehmann, J. 2011. Climate Change Impact of Biochar Cook Stoves in Western Kenyan Farm Households: System Dynamics Model Analysis. *Environmental Science & Technology* 45: 3687–3694.

Whitman, T., Scholz, S. M., and Lehmann, J. 2010. Biochar Projects for Mitigating Climate Change: An Investigation of Critical Methodology Issues for Carbon Accounting. *Carbon Management* 1 (1): 89–107.

Whitman, T., Enders, A., and Lehmann, J. 2014a. Pyrogenic Carbon Additions to Soil Counteract Positive Priming of Soil Carbon Mineralization by Plants. *Soil Biology and Biochemistry* 73 (June): 33–41.

Whitman, T., and Lehmann, J. 2015a. A Dual-Isotope Approach to Allow Conclusive Partitioning between Three Sources. *Nature Communications* 6: 8708.

Whitman, T., Pal Singh, B., Zimmerman, A. R., Lehmann, J., and Joseph, S. 2015b. Priming Effects in Biochar-Amended Soils: Implications of Biochar-Soil Organic Matter Interactions for Carbon Storage. In *Biochar for Environmental Management: Science, Technology and Implementation*, Lehmann and Joseph (Eds), Routledge (Oxon, UK): 455–488.

Whitman, T., Pepe-Ranney, C., Enders, A., Koechli, C., Campbell, A., Buckley, D. H., and Lehmann, J. 2016. Dynamics of Microbial Community Composition and Soil Organic Carbon Mineralization in Soil Following Addition of Pyrogenic and Fresh Organic Matter. *The ISME Journal* 10 (12): 2918–2930.

Whitman, T., Hanley, K., Enders, A., and Lehmann, J. 2013. Predicting Pyrogenic Organic Matter Mineralization from Its Initial Properties and Implications for Carbon Management. *Organic Geochemistry* 64: 76–83.

Whitman, T., Zhu, Z., and Lehmann, J. 2014b. Carbon Mineralizability Determines Interactive Effects on Mineralization of Pyrogenic Organic Matter and Soil Organic Carbon. *Environmental Science & Technology* 48 (23): 13727–13734.

Wicke, B., Verweij, P., van Meijl, H., van Vuuren, D. P., and Faaij, A. P. C. 2012. Indirect Land Use Change: Review of Existing Models and Strategies for Mitigation. *Biofuels* 3 (1): 87–100. doi.org/10.4155/bfs .11.154.

Wiedemeier, D. B., Abiven, S., Hockaday, W. C., Keiluweit, M., Kleber, M., Masiello, C. A., McBeath, A. V. et al. 2015. Aromaticity and Degree of Aromatic Condensation of Char. *Organic Geochemistry* 78: 135–143.

Wilson, M. F., and Henderson-Sellers, A. 1985. A Global Archive of Land Cover and Soils Data for Use in General Circulation Climate Models. *Journal of Climatology* 5 (2): 119–143.

Woolf, D. 2011. The Potential for Sustainable Biochar Systems to Mitigate Climate Change." Thesis for the Degree of Doctor of Philosophy, Swansea University.

Woolf, D. 2008. Biochar as a Soil Amendment: A Review of the Environmental Implications. *Organic Eprints* 13268.

Woolf, D., Amonette, J. E, Street-Perrott, F. A., Lehmann, J., and Joseph, S. 2010. Sustainable Biochar to Mitigate Global Climate Change. *Nature Communications* 1 (5): 1–9.

Woolf, D., and Lehmann, J. 2012. Modelling the Long-Term Response to Positive and Negative Priming of Soil Organic Carbon by Black Carbon. *Biogeochemistry* 111 (1–3): 83–95.

Woolf, D., Lehmann, J., and Lee, D. R. 2016. Optimal Bioenergy Power Generation for Climate Change Mitigation with or without Carbon Sequestration. *Nature Communications* 7: 13160.

Woolf, D., Lehmann, J., Fisher, E. M., and Angenent, L. T. 2014. Biofuels from Pyrolysis in Perspective: Trade-Offs between Energy Yields and Soil-Carbon Additions. *Environmental Science & Technology* 48 (11): 6492–6499.

Xu, H.-J., Wang, X.-H., Li, H., Yao, H.-Y., Su, J.-Q., and Zhu, Y.-G. 2014. Biochar Impacts Soil Microbial Community Composition and Nitrogen Cycling in an Acidic Soil Planted with Rape. *Environmental Science & Technology* 48 (16): 9391–9399.

Yang, W. H., Weber, K. A., and Silver, W. L. 2012. Nitrogen Loss from Soil through Anaerobic Ammonium Oxidation Coupled to Iron Reduction. *Nature Geoscience* 5: 538–541.

Zhang, Q., Wang, Y., Wu, Y., Wang, X., Du, Z., Liu, X., and Song, J. 2013. Effects of Biochar Amendment on Soil Thermal Conductivity, Reflectance, and Temperature. *Soil Science Society of America Journal* 77 (5): 1478–1487.

Zhang, Y., Hu, X., Zhang, D., Chen, W., and Zou, J. 2015. Effects of Biochar on Soil Surface Albedo, Temperature and Moisture in Agricultural Soil. *Research of Environmental Sciences* 8: 008.

Zheng, H., Wang, Z., Deng, X., Herbert, S., and Xing, B. 2013. Impacts of Adding Biochar on Nitrogen Retention and Bioavailability in Agricultural Soil. *Geoderma* 206: 32–39.

Zhu, X., Burger, M., Doane, T. A., and Horwath, W. R., 2013. Ammonia Oxidation Pathways and Nitrifier Denitrification Are Significant Sources of N_2O and NO under Low Oxygen Availability. *Proceedings of the National Academy of Science USA* 110: 6328–6333.

Zimmerman, A. R., Gao, B., and Ahn, M.-Y. 2011. Positive and Negative Carbon Mineralization Priming Effects among a Variety of Biochar-Amended Soils. *Soil Biology and Biochemistry* 43 (6): 1169–1179.

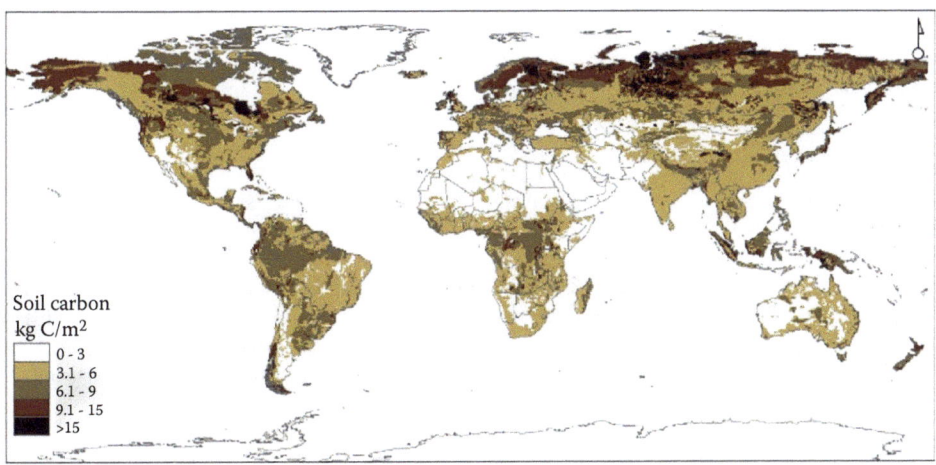

FIGURE 3.1 Global soil organic carbon density in kg Carbon/m² to 1 m depth (©IGBP-DIS (1998) Soil Data (V.0) A program for creating global soil-property databases, IGBP Global Soils Data Task, France), 0.5 degree resolution.

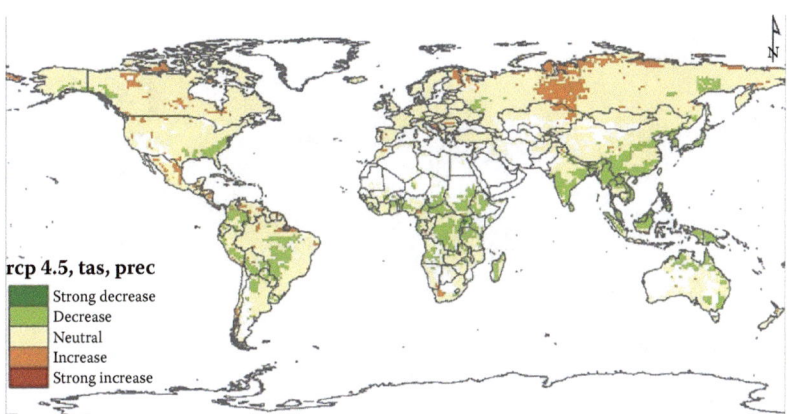

FIGURE 3.2 Hotspots of carbon release under changing climate conditions (based on rcp4.5 scenario for the period 2020 to 2039 and soil carbon stores). Red areas indicate increase of carbon release; green areas indicate decrease of carbon release.

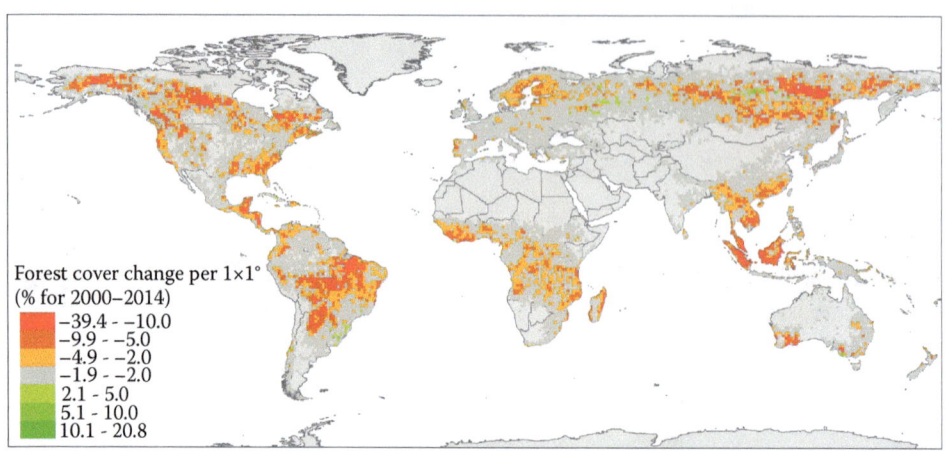

FIGURE 3.3 Forest cover change per 1 × 1° grid cell for the period 2000 to 2014 (in percent with regard to the baseline 2000). Calculations are based on the Global Forest Change maps by Hansen et al. (2013). (From Hansen, M.C., Potapov, P.V., Moore, R., Hancher, M., Turubanova, S.A., Tyukavina, A., Thau, D., Stehman, S.V., Goetz, J., Loveland, T.R., Kommareddy, A., Egorov, A., Chini, L., Justice, C.O., & Townshend, J.R.G., *Science*, 342 (6160), 850–853, 2013.)

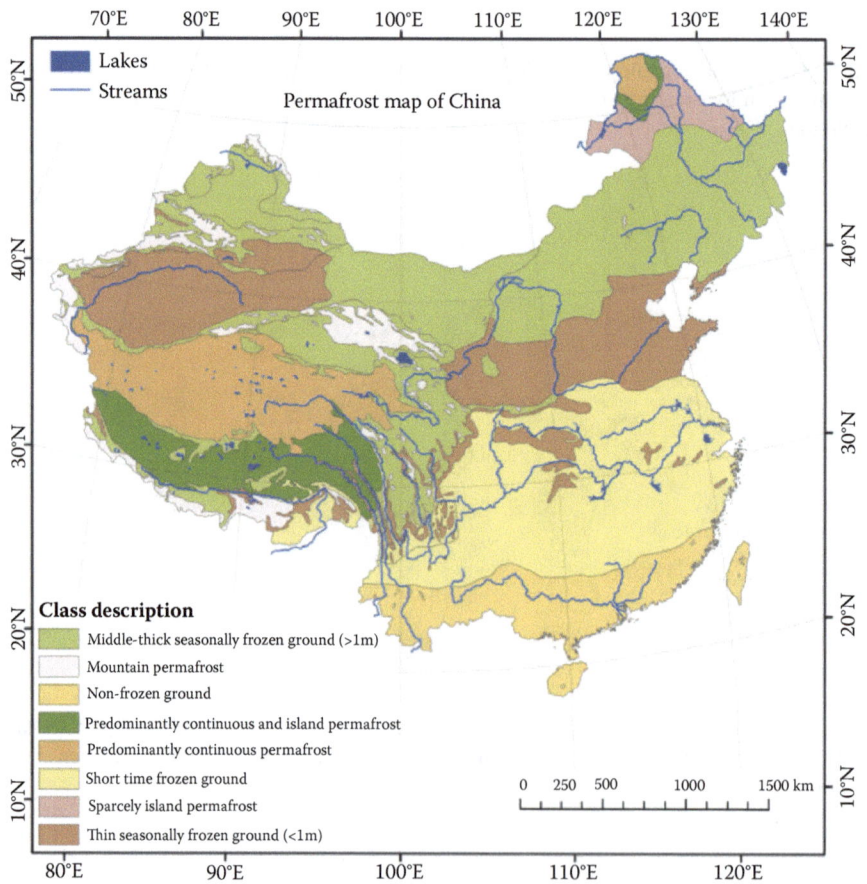

FIGURE 3.10 Permafrost distribution of China.(Modified from Zhou, Y., Guo, D., Qiu, G., Cheng, G., & Li, S., *China Permafrost*, Science Press, Beijing, pp. 145–151, 2005.)

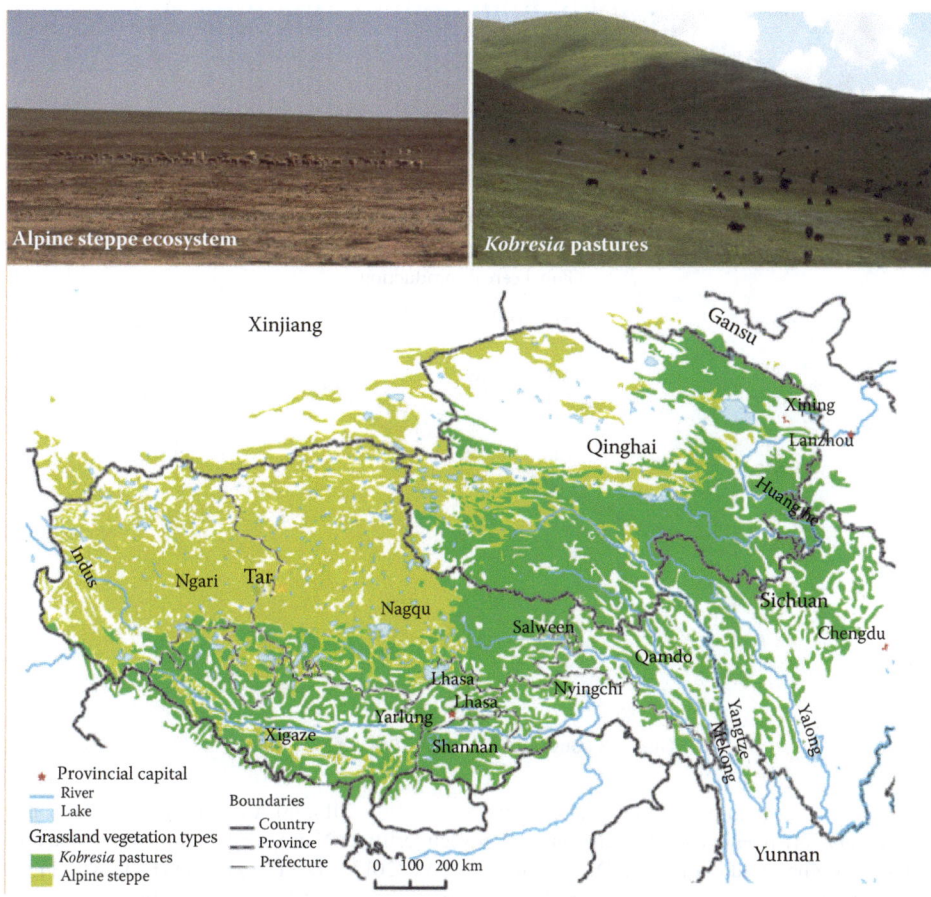

FIGURE 3.11 The distributions of the two predominate ecosystems on the Tibetan Plateau, the alpine steppes in the northwest and the *Kobresia* pastures in the southeastern part of the highlands. (Modified from Miehe, G., Miehe, S., Kaiser, K., Liu, J., & Zhao, X., *Ambio*, 37, 272–279, 2008; cartography done by L. Lehnert and C. Enderle.)

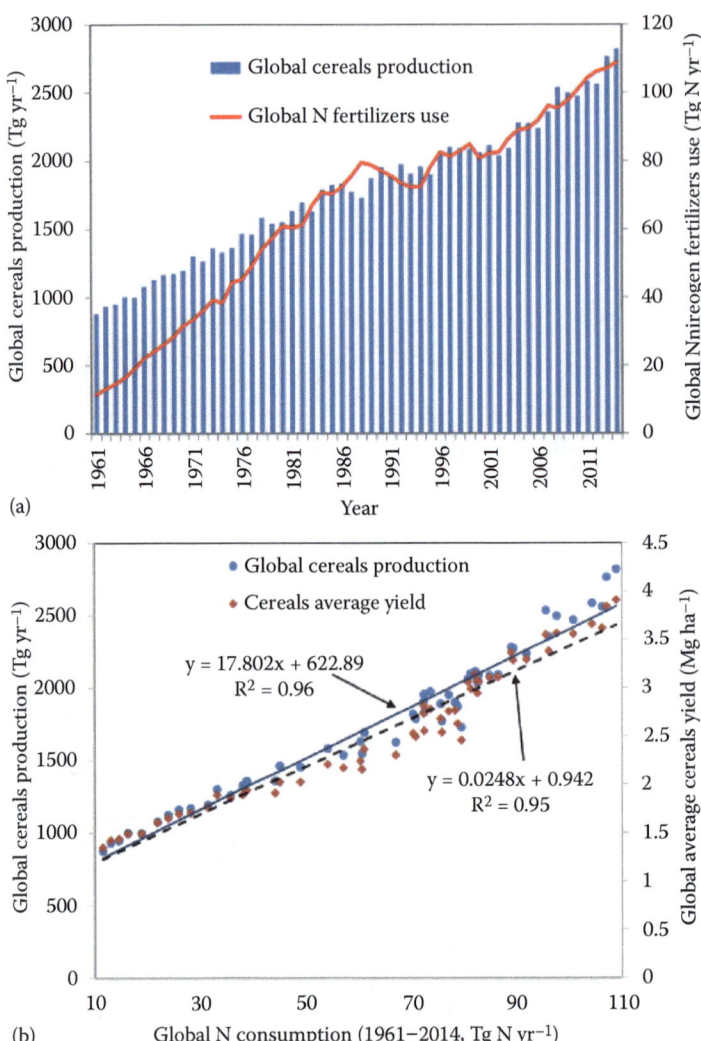

FIGURE 7.1 World cereal production and fertilizer N use, 1961–2014. (a) Annual cereal production and its relation to annual fertilizer N use. (b) Relationship between fertilizer N use, cereal production (blue dots) and cereal yield in Mg ha^{-1} (red diamonds). Data from FAOSTAT (2017). (Data from FAOSTAT, World Food and Agriculture Statistics. Food and Agriculture Organization of the United Nations, Statistical Division, Rome, Italy. Accessed April 2017, 2017.)

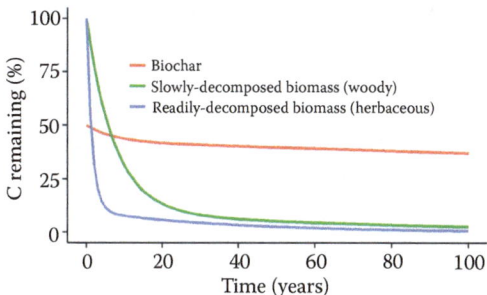

FIGURE 8.2 Conceptual comparison of un-mineralized biomass carbon (C) remaining from different grades of organic matter, as a function of time. The lines are modeled using a two pool exponential decay model, comparing slow-turnover woody biomass (green line), fast-turnover herbaceous biomass (blue line), and biochar (red line). Assumed representative fast and slow fraction half lives, respectively, were 4 and 25 years for woody biomass, 1 and 25 years for herbaceous biomass, and 5 and 500 years for biochar. It was also assumed that half of the initial biomass carbon is lost during biochar production, hence the carbon remaining in biochar starts at 50% at time equal to zero.

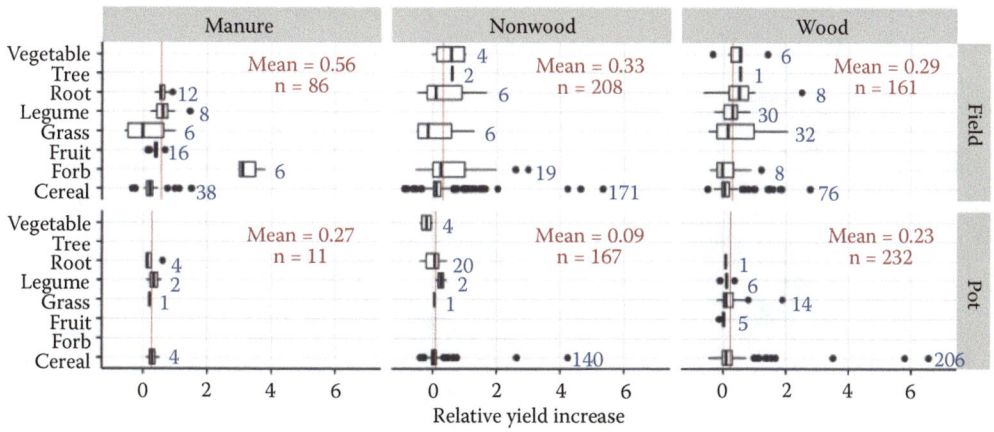

FIGURE 8.4 Yield response of crops to biochar additions (difference between biochar amended yield and control, expressed as a fraction of control yield). Data from 865 treatments from 74 published articles are broken down by field trial versus pot trial, by feedstock type (manure, wood or non-wood), and by crop type. Vertical red line on each panel shows the mean crop response. Mean response and number treatments for each panel are also given in red text. Numbers shown in blue adjacent to each box indicate the number of treatments in the sub-category.

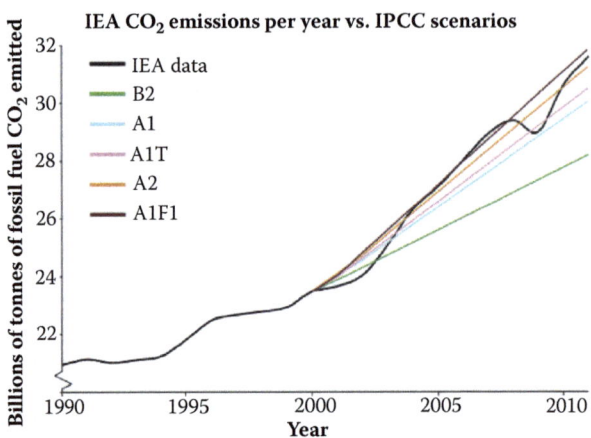

FIGURE 9.1 Observed vs IPCC model-predicted CO_2 emissions (https://static.skepticalscience.com/graphics/IEA_IPCC_2012.jpg with permission).

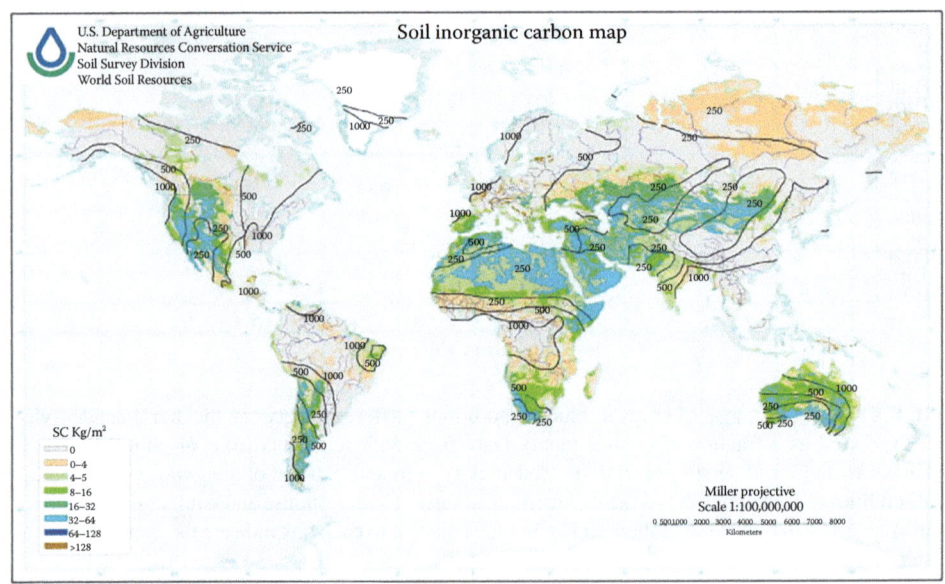

FIGURE 10.1 Worldwide SIC distribution. (From USDA-NRCS, 2000 (worldwide SIC map circa 2000), https://www.nrcs.usda.gov/wps/portal/nrcs/detail/soils/use/?cid=nrcs142p2_054016.)

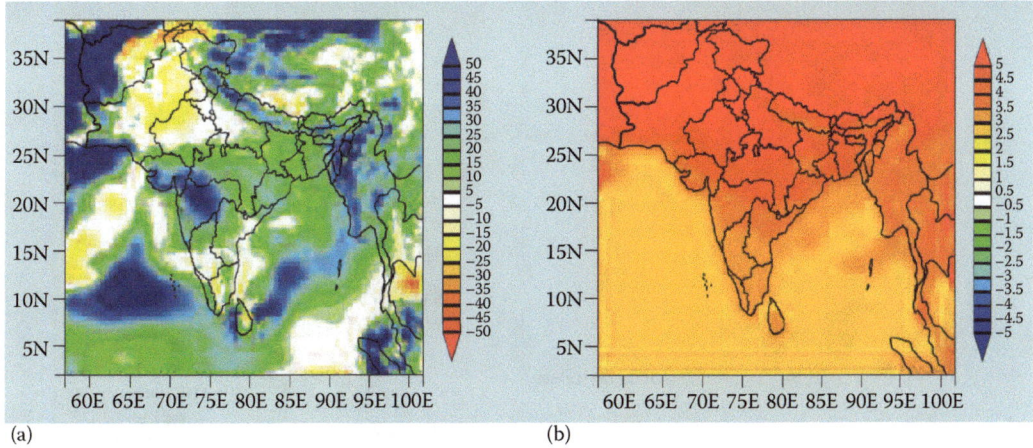

FIGURE 12.1 Spatial patterns of the changes in (a) summer monsoon rainfall (%) and (b) annual mean surface air temperature (°C) for the period 2071–2100 concerning the baseline of 1961–1990, under the A2 scenario; an output of PRECIS model version 2.0.0; www.metoffice.gov.uk/precis.

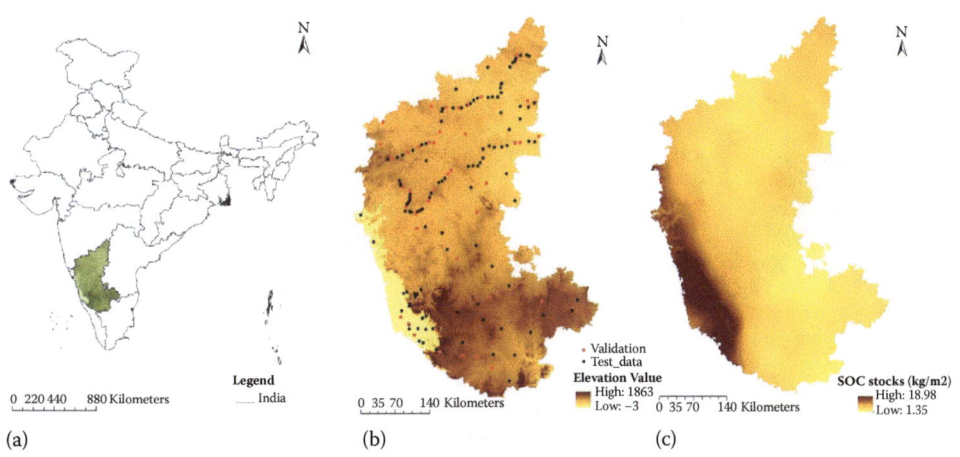

FIGURE 12.3 Location of the study site (a), sampling points (b), and current prediction of SOC using base model (c).

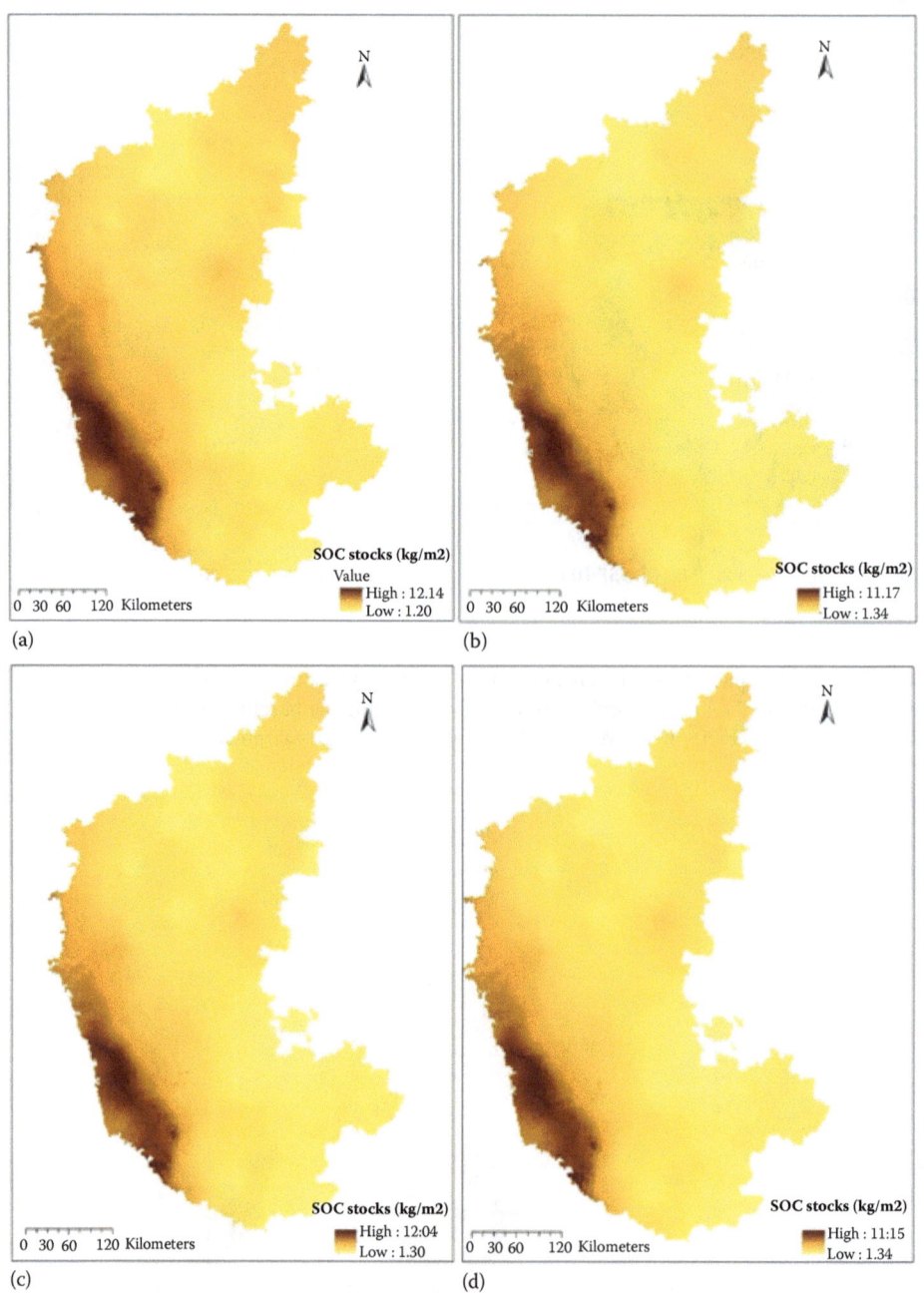

FIGURE 12.5 Future Prediction of SOC stocks under climate scenario CCM4_RCP2.6 (a), CCM4_RCP4.5 (b), CCM4_ RCP6.0 (c), and CCM4_RCP8.5 (d).

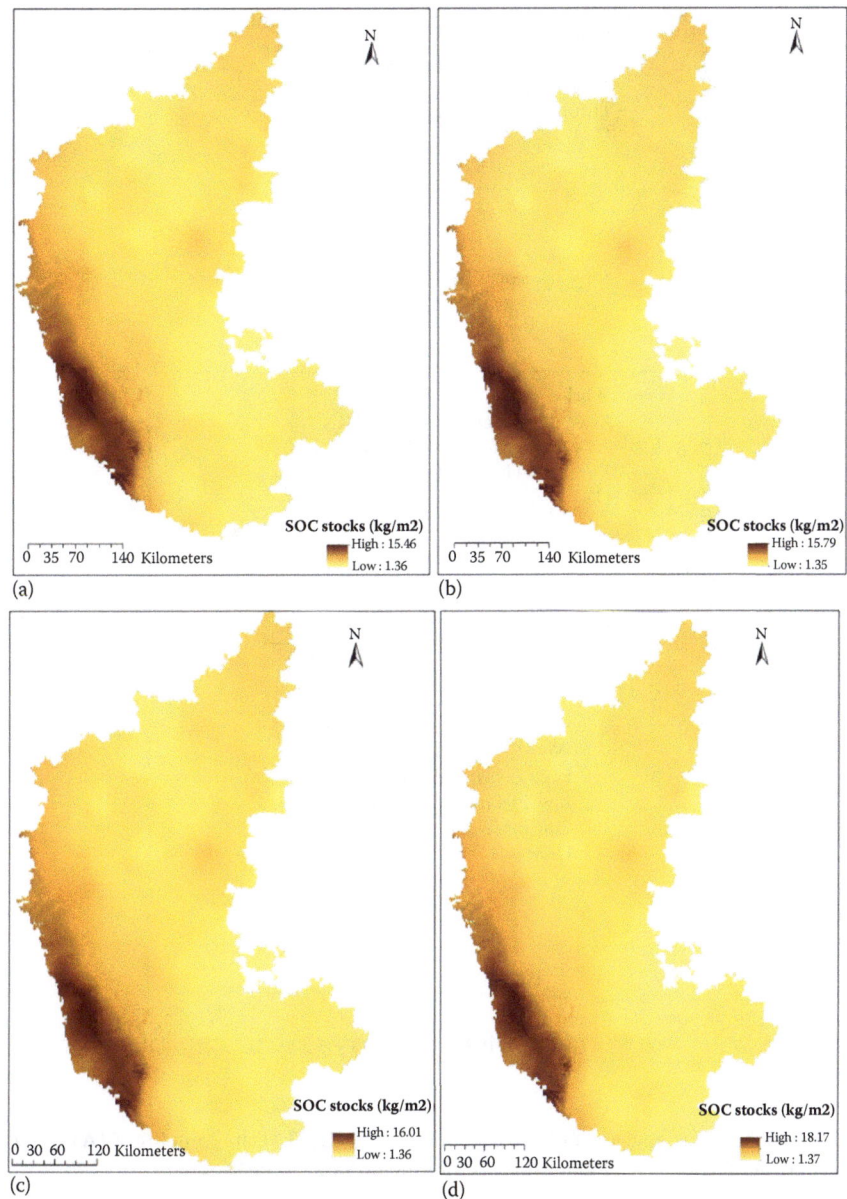

FIGURE 12.6 Future prediction of SOC stocks under climate scenario HadGEM2-AO_RCP2.6 (a), HadGEM2-AO_RCP4.5 (b), HadGEM2-AO_RCP6.0 (c), and HadGEM2-AO_RCP8.5 (d).

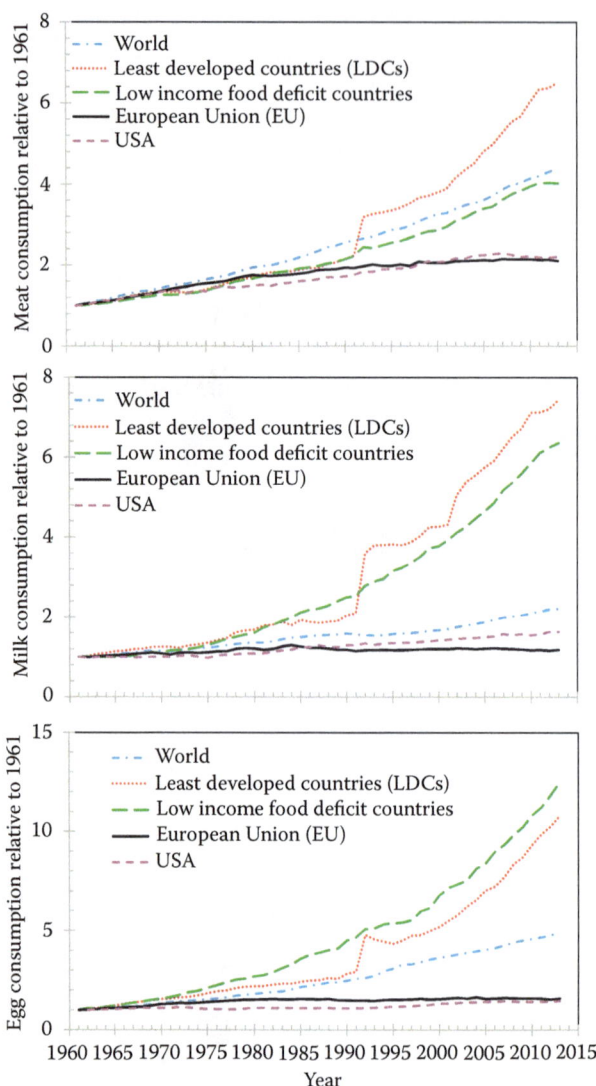

FIGURE 14.1 Global meat, egg, and milk consumption 1961 to 2013. (Data from FAOSTAT, *FAO Statistical Databases.* Food and Agriculture Organization of the United Nations (FAO), 2017.)

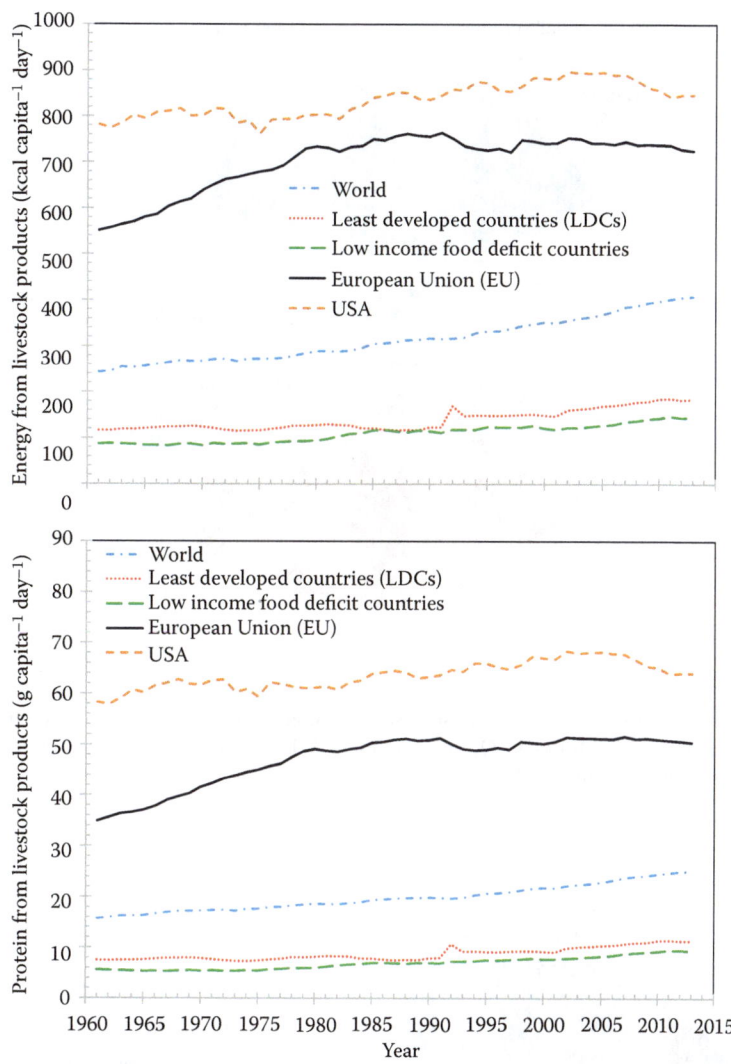

FIGURE 14.2 Global per capita energy and protein supply from meat, eggs, and milk. (Data from FAOSTAT, *FAO Statistical Databases.* Food and Agriculture Organization of the United Nations (FAO), 2017.)

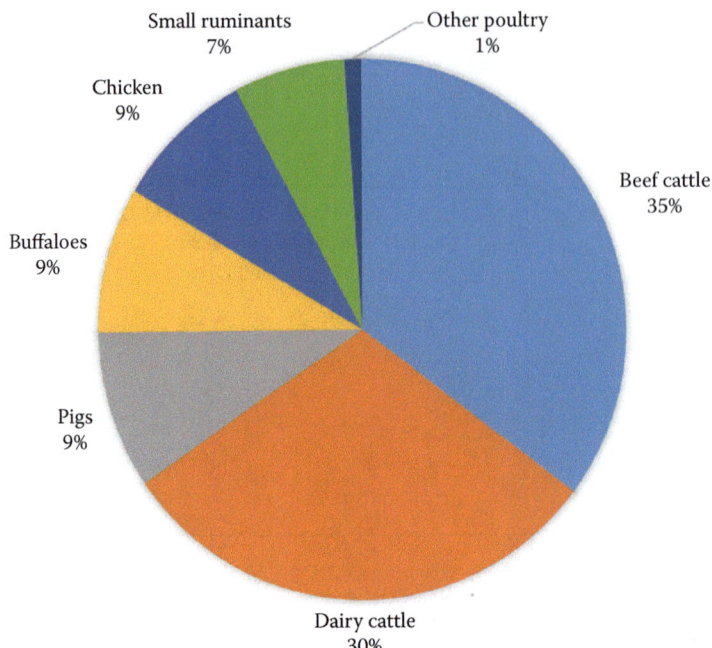

FIGURE 14.3 Global estimates of greenhouse gases emission by species. (Based on data from Gerber, P.J., H. Steinfeld, B. Henderson, A. Mottet, C. Opio, J. Dijkman, A. Falcucci, and G. Tempio, Food and Agriculture Organization of the United Nations (FAO), Rome, Italy, 2013.)

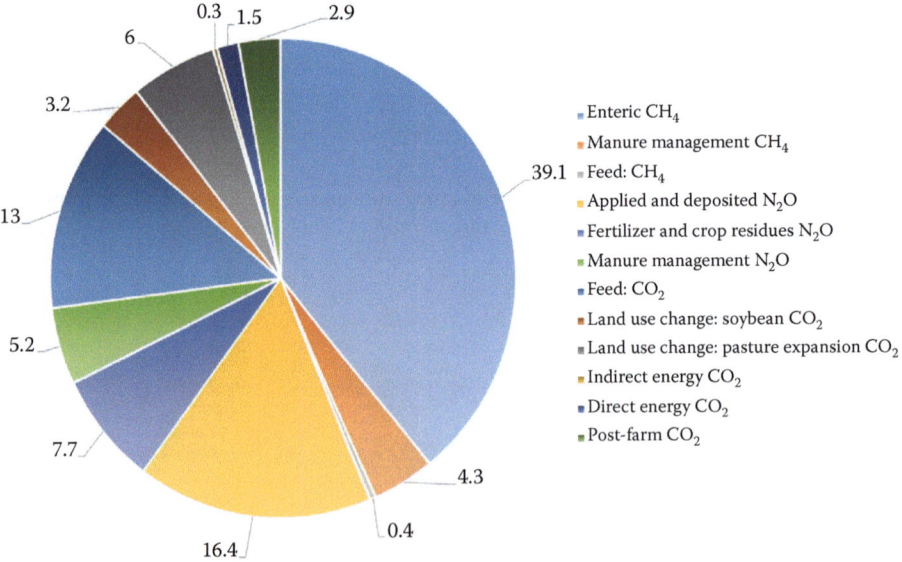

FIGURE 14.4 Global livestock sector supply chain categories greenhouse gases emissions percentage. (Data from Gerber, P.J., H. Steinfeld, B. Henderson, A. Mottet, C. Opio, J. Dijkman, A. Falcucci, and G. Tempio, Food and Agriculture Organization of the United Nations (FAO), Rome, Italy, 2013.)

9 Silicate Weathering to Mitigate Climate Change

Darryl D. Siemer

CONTENTS

9.1 INTRODUCTION

From 1870 to 2014, mankind's dependence upon fossil fuels generated anthropogenic carbon (C) emissions totaling about 545 Pg C (Pg = petagram = 10^{15}g = billion tonnes), all of which was dumped into the atmosphere. That "greenhouse gas" (GHG), ~1998 Pg of carbon dioxide (CO_2), partitioned itself between the atmosphere (approx. 230 PgC, 42%), ocean (approx. 155 PgC, 28%), and land (approx. 160 PgC, 29%) (these figures are based on "ICPP 2014"'s total CO_2 emission estimate and Sabine et al.'s (2004) partitioning factors). Its negative consequences include increasingly severe weather "events," ocean acidification (Orr et al. 2006), drought and biofuel production-driven food cost escalation, air pollution, deforestation, water shortages, oceanic shoreline erosion/flooding, onerous cost of living increases, and interminable squabbles/wars over resources (CNA 2014)—mostly "lebensraum," fresh water, and petroleum. During that same time the human population quintupled and ~80% of the cropland that feeds us experienced significant erosion along with loss of much of the soil organic matter (SOM), natural biota, and minerals that had originally rendered it more productive. Currently, a total of ~75 Pg of topsoil erodes per year, ~13–40 times faster than it did before humans invented agriculture. When topsoil thickness becomes less than mean food crop root zone depth (~15 cm), productivity drops off sharply (Sundquist 2005). Additionally, peak summer temperatures in the US's "corn belt" are now getting high enough to significantly reduce soybean (*Glycine max*) and maize (*Zea mays*) yields (Schauberger et al. 2017).

Approximately 20% of the world's human population is currently malnourished due to outright food shortages, prohibitive costs (to poor people), inequitable distribution, and in those areas most impacted by climate change (North Africa and the Middle East), desertification of formally arable soils—environmental refugees now outnumber people categorized as political refugees. In such a crowded world, the high productivity of modern intensive farming is both absolutely necessary and totally unsustainable because of its dependence upon cheap fossil fuels. Continued population growth (~210,000/day; UN 2017), land erosion, aquifer drawdown and the fact that we consume about ~1% of the world's remaining fossil fuels each year (Shafiee and Topal 2009) means that radical changes must take place well before 2100 AD. This chapter describes a technical fix for most of those problems that invokes the addition of mafic rock powder to farmland where it would serve to absorb excess atmospheric CO_2 and restore/maintain soil fertility. It will also explain why neither it nor any other realistic CO_2 removal (CDR) scheme could be implemented with the current suite of politically correct renewable energy sources.

9.2   BACKGROUND

Concerted international effort to address fossil fuel's environmental consequences began with the UN's 1997 Kyoto Protocol, to which a sizable number of mostly small nations signed up. Since circa 1992, many billions of dollars have been spent on research (mostly data gathering and computer-based modeling) that has generated thousands of peer reviewed papers/reports and hundreds of subsequent conferences and "topical sessions" both large and small. However, neither that science nor the policy changes favoring development/implementation of politically correct alternative/renewable energy sources (mostly wind turbines, solar panels, low-head hydro plants, and bio-fuels (Jacobson and Delucchi 2009; Jacobson et al. 2017; Evans 2016)) have apparently done much to reduce the rate that anthropogenic CO_2 enters the atmosphere (Figure 9.1). In spite of the US's and China's much ballyhooed decisions to "fully commit," the recent two such meetings, the United Nations Framework Convention on Climate Change (UNFCCC) "COP 21" (Paris 2015) and "COP 22" (Marrakech 2016) conferences did not really accomplish much because there is no legal mechanism ensuring that promised "Intended Nationally Determined Contributions" (INDCs; i.e., intended greenhouse gas emissions) or contributions to a proposed $100 billion/a "Green Climate Fund" (GCD 2018) are honored. Worse, most of the current climate models indicate that the sum of the INDCs of those conference's ~195 attendees would eventually cause global warming of about 3.4 degrees Celsius (°C), well above the 2°C figure that many of those model's creators consider a tipping point above which positive feedback mechanisms will render catastrophic consequences inevitable (Hansen et al. 2008). Probably the most useful outcomes of the climate science-related work performed to date is that global warming has definitely been proven to be man-caused and the fact that reasonably consistent/accurate estimates of global C fluxes, sources, sinks, etc., have joined the massive amount of other technical information freely available to anyone with internet access (CIA World Factbook 2017). Tables 9.1 and 9.2 pull together some of the numbers that will be used in subsequent arguments and calculations along with their sources.

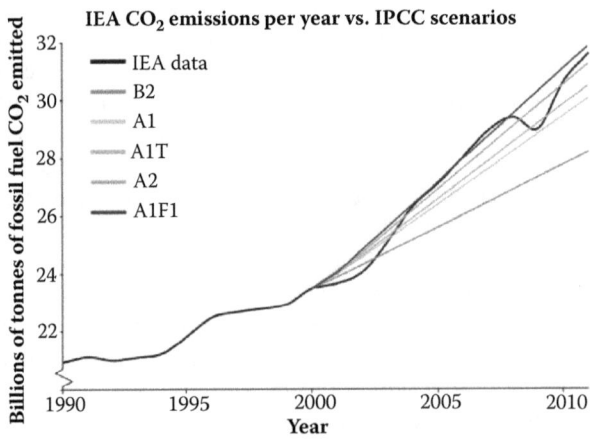

FIGURE 9.1 (See color insert.)   Observed vs IPCC model-predicted CO_2 emissions (https://static.skeptical science.com/graphics/IEA_IPCC_2012.jpg with permission).

TABLE 9.1

Some Especially Important Numbers

What?	Amount	Units	Reference
Total anthropogenic C emissions, 1870 to 2014	545	Pg	http://knoema
Total world grain production 2015	~2.5	Pg	http://www.statista
Total world bone-dry wood production 2015	~1.9	Pg/a	Globalwood 2016
Total C emissions fuel burning & cement-making (10 Pg C corresponds to 36 Pg CO_2)	~10	Pg/a	Le Quere et al. 2016
Anthropogenic C emissions from unsustainable land use	~1	Pg/a	Le Quere et al. 2016
Total soil organic carbon (SOC) to 1-m depth	~1550	Pg	Ontl 2012
Total soil inorganic carbon (SOI) to 1-m depth	~950	Pg	Ontl 2012c
Total soil CO_2 respiration rate (as elemental C)	60–77	Pg/a	Raich 1995
Total atmospheric C (about 99% CO_2)	~800	Pg	WMO 2016
C_1 oceans (97.6% inorganic (DIC) ≈ 0.0023 moles/liter)	~38,400	Pg	Falkowski 2000
Total C in remaining fossil fuel reserves	~1190	Pg	World 2016
Total mass of land-based flood basalts (assumes 3 g/cm^3)	~6.9E^{+7}	Pg	Ross 2005
Total accessible peridotites (high olivine basalt)	5E^{+5}	Pg	Kelemen 2011
C fixation potential of average earth mantle basalt	~750*	cmol$_+$/kg	Sanloup 2013,
Total water volume (~97.5% seawater)	~1.35E^{+9}	km^3	WIKIPEDIA
Mass of atmosphere (mean molecular wt. ≈29 g/mole)	5.15E^{+6}	Pg	WIKIPEDIA

TABLE 9.2

Chemical Compositions of Representative Mafic Rocks

	Average US Ultramafic[a]	Typical CA Dunite[a]	Typical US Serpentine[a]	Typical Snake River Plain Basalt[b]
SiO_2	44.9	39	38.1	47.04
TiO_2	0.65	0.02	0.1	2.7
Al_2O_3	1.8	0.04	0.7	15.11
FeO	4.85	7.52	2.79	13.51
MnO	0.22	0.11	–	0.2
MgO	37.3	46.1	40.3	7.65
CaO	0.74	0	0.3	10.06
Na_2O	0.15	0	0.1	2.54
K_2O	0.1	0.23	0.23	0.61
P_2O_5	0.02	0.03	0.03	0.55
NiO	± 0.3	0.35	–	–
Cr_2O_3	± 0.5	0.44	–	–
"Mg no."[c]	93.2	91.6	96.3	50.2
meq/g all bases[d]	0.0188	0.0229	0.0202	0.0076

[a] Compositions from Table 5.6 in Knuckleberg (2002).

[b] Average of 78 samples (Leeman 1982).

[c] "Mg no" = 100* mole% MgO/(mole%MgO + Mole%FeO).

[d] 1 meq/g = –0.01 times either cmol$_+$/kg or meq/100.

9.3   CARBON DIOXIDE REMOVAL (CDR)

Most of the world's leading climate scientists now believe that heading off the consequences of climate change will require more than just slashing CO_2 emissions—it is too late for that alone—excess CO_2 already in the atmosphere must be removed (Gasser et al. 2015).

Carbon dioxide removal (CDR, a.k.a., "negative emissions") is one of two ways to implement "geoengineering" (active transformation of climate via human intervention): the other, solar-radiation management (SRM, a.k.a., "albedo enhancement"), seeks to reflect sunlight back into space (and generate acid rain?) by injecting sulfur dioxide (SO_2) into the atmosphere, is now generally considered "too risky" (Caldiera et al. 2013).

Three ways to affect CDR have received a great deal of attention. The first, direct air separation, invokes huge "factories" that would first "scrub" CO_2 from the atmosphere with any of several theoretically possible ways and then sequester (permanently immobilize) it via an equally theoretically possible means. The second, biological CO_2 reduction, invokes either (1) adding some sort of fertilizer (usually iron is assumed) to the oceans to increase the rate at which atmospheric CO_2 ends up in abysmal sediments via increased "organic fallout"; (2) plowing purpose-raised crops directly back into the soil to sequester atmospheric C as soil organic carbon (SOC); (3) converting such crops to a less labile "biochar" (charcoal), before burying them; or (4) growing biofuels to produce energy, capturing the resulting CO_2 and then "sequestering" it (BECCS) in any of several theoretically possible ways.

The third approach, "enhanced weathering," invokes the scattering of powdered basic (mafic) rocks/minerals over the oceans (e.g., Taylor et al. 2016), seashores (Hangx and Spiers 2009), and/or land surfaces (Schuiling and Krijgsman 2006) to enhance their ability to absorb/sequester CO_2. This chapter discusses one such scenario invoking adding powdered basaltic rock (two types) to the world's farmland in order to simultaneously render agriculture sustainable and sequester atmospheric CO_2 as soil inorganic carbon (SIC) and groundwater "alkalinity" (bicarbonate). This in itself is not new—what's new is that it would be implemented with a politically incorrect energy source (Anti-nuclear 2018).

9.3.1   IMPRACTICAL (MOST) CDR PROPOSALS

Most of the proposed CDR schemes are impractical. For example, an internet search of "oceanic acidification mitigation" will bring up several invoking the addition of bicarbonate generated by contacting air (CO_2 is the anhydride of carbonic acid) with electrolytically generated hydroxide after which that solution would be dumped into the oceans where it would then also serve to enhance their atmospheric CO_2 absorption capacity due to "increased alkalinity." The latter contention is wrong because adding bicarbonate to water (e.g., seawater) that is primarily buffered by its carbonate to bicarbonate ratio $\left(K_{2a} = (CO_3 =)(H^+) / HCO_3^- \right)$ tends to lower, not raise, its pH. Lowering pH renders water able to absorb less, not more, CO_2 which is one of the reasons why the oceans are now less absorbant than they were before anthropogenic C lowered their surface-water pH by about 0.1 unit (the other reason is that such carbon has also raised mean surface seawater temperatures sufficiently (about 1.5°C) to reduce CO_2 solubility by ~6%).

However, in principle such scenarios could indeed sequester atmospheric CO_2. The problem is that a quantitative review of the numbers associated with implementing any electrochemistry-based CDR proposal will show that it would be impractical. For example, one such proposal (House et al. 2007) invokes giant chlor-alkali cells that would electrolyze aqueous solutions of pure NaCl (the alkaline earth metals, organic matter, and sulfate in natural seawater would quickly plug such cells) to generate sodium hydroxide that would be utilized in contactors to scrub CO_2 from the atmosphere after which the solution would be dumped into the oceans (see reactions below). The hydrogen and chlorine gas co-generated by the chlor alkali cells would be recombined in fuel cells to recover some of the power required to implement the process. The fuel cells' product, HCl (a strong acid), would then be neutralized via reaction with powdered mafic rock in huge "pressure cookers" (to accelerate reaction rates); this step would generate waste comprised of a slurry of decomposed rock sludge particles (mostly silica) in a chloride-salt brine.

Electrolysis: $2NaCl + 2H_2O \rightarrow 2NaOH + Cl_2 + H_2$
Air scrubbing: $NaOH$ +water+ CO_2 in air $\rightarrow NaHCO_{3\ aq}$
$H_2 + Cl_2$ (fuel cell) $\rightarrow 2HCl$ (in a water-based electrolyte)
HCl_{aq}+ a Mg/Ca-containing rock $\rightarrow CaCl_2 + MgCl_2+$ rock sludge.

Electrochemistry's drawbacks are electrode area-limited (slow) kinetics and high energy cost. Consequently, in today's world (expensive electricity) large-scale industrial processes are not implemented that way if there's another practical way to do it. While the above example was characterized as "energetically feasible," real chlor-alkali cells require about 3.9 volts to operate at reasonably productive rates (O'Brien et al. 2005) and real H_2/Cl_2 fuel cells generate only about one volt at similarly practical current densities ($\sim$0.5 A/cm^2; Huskinson et al. 2012). The consequence is that the net energy required to produce one mole (or equivalent) of hydroxide would be $\sim$2.8E^{+5} J [1 equivalent * (3.9-1) volts * 96,500 coulombs/equivalent) * 1J/(volt * coulomb)]. Implementing this example's process on a scale capable of capturing anthropogenic CO_2 as quickly as it is now being discarded ($\sim$36 Pg or 8.2^{+15}gram moles per year) would require the *continuous* input of $\sim$7260 [2.3E^{+20} J/1 E^{+9} J/s/3600 s/hr/24 hr/day/365 day/year] full-sized (i.e., 1 GW$_e$) power plants. Because the time-averaged power output of US-sited wind turbines is about 30% of their nominal ratings (i.e., capacity factor$_{wind}$ (CF$_w$) $\approx$ 0.3), if this scheme were to be wind powered, 24,200 (7260/0.3) GWs worth of such capacity would be required (e.g., 16.1 million, 1.5 MW$_e$ wind turbines). Additionally, in order to achieve the same time-averaged CO_2 capture rate, the process equipment itself would also have to be similarly scaled up. Since the capacity factors of photovoltaic solar panels are generally considerably lower than those of wind turbines even larger scale up factors would be required if it were to be solar-powered (e.g., require $\sim$72 billion one kilowatt-rated, rooftop-type, solar panels).

Figure 9.2 demonstrates why trying to implement any such scenario with politically correct 100% renewable energy (e.g., Jacobson and Delucchi 2009) would be virtually impossible.

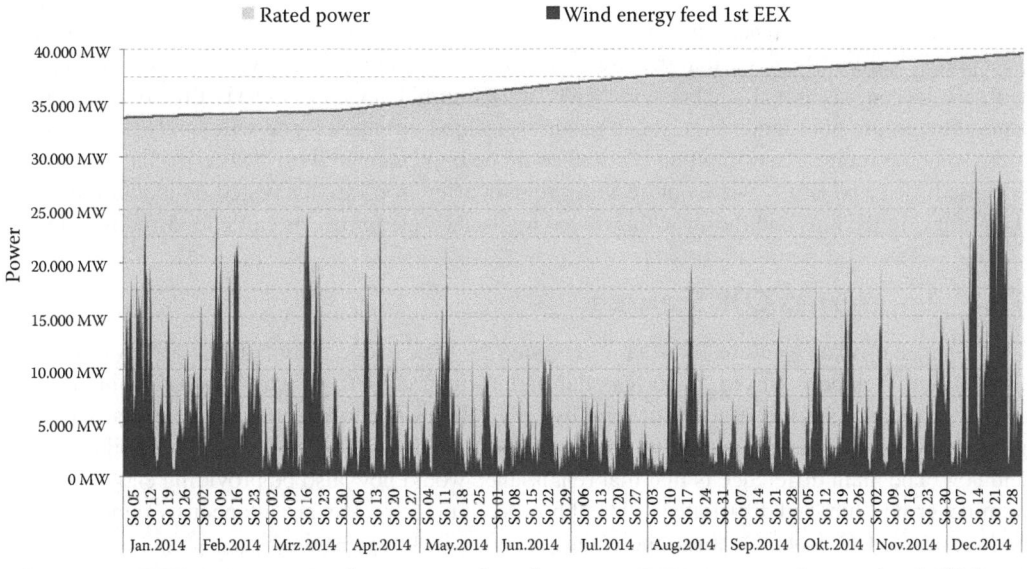

Data source: EEX-Leipzig   Resolution: quarter-hour data source: EEX-Leipzig   Presentation: Rolf Schuster

FIGURE 9.2   Real-time output of Germany's entire wind turbine fleet throughout 2014 (http://www.ver nunftkraft.de/85-prozent-fehlzeit-windkraftanlagen-sind-faulpelze/ (with permission)).

The plot's red line depicts the total nominal "capacity" of Germany's wind turbines throughout 2014 (its wind turbine fleet was still rapidly growing) and its blue spikes, real time power generation. That system's overall CF during that year was 0.163—about one half that of the US's (the US is windier). The several-year average CF of Germany's solar panels is 0.107 (Fraunhofer 2017).

Coal-fired power plants could not power such proposals either. In 2015, the US's coal fired power plants burned 0.69 Pg (740 million short tons) of coal to generate $1.509E^{+19}$ J (16.4 quads) of heat along with 1.364 Pg of CO_2 (EIA 2016). Assuming the USA's current mean coal-heat energy-to-electricity conversion efficiency (33.4%), these figures correspond to generating 62 Pg of "new" CO_2 [0.6907 * 1.364 *(2.3E+20 J/(1.509E+19J*0.334))] for every 36 Pg of atmospheric CO_2 so-sequestered.

Finally, inventors/champions of CDR sequestration technologies often suggest that they could be powered with "stranded" politically correct renewable energy (energy that could be generated but isn't because it can't be sold). Stranding is mostly caused by renewable energy's (esp. wind and solar) variability (unreliability), distribution grid limitations, and customer penny-pinching (aka "conservation") induced by policy-driven electricity cost escalation. In 2010, 25 TWh's worth of wind power was stranded in the USA (Stranded 2011). That figure, $9E^{+16}$ J, represents 7.1% of its wind turbine fleet's nominal capacity (then 40.27 GW) and is almost three orders of magnitude too small to power the above-described CDR scenario.

That example is by no means the most impractical electrochemical CDR scheme that has received a good deal of publicity and research funding. The "STEP" system (Lichte 2009) invoking the conversion of atmospheric CO_2 to graphite (electrochar?) with solar tower-generated heat and electricity would require at least four times that much energy to implement.

COP 21/22's rosy projections apparently assume that at least 10 Pg of CO_2 will be removed from the atmosphere each year by 2050 (Anderson and Peters 2016). The current front-running negative emissions technology, BECCS, is simultaneously eminently politically correct (both "biofueling" and "sequestering") and totally unrealistic. For instance, burning biofuel equivalent to 100% of humanity's current "total carbohydrate" production [all grains (food) plus all "bone dry wood"], ~4.4 Pg/a, in 40% Carnot efficient electrical power plants would generate $2.61E^{+19}$ J/a [0.4 * 4.4E+12 kg * 4.5 kWh/kg * 3.6E^{+6} J/kWh] or 43% of the electricity required to remove 10 Pg CO_2 via the above described approach. Doing so would also generate roughly 6.5 Pg of CO_2 [4.4 Pg*44 g CO_2/mole C/30g carbohydrate/mole C] that, if then quantitatively captured and permanently sequestered represents only 65% of the "negative emissions" that COP 21/22's modelers have assumed. The key assumption behind BECCS—that the world has lots of extra arable land available to grow massive biofuel crops on—is unrealistic because "good" agricultural land (and water) is already almost fully committed as are poor lands (e.g., sheep pastures) to food production and will surely become even more so in the future. Asserting that BECCS could be implemented with "waste" is also unrealistic because little or no BECCS-relevant residues remain after crops are harvested on modern farms—the grain is in the bins and the straw/hay baled up/trucked off for use as cattle feed and bedding.

9.3.2 This Chapter's CDR Scenarios

This chapter invokes a future like that envisioned by Oak Ridge National Laboratory's (ORNL) Goeller and Weinberg 40 years ago (Goeller and Weinberg 1976) in which a sustainable nuclear renaissance addresses the issues generated by civilization's dependence upon rare/expensive high-grade ores, petroleum, and fresh water that are responsible for most of the world's turmoil both then and now. The main difference is that that renaissance would now also be providing enough energy to mine, grind, ship, and distribute sufficient powdered mafic rock to simultaneously remove excess anthropogenic CO_2 from the atmosphere and render agriculture sustainable.

9.3.2.1 Its Whys

Why would adding freshly powdered (unweathered) mafic rock to the world's farmland simultaneously address the root cause of climate change and render agriculture sustainable? In 2014,

Gislason and Oelkers observed that, "all the C in the atmosphere, living creatures and dissolved in the oceans is derived from rocks and will (eventually) end up in (basic) rocks, the largest C reservoir on Earth" (Gislasen and Oelkers 2014). The gist of their observation first appeared in the scientific literature 172 years ago (Ebelman 1845) and has subsequently served as the rationale for all of the "enhanced weathering" C sequestration proposals described in Hartmann et al.'s (2013) comprehensive review of such literature (~375 citations). Rock weathering engendered by plate tectonic-driven exposure of mafic crustal rocks (mostly various sorts of basalts) to the atmosphere's carbon dioxide has stabilized its concentration within a "livable" range for over a billion years (Dunsmore 1992).

To date, almost all of the funding devoted to testing/demonstrating enhanced weathering-based CDR technologies has been spent upon those seeking to capture/sequester CO_2 (CCS) from point sources (gases from fossil fuel-fired (Lackner 2002) or geothermal (Gislason and Oelkers 2014) power plants, cement factories, refineries, steelworks, fertilizer and synthetic fuel plants, etc.) The driver for those exercises is that individual nations are incentivized to capture some of their own CO_2 to demonstrate commitment to achieving IPCC goals. However, doing so is inefficient because the atmosphere is a well-mixed reservoir which means that neither its origin nor where it is captured is relevant with respect to anthropogenic carbon's environmental impact. Another downside is that the cost of implementing high tech but small-scale CCS demonstrations severely limits the amount of R&D funding available to investigate more efficient proposals. The goal of any such commitment to R&D should be to develop something capable of capturing/sequestering as much CO_2 as possible, at the lowest possible cost, anywhere independent of origin, not to demonstrate patented "cutting-edge technologies" applicable only to special situations.

These are the reasons why this chapter's approach to implementing CDR would work:

1. The world's plant life transfers CO_2—the anhydride of carbonic acid—from the atmosphere into soils that then collectively "respire" about 7 times more of it (~275 Pg/a) back into the atmosphere as humanity is adding to it because the majority of the mineral matter within natural soils is already "weathered" (unreactive).
2. Biologically active soils including today's intensively-farmed land greatly accelerate rock weathering (Manning 2008; Manning and Renforth 2013; Schuiling and Krijgsman 2006).
3. Farmers are apt to cooperate because they already add lots of different "amendments" to their soils, any such scheme would surely be government subsidized, and doing so would increase their crop yields and land values.
4. It would be much cheaper than any sort of factory-based CDR scheme and therefore more likely to actually be implemented.
5. The nuclear renaissance powering it would also provide enough desalinated water to grow the crops responsible for pulling atmospheric CO_2 down into the amended soil.

9.3.2.2 Energetics

The correlation between per-capita energy consumption and human living standards, whether measured in terms of Gross Domestic Product (GDP), Quality of Life (Quality 2017), or Human Development (Development 2017), is high; i.e., the US's high per-capita energy consumption—about 5 times the global average—is largely responsible for its high standard of living. Because today's fossil fuel dominated energy system (currently ~85%) is both unsustainable and problematic, the single most important thing humanity must do well before the end of this century is to develop a replacement that is simultaneously reliable (not intermittent), sustainable (won't run out of fuel), environmentally clean, and "cheaper than coal" and then actually implement it (Pielke 2010). While further efficiency improvements and the construction of even more intermittent energy sources (mostly wind and solar) will somewhat mitigate CO_2 emissions, such measures alone cannot address the future's energy-related challenges because energy demand will outstrip their effects (Beckers 2016a; Pielke 2010; Clack et al. 2017).

Estimating the amount of power/energy that the world's new energy system must supply begins with the assumption that future world leaders would choose to size their "clean energy initiative" so that it could comfortably satisfy *everyone*'s power needs and thereby eliminate a root cause of human conflict and misery.

The world's raw/primary energy consumption is currently about 570 exajoules/a, ~18% of which is consumed by the US's relatively wealthy (energy-rich) 0.319 billion people (EIA 2016). If the human population eventually asymptotes at 9 billion (UN 2017) and everyone is to become as energy-rich as US citizens are now, the future's total primary/raw energy requirement would rise to ~2895 EJ/a [570E^{+18} * 0.18 * 9/0.319]. Since the efficiency with which technological civilizations convert raw heat energy to useful "energy services" (e.g., electricity) is about 40% (LLNL 2015), that figure corresponds to a useful future power requirement of ~3.67E^{+13}W [0.4 * 2895E^{+18} J/(3600 s/hr*24 hr/day*365 day/year)]. Next assumption: since the consequences of Germany's heroic experiment with wind and solar power ("Energiewende") suggest that "integration costs" renders electricity from unreliable sources prohibitively expensive above ~20% market penetration (Evans 2016; Beckers 2016b), and fission-based nuclear power is simultaneously reliable, "green," and *potentially* capable of satisfying mankind's requirements indefinitely, an appropriately implemented nuclear renaissance is to provide 80% of such services. Such a renaissance would require ~30,000 [5.67E^{+13} * 0.8/1E^{+9}/3600/24/365] full-sized (i.e., ~1 GW$_e$, not "small modular") nuclear reactors. Because today's ^{235}U-fueled light water reactors (LWRs) consume roughly 160 Mg (160 tonnes) of raw uranium per GW$_e$-year and the world has only about 18 Tg (1 Tg = 1 million metric tonnes) of "affordable uranium resources" (Redbook 2014; "affordable" means costing under $260/kg), implementing such a renaissance with them would be impossible because 18E^{+6} t/(30000 GW$_e$/a * 160 t/GW$_e$ year = (only) 3.75 years. Consequently, "appropriately implemented" means that its reactors would "burn" natural uranium and/or thorium, not just the ^{235}U isotope. It would be sustainable because breeder reactors close-coupled to efficient fuel recycling (reprocessing) systems would consume only about one Mg (tonne) of U and/or Th per GW$_e$-year which would render fuel extraction from "average crustal rock" affordable. Such rock has a density of ~3 g/cm^3 and contains ~15 ppm U+Th (Taylor 1964) which translates to ~7E^{+12} t of U+Th within the topmost kilometer of the Earth's crustal landmass (not under its oceans) —sufficient fuel to generate 30,000 GW$_e$ for 231 million years.

9.3.2.3 Materials

It is difficult to grasp the scale of these problems. Current annual anthropogenic CO_2 emission rates are now about 36 Pg (36 billion tonnes) per year, which means that any CDR technology implemented on an environmentally significant scale would require similarly huge amounts of reactants. Most of the reactants/materials that have been proposed for enhanced weathering CDR (and CSS) schemes are naturally occurring, intrinsically basic silicates containing high percentages of alkali and alkaline earth elements, although some industrial wastes and residues have also been suggested. The advantages of the latter (e.g., Portland cement concrete-based demolition wastes (Manning 2008; Renforth et al. 2009), coal and oil shale ash, metallurgical slags including those generated by the production of elemental phosphorus, steel, and aluminum) are that they are intrinsically highly reactive, readily available, and must be disposed of anyway. However, since their production rate is much lower than current anthropogenic CO_2 dump rates, natural rocks and minerals among which olivine-containing rocks (dunites) usually take precedence. The reason for this is that olivine (Forsterite—crystalline Mg_2SiO_4) is highly basic—contains up to ~57.1 wt% MgO) —meaning that relatively little of it would be needed to absorb any given amount of CO_2, weathers quickly, and is readily available (see Figure 9.3).

Other candidates include various sorts of basalts, basaltic tuffs, hydrated olivines (mixtures of serpentine—$Mg_3Si_2O_5(OH)_4$ and brucite $Mg(OH)_2$), and, possibly, high calcium plagioclase (e.g., anorthsite, $CaAl_2Si_2O_8$). Another commonly mentioned candidate is wollastonite ($CaSiO_3$). Wollastonite does indeed react (weather) quickly in wet soils but there is not enough of it to implement CDR on an environmentally relevant scale. In any case, initial priority would probably be

FIGURE 9.3   The world's major dunite outcrops (each dot may represent several closely situated deposits). (From Schuiling. R. D., Weathering approaches to CO_2 sequestration, in R. A. Meyers, Ed., *Encyclopedia of Sustainability Science and Technology*, Vol. 3, pp. 1909–1927, ISBN 978-0-387-89469-0, http://ecological aquaculture. org/Part2.pdf, 2013. With permission.)

given to materials that are already crushed and/or milled to minimize cost. Most chromite, nickel, magnesite, peridot (a semi-precious variety of olivine), and diamond ore deposits are located in olivine-rich host rocks (e.g., kimberlite), meaning that such mines are invariably surrounded by huge mine tailings heaps comprised of already mined & crushed, mafic rock.

However, it is important to ascertain that there really is enough of such materials available.

A report inspired by the recent surge of interest in advanced weathering CDR/CSS concepts (Krevor et al. 2009) characterizes the US's entire ultramafic rock (<45 wt% SiO_2) resource base. Unfortunately, with one exception, that report does not include estimates of either the quantities or chemical compositions of those resources either at individual sites or in toto. That exception characterizes Washington State's Twin Sisters "dunite" site as being one of the largest olivine deposits in the world (remaining reserves of unaltered olivine ~200 Pg) and the site of the USA's largest ongoing olivine production operation (~90,000 t/a). An earlier report inspired by the onset of WWII (Hunter 1941) —olivine was then used to line blast furnaces—characterized the ~ 0.23 Pg of "high grade olivine" in the relatively small outcrops scattered along the Blue Ridge Mountains as containing 48.1 wt% MgO. Table 9.2 lists the compositions of materials that could be utilized to implement enhanced weathering CDR in the US.

If we assume that Twin Sister site olivine possesses the same composition as that of the Blue Ridge Mountains and that the atmospheric CO_2 it sequesters ends up as bicarbonate ion, that site's 200 Pg of olivine alone could fix ~210 Pg of CO_2 [200 Pg rock * 2 eq/mole MgO * 0.481 g MgO/g rock * 44 g CO_2/mole/(24.31+16) g/mole MgO]. Since the US currently dumps about 5.2 Pg of CO_2 into the atmosphere each year (14.4% of the world's total), it certainly possesses enough olivine to deal with its contribution to that problem for as long as its remaining fossil fuels are cheap enough to burn. Since the world's land masses purportedly have roughly 5E^{+7} Pg of "high olivine basalt" on or near their surfaces, raw material availability won't be a problem anywhere if mankind should suddenly become both willing and (eventually) able to implement it. Realistically, that will depend upon its perceived cost-to-benefit ratio at such time which, in turn, will be largely determined by the cost of sufficiently reliable electricity.

9.3.2.4   The All-In CDR Option

Let's do some more ball parking. First, here is the first set of assumptions:

1. If achieving rapid ("emergency") CDR is deemed to be of paramount importance, the rock so-used should possess maximal acid/CO_2 neutralizing/fixing capability (i.e., high total meq/g of alkali plus alkaline earth elements—the example above shows how that figure is arrived at) and highly reactive (quickly weathered). Consequently, to meet that scenario's goals most efficiently, I shall assume the composition of Table 9.2's "typical Californian dunite" (second column).

2. I shall also assume that a crushing/grinding energy input of 100 kW/hr per Mg can achieve the degree of comminution required to render such rock fine/reactive enough to release 100% of its basic components (in this case, mostly magnesium) within one year under field conditions [Hartmann et al. 2013]—this figure is ~5 times greater than that required to reduce chunks of raw cement clinker to cement powder.

3. To be consistent with my electrochemistry-based CRD example, let us also assume that future decision makers choose to sequester anthropogenic CO_2 at the same rate that we are currently generating it, i.e., ~36 Pg/a (probably wildly optimistic).

With these assumptions, the amount of rock that must be mined, crushed, ground, shipped, and plowed-in across the world each year would be about 36 Pg [36 Pg CO_2 * 44 g CO_2/eq * 0.0229 eq/g olivine = 36 Pg olivine]. At 100 kWhr/t ($3.6E^{+8}$ J/t), mining and powdering it would require ~$1.30E^{+19}$ J/a, which figure translates to the continuous electrical power output of 411 full-sized (1 GW_e) nuclear reactors—about the same as that generated by the world's current utility nuclear reactor fleet. To continue, if this powder is to be shipped an average of 2000 km via electrified trains as efficient as those currently used to move coal (~185 km/L diesel fuel/short ton), this scenario's total power requirement would increase to ~141 kWh/t, thereby requiring the continuous output of ~563 full sized power plants to implement—well under 2% of the total required to meet the needs of this scenario's 9 billion uniformly rich people.

Because this approach to implementing CDR would require far less power than would any electrochemical alternative (e.g., 563 vs 7260 GW_e), it could, in principle at least, be temporarily implemented with fossil fuel burning power plants. However, the process would have to be significantly upsized to compensate for the fact that such fuel would generate additional CO_2. For example, in the UK, coal-generated electricity currently emits ~0.27 kg C per kWh (Renforth 2012), which translates to generating 0.14 Mg of new CO_2 for every Mg of old CO_2 removed by olivine-weathering based CDR—with "average Idaho basalt the new-to-old carbon ratio would be 0.42.

What soils would be "best"? Although tropical forests represent ideal sites (warm, wet, little seasonal change, and underlain with intrinsically acidic/lateritic (already leached-out) soils (Stallard and Edmond 1983; Manning and Renforth 2013), it is more realistic to assume that already-farmed, privately owned (and therefore deemed valuable enough to spend a good deal of money/effort upon) land would be treated. Since warm, well-watered, and already productive soils are especially biologically active, the world's ~400 million acres (1.61 million km^2) of rice (*Oryza sativa*) fields might be "best." Land devoted to corn and soybean production (combined ~2.8 million km^2) is apt to work pretty well too. Assuming 36 Pg/a and a soil bulk density of 1.21 g/cm^3 (Pimentel and Burgess 2013), if land currently devoted to all three of those crops were to be so-employed, this scenario's soil "rebuild rate" would be about 7.1 mm/year. If the rock powder were to be applied to the Earth's entire ~1.4 billion ha of arable land, that rate would be ~2 mm/a. In either case, soil rebuild rate would be similar to the rate at which modern agricultural practices tend to erode European and US top soils, ~17 Mg/ha/a ≈ 1.4 mm/a (Pimentel and Burgess 2013).

9.3.2.5 The Sustainable Agriculture Plus CDR Option

The all-in CDR scenario's main drawback is that the weathering of either "pure" periodite (a.k.a., dunite) or its alteration product, serpentine, generates poor quality soils (Kruckeberg 2002)—their extremely high Mg:Ca ratio is not suitable for some plants as is the fact that they don't contain appreciable amounts of other biologically important elements (K, P, trace minerals, etc.). This issue is compounded by the fact that high-Mg (more "basic" and therefore better for CDR) dunites also tend to have relatively high percentages of nickel (Ni) and chromium (Cr) in them (see Table 9.2 and Hertzberg et al. 2013) which raises toxicity concerns. Of course, such fears aren't *really* justified for two reasons. First, the amount of olivine to be added each year (about 86 t/ha if only rice, corn, and soybean fields were to be so treated) is too small to appreciably change the gross composition of such soils. For example, the depth of Illinois' topsoil is currently about 30 cm—at a bulk density of ~1.2 g/cm^3 (Pimentel 2013) that works out to 3600 Mg of topsoil/ha (in other words, the solution to pollution would be dilution). The second is that both nickel and chromium are apt to be essentially insoluble in soils experiencing the "liming" effects of powdered dunite (or basalt), which fact would likely keep the amounts of either ending up in foodstuffs grown/harvested from such soils well below harmful levels (Vágó et al. 1996). Unfortunately, as is also the case with nuclear power, fact-based arguments probably won't influence decision making in this arena.

The second implementation scenario assumes that "average Idaho basalt" (fourth column of Table 9.2) is used instead. One reason for this is that there is roughly two orders of magnitude more common basalt than ultramafic rock on or near the surface of the continents. A better reason is that the chemical compositions of common basalts closely approximate those of volcanic ashes and thereby provide the most compelling argument for why this scenario would render agriculture permanently sustainable.

Advanced life can exist on the Earth's surface only because of its partnership with other natural phenomena—heat from the sun and nutrients from rocks decomposed into soil via reactions with moisture and acidic atmospheric gases, primarily CO_2 plus the (accelerating) effects of a host of indigenous soil biota. Such weathering releases key elements from volcanic rocks, minerals, and glasses, rendering them readily accessible to anything grown in such soils (Hensel 1894; Fisher et al. 1997; Sheets and Grayson 1979; Van Straaten 2002). Volcanic soil is why humans insist upon living near volcanoes. Close to an erupting volcano, short-term destruction by pyroclastic flows, heavy ash falls, and lava flows can be complete, the extent of which depends upon the eruption magnitude. Crops, forests, orchards, and animals grazing/browsing on its slopes, downwind, or on surrounding lowlands may be leveled or buried. However, this is only a short-term effect. In the long run, volcanic deposits develop into some of the richest agricultural lands on earth.

Italy provides a good example. Except for the volcanic region around Naples, farming in southern Italy is difficult because its soil's "basement" rock is dolomitic limestone, which means that the soil made from it is quite poor. However, the Naples region, which includes Mount Vesuvius, is very rich because two eruptions 35,000 and 12,000 years ago left it blanketed with thick deposits of tephra, which has since weathered to form soil that has been intensively cultivated since well before the birth of Christ. Every square meter of it is so used. For example, even a small vineyard will have, in addition to the grapes, tomatoes and spring beans on its trellises, fava beans (*Vicia faba*), cauliflower (*Brassica oleracea*), and onions (*Allium cepa*) between the trellis rows, and its margin rimmed with orange and lemon trees, herbs, and flowers.

The verdant fertility of the New Zealand's North Island is also due to its numerous prehistoric volcanic ash falls. Volcanic loams developed on 4,000- and 40,000-year-old volcanic ash deposits in the Waikato and Bay of Plenty regions. With the help of ample rainfall, warm summers, and mild winters, those regions now produce abundant crops, including the kiwifruit shipped around the world. Its altered volcanic ash soils are rich in SOM, well-drained yet still hold ample water for vegetation, and are easily tilled. Its deep volcanic loams are particularly good for pastures (New Zealand has a large dairy industry), horticulture, and maize (Molloy 1993).

After the 1980 Mount St. Helens eruption, people living downwind of it were concerned that its ash would be detrimental to the rich agricultural farmlands of eastern Washington and western Idaho. That concern was quickly countered by that region's agronomists, who pointed out that such ash constitutes a time-release capsule for plant nutrients. The same rationale explains why the US's "corn belt" is so productive. It is directly downwind of where three of the world's most powerful "supervolcanos" each spewed out several thousands of times more volcanic ash than did Mount St. Helens.

In this case I will again assume that sufficient mafic rock is mined/ground/transported and scattered each year to convert 36 Pg of CO_2 to bicarbonate when it has completely weathered. If so, the amount required would be 108 Pg (36 Pg olivine * 0.0229 meq/g olivine/0.0076 meq/g basalt].

Assuming a bulk soil density of 1.21 g/cm³, application of that much powder to the world's 1.4 billion Ha of arable land (~77 Mg/ha) would rebuild it at a "negative erosion" rate of ~6.4 mm/year while adding 471 kg of K_2O and 425 kg of P_2O_5/ha per year. Since these figures represent about 1.5 and 7 times as much K_2O and P_2O_5 as in the harvested moiety of 1 ha/a's worth of US soybean or corn crops, most of today's artificial fertilizers would no longer be needed. The exception, either nitrate or ammonia-type nitrogen, would be cheap to produce with "cheaper than coal" nuclear power. If water limits "reaction rates," fields could be irrigated with desalinated seawater. Assuming that 20" (0.508 m) of water is added each season and a desalination energy demand of 3kWh/m³ (typical for modern reverse osmosis desalination systems), a single one GW_e nuclear reactor could irrigate ~575,000 ha of farm land.

9.4   PERSPECTIVES

9.4.1   How Big Is Big?

To put any worthwhile (big enough to make a significant difference) enhanced weathering CDR scenario's "numbers" into proper perspective, we should remind ourselves that mankind currently mines, treats (washes/screens/crushes, etc.), transports, etc., about 36 Pg of sand and gravel each year to make Portland cement-based concrete (estimate based upon the percentage of Portland cement in such concretes (about 11%) and that sort of cement's current consumption rate (~4.2 Pg/a)). That figure underestimates the total amounts mined/treated/moved/etc. because sand and aggregate are used for other things too. That total is probably dwarfed by the amount of overburden rock routinely blasted/mined/moved from strip-mined coal deposits each year—most notoriously by the US's mountaintop removal approach to accessing that fossil fuel.

9.4.2   What Should It Cost?

Cost estimating is often "trans scientific" (see Weinberg 1994; Pielke 2010) because the price of many things we buy doesn't have much to do with their intrinsic cost or value. Prices are instead usually determined by "what the market (or system) will bear," which depends more upon human behavior than upon anything that can be evaluated quantitatively (the current costs of getting a tooth filled or building a nuclear reactor are so-determined). To be consistent with my previous examples, I will base an educated guess upon the "should-cost" of the energy needed to implement these proposals. In 1970, a paper written by Bettis and Robertson (1970) included an analysis of what power generated by a full-sized molten salt breeder reactor should cost (Bettis 1970—that reactor was never built). Applying the US's subsequent *mean* 6.2× inflation factor to their conclusion ($0.0041/kWhr) generates a should-cost of $0.0254/kWh in today's money—about 40% of the US's current average wholesale electricity cost. At 100 kWh/t, rock comminution should cost about $2.54 per Mg (tonne). If that powder is to be transported an average of 2000 km from mine-to-farm by an electrified rail system as energy-efficient as is that currently used to move coal (~185 km/L diesel fuel/short ton), the energy cost of transporting each Mg would be about $1.10. Trebling the sum of those figures to provide a profit and defray other costs generates a total implementation cost of about $11/Mg. Again, assuming that 36 Pg of powdered olivine is to be applied each year, this sums up to a whole-world CDR cost of ~$563 billion/a—about 70% of the US's current annual military expenditures. If the USA decided to remove/sequester only its own CO_2 emissions (~5.2 Pg/a) via that all-in CDR scenario, it would have to build about 77 one GW_e reactors to power it (the US currently operates ~100 full sized LWRs).

Finally, if future decision makers decided to remove/sequester just 10 Pg of CO_2/a (COP 21 and 22's assumption), total implementation cost of this paper's olivine-based scenario would be about $100 billion/year—the same figure currently assumed for the IPCC leadership's "Green Climate (i.e., climate change mitigation/adaptation) Fund."

9.5 REMAINING ISSUES, RECOMMENDATIONS, AND CONCLUSIONS

Reaction rates constitute any enhanced weathering CDR scheme's big unknown. Any soil amendment-based proposal's absolute maximum rate would be that at which soil microorganisms, bugs, worms, and roots respire (emit) CO_2 into the soil's interstitial gas and liquid phases. Manning and Renforth (2013) reported that the rate at which the portlandite ($CaOH_2$) in "fresh" (not already carbonated) Portland cement-based demolition waste buried in grass-covered English soil was converted to calcite corresponds to a C fixation rate of ~25 t C/ha/year. That translates to a mean soil respiration rate of 6.6 umole $C/s/m^2$ which figure is consistent with measured respiration rates of temperate region soils during summer months (Curiel Yuste et al. 2007). It is also the same number obtained by dividing one-half of the world's current total soil respiration rate (60–77 Pg C/a depending upon reference) by its ~13.8 million km^2 of arable land; i.e., 21–27 t C/ha/a. The degree to which such maxima are approached would depend upon the rate at which weathering releases rock constituents (alkalis and alkaline earths) capable of converting soil-gas CO_2 to carbonate and bicarbonate ions.

Manning and Renforth's (2013) observations support a contention that soil containing enough of a sufficiently reactive CO_2 adsorbent (e.g., portlandite) would sequester/fix CO_2 as quickly as soil respiration generates it but doesn't tell us how closely more leach resistant (less reactive) materials would approach that figure.

Most (all?) of the factors likely to affect powdered rock weathering rates (particle size/mineralogy, temperature, biological vectors—worms, fungi, bacteria, plant roots/exudates, etc., the kind and amount of water, physical factors–cultivating/plowing/etc.) have been identified and the subject of much discussion (e.g., Needham et al. 2006; Schuilling and Krijgsman 2006; Renforth et al. 2009; Schuiling et al. 2011; Schuiling 2013; Hartmann et al. 2013). Unfortunately, to date, no reports containing data generated by realistic field tests have been published. The two semi-relevant tests performed (ten Berge et. al. 2012; Renforth et al. 2015) attempted to determine rock weathering rates by doing mass balances around magnesium (the amendment utilized was almost pure forsterite (Mg_2SiO_4)). The results were both inconclusive and of questionable relevance because carbon's behavior wasn't determined.

Consequently, dozens (hundreds?) of person-years of intensely scientific study have not yet rendered it possible to assign reasonably definite/accurate rates to proposals invoking the natural processes responsible for both soil formation and long term atmospheric CO_2 stabilization. This stubbornly persistent ignorance about how deliberate efforts to implement such technologies would perform under real world conditions could be relieved by 2–3 years of appropriately managed/funded research performed by agricultural experiment station personnel (not by a government or university "think tank's" modeling experts). Almost every US state has at least one such facility associated with its universities. It is likely that experiments utilizing cheap plastic bucket (or barrel) lysimeters (Llansadwrn 2017) as "grow pots" could quickly generate unambiguous data. Such experiments would involve the addition of the powdered rock to well-watered, noncalcerous (low pH) soils which may contain lots of organic carbon but not much "background" inorganic carbonate/bicarbonate. Sufficient powder would be added to convert a substantial fraction of soil gas CO_2 to carbonate/bicarbonate meaning that carbon fixation (not rock weathering) rates could be directly determined by comparing the sum of soil inorganic carbon and total bicarbonate (alkalinity) in collected sub-water from amended vs control grow buckets. Such tests would be relatively cheap and simple to do and quickly generate immediately useful/relevant results.

There's a pretty good chance that these uncertainties will soon be resolved by the University of Sheffield's newly established "Leverhulme Centre for Climate Change Mitigation" (Leverhulme 2017). It has recently formed a collaboration with the University of Illinois, Champaign-Urbana which will perform field-scale, basalt dust soil amendment experiments on soybean and maize crops in the heart of America's Corn Belt. Exactly how the degree of CO_2 sequestration so-achieved is to be determined hasn't been revealed yet.

The main reason why the leaders of the First World's industrialized democracies resist the changes needed to realize Weinberg/Goeller's "Age of Substitutability" is that once a particular way of doing/providing anything is established, the people that own, implement, regulate, support, or otherwise profit from that/their business model resist change. This is why unsustainable light water–type reactors continue to dominate the nuclear power industry and that very little effort/resources have been devoted to developing superior alternatives during the last two decades (Till and Chang 2011; Siemer 2015). It is also the reason that chronic problems like those discussed in this chapter currently only get "studied," not solved—in many respects, "Big Science" has become just another "Big Business" seeking to maintain its prerogatives, avoid risk, and keep the "study money" flowing in (Beckers 2016a,b and NAP 1996).

ACKNOWLEDGMENT

The author acknowledge Professor Olaf (R. D.) Schuiling's contributions to furthering the "advanced weathering" concept and to this paper in particular.

ABBREVIATIONS

BECCS burning $\underline{B}$iofuels to produce $\underline{E}$nergy, $\underline{C}$apturing the $\underline{C}O^2$ and then $\underline{S}$equestering it

CDR Carbon Dioxide Removal

CF Capacity Factor—the ratio of mean delivered power to nameplate capacity

COP Conference of the Parties

CSS Carbon (CO^2) Capture and Storage (or "Sequestration") —usually envisioned to apply to point sources such as power plants, cement kilns, or synthetic fuel production facilities

GHG $\underline{G}$reen$\underline{h}$ouse $\underline{G}$as – any gas that absorbs/reemits infrared radiation – water vapor, CO^2, CH^4...

INDC Intended Nationally Determined Contributions (each country's intended contribution to GHGs)

IPCC International Panel on Climate Change

Mg (megagram) $1E^{+6}$ grams = 1 metric tonne (t)

ORNL Oak Ridge National Laboratory (worked to develop thorium fueled molten salt reactors from circa 1950 to 1978)

Pg (petagram) $1E^{+15}$ grams = 1 billion metric tonnes

SIC Soil Inorganic Carbon (the carbon in soil present as either carbonate or bicarbonate)

SOC Soil Organic Carbon (total soil carbon—SIC)

Tg (teragram) 10^{+12}g = 1 million metric tonnes

WFP World Food Program (UN agency currently distributing food to ~ 80 million especially poor people in ~75 different countries)

REFERENCES

Anderson, K. and G. Peters, 2016, The trouble with negative emissions, *Science*, 354, 182–183, 14Oct16 (paywalled, contact author).

Anti-Nuclear, 2018, https://en.wikipedia.org/wiki/Anti-nuclear_movement.

Beckers, M., 2016a, *Science a la Carte—and the Cherry Picking*, CreateSpace, ISBN-13: 978-150880690.

Beckers, M., 2016b, *The Non Solutions Project, Dec 2016*, CreateSpace, ISBN-13: 978-1537673806.

Bettis, E. S. and R. C. Robertson, 1970, The design and performance features of a single fluid molten salt breeder reactor, *Nuclear Applications and Technology*, 8, 2.

Caldiera, K., Govindasamy, B., and Long, Y., 2013, The science of geoengineering, *Annual Review Of Earth and Planetary Science*, 41:231–256

CIA World Factbook 2017, https://www.cia.gov/library/publications/the-world-factbook/geos/xx.html (good source of statistics).

Clack, C. et al., 2017, Evaluation of a proposal for reliable low-cost grid power with 100% wind, water, and solar, *Proceedings of the National Academy of Science*, 114, 26:6722–6727, doi.org/10.1073/pnas. (http://www.pnas.org/content/114/26/6722.full).

CNA Military Advisory Board, National Security and the Accelerating Risks of Climate Change, May 2014, http://www.cna.org/sites/default/files/MAB_2014.pdf.

Curiel Yuste, J., Baldocchi, D. D., Gershenson, A., Goldstein, A., Mission, L., and Wong, S., 2007, Microbial soil respiration and its and its dependency on carbon inputs, soil temperature and moisture, *Global Change Biology* 13:1–18, doi.org/10.1111/j.1365-2486.2007.01415.x, https://nature.berkeley.edu/ahg /pubs/microbial.pdf.

Development 2017, https://en.wikipedia.org/wiki/Human_Development_Index.

Dunsmore, H. E., 1992, A Geological perspective on global warming and the possibility of carbon dioxide removal as calcium carbonate mineral, *Energy Convers Management*, 33:565–572.

Ebelmen, J. J., 1845, Sur les produits de la décomposition desespèces minérales de la famille des silicates, *Anna. Mines*, 7:3–66.

EIA, 2016, http://www.iea.org/publications/freepublications/publication/KeyWorld2016.pdf.

Evans, S., 2016, The history of the Energiewende, https://www.carbonbrief.org/timeline-past-present-future -germany-energiewende.

Falkowski, P., Scholes, R. J., Boyle, E., Canadell, J., Canfield, D. et al., 2000, The global carbon cycle: A test of our knowledge of earth as a system, *Science*, 290, 5490:291–296, doi.org/10.1126 /science.290.5490.291.

Fisher, R. V., Heiken, G., and Hulen, J. B., 1997, *Volcanoes; Crucibles of Change*, Princeton University Press.

Fraunhofer 2017, https://www.ise.fraunhofer.de/content/dam/ise/en/documents/publications/studies/recent -facts-about-photovoltaics-in-germany.pdf.

Gasser, T., Guivarch, C., Tachiiri, K., Jones, C. D., and Ciais, P., 2015, Negative emissions physically needed to keep global warming below 2°C, *Nature Communications*, 6, Article number: 7958, doi .org/10.1038/ncomms8958, 03 August, http://www.nature.com/ncomms/2015/150803/ncomms8958/abs /ncomms8958.html.

GCD 2018, https://www.greenclimate.fund/documents/20182/24868/Status_of_Pledges.pdf/eef538d3-2987 -4659-8c7c-5566ed6afd19.

Gislason, S. R, and Oelkers, E. H, 2014, Carbon storage in Basalt, *Science*, 344: 373–374 (paywalled, contact author).

Globalwood, 2016, "~1.9 Pg" is derived from a total global wood production figure (3.5E^{+9} m^3), https:// en.wikipedia.org/wiki/Wood_economy and wood characterization data from www.globalwood.org/tech /tech_wood_weights.htm.

Goeller, H. E. and Weinberg, A. M., 1976, The Age of Substitutability, *Science*, 191, 4228: 683, Feb 20, 1976. (paywalled, available gratis at http://www.osti.gov/scitech/servlets/purl/5045860).

Hansen, J., Sato, M., Kharecha, P., Beerling, D., Berner, R., Masson-Delmotte, V., Pagani, M., Raymo, M., Royer, D. L., and Zachos, J. C., 200, Target atmospheric CO2: Where should humanity aim?, *The Open Atmospheric Science Journal*, 2: 217–231. doi.org/10.2174/1874282300802010217.

Hartmann, J., West, A. J., Renforth, P., Köhler, P., De La Rocha, C. L. et al., 2013, Enhanced chemical weathering as a geoengineering strategy to reduce atmospheric carbon, supply nutrients, and mitigate ocean acidification, *Reviews of Geophysics*, 51, https://www.researchgate.net/publication/26040781 _Enhanced_chemical_weathering_as_a_geoengineering_strategy_to_reduce_atmospheric_carbon _dioxide_supply_nutrients_and_mitigate_ocean_acidification.

Hangx, S. J. T. and C. J. Spiers, 2009, Coastal spreading of olivineto control atmospheric CO2 concentrations: A critical analysis of viability, *International Journal of Greenhouse Gas Control*, 3, 6:757–767.

Hensel, J., 1894, *Bread from Stones*, ISBN: 0-932298-85-0 (free—GOOGLE it].

Hertzberg, C., Asimow, P. D., Ionov, D. A., Vidito, C., Jackson, M. G., and Geist, D., 2013, Nickel and helium evidence for melt above the core–mantle boundary, *Nature*, 493:393–397, 17 January, doi.org/10.1038 /nature11771, http://www.nature.com/nature/journal/v493/n7432/fig_tab/nature11771_F1.html.

House, K. Z., House, C. H., Schrag, D. P., and Aziz, M. J., 2007, Electrochemical acceleration of chemical weathering as an energetically feasible approach to mitigating anthropogenic climate change, *Environmental Science Technology*, 41:8464–8470.

Hunter, C. E., 1941, Forsterite olivine deposits of North Carolina and Georgia, Georgia Department of Natural Resources, Bulletin number 47 (survey prompted to ascertain whether there would be enough blast furnace refractory to "fight" WWII with), https://epd.georgia.gov/sites/epd.georgia.gov/files/related_files /site_page/B-47.pdf.

Huskinson, B., Rugolo, J., Mondal, S. K., and Aziz, M. J., 2012, A high power density, high efficiency hydrogen-chlorine regenerative fuel cell with low precious metal content catalyst, *Energy & Environmental Science*, Issue 9. http://arxiv.org/pdf/1206.2883.pdf.

Jacobson, M. Z. and Delucchi, M. A., 2009, A plan to power 100 percent of the planet with renewables, *Scientific American*, Nov. issue.

Jacobson, M. Z., Delucchi, M. A., Bauer, Z. A. F., Goodman, S. C., Chapman, W. E. et al., 2017, 100% Clean and renewable wind, water, and sunlight (WWS) all sector energy roadmaps for 139 countries of the world," January 27, 2017, http://web.stanford.edu/group/efmh/jacobson/Articles/I/CountriesWWS.pdf.

Kelemen, P. B., Matter, J., Streit, L., Rudge, J., Curry, B., and Blusztajn, J., 2011, Rates and mechanisms of mineral carbonation in peridotite: Natural processes and recipes for enhanced, in situ CO_2 capture & storage, *Annual Review of Earth and Planetary Sciences*, 39:545–576, doi.org/10.1146/annurev-earth-092010-152509.

Krevor, S. C., Graves, C. R., Van Gosen, B. S., and McCafferty, A. E., 2009, Mapping the Mineral Resource Base for Mineral Carbon-Dioxide Sequestration in the Conterminous United States, U.S. Geological Survey, Reston, Virginia, Data Series 414, http://pubs.usgs.gov/ds/414/downloads/DS414_text_508.pdf.

Kruckeberg, A. R., 2002, *Geology and Plant Life: The Effects of Landforms and Rock Types on Plants*, Washington University Press, Table 5.6, p 112.

Lackner, K. S., 2002, Carbonate chemistry for sequestering fossil carbon. *Annual Review of Energy and the Environment*, 27, 193–197.

Leeman, W. P., 1982, Olivine tholeiitic basalts of the Snake River Plain, Idaho, in B. Bonnichsen and R. M. Breckenridge, Eds., *Cenozoic Geology of Idaho*, Idaho Bureau of Mines and Geology Bulletin 26, pp. 181–191 (also characterizes other US basalts), http://geology.isu.edu/Digital_Geology_Idaho/papers/B-26Ch4-1.

Leverhulme, 2017, http://www.lc3m.org/.

Licht, S., 2009, STEP: A solar chemical process to end anthropogenic global warming, *Journal of Physical Chemistry*, C, 113:16283–16292. Also see http://phys.org/news/2010-07-solar-powered-decrease-carbon-dioxide-pre-industrial.html.

Llansadwrn, 2017, http://www.llansadwrn-wx.co.uk/evap/lysim.html#calcs.

LLNL 2015, Its annual "primary energy" to "energy services" conversion factor (consistently about 40%) is based upon DOE/EIA-0035 (2015-03), e.g., see https://flowcharts.llnl.gov/.

Manning, D. A., 2008, Biological enhancement of soil carbonate precipitation: Passive removal of atmospheric CO2, *Mineral Magazine*, 72:639–649.

Manning, D. A. and Renforth, P., 2013, Passive sequestration of atmospheric CO_2 through coupled plant-mineral reactions in urban soils, *Environmental Science Technology*, 47, 1:135–141 (paywalled, contact author).

Molloy, L., 1993, *Soils in the New Zealand Landscape—The Living Mantle*, New Zealand Society of Soil Science, Canterbury.

NAP, 1996, Barriers to Science: Technical Management of the Department of Energy Environmental Remediation Program, *National Academy of Science*, http://www.nap.edu/catalog/10229.html.

Needham, S. J., Worden, S. H., and Cuadros, J., 2006, Sediment ingestion by worms and the production of bio-clays: A study of macrobiologically enhanced weathering and early diagenetic processes. *Sedimentology*, 53:567–579.

O'Brien, T. F., Bommaraju, T. V., and Hine, F., 2005, History of the Chlor Alkali Industry, Chap. 2, Springer, http://www.springer.com/978-0-306-4.

Ontl, T. A. and Schulte, L. A., 2012, Soil carbon storage, *Nature Education Knowledge*, 3, 10:35.

Orr, J. C. et al., 2005, Anthropogenic ocean acidification over the twenty-first century and its impact on calcifying organisms, *Nature* 437, 7059:681–686, Bibcode:2005Natur.437..681O, doi.org/10.1038/nature04095.

Pielke, R. Jr., 2010, *The Climate Fix*, Basic Press, ISBN 978-0-465-02052-2.

Pimentel, D. and Burgess, M., 2013, Soil erosion threatens food production, *Agriculture* 3(3):443–463; doi:10.3390/agriculture3030443. http://www.mdpi.com/2077-0472/3/3/443.

Quality 2017, https://www.numbeo.com/quality-of-life/rankings_by_country.jsp.

Raich, J., and Potter, C., 1995, Global patterns of carbon dioxide emissions from soils, *Global Biogeochemical Cycles*, 9:23–36.

Redbook 2014, http://www.oecd-nea.org/ndd/pubs/2014/7209-uranium-2014.pdf.

Renforth, P., Manning, D. A. C., and Lopez-Capel, E., 2009, Carbonate precipitation in artificial soils as a sink for atmospheric carbon dioxide, *Applied Geochemistry*, doi.org/10.1016/j.apgeochem.2009.05.005.

Renforth, P., 2012, The potential of enhanced weathering in the UK, *International Journal of Greenhouse Gas Control*, 10:229–243. https://doi.org/10.1016/j.ijggc.2012.06.011.

Renforth, P., Pogge von Strandmannc, P. A. E., and Henderson, G. M., 2015, The dissolution of olivine added to soil: Implications for enhanced weathering, *Applied Geochemistry*, 61:109–118, http://www.science direct.com/science/article/pii/S0883292715001389.

Ross, P. S., Ukstins Peateb, I., McClintocka, M. K., Xuc, Y. G., Skillingd, I. P., Whitea, J. D. L., and Houghtone, B. F., 2005, Mafic volcaniclastic deposits in flood basalt provinces: A review, *Journal of Volcanology and Geothermal Research*, 145:281–314.

Sabine, C. L., Feely, R. A., Gruber, N., Key, R. M., Lee, K. et al., 2004, The Oceanic Sink for anthropogenic CO_2, *Science*, 305, 5682:367–371, http://www.pmel.noaa.gov/pubs/outstand/sabi2683/sabi2683.shtml.

Schauberger, B., Archontoulis, S., Arneth, A., Balkovic, J., Ciais, P., Deryng, D. et al., 2017, Consistent negative response of US crops to high temperatures in observations and crop models, *Nature Communications* 8, Article number: 13931, doi.org/10.1038/ncomms13931, http://www.nature.com/articles/ncomms13931.

Schuiling. R. D., 2013, Weathering approaches to CO_2 sequestration. In R. A. Meyers, Ed., *Encyclopedia of Sustainability Science and Technology*, Vol. 3, pp. 1909–1927, ISBN 978-0-387-89469-0, http://ecological aquaculture.org/Part2.pdf.

Schuiling, R. D. and Krijgsman, P., 2006, Enhanced weathering: An effective and cheap tool to sequester CO_2, *Climate Change*, 74(1–3):349–354, http://www.innovationconcepts.eu/res/literatuurSchuiling /enhanced.pdf.

Schuiling, R. D., Wilson, S. A., and Power, I. M., 2011, Enhanced silicate weathering is not limited by silicic acid saturation, *PNAS*, 108, 12:E41, doi.org/10.1073/pnas.1019024108, http://www.pnas.org/content /108/12/E41.full.pdf.

Shafiee, S., and Topal, E., 2009, When will fossil fuel reserves be diminished?, *Energy Policy*, 37, 1:181–189.

Sheets, P. D. and Grayson, D. K. (eds.), 1979, *Volcanic Activity and Human Ecology*, Academic Press, New York.

Siemer, D. D., 2015, Why the MSFR is the "Best" GEN IV Reactor, *Energy Science and Engineering*, 3, 2:83–97, http://onlinelibrary.wiley.com/doi/10.1002/ese3.59/full.

Stallard, R. F. and Edmond, J. M., 1983, Geochemistry of the Amazon: 2. The influence of geology and weathering environment on the dissolved-load, *Journal of Geophysical Research*, 88, C14:9671–9688.

Stranded 2011, http://cleantechnica.com/2011/03/27/25-twh-of-wind-power-idled-in-2010-in-us-grid-storage -needed/.

Sundquist, B., 2005, Topsoil Loss—Causes, Effects, and Implications: A Global Perspective, on-line literature review, edition 6 @ http://home.alltel.net/bsundquist1/se0.html.

Taylor, S. R., 1964, Abundance of chemical elements in the continental crust—A new table, *Geochim. Cosmochim. Acta*, 28:1273–1285.

Taylor, L. L., Quirk, J., Thorley, R. M. S., Kharecha, P. A., Hansen, J., Ridgwell, A., Lomas, M. R., Banwart, S. A., and Beerling, D. J., 2016, Enhanced weathering strategies for stabilizing climate and averting ocean acidification, *Nature Climate Change*, 6:402–406, doi.org/10.1038/nclimate2882 (free at http:// escholarship.org/uc/item/3hw1h419#page-1).

Ten Berge, H. F. M., van der Meer, H. G., Steenhuizen, J. W., Goedhart, P. W., Knops, P., and Verhagen, J., 2012, Olivine weathering in soil, and its effects on growth and nutrient uptake in ryegrass (Lolium perenne L.): A pot experiment, *PLoS ONE*, 7, 8:e42098, http://journals.plos.org/plosone/article?id=10.1371/journal.pone .0042098.

Till, C. E. and Y. I. Chang, 2011, *Plentiful Energy: The Story of the Integral Fast Reactor*, CreateSpace, p. 182.

UN 2017, https://esa.un.org/unpd/wpp/Graphs/Probabilistic/POP/TOT/ (predicted populations circa 2100 AD range from ~6 to 18 billion agrees with WIKIPEDIA).

Vágó, I., Győri, Z., and Loch, J., 1996, Comparison of chromium and nickel uptake of plants grown in different soils, Fresenius *Journal of Analytical Chemistry*, 354:714–717.

Van Straaten, P., 2002, *Rocks for Crops: Agrominerals of Sub-Saharan Africa*, ICRAF, Nairobi, Kenya, 338 pp. (much referenced).

Weinberg, A., 1994, The First Nuclear Era – The Life and times of a Nuclear Fixer, AIP Press.

World 2016: http://knoema.com/smsfgud/world-reserves-of-fossil-fuels.

10 Determination of Secondary Carbonates

Darryl D. Siemer

CONTENTS

10.1 INTRODUCTION

Soils currently contain almost three times as much carbon (to 1-m depth) as does the atmosphere (Table 10.1) in both organic (SOC) and inorganic (SIC) forms. The latter consists of poorly water-soluble carbonates along with some bicarbonate and is frequently disregarded both because it is partly derived from soil parent material (underlying rocks) and, under "natural" conditions, forms relatively slowly from/with atmospheric CO_2. However, it is vitally important because SIC links the Earth's currently strongly human impacted C cycle with the relatively fast biotic soil C cycle which means that its deliberate management could significantly reduce that impact (Wang et al. 2015). Based upon its origin, formation, and morphology, SIC is categorized as lithogenic carbonate (LIC—unmodified base rock detritus), biogenic carbonate (BIC—clamshells, etc.), or pedogenic carbonate (PIC). The last of these is formed by dissolution/re-precipitation of LIC or via dissolution of carbon dioxide (CO_2) to form HCO_3^- followed by its precipitation with Ca^{2+} and/or Mg^{2+} originating from non-LIC minerals (weathered silicate rocks, dusts and/or fertilizers) as calcite/aragonite/dolomite.

10.2 ORIGINS OF PEDOGENIC (SECONDARY) CARBON

Figure 10.1 is the USDA's world map of SIC distribution. The most striking feature of SIC distribution in soils is that much of the world's best farmland contains essentially no SIC within its agriculturally relevant soil horizons. The reasons for this are that they are situated within a udic moisture regime (rainfall exceeds transpiration) which means that the gaseous CO_2 within them (typically two orders of magnitude higher concentration than that of the atmosphere immediately above them) renders their "free" water sufficiently acidic to dissolve calcium and/or magnesium

TABLE 10.1
The Earth's Carbon Reservoirs

Where	Pg C	References
Crustal rocks (mostly limestone)	$1.23E^{+07}$	Earthfacts 2017 and Blatt 1996
Oceans	~40000	Waterencyclopedia 2017
Current atmosphere	852	CDIAC/ORNL 2016
Excess atmosphere (anthropogenic)	255	(400-280)/400*852)
Terrestrial biosphere (total)	2 157–2293	Batjes 1996
Terrestrial biosphere living	600–1000	Falkowski et al. 2000
Terrestrial biosphere dead	~1200	Falkowski et al. 2000
Aquatic biosphere	1–2	Falkowski et al. 2000
Fossil fuels	~4130	Falkowski et al. 2000
SOC total (100 cm)	1462–1548	Batjes 1996
SIC total (100 cm)	695–949	Batjes 1996 and Eswaran et al. 1995
SIC udic soils	6	Eswaran et al. 1995

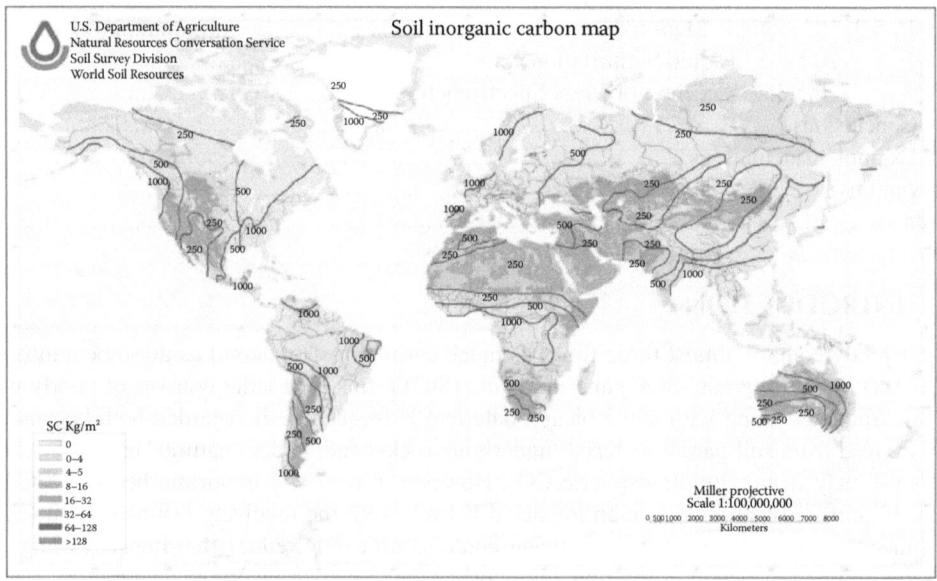

FIGURE 10.1   (See color insert.) Worldwide SIC distribution. (From USDA-NRCS, 2000 (worldwide SIC map circa 2000), https://www.nrcs.usda.gov/wps/portal/nrcs/detail/soils/use/?cid=nrcs142p2_054016.)

carbonates. That, in turn, converts insoluble carbonate minerals to bicarbonate ions which, in the presence of sufficient good quality water, quickly leaches out of the upper soil horizons where most biological activity occurs (Cerling 1984 is an excellent review of this subject). The majority of the US's, South America's, and Europe's most productive farmland fits this description. On the other hand, under arid to semihumid climatic conditions, the dissolution of primary carbonates followed by recrystallization with transpiration-derived CO_2 leads to the formation of pedogenic (secondary) carbonates that usually migrate downwards into the soil column (Zamanian 2016). In many respects, the process is the opposite of podzolization—calcification tends to concentrate calcium in the lower part of the B horizon, whereas podzolization leaches the entire soil column of calcium along with other basic components apt to form insoluble carbonates. Most such soils contain one or more carbonate accumulation horizons, the depth(s) of which depends upon climatic conditions and the characteristics and amounts of calcium/magnesium bearing parent materials. Pedogenic (secondary) carbonates exist in several forms (earthworm biospheroliths, rhizoliths and calcified roots, hypocoatings, nodules, clast coatings, calcretes, and laminar caps) and, in toto, comprise 30%–40% of the world's soil carbon inventory (Table 10.1). They can also serve as an archive of the paleoenvironmental conditions under which the soil had formed. Evidence of that history—paleovegetation, paleotemperatures, paleorainfall, and paleo-pCO_2—is provided by its morphology, stable carbon ($^{13}C/^{12}C$) and oxygen ($^{18}O/^{16}O$) isotopic ratios, unstable carbon (^{14}C) to stable carbon ratios, and the depths of any such carbonate mineral horizons/accumulations. The determination of stable carbon isotope ratios is especially useful because pedogenic carbonates form in isotopic equilibrium with soil-gas ("respired") CO_2 that possesses carbon ratios reflecting the photosynthetic pathway of the vegetation extant at that time. Similarly, $^{18}O/^{16}O$ ratios may reflect the origin of water present during the same period and $^{14}C/^{12}C$ ratios constitute a more or less direct measure of their age. No single isotope-based measurement/indicator is definitive because both the temporal and spatial variability in soil respiration is often high (Singh and Gupta 1977), and a great deal of isotopic exchange can take place (especially of the oxygen atoms) under some conditions that may "scramble" the evidence (Mills and Urey 1940).

10.3 CHEMICAL CHARACTERIZATION

Wang et al. (2012) have compared several of the methods utilized for determining SIC and SOC Both are often determined by difference; i.e., "total" carbon is first determined by analyzing an aliquot of the raw soil and then SOC is determined by analyzing another aliquot after it has been pretreated with a strong inorganic, non-oxidizing (usually hydrochloric) acid. Both often involve combustion of the sample with either oxygen or cupric oxide at a temperature high enough to quickly decompose lime (>900°C) after which the so-generated CO_2 is purged out of the combustion system with a carrier gas and then measured in some fashion. These "finish" measurements may be performed in several ways including infrared spectrometry, absorption/ titration, thermal conductivity, or mass spectrometry. SIC may also be determined "directly" by first pretreating the sample in a manner that removes most/all of the organic carbon before analysis. Such pretreatment may include moderate temperature digestion with concentrated hydrogen peroxide, low temperature combustion with air, elemental oxygen, singlet oxygen, or ozone, and/or extraction with a strongly basic aqueous solution followed by copious water rinses. A more direct approach would be to heat the sample aliquot with a concentrated, non-oxidizing mineral acid (usually phosphoric), and collect the so evolved CO_2 by either freezing it out of an inert carrier gas within a U-tube immersed in liquid nitrogen (LN2) or passing it through an ice cooled tube containing a small amount of molecular sieve (Brochure 2016) prior to the finish. In either case, the CO_2 is subsequently revolved into the measurement system/device by rapidly heating the collection tube. In many laboratories some or most of these operations are automated.

10.4 MORPHOLOGICAL CHARACTERIZATION

Determining the history of a soil begins with optical examination of its structure/morphology. While this starts with the use of a hand lens or eye loupe, much more can be learned by examining thin sections of it with the aid of a petrographic polarizing light microscope. To do so, the soil is first impregnated with an epoxy or polyester resin and then sliced and ground to a thickness of ~0.03 millimeter to render it partially translucent. The object of such examination is to determine the degree that individual features (1) consist of a coating or filler or in nodular forms (a calcrete); (2) appear to be micritic (calcitic mud), micro-spiritic, or sparitic (cemented); (3) appear to have affected the weathering of neighboring primary minerals; (4) possess a "fabric" similar to that of adjacent soil particles; or (5) are mixed with alluvial clays.

Such features are *non*-pedogenic if they (1) exhibit sharp boundaries with neighboring soil particles; (2) appear "pure"—not much inclusion of obviously "primary" minerals (e.g., quartz granules); (3) possess a coarse texture; and/or (4) appear to be free from other pedogenic features such as illuviated clay.

A review of the morphological characterization has been published (Durand et al. 2010) and at least one well-written example is freely available (Srivastava et al. 2016).

10.5 ISOTOPIC CHARACTERIZATION

The fundamental reason why stable isotopic ratios are of interest is that during chemical and/or physical reactions (e.g., diffusion through air, liquids, particulate matter, or biological membranes), lighter mass isotopic species almost always react/move more rapidly than do heavier ones. For example, all else being equal, $Ca^{12}CO_3$ tends to dissolve slightly (very slightly) faster in an acidic solution than does $Ca^{13}CO_3$ (Mook and Vries 2004). In any gas/liquid equilibration between two otherwise identical chemical compounds, the one containing the heavier isotope(s) tends to favor the liquid phase. For example, $^1H_2^{18}O$ and $^2HH^{16}O$ possesses lower vapor pressure than does $^1H_2^{16}O$ and therefore tend to stay in the oceans when higher temperatures shifts some of their water into the atmosphere. Another example is that water falling at the Earth's polar regions tends to be depleted in ^2H (D) and ^{18}O content with respect to equatorial rainfall—more evaporation/precipitation/re-evaporation, etc., translates to greater isotopic differentiation.

"Fractionation factor" or α is analogous to a chemical equilibrium constant, K, but applied to isotope exchange processes. It is a multiplication factor indicating how isotope ratios differ at any given temperature. The following is an example for the exchange of oxygen atoms between calcite and water:

$$\alpha CaCO_3 - H_2O = {}^{18}O/{}^{16}O\ CaCO_3/{}^{18}O/{}^{16}O\ HO = 1.031\ \text{at}\ 25°C\ (\text{Hoefs 2004})$$

Since fractionation factors are small in absolute terms, direct ratio comparisons between a standard and samples usually generate ratios close to unity. For example, significantly "different" specimens may generate raw $^{13}C/^{12}C$ ratios (e.g., 0.99 and 0.98) that are hard to discern because they differ only in the second/third decimal point. To make this difference more readily apparent, 1.000 is usually subtracted from the ratio of the sample's $^{13}C/^{12}C$ ratio to that of the standard, and that result multiplied by 1,000 to produce a "$\delta^{13}C$" or "delta" value.

In general,

$$R_A = A_{heavy}/A_{light}\ \text{where}\ A = \text{Element (C, O, S, H etc.)}$$
$$\delta A‰ = (R_{Asample}/R_{Astd} - 1) * 1000$$

Isotopic deltas are expressed in terms of parts per thousand (‰) which figure is analogous to the parts per million (ppm) or percent (%) figures convenient to other disciplines. For example, R's of 0.99 and 0.98 translate to delta values of −10‰ and −20‰, respectively. Most such figures are negative because the more or less universally accepted standard material, Vienna Pee Dee Belemnite (VPDB), is made from the shells of a Cretaceous marine fossil which possesses an anomalously high absolute $^{13}C/^{12}C$ (R = 0.0112372) defined as a $\delta^{13}C$ of zero. The $^{13}C/^{12}C$ ratios of other "working" standards currently used are referenced to it. Table 10.2 depicts $\delta^{13}C$ values (vs VPDB) obtained for a wide range of carbon-bearing materials. In soil-type carbon, the primary cause of isotopic differentiation is that during growing seasons, root and microorganism respiration rates are both high and represent the only readily mobile (gaseous) source of CO_2 within it (Cerling 1984). The relative abundance of C3 and C4 plants in local vegetation controls the $\delta^{13}C$ of PIC formed during that season and overall growing conditions determine the absolute amount so-generated. Due to the isotopic discrimination by the two photosynthetic pathways employed by advanced plant life forms, the $\delta^{13}C$ of CO_2 in/under/around C3 plants (−27‰ on average) differs from that of C4 species

TABLE 10.2
$^{13}C/^{12}C$ Isotopic Ratios of the Earth's Carbon Sources

Carbon Source	δ ^{13}C ‰	Reference
PDB standard (a marine mollusk)	0	ESRL/NOAA 2016
Modern marine plankton	–17 to –22	Maslan and Swann 2006
C3 plant life	–22 to –30	Cerling 1984
C4 plant life	–9 to –15	Cerling 1984
Atmosphere and ocean	–8	ESRL/NOAA 2016
Bitumen/coal	–25–32	Muehlenbachs et al. 1988
Bacterial methane[a]	–60	Washington State University 2016
Biomass methane	–20	Washington State University 2016
Fossil Fuel derived methane	–40	Washington State University 2016

[a] Bacterial = termite, ruminant, and wetlands, mean atmospheric CH_4 ^{13}C = ~–57, CH_4 lost from atmosphere/a = 520 Tg. Current atmospheric CH_4 conc. ~1.7 ppm by volume. Short-term, it's become almost as important as anthropogenic CO_2 (Washington State University 2016).

(–13‰ average). Additional isotopic C discrimination results from CO_2 diffusion in soil (~+4.4‰) and carbonate precipitation rates (~+11‰).

Because absolute differences of $^{13}C/^{12}C$ ratios are small and the less abundant isotopes (usually the heavier ones) of relevant elements are usually relatively rare (^{13}C represents about1% of natural carbon), the instruments utilized to determine these ratios must simultaneously exhibit excellent precision along with the ability to accurately determine both large and small signals generated by different analytes (isotopes) at essentially the same time (high degree of response linearity).

10.5.1 IRMS-Based $^{13}C/^{12}C$ Ratio Determination

To date the majority of soil/mineral δ^{13}C determinations have been performed with specialized magnetic sector-type—not time-of-flight or quadrupole—isotopic ratio mass spectrometers (IRMSs) featuring multiple (at least two) detectors. This dual (or more) channel approach to mass spectrometry dates from Harold Urey's conceptual breakthrough almost 7 decades ago (Urey 1948) which instrument was subsequently developed/described by McKinney et al. (1950) two years later. Despite significant improvements in electronics, vacuum system design and currently, automation, the principles embodied in that spectrometer still constitute the basis of modern stable isotope ratio mass spectrometry. That concept employs simultaneous (or almost simultaneous) detection of signals generated by ions for both a standard and samples that in turn eliminates most of the instrumental "drift" and non-linearity imperfections that would otherwise fatally compromise such determinations. The key feature of the McKinney et al. (1950) instrument was a "changeover valve" which repeatedly switched CO_2 fed to the spectrometer from that generated from samples to that from an isotopic standard. While the gas from one is introduced into the mass spectrometer, gas from the other goes to a waste line and vice versa. Such dual-inlet IRMS systems can routinely deliver precision levels of 0.01‰ for δ^{13}C. Continuous-flow IRMS (CFIRMS), an alternative configuration to dual-inlet IRMS, is capable of analyzing much smaller samples but may offer lower precision, typically 0.1‰ for δ^{13}C (Brenna et al. 1997).

Muccio and Jackson (2008) and Chapter 11 of Mook and Vries' (2004) book are well written, comprehensive, and readily obtained reviews of this subject. Unfortunately, no analytical method is perfect. Werner and Brand (2001) wrote the best review of IRMS's foibles. These issues mostly have to do with the fact that calculation of a δ^{13}C value via comparison of the responses of a mass spectrometer fed with CO_2 is complicated by the fact that oxygen consists of three different isotopes meaning that any given sample will contain non-zero quantities of 12 stable isotopologs having molecular weights ranging from 44 to 49. However, since more than 99.98% of such gas will consist of the four most abundant such isotopologs: $^{12}C^{16}O^{16}O$ (~98.42%, AMU = 44), $^{13}C^{16}O^{16}O$ (~1.09%, AMU = 45), $^{12}C^{16}O^{17}O$ (~0.07%, AMU = 45), and $^{12}C^{16}O^{18}O$ (~0.40%, AMU = 46), in practice, $\delta^{13}C$ figures are derived from instrumental responses generated by species possessing masses of 44, 45 *and* 46 AMU. In principle though, unless the spectrometer possesses a mass resolution capability of >52000 (i.e., ~45/(difference between the exact masses of 13CO$_2$ and C16O17O)) (which characteristic conventional IRMS instruments do not possess), accurate $\delta^{13}C$ values cannot be determined without knowing the sample's oxygen isotope ratios. The problem arises because natural $\delta^{17}O$ ratios vary for the same reasons that $\delta^{13}C$'s do and the materials that serve as $\delta^{13}C$ standards do not possess exactly the same $^{18}O/^{17}O/^{16}O$ ratios as did the original VPDB (or PDB) standard. Most of the resulting confusion is due to the fact that "deltas" rather than actual isotopic makeups ratios are usually discussed/reported (Werner and Brand's paper is "best" because it lists $^{16}O/^{18}O$ as well as the $^{13}C/^{12}C$ ratios of the common IRMS carbon standard materials; Werner and Brand 2001). However, there is still uncertainty due to the fact that while ^{18}O and ^{17}O concentrations tend to vary in the same directions for the same reasons, a correction based upon one may not be accurate for situations involving the other isotope. In any case, such confusion may generate uncertainties regarding the comparability of $\delta^{13}C$ values generated by different laboratories.

One way to reduce the uncertainty of an IRMS-determined $\delta^{13}C$ value due to such uncertainties is to shake the sample and standard gases with aliquots of the same water before analysis. Because oxygen atoms within CO_2 rapidly exchange with those of water (Urey 1948), as long as there is a far greater number of water molecules than of CO_2 present when such "normalization" is performed, the latter's $\delta^{17}O$ will then reflect that of the water regardless of what it may have been in the as-received material.

The Example section at the end of this chapter goes through some of these calculations. The "importance" of these errors depends upon what the person receiving the data chooses to do with it, which consideration is beyond the scope of this review. The bottom line is that while it is possible to achieve precisions of 0.01% for $\delta^{13}C$ determinations via IRMS, the mean of such values obtained during one day may differ "significantly" from those gotten at another time or generated by a different laboratory—precision isn't necessarily the same as accuracy.

IRMS's other drawback is that the instruments are rather expensive—typically about $140,000—and their operation often requires a good deal of rather expensive "consumables"—typically about $10,000 per year.

10.5.2 INFRARED SPECTROMETRIC ISOTOPIC DETERMINATION

A fundamentally different way to perform isotopic ratio determinations is based on the fact that infrared spectra are generated by the electrons within multiatomic species (molecules, ions, etc.) shifting between different quantized vibrational/rotational energy levels. Because the frequency (energy) of such transitions is determined by the "reduced mass" (m1 * m2/(m1 + m2) of atoms bonded to each other (e.g., the C and O within carbonates), absorption spectra generated by molecules containing different isotopes differ sufficiently that their magnitudes can be distinguished/determined with high resolution infrared spectrometers.

Vibrational energy = $1/2pi*(k/\mu)^{0.5}$ where μ = m1 * m2/(m1 + m2)
(e.g., for the $^{12}C^{16}O$ "stetch", m1 = 12 & m2 = 16)

These instruments are different from those utilized to generate the CO_2 absorption spectra familiar to people who have pondered the mechanisms responsible for anthropogenic global warming (Figure 10.2). The latter spectra are pressure-broadened, nearly "saturated" (almost all light absorbed), nearly structure-less, "bands" representing the attenuation of infrared radiation passed through the entire atmosphere. The former (Figure 10.3) focus on specific features within one band

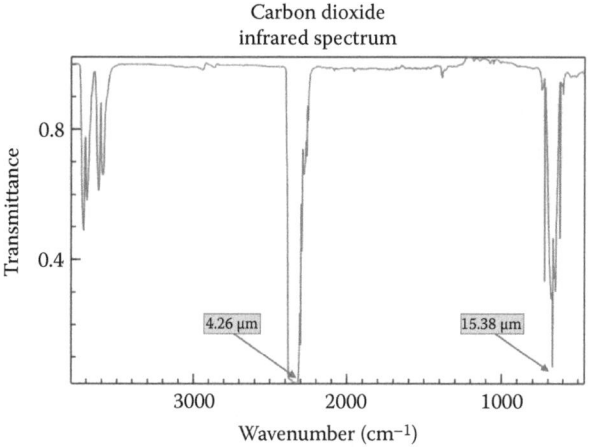

FIGURE 10.2 Low-resolution CO_2 absorption spectra (the 4.2 µm/2300 cm^{-1} line is saturated).

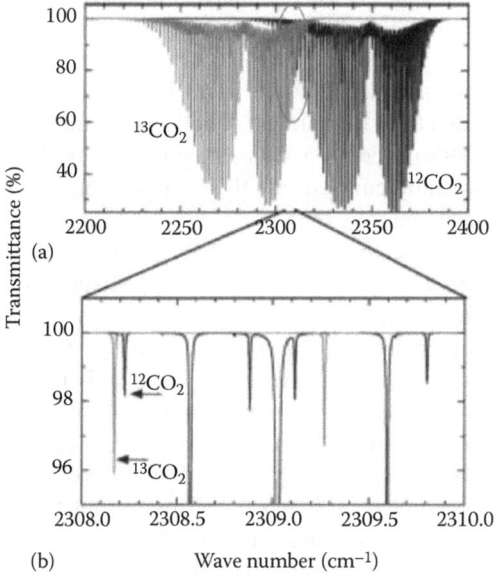

FIGURE 10.3 Medium (a) and high (b) resolution "4.2 µm CO_2 band" infrared spectra.

of that spectrum (usually the one situated at ~4.2 µm/2300 cm^{-1}) measured in a way that enables the resolution of individual "lines" and generates a linear (not saturated) analytical response.

Because isotopically generated energy (or wavelength) perturbations of molecular vibronic energy levels are small, these instruments must be capable of resolving/measuring spectral features that are both rather "weak" (possess relatively low absorption extinction coefficients) and are often closely situated to potentially interfering spectral features.

The key to achieving high sensitivity is to utilize a much longer absorption path length (typically 1–10 m) than is possible with conventional infrared spectrometers. Achieving sufficiently high resolution is done by either utilizing a light source that is simultaneously bright, inherently monochromatic, and capable of being shifted back and forth either to or over two or more adjacent wavelengths (i.e., tunable diode laser absorption spectrometry (TDLAS)—the most common approach) or high resolution (Fourier Transform-type) IR spectrometer (Griffith et al. 2012).

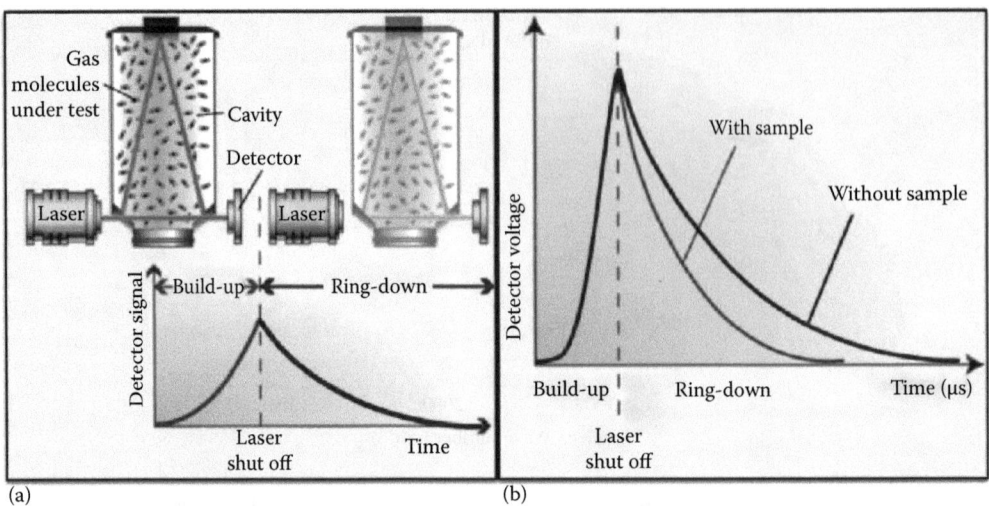

FIGURE 10.4 (a) Schematic of instrument and (b) difference in detector voltages with and without sample. (Courtesy of Picarro Inc., Santa Clara, CA, 2015.)

Cavity ring-down spectrometry (CRDS) was one of the first practical techniques to overcome conventional infrared spectrometry's limitations. A cell containing the sample gas is bounded by highly reflective mirrors (99.99%) to create a resonant optical cavity (Figure 10.4). A coherent (laser generated) light beam is resonantly coupled into this cell and bounced back and forth between its mirrors. The ring-down measurement is then taken by switching off the laser and measuring the time constant of the exponential decay of light intensity ("ringdown time") that is determined by the analyte's concentration. Its ability to distinguish different species is solely determined by its light source—a wavelength tunable pulsed laser that is repeatedly jumped back and forth between discrete wavelengths (e.g., the absorption peaks of neighboring $^{13}CO_2$ and $^{12}CO_2$ lines). Gas pressure within the gas cell/chamber is generally reduced (to typically under ~0.1 atmosphere) in order to facilitate the "sucking" of sample gas into/through it and reduce/normalize spectral line widths.

CRDS represented a major advance because it was the first optical technique able to match the sensitivity of conventional gas chromatography while also being able to distinguish between isotopomers. However, it is/was rather fussy to use because resonant wavelength coupling requires subnanometer optomechanical precision and stability which tends to generate "noise" and complicate field service. In addition, any resonant CRDS-based system requires a high degree of laser wavelength monitoring and control. These issues can be overcome but tend to increase cost and complexity and decrease robustness.

Figure 10.5 represents a fixed path length (1.5 meter) tunable diode laser absorption spectrometer (TDLAS—in this case, Campbell Scientific's TDL-100) that has been commercially available for almost two decades (see Bowling et al. 2003—another especially well-written paper). A tunable solid-state laser is directed through a low pressure gas sample (e.g., 400 ppmv CO_2 in air at 0.02–0.10 bar within a 1.5-m long cell) and focused onto a photodiode. The laser is frequency-tuned through nearby molecular absorption features ("lines") of both $^{13}CO_2$ and $^{12}CO_2$—a "stronger" spectral line is chosen for the former to compensate for the fact that it will always be present at ~100× lower concentration—and the change in light transmission over each peak is directly related to concentration via the Beer–Lambert–Bouguer Law, i.e., $\log_{10}(Io/I)$ or absorbance, which figure is directly proportional to an analyte's (e.g., $^{13}CO_2$) concentration. The light's intensity is continuously monitored, not a "ring-down" time after the source is shut off. As is also the case with CRDS, to insure both the linearity and magnitude of its analytical response, the light source's bandwidth must be considerably narrower than that of the molecule's absorption line—one of the reasons why

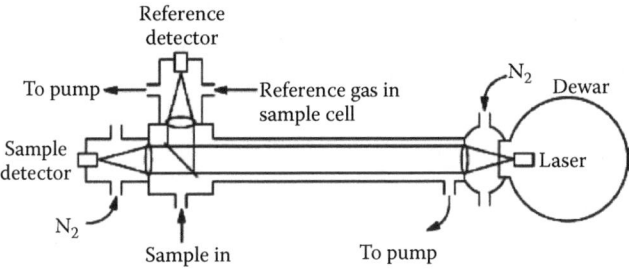

FIGURE 10.5 Tunable diode laser isotope spectrometer.

a tunable laser, not light generated by a hot wire or "globar" and then dispersed by a prism or grating serves as its light source. The current version of this instrument is easier/cheaper to maintain because a Peltier (thermoelectric) cooler rather than LN2 cools its tunable diode light source. Both versions simultaneously measure the isotope ratios of a standard and sample gas to minimize measurement errors.

An off-axis integrated cavity output spectrometer (OA-ICOS) has recently been developed to provide greater sensitivities for trace level atmospheric gas determinations (Figure 10.6). In this case the laser beam enters the optical cavity (measurement cell) at an off-axis angle, resulting in a multibounce light path between the mirrors and therefore a much longer measurement light path than is the case with the TDL (all else being equal that translates to higher sensitivity). Unlike the CDRC, its off-axis optical beam is non-resonant, meaning that alignment (light trajectory) is neither unique nor fussy. Since especially high sensitivities are not needed for measuring $\delta^{13}C$ ratios in air, water, or soils, the pair of lines is different than that utilized by the TDL; i.e., lines situated within the weak ~2.1-micron band instead of the much brighter/stronger fundamental lines making up CO_2's 4.2-micron absorption band (see Figure 10.2).

Any of these instruments can be rendered portable and/or "retuned" to measure the concentrations of other trace level atmospheric gases. Zellweger et al. (2012) have written a comparison of all currently available, IRIS-capable, instruments "tuned" to determine atmospheric CO rather than CO_2. Most of them can also simultaneously determine $\delta^{17}O$ and/or $\delta^{18}O$ in addition to $\delta^{13}C$ in normal environmental-type samples. (However, none are sensitive enough to determine $\delta^{14}C$ in such materials.)

In general, all IRIS-based isotope ratio spectrometers are relatively compact, rugged, and cost about half as much as state-of-the art IRMS instruments. They also generally require less attention and consumables. For soil carbon analyses, all of them would have to be coupled to something capable to converting it to CO_2, which means that they are often direct coupled to "total carbon" analyzers. However, there are many other ways to accomplish this limited only by the analyst's

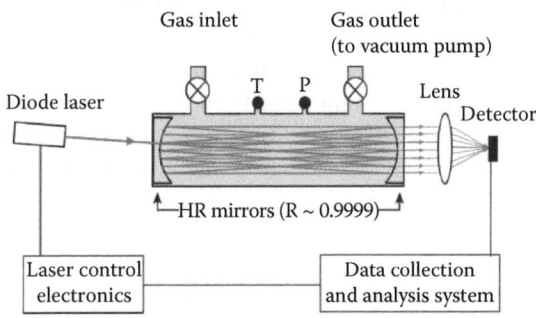

FIGURE 10.6 Schematic of off-axis integrated cavity output spectrometer. (From Gupta M. and Leen, J. B., American Laboratory, http://www.americanlaboratory.com/913-Technical-Articles/129080-Field-Portable -Analyzers-Based-on-Cavity-Enhanced-Laser-Absorption-Spectrometry/, 2013.)

time constraints and willingness to "fiddle." For example, SIC could be determined by reacting soil with sulfuric acid within a syringe, noting the volume of gas generated (total SIC) and that gas then introduced into the IRIS (or IRMS) for isotopic evaluation.

10.5.3 ^{14}C ISOTOPIC MEASUREMENTS

10.5.3.1 Liquid Scintillation

Currently, the two primary ways of determining ^{14}C are liquid scintillation counting (LSC) and accelerator mass spectrometry (AMS). The former technique detects (counts) beta particles emitted by the radioactive decay of ^{14}C (half-life = 5730 years) within a "scintillation cocktail" comprised of organic materials that convert ionizing radiation to light pulses detected by a photomultiplier tube. Signals/count rates generated by natural carbon-type samples are small because ^{14}C represents only about 1 in 10^{12} atoms in "new" carbon and even less in "old" carbon (Godwin 1962). For example, the following calculation derives the maximum possible ^{14}C LSC response/signal generated by 1 g of new natural carbon.

^{14}C Activity = disintegrations/s (or "Becquerel" Bc) = #^{14}C atoms * ln2/half-life in seconds
Number C atoms in 1 g carbon (mostly ^{12}C) $\approx$ 6.023E23 (N Avogadro)/12 = 5.02E22
Number ^{14}C atoms/g fresh carbon = 5.02E22/1E12 = 5.02E10
^{14}C activity/g C = 5.02E10 * 0.693/(5730 * 365 * 24 * 3600) = 0.192 Bc/g ($\approx$11.5 dpm/g)

In practice, only a fraction, typically 30%–60%, of the beta particles ("hot" electrons) emitted by decay of the sample's ^{14}C atoms generate light pulses "counted" by the instrument. The rest are "quenched" which means that real ^{14}C responses generated from natural soil carbon samples will be even smaller and therefore even more difficult to accurately characterize.

For example, if a soil chemist manages to quantitatively convert/transfer the carbon in 100 g of a 3% "new pedogenic $CaCO_3$"–containing soil (no SOC) to a form compatible with 20 cm^3 of liquid scintillation cocktail (no mean feat) the maximum possible response in terms of disintegrations (decays) per minute (dpm) would be

100 * 0.03 * 12/(40 + 12 + 3 * 16) * 11.5 = 4.03 dpm

If counting efficiency (1/quenching factor) = 0.5, net response = 2.015 dpm.

If that soil's $CaCO_3$ were formed from vegetative carbon 13,000 years ago, its quench-corrected net LSC count rate would be

2.015 * (1/2)^(13000/5730) = 0.446 dpm

Since natural background radiation typically generates LSC count rates of about 2 dpm, soil samples must be counted for long time periods to generate net responses accurately distinguishable from "noise" (standard deviation of #counts = #counts$^{1/2}$).

For example, if one were to count the immediately above-described sample for "only" one hour, the relative standard deviation of the analytical response would be determined as follows:

Since error propagation of n results generated by the addition or subtraction of two numbers (A & B) is

$\sigma R^2 = \sigma A^2 + \sigma B^2$ (Laitinen 1960)
$RSD_{60\,min} = \sigma_R/R = (60 * (0.446+2) + 60 * 2)^0.5/(60 * 0.446) = 0.61$ (terrible!)

a 6-day count would be required to lower the RSD of this determination to 5%.

This example demonstrates that while LSC represents a handy/cheap (a still-viable used LSC instrument costs only about $10k) way of tracking the behavior of carbon to which a ^{14}C tracer has been added/equilibrated, it is generally not very useful for experiments involving only natural carbon.

10.5.3.2 Accelerator Mass Spectrometry

Since age determinations are the main reason why $^{14}C/^{12}C$ ratio ($\delta^{14}C$) measurements are made, AMS has largely supplanted LSC (or any other radiochemical) approach. Since AMS counts/detects all of the ^{14}C atoms entering the instrument, not just those that happen to decay while within it, it is intrinsically much more sensitive. For this application it is superior to other forms of mass spectrometry due to its far more efficient suppression of interfering isobaric (same AMU) species such as ^{14}N, ^{13}CH and $^{12}CH_2$. However, it also requires a much greater investment of time and money. With very few exceptions samples must be sent to "outside" laboratories that specialize in this technique. The US currently has about 12 such facilities and costs quoted for a single such analysis currently range from $300 to $2400.

A typical ^{14}C SIC analysis might begin with the following:

1. Gently heat a gram of the sample with super concentrated H_3PO_4 (it retains water but releases CO_2) within a tube through which helium is slowly passing.
2. Sweep the so generated CO_2 into a U tube immersed in a LN2 dewar to freeze-concentrate it.
3. While still passing the helium carrier gas through the tube, remove it from the dewar to revolatilize the CO_2.
4. Pass that gas through a small quartz glass tube containing powdered iron within a ~500°C tube furnace (the carbon is converted to iron carbide).
5. Cool that powder, mix it with powdered aluminum or copper and press the mixture into a pellet.
6. Mount that pellet into the AMS instrument's sampling system and push the "start" button.

The finish analysis involves focusing a beam of high energy cesium ions (the periodic table's least electronegative element) onto that pellet, which serves to convert (volatilize/atomize/ionize) the sample's currently carbide-type carbon to a form (negatively charged carbon ions) that the spectrometer can subsequently extract, separate and (eventually) measure. A good description of what that involves is available at Purdue University's website (https://www.physics.purdue.edu /ams/introduction/ams.html). A brochure written by one of the companies that perform AMS determinations provides a more detailed description of how these spectrometers operate (BETA Analytic 2017).

10.6 CONCLUSIONS

The review presented is useful to anyone wishing to perform the experimentation required to assign numbers to "geoengineering" schemes like the one described in this volume's Chapter 9, "Silicate Weathering to Mitigate Climate Change." If properly designed experiments were to be performed, it would be neither difficult nor expensive to determine how such scenarios would actually work. To date none of the in-situ "enhanced weathering" based atmospheric carbon reduction schemes that have been proposed has been evaluated in a manner that affords information of much use to policy makers.

10.7   EXAMPLE CALCULATIONS

"DATA"

	Mean Earth Fraction	Range of Fractions	Fractional Variation
^{13}C	0.01108	0.0105–0.0113	~7%
^{12}C	0.98892	NA	NA
^{16}O	0.99757	0.99738–0.99776	0.04%
^{17}O	$3.8E^{-4}$	$3.7E^{-4}$–$4.0E^{-4}$	7.9%
^{18}O	$2.05E^{-3}$	$1.88E^{-3}$–$2.22E^{-3}$	16.6%
$^{13}C/^{13}C$ (V)PDE	0.011237	NA	0

$\delta^{13}C3$ plant life $= -27‰$
$\delta\ ^{13}C4$ plant life $= -13‰$
Volume of one mole of any gas at one bar pressure and $0°C \approx 22.4$ liters
"Natural" surface carbon $^{14}C/C_{tot}$ ratio $= 1.00E^{-12}$
https://en.wikipedia.org/wiki/Isotopes_of_oxygen & carbon
https://en.wikipedia.org/wiki/%CE%9413C#Reference_standard

Average earth carbon's $\delta^{13}C = ((0.01108/0.98892)/0.11237 - 1) * 1000 = -2.92‰$
Average earth $^{17}O/^{13}C$ in CO_2 ratio $= 2 * 3.8E^{-4}/0.01108 = 0.06859$
Range $^{17}O/^{13}C$ ratios $= 7.87\%$ of $0.06859 = 0.005398 = 0.54\%$
(translates to an uncertainty of 0.54% (0.005398/1000) in IRMS-determined $\delta^{13}C$ values if the sample's ^{17}O isotope ratio isn't co-determined—worse if the standard's $\delta^{17}O$ is also unknown)

Hypothetical soil example

Assume that 1.23 g of dry powdered soil reacted with sulfuric acid within a syringe at 25°C and 0.93 bar generated 11.2 cm³ of CO_2 gas having a $\delta^{13}C$ of $-15‰$
Wt% SIC = moles CO_2 * 44 g/mole/1.23 g = 100 * (11.2 * 273/(273 + 25)) * .93) * 44/22400 = 1.87%

Question: If another isotopic analysis indicated a soil gas CO_2 $\delta\ ^{13}C$ of $-22\ ‰$, what fraction of this soil sample's SIC is pedogenic?

Ans: Since >99% of "average earth carbon" is non pedogenic (limestone, etc.) and its $\delta13$ is $-2.92\ ‰$
Fraction pedogenic carbon (PIC) in this SIC $= (-15-(-2.92)/(-22-(-2.92)) = 0.633$

Next, if $^{14}C/C_{tot}$ of this sample's SIC were $0.23E^{-12}$, how "old" is its pedogenic carbon?

Ans: Since most non-pedogenic SIC is too old (>10 * 5730 years) to have "any" ^{14}C, the number fraction of ^{14}C in "new" pedogenic C would $= 1E^{-12} * 0.633 = 0.633E^{-12}$
Which means that it's "age" in terms of ^{14}C half-lives $= \log_{10}(0.23/0.633)/\log_{10}(1/2) = 1.46$
Which translates to $1.46 * t½ = 1.46 * 5730 = 8370$ years

Next, let's assume that this soil sample was taken immediately downwind of a bituminous coal briquette factory and that its organic carbon $^{14}C/^{12}C$ ratio were $0.25E^{-12}$. What fraction of its organic carbon originated from that coal?

Ans: Since coal is also too old to contribute much ^{14}C, virtually all of it must have come from other organic materials. If we assume that such carbon exchanges back and forth between the soil and atmosphere relatively quickly (t << 5730 years; see Simonson 1959), then its ^{14}C content should roughly approximate that of the atmosphere's. If so, fraction "coal" = 1– 0.25E[-12]/1E[-12] = 0.75

Finally, let's say that another sample came from a 28-cm diameter plastic pail grow bucket/ lysimeter (Llansadwrn 2017) like that invoked at the end of my other contribution to this book (it was amended with easily-weathered basaltic rock powder to convert its soil gas CO_2 to SIC). Let's also assume that the 20 kg of soil in that grow bucket originally possessed no SIC and that enough rainwater (no minerals) was added during the growing season to penetrate the entire soil column so that "runoff" could be drained from the bucket's bottom, its volume measured, and then analyzed. Let's also assume that 13 liters of such water was collected, its pH was 7.5 and its alkalinity (sum of its bases titratable with strong acid ($mHCO_3^-$ + $2mCO_3^{2-}$ to pH of 4.3) was 0.015 equivalents per liter). Finally, let's assume that at the end of the season, the soil ended up containing 0.32 wt% SIC. First question: How much carbon was in the runoff water?

Since pK_{a2} carbonic acid = 10.23 and pH ($-\log_{10}[H^+]$) = 7.5
$[CO_3^=]/[HCO_3^-]$ = $10^{(-10.23-7.5)}$ = 0 0019 (i.e., almost all of that water's carbon is bicarbonate)
If so, it contains 0.015 * 13 = 0.195 mole of originally atmospheric-type Ct

Next, how much atmospheric carbon was "sequestered"?

Its post-season SIC corresponds to 20 kg * 1000 g/kg * 0.0032/12 = 5.33 moles of carbonate-type C
Total carbon fixed/season = (5.33 + 0.195) = 5.55 g moles

Finally, assuming a six-month growing season, what was the C sequestration rate?

CDR rate = 5.55/(6 * 30 * 24 * 3600)/($\pi(0.28/2)^2$) = 5.8 μmole/s/m^2

ABBREVIATIONS

AMS   Accelerator mass spectrometry (or spectrometer)
AMU   Atomic mass unit (nearest whole number atomic/molecular mass)
Bc   Becquerel disintegrations (atom decays) per second
CDR   Carbon dioxide removal (from atmosphere)
CDRS   Cavity down-ring spectrometry (or spectrometer)
dpm   disintegrations (atom decays) per minute
IRIS   Infrared isotope spectrometry
IRMS   Isotope ratio mass spectrometry (or spectrometer)
LN2   liquid elemental nitrogen
LSC   Liquid scintillation counter (or counting)
MS   Mass spectrometry (or spectrometer)
OA-ICOS Off-axis integrated cavity output spectrometry (or spectrometer)
SIC   Soil inorganic carbon
SOC   Soil organic carbon
TDL   Tunable diode laser absorption spectrometry (or spectrometer)

REFERENCES

Batjes, N. H., 1996, Total carbon and nitrogen in the soils of the world, *European Journal of Soil Science*, 47, 151–163, http://library.wur.nl/WebQuery/file/isric/fulltext/isricu_t47d6414d_001.pdf.

BETA Analytic, 2017, Introduction to radiocarbon determination by the accelerator mass spectrometry method, https://www.radiocarbon.com/PDF/AMS-Methodology.pdf.

Bowling, D. R., Sargent, S. D., Tanner, B. D., and Ehleringer, J. R., 2003, Tunable diode laser absorption spectroscopy for stable isotope studies of ecosystem—Atmosphere CO_2 exchange, *Agricultural and Forest Meteorology*, 118, 1–19, http://www.sciencedirect.com/science/article/pii/S0168192303000741.

Brenna, J. T., Corso, T. N., Tobias, H. J., and Caimi, R. J., 1997, High-precision continuous-flow isotope ratio mass spectrometry, *Mass Spectrometry Reviews*, 16, 227–258.

Brochure, 2016, https://tools.thermofisher.com/content/sfs/brochures/BR30178-KIEL-IV-Carbonate-Device -EN.pdf.

Cerling T. E., 1984, The stable isotopic composition of modern soil carbonate and its relationship to climate, *Earth and Planetary Science Letters*, 71 (1984), 229–240.

Durand, N., Monger, H. C., and Canti, M. G., 2010, Calcium carbonate features. In *Interpretation of Micromorphological Features of Soils and Regoliths*, edited by G. Stoops, V. Marcelino, and F. Mees, pp. 149–194, Elsevier, ISBN: 978-0-444-53156-8.

Earthfacts, 2017, http://www.earthfacts.com/earth-dynamics/earthscrust/.

ESRL/NOAA, 2016, https://www.esrl.noaa.gov/gmd/outreach/isotopes/deltavalues.html.

Eswaran, H., Van den Berg, E., Reich, P., and Kimble, J., 1995, Global soil C resources. In *Soils and Global Change*, edited by R. Lal, J. Kimble, E. Levine, and B. A. Stewart, pp. 27–44. CRC/Lewis Publishers.

Falkowski, P., Scholes, R. J., Boyle, E., Canadell, J., Canfield, D. et al. 2000. The global carbon cycle: A test of our knowledge of earth as a system, *Science*, 290 (5490), 291–296, doi.org/10.1126/science.290.5490.291. PMID 11030643.

Godwin, H., 1962. Half-life of radiocarbon, *Nature*, 195 (4845), 984, doi.org/10.1038/195984a0.

Griffith, D. W. T., Deutscher, N. M., Caldow, C., Kettlewell, G., Riggenbach, M., and Hammer, S., 2012, A Fourier transform infrared trace gas and isotope analyser for atmospheric applications, *Atmospheric Measurement Technology*, 5, 2481–2498, doi.org/10.5194/amt-5-2481-2012,2012.

Gupta M. and Leen, J. B., 2013, Field Portable Analyzers Based on Cavity Enhanced Laser Absorption Spectrometry, American Laboratory, posted: January 18, http://www.americanlaboratory.com/913 -Technical-Articles/129080-Field-Portable-Analyzers-Based-on-Cavity-Enhanced-Laser-Absorption -Spectrometry/.

Hoefs, J., 2004, *Stable Isotope Geochemistry*, 5th ed., Springer.

Laitinen, H. A., 1960, *Chemical Analysis*, Table 26-2, McGraw-Hill.

Llansadwrn, 2017, http://www.llansadwrn-wx.co.uk/evap/lysim.html#calcs.

Maslin, M. A. and Swann, G. E. A., 2006, Isotopes in marine sediments. In *Isotopes in Palaeoenvironmental Research*, edited by Melanie J. Leng, pp. 227–290. Springer, ISBN 978-1-4020-2503-7.

McKinney, C. R., McCrea, J. M., Epstein, S., Allen, H. A., and Urey, H. C., 1950, Improvements in mass spectrometers for the measurement of small differences in isotope abundance ratios. *Review of Scientific Instruments*, 21, 724, doi.org/10.1063/1.1745698.

Mills, G. A. and Urey, H., 1940, The kinetics of isotopic exchange between carbon dioxide, bicarbonate ion, carbonate ion and water, *Journal of the American Chemical Society*, 6 (5), 1019–1026.

Mook, W. and Vries, J., 2004, *Environmental Isotopes in the Hydrological Cycle Principles and Applications. Volume 1: Introduction—Theory, Methods, Review*, edited by W. Mook, International Atomic Energy Agency, Vienna, pp. 1–271, http://www-naweb.iaea.org/napc/ih/documents/global _cycle/Environmental%20Isotopes%20in%20the%20Hydrological%20Cycle%20Vol%201.pdf.

Muccio, Z. and Jackson, G. P., 2009, Isotope ratio mass spectrometry, *Analyst*, 134, 213–222 (first published as an Advance Article on the web 14th November 2008), doi.org/10.1039/b808232.

Muehlenbachs, K., Steer, J. G., Hogg, A., Ohuchi, T., and Beaulieu, G., 1988, Natural Variations of [13]C Abundance in Coal and Bitumen as a Tool to Monitor Co-Processing, pp. 123–130, https://web.anl.gov /PCS/acsfuel/preprint%20archive/Files/33_1_TORONTO_06-88_0122.pdf.

Simonson, R. W., 1959, Outline of a generalized theory of soil genesis, *Soil Science Society of America Proceedings*, 23, 152–156.

Singh, J. S. and Gupta, S. R., 1977, Plant decomposition and soil respiration in terrestrial ecosystems, *Botanical Review*, 43 (4), 449–528, http://www.jstor.org/stable/4353928.

Srivastava, P., Aruche, M., Arya, A., Pal, D. K., and Singh, L. P., 2016, Micromorphological record of contemporary and relict pedogenic processes in soils of the Indo-Gangetic Plains: Implications for mineral weathering, provenance and climatic changes, *Earth Surface Processes and Landforms* 41, 771–790.

University of Washington, 2016, Problem Set #3, http://www.ocean.washington.edu/courses/geol330/PS3_2016 .pdf.

Urey, H. C., 1948, Oxygen isotopes in nature and in the laboratory, *Science*, 108, 489–96, PMID 17808921, doi.org/10.1126/science.108.2810.489.

USDA-NRCS, 2000 (worldwide SIC map circa 2000), https://www.nrcs.usda.gov/wps/portal/nrcs/detail/soils /use/?cid=nrcs142p2_054016.

Wang, X., Wang, J., and Zhang, J., 2012, Comparisons of three methods for organic and inorganic carbon in calcareous soils of northwestern China, *PLoS ONE*, 7 (8), e44334, doi.org/10.1371/journal.pone.0044334.

Wang, X., Wang, J., Xu, M., Zhang, W., Fan, T., and Zhang, J., 2015, Carbon accumulation in arid croplands of northwest China: Pedogenic Carbonate exceeding organic carbon, *Nature Scientific Reports*, 5, 11439, https://www.ncbi.nlm.nih.gov/pmc/articles/PMC4473677/pdf/srep11439.pdf.

Werner, R. A. and Brand, W. A., 2001, Referencing strategies and techniques in stable isotope ratio analysis, *Rapid Communications in Mass Spectrometry*, 15, 501–519.

Zamanian, Z. G., 2016, Pedogenic carbonates: Forms and formation processes, *Earth Science Reviews*, 157, 1–17.

Zellweger, C., Steinbacher, M., and Buchmann, B., 2012, Evaluation of new laser spectrometer techniques for in-situ carbon monoxide measurements, *Atmospheric Measurement Techniques*, 5, 2555–2567.

11 Effects of Plant Invasions on the Soil Carbon Storage in the Light of Climate Change

Constanze Buhk and Hermann F. Jungkunst

CONTENTS

11.1 INTRODUCTION

One crucial driver of environmental change that affects soils is biotic exchange leading to exotic plant invasions. This is exacerbated by human actions like disturbances and travel (Bradley, Blumenthal, Wilcove, & Ziska, 2010; van Kleunen et al., 2015), and increases its relevance under elevated temperatures, increased CO_2 (Liu et al., 2016), and increasing frequency of extreme weather events (Sorte et al., 2013). Wind storms and pathogens increase invaders' success (Bebber et al., 2013), as do fluctuations in resources typically found under more extreme weather conditions (Liu, van Kleunen, & Cornelissen, 2017).

As a feedback mechanism, invasions by certain plant species may influence the soil's capacity to serve as a carbon (C) sink (Tamura & Tharayil, 2014), and may, therefore, have a direct positive or negative effect on future global warming. Vegetation dominated by exotic species tends to build more biomass compared with native vegetation (Vila et al., 2011), which could indicate an increase in C storage through invasions and consecutively a decelerating effect on climate change. However, opposing processes indicate a loss of the function of the soil as a C sink. For example, certain changes in biogeochemical cycles could have an accelerating effect on climate change (Tamura & Tharayil, 2014).

The objective of this chapter is to evaluate the positive and negative aspects of massive plant invasions on soil C emissions and C sequestration in soils and to open the perspectives of future investigations.

Impacts on biogeochemical cycling in the soil are strong and complex under increasing rates of new plant invasions with usually high growth rates (Ehrenfeld, 2010; Sousa, Morais, Dias, & Antunes, 2011; Stricker, Hagan, & Flory, 2015). The impacts may be classified into three categories (Figure 11.1): **(A)** direct effect of invaders that **increases biomass production** and consequently changes C and N pools (Tamura & Tharayil, 2014) and potentially decreases soil-bound

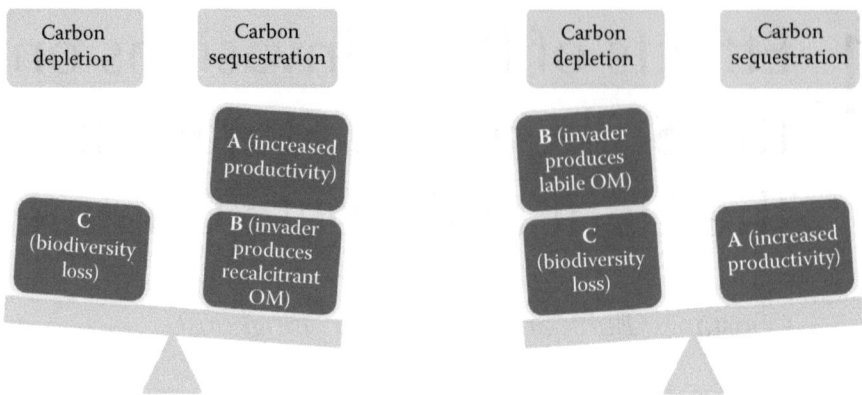

FIGURE 11.1 Simplified scheme to evaluate the factors leading to either carbon depletion or carbon sequestration after massive invasion concerning (**A**) direct effects, and (**B** and **C**) indirect effects. Sides represent
situation for invaders that build recalcitrant organic matter (left) and labile organic matter (right). See text for
explanation.

micronutrients (Marschner & Marschner, 2012), (**B**) indirect effect of invaders **influencing the
soil biota** (fauna and microbial) activity and decomposition via root exudates like allelopathic
agents or changes in litter quality and soil pH (Prescott & Zukswert, 2016; Tamura & Tharayil,
2014; Wolfe & Klironomos, 2005) leading to either sequestration or depletion of C pools in soils,
or (**C**) feedback from the indirect effects due to **biodiversity loss**, which is a typical consequence of plant invasions (Pyšek et al., 2012). Biodiversity loss leads to various effects on biotic
and abiotic soil characteristics due to less effective nutrient use and water retention, changes in
aggregate stability, leaching, or erosion (Fischer et al., 2014; Hooper et al., 2005; Soliveres et al.,
2016; Verheyen et al., 2008) and may consequently decrease the C storage capability (Lange et
al., 2015). See a summary of the potential characteristics and effects in Table 11.1 and detailed
explanations below.

11.1.1 Direct Effects of Plant Invasions on Carbon Cycling and Storage

Massive plant invasion commonly leads to a strong increase in biomass and litter production (Castro-
Diez, Godoy, Alonso, Gallardo, & Saldana, 2014; Liao et al., 2008; Vila et al., 2011). Accordingly,
one could expect generally increased C pools (Liao et al., 2008; Vila et al., 2011) and potentially
increased C sequestration in the soil (Lal, 2008). As most studies refer to overall C pools (soil C,
above ground biomass and litter) and not to soil C pools alone (Vila et al., 2011) and refer only to
the uppermost few centimeters of the soil (Liao et al., 2008; Osunkoya & Perrett, 2011), it remains
unclear if soils invaded by exotic species are C sinks or sources. For agricultural sites, it is known that
gains in the topsoil can be nullified by losses in the subsoil (cf. Chapter 13, Jungkunst et al.). There is
a lack of research data on soil-specific processes including soil C derived from root and soil microbial
(faunal) necromass also present in deeper soil layers. This seems especially relevant as many invaders tend to increase rooting depth in contrast to the native vegetation (Table 11.2). Context-specific
results are often contradictory (Hulme et al., 2013), possibly due to the large variability of traits
found in different groups of invaders (see Table 11.2) and as a consequence of different soil depths
studied. Such context specificity could be the similarity of the invader's traits in relation to traits of
the native vegetation. The higher the similarity, the less impact is to be expected (Levine et al., 2003).
Also, site conditions matter. Wet but not water logged soil conditions may lead to a heavy loss of soil
C, whereas dry conditions may lead to a gain in soil C stocks during invasions (Jackson, Banner,

TABLE 11.1
Summary of Mechanisms Potentially Effecting Soil C Dynamics Relevant after Plant Invasions

		Characteristic Changes with Plant Invasion	Effects Relevant for C Dynamics	Possible Effects on Soil C Pools	References
Direct effects	A	+ Biomass	+ Litter OM	C sequestration	Vila et al., 2011
		+ Rooting depth	+ Root OM	C sequestration	Schmidt et al., 2011
			+ Root OM in subsoil	C sequestration	Schmidt et al., 2011
Indirect effects	B	+ Leaf N content (e.g., in N fixers)	+ Microbial activity	(C depletion)	Liao et al., 2008; Tamura & Tharayil, 2014
		+ Recalcitrant litter	– Microbial activity or microbial community composition changes	(C sequestration)	Schmidt et al., 2011
		+ Soil acidification		C sequestration	Pyšek et al., 2012
		– Water availability		C sequestration	Jackson et al., 2002
		+ Allelopathic compounds		(C sequestration)	Bardon et al., 2016; Kim & Lee, 2011
		+ Recalcitrant root OM		(C sequestration)	Schmidt et al., 2011
		+ Labile root OM	+ Microbial activity (priming effect)	C depletion	Kuzyakov, 2010
		+ Subsoil OM (deeper roots, DOC leaching)	+ or – microbial activity	C sequestration or depletion	Strickland et al., 2010; Schmidt et al., 2011; Bonanomi et al., 2014
	C	+ Fire frequency		(C depletion)	Knicker, 2007
		– Biodiversity	Various interactions	(C depletion)	Tilman et al., 2014

Note: + indicates increase; – indicates a decrease; terms in brackets indicate especially uncertain information according to most recent literature; DOC, dissolved organic carbon.

TABLE 11.2

Collection of Characteristics of Several Highly Invasive Species Selected from the Lists of the 100 Worst Invasive Species of the World and in Europe

Plant Invader	Life Form	N-Fixing	Biomass (Above/Below Ground)	Decomposition	Choot C/N	Effects on Fire Frequency	Allelopathic	Decreases (Plant) Diversity	Further Reported Effects	References
Acacia mearnsii (Australian acacia), Fabaceae	Large shrub or tree, 5–15 m tall	Yes	+/~	(Fast)	~	~	Yes	Yes	+ Erosion, (– Soil C), – Soil water	1, 2
Arundo donax (giant reed), Poaceae	Graminoid, rhizomatous, 2–10 m tall, deep rooting	No	+/+	Slow	High	High	Yes	Yes	+ Erosion, – Soil water	1, 2, 3
Chromolaena odorata (*Siam weed*), Asteraceae	Herbaceous to woody perennial, thickets up to 2 m high	No	+/+	~	~	High	Yes	Yes		1, 2 (te Beest, Stevens, Olff, & van der Putten, 2009)
Clidemia hirta (Koster's curse), Melastomataceae	Shrub 0.5–3 m tall	No	+/~	~	~	~	~	Yes		1, 2
Euphorbia esula (leafy spurge), Euphorbiaceae	Herb 0.3–0.9 m tall, several meters deep taproots + horizontal roots	No	+/+	~	~	~	Yes	Yes		1, 2, 3
Fallopia japonica (Japanese knotweed), Polygonaceae	Herbaceous perennial, 2–3 m tall, rhizomatous roots system 5–7 m deep	No	+/+	Slow	~	~	Yes	Yes	+ Flooding + Litter	1, 2 (Kovářová et al., 2011)

(Continued)

TABLE 11.2 (CONTINUED)

Collection of Characteristics of Several Highly Invasive Species Selected from the Lists of the 100 Worst Invasive Species of the World and in Europe

Plant Invader	Life Form	N-Fixing	Biomass (Above/Below Ground)	Decomposition	Choot C/N	Effects on Fire Frequency	Allelopathic	Decreases (Plant) Diversity	Further Reported Effects	References
Hedychium gardnerianum (wild ginger), Zingiberaceae	Herbaceous perennial, 1–2 m tall, dense rhizomatous root system 1 m deep	No	+/+	Fast	~	~	Yes	Yes	+ Erosion	1, 2 (Funk, 2005)
Impatiens glandulifera (Himalayan balm), Balsaminaceae	Annual herb, 0.5–2.5 m tall, shallow and weak root system 0.15 cm deep	No	+/−	~	~	No	Yes	(Yes)	+ Erosion − Arbuscular mycorrhiza	1, 3 (DAISIE, 2008)
Imperata cylindrica (cogon grass), Poaceae	Graminoid, 0.3–1.5 m tall, rhizomatous root system	No	+/+	Fast	~	Yes	Yes	Yes	− Soil water + Habitat openness	1, 2 (Estrada & Flory, 2015; Holly, Ervin, Jackson, Diehl, & Kirker, 2009)
Lantana camara (lantana) Verbenaceae	Shrub, climbing 2–5 m tall	No	+/~	Slow~	~	Yes	Yes	Yes		1, 2 (Osunkoya & Perrett, 2011)

(Continued)

TABLE 11.2 (CONTINUED)

Collection of Characteristics of Several Highly Invasive Species Selected from the Lists of the 100 Worst Invasive Species of the World and in Europe

Plant Invader	Life Form	N-Fixing	Biomass (Above/Below Ground)	Decomposition	Choot C/N	Effects on Fire Frequency	Allelopathic	Decreases (Plant) Diversity	Further Reported Effects	References
Leucaena leucocephala (leucaena), Fabaceae	Small tree, 3–15 m tall, shallow root system	Yes	+/~	Fast	High	~	Yes	Yes		1, 3
Melaleuca quinquenervia (paperbark tree), Myrtaceae	Tree, 8–12 m tall	No	+/~	Slow	~	Yes	Yes	Yes	– Soil water – C and N in organic layer	1, 2
Mikania micrantha (bitter vine), Asteraceae	Herbaceous to woody vine, climbing up to 25 m in trees	No	+/~	~	~	No	Yes	Yes		1, 2
Mimosa pigra (catclaw mimosa), Fabaceae	Thorny shrub, 2 m tall, 5-year lifespan	Yes	+/~	~	~	~	~	Yes	– Habitat openness	1, 2
Pinus pinaster (maritime pine), Pinaceae	Large tree, 20–40 m tall, deep taproots + well-developed secondary roots	No	+/~	Slow	~	Yes	~	Yes	+ Soil erosion + Litter – Soil water	1, 2
Pueraria montana ssp. *lobata* (kudzu), Fabaceae	Densely growing vine, stems of 30 m length, root systems with tubers of 2 m length and 20–45 cm width	Yes	(+)/~	~	~	Yes	Yes	(Yes)	– Habitat openness – Trees	1, 2

(Continued)

TABLE 11.2 (CONTINUED)

Collection of Characteristics of Several Highly Invasive Species Selected from the Lists of the 100 Worst Invasive Species of the World and in Europe

Plant Invader	Life Form	N-Fixing	Biomass (Above/Below Ground)	Decomposition	Choot C/N	Effects on Fire Frequency	Allelopathic	Decreases (Plant) Diversity	Further Reported Effects	References
Robinia pseudoacacia (black locust), Fabaceae	Tree, 12–18 m tall, root system shallow, widespread + roots up to 7 m deep	Yes	(+)/~	Fast	Medium	~		Yes		1 (DAISIE, 2008)
Schinus terebinthifolius (Brazilian pepper tree), Anacardiaceae	Small tree, 3–10 m tall, shallow root system?	No	+/~	~	High	~	Yes	Yes		1, 2, 3
Tamarix ramosissima (saltcedar), Tamaricaceae	Shrub, 1–5 m tall, taproot reaching 30 m deep + 50 m lateral spread	No	+/+	~	~	Yes	~	Yes	– Soil water + Salinity + Sedimentation + Litter	1, 2
Ulex europaeus (gorse), Fabaceae	Spiny shrub, 2–2.5 m tall, central taproot + shallow lateral roots	Yes	+/~	~	~	Yes	~	~	– Soil water + Acidification	1, 2

Sources: IUCN, Global Invasive Species Database, retrieved from http://www.iucngisd.org/gisd/100_worst.php. 2017; DAISIE (Ed.), *Handbook of alien species in Europe*, Dordrecht: Springer, retrieved from http://www.europe-aliens.org/pdf/DAISIE_press_release_1.pdf, 2008.

Note: + indicates positive effects or elevated values, – indicates negative effects or reduced values, ~ indicates unknown or very uncertain information, data in brackets indicates partly uncertain or contrasting information. The table does not refer to soil characteristics met below the invaders as there is practically no information available.

Jobbagy, Pockman, & Wall, 2002). Gains can be expected because microbes cannot decompose the additional organic matter (OM) due to the lack of water. Accordingly, there is still no clear answer to the question if exotic species directly lead to C sequestration or depletion in the soils. An answer may be found in the differentiation of the C type—are leaf and particularly root litter of the invader labile or recalcitrant? (Suseela, Tharayil, Xing, & Dukes, 2013).This question is addressed in the next section describing indirect effects of the invasion via the changing decomposition by soil organisms.

11.1.2 Indirect Effects Changing Soil Microbial Activity

Evidence suggests that nutrient cycling is commonly accelerated, as litter quality is often enhanced due to massive plant invasion (Liao et al., 2008). This may prevent the storage of additional OM in the soil. Herbal plant invasions, in particular, apparently do not increase C storage in the soil even though their biomass and litter are higher compared to the original vegetation (Liao et al., 2008). Some authors even document depletion of soil C pools due to invasion (Strickland, Devore, Maerz, & Bradford, 2010; Tamura & Tharayil, 2014). Strickland et al. (2010) believe that the activation of deeper C pools or alternative nutrient pools by soil microorganisms is responsible for the reduction of overall soil C. This trend may be caused by the so-called priming effect—when labile C is added to stable C, the latter can be "activated" leading to a net loss in C (Kuzyakov, 2010). Examples of invaders adding such labile OM to the soil are N-fixing legumes. Tamura and Tharayil (2014) reported the N-fixing ability of their studied legume invader to be responsible for heavy decomposition and loss of stored soil C—even of C bound in recalcitrant OM (Bonanomi, Capodilupo, Incerti, & Mazzoleni, 2014). This happens probably because stabile organic compounds are decomposed more easily in the presence of high quality litter as the rapidly developing bacteria die after the labile C is used up and their necromass helps to feed microbial species specialized to mineralize recalcitrant parts of the soil organic matter (SOM)—this is the priming effect (Kuzyakov, 2010). However, in a meta-analysis, N-fixing invaders slightly increased soil C stocks, whereas non-N-fixing invaders did not show a clear trend (Liao et al., 2008). They compared soil C pools only up to 10 cm depths, which excluded large parts of the rhizosphere zone and therefore potential priming effects due to root exudates.

To estimate changes in C pools before and after invasion, the characteristics of the litter and root exudates of the invader have to be considered in relation to the characteristics of the native vegetation. As priming effects may be especially relevant for the mobilization of deep soil C (Schmidt et al., 2011), invaders reaching soil zones deeper than the native vegetation (see Table 11.2) might lead to a strong C release from the soil. Many invaders were found to increase decomposition rates (Pyšek et al., 2012; Vila et al., 2011) due to a lower C:N ratio of the litter of invaders compared with the litter of native plant species (Liao et al., 2008). However, this trend may be an atypical characteristic of all invaders (see Table 11.2), but due to a bias toward particularly influential invader species (Prescott & Zukswert, 2016). Further, the effects may depend on the ecosystem type (McCary, Mores, Farfan, & Wise, 2016). Jo et al. (2016) could not find systematic differences in decomposition rates between native and invasive woody plant forest species' litter under laboratory conditions. Any number of topics along these lines remain understudied: (1) the role of root OM and root exudates, as well as their quality for decomposition, and (2) effects on decomposition due to plant secondary compounds produced by the roots. Allelopathic effects of phenolic or other plant secondary metabolites in plant parts and root exudates may influence other plants (Bais, Vepachedu, Gilroy, Callaway, & Vivanco, 2003) and the microbial community (Bais, Weir, Perry, Gilroy, & Vivanco, 2006). Such root exudates play crucial roles in many biotic and abiotic processes (Rasmann & Turlings, 2016). For example, denitrification in soils below the invasive knotweeds *Fallopia japonica* s.l. is reduced by procyanidin produced in the roots, and it has antibacterial properties as it inhibits aerobial respiration (Bardon et al., 2016). Antimicrobial phenolic compounds in leaf material of several invasive plants may have antimicrobial effects on many different taxa of fungi, whereas native species or non-invasive alien species did not show these effects (Kim & Lee, 2011).

An impact on microbial communities and concurrent changes in the C storage may also be due to acidification processes triggered by many invaders (Pyšek et al., 2012), which may for example increase the fungi:bacterial ratio. Further, the massive water use of many invaders may also be responsible for changes in microbial community composition and decomposition. As priming effects are reduced under dry conditions (Kuzyakov, 2010), the availability of water for the microorganisms in the rhizosphere is of fundamental importance to C sequestration or depletion. Indirect effects of unclear consequences for carbon stocks are also found in the increase of the fire frequency and often also fire heat development due to plant invasions (see Table 11.2). Carbon loss in the vegetation and upper soil layers are obvious. However, recalcitrant char particles, hydrophobicity, and leaching of carbon to deeper soil layers occur and complicate the answer to the question how increased fire frequencies of invaded areas influence overall long-term carbon pools (Knicker, 2007).

11.1.3 Indirect Effects due to the Reduction of Biodiversity

A second pathway of indirect impact is the commonly found dramatic reduction of the plant diversity along with the invasion process (Pyšek et al., 2012). This is often followed by dramatic reductions of species diversity in other taxonomic groups (Pyšek et al., 2012; Soliveres et al., 2016). Changes in microbial community diversity lead to reduced functionality and probably also to consequences on C sequestration (Delgado-Baquerizo et al., 2016). Plant diversity is known to increase ecosystem functions (Cardinale et al., 2011) and to have a stabilizing effect under environmental change on several soil processes (Isbell et al., 2011). Biodiversity, as such, increases multifunctionality (Lefcheck et al., 2015). High plant diversity interacts with climate change by increasing nutrient cycling (Niklaus et al., 2016), and plant diversity has been shown to increase the soil functions to accumulate soil C (Tilman, Isbell, & Cowles, 2014). The mechanisms behind the effects of species diversity are expected to be found in the complementarity of different species growing next to each other (Cardinale et al., 2011). This functions to combine different resource use, litter quality, root exudates, and other effects, like soil acidification, to build up a well-buffered system of small-scale processes that does not lead to large nutrient losses or leaching and that fix the soil avoiding erosion. A massively invading plant tends to suppress other species, and the diversity in root architecture and plant-soil feedback mechanisms decreases dramatically. In addition to accumulation processes of litter or sedimentation, erosion processes can also be affected, as the simplified root system of the mono-species stand do not effectively stabilize the soil (Fei, Phillips, & Shouse, 2014).

It is difficult to distinguish between invader-specific effects on soil food webs and the effect of the simultaneously declining biodiversity. Several analyses have attributed the invader-specific effects to the impact of certain plant invaders. These effects, summarized by McCary et al. (2016), might be strongly context-dependent and are primarily a result of the loss in biodiversity.

11.2 UNCERTAINTIES TO BE RESOLVED HIGHLIGHTED BY SELECTED CASE STUDIES

Despite high uncertainties, one thing is clear: exotic plant invasions do influence soil biogeochemistry and therefore C- and N-stocks with feedback to the climate systems. However, there is no easy answer to the question—do plant invasions lead to soil C depletion or soil C sequestration? On the one hand, this is because invaders have no uniform effects. Effects are species-specific (Table 11.2), vary between ecosystems and environmental conditions and depend on the former native species composition (Hulme et al., 2013; Kumschick et al., 2015; McCary et al., 2016). On the other hand, the interactions between vegetation, soils, environmental conditions, microbial activity, and soil fauna are clearly understudied, especially considering deeper soil layers (Schmidt et al., 2011). Of course, several studies try to define the processes in the soils from the perspective of vegetation scientists (for example, Pyšek et al., 2012), plant physiologists (Rasmann & Turlings, 2016), soil scientists (Kuzyakov, 2010), and microbiologists (Bodelier, 2011). However, single expert groups commonly cannot provide detailed insights into the relevant processes in the plant-soil interface under natural conditions, as analysis techniques tend to be applicable only in small microcosms. Usually the expertises between all relevant subjects are not linked at all or address just two of the mentioned perspectives. Vegetation ecologists and nature conservationists tend to try and cover more than one discipline, but their analysis apparently was limited so far to the top soil, which decreases the ability to judge ecosystem processes (for example, Liao et al., 2008; Prescott, 2010; Stefanowicz, Stanek, Nobis, & Zubek, 2017). Microbiologists and soil scientists, in contrast, focus on fine-scale studies (Kuzyakov, 2010) and those incorporating biodiversity effects and mechanisms from soils to plants to microbes and their feedback effects. Such multiple interactions cannot be studied by summing up single quantified effects, but call for comprehensive hypotheses formulated by soil scientists, ecologists, and biologists and consecutively joint research and publication activity. The schematics in Figure 11.2 highlight some ideas for hypotheses that could help to answer questions about the relevancy of massive plant species invasions to soil or ecosystem C storage. Outlined below are some ideas about future research directions that arise from the review and synthesis of the published literature.

The simple separation into plants producing recalcitrant OM and plants producing labile OM no longer holds, as the storage of C in the soil is influenced by a large number of factors that may lead to easy decomposition of recalcitrant litter or long-time storage of labile litter (Schmidt et al., 2011). The role of N-fixing species on C storage, for example, is still under debate though they are expected to build easily decomposable OM and labile C pools (Tamura & Tharayil, 2014). On the basis of a meta-analysis, Prescot (2010) proposes establishing N-fixing plants in order to add N to the soil. Such additions should increase humification, and therefore C storage in the soil. However, most data are based on small-scale observations, and sugar-containing root exudates could simultaneously lead to increased decomposition instead of humification and C storage. In woody plants, roots of N-fixing Eleagnaceae species contain elevated acid-insoluble residues, like lignins, which tend to build recalcitrant OM even under the influence of additional N (Jo et al., 2016). This trend shows the relevance of the quality of the root biomass, as it might be more important than the above ground biomass for C storage. Decomposition of roots has a greater likelihood to release recalcitrant OM in the soil than does leaf litter (Rasse, Rumpel, & Dignac, 2005). Accordingly, the most obvious scientific gap to quantify soil C pool changes after massive

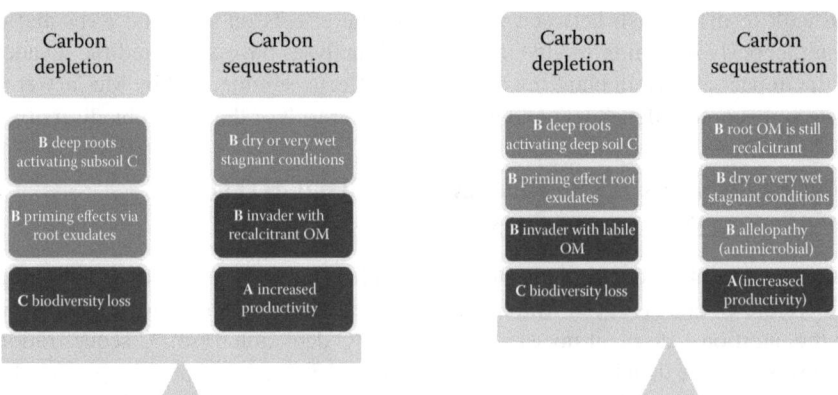

FIGURE 11.2 A more realistic overview on the factors affecting carbon depletion or sequestration after massive plant invasions. Invader has recalcitrant OM on the left or more labile OM on the right. Light gray boxes indicate factors that may occur and would tip the balance into their respective direction. See text for explanation. Effect sizes are not yet definable and are among researchable priorities.

plant invasions (above ground examples in Figure 11.3 through 11.6, however below ground extension is hardly known), the missing knowledge on below ground effects on biotic and abiotic processes in the rhizosphere. In meta-analyses, the great variability of decomposition under N-fixing plants or experimentally add N becomes obvious (Knorr, Frey, & Curtis, 2005; Liao et al., 2008). There might also be a possible time lag, whereby N addition might enhance decomposition during the first couple of years, but tends to reduce decomposition thereafter (Knorr et al., 2005). Therefore, as proposed in Figure 11.2, the role of above and below ground organic matter quality, the priming effect, and root exudates in the rhizosphere has to be urgently addressed at

FIGURE 11.3   The hybrid knotweed *Fallopia x bohemica* overgrows large areas, especially in Europe, North America, and Australia. It outperforms the invasive parental taxa Japanese and giant knotweed in terms of aggressiveness of growth. (From Buhk, C., & Thielsch, A., *Perspectives in Plant Ecology, Evolution and Systematics*, 17, 274–283, 2015.) Hardly any other plant species can grow below these thickets. Litter production is very large every year, root systems reach over 5 m in depth and allelopathic agents act in the soil. Biogeochemical cycling is therefore highly likely transformed from top soil down to subsoil in comparison to native vegetation.

FIGURE 11.4 Himalayan balm (*Impatiens glandulifera*) and Japanese knotweed (*Fallopia japonica*) in the foreground overgrow large riverside areas in Europe and parts of North America. The annual Himalayan balm produces tall stems of several meters every year but the root system is weaker than in the native vegetation. The decay of the roots during winter increases soil erosion. Allelopathic substances are found in all parts of the plant. Effects on biogeochemical cycles are therefore likely.

small, medium, and large scales to study the relevance of these existing and interrelated factors. Specific to the effects of plant invasions, there is a strong need for long-term research because time lag effects in short experiments do not provide realistic data (Vila et al., 2011). It is the long-term experiments that would provide the much-needed information on how relevant invasions are at the ecosystem, landscape, or even global scale (Cardinale et al., 2011).

The examples of invasive species in Table 11.2 demonstrate that high biomass production and reduced biodiversity are very common effects of invaders, but there is no clear pattern of characteristics concerning other traits. Every species has a specific set of traits, which makes general conclusions on effects of carbon sequestration and depletion unconvincing on the basis of today's knowledge.

Main deficits are that short-term carbon increase in the top soil may easily be outweighed by carbon losses in the subsoil. However, the urgently needed information of plant characteristics is by far not complete, as data on root characteristics like mass, distribution, or exudates and microbial/plant interactions in the soil and especially in deeper soil layers are understudied. Further, effects strongly depend on the characteristics and similarity of traits of the native vegetation that is replaced by the invader, but little is known about the (sub)soil activities and interactions between different native species to compare with the new situation under invasion.

As a consequence of our review, we therefore call for interdisciplinary research on plant-soil interrelation at small to large scales as neither biologists nor soil scientists alone can bring light into this highly complex system. We demonstrate that our understanding of the role of plant invasions for carbon dynamics is lacking and not specific, and also uncovers fundamental scientific gaps on how plant characteristics influence carbon dynamics. Along with Sousa et al. (2011), we highlight that **invasive species** may be excellent **model organisms** to **bring light** into these fundamental ecological **processes in soils** and help to understand overall carbon dynamics at local to regional scales. Due to the strong engineering character of many plant invaders and fast and easy reproduction, they allow easily replicated laboratory and field experiments as well as field studies whose effects are quantifiable even in reasonable time spans (Stricker et al., 2015). It's time to merge invasion ecology into fundamental ecological theories and studies on ecosystem functioning (Sousa et al., 2011), in particular soil functioning.

FIGURE 11.5 Not (yet) listed in the top 100 of the worst invaders is American pokeweed (*Phytolacca americana*). However, the potential of this plant species and related taxa to conquer large forested areas in temperate regions is large and transformation processes on biogeochemical cycles are to be expected. Today, the plants invade various types of forests in the upper Rhine valley in southern Germany and France and colonizes the adjacent low mountain ranges where it builds a dense shrub layer even in undisturbed forests, where shrubs usually do not occur (Buhk, 2016). The deep taproots develop already in an early stage. The plant is known for the large content of allelopathic agents. It is also a large problem in nature parks in Asia. (From Zhou, B. et al., *Ecology and Environmental Sciences*, 22(4), 567–574, 2013.)

11.3 CONCLUSIONS

- Invasive plants influence soil carbon storage and cycling but not in a unidirectional manner.
- Differing effects depend on the specific invasive species, suppressed native vegetation type, soil characteristics, and hydrological conditions.
- Effects were grouped into three major sections: (A) direct effects, (B) indirect effects induced by plant-soil-microbial interactions, and (C) indirect effects caused by the loss of plant diversity going along with the invasion process.
- Major knowledge gaps are (a) missing subsoil (below 20 cm) information on stock changes and biogeochemical processes, (b) rooting depth and structure as well as allelopathic characteristics of the invasive species compared to the native vegetation, (c) systematic studies under varying soil conditions, and (d) upscaling of processes from micro scale to stand level and landscape scale.

FIGURE 11.6   A garden escape in many areas in Europe, North America, and Australia with large trans-
former potential is the Himalayan blackberry (*Rubus armeniacus*). The stand in the photograph is about 3 m
high and was recently cut by a lateral mowing machine allowing the view inside the stand. The soil inside
is completely dry, hydrophobic, and bare of other vegetation. Long looping runner shoots of up to 12 m
length easily spread into the surrounding area overgrowing even adjacent low growing trees. (From Buhk, C.,
*Wissenschaftliches Jahrbuch des Grenzüberschreitenden Biosphärenreservates Pfälzerwald - Vosges du
Nord*, 18, 17–43, 2016.)

- Concerted joint studies of soil scientists, microbiologists, and ecologists are required to
  bring light into the high complexity of the processes involved.
- Plant invasion processes provide the unique opportunity to advance our understanding of
  fundamental ecological processes and theories—use it!

REFERENCES

Bais, H. P., Vepachedu, R., Gilroy, S., Callaway, R. M., & Vivanco, J. M. (2003). Allelopathy and exotic
plant invasion: From molecules and genes to species interactions. *Science (New York, N.Y.)*, *301*(5638),
1377–1380. https://doi.org/10.1126/science.1083245.

Bais, H. P., Weir, T. L., Perry, L. G., Gilroy, S., & Vivanco, J. M. (2006). The role of root exudates in rhizosphere interactions with plants and other organisms. *Annual Review of Plant Biology*, *57*, 233–266.

Bardon, C., Piola, F., Haichar, F. e. Z., Meiffren, G., Comte, G., Missery, B.,… Poly, F. (2016). Identification
of B-type procyanidins in Fallopia spp. involved in biological denitrification inhibition. *Environmental
Microbiology*, *18*(2), 644–655. https://doi.org/10.1111/1462-2920.13062.

Bebber, D. P., Ramotowski, M. A. T., & Gurr, S. J. (2013). Crop pests and pathogens move polewards in a
warming world. *Nature Climate Change*, *3*(11), 985–988. https://doi.org/10.1038/nclimate1990.

Bodelier, P. L. E. (2011). Toward understanding, managing, and protecting microbial ecosystems. *Frontiers in
Microbiology*, *2*, 80. https://doi.org/10.3389/fmicb.2011.00080.

Bonanomi, G., Capodilupo, M., Incerti, G., & Mazzoleni, S. (2014). Nitrogen transfer in litter mixture
enhances decomposition rate, temperature sensitivity, and C quality changes. *Plant and Soil*, *381*(1–2),
307–321. https://doi.org/10.1007/s11104-014-2119-4.

Bradley, B. A., Blumenthal, D. M., Wilcove, D. S., & Ziska, L. H. (2010). Predicting plant invasions in an era of
global change. *Trends in Ecology & Evolution*, *25*(5), 310–318. https://doi.org/10.1016/j.tree.2009.12.003.

Buhk, C. (2016). Invasive Neophyten auf dem Weg ins Biosphärenreservat Pfälzer Wald Nordvogesen:
Kurzer Abriss aus der Forschung. *Wissenschaftliches Jahrbuch des Grenzüberschreitenden
Biosphärenreservates Pfälzerwald - Vosges du Nord*, *18*, 17–43.

Buhk, C., & Thielsch, A. (2015). Hybridisation boosts the invasion of an alien species complex: Insights into
future invasiveness. *Perspectives in Plant Ecology, Evolution and Systematics*, *17*, 274–283.

CABI. (2017). Invasive Species Compendium. Retrieved from www.cabi.org/isc.

Cardinale, B. J., Matulich, K. L., Hooper, D. U., Byrnes, J. E., Duffy, E., Gamfeldt, L.,... Gonzalez, A. (2011). The functional role of producer diversity in ecosystems. *American Journal of Botany*, *98*(3), 572–592. https://doi.org/10.3732/ajb.1000364.

Castro-Diez, P., Godoy, O., Alonso, A., Gallardo, A., & Saldana, A. (2014). What explains variation in the impacts of exotic plant invasions on the nitrogen cycle? A meta-analysis. *Ecology Letters*, *17*(1), 1–12. https://doi.org/10.1111/ele.12197.

DAISIE (Ed.). (2008). *Handbook of alien species in Europe*. Dordrecht: Springer. Retrieved from http://www.europe-aliens.org/pdf/DAISIE_press_release_1.pdf.

Delgado-Baquerizo, M., Maestre, F. T., Reich, P. B., Jeffries, T. C., Gaitan, J. J., Encinar, D.,... Singh, B. K. (2016). Microbial diversity drives multifunctionality in terrestrial ecosystems. *Nature Communications*, *7*, 10541. https://doi.org/10.1038/ncomms10541.

Ehrenfeld, J. G. (2010). Ecosystem consequences of biological invasions. *Annual Review of Ecology, Evolution, and Systematics*, *41*, 59–80.

Estrada, J. A., & Flory, S. L. (2015). Cogongrass (Imperata cylindrica) invasions in the US: Mechanisms, impacts, and threats to biodiversity. *Global Ecology and Conservation*, *3*, 1–10. https://doi.org/10.1016/j.gecco.2014.10.014.

Fei, S., Phillips, J., & Shouse, M. (2014). Biogeomorphic impacts of invasive species. *Annual Review of Ecology, Evolution, and Systematics*, *45*(1), 69–87. https://doi.org/10.1146/annurev-ecolsys-120213-091928.

Fischer, C., Tischer, J., Roscher, C., Eisenhauer, N., Ravenek, J., Gleixner, G.,... Hildebrandt, A. (2014). Plant species diversity affects infiltration capacity in an experimental grassland through changes in soil properties. *Plant and Soil*. Advance online publication. https://doi.org/10.1007/s11104-014-2373-5.

Funk, J. L. (2005). *Hedychium gardnerianum* invasion into Hawaiian montane rainforest: Interactions among litter quality, decomposition rate, and soil nitrogen availability. *Biogeochemistry*, *76*(3), 441–451. https://doi.org/10.1007/s10533-005-7657-7.

Holly, D. C., Ervin, G. N., Jackson, C. R., Diehl, S. V., & Kirker, G. T. (2009). Effect of an invasive grass on ambient rates of decomposition and microbial community structure: A search for causality. *Biological Invasions*, *11*(8), 1855–1868. https://doi.org/10.1007/s10530-008-9364-5.

Hooper, D. U., Chapin Iii, F. S., Ewel, J. J., Hector, A., Inchausti, P., Lavorel, S.,... others. (2005). Effects of biodiversity on ecosystem functioning: A consensus of current knowledge. *Ecological Monographs*, *75*(1), 3–35.

Hulme, P. E., Pyšek, P., Jarošík, V., Pergl, J., Schaffner, U., & Vilà, M. (2013). Bias and error in understanding plant invasion impacts. *Trends in Ecology & Evolution*, *28*(4), 212–218. https://doi.org/10.1016/j.tree.2012.10.010.

Isbell, F., Calcagno, V., Hector, A., Connolly, J., Harpole, W. S., Reich, P. B.,... others. (2011). High plant diversity is needed to maintain ecosystem services. *Nature*, *477*(7363), 199–202.

IUCN. (2017). Global Invasive Species Database. Retrieved from http://www.iucngisd.org/gisd/100_worst.php.

Jackson, R. B., Banner, J. L., Jobbagy, E. G., Pockman, W. T., & Wall, D. H. (2002). Ecosystem carbon loss with woody plant invasion of grasslands. *Nature*, *418*(6898), 623–626. https://doi.org/10.1038/nature00910.

Jo, I., Fridley, J. D., & Frank, D. A. (2016). More of the same? In situ leaf and root decomposition rates do not vary between 80 native and nonnative deciduous forest species. *New Phytologist*, *209*(1), 115–122. https://doi.org/10.1111/nph.13619.

Kattge, J., Diaz, S., Lavorel, S., Prentice, I. C., Leadley, P., Bönisch, G., others. (2011). TRY—A global database of plant traits. *Global Change Biology*, *17*(9), 2905–2935.

Kim, Y. O., & Lee, E. J. (2011). Comparison of phenolic compounds and the effects of invasive and native species in East Asia: Support for the novel weapons hypothesis. *Ecological Research*, *26*(1), 87–94. https://doi.org/10.1007/s11284-010-0762-7.

Knicker, H. (2007). How does fire affect the nature and stability of soil organic nitrogen and carbon? A review. *Biogeochemistry*, *85*(1), 91–118. https://doi.org/10.1007/s10533-007-9104-4.

Knorr, M., Frey, S. D., & Curtis, P. S. (2005). Nitrogen additions and litter decomposition: A meta-analysis. *Ecology*, *86*(12), 3252–3257. https://doi.org/10.1890/05-0150.

Kovářová, M., Frantík, T., Koblihová, H., Bartůňková, K., Nývltová, Z., & Vosátka, M. (2011). Effect of clone selection, nitrogen supply, leaf damage and mycorrhizal fungi on stilbene and emodin production in knotweed. *BMC Plant Biology*, *11*, 98. https://doi.org/10.1186/1471-2229-11-98.

Kumschick, S., Gaertner, M., Vila, M., Essl, F., Jeschke, J. M., Pyšek, P.,... Winter, M. (2015). Ecological impacts of alien species: Quantification, scope, caveats, and recommendations. *BioScience*, *65*(1), 55–63. https://doi.org/10.1093/biosci/biu193.

Kuzyakov, Y. (2010). Priming effects: Interactions between living and dead organic matter. *Soil Biology and Biochemistry, 42*(9), 1363–1371. https://doi.org/10.1016/j.soilbio.2010.04.003.

Lal, R. (2008). Carbon sequestration. *Philosophical Transactions of the Royal Society of London. Series B, Biological Sciences, 363*(1492), 815–830. https://doi.org/10.1098/rstb.2007.2185.

Lange, M., Eisenhauer, N., Sierra, C. A., Bessler, H., Engels, C., Griffiths, R. I.,... Gleixner, G. (2015). Plant diversity increases soil microbial activity and soil carbon storage. *Nature Communications, 6*, 6707. https://doi.org/10.1038/ncomms7707.

Lefcheck, J. S., Byrnes, J. E. K., Isbell, F., Gamfeldt, L., Griffin, J. N., Eisenhauer, N.,... Duffy, J. E. (2015). Biodiversity enhances ecosystem multifunctionality across trophic levels and habitats. *Nature Communications, 6*, 6936. https://doi.org/10.1038/ncomms7936.

Levine, J. M., Vila, M., D'Antonio, C. M., Dukes, J. S., Grigulis, K., & Lavorel, S. (2003). Mechanisms underlying the impacts of exotic plant invasions. *Proceedings. Biological Sciences, 270*(1517), 775–781. https://doi.org/10.1098/rspb.2003.2327.

Liao, C., Peng, R., Luo, Y., Zhou, X., Wu, X., Fang, C.,... Li, B. (2008). Altered ecosystem carbon and nitrogen cycles by plant invasion: A meta-analysis. *The New Phytologist, 177*(3), 706–714. https://doi.org/10.1111/j.1469-8137.2007.02290.x.

Liu, Y., Oduor, A. M. O., Zhang, Z., Manea, A., Tooth, I. M., Leishman, M. R.,... van Kleunen, M. (2016). Do invasive alien plants benefit more from global environmental change than native plants? *Global Change Biology*. Advance online publication. https://doi.org/10.1111/gcb.13579.

Liu, Y., van Kleunen, M., & Cornelissen, H. (2017). Responses of common and rare aliens and natives to nutrient availability and fluctuations. *Journal of Ecology*. Advance online publication. https://doi.org/10.1111/1365-2745.12733.

Marschner, H., & Marschner, P. (Eds.). (2012). *Marschner's mineral nutrition of higher plants* (3rd ed.). Amsterdam, Boston, MA: Academic Press. Retrieved from http://search.ebscohost.com/login.aspx?direct=true&scope=site&db=nlebk&db=nlabk&AN=453799.

McCary, M. A., Mores, R., Farfan, M. A., & Wise, D. H. (2016). Invasive plants have different effects on trophic structure of green and brown food webs in terrestrial ecosystems: A meta-analysis. *Ecology Letters, 19*(3), 328–335. https://doi.org/10.1111/ele.12562.

Niklaus, P. A., Le Roux, X., Poly, F., Buchmann, N., Scherer-Lorenzen, M., Weigelt, A., & Barnard, R. L. (2016). Plant species diversity affects soil-atmosphere fluxes of methane and nitrous oxide. *Oecologia, 181*(3), 919–930. https://doi.org/10.1007/s00442-016-3611-8.

Osunkoya, O. O., & Perrett, C. (2011). *Lantana camara* L. (Verbenaceae) invasion effects on soil physicochemical properties. *Biology and Fertility of Soils, 47*(3), 349–355. https://doi.org/10.1007/s00374-010-0513-5.

Prescott, C. E. (2010). Litter decomposition: What controls it and how can we alter it to sequester more carbon in forest soils? *Biogeochemistry, 101*(1–3), 133–149. https://doi.org/10.1007/s10533-010-9439-0.

Prescott, C. E., & Zukswert, J. M. (2016). Invasive plant species and litter decomposition: Time to challenge assumptions. *New Phytologist, 209*(1), 5–7. https://doi.org/10.1111/nph.13741.

Pyšek, P., Jarošík, V., Hulme, P. E., Pergl, J., Hejda, M., Schaffner, U., & Vilà, M. (2012). A global assessment of invasive plant impacts on resident species, communities and ecosystems: The interaction of impact measures, invading species' traits and environment. *Global Change Biology, 18*(5), 1725–1737. https://doi.org/10.1111/j.1365-2486.2011.02636.x.

Rasmann, S., & Turlings, T. C. (2016). Root signals that mediate mutualistic interactions in the rhizosphere. *Current Opinion in Plant Biology, 32*, 62–68. https://doi.org/10.1016/j.pbi.2016.06.017.

Rasse, D. P., Rumpel, C., & Dignac, M.-F. (2005). Is soil carbon mostly root carbon? Mechanisms for a specific stabilisation. *Plant and Soil, 269*(1–2), 341–356. https://doi.org/10.1007/s11104-004-0907-y.

Schmidt, M. W. I., Torn, M. S., Abiven, S., Dittmar, T., Guggenberger, G., Janssens, I. A.,... Trumbore, S. E. (2011). Persistence of soil organic matter as an ecosystem property. *Nature, 478*(7367), 49–56. https://doi.org/10.1038/nature10386.

Soliveres, S., van der Plas, F., Manning, P., Prati, D., Gossner, M. M., Renner, S. C.,... Allan, E. (2016). Biodiversity at multiple trophic levels is needed for ecosystem multifunctionality. *Nature, 536*(7617), 456–459. https://doi.org/10.1038/nature19092.

Sorte, C. J. B., Ibanez, I., Blumenthal, D. M., Molinari, N. A., Miller, L. P., Grosholz, E. D.,... Dukes, J. S. (2013). Poised to prosper? A cross-system comparison of climate change effects on native and non-native species performance. *Ecology Letters, 16*(2), 261–270. https://doi.org/10.1111/ele.12017.

Sousa, R., Morais, P., Dias, E., & Antunes, C. (2011). Biological invasions and ecosystem functioning: Time to merge. *Biological Invasions, 13*(5), 1055–1058. https://doi.org/10.1007/s10530-011-9947-4.

Stefanowicz, A. M., Stanek, M., Nobis, M., & Zubek, S. (2017). Few effects of invasive plants Reynoutria japonica, Rudbeckia laciniata and Solidago gigantea on soil physical and chemical properties. *Science of The Total Environment*, *574*, 938–946. https://doi.org/10.1016/j.scitotenv.2016.09.120.

Stricker, K. B., Hagan, D., & Flory, S. L. (2015). Improving methods to evaluate the impacts of plant invasions: lessons from 40 years of research. *AoB PLANTS*, *7*. https://doi.org/10.1093/aobpla/plv028.

Strickland, M. S., Devore, J. L., Maerz, J. C., & Bradford, M. A. (2010). Grass invasion of a hardwood forest is associated with declines in belowground carbon pools. *Global Change Biology*, *16*(4), 1338–1350. https://doi.org/10.1111/j.1365-2486.2009.02042.x.

Suseela, V., Tharayil, N., Xing, B., & Dukes, J. S. (2013). Labile compounds in plant litter reduce the sensitivity of decomposition to warming and altered precipitation. *The New Phytologist*, *200*(1), 122–133. https://doi.org/10.1111/nph.12376.

Tamura, M., & Tharayil, N. (2014). Plant litter chemistry and microbial priming regulate the accrual, composition and stability of soil carbon in invaded ecosystems. *The New Phytologist*, *203*(1), 110–124. https://doi.org/10.1111/nph.12795.

te Beest, M., Stevens, N., Olff, H., & van der Putten, W. H. (2009). Plant-soil feedback induces shifts in biomass allocation in the invasive plant Chromolaena odorata. *Journal of Ecology*, *97*(6), 1281–1290. https://doi.org/10.1111/j.1365-2745.2009.01574.x.

Tilman, D., Isbell, F., & Cowles, J. M. (2014). Biodiversity and ecosystem functioning. *Annual Review of Ecology, Evolution, and Systematics*, *45*(1), 471–493. https://doi.org/10.1146/annurev-ecolsys-120213-091917.

van Kleunen, M., Dawson, W., Essl, F., Pergl, J., Winter, M., Weber, E.,... Pyšek, P. (2015). Global exchange and accumulation of non-native plants. *Nature*, *525*(7567), 100–103. https://doi.org/10.1038/nature14910.

Verheyen, K., Bulteel, H., Palmborg, C., Olivié, B., Nijs, I., Raes, D., & Muys, B. (2008). Can complementarity in water use help to explain diversity-productivity relationships in experimental grassland plots? *Oecologia*, *156*(2), 351–361. https://doi.org/10.1007/s00442-008-0998-x.

Vila, M., Espinar, J. L., Hejda, M., Hulme, P. E., Jarosik, V., Maron, J. L.,... Pyšek, P. (2011). Ecological impacts of invasive alien plants: A meta-analysis of their effects on species, communities and ecosystems. *Ecology Letters*, *14*(7), 702–708. https://doi.org/10.1111/j.1461-0248.2011.01628.x.

Wolfe, B. E., & Klironomos, J. (2005). Breaking new ground: Soil communities and exotic plant invasion. *BioScience*, *55*(6), 477. https://doi.org/10.1641/0006-3568(2005)055[0477:BNGSCA]2.0.CO;2.

Zhou, B., Yan, Xiaohong, Xiao, Yian, Thang, Z., Li, X., & Yang, J. (2013). Traits of reproductive biology associated with invasiveness in alien invasive plant *Phytolacca americana*. *Ecology and Environmental Sciences*, *22*(4), 567–574.

12 Climate Change Impact on Soil Carbon Stocks in India

Tarik Mitran, Rattan Lal, Umakant Mishra,
Ram Swaroop Meena, T. Ravisankar, and K. Sreenivas

CONTENTS

12.1   INTRODUCTION

Soils are directly related to the atmospheric and climate systems via nitrogen (N), carbon (C), and hydrologic cycles. Therefore, any changes in atmosphere and climate may influence soils by altering the functioning of such cycles specifically that of C. The top 1 m of world soil stores approximately 1408 Pg as organic C (Batjes 2016) and exchanges 60 Pg C yr^{-1} with the atmosphere globally, which at present contains ~820 Pg C as carbon dioxide (CO_2) (Le Quéré et al. 2015). The annual flux of C passing through the soil and the magnitude of the soil C stock are two of the reasons that soil organic carbon (SOC) plays a significant role in the global C cycle (GCC), and its judicious management can offset some of the anthropogenic emissions. Soil C also plays a significant role in soil fertility and environmental quality, and there is a large potential of C sequestration in the biota and soil. The quantity of organic C stored in different soil pools at a specific time reflects the balance between the rate of input of biomass C and that of its mineralization in each of the organic C pools (Post and Kwon 2000). There are several factors that influence active soil C stocks in magnitude and dimension, including climate, vegetation, soil characteristics, land surface characteristics, and management practices (Gray et al. 2014; Hobley et al. 2013; Minasny et al. 2013; Willaarts et al. 2015). Among these factors, climate, and specifically the rainfall and temperature, are the most important determinants of SOC contents (Homann et al. 1995; Alvarez and Lavado 1998). This trend might be due to the effects of rainfall and temperature on the quality and quantity of organic residues inputs and on the rates of soil organic matter (SOM) mineralization and litter decomposition (Quideau et al. 2001; Heviaa et al. 2003). Baldock (2007) reported that the amount of C in the soil could be influenced by climate attributes because biological processes such as the amount of SOM input and the rate of decomposition are affected by soil temperature, oxygen, and moisture content. A large amount of SOC can be lost via respiration due to faster decomposition of SOM because of specific soil moisture and temperature conditions which in turn lead to less storage of C in the slow and passive stocks. Knorr et al. (2005) observed that SOC stocks with longer turnover times are sensitive to temperature. Thus, SOC content is lower in soils of warmer climate than those in colder regions (Lal et al. 2007). However, the quantitative relationship between the temperature, decomposability, and decomposition rate as well as the fate of soil C in a warmer world remain unresolved, and the issue has been addressed differently in leading climate–carbon models (Huntingford et al. 2009). Precipitation influences SOC content by promoting higher plant growth. The soils prevailing in the humid regions contain higher organic C as compared to those in dry regions (Lal 2007). The storage of organic C in the soil depends on the balance between gains and losses of C, which are governed by the changing climate. Therefore, projected climate change may accentuate the mineralization, and deplete the soil C stocks and affect its dynamics. However, the potential effects of climate change on SOC dynamics are still largely uncertain (Zaehle et al. 2007; Álvaro-Fuentes and Paustian 2011). Although several studies have been conducted globally and in India on soil C inventories, the quantitative data on the relationship between SOC concentration and the controlling factors do not exist at a large scale (Lal 2004). Further, procuring data on soil profile C distribution in different landscapes is complex and challenging. Thus, an understanding of the C status in soils is needed because it is an important indicator of soil and environmental quality and integral part of GCC. Anthropogenic emissions of greenhouse gases (GHGs) and especially of CO_2 affect changes in temperature, rainfall patterns, net primary production (NPP), C inputs into the soil, decomposition rate of soil organic matter (SOM), and terrestrial C storage (Falloon 2007). The magnitude of change in soil C storage due to climate change depends upon specific processes that dominate land use and the soil type (Fang et al. 2005; Knorr et al. 2005; Davidson and Janssens 2006; Eglin et al. 2010). The results of the general circulation models (GCMs) on climate change indicate that the global average surface temperature will be increased by 1.5–4.5°C over the next 100 years due to rising levels of GHGs. Similarly, precipitation patterns also will be altered in the future (Brevik 2013). Khan (2009) reported that the average temperature in India is projected to increase by 0.4–2.0°C and 1.1–4.5°C by 2070 during summer and winter seasons, respectively. Therefore, it is important to know how the projected changes in climate will affect SOC dynamics and stocks in India. Estimation of soil C stock and changes at a national level are also required to develop appropriate management strategies for land use, protect soil resources and biodiversity, and develop an action plan to mitigate the climate change. The objectives of this chapter are to provide estimates of existing soil C stocks and document the projected climate change impacts on soil C stocks in India.

12.2 CLIMATE CHANGE SCENARIO IN INDIA

The Intergovernmental Panel on Climate Change (IPCC) refers to climate change as a change in climatic factors over time, whether due to natural variability or as a result of human activity. Over the last few decades, the increasing human activities have contributed significantly to climate change by adding CO_2 and other heat-trapping gases like CH_4 and N_2O to the atmosphere. The increase in the concentration of such gases in the atmosphere has increased the greenhouse effect, which leads to the rise in Earth's surface temperature and altered precipitation pattern. A number of researchers across India have reported an increasing trend in surface temperature (Kothawale et al. 2010; Jain and Kumar 2012; Pai et al. 2013; Rohini et al. 2016), but without any significant trend in rainfall on an all-India basis (Thapliyal and Kulshrestha 1991; Kumar et al. 2010) and decreasing/increasing trends in rainfall on a regional basis (Kripalani et al. 2003; Jain and Kumar 2012) over the 20th century. The World Bank (2017) reported that mean annual temperature in India has increased by 0.56°C between 1901 and 2007 and is expected to increase by 1.7°C to 2°C by the 2030s and 3.5°C to 4.3°C by 2100 (Table 12.1). Kumar and Gautam (2014) observed that throughout the 21st century, India is projected to experience

TABLE 12.1

Climate Change Scenario in India

Climate Parameter	Baseline	Future
Mean annual temperature	Increases 0.56°C/100 years (1901–2007)	By 1.7°C to 2°C by the 2030s; 3.5°C to 4.3°C by the end of the century
Seasonal mean rainfall	Decreases over the 20th century	Projected to increase during the 21st century
Extreme rainfall event has become more frequent	Over the 20th century in northern India	By 30–45 days under RCP2.6; by 15–-200 days under RCP8.5
Sea level has increased	By 0.21 m as of 2009	0.6 and 1.1 m by 2081–2100

Source: Climate Change Knowledge Portal, World Bank Group (2017) http://sdwebx.world bank.org

warming above the global level and the average temperature change is predicted to be 2.33°C–4.78°C with a doubling of CO_2 concentrations. The PRECIS climate models including two regional models, namely HadRM2 and HadRM3H developed by The Hadley Centre for Climate Prediction, UK (www .metoffice.gov.uk), have depicted the real change scenario on surface climate over the Indian region. The model simulation results under the scenario of increase in GHG concentration and sulfate aerosols indicate considerable spatial differences in the projected rainfall changes across India. The maximum increment of 10%–30% in rainfall is expected to occur over central India. The temperatures are projected to increase by as much as 3°C to 4°C towards the end of the 21st century. The data in Figure 12.1 show that the warming is widespread over the country, and the effect will be more pronounced over the northern parts of India. The changing climatic factors and their impact on soil C could decline agricultural production and threaten the nation's food security. The effect of climate change on agricultural productivity can be described in two ways: first, direct effects due to deviations in rainfall pattern, temperature, and/or CO_2 concentration, and, second, indirect effects, through changes in soil processes

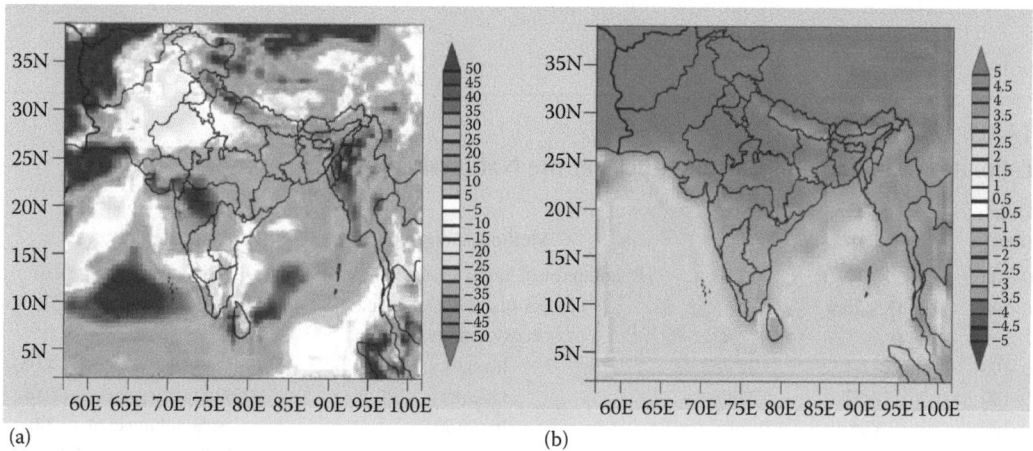

FIGURE 12.1 **(See color insert.)** Spatial patterns of the changes in (a) summer monsoon rainfall (%) and (b) annual mean surface air temperature (°C) for the period 2071–2100 concerning the baseline of 1961–1990, under the A2 scenario; an output of PRECIS model version 2.0.0; www.metoffice.gov.uk/precis.

particularly reduction in SOC stock. Few experiments have been conducted on the second aspect to understand how climate change can affect soil processes, mainly its impact on SOM decomposition, which has a direct link to soil fertility. The estimates and proper understanding of soil C stock at various depths is a prerequisite to assess its dynamics under climate change.

12.3   ESTIMATES OF SOIL CARBON STOCKS IN INDIA

The stock of SOC up to 1-m depth is about 1500 to 1600 Pg on a global scale (Batjes 1996; Lal 2004). But the data of SOC and soil inorganic carbon (SIC) stocks at a national scale are still scanty for many countries including India. Existing estimates on SOC and soil inorganic carbon (SIC) stocks based on available information in India are discussed in the following sections.

12.3.1   SOIL ORGANIC CARBON

Several researchers in India have estimated the SOC stock up to 1-m depth using various techniques, namely soil survey and inventory, digital soil mapping, process-based modeling, etc. (Dadhwal and Nayek 1993; Bhattacharyya et al. 2000; Banger et al. 2015; Sreenivas et al. 2016). These studies have reported an estimated SOC stock between 21 and 27 Pg C in India (Table 12.2). However, the first comprehensive study of SOC in Indian soils was not on the SOC stock estimation; rather, it was carried out by collecting data from the selective agricultural fields and forests representing variable rainfall and temperature patterns (Jenny and Raychaudhuri 1960). The first study showed the possible effects of climate on C storage in soil. The first ever estimate of SOC stock in soils of India was 24.3 Pg based on 48 soil series taking into account a few major soils ranging from the surface to an average subsurface depth of 44 to 186 cm (Gupta and Rao 1994). Although the first comprehensive report of SOC, SIC, as well as total C (TC) was conducted by Velayutham et al. (2000) and Bhattacharyya et al. (2000). These estimates were useful for various mapping schemes (Bhattacharyya et al. 2008). In contrast, Falloon et al. (2007) estimated soil C stock in India at 6.5–8.5 Pg using a coupled climate-C cycle global circulation model. Bhattacharyya et al. (2008) reported that the SOC stocks in soils of India were 24.04 and 29.92 Pg up to 100 and 150 cm depth, respectively, with a comprehensive soil survey database and mapping. By using the process-based dynamic land ecosystem modeling technique, Banger et al. (2015) estimated the SOC stock in India at 20.5–23.4 Pg, a large

---

TABLE 12.2
Estimates of Indian Soil Carbon Stocks at the National Level

Depth (cm)	SOC (Pg)	SIC (Pg)	Methods Adopted	References
0–100	22.72	12.83	Random forest-based spatial interpolation	Sreenivas et al. (2016)
0–100	20.5–23.4	–	Process-based dynamic land ecosystem model	Banger et al. (2015)
0–100	24.04	22.46	Inventory	Bhattacharyya et al. (2008)
0–150	29.92	33.98	Inventory	Bhattacharyya et al. (2008)
0–100	47.5	–	Inventory	Velayuthum et al. (2000)
–	23.4–27.1	–	Inventory	Dadhwal and Nayak (1993)
–	24.4–26.5	–	Inventory	Dadhwal and Nayak (1993)
44–186	24.3	–	Inventory	Gupta and Rao (1994)

TABLE 12.3
Soil Carbon Stock (in Pg) in Different Agro-Ecological Regions of India

		Bhattacharyya et al. (2008)		Sreenivas et al. (2016)	
		Up to 0–0.3-m Depth		Up to 1-m Depth	
Agro-Ecological Zones		SOC Stock (Pg)	SIC Stock (Pg)	SOC Stock (Pg)	SIC Stock (Pg)
1	Western Himalayas—skeletal soils—arid region	0.586	0.669	0.06	0.14
2	Western Plains—kachchh peninsula—arid region	0.266	0.891	1.13	3.10
3	Deccan plateau—mixed red and black soils—arid (typic) regions	0.121	0.05	0.19	0.23
4	Northern plain—Gujarat plain—central highlands—mixed red black and brown and alluvial soils—semi-arid regions	0.666	0.219	1.75	2.11
5	Kachchh peninsula—central highlands—semi-arid (dry to moist) regions	0.609	0.348	1.04	0.92
6	Deccan plateau—shallow to deep black soils—semi-arid to sub-humid	0.648	0.827	1.83	1.30
7	Deccan plateau—eastern ghats-mixed red and black soils—semi-arid (dry to moist) regions	0.554	0.366	1.07	0.52
8	Eastern ghats and Tamil Nadu uplands to Deccan plateau—mixed red (loamy) and black soils—semi-arid (dry to moist) regions	0.351	0.298	1.28	0.70
9	Northern plain—alluvial soils—semi-arid to sub-humid regions	0.15	0.006	0.94	0.55
10	Central Highlands to Deccan plateau—mixed red soils and shallow to deep black soils—sub-humid (dry to moist) regions	0.616	0.053	1.95	0.85
11	Eastern plateau—red and yellow soils—sub-humid regions	0.141	0	1.33	0.38
12	Eastern plateau—Eastern ghats and TN uplands—red and lateritic soils—sub-humid (dry to moist) regions	0.765	0	2.54	0.51
13	Eastern plain to central Himalayas—alluvial derived to Tarai soils—sub-humid (dry to moist) regions	0.172	0.153	0.66	0.64
14	Western Himalayas—brown forest and podzolic soils to red and yellow soils to tarai soils—sub-humid to humid regions	0.747	0.109	1.34	0.39
15	Assam and Teesta valley—alluvial derived soils—sub-humid to per-humid regions	0.403	0.041	1.26	0.17
16	Eastern Himalayas—Tarai soils to brown and red hill and loamy soils—per-humid region	0.787	0	0.93	0.02
17	North eastern hills—red loamy and lateritic soils—per-humid region	0.771	0	1.28	0.06
18	East coastal plain—coastal and deltaic alluvium derived soils—semi-arid to sub-humid regions	0.238	0	0.56	0.13
19	The Western Ghats to west coastal plains—mixed red and lateritic and black soils—sub-humid to humid to per-humid regions	0.843	0.063	1.60	0.10

(Continued)

TABLE 12.3 (CONTINUED)
Soil Carbon Stock (in Pg) in Different Agro-Ecological Regions of India

Agro-Ecological Zones		Bhattacharyya et al. (2008)		Sreenivas et al. (2016)	
		Up to 0–0.3-m Depth		Up to 1-m Depth	
		SOC Stock (Pg)	SIC Stock (Pg)	SOC Stock (Pg)	SIC Stock (Pg)
20	Islands—sandy and littoral soils to red and yellow soils—sub-humid to humid regions	0.121	0.014	0	0
	Total	9.555	4.107	22.74	12.82

proportion of which is stored in the forest soils of the north east, north and few dispersed regions in the southern India (Table 12.2).

By using the random forest based digital mapping technique, Sreenivas et al. (2016) estimated the SOC stock at 22.72 ± 0.93 in the top 100-cm depth of soil. This study used 1198 soil samples collected from various parts of India considering different land use, soil, topography, and agro-ecological regions based on a stratified random sampling method. They made a spatially explicit mapping of soil organic and inorganic C stocks at 250 m resolution. On the basis of soil data collected during 1980, total SOC stock in soils of India was estimated as 9.55 Pg to 30 cm depth (Bhattacharyya et al. 2005). Regional estimates of SOC stock in India also have been carried out by some researchers (Table 12.3). Bhattacharyya et al. (2008) estimated that the semi-arid (116.4 Mha) and sub-humid (105.0 Mha) regions contribute about 56% of total SOC stock in India to 30-cm depth. The Himalayan region (western and eastern zones) constitutes 19% of total geographical area and contributes 33% of SOC stocks of the country. Sreenivas et al. (2016) observed that soils of Western Ghat and coastal plains of India, under prevailing hot humid to per-humid climates, have higher SOC stock up to 1-m depth compared with lower values in arid and semi-arid regions (Table 12.4). This trend might be due to the humid tropical climate with a hyperthermic temperature regime having cool winter months and high rainfall which promote soil C sequestration. Thus, climate is a significant determinant of the magnitude of soil C stock. Venkanna et al. (2014) and Sreenivas et al. (2016) also reported variations in SOC density and C stock of soils in the sub-humid region of India. They emphasized the impact of land management practices (balanced fertilization) rather than climate change per se. Balanced fertilization with supplementary irrigation enhances root and crop biomass production, which in turn influence SOC storage of soils in such regions. The wide variability in SOC stock also is reported within diverse land uses and land covers throughout the country by Chhabra et al. (2003), Ravindranath et al. (1997), Banger et al. (2015), and Sreenivas et al. (2016). In India, SOC stock for diverse forest types has been estimated as 4.13 Pg (Chhabra et al. 2003) and 5.25 Pg (Velmurugan et al. 2014) to 50-cm depth. Using SOC densities and remote sensing–based area of forest types, estimates of 9.38 and 6.81 Pg were reported by Sreenivas et al. (2016) and Chhabra et al. (2003) respectively for forest soil to 100-cm depth. Sreenivas et al. (2016) reported the highest mean SOC density to 1-m depth in soils under plantation (25.3 kg m^{-2}) followed by those under forest (13.99 kgm^{-2}) and agricultural land use (5.85–6.74 kg m^{-2}). They also observed that almost 80% of the total SOC stock was contributed from forest lands (9.38 Pg) and mono- and double-cropped lands (8.81 Pg). These data were corroborated by Banger et al. (2015), who reported SOC stock of 3.9–7.8 and 11.0–11.8 Pg for cropland and forest land, respectively (Table 12.4).

TABLE 12.4

Distribution of Soil Carbon Stocks in India in Relation to Land Cover

Method	Depth (cm)	SOC Stock (Pg)	SIC Stock (Pg)	References
		Cropland		
Random forest-based spatial interpolation	0–100	8.81	8.30	Sreenivas et al. (2016)
Process-based dynamic land Ecosystem Model		3.90–7.80	–	Banger et al. (2015)
		Plantations		
Random forest-based spatial interpolation	0–100	0.06	0.06	Sreenivas et al. (2016)
		Forests		
Random forest-based spatial interpolation	0–100	9.38	1.41	Sreenivas et al. (2016)
Inventory	0–100	6.81	–	Chhabra et al. (2003)
Inventory		5.40	–	Rabindranath et al. (1997)
Inventory		5.90–6.20	–	Dadhwal et al. (1998)
Process-based dynamic land Ecosystem Model		11.0–11.8	–	
		Grassland		
Random forest-based spatial interpolation	0–100	0.31	0.15	Sreenivas et al. (2016)
Process-based dynamic land Ecosystem Model		0.41–0.47[a]	–	Banger et al. (2015)
Waste and scrubland				
Random forest-based spatial interpolation	0–100	1.29	2.69	Sreenivas et al. (2016)

[a] (approx.) including shrub land.

12.3.2 Soil Inorganic Carbon

The estimations of SIC stocks are poorly documented compared to those of SOC at a national or global scale (Lal et al. 1998; Batjes 2006; Guo et al. 2006). In India also, the primary focus has been on estimating the average SOC and mean density of C to various depths. Thus, few studies have been conducted on SIC stock estimation and their patterns, controls, and dynamics in the soil. Using a comprehensive soil survey database and mapping, Bhattacharyya et al. (2008) reported that the SIC stocks in soils of India were 22.46 and 33.98 Pg to 100 and 150-cm depth, respectively. With random forest-based digital mapping techniques and intensive field and remote sensing–based data, Sreenivas et al. (2016) estimated the SIC stocks at 12.83 ± 1.35 Pg to 100-cm depth. This is the first study in India which used explicit mapping of SIC stock at 250 m resolution.

They reported high SIC stock in the western Rajasthan and Rann areas of Gujarat, particularly in the Western plains–Kachchh peninsula arid region (3.10 Pg), and no accumulation of SIC was observed in northeastern India. Bhattacharyya et al. (2008) reported that the Himalayan region (western and eastern zones) constitutes 19% of the total geographical area (TGA) and contributes 20% of SIC stocks of the country. A wide variability in SIC density and stock was reported within different land use and land cover by Sreenivas et al. (2016). They reported that 65% of total SIC stock to 1-m depth is located in cultivated soils (8.3 Pg), followed by that in scrub and wasteland (2.69 Pg), and forest and grassland (1.56 Pg) (Table 12.5).

TABLE 12.5
Changes in SOC Stock in India Estimated Using Various Modeling Approaches and Climate Change Scenario

Site	Time Frame	SOC Stock Change Scenario	Modeling	References
India	1860–2100	Decline in SOC stock by 2.07 Pg by 2000 whereas the value predicted to be reduced by 0.11 Pg by 2100 from the baseline data of 8.62 Pg in 1860	HadCM3LC model (IS92a emissions scenario)	Falloon et al. (2007)
India	1860–2100	Decline in SOC stock by 1.22 Pg by 2000 whereas the value predicted to be reduced by 0.63 Pg by 2100 from the baseline data of 8.0 Pg in 1860	Predicted by RothC model driven with HadCM3LC model (IS92a emissions scenario)	Falloon et al. (2007)
India	1901–2010	Decline in SOC stock by 0.78 Pg by 2010 from the baseline data of 1901	Process-based dynamic land ecosystem model	Banger et al. (2015)
Indo-Gangetic plains of India	1967–2030	Decline in SOC stock by 0.283 Pg 000 whereas the value predicted to be reduced by 0.35 Pg by 2030 from the baseline data of 1.61 Pg in 1967	GEFSOC modeling (century)	Bhattacharyya et al. (2007)
Indo-Gangetic plains of India	1967–2030	Dynamic equilibrium reached by 2000 whereas the value predicted to be reduced by 0.05 Pg by 2030 from the baseline data of 0.96 Pg in 1967	GEFSOC modeling (empirical IPCC method)	Bhattacharyya et al. (2007)

12.4 ASSESSMENT OF CLIMATE CHANGE IMPACT ON SOIL CARBON STOCKS: RELIEVANCE

The relations between the atmosphere and soils in a climate change scenario are important to comprehend altered climate and its possible influence on soil C stock. Soil C is the integral component of SOM (Brady and Weil 2008) and vital from the perspective of soils and climate change interactions. Climate change affects changes in temperature and precipitation. In addition, changes in atmospheric CO_2 concentration affect NPP, C inputs to soil, the decomposition rate of SOC, and the terrestrial C stock (Falloon 2007). Continuous rises in CO_2 concentration increase photosynthetic rates as well as water-use efficiency (WUE) of plants and thus increase in organic matter supplies to soils (Brinkman and Sombroek 1996). Therefore, SOM is both a driver as well as an effect of climate change, and can be both a source and sink of atmospheric C. Some researchers have reported that rises in temperature reduce soil C storage due to increased decomposition of SOM and CO_2 emission (Nuttall 2007; Allen et al. 2011). Although all these effects are highly region specific and depend on the magnitude of the climate change, soil properties, and climate. Both the C-to-N ratio and SOM content diminish in a warmer soil temperature regime, compared to those in cooler regions. Davidson and Janssens (2006) reported that increases in temperature due to global warming might enhance the release of CO_2 to the atmosphere from the decomposition of SOM leading to higher CO_2 levels and aggravated global warming, which in turn may significantly affect the SOC stock. Depletion of the SOC stock has a direct influence on the environmental quality (emission of GHGs) as well as on soil productivity (soil degradation and nutrient depletion) which can aggravate food insecurity (Figure 12.2). On the contrary, increases in SOC can enhance biodiversity by reducing risks of soil erosion and desertification. Through its strong effect on the GCC, management of SOC stock is of global importance to the mitigation or worsening of atmospheric levels of GHGs. However, the understanding regarding how climate change will influence the soil C stock is incomplete and relatively

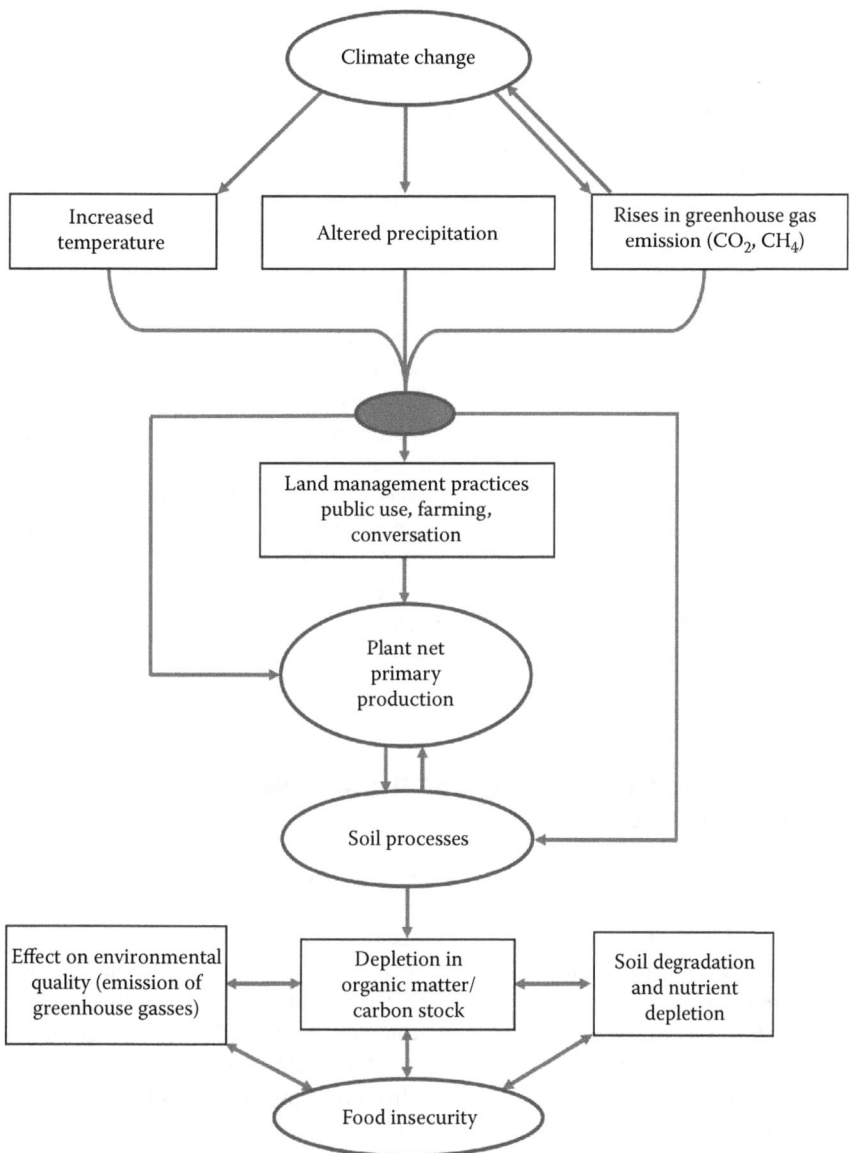

FIGURE 12.2 Schematic diagram showing direct and indirect effects of climate change in relation to soil carbon stock and food security.

unexplored in India, and additional research is needed on this aspect. There is a need to develop some methods or approaches to predict how SOC stocks will change as a function of changes in land use and climate in order to make appropriate management decisions. Periodic estimation of SOC stock nationally could be a possible option to assess its dynamics under the changing climate.

12.5 APPROACHES TO ASSESS CLIMATE CHANGE IMPACT ON SOIL CARBON STOCKS

Several approaches have been proposed to evaluate climate change impacts on SOC stocks, dynamics, and distribution and to predict trends under climate change scenarios from regional to global scale (Link et al. 2003; Wan et al. 2011; Follett et al. 2012; Brevik 2013; Banger et al. 2015).

The empirical and computational methods outlined by the IPCC (1997) have been used initially to estimate SOC stock and changes at the regional scale. But such methods are not able to predict future changes. Later, the regression approaches based on a long-term experimental database and a spatially explicit soil database have been used to predict changes in SOC stocks in the future (Gupta and Rao 1994). But all these approaches are not able to address the dynamic changes and assume a constant rate of change in SOC, which is unrealistic (Falloon et al. 2002). Dealing with dynamic changes in SOC stocks requires models that are linked with spatial soil data. Such modeling approaches also would be able to identify hot spots of SOC stock change (Falloon and Smith 2003). However, the absence of adequate models that are validated and calibrated with field data and monitoring is the real challenge for climate change impact assessment. Climate change modeling using long-term intensive spatial databases could provide a better understanding of climate change impact on soil C dynamics. In India, where an estimation of SOC and SIC stocks at the national scale is a big concern, assessing the climate change impact on C stock is a challenging task. However, an attempt has been made by several researchers to assess the impacts of climate change on soil C stock (mostly for organic carbon) using various climate change modeling approaches (Bhattacharyya et al. 2007; Falloon et al. 2007; Banger et al. 2015). The most commonly used models are process-oriented models in which the decomposition rate is based on the first order kinetics. Most frequently used models (namely Century, Roth-C, DNDC) simulate SOC dynamics (Viaud et al. 2010) under various emission scenarios. Further, these models are coupled with the GCM model, which is a powerful tool for estimating the climate change impact (Ghosh and Majumder 2008). But the impact of climate change on soil SIC stock has not been explored properly at a national level in India.

12.6 CLIMATE CHANGE IMPACT ON SOIL CARBON STOCKS IN INDIA: A REVIEW

The impact of climate change on SOC stock and its future trends under different emission scenario over India are presented in the following.

Using the HadCM3LC model under the IS92a emissions scenario, Falloon et al. (2007) reported that the SOC stock (whole profile based) in India decreased by 2.07 Pg by the end of 2000 and was predicted to be reduced by 0.11 Pg by the end of the 21st century from the baseline value of 8.62 Pg in 1860 (Table 12.5). Use of the RothC model driven with the HadCM3LC model under the same emissions scenario estimated a reduction in SOC stock by 1.22 Pg by the end of the 20th century and predicted it to be reduced by 0.63 Pg by 2100 from the baseline data of 8.0 Pg in 1860. Using the process-based dynamic land ecosystem model, Banger et al. (2015) reported that climate change and variability reduced the total SOC stock of India by 0.78 Pg C during 1901–2010. They analyzed that the majority of the climate-induced reduction of SOC stock occurred after the 1950s due to increases in temperature and frequent occurrences of severe drought years that affected the NPP. Using global environment facility soil organic carbon (GEFSOC) modeling coupled with an empirical IPCC method in a study over the Indo-Gangetic plains of India, Bhattacharyya et al. (2007) observed a dynamic equilibrium in SOC stock (0–30 cm depth) in the year 2000, and the SOC stock was predicted to be decreased by 0.05 Pg by 2030 due to climate and land use changes from the baseline data of 0.96 Pg in 1967. However, GEFSOC modeling coupled with the Century model predicted the decline in SOC stock of 0.283 Pg by 2000 and an additional decline of 0.35 Pg by 2030 from the baseline data of 1.61 Pg in 1967. The above-mentioned findings demonstrate that climate change has negatively affected the SOC stocks in India. Thus, most soils in India are characterized by low contents of the slow C stock (Banger et al. 2010) which is more sensitive to temperature (Davidson and Janssens 2006). Hence higher temperatures may have long-term implications to soil C storage. Link et al. (2003) observed 32% decreases in SOC content over a 5-year period due to soil warming and drying, which is very rapid and pronounced as compared to reduction caused by increased tillage. Gupta (2015) reported a reduction of soil C content by 11.6–19.2% (under the A2 climate change scenario) and 9.62%–16.9% (under the B2 climate change scenario) by 2099 due to climate change using the Century model as compared to the baseline data of 2010 in the Indian Himalayan region. The impact of climate change on SOC storage in India in the future could be visualized by the fact that the SOM content, which is already low in the soils of India, would still get lower. The changing climate scenario and its impact could decline agricultural production and threaten the food security. So there is a need to develop some appropriate management decisions in order to offset the changes in SOC stocks because of the changing climate for better productivity.

12.7 IMPACT OF CLIMATE CHANGE ON SOIL ORGANIC CARBON STOCKS: A CASE STUDY

12.7.1 STUDY AREA AND SOIL SAMPLING AND CHEMICAL ANALYSIS

The study was conducted in the Indian state of Karnataka in southern India which is located between 12.97°N latitude and 77.50°E longitude (Figure 12.3). The geographical area of the state is 191,976 km². A total number of 166 soil samples (GPS based, position accuracy of 10 m) were collected from the surface depth (0–30 cm) at randomly selected sites in the study region during the period of August 2016 to October 2016. One kilogram of mixed soil sample was collected from each site and brought to the laboratory for chemical analysis. The oxidizable organic carbon content of soil samples was determined by the wet digestion method (Walkley and Black, 1934). Another set of soil samples from the same locations was collected by using a core sampler of known volume for bulk density measurement (Grossmann and Reinsch 2002). The organic carbon stock was calculated by multiplying the measured organic carbon content with the soil depth and the bulk density. All soil samples were randomly divided into calibration (75%) and validation dataset (25% of total) for modeling (Table 12.6).

12.7.2 ENVIRONMENTAL COVARIATES AND CLIMATE SCENARIOS

The Shuttle Radar Topography Mission's (SRTM) digital elevation model (DEM) (90 m) dataset was extracted for the study region from the U.S. Geological survey site (http://earthexplorer.usgs .gov). The slope angle was calculated from the DEM using the spatial analyst function of ArcGIS software version 10.2.1. The 10-year mean NDVI data (250m resolution), available from MODIS as

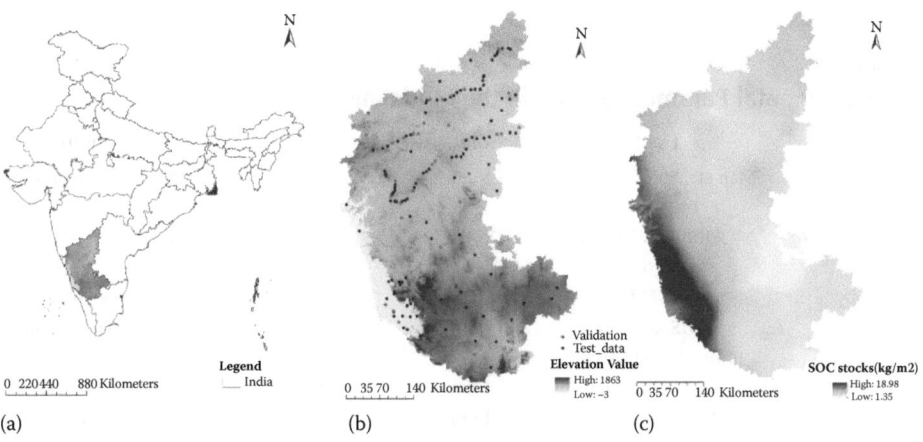

FIGURE 12.3 **(See color insert.)** Location of the study site (a), sampling points (b), and current prediction of SOC using base model (c).

TABLE 12.6

Summary Statistics of the Soil Carbon Database

Fitting Database (n = 124)			Validation Database (n = 42)		
Mean SOC Stocks (kg m^{-2})	StD SOC Stocks (kg m^{-2})	CV	Mean SOC stocks (kg m^{-2})	StD SOC Stocks (kg m^{-2})	CV
4.79	3.65	0.76	4.40	3.51	0.79

Abbreviations: CV—coefficient of variance; StD—standard deviation.

16-day composite, was collected (http://modis.gsfc.nasa.gov). The mean NPP data (1 km resolution) of the last 10 years was extracted from the latest available MODIS data as an annual composite (http://modis.gsfc.nasa.gov) for the study. The current monthly composited mean, the maximum and minimum air temperatures, and precipitation data were collected from the WorldClim data source (Fick and Hijmans 2017). The land use land cover map at 56 m resolution published by the National Remote sensing Centre, ISRO, for the year 2015–2016 was used to extract land cover information of the study region (http://bhuvan.nrsc.gov.in).

The HadGM2-AO and CCM4 GCM climate projections for the year 2050 used in the study were taken from global climate models (GCMs) for four representative concentration pathways (RCP 2.6, RCP 4.5, RCP 6.0, and RCP 8.5) which are available on the WorldClim data portal (Table 12.7). These are the most recent GCM climate projections that were used in the Fifth Assessment IPCC report (IPCC 2013). The spatial resolution of all the environmental covariates was finally resampled to 90 m using the resampling function of ArcGIS 10.2.1.

12.7.3 METHODS

A geospatial modeling approach was evaluated in this study for spatiotemporal predictions of SOC stocks. This method consists of two main steps (Figure 12.4). First, the uncorrelated linear predictors were identified for the SOC stocks for the present climate. The identified linear predictors were used in a geographically weighted regression kriging (GWRK) approach to predict the SOC stocks at a regional scale. Second, the climatic predictors were replaced using

TABLE 12.7

List of Environmental Parameters Used in the Modeling

Variables	Data	Resolution	Resample to (m)	Source
LULC	Map (1:250K) prepared by NRSC	56 m	90	http://bhuvan.nrsc.gov.in
NDVI	MODIS NDVI 16 days composite	250 m	90	http://modis.gsfc.nasa.gov
DEM	SRTM	90 m	–	http://earthexplorer.usgs.gov
NPP	MODIS annual net primary productivity	1 km	90	http://modis.gsfc.nasa.gov
Climate (current)	WorldClim data	1 km	90	Fick and Hijmans (2017)
Climate (Future 2050)	WorldClim data	1 km	90	Fick and Hijmans (2017)

Abbreviations: DEM—digital elevation model; LULC—land use land cover; NDVI—Normalized Difference Vegetation Index; NPP—net primary productivity.

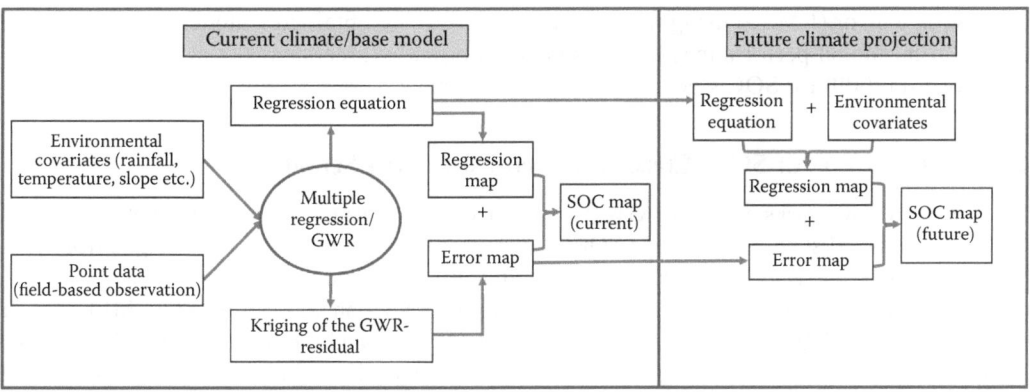

FIGURE 12.4 Soil organic carbon prediction workflow: Current and future climate scenarios.

the projected climatic data for different IPCC emission scenarios up to 2050, and the GWRK approach was used to predict the SOC stocks for the future climatic scenarios. The environmental factors of the soil sampling sites were extracted and correlation analysis was performed using the SPSS software (IBM, version 24) to assess the linear relationship between SOC stocks and environmental covariates. The significant correlation of SOC stocks was observed with NDVI, slope, elevation, precipitation, NPP, mean, and maximum surface air temperature. Stepwise multiple linear regression analysis was carried out to identify the uncorrelated environmental predictors of the SOC stocks. Based on the regression coefficient and probability values (Table 12.8) obtained from the stepwise regression, we used precipitation and NDVI in a geographically weighted regression (GWR) model to predict the SOC stocks for the present climate (Table 12.9). The GWR–kriging (GWRK) technique, which combines GWR with ordinary kriging of the spatially autocorrelated residuals (Kumar et al. 2012), was used to predict the SOC stocks under current and future climates (2050). The performance and prediction accuracy of the SOC stock predictions under the current climate was evaluated by calculating the mean estimation error (MEE), the mean absolute estimation error (MAEE), and the root

TABLE 12.8

Descriptive Statistics of Selected Environmental Variables from Stepwise Regression

Covariates	Model Coefficient	t Value	p Value
Intercept	0.26	1.47	0.00403
Precipitation	0.002	4.07	0.00001
NDVI	1.71	3.84	0.00004

TABLE 12.9

Coefficients Used in the GWRK Modeling

Covariate	Minimum	Maximum	Median	Probability
Intercept	0.07	0.37	0.16	0.006
Precipitation	0.001	0.002	0.002	0.005
NDVI	1.32	2.11	1.97	0.002

mean square error (RMSE) using the validation dataset. We obtained a model validation R^2 of 0.61. Similar model performance also has been reported by Meersmans et al. (2008) and Bell and Worrall (2009) for SOC stock estimation at regional scales.

12.7.4 PREDICTION OF SOC: CURRENT AND FUTURE CLIMATE SCENARIO

The predicted SOC stocks map using GWRK approach for the current climate is presented in Figure 12.3. The predicted state-wise average SOC stock was 4 kg m^{-2}, which spatially ranged from 1.35 to 19 kg m^{-2}. The current SOC stock of the study area was estimated as 732.2Tg. The model prediction errors are presented in Table 12.10. The bias of the prediction was estimated by MEE, which shows a value of −0.19 kg m^{-2}. The estimation error (MAEE) and the minimum prediction error (RMSE) values of 0.89 and 1.16 kg m^{-2} were observed, respectively, in the SOC stocks under the current climate. The predicted SOC stocks maps using the GWRK approach for future climate scenarios (2050) are presented in Figures 12.5 and 12.6. The SOC stock maps predicted under CCM4 and HadGEM2-AO climate scenarios with four representative concentration pathways show the similar spatial distribution patterns as the current SOC stock map. However, slightly lower values were observed as compared to current predictions (Table 12.11). The highest mean SOC stock of 4.02 kg m^{-2} was predicted by HadGEM2-AO (RCP8.5) in 2050, whereasthe lowest mean SOC stock of 3.74 kg m^{-2} was predicted by CCMP (RCP2.6) in 2050. The result indicates a reduction in SOC stocks in all future climate scenarios in comparison to the current SOC stock. In the present study, the projected decrease in SOC stocks is 3.10–53.5 Tg by 2050. The highest decrease was predicted by CCMP (RCP2.6) by 53.5 Tg, and the lowest decrease was by HadGEM2-AO (RCP8.5) by 3.10 Tg. The findings of the present study indicate a decreasing trend in SOC stocks at 0–30-cm depth in near future due to changes in climate, which are corroborated with the findings of Bhattacharyya et al. (2007), Falloon et al. (2007), and Banger et al. (2015). However, our prediction accuracy of current and projected SOC stocks was constrained by the limited soil samples, their uneven distribution across the study sites, variation in soil C measurement and stock estimation, effect of data derived from different environments on model parameterization thereby model outputs, and use of the coarse resolution environmental database.

Because of such limitations, we consider the model-based prediction of the current and future SOC stocks presented here to be initial estimates that will require future refinements. The current and projected SOC stocks (2050) across various land covers were also assessed and are presented in Table 12.12. The major decrease in SOC stocks in the near future was observed in forest (4 to 33 Tg) followed by that in crop land (6 to 10 Tg) under diverse climate scenarios. The results presented indicate that the soil C stock of forest and cropland in the study area may be adversely affected under the projected climate change. This trend might be due to changes in precipitation patterns which are predicted to create exaggerated periods of both drought and extreme precipitation leading to either increased erosion of stabilized mineral soil carbon or decreased decomposition linked to low oxygen availability in wet soils (Wuebbles et al. 2014).

So there is an urgent need to develop suitable interventions for improving soil C storage. Soil C sequestration can be accomplished by a range of management practices that can add biomass

TABLE 12.10
Validation Indices of the Base Model Predicting Current Conditions

Model	R^2	RMSE (kg m^{-2})	MEE (kg m^{-2})	MAEE (kg m^{-2})
Base model	0.61	1.16	−0.19	0.89

Abbreviation: MAEE, mean absolute estimation error; MEE, mean estimation error; RMSE, root mean square error.

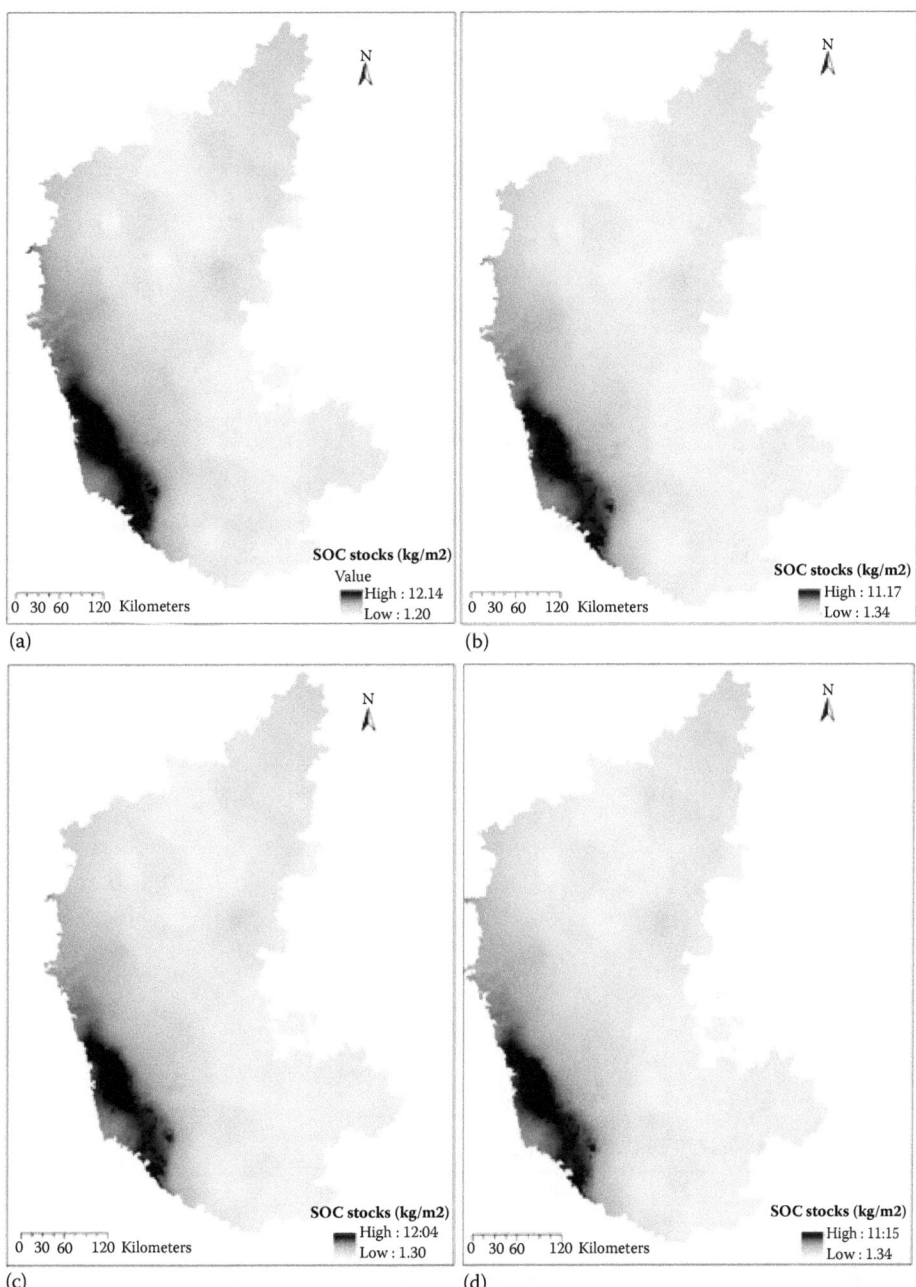

FIGURE 12.5 **(See color insert.)** Future Prediction of SOC stocks under climate scenario CCM4_RCP2.6 (a), CCM4_RCP4.5 (b), CCM4_ RCP6.0 (c), and CCM4_RCP8.5 (d).

to soil in a high quantity, conserve soil and moisture by improving soil structure by minimal soil disturbance and enhance soil faunal activity. The maintenance of forest cover, optimizing tree biomass, establishing forests where they did not exist previously (afforestation) and avoiding drainage of ecosystems with deep organic soils (which contain substantial carbon stores) are likely to have the best results for protecting carbon in the forest soils (D'Amore and Kane 2016). Increasing area under forests including agroforestry systems is one of the important aspects to

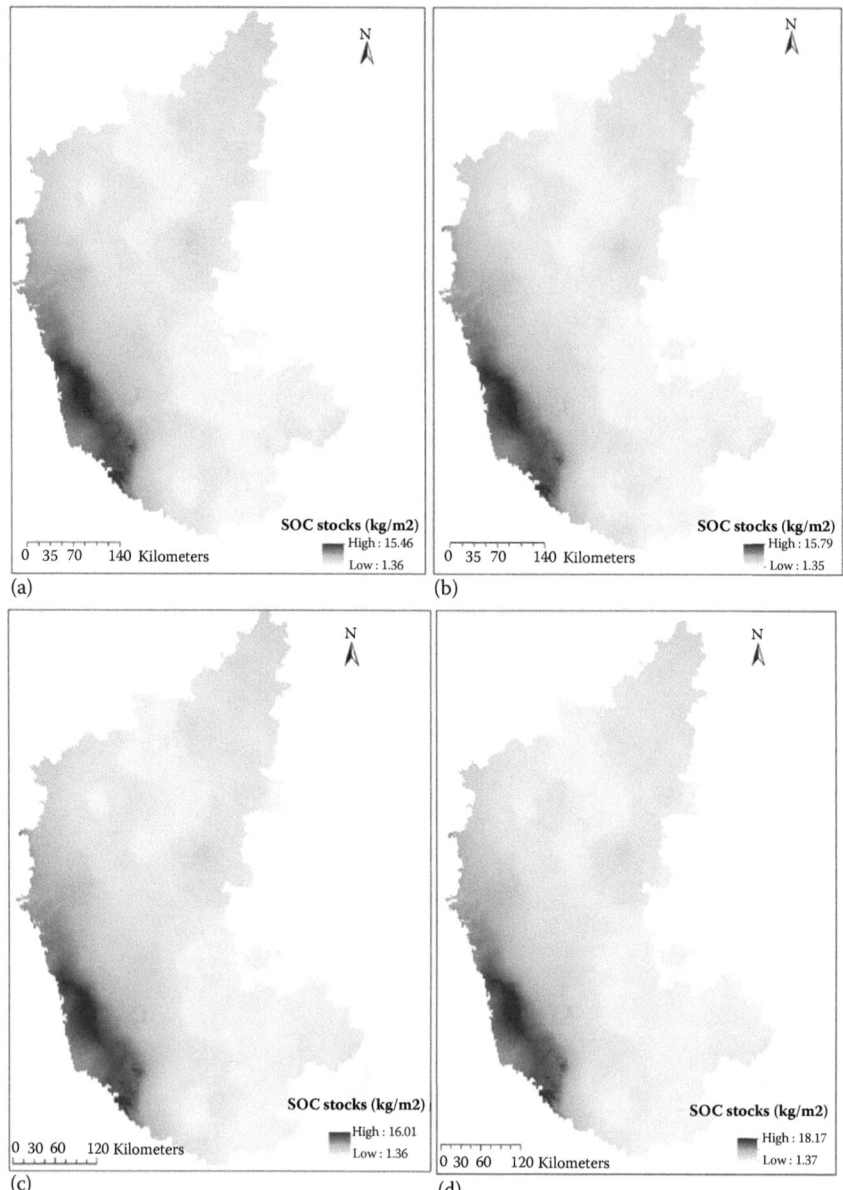

FIGURE 12.6   (See color insert.) Future prediction of SOC stocks under climate scenario HadGEM2-AO_RCP2.6 (a), HadGEM2-AO_RCP4.5 (b), HadGEM2-AO_RCP6.0 (c), and HadGEM2-AO_RCP8.5 (d).

improve SOC storage and offset the impact of climate change (Vasievich and Alig 1996; Nelson et al. 1999). This sector plays an important role in regulating the emission profile from the agriculture sector and provide avenues for increasing the C sink capacity. An increment in C stock in agricultural soils also can be achieved through the adoption of suitable cropping sequences (Wright and Hons 2005), integrated soil fertility management (Lal 2010), precise use of fertilizers and organic inputs (Mandal et al. 2007; Majumder et al. 2008), and adoption of conservation agriculture (Lal 2009).

TABLE 12.11

Present and Projected Soil Organic Carbon Stocks

Model	Minimum (kg m^{-2})	Maximum (kg m^{-2})	Mean (kg m^{-2})	Standard Deviation (kg m^{-2})	Stock (Tg)	Change Compared to Base (Tg)
			Current Prediction (Base)			
Base model	1.35	18.97	4.03	2.36	732.2	
			Future Prediction (2050)			
CCM4						
RCP 8.5	1.34	17.10	3.97	2.19	714.0	−18.2
RCP6.0	1.35	17.09	3.96	2.20	715.6	−16.6
RCP4.5	1.34	17.17	3.97	2.22	717.1	−15.1
RCP2.6	1.33	13.14	3.74	1.76	678.7	−53.5
HadGEM2AO						
RCP 8.5	1.37	18.17	4.02	2.27	729.1	−3.10
RCP6.0	1.36	16.01	3.87	1.99	699.6	−32.6
RCP4.5	1.35	15.79	3.88	2.03	700.9	−31.3
RCP2.6	1.36	15.46	3.88	1.99	702.0	−30.2

TABLE 12.12

Present and Projected Soil Organic Carbon Stocks (Tg) across Land Cover Types

Land Cover Type	Current Prediction Base Model	Future Prediction (2050) HadGEM2AO				Future Prediction (2050) CCM4			
		RCP8.5	RCP6.0	RCP4.5	RCP2.6	RCP8.5	RCP6.0	RCP4.5	RCP2.6
Developed	13.53	13.36	13.38	13.40	13.70	13.08	13.45	13.45	13.34
Crop Land	409.0	401.2	399.44	399.65	409.87	395.19	402.77	402.41	401.86
Plantation	69.59	66.94	65.97	66.17	69.67	63.43	68.96	68.79	68.92
Forest	223.1	203.7	205.53	203.84	218.89	190.60	215.21	214.28	212.67
Waste Land	11.20	11.20	11.08	11.09	11.40	11.00	11.16	11.17	11.18
Grassland	1.03	0.86	0.87	0.87	0.91	0.83	0.90	0.90	0.90
Scrub land	4.66	4.62	4.60	4.60	4.69	4.58	4.65	4.65	4.63
Total	**732.2**	**702.0**	**700.9**	**699.6**	**729.1**	**678.7**	**717.1**	**715.6**	**713.5**

12.8 CONCLUSION

SOC stock in India is vulnerable as the mean annual temperature and precipitation are projected to change due to the increased concentration of GHGs in the atmosphere. The current estimated SOC stock size of India is about 25 Pg. Different modelling approaches under various climate changes or emission scenarios show a declining trend in soil C stock in the future. Depletion in soil C stock has the direct influence on the environmental quality by the emission of GHGs as well as on soil fertility by soil degradation and nutrient depletion which also aggravate food insecurity. Thus, there is an urgent need to develop methodologies to monitor periodically how multiple global change factors control SOC dynamics specifically the SOC stocks in India. An understanding of SOC stocks and changes at the national and regional scales would be important to policy makers in making prudent land use/management decisions. Formulation of suitable management plans including the adoption of appropriate crop rotations, integrated soil fertility management, precise use of fertilizers and organic amendments, adoption of conservation agriculture, promotion of agroforestry could offset the effects of climate change on soil C stock through C sequestration. The SOC stocks under these management options can be estimated by using GWRK technique and validated by the ground truthing data.

12.9   ACKNOWLEDGMENT

The first author is greatly thankful to the Science and Engineering Research Board and the Indo-US Science and Technology Forum of India and the Carbon Management and Sequestration Center, The Ohio State University, for necessary help and support.

REFERENCES

Allen, D. E., B. P. Singh, and R. C. Dalal. 2011. Soil health indicators under climate change: A review of current knowledge. In B.P. Singh et al. (eds.), *Soil Health and Climate Change*, Soil Biology 29, Springer-Verlag, Berlin, Heidelberg, 25–45.

Alvarez, R., and R. S. Lavado. 1998. Climate, organic matter and clay content relationships in the Pampa and Chaco soils, Argentina. *Geoderma* 83:127–141.

Álvaro-Fuentes, J., and K. Paustian. 2011. Potential soil carbon sequestration in a semiarid Mediterranean agro-ecosystem under climate change: Quantifying management and climate effects. *Plant Soil* 338:261–272.

Baldock, J. 2007. Composition and cycling of organic carbon. In: P. Marschner and Rengel, Z. (eds.), *Soil Nutrient Cycling in Terrestrial Ecosystems*. Springer, Berlin, Heidelberg, 1–35.

Banger, K., H. Tian, B. Tao, C. Lu, W. Ren, and J. Yang. 2015. Magnitude, spatiotemporal patterns, and controls for SOC stocks in India during 1901–2010. *Soil Sci Soc Am J* 79:864–875.

Banger, K., G. S. Toor, A. Biswas, S. S. Sidhu, and K. Sudhir. 2010. SOC fractions after 16-years of applications of fertilizers and organic manure in a Typic Rhodalfs in semi-arid tropics. *Nutr Cycl Agroecosystem* 86:391–399.

Batjes, N. H. 1996. Total carbon and nitrogen in the soils of the world. *European J Soil Sci* 47:151–156.

Batjes, N. H. 2006. Soil carbon stocks of Jordan and projected changes upon improved management of croplands. *Geoderma* 132:361–371.

Batjes, N. H. 2016. Harmonized soil property values for broad-scale modelling (WISE30sec) with estimates of global soil carbon stocks. *Geoderma* 269:61–68.

Bell, M. J., and F. Worrall. 2009. Estimating a region's soil organic carbon baseline: The undervalued role of land-management. *Geoderma* 152(1–2):74–84.

Bhattacharyya, T., M. Easter, K. Paustian, K. Killian, S. Williams, S. K. Ray, P. Chandran, S. L. Durge, D. K. Pal, K. S. Gajbhiye, E. Milne, and B. Singh. 2005. Description of newly added crop database and modified crop files: Indo-Gangetic plains, India case study. Special publication for: Assessment of SOC Stocks and Change at National Scale. NBSS & LUP, Nagpur, India, 196 pp.

Bhattacharyya, T., D. K. Pal, P. Chandran, S. K. Ray, C. Mandal, and B. Telpande. 2008. Soil carbon storage capacity as a tool to prioritize areas for carbon sequestration. *Curr Sci* 95:482–484.

Bhattacharyya, T., D. K. Pal, M. Easter, N. H. Batjes, E. Milne, K. S. Gajbhiye, P. Chandran, S. K. Ray, C. Mandal, K. Paustian, S. Williams, K. Killian, K. Coleman, P. Falloon, D. S. Powlson. 2007. Modelled SOC stocks and changes in the Indo-Gangetic Plains, India from 1980 to 2030. *Agric Ecosyst Environ* 122:84–94.

Bhattacharyya, T., D. K. Pal, M. Velayutham, P. Chandran, and C. Mandal. 2000. Total carbon stock in Indian soils: Issues, priorities and management. In: *Special Publication of the International Seminar on Land Resource Management for Food, Employment and Environment Security (ICLRM)*. Soil Conservation Society of India, New Delhi, pp. 1–46.

Brady, N. C., and R. R. Weil. 2008. *The Nature and Properties of Soils*, 14th ed. Pearson Prentice Hall, Upper Saddle River, NJ.

Brevik, E. C. 2013. The potential impact of climate change on soil properties and processes and corresponding influence on food security. *Agriculture* 3(3):398–417.

Brinkman, R., and W. G. Sombroek. 1996. The effects of global change on soil conditions in relation to plant growth and food production. In F. Bazzaz and W. Sombroek, eds., *Global Change and Agricultural Production*. Food and Agricultural Organization of the United Nations and John Wiley & Sons, New York, 345 pp.

Chhabra, A., S. Patria, and V. K. Dadhwal. 2003. SOC pool in Indian forests. *Forest Ecol Manag* 173:187–199.

Dadhwal, V. K., and S. R., Nayak. 1993. A preliminary estimate of biogeochemical cycle of carbon for India. *Sci Cult* 59:9–13.

Dadhwal, V. K., N. Shukla, and A. B. Vora. 1998. Carbon cycle for Indian forest ecosystem: A preliminary estimate In: B.H. Subbaraya et al. eds., *Global Change Studies: Scientific Results from ISRO-GBP*. ISRO, Bangalore, 411–430.

D'Amore, D., and E. Kane. 2016. Climate Change and Forest Soil Carbon. U.S. Department of Agriculture, Forest Service, Climate Change Resource Center. www.fs.usda.gov/ccrc/topics/forest-soil-carbon.

Davidson, E. A., and I. A. Janssens. 2006. Temperature sensitivity of soil carbon decomposition and feedbacks to climate change. *Nature* 440:165–173.

Eglin, T., P. Ciais, S. L. Piao, P. Barre, V. Bellassen, P. Cadule, C. Chenu, T. Gasser, C. Koven, M. Reichstein, and P. Smith. 2010. Historical and future perspectives of global soil carbon response to climate and land-use changes. *Tellus* B 62:700–718.

Falloon, P., and P. Smith. 2003, Accounting for changes in soil carbon under the Kyoto Protocol: Need for improved long-term data sets to reduce uncertainty in model projections. *Soil Use and Manage* 19:265–269.

Falloon, P., C. D. Jones, C. E. Cerri, R. Al-Adamat, P. Kamoni, T. Bhattacharyya, M. Easter, K. Paustian, K. Killian, K. Coleman, and E. Milne. 2007. Climate change and its impact on soil and vegetation carbon storage in Kenya, Jordan, India and Brazil. *Agric Ecosyst Environ* 122:114–124.

Falloon, P., P. Smith, J. Szabo, and L. Pasztor. 2002. Comparison of approaches for estimating carbon sequestration at the regional scale. *Soil Use Manage* 18:164–174.

Fang, C., P. Smith, J. B. Moncrieff, and J. U. Smith. 2005. Similar response of labile and resistant soil organic matter pools to changes in temperature. *Nature* 433:57–59.

Fick, S. E., and R. J. Hijmans. 2017. WorldClim 2: New 1-km spatial resolution climate surfaces for global land areas. *Int J Climatol* 37(12):4302–4315.

Follett, R. F., C. E. Stewart, E. G. Pruessner, and J. M. Kimble. 2012. Effects of climate change on soil carbon and nitrogen storage in the US Great Plains. *J Soil Water Conserv* 67(5):331–342.

Ghosh, S., and P. P. Majumder. 2008. Statistical downscaling of GCM simulation to streamflow using relevance vector machine. *Adv Water Resour* 31(1):132–146.

Gray, J. M. et al. 2014. Direct human influence on atmospheric CO_2 seasonality from increased cropland productivity. *Nature* 515:398–401.

Grossmann, R. B., and T. G. Reinsch. 2002. Bulk density and linear extensibility. In: J. H. Dane and G. C. Topp, eds. *Methods of Soil Analysis*. Part 4. SSSA Book Ser. 5. SSSA, Madison, WI, 201–228.

Guo, Y. Y., R. Amundson, P. Gong, and Q. Yu. 2006. Quantity and spatial variability of soil carbon in the conterminous United States. *Soil Sci Soc Am J* 70:590–600.

Gupta, R. K., and D. L. N. Rao. 1994. Potential of wastelands for sequestering carbon by reforestation. *Curr Sci* 66(5):378–380.

Gupta, S. 2015. Simulating Climate Change Impact on Soil Erosion and Carbon Sequestration. Ph.D thesis. Andhra University, Visakhapatnam, India.

Heviaa, G. G., D. E. Buschiazzoa, and E. N. Heppera, 2003. Organic matter in size fractions of soils of the semiarid Argentina. Effects of climate, soil texture and management. *Geoderma* 116:265–277.

Hobley, E., G. R. Willgoose, S. Frisia, and G. Jacobsen. 2013. Environmental and site factors controlling the vertical distribution and radiocarbon ages of organic carbon in a sandy soil. *Biol Fert Soils* 49:1015–1026.

Homann, P. S., P. Sollins, H. N. Chappell, and A. G. Stangenberger. 1995. SOC in a mountainous, forested region: Relation to site characteristics. *Soil Sci Soc Am J* 59:1468–1475.

http://bhuvan.nrsc.gov.in

http://earthexplorer.usgs.gov

http://modis.gsfc.nasa.gov

http://sdwebx.worldbank.org

Huntingford, C., J. Lowe, B. Booth, C. Jones, G. Harris, L. Gohar, and P. Meir. 2009. Contributions of carbon cycle uncertainty to future climate projection spread. *Tellus* 61B:355–360.

IPCC. 1997. Revised 1996 IPCC Guidelines for National Greenhouse Gas Inventories: Reporting Instructions (vol. 1); Workbook (vol. 2); Reference Manual (vol. 3). Intergovernmental Panel on Climate Change, United Nations Environment Programme, Organization for Economic Co-Operation and Development, International Energy Agency, Paris.

IPCC. 2013. Summary for policy-makers. In: T.F. Stocker, D. Qin, G.-K. Plattner, M. Tignor, S. K., Allen, J. Boschung, A. Nauels, Y. Xia, V. Bex, and P. M. Midgley, eds. *Climate Change 2013: The Physical Science Basis*. Contribution of Working Group I to the Fifth Assessment Report of the Intergovernmental Panel on Climate Change. Cambridge University Press, Cambridge, United Kingdom and New York, NY.

Jain, S. K., and V. Kumar. 2012. Trend analysis of rainfall and temperature data for India *Current Sci* 102(1):37–49.

Jenny, H., and S. P. Raychaudhuri. 1960. *Effect of Climate and Cultivation on Nitrogen and Organic Matter Reserves in Indian Soils*. ICAR, New Delhi, India, 126 pp.

Khan, S. A., S. Kumar, M. Z. Hussain, and N. Kalra. 2009. Climate change, climate variability and indian agriculture: Impacts vulnerability and adaptation strategies . In: S. N. Singh, ed. *Climate Change and Crops, Environmental Science and Engineering*. Springer-Verlag, Berlin, Heidelberg. doi.org/10.1007/978-3 -540-88246-6 2.

Knorr, W., I. C. Prentice, J. I. House, and E. A. Holland. 2005. Long-term sensitivity of soil carbon turnover to warming. *Nature* 433:298–301.

Kothawale, D. R., A. A. Munot, and K. Krishna Kumar. 2010. Surface air temperature variability over India during 1901–2007, and its association with ENSO. *Climate Res* 42:89–104.

Kripalani, R. H., A. Kulkarni, S. S. Sabade, and M. L. Khandekar. 2003. Indian monsoon variability in a global warming scenarios. *Natural Hazards* 29(2):189–206.

Kumar, R., and J. Gautam. 2014. Climate change and its impact on agricultural productivity in India. *Climatol Weather Forecasting* 2(1):1–3.

Kumar, S., R. Lal, and L. Desheng. A geographically weighted regression kriging approach for mapping soil organic carbon stock. *Geoderma* 189(2012):627–634.

Kumar, V., S. K. Jain, and Y. Singh. 2010. Analysis of long-term rainfall trends in India. *Hydrol Sci J* 55:484–496.

Lal, R. 2004. Soil carbon sequestration to mitigate climate change. *Geoderma* 123:1–22.

Lal, R. 2009. Soil carbon sequestration for climate change mitigation and food security. In: *Souvenir, Platinum Jubilee Symposium on Soil Science in Meeting the Challenges to Food Security and Environmental Quality, Indian Society of Soil Science*, New Delhi, 39–46.

Lal, R. 2010. Carbon sequestration potential of rainfed agriculture. *Indian J Dry land Agri Res and Develop* 25(1):1–16.

Lal R., R. F. Follett, B. A. Stewart, and J. M. Kimble. 2007. Soil carbon sequestration to mitigate climate change and advance food security. *Soil Sci* 172(12):943–956.

Lal, R., J. Kimble, J. M. Follett, R. B. A. Stewart. 1998. *Soil Processes and the Carbon Cycle*. CRC Press, Boca Raton, FL.

Le Quere, C. et al. 2015. Global carbon budget 2014. *Earth Syst. Sci. Data* 7:47–85.

Link, S. O., J. L. Smith, J. J. Halverson, and H. Bolton. 2003. A reciprocal transplant experiment within a climatic gradient in a semiarid shrub-steppe ecosystem: Effects on bunchgrass growth and reproduction, soil. *Glob Change Biol* 9:1097–1105.

Majumder, B., B. Mandal, P. K. Bandyopadhyay, A. Gangopadhyay, P. K. Mani, A. L. Kundu, and D. Majumder. 2008. Organic amendments influence SOC pools and crop productivity in 19 years old rice-wheat agroecosystems. *Soil Sci Soc of America J* 72:775–785.

Mandal, B., B. Majumder, P. K. Bandyopadhyay, G. C. Hazra, A. Gangopadhyay, R. N. Samantaray, A. K. Mishra, J. Chaudhury, M. N. Saha, and S. Kundu. 2007. The potential of cropping systems and soil amendments for carbon sequestration in soils under long-term experiments in subtropical India. *Glob Change Biol* 13:1–13.

Meersmans, J., F. De Ridder, F. Canters, S. De Baets, and M. Van Molle. 2008. A multiple regression approach to assess the spatial distribution of soil organic carbon (SOC) at the regional scale (Flanders, Belgium). *Geoderma* 143:1–13.

Minasny, B., A. B. McBratney, B. P. Malone, and I. Wheeler. 2013. *Digital Mapping of Soil Carbon. Advances in Agronomy*, Volume 118. Academic Press, Newark, NJ.

Nelson, F. E., N. I. Shiklomanov, and G. R. Mueller. 1999. Variability of active-layer thickness at multiple spatial scales, north central Alaska, U.S.A. *Arctic, Antarctic, and Alpine Research* 31(2):158–165.

Nuttall, J. G. 2007. Climate change—Identifying the impacts on soil and soil health. Department of Primary Industries, Future Farming Systems Research Division, Victoria.

Pai, D. S., A. Smitha, and A. N. Ramanathan. 2013. Long term climatology and trends of heat waves over India during the recent 50 years (1961–2010). *Mausam* 64(4):585–604.

Post, W. M., and K. C. Kwon. 2000. Soil carbon sequestration and land-use change: Processes and potentials. *Glob Change Biol* 6:317–328.

Quideau, S. A., Q. A. Chadwick, A. Benesi, R. C. Graham, and M. A. Anderson. 2001. A direct link between forest vegetation type and soil organic matter composition. *Geoderma* 104:41–60.

Ravindranath, N., B. S. Somanshekhar, and M. Gadgil. 1997. Carbon flows in Indian forests. *Climatic Change* 35:297–320.

Rohini, P., M. Rajeevan and A. K. Srivastava. 2016. On the variability and increasing trends of heat waves over India. *Scientific Reports* 6:26153, doi.org/10.1038/srep26153.

Sreenivas, K., V. K. Dadhwal, K. Suresh, S. G. Harsha, T. Mitran, G. Sujatha, J. R. G. Suresh, M. A. Fyzee, and T. Ravisankar. 2016. Digital organic and inorganic carbon mapping of India. *Geoderma* 269:160–173.

Thapliyal, V., and, S. M. Kulshrestha. 1991. Climate changes and trends over India. *Mausam* 42:333–338.

Vasievich, M., and R. Alig. 1996. Opportunities to increase timber growth and carbon storage on timberlands in the contiguous United States. In: R. N. Sampson and D. Hair, eds. *Forests and Global Change, Vol. 2: Forest Management Opportunities*. American Forests, Washington, DC, 91–104.

Velayutham, M., D. K. Pal, and T. Bhattacharyya. 2000. Organic carbon stock in soils of India. In: R., Lal, J. M. Kimble, and B. A. Stewart, eds. *Global Climate Change and Tropical Ecosystems*. Lewis Publishers, Boca Raton, FL, 71–96.

Velmurugan, A., S. Kumar, V. K. Dadhwal, and M. K. Gupta. 2014. SOC status of Indian Forests. *The Indian Forester* 140:468–477.

Venkanna, K., U. K. Mandal, A.S. Raju, K. L. Sharma, R. V. Adake, R. B. Pushpanjali, R. N. Masane, K. Venkatravamma, and B. P. Babu. 2014. Carbon stocks in major soil types and land-use systems in semiarid tropical region of southern India. *Current Sci* 106(4):604–611.

Viaud, V., D. A. Angers, and C. Walter. 2010. Towards land scape-scale modelling of soil organic matter dynamics in agro ecosystems. *Soil Sci Soc of America J* 74(6):1–14.

Walkley, A., and I. A. Black. 1934. An examination of Degtjareff method for determining soil organic matter and a proposed modification of the chromic acid titration method. *Soil Sci* 37:29–37.

Wan, Y., E. Lina, W. Xionga, Y. Lia, and L. Guo. 2011. Modeling the impact of climate change on SOC stock in upland soils in the 21st century in China. *Agril Ecosyst Environ* 141(1–2):23–31.

Willaarts, B. A., C. Oyonarte, M. Muñoz-Rojas, J. J. Ibáñez, and P. A. Aguilera. 2015. Environmental factors controlling SOC stocks in two contrasting Mediterranean climatic areas of southern Spain. *Land Degrad Dev* (on-line). doi.org/10.1002/ldr.2417.

World Bank. 2017. Climate Change Knowledge Portal Data Source: http://sdwebx.worldbank.org.

Wright, A. L., and F. M. Hons. 2005. Tillage impacts on soil aggregation and carbon and nitrogen sequestration under wheat cropping sequences. *Soil Tillage Res* 84:67–75.

Wuebbles, D., G. Meehl, K. Hayhoe et al. 2014. CMIP5 Climate Model Analyses: Climate Extremes in the United States. *Bull Amer Meteor Soc* 95(4):571–583.

Zaehle, S., A. Bondeau, T. R. Carter, W. Cramer, M. Erhard, I. C. Prentice, I. Reginster, M. D. A. Rounsevell, S. Smith, B. Smith, P. C. Smith, and M. Sykes. 2007. Projected changes in terrestrial carbon storage in Europe under climate and land-use change, 1990–2100. *Ecosyst* 10:380–401.

13 Soil Degradation and Climate Change in South Asia

Muhammad Farooq, Muhammad Sanaullah, Faisal Nadeem, Nirmali Gogoi, Muhammad Shakeel Arshad, and Rattan Lal

CONTENTS

13.1 INTRODUCTION

The 68th United Nations General Assembly affirmed 2015 as the International Year of Soils and proclaimed that "soil is vital for ecosystem functioning and agriculture development to achieve food security and is the foundation for life sustainability on Earth" (United Nations 2015). The whole terrestrial ecosystem is vulnerable due to the increased climatic variability, resulting in soil degradation under adverse climatic changes in most eco-regions. The Food and Agriculture Organization (FAO 2015) described soil degradation as a phenomenon of "change in the soil health status resulting in a diminished capacity of the ecosystem to provide goods and services for its beneficiaries."

The broader term, land degradation, includes all adverse changes in an ecosystem's capacity to provide services and goods (FAO 2014). Soils with poor health do not have the capacity to support their ecosystems by providing food and other ecosystem services for human wellbeing and nature conservancy. DeLong et al. (2015) indicated that soil degradation is a global pandemic. It is induced by several drivers such as deforestation, excessive removal of vegetation and tillage practices, inappropriate crop rotations, industrialization, and urban sprawl (Karlen and Rice 2015). Therefore, degradation of soil limits crop and animal performance and productivity, resulting in economic loss and causing threats to farmers' livelihood and global food security (Bhattacharyya et al. 2015).

Soil degradation is intense and severe in most regions around the globe, and it affects about 33%, 20%, and 10% of forests, cultivated lands, and grasslands, respectively (Bai et al. 2008). Adams and Eswaran (2000) reported that 2.6 billion people (from >100 countries) and 33% of the Earth's surface are affected by desertification and land degradation. South Asia covers a total land area of 5.1 million Km², which is 11.5% of the Asian continent and 3.4% of the world's land surface area. South Asia is also the home to 1.88 billion people or about one fourth (24.7%) of the world's population (Table 13.1). Therefore, about one fourth of the world population lives in 3.4% of the world land area. The scarcity of soil and water resources is being exacerbated by the projected climate change and is the attendant issue of soil degradation and water depletion as well as pollution. While the soils of South Asia are diverse and fertile and are also prone to a wide range of degradation processes. For instance, in India alone, an area of about 114 and 147 million ha (Mha) is affected by land degradation, causing severe threat to food security (ICAR/NAAS 2010; Bhattacharyya et al. 2015). Soil degradation is, therefore, a serious threat to the South Asian region (Shah and Arshad 2012).

Soil degradation may be categorized into: (i) physical degradation (including soil compaction and erosion), (ii) chemical degradation (including salinization, acidification, and heavy metal pollution), (iii) biological degradation (including biodiversity losses and soil organic matter depletion), and (iv) ecological (disruption in elemental cycling and changes in water and energy balance). Various factors that affect the type of degradation include climate (temperature and precipitation), vegetation (biodiversity and biomass), inherent soil abilities (chemical and physical

TABLE 13.1
Population and Land Distribution Statistics in South Asian Countries

Countries	Population (millions)	Total Area (km²)	Degraded Area (km²)	Degraded Area (%)	Percent of Global Degraded Area	Cropland Area (000′ ha)	Forest Land (000′ ha)
Afghanistan	034	38,394	7658	33	0.025	7771	1350
Bangladesh	164	147,610	68,422	75	0.199	7669	1431.6
Bhutan	0.80	38,394	27,011	10	0.073	100	2745
India	1339	3,287,263	592	25	1.751	156360	70503
Nepal	029	147,181	54,704	26	0.182	2114	3636
Pakistan	207	881,913	20,644	61	0.073	30440	1515
Sri Lanka	021	65,610	21, 057	44	0.060	1300	2076
Total	**1794.8**	**4,606,365**	**179031**	**274**	**2.363**	**205754**	**83256.6**

Sources: Bai ZG, Dent DL, Olsson L, Schaepman ME, Report 2008/01, ISRIC, Wageningen, the Netherlands, 2008; Shah ZH, Arshad M, 2012, ECO Services International, 2012; Anonymous, The World Factbook, https://www.cia.gov /library/publications/the-world-factbook/fields/2147.html, 2017; FAOSTAT, Food and Agriculture Data. http://fao stat.fao.org, 2017.

properties and processes), and terrain characteristics (drainage and slope) (Lal 1997, 2015). In physical degradation, the soil structure is adversely affected by compaction, blocked pore spaces, and reduced air permeability and water holding capacity, which restricts root development and biological activities in soil (Van Den Akker et al. 2003). Physical degradation mainly occurs due to inadequate soil management practices and pressure under machinery and/or animals on soil (Houskova and Montanarella 2008). In chemical degradation, soil has excessive amounts of toxic substances, which alter its functions. These substances impede the uptake, availability, and transformation of nutrients (Suraj et al. 2001). Approximately 240 Mha of soil area worldwide are affected by chemical degradation (Oldeman 1992). Atmospheric deposition of eutrophying and acidifying substances, contaminations from eroded soil, and excessive use of fertilizers, manures, and sewage sludge add high amounts of heavy metals, which are major determinants of chemical degradation (Kavvadias 2014). Additionally, the major concerns are contamination from native sources, chemicals and effluents from heavy industry, and wastes from densely populated urban centres (EEA-UNEP 2000). Industrial contaminants include mineral oils, chlorinated hydrocarbons, organic compounds, and heavy metals (EEA 2014).

Soil productivity is also degraded by salinization, a type of chemical degradation. Salinization is the build-up of water soluble cations and/or anions above or just below the soil surface. Similarly, sodification is a condition caused by an excessive accumulation of salts, particularly exchangeable Na^+, which results in the complete loss of soil productivity (Kavvadias 2014). Several factors contribute to accumulation of salts in the soil, and the process is aggravated and confounded by anthropogenic activities. The principal cause of salt accumulation in agricultural soils is the poor quality of underground irrigation water, which aggravates the effects of other factors including weathering process, artificial nutrient application, encroachment of saline water from sea, and transfer of salt water from sea through inundation and volatilization (Hedge et al. 2011).

Acidification, which leads to the modification of soil chemical properties, causes movement of toxic metals in the soil profile (Nagle 2006). Soil acidification leads to decline the soil pH due to the acid deposition in soil, which negatively influences soil water chemistry and soil ecosystems. Acidification is caused by the input of sulphates from areal deposition, contamination from industrial effluents, and non-judicious use of nitrogenous fertilizers (i.e., ammonium sulphate, urea, etc.). Dent (1986) reported that drainage of parent material rich in pyrite may oxidize and can cause increase in land area under acidic soils. Pyrite accumulation also takes place under waterlogged conditions due to high organic matter and dissolution of sulphates.

The loss of soil humus by mineralization, accelerated erosion, and leaching is a type of biological degradation (Solomon 1994). Although, SOM contents are higher in topsoil horizon due to plant biomass additions but still subsoil horizons have more than half of the total SOM stock is in deep soil horizons (Batjes 1996). This important SOM pool is directly affected by the extensive tillage practices and accelerated erosion, which expose subsoil stabilized organic matter to microbial decomposition (Guillaume et al. 2015).

Global warming is also inducing changes in the hydrological cycle and thermal regimes including higher intensity as well as higher frequency of rainfalls. Variable temperatures and carbon dioxide (CO_2) concentrations in the atmosphere can substantially influence the rate of soil erosion and degradation (Nearing et al. 2004). South Asia is one of the most vulnerable regions to the impacts of climate change. According to a recent estimate, with increasing CO_2 emissions from South Asian countries (Table 13.2), there is prediction of increasing air temperatures as well as changes in the precipitation patterns (Table 13.3; IPCC 2014). In some regions, change in soil temperature can reduce soil C by more than 32%, which is a more rapid rate of decrease as compared to increased tillage intensity during a period of 5 years (Link et al. 2003). Climate change and anthropogenic activities are increasing the risks of soil erosion by wind and water (Figure 13.1; Zhang et al. 2004; Sivakumar 2011).

TABLE 13.2

Terrestrial Carbon Fluxes and Average Fossil Fuel CO$_2$ Emissions for the Countries in South Asia

Countries	E$_{LULCC}$ (Tg C yr^{-1})	E$_{FIRE}$ (Tg C yr^{-1})	NEP (Tg C yr^{-1})	NBP (Tg C yr^{-1})	Average Fossil Fuel CO$_2$ Emissions (Tg C yr^{-1}) 1990–1999	Average Fossil Fuel CO$_2$ Emissions (Tg C yr^{-1}) 1990–2009
Bangladesh	−1.4 ± 4.6	−0.0 ± 0.0	10.6 ± 8.8	9.3 ± 9.2	−5.638	−8.207
Bhutan	−0.4 ± 1.1	−0.5 ± 0.0	2.2 ± 1.9	1.2 ± 2.3	−0.072	−0.113
India	−6.2 ± 44.4	−8.5 ± 0.6	200.6 ± 137.7	185.9 ± 145.6	−247.44	−319.81
Nepal	−0.2 ± 2.8	−1.0 ± 0.1	9.2 ± 7.6	8.0 ± 8.4	−0.514	−0.70
Pakistan	−1.1 ± 2.8	−0.4 ± 0.0	14.2 ± 7.7	12.7 ± 10.9	−23.019	−29.99
Sri Lanka	−3.1 ± 2.2	−0.2 ± 0.0	3.6 ± 1.6	0.4 ± 3.2	−1.629	−02.37
	−12.4 ± 44.8	−10.6 ± 0.6	240.6 ± 138.4	217.5 ± 146.6	278.312	361.19

Sources: Patra PK, Canadell JG, Houghton RA, Piao SL, Oh N-H, Ciais P, Manjunath KR, Chhabra A, Wang T, Bhattacharya T, Bousquet P, Hartman J, Ito A, Mayorga E, Niwa YP, Raymond A, Sarma VVSS, Lasco R, *Biogeosciences*, 10, 513 -527, 2013; Cervarich M, Shu S, Jain AK, Arneth A, Canadell J, Friedlingstein P, Houghton RA, Kato E, Koven C, Patra P, Poulter B, Sitch S, Stocker B, Viovy N, Wiltshire A, Zeng N, *Environmental Research Letters*, 11, 105006, 2016.

Note: E$_{LULCC}$ = Emission due to land use and land cover change; E$_{FIRE}$ = Emissions due to non-land use change activities; NEP = Net ecosystem productivity; NBP = Net biome productivity (NBP = NEP − E$_{LUC}$ − E$_{FIRE}$). Positive and negative values are sink and source of carbon, respectively for the years 2000–2013.

TABLE 13.3

Projected Rate of Climate Change (Temperature and Precipitation) in South Asia under High Emission Rates of Greenhouse Gases

Countries	Temperature (°C) 2030	Temperature (°C) 2050	Temperature (°C) 2080	Precipitation (%) 2030	Precipitation (%) 2050	Precipitation (%) 2080
Afghanistan	1.6	2.5	4.5	−10.0	−1.5	−0.9
Bangladesh	0.9	1.6	4.2	12.9	26.1	16.5
Bhutan	1.5	2.3	4.5	−3.3	6.4	0.2
India	1.8	2.9	5.6	16.3	27.5	32.1
Maldives	0.9	1.6	3.3	11.8	15.5	28.8
Nepal	1.6	2.5	4.8	−5.0	−1.7	−0.9
Pakistan	1.7	2.4	4.5	15.0	30.0	40.0
Sri Lanka	1.0	1.8	3.6	7.4	15.8	39.6

Source: IPCC (Intergovernmental Panel on Climate Change), *Proceedings of the Contribution of Working Groups I, II and III to the Fifth Assessment Report of the Intergovernmental Panel on Climate Change*, Cambridge University Press, Cambridge, 2014.

Readers are referred to other reviews on soil degradation, land scarcity and food security, soil degradation and soil quality, and restoring soil quality to mitigate soil degradation in Europe (Virto et al. 2015; Lal 2015; Gomiero 2016). However, there are few, if any, syntheses of literature on soil degradation and climate change for South Asia. Thus, the objective of this chapter is to deliberate the rate of soil degradation, types of soil degradation, effect of climate change on soil degradation, and different strategies and options to combat soil degradation under changing climate with a specific focus on South Asia.

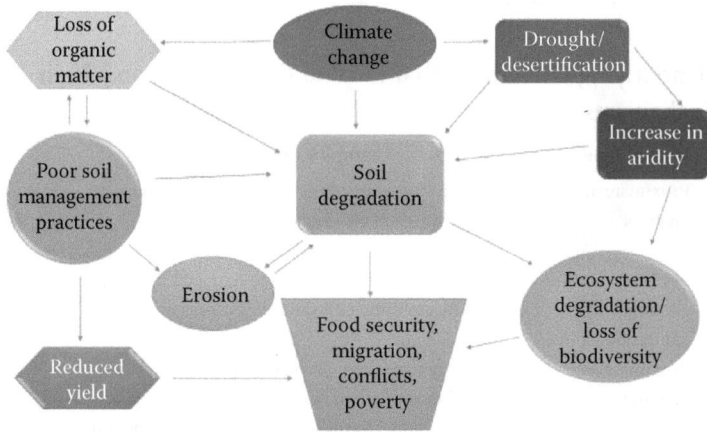

FIGURE 13.1 Impact of climate change on soil degradation and its associated factors.

13.2   DRIVERS OF SOIL DEGRADATION (FACTORS VS CAUSES)

Soil degradation is driven by a range of factors, which are interrelated and are complex, frequently having context-dependent features (Tables 13.4 and 13.5; Von Braun et al. 2013). Identification of these drivers is essential to restrict and mitigate the losses caused by soil degradation, and for the improvement and restoration to harness the authentic and long-term benefits. Important among

TABLE 13.4
Land Degradation in South Asian Countries by Different Drivers

Countries	Salinization	Water Erosion	Wind Erosion	Soil Organic Matter/Fertility Depletion	Deforestation	Reference
Area (000 ha) as per Old Estimate						
Afghanistan	1271	11156	2082	–	–	FAO (1994)
Bangladesh	–	1504	–	6367	–	FAO (1994)
Bhutan	–	40	–	–	–	FAO (1994)
India	7000	32773	10796	29383	–	FAO (1994)
Nepal	–	1592	–	–	–	FAO (1994)
Pakistan	4200	7204	10740	5200	–	FAO (1994)
Sri Lanka	47	1074	–	1425	–	FAO (1994)
Area (000 ha) as per New Estimate						
Afghanistan	–	–	–	–	–	
Bangladesh	900	3400	–	7500	15500	Hossain (2007)
Bhutan	–	–	–	–	–	
India	5900	93700	9500	3700	–	Aulakh and Sidhu (2015); Bhattacharyya et al. (2015)
Nepal	–	106	–	–	2290	CBS (2005); MoEST (2006)
Pakistan	6281	13050.2	6173.5	–		Shah and Arshad (2012)
Sri Lanka	–	–	–	–	–	

TABLE 13.5

Proximate and Underlying Drivers Related to Land Degradation and Their Potential Cause-Effect Mechanisms

Drivers	Type	Examples of Degradation	References
Topography	Proximate and natural	Steep slopes are vulnerable to severe water-induced soil erosion	Wischmeier (1978), Voortman et al. (2000)
Land cover change	Proximate and natural/ anthropogenic	Conversion of rangelands to irrigated farming with resulting soil salinity. Deforestation	Gao and Liu (2010), Lu et al. (2007)
Climate	Proximate and natural	Dry, hot areas are prone to naturally occurring wildfires, which, in turn, lead to soil erosion. Strong rainstorms lead to flooding and erosion. Low and infrequent rainfall and erratic and erosive rainfall (monsoon areas) lead to erosion and salinization	Safriel and Adeel (2005), Barrow (1991)
Soil erodibility	Proximate and natural	Some soils, for example those with high silt content, could be naturally more prone to erosion	Bonilla and Johnson (2012)
Pest and diseases	Proximate and natural	Pests and diseases lead to loss of biodiversity, loss of crop and livestock productivity, and other forms of land degradation	Sternberg (2008)
Unsustainable land management	Proximate and anthropogenic	Land clearing, overgrazing, cultivation on steep slopes, bush burning, pollution of land and water sources, and soil nutrient mining are among the major causes of land degradation	Nkonya et al. (2008, 2011), Pender and Kerr (1998)
Infrastructure development	Proximate and anthropogenic	Transport and earthmoving techniques, such as trucks and tractors, as well as new processing and storage technologies, could lead to increased production and foster land degradation if not properly planned	Geist and Lambin (2004)
Population density	Underlying	No definite answer. Population density leads to land improvement	Bai et al. (2008), Tiffen et al. (1994), Boserup (1965),
Population density	Underlying	Population density leads to land degradation	Grepperud (1996)
Market access	Underlying	No definite answer. Land users in areas with good market access have more incentives to invest in sustainable land management	Pender et al. (2006)
Market access	Underlying	High market access raises opportunity cost of labor, making households less likely to adopt labor-intensive sustainable land management practices	Scherr and Hazell (1994)
Land tenure	Underlying	No definite answer. Insecure land tenure can lead to the adoption of unsustainable land management practices	Kabubo-Mariara (2007)
Land tenure	Underlying	Insecure land rights do not deter farmers from making investments in sustainable land management	Besley (1995), Brasselle et al. (2002)

(Continued)

TABLE 13.5 (CONTINUED)
Proximate and Underlying Drivers Related to Land Degradation and Their Potential Cause-Effect Mechanisms

Drivers	Type	Examples of Degradation	References
Poverty	Underlying	No definite answer. There is a vicious cycle between poverty and land degradation. Poverty could lead to land degradation while land degradation could lead to poverty	Way (2006), Cleaver and Schreiber (1994), Scherr (2000)
Poverty	Underlying	The poor heavily depend on the land, and thus, have a strong incentive to invest their limited capital into preventing or mitigating land degradation if market conditions allow them to allocate their resources efficiently	De Janvry et al. (1991), Nkonya et al. (2008)
Access to agricultural extension services	Underlying	No definite answer. Access to agricultural extension services enhances the adoption of land management practices	Clay et al. (1996), Paudel and Thapa (2004)
Access to agricultural extension services	Underlying	Depending on the capacity and orientation of the extension providers, access to extension services could also lead to land-degrading practices	Benin et al. (2007), Nkonya et al. (2010)
Decentralization	Underlying	Strong local institutions with a capacity for land management are likely to enact bylaws and other regulations that could enhance sustainable land management practices	FAO (2011)
International policies	Underlying	International policies through the United Nations and other organizations have influenced policy formulation and land management	Sanwal (2004)
Non-farm employment	Underlying	Alternative livelihoods could also allow farmers to rest their lands or to use nonfarm income to invest in land improvement	Nkonya et al. (2008)

these drivers are soil properties, topography, and climate, an understanding of which is essential because of their vital role in soil degradation and causal mechanisms (Tables 13.4 and 13.5).

13.2.1 WEATHER VARIABILITY

Changing climate (e.g., temperature fluctuations and uncertain precipitation) is supposed to melt the glaciers and ice caps (Malla 2007; IPCC 2014), leading to unexpected and erratic rainfalls, increasing surface runoff and flood waves, causing fluctuations in water flow level and rate, raising sea levels, increasing salt water intrusion into aquifers, aggravating erosion, and degrading agricultural and other managed lands.

Natural flora and practices involved in land use change dramatically with the climatic-extremes. Other consequences in response to these changes include imbalance between energy and heat in the Earth's atmosphere, change in micro-climatic events and precipitation, increase in soil roughness and temperature regimes (Varallyay 2010). The rise in global ocean temperature is expected to raise sea levels (IPCC 2007), increase salt contents in the groundwater and soil, and aggravate the risks of salinization (Varallyay 2010). All these changing climatic events increase the risks of soil degradation and undermine sustainable development. Erratic rainfall patterns and monsoon precipitation in Pakistan have caused soil degradation and destroyed rangeland areas by erosion (water and wind) and salinity (Rasul et al. 2012). Natural resource deterioration due to climatic variability is expected to make the South Asian region more vulnerable to the soil degradation. Soil degradation increases by changing land use patterns and pollution, putting ecosystems at risk because these factors operate independently or in interaction with one another (Loper et al. 2005; Easterling et al. 2007). Climatic models have predicted that due to weather variability, aridity conditions will intensify more during the current century over most of Asia and Africa (Thornton et al. 2014) because precipitation patterns are becoming irregular (intense and longer dry episodes), and this situation increases floods and runoff with the attendant increase in soil erosion (Dai 2011). Therefore, there is an urgent need to manage soil resources by an integrated approach and to highlight the critical factors responsible for unsustainable land use rather than merely discussing the negative impacts of climate change and soil degradation (Kumar and Das 2014).

13.2.2 WATER DEFICIT

There exists a close link between water and soil, both of which are important natural resources. Unavailability of fresh water to meet the domestic, commercial, social, and energy requirement/demands is known as water shortage/scarcity. Water shortage is one of the important factors responsible for soil degradation, as soil health depends on an adequate availability of water and its management. Desertification, soil degradation in dryland areas (Gnacadja 2013), is aggravated by poor water management policies (Gnacadja 2013, UNCCD; United Nations Convention to Combat Desertification). Long-term and sustainable efforts are urgently required for the curtailment of losses induced by soil degradation, for which policies are needed for better management of water domestically and commercially. Important among these policies are those that minimize the losses, stabilize the ecosystem, restore the soil environment, and build resilience against climate change vulnerability (Gnacadja 2013).

In Pakistan, water scarcity and frequent drought episodes have exacerbated the problems of soil degradation resulting in reduced soil productivity (Khan et al. 2012). Underground water resources in Baluchistan are also shrinking due to overexploitation and poor recharge. In Pakistan, three provinces (Sindh, Southern Punjab, and Baluchistan) have been severely affected by long drought spells for the last 3 years (Khan et al. 2012), resulting in deterioration of the soil quality. Uncertain and decreasing precipitation due to climatic variability has accentuated the aridity in the Central Asian region particularly in the western regions of Kazakhstan, Uzbekistan, and Turkmenistan (Lioubimtseva and Henebry 2009); therefore, lack of a judicious irrigation system has made soil degradation pandemic (Lioubimtseva and Henebry 2009). Large areas in South Asia are under threat of degradation because of excessive withdrawal of water from the aquifer (Lal 2007).

Unavailability of water over long periods of time, coupled with imbalance between demand and supply through anthropogenic activities or natural phenomena, trigger desertification by adversely affecting soil health. Strong variations in precipitation and prolonged drought spells have damaged the freshwater pools, and the poor water management makes soil rough and susceptible to accelerated erosion and degradation.

13.2.3 SOIL EROSION

Archaeologists have highlighted that soil erosion has been a serious threat for the last 8000 years and was a cause of extinction of many ancient civilizations (Montgomery 2007). Several factors, including biophysical and human induced, exacerbate soil erosion by destroying soil structures and resources, changing climate and polluting the environment and crop production (Lal 1994). Oldeman et al. (1991) reported that 550 and 1094 Mha of land is prone to wind and water erosion worldwide, respectively. Moreover, prolonged drought spells and inadequate soil management practices aggravated loss of soil in the form of dust emission in 1930s, also known as the "dust bowl" period (Worster 1979). Desertification may also be referred as an effect of accelerated soil erosion (Lal 2001; MEA 2005), which results in loss of biodiversity and poor soil health. Accelerated soil erosion due to climatic extremes, such as temperature and fluctuations in precipitation, cause loss of fertile soil and biodiversity by stimulating aridity; this results in severe loss of resources such as carbon capturing processes and primary productivity particularly in dry regions (drylands) (Chapin et al. 1997). Loss of biodiversity causes reduction in the resilience ability of the ecosystem, i.e., the capacity and strength of the ecosystem to retain its natural structure after disturbance (Elmqvist et al. 2003). Ecosystems with less diversity are susceptible to irreversible change that may occur in response to climatic, human induced, and/or environment fluctuations. Arid and semi-arid regions are mainly threatened by soil erosion due to unsustainable agricultural practices (Lal 2001) which reduce soil productivity, vegetation, and other resources (Puigdefabregas 2005). Li et al. (2008) reported that soil fertility and productivity is greatly reduced by soil erosion due to the removal of nutrients present in the topsoil layer resulting in drastic reductions in crop yields. The loss of SOM contents by erosion disturbs the carbon budget globally (Li et al. 2008) and thus reduces the ability of soil to retain nutrients, water, and support for successful crop production or vegetation. Due to accelerated soil erosion, water reserves are depleted and polluted because of deposition of soil, sediments, and pollutants (Lal 2001). For instance, wind erosion can transport the dust particles from nearby or long distances and can deposit it in oceans (Swap et al. 1996). Yu et al. (2015) highlighted that the emission of dust through wind erosion causes aerosol deposition in the troposphere and contaminates the air quality, and disrupts the hydro-biogeochemical cycle, and the consequences extend beyond and toward a wide range of dry regions and geographic domains. Various factors such as soil water contents, soil physical and chemical properties, slope, vegetation cover, and climatic variability determine the susceptibility of soil to wind erosion (Lal 1994). Soil erosion due to intensive agricultural practices perturbs the global carbon cycle; however, its effect ranges from a source of 1 Pg C/year (Lal 2003) to a sink of the same magnitude (Van Oost et al. 2007). Moreover, erosion induced sink of atmospheric carbon is reportedly equivalent to about 26% of carbon transported by erosion (Van Oost et al. 2007). Another study by Lal (2007) suggested that carbon emission into the environment due to erosion remains an unquantified and misunderstood component of the global carbon budget. Soil erosion is a four–step process that includes detachment, breakdown of soil particles, transport, and deposition; and the soil organic pool is influenced by all these processes. Therefore, erosion causes depletion of SOM, which is a light organic fraction of a low density of 0.6 Mg/m^3 (Lal 2007).

Land cover is an important biophysical factor which may help in preventing water and wind erosion under dry-land ecosystems as it provides protection and cover to soil against water and wind and reduces the loss of nutrients and soil fertility. Soil cover also enhances the soil microbial activity, and affects soil physical and biological properties (Raupach 1992; Singer and Shainberg 2004). Meanwhile, soil water contents influence the bonding between soil particles and determines its susceptibility to erosion (Ravi et al. 2006). Poor agricultural practices, excessive tillage, in-field burning of crop residues, overgrazing, urbanization, and industrialization aggravate the soil erosion hazard (Ravi et al. 2009). Arid regions are more prone to wind erosion due to more wind intensity, whereas humid regions are more susceptible to water erosion (Field et al. 2009), and semi-arid regions are prone to both wind and water erosion. Erosion begins with the loosening of soil particles (detachment) due to intense rainfall followed by runoff and results in heavy loss of soil nutrients compounded with saltation of particles by wind. The hydrological cycle is closely linked with water erosion followed by sheet, rill, and gully formation by runoff water from the same land. Similarly, rainfall directly causes soil erosion by loosening the soil aggregates due to excessive kinetic energy and indirectly by enhancing the surface runoff into gully formation (Lal 1994). Loss in soil health by water erosion is less as compared to wind erosion, which results in loss of soil due to the detachment and transportation of large size particles over long distances (Field et al. 2009). The transportation, however, depends upon the slope and fluvial movement of soil particles sometimes unidirectional or irreversible (Ravi et al. 2010). Wind erosion is more common and two dimensional, horizontal, and vertical, than the water erosion and occurrence is accelerated by winds with more velocity (Breshears et al. 2003; Field et al. 2009). The force at which wind erosion takes place may be parallel and omni-directional to the direction of wind flow, whereas in water erosion it is parallel. Therefore, during wind erosion, the direction of the particles' movement at one time is one way, and afterwards they move in the opposite direction. Thus, the shift in particle movement from one direction to the opposite direction is linked with the change in wind direction.

Wind erosion is reversible and less affected through soil slope as compared with water erosion. The particles displaced and eroded by wind may be re-deposited on their place of origin (Field et al. 2009). Wind erosion is, however, affected by sediment/soil amount, size, and sorting (Ravi et al. 2010). Particles detached and transported by water and wind erosion can be quantified by estimating the detached and transported particles that move per unit length of line perpendicular to the erosion force (Breshears et al. 2003).

The relative loss and importance of water and wind erosion is related to climate variability, vegetation, slope/relief, hydrological cycle, and aeolian dominance of both wet and dry regions (Ravi et al. 2010). Temporal separation may exist between the interaction of hydrological cycle and aeolian, as aeolian processes are also affected by hydrological process legacy from preceding wet or dry periods. The removal of sediments is activated by the drying of lakes through aeolian cycles. The intrusion of sand dunes into ephemeral river beds indicates a close linkage between fluvial and aeolian landforms (Nanson et al. 2002; Goudie and Middleton 2006). Most of the case studies have indicated that the aeolian process and hydrological cycle operate simultaneously and there exists a relationship between both these factors, which also have implications to soil degradation, land cover pattern, and landscape evolution (Ravi et al. 2007). Although interactions between aeolian and hydrological processes can occur at all moisture levels, they are particularly significant in areas where neither wind nor water erosion dominate, as is the case in semi-arid regions (Bullard and Livingstone 2002).

Hilly regions, with terraced agriculture, are also susceptible to erosion due to (i) increased run-off, (ii) overgrazing, (iii) declining soil fertility, and (iv) sparse vegetation cover, steepness of terrace, ditches for drainage purpose because of poor maintenance (Lasanta et al. 2001; Van Dijk et al. 2003; Chapagain and Raizada 2017).

In Pakistan, about 3–5 and 11 Mha of soil are degraded by wind and water erosion respectively, and the amount of soil removed by wind is about 28% of total soil loss (Khan et al. 2012). Soil erosion is a major contributor of soil degradation. In Pakistan, loose soils of the Pothowar plateau are degraded by water erosion, and the gully formation induces severe loss of productive topsoil (Zia et al. 2004). In India, approximately 94 Mha and 9 Mha soil are affected by water and wind erosion, respectively (Bhattacharyya et al. 2015) Another study reported that about 30 and 10 Mha of land have been degraded due to water and wind erosion in India, respectively (Samuel and Al Mahmud Titumir 2011). In South Asia, soil productivity has declined to about 50% due to erosion. The annual loss of productivity in South Asia is approximately 36 million tons of cereals which are equivalently valued at US\$ 5 million, US\$ 400 (Eswaran et al. 2001).

Poor agricultural practices, salinization, and sedimentation are important drivers of soil degradation by erosion leading to desertification of the crop land areas (Vincini 1999). Sloping soils are highly susceptible to water erosion, whereas soils having high silt content are susceptible to natural degradation (Voortman et al. 2000; Bonilla and Johnson 2012). About 87% of land area is degraded by water and wind erosion (Middleton and Thomas 1997). Poor agricultural practices, removal of vegetation, and ignorance of conservation management options make soil prone to erosion and desertification (Nicholson et al. 1998). Information regarding the relationship between aeolian and hydrological cycles and removal of vegetation toward accelerated soil degradation is known, but how these factors and their inter-relationship cause desertification is still unknown.

13.2.4 NUTRIENT MINING

The deficiency of nutrients in drylands is also one of the main reasons for low yield output as fertilizers are not applied after sowing in dryland agriculture and the rate of fertilizer application is quite low due to weather vulnerability. Crop residues are removed from the fields for other domestic uses and are not incorporated in soil, resulting in depletion of SOM and widespread nutrient deficiencies in dryland agricultural soils (Rojas et al. 2016).

Nutrients mining and negative balance between nutrient input and output occurs when crop nutrients' removal and their losses to other sinks become higher than the soil inherent supply (Majumdar et al. 2016). When the nutrient removal by crops exceeds the combined applied and indigenous sources soil nutrient reserves get depleted (Majumdar et al. 2016). Over time, developing countries face the net depletion of macro nutrient stocks while more developed areas that replenish soils with macronutrients unintentionally cause excessive micronutrient mining (Emmett et al. 1997). The situation in drylands with marginal soils is even more alarming, because they are already under pressure due to various environmental impacts and increased human intrusions (Rojas et al. 2016). Intensive cropping systems in Bangladesh have deteriorated soil health and nutrients status by mining out scarce native soil nutrients (Quamruzzaman 2006). Volatilization, leaching, and soil erosion accelerate nutrients loss from soil; for instance, 50% of applied nitrogen (N) fertilizers is lost from soils of agro-eco systems (Ladha et al. 2005). Changing climate, such as elevated temperatures and frequent drought stress, directly affects microbial community composition as well as activities which directly or indirectly influence nutrient cycling (Hawkes et al. 2008).

An area as large as 11 Mha in South Asia has been degraded by severe nutrients depletion (Lal 2007). In India, about one third of soil suffers from loss of nutrients due to erosion that removed the topsoil layer and resulted in low productivity (State of the Environment 2001). Moreover, the State of the Environment (2001) in India reported an annual loss of 0.8 million ton of nitrogen, 1.8 million tons of phosphorus, and 26.3 million tons of potassium. In Pakistan, due to continuous monocropping and inadequate agricultural practices have caused substantial nutrients loss by making the soil more vulnerable to erosion (wind or water) (Ali et al. 2007).

13.2.5 INTENSIVE TILLAGE

Intensive tillage together with the use of heavy machinery for various agricultural practices (e.g., sowing, harvesting) and inadequate conservation measures exacerbate soil degradation. For instance, puddling of soil for rice destroys the soil structure, increases bulk density, and limits soil biological activity (Hobbs et al. 2008).

Soil structure determines nutrient and water retention ability, drainage capacity, and nutrient transformation in soil; however, the changing climate is deteriorating the soil health and decreasing its productivity (Tables 13.4 and 13.5; Lal 1994; Nordstrom 2009). Kumar and Das (2014) described that it is difficult to quantify changes in soil structure because of various land use and management practices, and hence focused research is needed for better understanding of the effects of changing climate on soil structure. Soils with high clay contents shrink more upon drying and, therefore, dry climatic conditions increase crack formation in soils particularly in temperate regions (Boden and Driscoll 1987). Moreover, acidic clayey soils of high clay content also experience cracks because of high aluminium, iron, manganese, and heavy metals when subjected to changing climate (Wild 1993; Kumar and Das 2014).

Intensive tillage practices in hilly areas accelerate soil degradation as such practices accelerate soil erosion (Govers et al. 1994). Erosion due to excessive tillage is affected by soil cover and soil structure, and field boundaries act as lines of zero flux, leading to sedimentation at the upslope side and erosion at the downslope side (Govers et al. 1994). Neglecting tillage induced erosion, the impact of climatic variability on soil erosion cannot be assessed accurately because the effects of climate changes on tillage induced erosion (rate and patterns) have not been clearly studied. In India, excessive tillage coupled with the use of heavy machinery for various agricultural operations induces loss in SOM, poor soil structure, and ultimately resulting soil degradation (Bhattacharyya et al. 2015).

13.2.6 OVERGRAZING/DEFORESTATION

Overgrazing and deforestation are major drivers of land degradation in grassland and forest ecosystems, respectively. Removal of vegetation beyond the permissible limits for other purposes such as fodder extraction, encroachment by agriculture into forest lands, forest fires, and overgrazing all contribute toward soil degradation. Moreover, impoverishment of the natural woody cover of trees and shrubs is a major factor responsible for wind and water erosion.

Removal of vegetation and depleting natural resources have been serious threats to the ecosystem in South Asia since the 5th century BC. Several reports have documented the adverse effects of deforestation on soil degradation (Elvin 2004). The effects of vegetative cover to protect soil against erosion also depend on the rooting system (e.g., *Agathis spp.* [resistant against erosion]; Scatena et al. 2005). Keppel et al. (2010) reported that trees with small leaves and larger roots resist soil erosion. Forests are also adversely affected by storm destructiveness and its effects can persist for several years. For instance, in Sarawak in 1880, storm bombardments destroyed large areas of forest. The regions having forests of vines canopy are frequently affected by storms. In Asia, removal of vegetation by overgrazing, deforestation, and unsustainable agricultural practices, reduce the soil's capacity to support life and it becomes vulnerable to degradation (Ma and Ju 2007). In India, eight states are suffering from land degradation due to overgrazing and deforestation and have >20% wastelands. Overgrazing due to high livestock density in arid areas makes soil more susceptible to runoff and erosion (Bhattacharyya et al. 2015). A cattle population of 467 million grazes on 11 Mha of pastures, implying an average cattle density of 42 animals/ha of land as compared to a sustainable threshold level of 5 animals/ha (Sahay 2000). Forest degradation is a major problem confronting the South Asia region (FAO 2010a). For Instance, in Pakistan, approximately 7000–9000 hectares are deforested every year with higher rates in the northern areas where the consumption of fuel wood is 10 times higher than that in the southern regions of Pakistan (Shah and Arshad 2012).

13.2.7 BIODIVERSITY LOSS

Intensive agricultural practices may lead to soil degradation by deteriorating soil health coupled with environmental pollution, resulting in loss of biodiversity, reduced ecosystem services, and increased erosion (Weiss and Fox 2003; Hendrickson 2003). In arid and semi-arid regions, the interaction between soil cover and erosion is considered to be an important determinant of vegetation pattern which may include rings, stripes, and bands (Borgogno et al. 2009). The spatial distributions of nutrients, moisture contents, and sediments are determinants of species composition, and crop growth is also affected by the geometric pattern of the vegetated cover (Puigdefabregas 2005). Moisture contents of soil and vegetation cover interact positively and improve heterogeneities in soil and vegetation distribution contributing toward stable regions in drylands vegetation (D'Odorico et al. 2007). Soils having high moisture retention capacity produce more vegetation than drier soils (with less moisture). Therefore, the vegetation pattern is a good biological indicator of abiotic stress (including infiltration and runoff) and of the source–sink region for sedimentation in dryland landscapes (Ludwig et al. 2005). In arid regions, environmental variations (e.g., desertification) can be readily observed by sediment transportation and changes in the vegetation patterns (Van de Koppel et al. 2002). Thus, understanding the relationship between vegetation and soil erosion can provide useful information about the occurrence of some vital processes in dryland ecosystems and the response of ecosystems to changing climate along with complementary management options. Grassland areas in dryland regions of South Asia consist of various species that are highly vulnerable to increasing carbon dioxide (CO_2) concentration and climatic variability with its impact related to their stability and resilience (Mitchell and Csillag 2001), because the climatic variability can increase the risks of vegetation degeneration that may also aggravate soil degradation (Zheng et al. 2002).

13.2.8 SOIL SALINIZATION

Soil salinization is a major environmental problem in arid and semi-arid regions (Wang and Li 2013) and has been a concern for millennia (Bennett et al. 2009). It adversely affects soil quality and its functioning along with reduction in crop productivity (Tilman et al. 2002). However, huge efforts and vast resources are needed to reclaim salt-affected soils (Rengasamy 2006). Ghassemi et al. (1995) reported, about 955 Mha of land is affected by salinization worldwide, whereas approximately 80 Mha is affected by secondary salinization particularly in dryland regions. Land-use changes (e.g., conversion of natural landscape into agricultural land) resulted in more salinization (Qureshi et al. 2008). Decreased evapo-transpiration, high water table level, land exploitation for agriculture, and changing land pattern particularly in arid areas induce salinization (Williams et al. 1997). Ritzema et al. (2008) observed that poor water management practices and more groundwater recharge resulted in soluble salts accumulation at the soil surface. In addition, excessive use of agricultural inputs by farmers had negative impacts on soil composition and properties resulting in salinity problems (Sveistrup et al. 2005). Plant growth is restricted in soils having high salt contents because of poor soil structure. In addition, low soil fertility also causes ecosystem instability and decreases agriculture productivity.

In Pakistan, about 5.7 Mha of irrigated land is severely affected by salinization, of which 55.4% is saline sodic, 0.5% is sodic, and 44.1% is saline (Zia et al. 2004). A relatively large salt affected area is in Punjab (2.5 Mha), followed by that in Sindh (2.3 Mha) (Zia et al. 2004). In Turkmenistan, 96% of the land is degraded by salinization resulting in fragile soil health and contraction of cultivated land (Gupta et al. 2009). Salinity intrusion is one of the major causes of changes in agricultural land use. In Bangladesh, as elsewhere in South Asia, the adverse effects of salinization are increased dramatically in the dry season due to increased temperature, low rainfall, and high evaporation (Alam 2013).

13.3 SOIL HEALTH UNDER CHANGING CLIMATE

Soil is a vital component of terrestrial ecosystem and is a buffer against any change in biogeochemical processes. It also affects global climate change by being a source or a sink for greenhouse gases (GHGs). Soil health is the ability of soil to sustain and support vital living ecosystem functions related to plants, animals, and humans on a long-term basis (Singh et al. 2011). Therefore, as a living entity, soil responds to any external changes (Kibblewhite et al. 2008). Changing climate and global warming have prominent effects on soil health components, i.e., physical, chemical, and/or biological (Table 13.6; Brevik 2013). It is important to evaluate the impact of climatic variability on these components as they have a vital role in the sustainable management of croplands, forest lands, and grazing/pasture lands. The impact of elevated carbon dioxide (CO_2) level, elevated temperature, and uncertain rainfall patterns because of climate change must be considered for managing soil health (French et al. 2009). Changing climate (e.g., temperature) affects soil functions directly by SOM decomposition and indirectly by altering soil moisture by changing evapotranspiration. However, soil can be a source of GHGs and thus contributes to climate change. Major soil physical, chemical, and biological properties which may indicate the status of soil health in relation to climate change impacts are discussed in the following sections.

TABLE 13.6
Impact of Climate Change on Soil Health Parameters

Climate Change Indicator	Soil Health Parameters	Decrease (–)/ Increase (+)	References
Elevated temperature	Aggregate stability	–	Garten et al. (2009)
	Soil pH	–	Ultra and Han 2015
	Soil EC	+	Ultra and Han 2015
	Microbial activities	+	Singh et al. (2011)
	Soil organic matter	–	Sanaullah et al. (2014)
Elevated CO_2	Microbial activities	+	Sanaullah et al. (2011a, 2016)
	Soil pH	–	Ultra and Han 2015
	Soil EC	+	Ultra and Han 2015
	Soil organic matter	+	Dalal et al. (2005)
Uncertain rainfall-low	Aggregate stability	–	Chan and Mullins (1994)
	Soil pH	+	Rengel 2002
	Soil EC	+	Rengel 2002
	Microbial activities	–	Sanaullah et al. (2011a, 2016)
Uncertain rainfall-high	Soil pH	–	Rengel 2002
	Soil organic matter	–	Singh et al. (2011)

13.3.1 Soil Physical Health

Climate change directly affects soil physical health by changing its physical properties including soil water, air, and temperature and soil structure/soil aggregation as well. Fluctuating soil temperature is an important indicator of changing climate (Karmakar et al. 2016). Soil physical properties are considered as the foundation of chemical and biological properties which further contribute toward climatic variability (Karmakar et al. 2016).

Erratic and uncertain rainfall patterns due to climatic variations cause fluctuations in soil water which particularly affect SOM content, aggregate formation, and their stability. Soil structure/ aggregate stability is an important physical indicator directly affected by external changes in moisture and temperature due to climate change (Dalal and Moloney 2000; Moebius et al. 2007). Climate change also impacts SOM accumulation, water infiltration, water holding capacity, and rhizo-microbial activities (Moebius et al. 2007; Rimal and Lal 2009). Periodic extreme climatic conditions such as drying and rewetting of soil results in the slaking of soil aggregates causing soil structure deterioration (Chan and Mullins 1994).Prolonged exposure of soil aggregates to extreme climatic conditions (temperature) can even cause dispersion of soil aggregates and accelerate the risks of soil degradation. Therefore, such climatic conditions increase soil erosion risks due to deterioration of soil structural/aggregate stability (Singh et al. 2011). Moreover, changes in soil water and temperature have indirect impacts on soil structural stability as soil biota and SOM contents involved in aggregate formation may alter under changing climate (Lavelle et al. 1997). Soil physical degradation has become a serious problem in rainfed and irrigated areas of India; about 90 Mha of soil is experiencing physical degradation (Indoria et al. 2017), while most of the rainfed areas are on the verge of land degradation due to low fertility status, erosion, intensive tillage practices, poor physical health, and subsurface compactness (Indoria et al. 2017).

13.3.2 Soil Chemical Health

Soil pH is an important indicator of soil chemical health. Several soil chemical properties are directly affected by soil pH. Cation exchange capacity (CEC) is also a vital component of soil chemical health as it is involved in the retention of nutrient cations and potentially toxic cations (heavy metals) through immobilization (Ross et al. 2008). Similarly, electrical conductivity (EC), the measurement of total soluble salts, is an important indicator of soil chemical health because it indicates salinity, sodicity, and nutrient cycling (Arnold et al. 2005). There may be no direct long-term impact of climate change on soil pH. However, change in precipitation patterns may change soil pH over a short time. In dry seasons, with low precipitation, accumulating salts can lower soil pH by replacing H^+ ions from exchange sites into the soil solution and can increase soil EC. Conversely, heavy rainfalls dilute salt concentrations, resulting in increased soil pH and decreased EC (Rengel 2002). Indirectly, the climate change drivers including elevated CO_2, temperature, and precipitation pattern alter soil organic C (SOC) and nutrient cycling which can affect soil pH and CEC (Brinkman and Sombroek 1999; Reth et al. 2005). Ultra and Han (2015) indicated that elevated temperature and CO_2 decreased soil pH but increased EC. In Bangladesh, about 27% of the land area is prone to degradation due to high concentrations of phytotoxic aluminium, iron, and manganese. Al⁻ ions on the soil surface generate acidic conditions, and a higher use of nitrogenous fertilizer also induces acidification due to more nitrification (hydrogen ions) (Reth et al. 2005).

13.3.3 SOIL BIOLOGICAL HEALTH

Soil biological health (e.g., soil biota) is a reliable indicator of any external perturbation including climatic changes. It significantly affects SOM dynamics, nutrient cycling, and soil physical properties such as aggregate formation (Kibblewhite et al. 2008). Soil biota has both casual and mitigation roles in climate change because they can either produce or consume GHGs. Increasing air and soil temperatures, changing rainfall distribution, and intensification directly affect soil biology. Climate variability can alter microbial species distributions and their interspecific interactions (Van der Putten 2012). Increasing temperature has no direct impact on microbial community composition because soil biota can function at broad temperature range. However, increased CO_2 in the atmosphere results in a shift of the microbial community because elevated CO_2 can enhance photosynthate production as well as exudation through roots which induces modification in the composition of the microbial communities in the rhizosphere (Sanaullah et al. 2011a, 2016).

The SOM is a vital soil component because almost all soil properties are directly or indirectly affected by SOM contents as well as its turnover. SOM is the prime habitat of soil micro-organisms and has a direct role in the soil health and productivity, and thus is a strong indicator of soil functions. Global climate change scenarios by 2100 may increase global temperature by 1.6–6.4°C, atmospheric CO_2 up to 550 ppm, and 20% change in precipitation (Denman et al. 2007) which may directly affect SOM decomposition. However, the response of SOM dynamics to increase in temperature is not clear due to wide temperature adaptation of SOM decomposers. Thus, SOM availability and accessibility to microorganism is more important than any change in climatic factors such as temperature (Kuzyakov and Gavrichkova 2010). With increasing temperature and moisture above the threshold level, respiration rate and SOM decomposition are mainly accelerated (Singh et al. 2011). In India, large areas have been degraded due to sever depletion of SOM (Aulakh and Sidhu 2015), because its depletion deteriorates soil structure and decreases the soil's ability to support life (Bhattacharyya et al. 2015).

In general, soil moisture is more important than soil temperature particularly under moisture-limited conditions (Garten et al. 2009). However, other factors such as physical protection of SOM in aggregates and its chemical recalcitrance affect the temperature dependence of SOM decomposition (Davidson and Janssens 2006; Sanaullah et al. 2011b). The decomposition of SOM is an enzymatic process, and enzyme activities responsible for SOM decomposition are affected by climatic variability. Enzymes present in soil are highly temperature-dependent; their activities decrease with increase in temperature and enzymes become denatured when the temperature exceeds the optimum range. Enzymes work efficiently at optimum temperature because of more available kinetic energy, which stimulates the enzyme reaction rate. However, if enzymes are denatured, their activity to function at a normal rate decreases because of loosening of the bond which holds the molecules together. Furthermore, active sites are denatured resulting in slow rate of SOM decomposition. Similarly, the observed 40% increase in CO_2 concentration since the industrial revolution may have indirectly affected the SOM pool because elevated CO_2 enhances total fine root production which ultimately helps in SOM accumulation, especially in soil aggregates (Dalal et al. 2005).

13.4 MONITORING AND EVALUATING THE EFFECTS OF CLIMATE CHANGE ON SOIL DEGRADATION

Climatic variability is a major factor responsible for soil degradation and dramatic changes in soil cover particularly in dryland regions causing rapid degradation of land. Climatic models have been used to simulate increase in aridity in most parts of dryland worldwide. Ravi et al. (2010) reported that, with climatic extremes abiotic factors prevail in ecosystems, and the soil becomes more sensitive to water and wind erosion. This variability causes a shift of vegetation at a faster rate in arid and semi-arid areas, and the erosion rates fluctuate. From the last century, human induced activities have affected the hydrological and aeolian processes. Furthermore, exploitation of resources and excessive anthropogenic activities based on poor management have disturbed ecosystem balance with deteriorated soil health and perturbed the nutrient and water dynamics in soil. Thus, an integrated approach is needed to manage the hydro-aeloian cycle and to maintain the soil cover which can reduce the risks of soil degradation by climate change. This holistic approach assists in maintaining the soil health status and its quality rather than improving it. Cooperation among social, biological, and physical scientists is useful in understanding the bio-physical processes and human induced activities causing soil degradation (Ravi et al. 2010).

13.4.1 DECOMPOSITION OF ORGANIC MATTER

The SOM is the most important component of soil; it builds the basic soil infrastructure by improving nutrients and water retention in soil, improves soil aggregates, and provides the medium for the growth and nourishment of soil micro-organisms. It enables soil to function as a living system. With changing climate, however, the composition and properties of SOM are also likely to change. For example, increased temperature increases the decomposition rate of SOM (Kumar and Das 2014). Intensive agriculture and associated practices deplete the SOM level. SOM is lost to the atmosphere as CO_2 or in the form of other GHGs such as methane (CH_4) by poor management practices (Tate 1992; Kumar and Das 2014). Continuous ploughing of soil breaks micro- and macro-soil aggregates, disperses clay and silt soil particles, and exposes the protected SOC present in aggregates for microbial decomposition (Six et al. 2000; Bronick and Lal 2005). The SOC content in tropical regions of rice cultivation is low due to high temperature induced mineralization of SOM (Ghimire et al. 2017). Removal of crop residues for fodder and fuel purposes in South Asia depletes SOM with adverse effects on soil fertility, productivity, and environmental quality (Lal 2004). Modeling of the global earth system has indicated variations in the global C flux between soil and atmosphere that reduces C input to soil and exacerbates decomposition of SOC (McGuire et al. 2001; Bondeau et al. 2007).

Decreased levels of SOC due to changing climate have been observed in Tundra, Boreal, and polar areas/regions. Soils of cold areas (Cryosols) and peat are a large sink of net carbon, but the sink capacity of these soil may decrease with increasing temperature (Lal et al. 2000). Some studies have been conducted to assess the impact of climatic variability on ecosystems and SOM. In the Mediterranean Basin, a rise in temperature (3°C) may cause a shift in the vegetation belt by 500 m (Bottner et al. 1995), whereas increasing the CO_2 level (500 ppm) together with a 2°C rise in temperature and a 30% decrease in precipitation level may substantially change the vegetation flora (Lal 2004). Further, change in climate may deplete the SOC pool by 28%, 20%, and 15% in humid, sub-humid, and arid zones, respectively (Cheddadi et al. 2001). Therefore, the SOC pools in the highland soils of South Asia (the Himalayan regions) may also be deceased by the projected change in climate. The melting of glaciers in the Himalayans (Laghari 2013) may also affect the water resources and the river flow with adverse impacts on irrigation and the groundwater recharge (Laghari 2013).

13.4.2 Soil Erosion

Soil degradation is frequently attributed to erosion by water or wind. Wind erosion is a major threat in arid and semi-arid regions (Kumar and Dias 2014). Soils with more vegetative cover have more infiltration rate and evapotranspiration which increases soil moisture contents and decreases run-off (Bruijnzeel 2004). For example, in the Sahel African region, integration between climatic fluctuations and vegetation develop two stable zones as alternatives: first a vegetated zone (moisture zone) and second a desert zone (dryland zone) (Zeng and Neelin 2000). Decreased soil cover and moisture regimes can cause climate shifts into drier regimes, where rainfall is not enough to promote vegetation and plant growth (Brovkin et al. 1998). In Pakistan, erosion is one of the most complex and detrimental types of soil degradation that deteriorates soil quality and results in loss of soil productivity (Shaheen 2016). In addition, nutrient loss through soil erosion is another reason for soil fertility depletion as it causes loss of 8 Mt of nutrients and 5.3 billion tons of soil (Prasad et al. 2000).

Inappropriate agricultural practices and excessive grazing in addition to climatic variability and urbanization cause degradation of about 70% of dryland areas. In dryland regions, soil degradation is increasing at a faster rate because of uncertain climatic variability, resulting in dryness and causing a threat to food security as well as environmental pollution (MEA 2005). Seager et al. (2007) reported that soil degradation is the most obvious due to soil erosion, especially under changing climate worldwide, particularly in arid and semi-arid areas. Moreover, chances to prevail abiotic components in response to increased degree of aridity results in more soil deterioration due to alteration of hydro-aeolian processes.

13.4.3 Leaching

Leaching refers to the downward movement of water enriched with nutrients. For example, soluble nitrates can leach down due to high rainfall causing acidification (Fenton and Helyar 2007). Leaching is a serious threat globally and is worsened by poor agricultural practices (Charman and Murphy 2007). Climatic events (e.g., rainfall, temperature) regulate this phenomenon. In areas of high rainfall, acidification is aggravated by leaching, and vice versa (Wild 1993; Kumar and Das 2014). Soils become acidified with increases in sulphate contents; moreover, pyrite oxidation turn soils into more acidic upon drainage (Rounsevell and Loveland 1992).

Soil nutrients are important determinants of soil fertility. The soil fertility is affected by environmental factors, and the nutrient concentration varies from soil to soil. For example, as temperature increases, the amounts of nutrients retained in soil are affected by reduced leaching and evaporative losses (Dent 1986; Kumar and Das 2014). Faster movement of water leads to nutrient depletion due to runoff and affects nutrient status in soil. Furthermore, low rainfall creates dryness and favors evaporation, which enhances the upward movement of nutrients resulting in salinization, which is a major problem in dry regions (Kumar and Das 2014).

13.5 COMBATING SOIL DEGRADATION UNDER CHANGING CLIMATE

Soil degradation due to climate change intensifies agricultural economic losses, disorganizes local and regional food markets, and causes social and ecosystem instability (Figure 13.1). Moreover, changing climate affects soil functions by accelerated erosion, salinization, elemental imbalance, and depletion of SOM. Restoration of degraded and desertified soils, converting marginal lands to agricultural productive lands, and adoption of adequate management practices have the potential to mitigate adverse impacts of climate change on soil degradation. The most important of these management practices are discussed below.

13.5.1 RAINWATER HARVESTING AND THE EFFICIENT USE OF WATER

Rainwater harvesting has been practiced for thousands of years to collect, store, and manage the runoff water of rain or flood in areas of erratic precipitation distribution to increase the water availability in periods of drought for irrigating the field crops, trees, pastures, rearing livestock, household consumption, and increase water availability for ecosystem sustenance (Reij et al. 1988; Grewal et al. 1989; Gupta 1989). In arid and semi-arid regions of the world, rainwater harvesting provides pragmatic solution to overcome problems of agricultural soil degradation and climate change induced changes in agricultural ecosystems (Wallace 2000; Lal 2001). Techniques or agricultural practices used to reduce runoff, enhance infiltration, and increase water storage in root zones of crops are called in situ rainwater harvesting (Ngigi 2003).

13.5.2 CROP DIVERSIFICATION

The impact of global climate change on small landholder farmers of semi-arid regions can be reduced by diversification of crops. Crop diversification provides opportunities of increased diversity in marketable produce, enhanced functioning of agricultural systems, and development of innovative approaches to mitigate climate change (McCord et al. 2015). However, the adoption of crop diversification strategies is challenging for farmers of semi-arid regions due to the fluctuations in total annual rainfall and occurrence of drought periods during the growing season (McCord et al. 2015). Crop diversification would be a viable option to increase the resilience of agro-ecosystems against climatic variability (Ewel 1999). On poor quality soils, intercropping with leguminous crops and trees can increase root proliferation, nutrient uptake, and efficient utilization of water resources (Morris and Garrity 1993; Lithourgidis et al. 2011). Crop rotation, multiple cropping, and green manuring enhance soil fertility (Tonitto et al. 2007; Grant et al. 2002). Soil fertility management is the key for sustainable management of agricultural systems to improved biodiversity and agricultural yields (Ponisio et al. 2015; Garbach et al. 2016). Balance and diversified crop rotation can improve soil health, enhance SOC sequestration (Omonode et al. 2007), reduce pest infestation and develop resistance to pests, and increase crop productivity (Katsvairo and Cox 2000; Krupinsky et al. 2006). In India, crop diversification has been shown to be a better option to mitigate the negative impacts of climate change (Lal et al. 2017). Moreover, diversified cropping, such as the incorporation of legumes in rice–wheat systems, increases the total SOC and, being a passive pool of C, can maintain soil health and food security throughout the Indo-Gangetic plains of South Asia (Samal et al. 2017). Growing cover crops, like mung bean (*Vigna radiata*) during the fallow period between the rice–wheat system, can protect the soil structure from deterioration and improve moisture reserves (Bhatt et al. 2016). Crop diversification practiced in South Asia results in improved soil health and fertility status through its positive effects on soil physical and chemical functions and nutrients conservation which result in better farm profitability (Aggarwal et al. 2004). Several researchers (Sapkota et al. 2015; Schwab et al. 2015; Ghimire et al. 2017) have reported that conservation practices and diverse cropping improve soil quality, fertility, and agricultural stability in South Asian regions. Inclusion of berseem (*Trifolium alexandrium*) crop in the rice–wheat rotation, once in at least 2–3 years, helps in improving the soil fertility (Gautam and Sharma 2004). Strategy of diversified farming systems (DFS) constitutes genetic variation of a crop and multiple cropping of a number of crops which will ultimately enhance the biodiversity in soil profile by the addition of manure. Adaptive measures must be taken by the farming community to mitigate impacts of climate variability which includes selection of climate resilient, short duration and stress tolerant crop cultivars, balanced crop diversification, inclusion of agroforestry into agroecosystems, and mixed cropping techniques (Altieri and Koohafkan 2008).

13.5.3 Conservation Agriculture

Conservation agriculture (CA) is a suite of technologies used to conserve soil and water and increase crop yield on a sustainable basis (Hobbs et al. 2008; Farooq et al. 2011, 2015; Powlson et al. 2016). The FAO (2010b) described that CA is a resource saving agricultural system to achieve higher profit along with environmental sustainability. All modern technologies that improve the quality and ecological integrity of the soil are employed in CA. In CA, soil disturbance is minimized along with improved water use efficiency (WUE) to reduce soil erosion and increase crop yield. There are four main principles of conservation agriculture: (1) continuous minimal soil disturbance, (2) year-round soil cover, (3) crop rotations including cover cropping, and (4) integrated nutrient management (Lal 2016). The world population is increasing continuously, and it is estimated to reach 9.8 billion by 2050 (United Nations 2017). To meet the growing demand of food in a limited amount of land, the CA plays an important role. Due to continual agricultural use, the land resources are degrading rapidly. In this scenario, risks of soil degradation can be reduced by adopting CA. Among numerous benefits, CA reduces moisture deficits, breaks up pest and disease cycles, minimizes weeds by improving the diversity of soil biota, increases soil fertility, enhances SOC sequestration, and reduces soil erosion (Kassam et al. 2009; Palm et al. 2014). It increases micro-pores in soil, improves water holding capacity, and minimizes evaporation of water from soil surface and topsoil (Kassam et al. 2009; Jemai et al. 2012). Poor aeration under CA reduces the rate of breakdown of SOM but can also initiate some chemical reactions that are toxic to plant roots (Shepherd et al. 2008). CA is a mitigation strategy against climate change, late and variable rainfall, and delivers advantage to smallholder farmers by increasing rain water harvesting (Thierfelder and Wall 2010; Pittelkow et al. 2014). Thus, use of CA can improve the livelihoods with sustainable agriculture. While comparing CA to conventional agriculture, it causes lesser disturbances to soil. In some of the CA experiments, the incidences of crop failure are basically due to extreme weather events (Schulze 2011).

In South Asia, CA is practiced over >5 Mha in the Indo-Gangetic plains (Rehman et al. 2015). The adoption of CA in South Asia is associated with integrated efforts of several factors including (i) availability of modern farm equipment and machinery, (ii) availability of herbicides and biocides with new chemistry to control weeds, insects and pests, (iii) decrease in labor supply and increase in labour cost, and (iv) increase in cost of production, energy shortage issues, and erosion losses (Jat et al. 2011). The use of CA in wheat crops was introduced early in the 1980s in South Asia, by using New Zealand imported zero tillage dill for first time in Punjab, Pakistan. In India, wheat cultivation is practiced on large areas through conservation tillage in double rice–wheat cropping systems (Friedrich et al. 2012). The conservation-based technologies have been widely adopted in South Asia rice–wheat, maize–sorghum, and barley cropping systems (Rehman et al. 2015). Adoption of CA practices improves soil quality and enhances soil fertility status as well as soil physical and biological health (Mohanty et al. 2007), by build-up of SOM content (Hobbs 2007), total C, total N, and aggregate stability (Aziz et al. 2013). Moreover, CA reduces the use of diesel that minimizes CO_2 emission to atmosphere and enhances C sequestration in soil (Hobbs and Govaerts 2010). Use of improved and site-specific CA practices in South Asia (i.e., no tillage and reduced tillage, crop residue retention as mulch) can enhance SOC accumulation over time (Ghimire et al. 2017).

Conservation agriculture is often practiced rescuing crops from failure and improving crop productivity and soil quality. In CA, use of cover crops and crop residues limits the weed infestation and soil erosion throughout the year. Mixed cropping and crop rotation systems are practiced in CA to maintain soil fertility and control pests, diseases, and weeds. By applying the principles of CA, farmers can save labor, reduce cost, prevent the formation of hardpan in soil, increase soil fertility, and improve water holding capacity, and thus soil achieve the power to stabilize yields and improve production for a long time (SUSTAINET EA 2010). CA can be practiced in different types of farms such as hoe farming, smallholder farming with draught animals, and even in mechanized farms by replacing mouldboard ploughs, disks and harrows with rippers, subsoilers and direct-drill planters pulled by tractors (Sims et al. 2009). Reduced tillage and residue retention can improve soil moisture and SOM contents and reduce soil erosion and runoff (Brouder and Gomez-Macpherson 2014; Palm et al. 2014). To limit soil disturbance to the narrow zone in CA, planting furrows are made with rippers or hoes and sowing is done directly into the ripped line or planting holes made by hoes or seeds are drilled into the undisturbed soil with the help of direct planters. In CA, crop residues are left in the field to improve SOM and stimulate the enzyme phosphatase activity even in highly weathered soils (Cui et al. 2015; Senwo et al. 2007).

Weeds are controlled by using techniques which disturb soil as little as possible, such as slashers and herbicides. Mulches and cover crops also smother the weeds and prevent quick growth of weeds (Stagnari et al. 2009). Clean seedbeds are a part of the cause of soil degradation and yield reduction because of the exposure of soil to wind and water. Cover crops are grown to protect soil from erosion; the use of legumes as cover crops increases the soil fertility by adding nitrogen to the topsoil (Lehmann et al. 2000). Cover crops and mulch shade the soil surface, which reduces the water evaporation while protecting the soil from erosion by heavy rain. Grazing in agricultural fields is also restricted to keep the soil covered. Smallholder farmers practice crop rotation with grain legumes, green manure cover crops, or cash crops (Thierfelder et al. 2014).

Legumes in crop rotation increase soil N content, improve fertilizer use efficiency, and enhance agricultural productivity (Reckling et al. 2014). Including a legume in the rotation plays an important role in CA as it can break up the hardpan soil with its deep roots. If there is no hardpan, water can percolate deep into the soil, and no waterlogging occurs. Crop diversification in CA occurs through mixed cropping, crop rotation, and cover cropping, all of which increase the sustainability of agricultural system. Crop residue and cover crop maintain the SOM at a high level and improve soil water holding capacity (Stagnari et al. 2009). Grain crops can utilize this stored water from soil, and crop yield under CA is resilient to seasonal rainfall variability compared to that under conventional agriculture due to water conservation. CA can increase resilience to soil degradation, drought and increase the WUE (FAO 2009).

Globally, CA is not consistent in increasing crop yield compared to conventional tillage, but it is effective when all the four principles are implemented in dry agro-ecosystems using a system-based approach (Rusinamhodzi et al. 2011; Pittelkow et al. 2014). Labor requirements for weeding and most of the other farm activities are less in CA. With the adoption of CA, farmers can cultivate a large area at low cost and labor compared to conventional agriculture. Crop rotation and mixed cropping in CA increase the crop diversity (Kassam and Friedrich 2009), and if one crop fails, another is there to replace it. Thus, CA delivers more a diverse diet compared to conventional agricultural systems. It can also reduce global warming as it increases the quantity of SOM, which is a "sink through carbon sequestration," stabilizes the CO_2 concentration, and decreases global temperature (Bernoux et al. 2006). Some benefits of CA include increased soil moisture retention, reduced runoff, reduced surface soil temperature, and decreased soil erosion (Findlater 2013). Practicing of CA decreases crop sensitivity to weather variability by reducing waterlogging and improving water holding capacity (Thierfelder and Wall 2010). Thus, CA is preferable in changing climate for sustainable agriculture and soil management.

13.5.4 Mixed Crop–Livestock Systems

The intensity of land utilization and efficiency of mixed farming systems is better as compared to specialized livestock or crop production systems (Mcintire et al. 1992). Soil fertility improves under mixed farming systems by the addition of organic residues in the form of manure resulting improved nutrient status and soil water retention (De Haan et al. 1997). Moreover, inclusion of legumes into the crop rotation reduces soil erosion and stabilizes soil nutrient status. Integrated crop and livestock production systems reduce soil loss by erosion, increase soil biological health, replenish nutrients, intensify use of land resources, and increase profitability, which ultimately contribute to culminating poverty, ending malnutrition, and enhancing environmental sustainability (Gupta et al. 2012). In Asia, the transfer of nutrients from grazing land to crop land through manure substantially contributes to the maintenance of soil fertility and farm productivity. Livestock provide the least cost, labor efficient route to intensification through their role in nutrient cycling (Devendra and Thomas 2002).

Inclusion of animals into agricultural crop production systems is an economical method for enhancing soil fertility as animals excrete nutrients particularly N, P, and K in variable quantities depending on the quality of the fodder (Watson et al. 2005). The low productivity of mixed-crop livestock systems in South Asia is due to the low adoption of available technologies or their uptake has not been sustainable, because they were improperly targeted into the farming systems (Parthasarathy Rao and Birthal 2008). Therefore, there exists a need for better understanding of the nature of small scale mixed farming systems in South Asia, and the recognition of the strong nexus between crop and animal production. There is a striking variation in systems over time and space which gives rise to the need for differential intervention. Thus, a crop–livestock system topology has been constructed that delineates the regions of each country into homogeneous crop–livestock zones with similar responses to technology uptake and development initiatives (Parthasarathy Rao and Birthal 2008).

13.5.5 Carbon Sequestration

Carbon sequestration in soil for longer periods could be the best mitigation strategy to minimize climate change because it enhances ecosystem resilience and adaptation to climate change. Agricultural soils usually have low SOM content because of low return of plant biomass and intensive cultural practices. Due to anthropogenic activities, the global historic C losses can be 78 Pg (Lal et al. 2007) or as much as 132 Pg, which must be addressed by enhancing C sequestration in soil It requires a range of improved management practices which help in increasing SOM, including CA and the use of organic inputs (plant residues, animal manures, and biosolids) as amendments. Manure application or residue incorporation in rice-wheat system in combination with nitrogen and phosphorus increased SOC by 0.30 Mg ha^{-1} yr^{-1} (Fang et al. 2005). Another study by Majumder et al. (2008) indicated a 24% increase in SOC through organic amendments (residue incorporation, green manuring, farmyard manure) in combination with NPK application in rice–wheat rotations in India. Moreover, SOC sequestration rates vary from 0.08 to 0.98 Mg ha^{-1} yr^{-1} under inorganic and organic amendments in Indo-Gangetic plains of India and Nepal (Duxbury 2001). Although enhancing crop productivity is the key element to meet the future challenges of food security but sustainability and maintaining soil health are still very important factors to mitigate climate change impacts. So, SOC sequestration requires such management strategies which enhance plant input in soil and reduce soil respiration losses. Planting trees and shrubs increases the amount of biomass-C sequestered aboveground. Agroforestry may also reduce SOC losses stemming from erosion, thus improving the SOC pool (Paustian et al. 1997; Lal and Bruce 1999; Lal 2003, 2004; Verchot et al. 2007). These management practices as well as strategies to improve carbon sequestration are given in Table 13.7.

TABLE 13.7

Management Strategies and Practices (to Discourage) to Sequester Carbon in Soil under Changing Climate

Strategies	Practices
• *In situ* plant reside input in soil	• Biomass burning
• Crop–pasture rotations	• Traditional heavy tillage practices
• Use of cover crops	• Residue removal for livestock feed
• Minimizing land-use changes	• Open grazing
• Reduced tillage and adopt conservation tillage	• Land-use change from forest to arable ecosystem
• Integrated nutrient management	
• Use of complexed organic fertilizers	
• Use of degraded soils for bioenergy feedstock production	

13.5.6 POLICY OPTIONS

Soil is the foundation for agricultural growth as it provides the nutrient support necessary for plant growth. It also has a major role in the proper functioning of ecosystems. With changing climate, soil degradation increases and becomes a serious threat for the sustainability of ecosystems and increases risks of food insecurity. Effective policy is needed to protect the soil from degradation and to sustain its healthy state. Building awareness of farmers by providing them knowledge about keeping soil healthy in a sustainable way and provision of incentives to the farming community to easily approach the inputs necessary for the improvement of the soil and protect it from loss is a good policy option. Moreover, encouraging farmers for tree plantation, adopting practices which build SOM content, good incentive packages to financially weak famers, developing small irrigation structures for farmers having low acreage, and investment for land improvement are fruitful policies to mitigate soil degradation. Enhancing the use of natural stock (resources) in efficient ways and delivery of land improvement (long term) right to poor farmers' communities by safeguarding their property right is also a good option (Scherr and Yadav 1996). Rehabilitation and restoring of marginal lands can be possible by improving and implementing agriculture support initiatives and ensuring attractive investments to build infrastructure. Furthermore, to alleviate soil degradation, agro-forestry is a competent and effective method as it helps in restoring soil health, improving agricultural input productivity, and enhancing energy security in a short time by providing fuelwood (Garrity et al. 2010). Intercropping between agro-forestry helps in maintaining the green cover on soil, improves nutrient cycling and their availability by nitrogen fixation, enhances SOM, increases soil aggregates and structure, improves water infiltration, and ultimately results in increased production of food and fiber (Garrity et al. 2010). Deforestation should be discouraged by implementing policy framework; it should be deliberated within a land rent: (i) the first policy should focus on the restructuring land tenure system, improving agriculture productivity and creating alternative income options; (ii) the second policy focus must be to build institutions (community forest management), safeguarding forest rent, developing markets to ensure more environmental services, which make possible for stakeholders to capture forest rent (Angelsen 2009).

Climate change agenda discussed at the 21st Conference of Parties to the United Nations Framework Convention on Climate Change in Paris (COP21; November 30–December 11, 2015) was unique regarding SOC and agriculture. It focussed on global negotiations/agreement to reduce climate change impact and limiting global warming to less than $2°C$ (Lal 2016; Minasny et al. 2017). The "4 per mil" proposal launched for voluntary action to increase SOC globally to 40 cm depth at the rate of 0.4% per year. The strategy is to promote SOC sequestration through adoption of adequate agricultural management practices including CA, agroforestry, cover cropping to restore the degraded soils (Lal 2016). COP21 put enormous responsibility on South Asian countries, as this region has major contribution in CO_2 emission accentuating global warming in the region. Therefore, South Asian countries signed the COP21 agreement and pledged to fulfil the global climate change policy by implementing COP21 objectives such as reducing CO_2 emission, limiting desertification and deforestation, and promoting decarbonisation on long-term bases (Lane 2016). To cope with climate change, farmers took the initiative and switched to CA and practiced soil and water conservation techniques including mulching and farm ponds similar to those used in African agriculture (Shivamurthy et al. 2015). Similarly, COP22 in Marrakech, Morocco (November 2016) adopted the program of Adapting African Agriculture (AAA) through improvement in soil health by restoration of SOC stocks. Specifically, the COP22 action plan targeted improvement in the soil and water management practices to reduce the emission of GHGs through soil carbon sequestration and afforestation (Anonymous 2016).

Agricultural policies framework should focus on striving to reduce erosion caused by human-induced activities and poor management practices leading to loss of soil cover, developing soil conservation techniques and new farming methods, and reducing the pollution induced by agricultural practices including that from animal husbandry and aquaculture.

13.5.7 Sustainable Land Management: The Multiple Win Option

Improving and restoring soils affected by degradation is a challenging task particularly in the areas characterized by low acreage and financially weak farmers. Restoring SOC as a natural resource regular input of biomass and of essential nutrients (Lal 2014). A sustainable development task of the United Nations has focussed on improving soil health to end poverty and hunger while improving economic growth and resolving health issues (Lal 2016). Therefore, rehabilitation of soil structure and quality requires a harmonized mindset and approach as this is national, international, and a global issue. Restoring and improving soil health requires the implementation of some basic approaches such as: (i) reducing losses of soil and nutrients, (ii) improving biodiversity and building SOC reserves, and (iii) strengthening nutrients and water cycling. Soil flora and fauna (macro, micro, and meso) have an important role in improving the soil health status by stabilizing soil structure and minimizing desertification and degradation. Microbiological quality or degradation indexes can be used as decision making tools to improve the management strategies (Bastida et al. 2006; Moreno et al. 2008).

Converting from conventional practices to a system-based CA by using cover crops and crop residues can enhance earthworm activities and improve soil structure (Lal 1987; Edwards et al. 1988). These practices also improve SOM content, aggregate stability, water holding capacity, SOC sequestration, and soil quality and conserve soil (Karlen and Rice 2015). Use of site-specific organic amendments offers sustainable and long-term solutions for reducing soil degradation while improving human livelihood and health (Horrigan et al. 2002). For example, using biochar—a carbon enriched product prepared after pyrolysis—enhances soil fertility and nutrients status and retention under tropical conditions (Lal 1997; Breulaman et al. 2015).

13.6 CONCLUSIONS, RESEARCHABLE PRIORITIES, AND POLICY INTERVENTIONS

South Asia is a world within the world. Its population (~1.9 billion) is increasing rapidly. Meeting the demands of the growing and increasingly affluent population is and will be a major challenge with limited soil resources, which are prone to degradation by the extreme climate events.

In South Asia, as anywhere else, soil is the most basic resource. In the densely populated region of South Asia, soil is vital for the sustainability, ecosystem functions, and sustainable development of the entire region. However, reductionism approaches and poor agricultural management have deteriorated soil health and aggravated degradation in South Asia. There are four types of soil degradation: (i) physical (poor structure and compactness), (ii) chemical (elemental toxicity and acidification), (iii) biological (organic matter loss and biodiversity loss), and (iv) ecological (reduced carbon sink and disturbed nutrient cycling), and all of these are rampant throughout the South Asia region Thus, there is a need to develop an interactive and integrated approach that can minimize soil losses and restrict, revert, and arrest the factors responsible for degradation and maintenance of ecosystem balance and functions in the South Asia region.

There is a dire need to develop researchable priorities and focus the attention of the scientific community toward practical and feasible options of reducing soil degradation and restoring degraded and desertified soils of South Asia and elsewhere. These priorities must include the use of soil according to its capability, use of water and soil resources in sustainable and judicious ways, and use of alternative and efficient options which restore soils. Farmers should be compensated through payments for ecosystem services towards human wellbeing and nature conservancy. Furthermore, site-specific approaches are needed for operationalizing and improving urban agricultural infrastructure, introduction of improved and system-specific agronomic suites, adopting precision agriculture technology, promoting carbon trading, promoting CA, and making agriculture a solution to environmental issues including mitigation of climate change. In this regards the importance of close cooperation among the scientific community from South Asia (covering all seven countries of the region) cannot be overemphasized. Such a cooperation is needed in restoring degraded soils, improving the use and management of water resources, reducing emission of greenhouse gases into the atmosphere, afforestation of the lower Himalayas spanning from Afghanistan in the west to the northern border of Cambodia in the east for re-carbonization of the biosphere and mitigation of the flood-drought syndrome, and making agriculture a solution to climate change, water scarcity, reduced biodiversity and ever degrading soil and water resources.

REFERENCES

Adams CR, Eswaran H. 2000. Global land resources in the context of food and environmental security. In *Advances in Land Resources Management for the 20th Century*, Gawande SP (ed). Soil Conservation Society of India: New Delhi, India, pp. 35–50.

Aggarwal PK, Joshi PK, Ingram JSI, Gupta RK. 2004. Adapting food systems of the Indo-Gangetic plains to global environmental change: Key information needs to improve policy formulation. *Environmental Science and Policy* 7: 487–498.

Alam MS. 2013. Climate change and land use change in the eastern coastal belt of Bangladesh, elucidated by analyzing rice production area in the past and future. *Journal of Life and Earth Science* 8: 83–92.

Ali I, Khan F, Bhatti AU. 2007. Soil and nutrient losses by water erosion under mono-cropping and legume intercropping on slopping land. *Pakistan Journal of Agricultural Research* 20: 161–166.

Altieri MA, Koohafkan P. 2008. *Enduring farms: Climate change, smallholders and traditional farming communities*. Third World Network, Penang, Malaysia.

Angelsen A. (ed.) 2009. *Realising REDD+. National Strategy and Policy Options*, CIFOR, Bogor.

Anonymous. 2016. Initiative for the Adaptation of African Agriculture to Climate Change. http://www.aaaini tiative.org/cp-cop22 (Accessed on October 30, 2017).

Anonymous. 2017. The World Factbook. https://www.cia.gov/library/publications/the-world-factbook/fields /2147.html (Accessed on October 27, 2017).

Arnold SL, Doran JW, Schepers J, Wienhold B. 2005. Portable probes to measure electrical conductivity and soil quality in the field. *Communication in Soil Science and Plant Analysis* 36: 2271–2287.

Aulakh MS, Sidhu GS. 2015. Soil degradation in India: Causes major threats, and management options. Presented in the MARCO Symposium 2015 on Next Challenges of Agro-Environmental research in Monsoon Asia. National Institute for Agro-Environmental Sciences (NIAES), 26–28 August 2015, Tsukuba, Japan, pp. 151–156.

Aziz I, Mahmood T, Islam KR. 2013. Effect of long term no-till and conventional tillage practices on soil quality. *Soil and Tillage Research* 131: 28–35.

Bai ZG, Dent DL, Olsson L, Schaepman ME. 2008. Global assessment of land degradation and improvement 1. Identification by remote sensing. Report 2008/01, ISRIC, Wageningen, the Netherlands.

Barrow C. 1991. *Land Degradation: Development and Breakdown of Terrestrial Environments*. Cambridge University Press: Cambridge, UK.

Bastida F, Moreno J, Hernandez T, Garcia C. 2006. Microbiological degradation index of soils in a semiarid climate. *Soil Biology and Biochemistry* 38: 3463–3473.

Batjes NH. 1996. Total carbon and nitrogen in the soils of the world. *European Journal of Soil Science*. 47: 151–163.

Benin S, Nkonya E, Okecho G, Pender J, Nahdy S, Mugarura S, Kato E, Kayobyo G. 2007. Assessing the Impact of the National Agricultural Advisory Services (NAADS) in the Uganda Rural Livelihoods. IFPRI Discussion Paper 00724. International Food Policy Research Institute: Washington, DC.

Bennett SJ, Barrett-Lennard EG, Colmer TD. 2009. Salinity and waterlogging as constraints to saltland pasture production: A review. *Agriculture, Ecosystems and Environment* 123: 349–360.

Bernoux M, Cerri C, Cerri CEP, Siqueira Neto M, Metay A, Perrin AS, Scopel E, Razafimbelo T, Blavet D, Piccolo MDC, Pavei M, Milne E. 2006. Cropping systems, carbon sequestration and erosion in Brazil: A review, *Agronomy for Sustainable Development* 26: 1–8.

Besley T. 1995. Property rights and investment incentives: Theory and evidence from Ghana. *The Journal of Political Economy* 103: 903–937.

Bhatt R, Kukal SS, Busari MA, Arorac S, Yadav M. 2016. Sustainability issues on rice-wheat cropping system. *International Soil and Water Conservation Research* 4: 68-83.

Bhattacharyya R, Ghosh BN, Mishra PK, Mandal B, Rao CS, Sarkar D, Das K, Anil KS, Lalitha M, Hati KM, Franzluebbers AJ. 2015. Soil degradation in India: Challenges and potential solutions. *Sustainability* 7: 3528–3570.

Boden JB, Driscoll RMC. 1987. House foundations-a review of the effect of clay soil volume change on design and performance. *Municipal Engineer* **4:** 181–213.

Bondeau A, Smith PC, Zaehle S, Schaphoff S, Lucht W, Cramer W, Gerten D, Lotze-Campen H, Müller C, Reichstein M, Smith B. 2007. Modelling the role of agriculture for the 20th century global terrestrial carbon balance. *Global Change Biology* **13:** 679–706.

Bonilla CA, Johnson OI. 2012. Soil erodibility mapping and its correlation with soil properties in Central Chile. *Geoderma* **189:** 116–123.

Borgogno F, D'Odorico P, Laio F, Ridolfi L. 2009. Mathematical models of vegetation pattern formation in ecohydrology. *Reviews of Geophysics* **47:** RG1005. doi.org/10.1029/2007RG000256.

Boserup E. 1965. *The Conditions of Agricultural Growth: The Economics of Agrarian Change under Population Pressure*. Aldine Press: New York.

Bottner P, Couˆteaux MM, Vallejo VR. 1995. Soil organic matter in Mediterranean-type ecosystems and global climate changes: A case study—The soils of the Mediterranean Basin. In *Global Change and Mediterranean Type Ecosystems*, Moreno JM, Oechel WC (eds). Springer-Verlag: New York, pp. 306–325.

Brasselle F, Brasselle A, Gaspart F, Platteau JP. 2002. Land tenure security and investment incentives: Puzzling evidence from Burkina Faso. *Journal of Development Economics* **67:** 373–418.

Breshears DD, Whicker JJ, Johansen MP, Pinder JE. 2003. Wind and water erosion and transport in semi-arid shrubland, grassland and forest ecosystems: Quantifying dominance of horizontal wind-driven transport. *Earth Surface Processes and Landforms* **28:** 1189–1209.

Brevik EC. 2013. The potential impact of climate change on soil properties and processes and corresponding influence on food security. *Agriculture* **3:** 398–417.

Brinkman R, Sombroek W. 1999. The effects of global change on soil conditions in relation to plant growth and food production. In *Global Climate Change and Agricultural Production*, Bazzaz F, Sombroek W (eds). Food and Agriculture Organization of the United Nations, John Wiley and Sons: Rome, Italy, pp. 49–63.

Bronick CJ, Lal R. 2005. Soil structure and management: A review. *Geoderma* **124:** 3–22.

Brouder SM, Gomez-Macpherson H. 2014. The impact of conservation agriculture on smallholder agricultural yields: A scoping review of the evidence. *Agriculture, Ecosystem and Environment* **187:** 11–32.

Brovkin V, Claussen M, Petoukhov V, Ganopolski A. 1998. On the stability of the atmosphere–vegetation system in the Sahara/Sahel region. *Journal of Geophysical Research* **103:** 613–631.

Bruijnzeel LA. 2004. Hydrological functions of tropical forests: Not seeing the soil for the trees? *Agriculture, Ecosystems and Environment* **104:** 185–228.

Bullard JE, Livingstone I. 2002. Interactions between aeolian and fluvial systems in dryland environments. *Area* **34:** 8–16.

CBS. 2005. *Statistical Year Book of Nepal*. Central Bureau of Statistics, National Planning Commission Secretariat, Government of Nepal, Kathmandu.

Cervarich M, Shu S, Jain AK, Arneth A, Canadell J, Friedlingstein P, Houghton RA, Kato E, Koven C, Patra P, Poulter B, Sitch S, Stocker B, Viovy N, Wiltshire A, Zeng N. 2016. The terrestrial carbon budget of South and Southeast Asia. *Environmental Research Letters* **11:** 105006.

Chan KY, Mullins CE. 1994. Slaking characteristics of some Australian and British soils. *European Journal of Soil Science* **45:** 273–283.

Chapagain T, Raizada MN. 2017. Agronomic challenges and opportunities for smallholder terrace agriculture in developing countries. *Frontiers in Plant Science*, p. 8.

Chapin FS, Walker BH, Hobbs RJ, Hooper DU, Lawton JH, Sala OE, Tilman D. 1997. Biotic controls of the functioning of ecosystems. *Science* **277:** 500–504.

Charman PEV, Murphy BW. 2007. *Soils: Their Properties and Management*. Oxford University Press Melbourne.

Cheddadi R, Guiot J, Jolly D. 2001. The Mediterranean vegetation: What if the atmospheric CO_2 increased? *Landscape Ecology* **16:** 667– 675.

Clay DC, Byiringiro FU, Kangasniemi J, Reardon T, Sibomana B, Uwamariya L, Tardif-Douglin D. 1996. Promoting Food Security in Rwanda through Sustainable Agricultural Productivity: Meeting the Challenges of Population Pressure. Department of Agricultural Economics/Department of Economics Michigan State University. Technical paper No. 28.

Cleaver KM, Schreiber GA. 1994. *Reversing the Spiral: The Population, Agriculture, and Environment Nexus in Sub-Saharan Africa*. The World Bank: Washington, DC.

Cui H, Zhou Y, Gu Z, Zhu H, Fu S, Yao Q. 2015. The combined effects of cover crops and symbiotic microbes on phosphatase gene and organic phosphorus hydrolysis in subtropical orchard soils. *Soil Biology and Biochemistry* **82:** 119–126.

Dai A. 2011. Drought under global warming: A review. *WIREs Climate Change* **45–65**.

Dalal RC, Harms, B, Krull E, Wang W. 2005. Total soil organic matter and its labile pools following Mulga (*Acacia ane*ura) clearing for pasture development and cropping 1. Total and labile carbon. *Soil Research* **43**: 13–20.

Dalal RC, Moloney D. 2000. Sustainability indicators of soil health and biodiversity. In *Management for Sustainable Ecosystems*, Hale P, Petrie A, Moloney D, Sattler P. (eds). Centre for Conservation Biology: Brisbane, pp. 101–108.

Davidson EA, Janssens IA. 2006. Temperature sensitivity of soil carbon decomposition and feedbacks on climate change. *Nature* **440**: 165–173.

De Haan C, Steinfield H, Blackburn H. 1997. Livestock and the Environment. Finding a Balance. European Commission Directorate-General for Development, Development Policy Sustainable Development and Natural Resources, Rome, Italy, p. 115.

De Janvry A, Fafchamps M, Sadoulet E. 1991. Peasant household behaviour with missing markets: Some paradoxes explained. *The Economic Journal* **101**: 1400–1417.

DeLong C, Cruse R, Wieneret J. 2015. The soil degradation paradox: Compromising our resources when we need them the most. *Sustainability* **7**: 866–879.

Denman KL, Brasseur G, Chidthaisong A, Clais P, Cox PM, Dickinson RE, Hauglustaine D, Heinze C, Holland E, Jacob D, Lohmann U, Ramachandran S, Da Silva Dias PL, Wofsy SC, Zhang X. 2007. Coupling between changes in the climate system and biogeochemistry. In *Climate Change 2007: The Physical Science Basis*, Solomon S, Qin D, Chen Z, Marquis M, Averyt KB, Tignor M, Miller HL (eds). Contribution of Working Group I to the Fourth Assessment Report of the Intergovernmental Panel on Climate Change. Cambridge University Press, Cambridge, pp. 499–587.

Dent D. 1986. Acid Sulphate Soils: A baseline for research and development. *International Institute for Land Reclamation and Improvement*, Wageningen **39**: 204.

Devendra C, Thomas D. 2002. Crop–animal interactions in mixed farming systems in Asia. *Agricultural Systems* **71**: 27–40.

D'Odorico P, Caylor K, Okin GS, Scanlon TM. 2007. On soil moisture–vegetation feedbacks and their possible effects on the dynamics of dryland ecosystems. *Journal of Geophysical Research* **112**: 4.

Duxbury JM. 2001. Long-term yield trends in the rice-wheat cropping system: Results from experiments and northwest India. *Journal of Crop Production* **3**: 27–52.

Easterling W, Aggarwal PK, Batima P, Brander KM, Erda L, Howden SM, Kirilenko A, Morton J, Soussana JF, Schmidhuber J, Tubiello FN. 2007. Food, fibre and forest products. In *Climate Change 2007: Impacts, Adaptation and Vulnerability*, Parry ML, Canziani OF, Palutikof JP (eds). Contribution of Working Group II to the Fourth Assessment Report of the Intergovernmental Panel on Climate Change. Cambridge University Press: Cambridge, UK, pp. 273–313.

Edwards WM, Shipitalo MJ, Norton LD. 1988. Contribution of macropososity to infiltration into a continuous corn no-tilled watershed: Implications for contaminant movement. *Journal of Contaminant Hydrology* **3**: 193–205.

EEA (European Environment Agency). 2014. Soil degradation. Available at: www.eea.europa.eu/publications/92-9157-202-0/page306.html (Accessed September 09, 2017).

EEA-UNEP. 2000. Down to earth. Soil degradation and sustainable development in Europe. A challenge for the 21st century. Environmental issues, series EEA, UNEP, Luxembourg, p. 6.

Elmqvist T, Folke C, Nystrom M, Peterson G, Bengtsson L, Walker B, Norberg J. 2003. Response diversity and ecosystem resilience. *Frontiers in Ecology and the Environment* **1**: 488–494.

Elvin M. 2004. *The Retreat of the Elephants: An Environmental History of China*. Yale University Press: New Haven and London.

Emmett BA, Cosby BJ, Ferrier RC, Jenkins A, Tietema A, Wright RF. 1997. Modelling the ecosystem effects of nitrogen deposition: Simulation of nitrogen saturation in a Sitka spruce forest, Aber, Wales, UK. *Biogeochemistry* **38**: 129–148.

Eswaran H, Lal R, Reich PF. 2001. Land degradation: An overview. *Responses to Land degradation*, pp. 20–35.

Ewel JJ. 1999. Natural systems as models for the design of sustainable systems of land use. *Agroforestry Systems* **45**: 1–21.

Fang C, Smith P, Moncrieff JB, Smith JU. 2005. Similar response of labile and resistant soil organic matter pools to changes in temperature. *Nature* **433**: 57–59.

FAO (The Food and Agriculture Organization of the United Nations). 1994. Land degradation in South Asia: Its severity, causes and effects upon the people. World Soil Resources 78, Rome, Italy, p. 100.

FAO (The Food and Agriculture Organization of the United Nations). 2009. Global Agriculture Towards 2050. Rome, Italy.

FAO (The Food and Agriculture Organization of the United Nations). 2010a. Global Forest Resources Assessment 2010: Main report. FAO forestry Paper 163. Rome, Italy.

FAO (The Food and Agriculture Organization of the United Nations). 2010b. Conservation Agriculture and Sustainable Crop Intensification in Lesotho. Integrated crop management. Rome, Italy.

FAO (The Food and Agriculture Organization of the United Nations). 2011. State of the World's Forests. Rome, Italy.

FAO (The Food and Agriculture Organization of the United Nations). 2014. International Year of Soils 2015: Healthy Soils for a Healthy Life. [Online] Available at: http://www.fao.org/resources/infographics /infographics-details/en/c/271187/ (Accessed on January 10, 2017).

FAO (The Food and Agriculture Organization of the United Nations). 2015. Status of the World's Soil Resources (SWSR)–Main Report. Food and Agriculture Organization of the United Nations and Intergovernmental Technical Panel on Soils. Rome, Italy, 650.

FAOSTAT. 2017. Food and Agriculture Data. http://faostat.fao.org (Accessed on October 25, 2017).

Farooq M, Flower K, Jabran K, Wahid A, Siddique KHM. 2011. Crop yield and weed management in rainfed conservation agriculture. *Soil and Tillage Research* **117**: 172–183.

Farooq M, Siddique KHM. 2015. *Conservation Agriculture*. Springer International Publishing: Switzerland.

Fenton G, Helyar K. 2007. Soil acidification. In *Soils: Their Properties and Management*, Charman PEV, Murphy BW (eds). Oxford University Press: Melbourne.

Field JP, Breshears DD, Whicker JJ. 2009. Toward a more holistic perspective of soil erosion: Why aeolian research needs to explicitly consider fluvial processes and interactions. *Aeolian Research* **1**: 9–17.

Findlater K. 2013. Conservation agriculture: South Africa's new green revolution? *The Africa Potal Backgrounder*, p. 61.

French S, Levy-Booth D, Samarajeewa A, Shannon KE, Smith J, Trevors JT. 2009. Elevated temperatures and carbon dioxide concentrations: Effects on selected microbial activities in temperate agricultural soils. *World Journal of Microbiology and Biotechnology* **25**: 1887–1900.

Friedrich T, Derpsch R, Kassam A. 2012. Overview of the global spread of the conservation agriculture. *Field Actions Science Reports* **6**: 1–7.

Gao J, Liu Y. 2010. Determination of land degradation causes in Tongyu County, Northeast China via land cover change detection. *International Journal of Applied Earth Observation and Geoinformation* **12**: 9–16.

Garbach K, Milder JC, DeClerck FA, de Wit MM, Driscoll L, Gemmill-Herren B. 2016. Examining multi-functionality for crop yield and ecosystem services in five systems of agroecological intensification. *International Journal of Agriculture Sustainability*. doi.org/10.1080/14735903.2016.1174810.

Garrity DP, Akinnifesi FK, Ajayi OC, Weldesemayat SG, Mow JG, Kalinganire A, Larwanou M, Bayala J. 2010. Evergreen agriculture: A robust approach to sustainable food security in Africa. *Food Security* **2**: 197–214.

Garten CT Jr, Classen AT, Norby RJ. 2009. Soil moisture surpasses elevated CO_2 and temperature as a control on soil carbon dynamics in a multi-factor climate change experiment. *Plant and Soil* **319**: 85–94.

Gautam RC, Sharma AR. 2004. Diversification in cercal based cropping systems for sustained productivity and food security. *Indian Farming* **54**: 3–8.

Geist HJ, Lambin EF. 2004. Dynamical causal patterns of desertification. *AIBS Bulletin* **54**: 817–829.

Ghassemi F, Jakeman AJ, Nix HA. 1995. Salinisation of Land and Water Resources: Human Causes, Extent, Management and Case Studies. The Australian National University/CAB International, Canberra, Australia/Wallingford, Oxon, UK.

Ghimire R, Lamichhane S, Acharya BS, Bista P, Sainju UM. 2017. Tillage, crop residue, and nutrient management effects on soil organic carbon sequestration in rice-based cropping systems in South Asia: A review. *Journal Integrative Agricultural* **15**: 60345–60347.

Gnacadja L. 2013. The hidden face of water scarcity. *Harvard International Review* **35**: 50–54.

Gomiero, T. 2016. Soil degradation, land scarcity and food security: Reviewing a complex challenge. *Sustainability* **8**: 281.

Goudie A, Middleton N. 2006. *Desert Dust in the Global System*. Springer, Heidelberg.

Govers G, Vandaele K, Desmet PJJ, Poesen J, Bunte K. 1994. The role of soil tillage in soil redistribution on hillslopes. *European Journal of Soil Science* **45**: 469–478.

Grant CA, Peterson GA, Campbell CA. 2002. Nutrient considerations for diversified cropping systems in the northern Great Plains. *Agronomy Journal* **94**: 186–198.

Grepperud S. 1996. Soil conservation as an investment in land. Discussion Paper 163. Oslo, Statistiques Norvège.

Grewal SS, Mittal SP, Agnihotri Y, Dubey LN. 1989. Rainwater harvesting for the management of agricultural droughts in the foothills of northern India. *Agricultural Water Management* **16**: 309–322.

Guillaume T, Damris M, Kuzyakov Y. 2015. Losses of soil carbon by converting tropical forest to plantations: Erosion and decomposition estimated by delta ^{13}C. *Global Change Biology* **21**: 3548–3560.

Gupta JP. 1989. Integrated effects of water harvesting, manuring and mulching on soil properties, growth and yield of crops in pearl millet-mungbean rotation. *Tropical Agriculture* **66**: 233–239.

Gupta R, Kienzler K, Mirzabaev A, Martius C, de Pauw E, Shideed K, Oweis T, Thomas R, Qadir M, Sayre K, Carli C, Saparov A, Bekenov M, Sanginov S, Nepesov M, Ikramov R. 2009. Research prospectus: A vision for sustainable land management research in Central Asia. ICARDA Central Asia and Caucasus Program, Sustainable Agriculture in Central Asia and the Caucasus series no 1. CGIAR-PFU, Tashkent; p. 84

Gupta V, Rai PK, Risam KS. 2012. Integrated crop-livestock farming systems: A strategy for resource conservation and environmental sustainability. *Indian Research Journal of Extension Education* **2**: 49–54.

Hawkes CV, Hartley IP, Ineson P, Fitter AH. 2008 Soil temperature affects carbon allocation within arbuscular mycorrhizal networks and carbon transport from plant to fungus. *Global Change Biology* **14**: 1181–1190.

Hedge R, Natarajan A, Naidu LGK, Dipak S. 2011. Soil Degradation, Soil Erosion Issues in Agriculture. INTECH, https://www.intechopen.com/books/soil-erosion-issues-in-agriculture (Accessed on August 21, 2017).

Hendrickson O. 2003. Influences of global change on carbon sequestration by agricultural and forest soils. *Environmental Reviews* **11**: 161–92.

Hobbs PR, Govaerts B. 2010. How conservation agriculture can contribute to buffering climate change. In *Climate Change and Crop Production*. Reynolds M. (ed). CABI Climate Change Series. CABI: Cambridge, pp. 177–199.

Hobbs P, Sayre K, Gupta R. 2008. The role of conservation agriculture in sustainable agriculture. *Philosophical Transactions of the Royal Society of London* **363**: 543–555.

Hobbs PR. 2007. Conservation agriculture: What is it and why is it important for future sustainable food production? *The Journal of Agricultural Sciences* **145**: 127–137.

Horrigan L, Lawrence RS, Walker P. 2002. How sustainable agriculture can address the environmental and human health harms of industrial agriculture. *Environmental Health Perspectives* **110**: 445.

Hossain MS. 2007. The Bengal Delta: An Assessment of Desertification, Dhaka University.

Houšková B, Montanarella L. 2008. The natural susceptibility of european soils to compaction. In *Threats to Soil Quality in Europe*, Tóth G, Montanarella L, Rusco E. (eds). EUR 23438. European Commission-Institute for Environment and Sustainability, pp. 23–36.

ICAR and NAAS. 2010. Degraded and Wastelands of India, Status and Spatial Distribution. Indian Council of Agricultural Research and National Academy of Agricultural Science, New Delhi, p. 158.

Indoria AK, Sharma KL, Reddy KS, Rao CS. 2017. Role of soil physical properties in soil health management and crop productivity in rainfed systems-I: Soil physical constraints and scope. *Current Science* **112**: 2405–2414.

IPCC (Intergovernmental Panel on Climate Change). 2007. Intergovernmental Panel on Climate Change Fourth Assessment Report: Climate Change 2007. Synthesis Report. World Meteorological Organization, Geneva, Switzerland.

IPCC (Intergovernmental Panel on Climate Change). 2014. Climate Change 2014: Synthesis Report. In *Proceedings of the Contribution of Working Groups I, II and III to the Fifth Assessment Report of the Intergovernmental Panel on Climate Change*, Pachauri RK, Meyer LA (eds). Cambridge University Press: Cambridge.

Jat ML, Saharawat YS, Gupta RK. 2011. Conservation agriculture in cereal systems of South Asia: Nutrient management perspectives. *Karnataka Journal of Agricultural Sciences* **24**: 100–105.

Jemai I, Ben Aissa N, Ben Guirat S, Ben-Hammouda M, Gallali T. 2012. Impact of three and seven years of no-tillage on the soil water storage, in the plant root zone, under a dry subhumid Tunisian climate. *Soil and Tillage Research* **126**: 26–33.

Kabubo-Mariara J. 2007. Land conservation and tenure security in Kenya: Boserup's Hypothesis revisited. *Ecological Economics* **64**: 25–35.

Karlen DL, CW Rice. 2015. Soil degradation: Will humankind ever learn? **7**: 12490–12501.

Karmakar R, Das I, Dutta D, Rakshit A. 2016. Potential effects of climate change on soil properties: A review. *Science International* **4**: 51–73.

Kassam AH, Friedrich T. 2009. Nutrient management in conservation agriculture: A biologically-based approach to sustainable production intensification. 7th Conservation Agriculture Conference, Dnipropetrovsk, Ukraine.

Kassam A, Friedrich T, Shaxson F, Pretty J. 2009. The spread of conservation agriculture: Justification, sustainability, and uptake. *International Journal of Agricultural Sustainability* **7:** 292– 320.

Katsvairo T, Cox WJ. 2000. Economics of cropping systems featuring different rotations, tillage, and management. *Agronomy Journal* **92:** 485–493.

Kavvadias V. 2014. Soil degradation. Soil science Institute of Athens-National Agricultural Research Foundation. Available at: www.prosodol.gr/sites/ (Accessed on September 20, 2017).

Keppel G, Buckley Y, Possingham H. 2010. Drivers of lowland rainforest community assembly, species diversity and forest structure in islands in the tropical South Pacific. *Journal of Ecology* **98:** 87–95.

Khan MA, Ahmed M, Hashmi HS. 2012. Review of available knowledge on land degradation in Pakistan. OASIS Country Report 3, ICARDA.

Kibblewhite MG, Ritz K, Swift MJ. 2008. Soil health in agricultural systems. *Philosophical Transactions of the Royal Society of London* **363:** 685–701.

Krupinsky JM, Tanaka DL, Merrill SD, Liebig MA, Hanson JD. 2006. Crop sequence effects of 10 crops in the northern Great Plains. *Agricultural Systems* **88:** 227–254.

Kumar R, Das AJ. 2014. Climate change and its impact on land degradation: Imperative need to focus. *Journal of Climatology and Weather Forecasting* **2:** 108.

Kuzyakov Y, Gavrichkova O. 2010. Time lag between photosynthesis and carbon dioxide efflux from soil: A review of mechanisms and controls. *Global Change Biology* **16:** 3386–3406.

Ladha JK, Pathak H, Krupnik TJ, Six J, Van Kessel C. 2005. Efficiency of fertilizer nitrogen in cereal production: Retrospects and prospects. *Advances in Agronomy* **87:** 85–156.

Laghari, J. 2013. Climate change: Melting glaciers bring energy uncertainty. Nature **502:** 617–618.

Lal R. 1987. *Tropical Ecology and Physical Edaphology*. John Wiley Sons, Chichester, UK.

Lal R. 1994. Sustainable land use systems and soil resilience. In *Soil Resilience and Sustainable Land Use*, Greenland DJ, Szabolcs I. (eds). CAB International, Wallingford, UK.

Lal R. 1997. Degradation and resilience of soils. *Philosophical Transactions of the Royal Society of London.* **352:** 997–1010.

Lal R. 2001. Soil degradation by erosion. *Land Degradation and Development* **12:** 519–539.

Lal R. 2003. Global potential of soil carbon sequestration to mitigate the greenhouse effect. *Critical Review in Plant Sciences* **22:** 151–184.

Lal R. 2004. Soil carbon sequestration to mitigate climate change. *Geoderma* **123:** 1–22.

Lal R. 2007. Soil degradation and environment quality in South Asia. *International Journal of Ecology and Environmental Science* **33:** 91–103.

Lal R. 2014. Climate strategic soil management. *Challenges* **5:** 43–74.

Lal R. 2015. Restoring soil quality to mitigate soil degradation. *Sustainability* **7:** 5875–5895.

Lal R. 2016. Beyond COP 21: Potential and challenges of the "4 per Thousand" initiative. *Journal of Soil and Water Conservation* **71:** 20A–25A.

Lal R, Bruce JP. 1999. The potential of world cropland soils to sequester C and mitigate the greenhouse effect. *Environmental Science and Policy* **2:** 177–186.

Lal R, Follett RF, Stewart BA, Kimble JM. 2007. Soil carbon sequestration to mitigate climate change and advance food security. *Soil Science.* **172:** 943–956.

Lal B, Gautam P, Panda BB, Raja R, Singh T, Tripathi R, Shahid M, Nayak AK. 2017. Crop and varietal diversification of rainfed rice based cropping systems for higher productivity and profitability in Eastern India. *PloS One* **12:** 0175709.

Lal R, Kimble JM, Stewart BA. 2000. *Global Climate Change and Cold Regions Ecosystems*. CRC/Lewis, Boca Raton, FL. p. 265.

Lane JE. 2016. The Asian miracles: Implementing the COP21 agreement. *Asia Pacific Journal of Public Administration* **38:** 75–86.

Lasanta T, Arnaez J, Oserin M, Ortigosa LM. 2001. Marginal lands and erosion in terraced fields in the Mediterranean mountains. *Mountain Research and Development* **21:** 69–76.

Lavelle P. 1997. Faunal activities and soil processes: Adaptive strategies that determine ecosystem function. *Advances in Ecological Research* **27:** 93–132.

Lehmann J, Silva Jr. JP, Schroth G, Da Silva LF. 2000. Nitrogen use in mixed tree crop plantations with a legume cover crop. *Plant and Soil* **225:** 63–72.

Li J, Okin GS, Alvarez L, Epstein H. 2008. Effects of wind erosion on the spatial heterogeneity of soil nutrients in two desert grassland communities. *Biogeochemistry* **88:** 73–88.

Link SO, Smith JL, Halverson JJ, Bolton H Jr. 2003. A reciprocal transplant experiment within a climatic gradient in a semiarid shrub-steppe ecosystem: Effects on bunchgrass growth and reproduction, soil carbon, and soil nitrogen. *Global Change Biology* **9**: 1097–1105.

Lioubimtseva E, Henebry GM. 2009. Climate and environmental change in arid Central Asia: Impacts, vulnerability, and adaptations. *Journal of Arid Environments* **73**: 963–977.

Lithourgidis AS, Vlachostergios DN, Dordas CA, Damalas CA. 2011. Dry matter yield, nitrogen content, and competition in pea-cereal intercropping systems. *European Journal of Agronomy* **34**: 287–294.

Loper CE, Balgos M, Brown J, Cicin-Sain B, Edwards P, Jarvis C, Lilley J, Torres de Noronha I, Skarke A, Tavares JF, Walker L. 2005. Small Islands, Large Ocean States: A Review of Ocean and Coastal Management in Small Island Developing States since the 1994 Barbados Programme of Action for the Sustainable Development of Small Island Developing States (SIDS). Papers Series No. 2005-1. UNEP/GPA and the Global Forum on Oceans, Coasts, and Islands.

Lu D, Batistella M, Mausel P, Moran E. 2007. Mapping and monitoring land degradation risks in the Western Brazilian Amazon using Multitemporal Landsat TM/ETM+ Images. *Land Degradation and Development* **18**: 41–54.

Ludwig JA, Wilcox BP, Breshears DD, Tongway DJ, Imeson AC. 2005. Vegetation patches and runoff-erosion as interacting ecohydrological processes in semiarid landscapes. *Ecology* **86**: 288–297.

Ma H, Ju H. 2007. Status and trends in land degradation in Asia. In *Climate and Land Degradation*, Sivakumar MVK, Ndiang'ui N (eds). Springer, Berlin Heidelberg, pp. 55–64.

Majumder B, Mandal B, Bandyopadhyay PK, Gangopadhyay A, Mani PK, Kundu AL, Mazumdar D. 2008. Organic amendments influence soil organic carbon pools and rice wheat productivity. *Soil Science Society of America Journal* **72**: 775–785.

Majumdar K, Sanyal SK, Dutta SK, Satyanarayana T, Singh VK. 2016. Nutrient mining: Addressing the challenges to soil resources and food security. In *Biofortification of Food Crops*, Singh U, Praharaj CS, Singh SS, Singh NP (eds). Springer, India, pp. 177–198.

Malla GS. 2007. Melting ice: "Warning Signs." *Journal of Agriculture and Environment* **8**: 66–73.

McCord PF, Cox M, Schmitt-Harsh M, Evans T. 2015. Crop diversification as a smallholder livelihood strategy within semi-arid agricultural systems near Mount Kenya. *Land Use Policy* **42**: 738–750.

McGuire AD, Sitch S, Clein JS, Dargaville R, Esser G, Foley J, Heimann M, Joos F, Kaplan J, Kicklighter DW, Meier RA. 2001. Carbon balance of the terrestrial biosphere in the twentieth century: Analyses of CO_2, climate and land use effects with four process-based ecosystem models. *Global Biogeochemical Cycles* **15**: 183–206.

Mcintire J, Bougat D, Pingali P. 1992. *Crop-Livestock Interaction in SubSaharan Africa*. The World Bank, Washington DC.

MEA (Millennium Ecosystem Assesment). 2005. *Ecosystems and Human Well-Being: Desertification Synthesis*. World Resource Institute, Washington DC.

Middleton N, Thomas D. 1997. *World Atlas of Desertification*. Arnold, London.

Minasny B, Malone BP, McBratney AB, Angers DA, Arrouays D, Chambers A, Chaplot V, Chen ZS, Cheng K, Das BS, Field DJ, Gimona A, Hedley CB, Hong SY, Mandal B, Marchant BP, Martin M, McConkey BG, Mulder VL, O'Rourke S, Richer-de-Forges AC, Odeh I, Padarian J, Paustian K, Pan G, Poggio L, Savin I, Stolbovoy V, Stockmann U, Sulaeman Y, Tsui CC, Vågen TG, Wesemael BV, Winowiecki L. 2017. Soil carbon 4 per mile. *Geoderma* **292**: 59–86.

Mitchell SW, Csillag F. 2001. Assessing the stability and uncertainty of predicted vegetation growth under climatic variability: Northern mixed grass prairie. *Ecological Modelling* **139**: 101–121.

Moebius BN, Van Es HZ, Schindelbeck PR, Idowu OJ, Clune DJ, Thies JE. 2007. Evaluation of laboratory-measured soil properties as indicators of soil physical quality. *Soil Science* **172**: 895–912.

MoEST. 2006. Rural Energy Policy. Ministry of Environment Science and Technology, Kathmandu.

Mohanty M, Painuli DK, Misra AK, Ghosh PK. 2007. Soil quality effects of tillage and residue under rice-wheat cropping on a Vertisol in India. *Soil and Tillage Research* **92**: 243–250.

Montgomery DR. 2007. *Dirt: The Erosion of Civilization*. University of California Press, Berkeley.

Moreno JL, Bastida F, Hernández T, García C. 2008. Relationship between the agricultural management of a semi-arid soil and microbiological quality. *Communications in Soil Science and Plant Analysis* **39**: 421–39.

Morris RA, Garrity DP. 1993. Resource capture and utilization in intercropping water. *Field Crops Research* **34**: 303–317.

Nagle G. 2006. Soil degradation—A creeping concern. Geo Factsheet. Curriculum Press, Bank House, 105 King Street, Wellington. Available: www.curriculum-press.co.UK (Accessed on November 20, 2014).

Nanson GC, Tooth S, Knighton AD. 2002. A global perspective on dryland rivers: Perceptions, misconceptions, and distinctions. In *Dryland Rivers: Hydrology and Geomorphology of Semiarid Channels*, Bull LJ, Kirkby MJ. (eds). John Wiley and Sons, New York, pp. 17–54.

Nearing MA, Pruski FF, O'neal MR. 2004. Expected climate change impacts on soil erosion rates: A review. *Journal of Soil and Water Conservation* **59**: 43–50.

Nicholson SE, Tucker CJ, Ba MB. 1998. Desertification, drought, and surface vegetation: An example from the West African Sahel. *Bulletin of the American Meteorological Society* **79**: 815–829.

Ngigi SN. 2003. What is the limit of up-scaling rainwater harvesting in a river basin? *Physics and Chemistry of the Earth, Parts A/B/C* **28**: 943–956.

Nkonya E, Gerber N, Baumgartner P, von Braun J, De Pinto A, Graw V, Kato E, Kloos J, Walter T. 2011. *The Economics of Land Degradation—Towards an Integrated Global Assessment*. Peter Lang, New York.

Nkonya E, Pender J, Kaizzi K, Kato E, Mugarura S, Ssali H, Muwonge J. 2008. Linkages between land management, land degradation, and poverty in Sub Saharan Africa: The case of Uganda. IFPRI Research Report 159, Washington DC.

Nkonya E, Phillip D, Mogues T, Pender J, Kato E. 2010. *From the Ground Up: Impacts of a Pro-Poor Community-Driven Development Project in Nigeria*. IFPRI Research Monograph. Washington, DC.

Nordstrom DK. 2009. Acid rock drainage and climate change. *Journal of Geochemical Exploration* **100**: 97–104.

Oldeman LR. 1992. Global extent of soil degradation. ISRIC Bi-Annual Report. Wageningen, The Netherlands, pp. 19–36.

Oldeman LR, Hakkeling RTA, Sombroek WG. 1991. *World map of human induced-soil degradation. An explanatory note*. ISRIC, Wageningen.

Omonode RA, Vyn TJ, Smith DR, Hegymegi P, Ga'l A. 2007. Soil carbon dioxide and methane fluxes from long-term tillage systems in continuous corn and corn–soybean rotations. *Soil and Tillage Research* **95**: 182–195.

Palm C, Blanco-Canqui H, DeClerck F, Gatere L. 2014. Conservation agriculture and ecosystem services: An overview. *Agriculture Ecosystems and Environment* **187**: 87–105.

Parthasarathy Rao P, Birthal PS. 2008. *Livestock in mixed farming systems in South Asia*. Patancheru, India: ICRISAT.

Patra PK, Canadell JG, Houghton RA, Piao SL, Oh N-H, Ciais P, Manjunath KR, Chhabra A, Wang T, Bhattacharya T, Bousquet P, Hartman J, Ito A, Mayorga E, Niwa Y, Raymond PA, Sarma VVSS, Lasco R. 2013. The carbon budget of South Asia. *Biogeosciences* **10**: 513–527.

Paudel GS, Thapa GB. 2004. Impact of social, institutional, and ecological factors on land management practices in mountain watersheds of Nepal. *Applied Geography* **24**: 35–55.

Paustian K, Andr'en O, Janzen HH, Lal R, Smith P, Tian G, Tiessen H, Noordwijk MV, Woomer PL. 1997. Agricultural soils as a sink to mitigate CO_2 emissions. *Soil Use and Management* **13**: 1–5.

Pender J, Kerr J. 1998. Determinants of farmers' indigenous soil and water conservation investments in semi-arid India. *Agricultural Economics* **19**: 113–125.

Pender J, Nkonya E, Jagger P, Sserunkuuma D, Ssali H. 2006. Strategies to increase agricultural productivity and reduce land degradation in Uganda: An econometric analysis. In *Strategies for Sustainable Land Management in the East African Highlands*, Pender J, Ehui S. (eds). International Food Policy Research Institute, Washington, DC, pp 165–190.

Pittelkow CM, Liang X, Linquist BA, Van Groenigen KJ, Lee J, Lundy ME, Van Gestel N, Six J, Venterea RT, Van Kessel C. 2014. Productivity limits and potentials of the principles of conservation agriculture. *Nature* **517**: 365–368.

Ponisio LC, M'Gonigle LK, Mace KC, Palomino J, De Valpine P, Kremen C. 2015. Diversification practices reduce organic to conventional yield gap. *Proceeding of the Royal Society of London* **282**: 20141396.

Powlson DS, Stirling CM, Thierfelder C, White RP, Jat ML. 2016. Does conservation agriculture deliver climate change mitigation through soil carbon sequestration in tropical agro-ecosystems? *Agriculture, Ecosystems and Environment* **220**: 164–174.

Prasad RN, Biswas PP. 2000. Soil Resources of India. In *50 Years of Natural Resource Management*, Singh GB, Sharma BR. (eds). Indian Council of Agricultural Research, New Delhi, India.

Puigdefabregas J. 2005. The role of vegetation patterns in structuring runoff and sediment fluxes in drylands. *Earth Surface Processes and Landforms* **30**: 133–147.

Quamruzzaman M. 2006. Integrated nutrient management for sustaining crop productivity and improvement of soil fertility in Bangladesh agriculture. Food and Agriculture Organization, Improving Plant Nutrient Management for Better Farmer Livelihoods, Food Security and Environmental Sustainability. FAO, Rome, Italy, p. 257.

Qureshi AS, McCornick PG, Qadir M, Aslam Z. 2008. Managing salinity and waterlogging in the Indus Basin of Pakistan. *Agricultural Water Management* **95:** 179–189.

Rasul G, Mahmood A, Sadiq A, Khan SI. 2012. Vulnerability of the Indus delta to climate change in Pakistan. *Pakistan Journal of Meteorology* **8:** 89–107.

Raupach MR. 1992. Drag and drag partitioning on rough surfaces. *Boundary-Layer Meteorology* **60:** 375–395.

Ravi S, Breshears DD, Huxman TE, D'Odorico P. 2010. Land degradation in drylands: Interactions among hydrologic–aeolian erosion and vegetation dynamics. *Geomorphology* **116:** 236–245.

Ravi S, D'Odorico P, Okin GS. 2007. Hydrologic and aeolian controls on vegetation patterns in arid land-scapes. *Geophysical Research Letters* **34:** L24–S23.

Ravi S, D'Odorico P, Zobeck TM, Over TM. 2009. The effect of fire-induced soil hydrophobicity on wind erosion in a semiarid grassland: Experimental observations and theoretical framework. *Geomorphology* **105:** 80–86.

Ravi S, Zobeck TM, Over TM, Okin GS, D'Odorico P. 2006. On the effect of wet bonding forces in air-dry soils on threshold friction velocity of wind erosion. *Sedimentology* **53:** 597–609.

Reckling M, Hecker JM, Schläfke N, Bachinger J, Zander P, Bergkvist G,Walker R, Maire J, Eory V, Topp CFA, Rees RA, Toncea I, Pristeri A, Stoddard FL. 2014. Agronomic analysis of cropping strategies for each agroclimatic region. Legume Futures Report 1.4, pp. 75.

Rehman H, Nawaz A, Wakeel A, Saharawat YS, Farooq M. 2015. *Conservation Agriculture in South Asia.* Conservation Agriculture. Springer, Dordrecht, pp. 249–283.

Reij C, Mulder P, Begeman L. 1988. Water harvesting for plant production. World Bank Technical paper 91. World Bank, Washington, p. 123.

Rengasamy P. 2006. World salinization with emphasis on Australia. *Journal of Experimental Botany* **57:** 1017–1023.

Rengel Z. 2002. Role of pH in availability of ions in soil. In *Handbook of Plant Growth. pH as a Master Variable in Plant Growth*, Rengel Z. (ed). Marcel Dekker, New York, pp. 323–350.

Reth S, Reichstein M, Falge E. 2005. The effect of soil water content, soil temperature, soil pH value and root mass on soil CO_2 efflux. *Plant and Soil* **268:** 21–33.

Rimal BK, Lal R. 2009. Soil and carbon losses from five different land management areas under simulated rainfall. *Soil and Tillage Research* **106:** 62–70.

Ritzema HP, Satyanarayana TV, Raman S, Boonstra J. 2008. Subsurface drainage to combat waterlogging and salinity in irrigated lands in India: Lessons learned in farmers' fields. *Agricultural Water Management* **95:** 179–189.

Rojas RV, Achouri M, Maroulis J, Lucrezia C. 2016. Healthy soils: A prerequisite for sustainable food secu-rity. *Environmental Earth Sciences* **75:** 180.

Ross DS, Matschonat G, Skyllberg U. 2008. Cation exchange in forest soils: The need for a new perspective. *European Journal of Soil Science* **59:** 1141–1159.

Rounsevell MDA, Loveland PJ. 1992. An overview of hydrologically controlled soil responses to climate change in temperate regions. *SEESOIL* **8:** 69–78.

Rusinamhodzi L, Corbeels M, Van Wijk MT, Rufino MC, Nyamangara J, Giller KE. 2011. A meta-analysis of long-term effects of conservation agriculture on maize grain yield under rain-fed conditions. *Agronomy for Sustainable Development* **31:** 657–673.

Safriel UN, Adeel Z. 2005. Dryland Systems. In *Ecosystems and Human Well-Being: Current State and Trends*, Hassan R, Scholes R, Ash N. (eds). Island Press, Washington, DC, pp. 623–662.

Sahay KB. 2000. Problems of livestock population. Available online: http://www.tribuneindia.com /2000/20000411/edit.htm (Accessed on September 20, 2017).

Samal SK, Rao KK, Poonia SP, Kumar R, Mishra JS, Prakash V, Mondal S, Dwivedi SK, Bhatt BP, Naik SK, Choubey AK. 2017. Evaluation of long-term conservation agriculture and crop intensification in rice-wheat rotation of Indo-Gangetic Plains of South Asia: Carbon dynamics and productivity. *European Journal of Agronomy* **90:** 198–208.

Samuel, J., Al Mahmud Titumir R. 2011. Land. In: *Natural Resource Management in South Asia.* Dorling Kindersley, Chennai, India.

Sanaullah M, Blagodatskaya E, Chabbi A, Rumpel C, Kuzyakov Y. 2011a. Drought effects on microbial bio-mass and enzyme activities in the rhizosphere of grasses depending on plant community composition. *Applied Soil Ecology* **48:** 38–44.

Sanaullah M, Chabbi A, Leifeld J, Bardoux G, Rumpel C. 2011b. Decomposition and stabilization of root litter in top-and subsoil horizons: What is the difference? *Plant and Soil* **338:** 127–141.

Sanaullah M, Chabbi A, Girardin C, Durand JL, Rumpel C. 2014. Effects of elevated temperature and drought on biochemical composition of forage plants and their mineralization potential in grassland soil. *Plant and Soil* **374:** 767–778.

Sanaullah M, Chabbi A, Maron PA, Baumann K, Tardy V, Blagodatskaya E, Kuzyakov Y, Rumpel C. 2016. How do microbial communities in top- and subsoil respond to root litter addition under field conditions? *Soil Biology and Biochemistry* **103:** 28–38.

Sanwal M. 2004. Trends in global environmental governance: The emergence of a mutual supportiveness approach to achieve sustainable development. *Global Environmental Politics* **4:** 16–22.

Sapkota TB, Jat ML, Aryal JP, Jat RK, Khatri-Chhetri A. 2015. Climate change adaptation, greenhouse gas mitigation and economic profitability of conservation agriculture: Some examples from cereal systems of Indo-Gangetic Plains. *Journal of Integrative Agriculture* **14:** 1524–33.

Scatena FN, Planos-Gutierrez EO, Schellekens J. 2005. Natural disturbances and the hydrology of humid tropical forests. In *Forests, Water and People in the Humid Tropics*. Bonell M, Bruijnzeel LA. (eds). Cambridge University Press and UNESCO, Cambridge, pp. 489–512.

Scherr S. 2000. Downward spiral? Research evidence on the relationship between poverty and natural resource degradation. *Food Policy* **25:** 479–498.

Scherr S, Hazell P. 1994. Sustainable Agricultural Development Strategies in Fragile Lands. Environment and Production Technology Division Discussion Paper, no. 1. *International Food Policy Research Institute*, Washington, DC, USA.

Scherr SJ, Yadav S. 1996. Land degradation in the developing world. Issues and policy options, 2020 vision policy Brief No. 44. Washington, DC. IFPRI.

Schulze E. 2011. Approaches towards practical adaptive management options for selected water-related sectors in South Africa in a context of climate change. *Water SA WRC 40-Year Celebration Special Edition* **37:** 5.

Schwab N, Schickhoff U, Fischer E. 2015. Transition to agroforestry significantly improves soil quality: A case study in the central mid-hills of Nepal. *Agriculture, Ecosystems and Environment* **205:** 57–69.

Seager R, Ting M, Held MI, Kushnir Y, Lu J, Vecchi G, Huang HP, Harnik N, Leetmaa A, Lau NC, Li C. 2007. Model projections of an imminent transition to a more arid climate in southwestern North America. *Science* **316:** 1181–1184.

Senwo ZN, Ranatunga TD, Tazisong IA, Taylor RW, He Z. 2007. Phosphatase activity of Ultisols and relationship to soil fertility indices. *Journal of Food Agriculture and Environment* **5:** 262–266.

Shah ZH, Arshad M. 2012. Land degradation in Pakistan: A serious threat to environments and economic sustainability. ECO Services International. Available at: www.ecoweb.com/edi/060715.html (Accessed on September 22, 2017).

Shaheen A. 2016. Characterization of eroded lands of Pothwar Plateau, Punjab, Pakistan. *Sarhad Journal of Agriculture* **32:** 192–201.

Shepherd TG, Stagnari F, Pisante M, Benites J. 2008. Visual soil assesment—Field guide for annual crops. FAO, Rome, Italy, VIII, 26.

Shivamurthy M, Shankara MH, Radhakrishna R, Chandrakanth MG. 2015. Impact of climate change and adaptation measures initiated by farmers. In *Adapting African Agriculture to Climate Change*, Springer International Publishing, Heidelberg, pp. 119–126.

Sims B, Friedrich T, Kassam A, Kienzle J. 2009. Agroforestry and conservation agriculture: Complementary practices for sustainable development. *Agriculture for Development* **8:** 13–18.

Singer MJ, Shainberg I. 2004. Mineral soil surface crusts and wind and water erosion. *Earth Surface Processes and Landforms* **29:** 1065–1075.

Singh BP, Cowie AL, Chan KY. 2011. *Soil Health and Climate Change*. Springer-Verlag Berlin, Germany.

Sivakumar MVK. 2011. Climate and land degradation. In *Sustaining Soil Productivity in Response to Global Climate Change: Science, Policy, and Ethics*, Sauer TJ, Norman JM, Sivakumar MVK. (eds). Oxford University Press, Oxford, UK, pp. 141–154.

Six J, Paustian K, Elliott ET, Combrink C. 2000. Soil structure and organic matter I. Distribution of aggregate-size classes and aggregate-associated carbon. *Soil Science Society of America Journal* **64:** 681–689.

Solomon A. 1994. Land use dynamics, soil degradation and potential for sustainable use in Metu area, Illubabor Region, Ethiopia. University of Berne, Switzerland, p. 135.

Stagnari F, Ramazzotti S, Pisante M, Lichtfouse E. (eds). 2009. *Organic Farming, Pest Control and Remediation of Soil Pollutants, Sustainable Agriculture Reviews* 1, Springer Science+Business, Media BV.

State of the Environment India. 2001. Land degradation, Part III http://envfor.nic.in/sites/default/files /soer/2001/soer.html (Accessed September 2017).

Sternberg T. 2008. Environmental challenges in Mongolia's dryland pastoral landscape. *Journal of Arid Environments* **72:** 1294–1304.

Suraj B, Reddy CP, and Manda, Verma, 2001. Scientific management and development of saline land. Addressing the challenges of land degradation: An overview and Indian perspective. *Proceedings of the National Seminar Land Resource Management for Food and Environmental Security, Soil* Conservation Society of India, pp. 73–83.

SUSTAINET EA. 2010. *Technical Manual for Farmers and Field Extension Service Providers: Conservation Agriculture.* Sustainable Agriculture Information Initiative, Nairobi. ISBN 978-9966-1533-0-2.

Sveistrup TE, Haraldsen TK, Langohr R, Marcelino V, Kværner J. 2005. Impact of land use and seasonal freezing on morphological and physical properties of silty Norwegian soils. *Soil and Tillage Research* **81:** 39–56.

Swap R, Garstang M, Macko SA, Tyson PD, Maenhaut W, Artaxo P, Kållberg P, Talbot R. 1996. The long-range transport of southern African aerosols to the tropical South Atlantic. *Journal of Geophysical Research Atmospheres* **101:** 23777–23791.

Tate KR. 1992. Assessment, based on a climosequence of soils in tussock grasslands, of soil carbon storage and release in response to global warming. *Journal of Soil Science* **43:** 697–707.

Thierfelder C, Wall PC. 2010. Investigating conservation agriculture (CA) systems in Zambia and Zimbabwe to mitigate future effects of climate change. *Journal of Crop Improvement* **24:** 113–121.

Thierfelder C, Rusinamhodzi L, Ngwira AR, Mupangwa W, Nyagumbo I, Kassie GT, Cairns JE. 2014. Conservation agriculture in Southern Africa: Advances in knowledge. *Renewable Agriculture and Food System* **30:** 328–348.

Thornton PK, Ericksen PJ, Herrero M, Challinor AJ. 2014. Climate variability and vulnerability to climate change: A review. *Global Change Biology* **20:** 3313–3328.

Tiffen M, Mortimore M, Gichuki F. 1994. *More People, Less Erosion: Environmental Recovery in Kenya.* Wiley and Sons, London, UK.

Tilman D, Cassman KG, Matson PA, Naylor R, Polasky S. 2002. Agricultural sustainability and intensive production practices. *Nature* **418:** 671–677.

Tonitto C, David MB, Drinkwater LE, Li C. 2007. Application of the DNDC model to tile-drained Illinois agroecosystems: Model calibration, validation, and uncertainty analysis. *Nutrient Cycling in Agroecosystems* **78:** 51–63.

Ultra VU and Han SH. 2015. Elevated atmospheric temperature and CO_2 altered the growth, carbon, and nitrogen distribution and the rhizosphere properties of *Platanus occidentalis* L. seedlings. *Turkish Journal of Agriculture and Forestry* **39:** 679–691.

United Nations. 2015. Transforming our world: The 2030 Agenda for Sustainable Development. United Nations General Assembly Resolution, 2015 September 18. Available at: https://sustainabledevelop ment.un.org/post2015/transformingourworld (Accessed on May 13, 2017).

United Nations. 2017. News. Available at: https://www.un.org/development/desa/en/news/population/world -population prospects-2017.html (Accessed on August 23, 2017).

Van de Koppel J, Rietkerk M, Van Langevelde F, Kumar L, Klausmeier CA, Fryxell JM, Hearne JW, Van Andel J, De Ridder N, Skidmore A, Stroosnijder L, Prins HHT. 2002. Spatial heterogeneity and irreversible vegetation change in semiarid grazing systems. *American Naturalist* **159:** 209–218.

Van den Akker JJH, Arvidsson J, Horn R. 2003. Introduction to the special issue on experiences with the impact and prevention of subsoil compaction in the European Union. *Soil and Tillage Research* **73:** 1–8.

Van der Putten WH. 2012. Climate change, aboveground-belowground interactions and species range shifts. *Annual Review of Ecology, Evolution, and Systematics* **43:** 365–383.

Van Dijk AIJM, Bruijnzeel LA. 2003. Terrace erosion and sediment transport model: A new tool for soil conservation planning in bench-terraced steeplands. *Environmental Modelling and Software* **18:** 839–850.

Van Oost K, Quine TA, Govers G, De Gryze S, Six J, Harden JW, Ritchie JC, McCarty GW, Heckrath G, Kosmas C, Giraldez JV. 2007. The impact of agricultural soil erosion on the global carbon cycle. *Science* **318:** 626–629.

Várallyay G. 2010. The impact of climate change on soils and on their water management. *Agronomy Research* **8:** 385–396.

Verchot L, Van Noordwijk M, Kandji S, Tomich TP, Ong C, Albrecht A, Mackensen J, Bantilan C, Anupama KV, Palm CA. 2007. Climate change: Linking adaptation and mitigation through agroforestry. *Mitigation and Adaptation Strategies for Global Change* **12:** 901–918.

Vincini M. 1999. The effect of abandoning agricultural activity on the LS factors of the erosion models in the Perino Valley: Simulation using GIS technique. *Genio Rurale* **62:** 58–64.

Virto I, Imaz MJ, Fernandez-Ugalde O, Gartzia-Bengoetxea N, Enrique A, Bescansa P. 2015. Soil degradation and soil quality in Western Europe: Current situation and future perspectives. *Sustainability* **7:** 313–365.

Von Braun J, Gerber N, Mirzabaev A, Nkonya E. 2013. The economics of land degradation. ZEF Working Papers 109. Bonn, Germany.

Voortman RL, Sonneveld BG, Keyzer MA. 2000. African land ecology: Opportunities and constraints for agricultural development. Center for International Development Working Paper 37. Harvard University, Cambridge, MA.

Wallace J. 2000. Increasing agricultural water use efficiency to meet future food production. *Agriculture, Ecosystems and Environment* **82:** 105–119.

Wang Y, Li Y. 2013. Land exploitation resulting in soil salinization in a desert–oasis ecotone. *Catena* **100:** 50–56.

Watson CA, Oborn I, Eriksen J, Edwards AC. 2005. Perspectives on nutrient management in mixed farming systems. *Soil Use and Management* **21:** 132–140.

Way SA. 2006. Examining the links between poverty and land degradation: From blaming the poor toward recognizing the rights of the poor. In *Governing Global Desertification: Linking Environmental Degradation, Poverty, and Participation*, Johnson P, Mayrand K, Paquin M. (eds). Ashgate, Burlington, VT, pp. 27–41.

Weiss CM, Fox K. 2003. European food supply chains—Are they sustainable? *Executive Outlook* **3:** 54–63.

Wild A. 1993. *Soils and the Environment*. Cambridge University Press, Cambridge, UK.

Williams J, Bui EN, Gardner EA, Littleboy M, Probert ME. 1997. Tree clearing and dryland salinity hazard in the upper Burdekin catchment of north Queensland. *Australian Journal of Soil Research* **35:** 785–801.

Wischmeier W, Smith D. 1978. *Predicting Rainfall Erosion Losses: A Guide to Conservation Planning*. U.S. Department of Agriculture, Washington, DC.

Worster D. 1979. *Dust Bowl: The Southern Plains of 1930s*. Oxford University Press, New York.

Yu K, D'Odorico P, Bhattachan A, Okin GS, Evan AT. 2015. Dust-rainfall feedback in West African Sahel. *Geophysical Research Letters* **42:** 7563–7571.

Zeng N, Neelin JD. 2000. The role of vegetation–climate interactions and interannual variability in shaping the African savanna. *Journal of Climate* **13:** 2665–2670.

Zhang XC, Nearing MA, Garbrecht JD, Steiner JL. 2004. Downscaling monthly forecasts to simulate impacts of climate change on soil erosion and wheat production. *Soil Science Society of America Journal* **68:** 1376–1385.

Zheng YQ, Yu G, Qian YF, Miao M, Zeng X, Liu H. 2002. Simulations of regional climatic effects of vegetation change in China. *Quarterly Journal of the Royal Meteorological Society* **128:** 2089–2114.

Zia MS, Muhmood T, Baig MB, Aslam M. 2004. Land and environmental degradation and its amelioration for sustainable agriculture in Pakistan. *Science Vision* **9:** 21–25.

14 The Soil–Livestock–Climate Nexus

David A.N. Ussiri and Rattan Lal

CONTENTS

14.1 INTRODUCTION

Livestock production is one of the fastest growing subsectors of the global agricultural economy, with the annual growth rates up to 5% since the 1990s (FAO 2009). Currently it contributes 40% of the agricultural gross domestic product (GDP) (FAO 2009; Herrero et al. 2013b) and accounts for 1.4% of the world's GDP. Livestock also play a crucial role in food security, livelihoods and development at large (Herrero et al. 2013b). They contribute to human nutrition and health by providing protein, while also acting as a buffer against grain shortage and assuring food security to the human population (Smith et al. 2013a). Livestock are also providers of income and employment for producers and others working in its value chain, while also being an important asset and safety net for larger part of the world's poor, especially women and pastoralist groups in dry areas (McDermott et al. 2010). Globally, more than 1.3 billion people are employed in the livestock sector (including poultry), both producers and retailers (Thornton 2010; Herrero et al. 2009; FAO 2009). Livestock also provides nourishment and financial security for 815 million food-insecure people in the world (Thornton 2010; FAO 2015; FAOSTAT 2017).

Based on the projected increase in the human population, improvement in the worldwide standard of living and shift of diets toward animal-based food such as meat and dairy products, the gross increase in meat and milk demand is estimated to increase by 70% and 80% of current levels, respectively, by 2050 (Alexandratos and Bruinsma 2012; IAASTD 2009). The projected growth in demand for meat, milk, and dairy products during 2010–2050 ranges from 0.9% to 1.8% and 1.1% to 1.6% per year, respectively (Rosegrant et al. 2009; Alexandratos and Bruinsma 2012). Global per capita consumption of livestock products has more than doubled in the past 50 years (FAO 2009; Alexandratos and Bruinsma 2012). Demand for food, climate change, and water insecurity are among the major challenges for humankind in the 21st century, and they will dominate development policy priorities. The global climate change is causing shifts in local weather patterns, especially changes in temperature and precipitation that will result in changes in land and water regimes with significant impacts to local and global agriculture productivity. To provide food security and the socio-economic benefits, the livestock sector uses a significant amount of land, water, biomass, and other resources and leaves a significant imprint on the environment (Steinfeld et al. 2006). There is increasing concern on how to manage the sector's growth so that livestock benefits can be attained at a lower environment and water footprint. More than 17 billion animals in the world are using a substantial amount of natural resources, mostly in the developing countries, where most of the growth of the sector is occurring. Livestock production occurs in a wide range of production systems including pastoral–agro-pastoral and mixed crop–livestock systems at different intensities. In developing countries, pastoral/grassland-based systems are generally practiced in the land areas with low human population densities. Mixed crop–livestock systems occur in areas suitable for both arable crop and livestock production where the bulk of rural population lives and the intensive systems are generally in peri-urban/urban areas (Herrero et al. 2009; Herrero et al. 2013a). Landless livestock production systems are also found in urban areas (Herrero et al. 2013a). All these systems in developing countries produce about 50% of the beef, 41% of the milk, 72% of the lamb, 59% of the pork, and 53% of the poultry for the global demand (Herrero et al. 2009; Herrero et al. 2013a). These shares are likely to increase because most of the future growth in livestock production is projected to occur in the developing world (Rosegrant et al. 2009).

Livestock, which has shaped agriculture to a large extent through their demand for feed, have an important impact on nearly all aspects of environment, including climate change, air quality, deforestation, soil degradation, and water quality deterioration. The sector is also one of the leading causes of the global biodiversity loss. At each step in the production process, livestock contribute to air and water pollution through release of pollutants and excessive nutrients to the environment or climate change by (greenhouse gas) GHG emissions to the atmosphere. Their environmental impacts may be either direct through soil degradation by overgrazing and pollution by animal waste or indirect by the activities such as deforestation for expansion of pastures, ranching and livestock feed production (Arima et al. 2011; Nepstad et al. 2014) as well as conversion of cropland to feed production. Livestock activities also emit considerable amounts of carbon dioxide (CO_2), methane (CH_4), and nitrous oxide (N_2O) both directly and indirectly. Direct emission comes from the respiration process of all animals in the form of CO_2, digestive process of ruminants, and to some extent monogastrics that emit CH_4, and animal manures that emits CH_4, N_2O, ammonia (NH_3), and CO_2 depending on the way it is managed and stored. Livestock also affect the carbon (C) balance of land used for pasture or feed crops, thereby releasing CO_2 to the atmosphere through organic matter (OM) decomposition, especially when forests are cleared for pastures. The changing food consumption patterns over the past 50 years toward increased consumption of animal protein, especially in developing countries and emerging economies, have increased the role of the livestock sector in environment and GHG emissions significantly in recent years (Steinfeld et al. 2006; Gerber et al. 2013).

Global assessments indicate that between 2000 and 2050, global livestock production will increase by 115% leading to 23% and 54% increases in global N and P surpluses, respectively (Bouwman et al. 2013) as well as changes in C cycling. These elements are essential for plant growth and soil fertility but excess supply results in harmful effects to the environment. Most of the excess N will be lost to the environment as NO_3^- leaching and runoff which leads to ground and surface water pollution, respectively (Bouwman et al. 2013). In addition, volatilization, nitrification, and denitrification of N compounds lead to air pollution and GHG emissions. The surplus P lost through runoff to waterways causes eutrophication (i.e., nutrient over-enrichment) and exacerbates water quality deterioration (O'Higgins and Gilbert 2014). Anthropogenic induced eutrophication is a rapidly growing environmental crisis in freshwater and marine systems worldwide (Diaz and Rosenberg 2008). Nutrients that cause eutrophication are nitrogen (N) and phosphorus (P). Phosphorus is the main freshwater pollutant, while N is generally linked with the impairment of coastal and marine waters (Selman and Greenhalgh 2010). In addition to eutrophication, N pollution also contributes to other environmental impacts such as acid rain, climate change, and local air pollution. GHG emissions are also likely to increase in the coming decades as a response to increased demand for animal source food. The increased crop and livestock production at the global scale is the major cause alteration of global N and P cycles.

The livestock sector is the largest land use system on Earth and is increasingly competing for resources while causing widespread pressures in many parts of the world (Gerber et al. 2013). Globally, livestock-based systems occupy 30% of ice-free land surface or 3,900 million hectares (Mha) of land (Steinfeld et al. 2006; Herrero et al. 2009), and account for approximately 70% of the global agricultural use (Steinfeld et al. 2006). Overall, pastoral systems, both wildlife and domestic animals, occupy 45% of the global land area (Reid et al. 2008). Likewise, about 34% of the global cropland is used for animal feed production, while livestock also consume about 32% of freshwater. Livestock systems have both positive and negative effects on natural resource base, public health, social equity, and economic growth. Due to their fast growth rates, especially in the developing world (Herrero et al. 2012), livestock systems have been linked to deforestation and have been a topic of considerable research (Fearnside 2005; Nepstad et al. 2006; Morton et al. 2006; Fehlenberg et al. 2017). Conversion of forests to cattle ranches is the largest cause of deforestation in the Brazilian Amazon (Fearnside 2008) and 70% of deforestation in the Brazilian Amazon is used for medium to large sized cattle ranches (Fearnside 2005). It is estimated that the extensive cattle enterprises have been responsible for 65%–80% of the total deforestation of Amazon (Nepstad et al. 2006; Morton et al. 2006; Wassenaar et al. 2007), which puts considerable stress on global forest ecosystems (Fearnside 2008). Forests are also cleared for growing crops such as soybean and cereals mostly to feed pigs and poultry in industrial livestock production systems and to provide a high protein source for concentrates of dairy cattle feeds (Nepstad et al. 2006; Wassenaar et al. 2007; Fearnside 2005). Current rates of forest loss to animal feed production are estimated at 0.4–0.6 Mha yr^{-1} (Wassenaar et al. 2007) and are projected to increase as the demand for pig and poultry meat increases at a faster rate than consumption of red meat (Steinfeld et al. 2006; Herrero et al. 2009). It is estimated that the combined forest loss from cattle and feedstock production accounts for 2.4 Pg CO_2 emissions globally (IPCC 2014b).

In the US, livestock production accounts for 55% of soil and sediment erosion and about 30% of total N and P loading to national drinking water resources (Selman and Greenhalgh 2010). Livestock operation remains a primary accelerator of nutrient cycling (Bouwman et al. 2009). Furthermore, nutrient surpluses and emissions of NH_4 and NO_x are associated with intensive livestock production in many regions. Meeting the increased demand for livestock sourced food will put substantial pressure on land and water resources. Similarly, climate variability and biodiversity conservation demands add intense pressure on resource use and poses development challenges for both developed and developing countries. As resources required for sustaining livestock production are strained, future increases in livestock production must be accommodated within the existing resources—land, water, and nutrients (Alexandratos and Bruinsma 2012). Improving resource use efficiency and reducing environmental impacts of livestock production are important for the sustainability of the livestock production sector.

The total global cultivated land area has not changed since 1991 (O'Mara 2012), reflecting increased productivity and intensification efforts. However, agricultural production needs to expand by 70% globally and by about 100% in developing countries by 2050 to meet the demand for the expected increase in the global population. Similarly, annual meat consumption will increase from 42 kg per person in 2013 to 52 kg per person in 2050 (28 to 44 kg $person^{-1}yr^{-1}$ in the developing countries) (FAOSTAT 2017; Bruinsma 2009). Increasing food and livestock production implies that additional land will be required for food and feeding purposes. Arable land expansion will be an important factor for crop and livestock production growth, especially in sub-Saharan Africa and Latin America.

The continuing decline of arable land area per person, increasing average food consumption, and changing climate are perceived as indicators of the impending crisis (Bruinsma 2009). Interests in reducing agriculture's negative impacts and improving the natural environment to provide or maintain ecosystem services (e.g., water, air, carbon sequestration, and soil health) are creating new agricultural paradigms. Finite natural resources including land and water, as well as competing demands for land resulting from urbanization, industrial uses, biofuel production, as well as the need to preserve natural resources for future generations are also adding more stress to agricultural production.

Implications of climate change to food security have highlighted the need for new metrics for measuring food security across local and regional contexts, leading to Expert Meetings under the IPCC in Dublin, Ireland in May 2015, highlighting the need for global change researchers to also identify the link between diminishing food security and climate change (IPCC 2015). Progress toward food security requires the availability and accessibility of food, and of sufficient quantity and quality to guarantee good nutritional outcomes. This chapter reviews the environmental impacts of livestock production with special emphasis on soil quality, nutrient losses as influenced by both grazing and manure management, and GHG emissions. In addition, the sustainable livestock management and production as well as practices that mitigate GHG emissions under livestock production systems are identified. The objectives of this review are to highlight the role of global livestock production in GHG emissions and the potential benefits of minimizing environmental and ecological footprints associated with global intensive livestock production sector.

14.2 LIVESTOCK AND GLOBAL FOOD PRODUCTION

Livestock contribute significantly to global agricultural production, food supply, nutrition, rural employment, and soil fertility, while providing income, food security, poverty reduction, and livelihoods among the rural population in developing countries (Randolph et al. 2007; Moyo and Swanepoel 2010; Sakadevan and Nguyen 2017). Its current share of the agricultural GDP is 40% globally and about 30% for the developing countries (World Bank 2009). It is one of the fastest growing agricultural subsectors with global annual meat and milk production growth rates of 2.5 and 1.6%, respectively, mostly occurring in developing countries. Some key drivers of increased animal source food are global population growth, rapid urbanization, and increasing incomes in some developing countries. The global population is projected to increase from the current 7.6 billion to 9.8 billion by 2050 (U.N. 2017). Most of this increase will occur in Asia and Africa. Rapid population growth may impede the achievement of food security in some countries even when global population ceases to grow. Urbanization is also increasing faster in the developing countries than the developed countries. For example, between 1980 and 2003, the urban population in developing countries increased at an average annual rate of 4.9% and 2.6% in sub-Saharan Africa and in Latin America, respectively, compared to 0.8% in developed countries (FAO 2009). Currently, more people live in urban areas than in rural settings, with urbanization rates varying from less than 30% in South Asia to nearly 80% in developed countries and Latin America (U.N. 2007). It is projected that by 2030, about 5 billion people will live in urban areas, with the largest growth in Africa and Asia (U.N. 2007). Urbanization is generally associated with higher average household incomes and changing lifestyles. It has considerable impact on patterns of food consumption in general and demands for livestock products in particular. Urban consumers demand more processed food—which increases the role of agribusiness (Rosegrant et al. 2009). The past trends in meat and milk consumption and future estimates are shown in Table 14.1. The trends in demand for food in developing countries have been shifting in favor of both increased quantity and improved quality.

14.2.1 LIVESTOCK AND FOOD SECURITY

A widely used definition indicates that food security is achieved when all people, at all times, have physical, social, and economic access to sufficient, safe, and nutritious food that meets their dietary

TABLE 14.1
Past and Projected Meat and Milk Consumption Trends

Category	Year	Annual per Capita Consumption (kg)		Total Consumption (Tg)	
		Meat	Milk	Meat	Milk
World	1990	30	77	175.4	404.5
	2000	33	78	285	586
	2010	41	83	296	719
	2030	45	92	319	900
	2050	49	99	460	997
Developing	1990	18	38	73	152
	2000	28	44	126	222
	2010	32	55	184	323
	2030	38	67	252	452
	2050	44	78	326	585
Developed	1990	80	200	100	251
	2000	78	202	102	265
	2010	83	203	112	273
	2030	89	209	121	284
	2050	94	218	126	295

Source: Historical data from FAOSTAT, *FAO Statistical Databases*. Food and Agriculture Organization of the United Nations (FAO), 2017; projection data from Alexandratos, N., and J. Bruinsma, Agricultural Development Economics Division, Food and Agricultural Organization of the United Nations (FAO), Rome, Italy, 2012.

needs and food preferences for an active and healthy life. As commonly used in development, food security emphasizes food quantity more than quality. Nutrition security is generally used to capture the quality dimension, since it incorporates the need for micronutrient as well as energy and protein adequacy. Livestock contribute to food security through: (i) increasing direct access to animal source food; (ii) providing cash income from the sale of livestock and livestock products which can be used to purchase food during times of food deficit; (iii) providing manure and traction needed for increasing aggregate cereal production; and (iv) improving livestock production, lowering prices of livestock products, and hence increasing access to livestock products by the poor. At the global level, livestock provide 15% of total food energy and 25% of dietary protein, while in developed countries, milk, meat, and eggs currently provide 20% and 48% of energy and protein, respectively (FAO 2009).

Many poor people depend on animal-source food products, especially dairy products, to ensure that their diets deliver the nutrients necessary for cognitive and physical development. Products from livestock provide essential micronutrients that are not easily obtained from plant-based food. Animals convert low-biological value protein foods that are less palatable and low nutrient to high biological value foods that are palatable and nutrient dense. Animal source foods are nutritionally dense sources of energy, protein, and various micronutrients, and match particularly well with nutrients needed by people to support normal development, physiological functioning, and good health in general. With nearly one billion of the poorest world population relying on livestock for their livelihoods (FAO 2012), it can be assumed that livestock-keeping households consume animal source foods they produce, and that increased productivity will have a positive impact on their household nutrition. Despite their nutrition security, overconsumption of animal source foods can harm human health and well-being through increased obesity epidemic, cardiovascular disease, and other non-communicable diseases (WHO/FAO 2003; Larson and Wolk 2012). In addition, about half of the world's production of grain is fed to animals, especially monogastrics (Ericksen et al. 2009; Ingram et al. 2010). By consuming feeds that could be consumed by people directly such as grains and legumes, livestock also reduce the amount of available food, and part of the cropland is allocated to crops used for livestock feeding purposes. Estimates indicate that mixed crop–livestock systems contribute 46% of meat, 88% of milk, and 50% of cereals, while intensive systems provide 45% of meat (Thornton and Herrero 2010).

14.2.2 DEMAND FOR LIVESTOCK PRODUCTS

The global demand and consumption of livestock products has increased steadily since the 1980s, particularly in the developing countries where the per capita consumption has outpaced growth in consumption of other major food commodity groups (Figure 14.1; FAO 2009; FAOSTAT 2017). Whereas the contribution of animal products to calorie intake per capita is much lower in developing

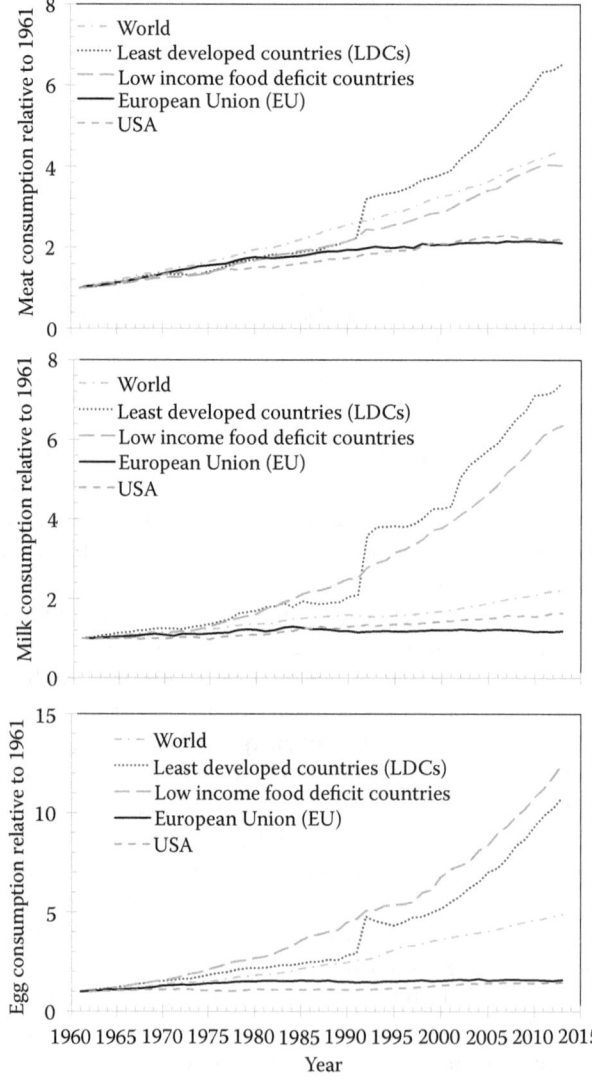

FIGURE 14.1 (See color insert.) Global meat, egg, and milk consumption 1961 to 2013. (Data from FAOSTAT, *FAO Statistical Databases*. Food and Agriculture Organization of the United Nations (FAO), 2017.)

countries compared to that of developed countries—i.e., 82 kcal person^{-1} day^{-1} (3.1% of total calorie consumption) versus 282 kcal person^{-1} day^{-1} (8.4% of total calorie consumption) in 2007 (Alexandratos and Bruinsma 2012), the trend has been an increasing one while that of developed countries is largely static (Figure 14.2; FAOSTAT 2017). Such a trend implies that consumers in developing countries have been increasing their consumption of animal products more rapidly than they have other sources such as cereal and root crops.

The global food demand and food consumption patterns, especially that of developing countries, have been shifting toward diets that favor livestock products in the past few decades (Alexandratos and Bruinsma 2012). This trend is sometimes referred to as the "livestock revolution" (Delgado 2003) due to rapidly changing global livestock production. In response to changing demand trends, the global livestock sector has increased production significantly. Beef and milk production have more than doubled since the 1970s, and monogastric production—pigs and poultry—has increased

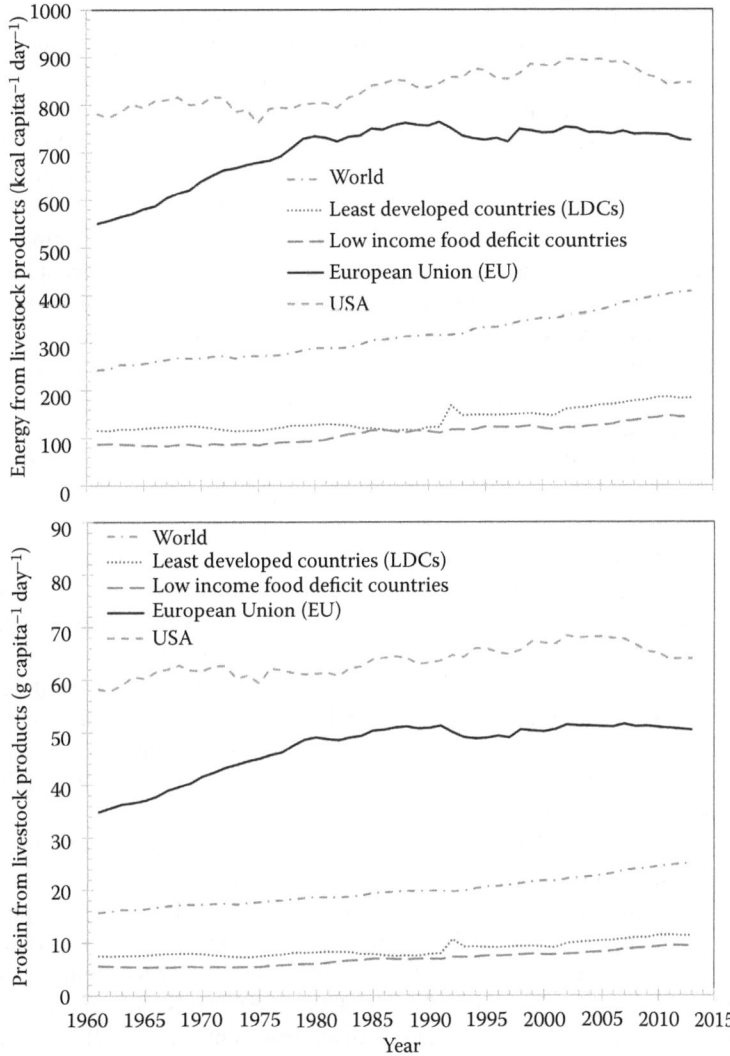

FIGURE 14.2 **(See color insert.)** Global per capita energy and protein supply from meat, eggs, and milk. (Data from FAOSTAT, *FAO Statistical Databases*. Food and Agriculture Organization of the United Nations (FAO), 2017.)

by a factor of five or more over the same period (Thornton 2010). In developing countries where nearly all world population increases take place, consumption of meat and dairy products has been increasing at an average of 5.1% and 3.6% per year, respectively, since the 1970s (Alexandratos and Bruinsma 2012).

The meat and milk consumption in developing countries increased by 3.3% and 3.9% per annum between 1990 and 2007, and is projected to increase by 2.7% and 2.1% per annum, respectively, between 2007 and 2030 (Alexandratos and Bruinsma 2012). Such an increased demand for livestock products will cause significant challenges for the sustainability of the industry including protection of land and water resources, management of manures, and reducing GHG emissions from the production chain. The csonversion of extensive livestock system to intensive and sustainable production systems will play an important role in reducing the negative impacts of livestock production to the environment (Havlik et al. 2014). In the second half of the last century, agriculture has become more intensive and shifted toward an industrialized production model, where crops and livestock production systems became increasingly specialized, potentially in the developed world (Entz et al. 2005), with more emphasis on technical efficiency, leading to significant increase in land productivity. Farms have been transformed into large-scale, highly specialized, energy-intensive operations, simplified, and concentrated to address the increasing demand for food, feed, and fiber (Kirschenmann 2007). Specialization has led to farm-level separation of crops and livestock production enterprises and specialized intensive livestock production (MacDonald and McBride 2009). This specialization also occurred in research to support technical developments in agricultural production systems (Lemaire et al. 2005). Within this model, concentrate feeds for livestock have been utilized to achieve production goals, often at the expense of environmental quality and important ecosystem services (Matson et al. 1997). Intensification of production has a played significant role in raising the output per unit of land and also per animal (Steinfeld et al. 2006). For example, in the US, 60% more milk is currently produced than in the 1940s with about 80% fewer dairy cows (Capper et al. 2009). Agricultural land expansion has also been an important component of increasing production in some areas such as Latin America and Africa, and land expansion is the main cause of deforestation and significant increase in GHG emissions as well as loss of biodiversity and other negative impacts on environment. Conventional economic wisdom suggests that the specialized intensive production systems have strong economic rewards. In contrast, challenges associated with diversification include higher labor costs, capital, and machinery requirements (Hendrickson et al. 2008; Wilkins 2008). As a result, large specialized livestock facilities focus more on producing animals and purchase most livestock feed off the farming systems. Consequently, this trend has led to a decline in available land for recycling livestock waste through cropping enterprises. For example, Gollehon et al. (2001) reported a 40% decrease in available farmland per animal unit in the US between 1982 and 1997. The short-term consequences of intensification were positive with significant increases in livestock production in most of the developed world. However, long-term consequences of intensive livestock production have not been positive, and some of the consequences include lack of sustainability and pollution through inefficient nutrients cycling and management (Lemaire et al. 2005) and increased livestock production-related GHG emissions.

The tendency of specialized livestock operations involving the purchase of a higher percentage of livestock feed requirements has led to growing imbalances in the supply of nutrients in livestock manure relative to the crop nutrient requirements in fields surrounding livestock production at the farm, watershed, and regional levels (Ribaudo et al. 2003). Intensification of agriculture has been driven by a large use of non-renewable resources that often impair environmental sustainability and simplification of production systems at all levels—field, farm, landscape, and region. Specialized livestock systems lead to negative manifestations in environment such as: (i) increasing water contamination with animal waste and excessive nutrients in the environment, (ii) decreasing groundwater levels due to high demand and competition from varieties of stakeholders; (iii) increasing GHG emissions; (iv) depleting soil organic matter (SOM); (v) degraded soils as a result of excessive tillage, salt accumulation, and pesticide inputs; and (vi) loss of biodiversity (Russelle et al. 2007).

Intensive livestock production systems have recently become the center of debate because of negative environmental effects since currently the increase in production is no longer the sole objective of agriculture production systems. Environmental regulations are becoming crucial aspects of production systems and trade markets are responding to these demands from the general public. Concerns over the environmental quality have led to renewed interest in livestock production systems with potential for protecting the environment through increased nutrients recycling and greater energy efficiency (Entz et al. 2005).

In many developing countries, particularly those in the tropics, mixed crop–livestock systems represent the backbone of smallholder production, and these systems provide 50% of the world's meat, over 90% of milk in developing countries, and 50% of cereals (Thornton and Herrero 2014). While the livestock revolution is fueling production and economic growth by providing growth opportunities for many poor rural populations involved in livestock sector (Delgado et al. 1999), it is also stretching the capacity of existing production and exacerbating environmental problems. In marginal semiarid rangelands, the impacts of climate change on agriculture—i.e., increased temperature and unpredicted rainfall—lead to frequent droughts and reduce livestock productivity. This is notably true in southern Africa and Central Asia, where droughts and frequent water deficits are common. The sustainability of using large volumes of groundwater to sustain intensification of livestock and crop production in these areas is questionable. While it is necessary to satisfy consumer demand, improve nutrition, and direct income growth opportunities to rural populations who need them most, it is also necessary to alleviate the associated environmental stress. Conventional agriculture is known to cause soil and pasture degradation because it involves intensive tillage, particularly when practiced on marginal lands. Management schemes and technologies that can enhance productivity while also preserving natural resource base are needed.

14.3 LIVESTOCK PRODUCTION SYSTEMS AND THEIR IMPACTS ON THE ECOSYSTEM

The current global demand for livestock product is met by heterogeneous production systems characterized by large variations in the scale and intensity of production, in varied agro-ecological zones, and in most cases with different production objectives. Livestock systems can be classified according to the extent of integration with crop production, the animal type, feed source, or agroecoregion. Based on feed source, livestock systems have been grouped into three main categories, namely: grazing, mixed crop–livestock, and industrial or landless systems (Sere and Steinfeld 1996; Herrero et al. 2009). Grazing systems are specific to ruminants only. In grazing systems more than 90% of the dry matter fed to animals comes from rangelands, pastures, and annual forages and less than 10% of feed comes from non-livestock farming activities. In mixed systems less than 90% of the dry matter fed to animal comes from grass, and the remainder comes from crop by-products, residues, lay crops, and feed grains (Sere and Steinfeld 1996). Mixed systems occur in all scales of production and range from extensive to intensive production, and as a category it is the least precise. Landless systems—often referred to as industrialized, confined, intensive, or grain-fed systems— are "livestock only" systems where less than 10% of the dry matter fed to animals is farm-produced, with the stocking rate exceeding 10 livestock units per hectare. Feed for landless systems are prepared commercially, consisting of cereals and oilseed proteins.

Ruminants (cattle, buffalo, sheep, and goats) produce 26% of global meat which contributes 13 g capita^{-1} day^{-1} global protein supply. Pig and poultry produce the three quarters of global meat in nearly equal amounts (FAOSTAT 2017). Animal source food contributes ~40% of all human protein intakes. Recent global livestock production data suggest that as much as 79% of global production of poultry and pig meat is produced in industrial systems, with smallholder farmers contributing a large share in areas such as Africa, South Asia, and parts of Southeastern Asia, and its share ranges between 40% and 55% (Herrero et al. 2013b). Mixed crop–livestock systems are important for ruminant production in both developed and developing regions. They produce most of the global ruminant meat and milk producing 69% and 61% of global milk and ruminant meat, respectively (Herrero et al. 2013b). Extensive grazing systems contribute significantly to the production of beef and small ruminant meat (Herrero et al. 2013b; Gerber et al. 2013). Grazing systems contribute significantly to beef in Latin America, Oceania, and North and Sub-Sahara Africa.

Traditional livestock production systems are generally resource-driven, and therefore they make use of locally available resources with limited alternative uses. For example, extensive grazing systems and mixed crop–livestock systems use land that is not suitable for crops or other uses and crop residues, respectively. In addition, traditionally managed mix production systems often provide valuable organic fertilizer inputs to crop production and maintain close integration between crop and livestock production systems. In contrast, modern industrial production systems have no direct link to local resource bases, since most of them are based on purchased feed rather than grown onsite.

Three major trends relating to global pastures are: (i) valuable ecosystems are being converted to pastureland (e.g., forest clearing); (ii) grassland are also being converted to other uses such as cropland, urban areas, and forest; and (iii) pasturelands are prone to degradation. Ranching remains a main cause of deforestation in Central and South America (Wassenaar et al. 2007). Grasslands are also becoming increasingly fragmented and encroached by cropland and urban settlements. It is estimated that more than 90% of North American tall grass prairie and about 80% of South America cerrado have been converted to cropland and urban land uses (Wassenaar et al. 2007). Globally, about 10%–20% of global pastures and rangelands have been degraded to some extent by overgrazing, and the proportion may be as high as 70% in dry areas, threatening food security and environmental quality in dryland (MEA 2005; Ravi et al. 2010). Some environmental consequences of pasture and grassland degradation include soil erosion, loss of quality and quantity of vegetation, release of C from OM storage in soils, reduced biodiversity, and interference with nutrients and water cycles. In addition, intensive feed crop production leads to severe land degradation, water pollution, and loss of biodiversity. Similarly, expanding arable land into natural ecosystems results in severe ecological consequences such as loss of biodiversity and ecosystem services such as regulation of water and erosion control. Although much of the increase in grain production has been achieved through intensification on existing arable land, most of the rapid increase in soybean production to meet the demand for intensive livestock production has been achieved through the expansion of cropland into natural habitats.

14.4 MANAGING THE IMPACTS OF INTENSIFICATION AND DEMAND FOR LIVESTOCK PRODUCTS TO THE ECOSYSTEMS

Technical options that could lessen the impact of intensive livestock production include: (i) adoption of agricultural practices that can reduce fertilizers and pesticide use in feed cropping and intensive pasture management; (ii) integration of ecological production systems and technologies to restore important soil habitats and reduce land degradation; (iii) improvement of extensive livestock production systems to contribute to biodiversity conservation—including silvipastoral and flexible grazing management systems that can increase biodiversity, quality of forage, soil cover, and SOC storage, thereby reducing water loss by runoff, soil erosion, and drought impacts and increase CO_2 sequestration in pasture and rangelands; (iv) conservation and restoration of ecological infrastructures at watershed levels as a way to reconcile the conservation of ecosystem function with expansion of agriculture; and (v) shifting toward pig and poultry that has better feed conversion ratio than ruminants. Under both industrial and mixed production systems there is a large gap between the current levels that are technically attainable, indicating that considerable efficiency gains can be realized through the adoption of best management practices (BMPs). However, meeting future demand for livestock products will require further improvements in livestock and land productivity as well as expanding feed production area, mostly at the expense of grassland and natural habitats (FAO 2009).

14.5 LIVESTOCK AND THE ENVIRONMENT

The past decade has witnessed significant research on livestock and the environment, generally driven by the recognition that the livestock sector is large, and has been growing at an accelerated rate, which is expected to continue for the foreseeable future. As a result, livestock production increasingly competes for resources while impacting the widespread range of environmental parameters in large parts of the world (Steinfeld et al. 2006; Herrero and Thornton 2013; Herrero et al. 2011, 2015; Gerber et al. 2013). Much of the effort in traditional production concentrated on increasing productivity per animal and/or per hectare. But the rising demand for livestock products is changing the relationship between livestock and natural resources. Modern industrial production systems are losing direct links with the local resource base, while continued growth in the livestock sector will exacerbate pressures on environment and natural resources, calling for approaches that allow for increased production while also lowering the environmental burden. The goal is currently shifting to sustainable intensification—i.e., to increase production from existing farmland while placing less pressure on the environment and also without undermining future production capacity (Garnett et al. 2013). In livestock production systems, the aim is to reduce impacts per unit of animal product generated. Livestock production has often been subjected to substantial scrutiny due to its impacts on environmental quality. In the process of providing social and economic benefits, livestock production uses large land areas, while also consuming significant amount of natural resources.

14.5.1 RESOURCES DEMANDS AND USE

Global grasslands cover 61.2 million km^2 or 45% of ice-free land which occurs across all biomes from arid to humid and with varying soil fertility. It encompasses many regional variations and includes savannahs, prairies, cerrado, steppe, llanos campos, pampas and shrub lands (Asner et al. 2004; Reid et al. 2008) and particularly dominates the lands outside the tropics. Grasslands therefore cover more land surface than other land types, occupying about 1.5 and 2.8 times more than that under global forests and cropland areas, respectively, and 17 times more than global settlement area. About two-thirds of all grasslands are in developing countries.

Livestock production is the largest land use system on the Earth, occupying 30% of the world's ice-free land surface (directly and indirectly). The land area utilized for livestock grazing is 33.8 million km^2 or about 26% of ice-free terrestrial land surface (Table 14.2; Steinfeld et al. 2006; FAO 2009). In addition, 500 Mha or about 33% of arable land is dedicated to feed and fodder crop production (Table 14.3; FAO 2009; Herrero et al. 2013b). The quality of this land and intensity of its use vary significantly, but most of it is used by extensive ruminant grazing systems. Some of the area used for grazing is too dry for cropping and is sparsely inhabited. Although extensive-grazing based systems have less contribution to the global food supply, they make important contributions

---

TABLE 14.2
Land Use Area by Country Group in 2010–2015

Countries Grouping	Pasture Area (Mha)	Pasture Share of Total Land Area (%)	Arable Land Area (Mha)	Arable Land Share of Total Land Area (%)	Forest Area (Mha)	Forest Share of Total Land Area (%)
Developed	1083.4	20.5	649.2	10.9	1829.0	34.7
Developing	2296.8	29.7	834.9	10.8	2170.2	27.3
World	3378.2	26.0	1484.1	11.1	3999.1	30.3

Sources: FAO, Food and Agriculture Organization of the United Nations (FAO), Rome, Italy, 2009; FAOSTAT, *FAO Statistical Databases*, Food and Agriculture Organization of the United Nations (FAO), 2017.

TABLE 14.3

Estimated Use of Important Land Physical Resources by Livestock Production in 2000

| | Land Use Type | | | |
| | Cropland | | Permanent Grassland | |
Livestock Group/Productivity	Grazing	Mixed	Grazing	Mixed
Dairy (million hectares)	2.0	130	560	400
• Productivity (ha Mg^{-1} protein)	6		50	
Ruminant meat (million hectares)	8.0	80	1,610	810
• Productivity (ha Mg^{-1} protein)	9		40	
Pigs and poultry (million hectares)	280	NA	NA	NA
• Productivity (ha Mg^{-1} protein)	10		NA	
Total land use area (million hectares)	500		3,380	

Sources: FAO, Food and Agriculture Organization of the United Nations (FAO), Rome, Italy, 2009; Herrero, M., P. Havlik, H. Valin, A. Notenbaert, M.C. Rufino, P.K. Thornton, M. Blummel, F. Weiss, D. Grace, and M. Obersteiner, *Proc. Natl. Acad. Sci. U.S.A.*, 110 (52), 20888–20893, 2013; Herrero, M., S. Wirsenius, B. Henderson, C. Rigolot, P. Thornton, P. Havlik, I. de Boer, and P. Gerber, *Annu. Rev. Environ. Resour.*, 40:177–202, 2015.

to livelihoods and sociocultural interactions. They also play a significant role in pasture degradation and land use change. The poor feed quality in developing countries where extensive livestock production is dominant means that livestock are mainly fed with nutrient deficient grasses and crop residues. As a result, the amount of feeds consumed by livestock in these resource poor countries can be as high as 10 times more than those in developed countries to produce the same amount of protein. Most of the global ruminants (beef, lamb, goat, and dairy) are produced from mixed systems. For example, in 2000, mixed crop–livestock systems produced 69% of global milk and 61% of global ruminant meat (Herrero et al. 2013b). In some countries, up to 85% of agricultural land is used for livestock production and contributes to significant proportion of GDP. Subsistence farming land in many developing countries across the world is part of land under livestock production (Steinfeld et al. 2006; Erb et al. 2007; Ramankutty et al. 2008; Herrero et al. 2013a). Management practices, productivity per hectare, and use of pastureland vary widely. In arid and semiarid rangelands where most of the global grasslands occur, intensification of pastures is generally not feasible or profitable (FAO 2009).

The global expansion of the livestock sector is one of the major drivers of the global land use change (Geist and Lambin 2002). Livestock production is the main contributor to deforestation, especially in Latin America, sub-Saharan Africa, and Southeast Asia, where the greatest deforestation has been occurring (Steinfeld et al. 2006; Wassenaar et al. 2007; Pan et al. 2007). As much as 70% of the previously deforested land in the Amazon is used primarily for cattle ranching, while feed crops and regeneration of secondary forest and subsistence farming cover large parts of the remainder (Fearnside 2008; de Toledo et al. 2017). For example, from 2010 to 2016, an estimated 43,000 km^2 of the Amazonian forest was cleared, and most of the cleared land is used for agriculture, ranching, and soybean cultivation for livestock feed (INPE 2017). In tropical Latin America, there is rapid expansion of pastures into some of the most vulnerable ecosystems, and 0.3% to 0.4% of forest loss to pastures is occurring annually. In the Amazon, cattle ranching and soybean production have been the major causes of deforestation for the past two decades (Steinfeld et al. 2006; Nepstad et al. 2014). Between 1990 and 2014, land area devoted to soybeans in South America tripled from 18 to 56 Mha, making it the largest area for asingle crop, far more than corn which ranks the second at 24 Mha (FAOSTAT 2017). Similarly, in Indonesia about 1.0 Mha of tropical forest has been cleared for agriculture (Hansen et al. 2009). It is estimated that throughout the tropical areas, about 12.0 and 9.0 Mha of forest were cleared in 2010 and 2015, respectively (Houghton et al.

2012; FAOSTAT 2017). Expansion of pastures into forest ecosystems has dramatic environmental consequences including biodiversity loss, stored soil C release to the atmosphere fueling climate change, impact on water cycles through reduced infiltration and water storage, and increased run-off. Over the past 300 years, grazing land expanded six-fold globally, including in North America, South America, and Australia, the continents where there had been little or no livestock grazing previously (Asner et al. 2004). Livestock production is projected to be the main land use replacing forest in the tropics after land clearing, and it is estimated that the expansion of pasture into forest is greater than that of cropland (Wassenaar et al. 2007). Therefore, the global livestock sector plays a significant role in human modification of Earth's surface with substantial demands being placed on food production worldwide as a result of recent growth in demand for animal feed due to livestock intensification and commodification (Steinfeld et al. 2006).

Feed links livestock to land use, both directly by grazing and indirectly through the purchase of grain or forage. Livestock feed can be grouped into grasses—either as directly grazing or hay and silage, grains—generally fed as concentrates, and other feed such as cut-and-carry forages, legumes, and crop stover. Globally, livestock consumed 5100 Tg yr^{-1} of feed biomass, with ruminants consuming bulk of feed biomass (~4,100 Tg DM) compared to pigs and poultry (~1000 Tg DM) (Table 14.4, Herrero et al. 2013b). Livestock consumed 1.3 Tg of grains in 2000. As much as 78% of feed grain is fed to pig and poultry (Table 14.4), mostly in regions where industrial systems are dominant. Overall, 3400 Tg DM yr^{-1} of biomass utilized for ruminants feed (more than 80% of annual dry matter used by ruminants feed) is currently produced under mixed crop–livestock production systems (Table 14.4, Herrero et al. 2015). Similarly, most of the global ruminant output comes from mixed systems which involve significant use of cropland-produced feed such as grains, hay, and silage. Averaged globally, the cropland use per unit protein output from ruminant meat under mixed crop–livestock systems is similar to that of pork and poultry (Table 14.3). Furthermore, intensive livestock production systems such as those in the US and Europe and cropland use per unit of

---

TABLE 14.4

Global Feed Consumption (Tg Dry Matter yr^{-1}) by Livestock Groups

Livestock Group	Grass and Legumes	Stover and Straw	Grains	Other Feed	Total Feed	Protein (kg DM kg^{-1})
Dairy cattle						60
• Grazing	170	1.6	4.6	12	188	–
• Mixed	620	230	170	140	1,160	–
Other ruminants						280
• Grazing	500	6.7	24	81	612	–
• Mixed	1,200	400	130	410	2,140	–
Total Ruminants	2,490	638	329	643	4,100	–
Pigs (total)	NA	NA	537	NA	537	
• Large scale (industrial)	–	–	469	–	469	–
• Smallholder	–	–	68	–	68	–
Poultry (total)	NA	NA	476	NA	476	
• Large scale (industrial)	–	–	400	–	400	–
• Smallholder	–	–	76	–	76	–
Total Pigs and Poultry	NA	NA	1013	NA	1,013	30
Total Livestock Feed	2,262	572	1,299	559	5,113	–

Sources: Bouwman, A.F., K.W. Van der Hoek, B. Eickhout, and I. Soenario, *Agric. Syst.*, 84 (2), 121–153, 2005; Herrero, M., P. Havlik, H. Valin, A. Notenbaert, M.C. Rufino, P.K. Thornton, M. Blummel, F. Weiss, D. Grace, and M. Obersteiner, *Proc. Natl. Acad. Sci. U.S.A.*, 110 (52), 20888–20893, 2013; Herrero, M., S., Wirsenius, B. Henderson, C. Rigolot, P. Thornton, P. Havlik, I. de Boer, and P. Gerber, *Annu. Rev. Environ. Resour.*, 40:177–202, 2015.

protein produced by beef systems are several times higher than that of pork and poultry (Wirsenius 2003). This contradicts the commonly held traditional notion (Peralta et al. 2014; Bradford 1999) that beef does not have an impact on the global food supply because it uses only inedible feed and/ or land resources with little or no food production value.

Land degradation is widely recognized as a global problem with implications to agronomic productivity and environment as well as its effects on food security and quality of life (Blaikie and Brookfield 2015). The impact of land degradation is manifested in chemical, physical, and biological properties of soil. Agricultural land degradation reduces productivity leading to expansion of agricultural land into natural habitats. While livestock contribute valuable nutrients to crops, both extensive and intensive livestock production systems contribute to soil degradation, and the resulting lower productivity of the soil causes natural habitats to convert to pastures and cropland and also extensive use of fertilizers. Poor and uncontrolled grazing can cause loss of ground cover due to trampling. Overgrazing and nutrient mining are among the major causes of grassland degradation. Overgrazing reduces infiltration, soil moisture and fertility, and accelerates runoff and soil erosion, which is a result of the mismatch between the capacity of the pasture to be grazed and trampled and the livestock density. Grazing livestock play significant role in desertification—in arid climates, increased woody plant cover—semiarid, subtropical rangelands, and deforestation—in humid climates. Globally, ~21% of permanent grassland is affected by human-induced soil degradation. Relatively large proportion of grasslands in Europe (36%) and Africa (31%) are degraded (Oldeman 1994). Degradation of grasslands particularly in the arid and semiarid environments of Africa and Asia and the sub-humid zones of Latin America remains an ongoing process. The livestock production sector can impact the environment through land use change and land management—deforestation, plowing grassland, or land degradation (i.e., from overgrazing), as well as water resources. However, full analysis and the extent of water resources use under the livestock sector remain scanty.

Other important environmental resources for livestock production include water resources and N, and ruminant meat production places approximately the same equivalent pressure to the environmental resources and crop production as pork and poultry (Herrero et al. 2015). Global livestock is a leading consumer of water, consuming about one-third of freshwater (Herrero et al. 2013b). Depending on the production method and climatic conditions, 1–5 m^3 are needed to produce 1.0 kg of grain, while 5–20 m^3 will produce 1.0 kg of livestock products (Chapagain and Hoekstra 2003). Ruminant production uses water intensively for drinking, forage and feed crop production, waste disposal, cleaning, and processing of animal products. Water is also essential for physiological and biochemical processes of the animals—such as thermal regulation, growth, reproduction, and lactation. Estimated annual consumption of blue water (surface and groundwater evaporated or used directly to make a product), green water (rain water evaporated or used directly to make a product), and N under different livestock management systems is summarized in Table 14.5. The global livestock sector uses 10% of anthropogenic freshwater for drinking, servicing, processing such as making meat and milk products, and irrigation, while an additional 15% is lost by the feed crops through transpiration (Steinfeld et al. 2006; Doreau et al. 2012). At the global scale, the livestock sector is estimated to utilize 10% of the global annual rainfall, which is 25%–32% of the total agricultural water use (Steinfeld et al. 2006; Hoekstra 2009). Overall, irrigation of feed crops is currently considered to be of minor importance at the global scale (Steinfeld et al. 2006). The water demand for livestock production is influenced by several factors including animal type, animal activity, the feed, and the quality of water available for livestock. For example, grain-fed beef production takes about 50 times the water required to produce the grain; raising broiler chicken takes 3500 L of water to produce 1 kg of meat. Unsustainable use of freshwater for feed production, animal care, and slaughterhouses contributes to water scarcity while also depleting water resources in many agricultural landscapes (Burkholder et al. 2007; Kraham 2017). In general, the water-use footprint of livestock is greater than that of crops with equivalent nutritional value (Table 14.5, Mekonnen and Hoekstra 2012; Legesse et al. 2017).

Resource use efficiency is often linked to environmental impacts. For example, low N use efficiency is closely linked with high N_2O emissions, NH_3 volatilization, and NO_3^- leaching

TABLE 14.5

The Global Estimated Consumption of Water and Nitrogen by Livestock Sector under Different Management Systems and Water Footprint for Selected Food Products

	Water Use			
Livestock and Production System	Green Water	Blue Water	Protein (m³ kg⁻¹ Protein)	Nitrogen in Feed (Tg N yr⁻¹)
Dairy (km³ yr⁻¹)			23	
• Grazing	2.3	45		3.6
• Mixed	21	450		21
Ruminant meat (km³ yr⁻¹)			84	
• Grazing	5.1	220		12
• Mixed	33	620		38
Pigs and poultry (km³ yr⁻¹)	65	590	23	26
Livestock products (m³ Mg⁻¹)				
• Beef	14,414	550	112	
• Pig meat	4,907	459	57	
• Poultry	3545	313	34	
• Sheep & goat meat	8253	457	63	
• Milk	863	86	31	
Cereals	1232	228	21	
Pulses	3180	141	19	

Sources: Herrero, M., S. Wirsenius, B. Henderson, C. Rigolot, P. Thornton, P. Havlik, I. de Boer, and P. Gerber, *Annu. Rev. Environ. Resour.*, 40, 177–202, 2015; Legesse, G., K.H. Ominski, K.A. Beauchemin, S. Pfister, M. Martel, E.J. McGeough, A.Y. Hoekstra, R. Kroebel, M.R.C. Cordeiro, and T.A. McAllister, *J. Anim. Sci.,* 95 (5), 2001–2018, 2017; Mekonnen, M.M., and A.Y. Hoekstra, *Ecosystems*, 15 (3), 401–415, 2012.

(Bouwman et al. 2013). It is estimated that animal feed production also uses a similar fraction of fertilizer use as that used for crops. Additionally, N in animal manure generated by livestock production is generally in excess of fertilizer use, and therefore global livestock production generally drives the nutrient cycling in total agroecosystems (van Horn et al. 1996; Wilkerson et al. 1997). Livestock manure is a valuable source of nutrients for improving soil fertility and soil quality through enhancing SOC storage. It is estimated that global animal manure produces an equivalent of 128 Tg N yr⁻¹ and 24 Tg P yr⁻¹ (Potter et al. 2010). In the United States, nearly 7.5 Tg N yr⁻¹ and 2.3 Tg P yr⁻¹ are generated from manure compared to 9 Tg N yr⁻¹ and 1.6 Tg P yr⁻¹ applied to agricultural land in the form of fertilizers (UF/IFAS 2017). Manure is also responsible for nutrient pollution in both surface and groundwater. Animal diet composition and quality are the main determinants of productivity and feed-use efficiency, and together with the animal characteristics, such as body weight and physiological state, they largely regulate feed intake, animal productivity, CH_4 emissions, and manure and urine output and composition (Herrero et al. 2013b). Globally estimated average feed conversion efficiency is currently estimated at 1%, 7%, and 10% for ruminant meat, dairy, and pork and poultry, respectively (Herrero et al. 2013b). The inherently low feed efficacy of ruminant meat and dairy is mainly due to comparatively low reproductive rates of cattle, sheep and goats.

Grazing animals also affect soil physical and chemical properties. For example, in both temperate and tropical regions, grazing animals significantly increase soil erosion under pastures (Pilon et al. 2017). In addition, poor and uncontrolled grazing increases the loss of vegetative cover due to trampling and grazing plants too close to the soil, which causes soil compaction and reduces soil quality. Overgrazing in marginal lands of Africa is the main cause of soil erosion and degradation (Lal 1990). Therefore, the livestock sector has many positive and negative qualities, and the roles played by livestock change depending on location and existing circumstances. There has been growing recognition that improving environmental performance and reducing livestock environmental footprint through establishing sustainable management are essential for the sustainability of global food system (Foley et al. 2011; McMichael et al. 2007; Herrero et al. 2013a).

14.5.2 IMPACT OF CLIMATE CHANGE ON LIVESTOCK SECTOR

The radiative forcing of GHG is likely increasing at a faster rate during the industrial era than any other time in the past 10,000 years, because of the increase in global abundance of CO_2, CH_4, and N_2O resulting in current and ongoing global and regional climate change with potential impact in various economic sectors including agriculture and livestock sectors (IPCC 2013, 2014a). Significant changes in physical and biological systems have already occurred in all continents and in most oceans, and most of these changes are in the direction expected with warming temperatures (Rosenzweig et al. 2008). It is estimated that changes in climate over the last 30 years have already reduced global agricultural production by 1% to 5% per decade (Thornton et al. 2015). However, unlike cropping systems, currently there is only limited evidence for recent impacts on livestock production systems (Porter et al. 2014). Climate change has negative impacts on forage quality and hence on livestock productivity in both high and low latitudes. For example, in much of Africa where millions of smallholder farmers depend on livestock based systems, climate change has cascading impacts on incomes and food security. Livestock are the most important risk management resource for about 170 million people in sub-Saharan Africa (Robinson et al. 2011). The potential impacts are primarily due to increases in atmospheric CO_2 and global temperature, climate variability and the associated precipitation variation, and a combination of these factors. Future projections indicate widespread challenges and negative impacts on animal agriculture associated with climate change. Generally, broad trends will be overshadowed by local differences, since the impacts of climate change are likely to be highly variable (Thornton et al. 2009). Climate change will alter the distribution of hungry people with particularly large negative effects in sub-Saharan Africa and South Asia. Smallholder and subsistence farmers and pastoralists will suffer complex localized impacts of climate change due to both constrained adaptive capacity in many places and additional climate related processes such as sea level rise and snow-pack decrease in areas such as Indo-Gangetic Plains (IPCC 2013).

Although there may be some benefits arising from global climate change in some regions such as potential increase in crop yields in northern Europe and lengthening of forage growing season in in North American cattle systems (Olesen and Bindi 2002; Gornall et al. 2010; Hatfield et al. 2011),

TABLE 14.6
Direct and Indirect Impacts of Climate Change on Livestock Production Systems

Impact	Grazing	Non-Grazing
Direct impacts	- Increased frequency of extreme weather events	- Change in water availability
	- Increased frequency and magnitude of drought and floods	- Increased frequency of extreme weather events
	- Productivity losses due to temperature increase (physiological stress)	
	- Change in water availability	
Indirect impacts	• Agroecological changes and ecosystem shifts:	• Increased resource prices
	• Alteration of fodder quality and quantity	• Disease epidemics
	• Diseases epidemics	
	• Changes in host-pathogen interactions	• Increased animal housing costs—e.g., cost of cooling systems
	• Increased incidence of emerging diseases	

most livestock producers will face serious problems associated with global warming (Thornton et al. 2009). Climate change impacts on livestock can be either direct or indirect (Table 14.6), and include effects on forage and feed, direct impacts of changes in temperature and water availability on animals, and indirect effects associated with livestock diseases. Impacts on feed crops and grazing systems include (i) changes in herbage growth due to changes in atmospheric CO_2 concentration and temperature, (ii) changes in composition of pastures (e.g., ratio of grasses to legumes), (iii) changes in herbage quality including water-soluble carbohydrates and N at a given dry matter yields, (iv) increased frequencies of drought that may offset any DM yield increases, and (v) greater intensities of rainfall which increase N leaching in some ecosystems (Thornton et al. 2009; Hopkins and Del Prado 2007; Chapman et al. 2012; Polley et al. 2013; Weindl et al. 2015). Increases in the maximum temperatures in tropical dry regions can lead to severe yield reductions and reproductive failure in crops such as maize; for example, each degree day spent above 30°C reduces yield by 1.7% under drought conditions (Lobell et al. 2011; Schauberger et al. 2017). Temperature also affects most of the critical factors of livestock production, such as water availability, animal production, reproduction, and health. Global agriculture is the largest water consumer at 70% of fresh water resources (Thornton et al. 2009). Due to water scarcity and depletion, it is projected that 64% of the world population may live under water-stressful conditions by 2025 (Rosegrant et al. 2009; Ercin and Hoekstra 2016). The livestock sector uses water for animal drinking, feed crops, and product processes, and accounts for 8% of global human water use. Increases in temperature will increase animal water consumption by a factor of 2 to 3 (Nardone et al. 2010). In addition, water availability will impact livestock production through pasture and forage crop quantity and quality, feed-grain production price, and pest and disease distributions (Henry et al. 2012).

The negative effects of increased temperature on feed intake, reproduction, and range of different livestock species performance is well documented (Porter et al. 2014; Wall et al. 2010). All animals have a range of ambient environmental temperatures that are beneficial to physiological function generally termed as the thermal comfort zone. Above the critical range of the temperature, animals suffer heat stress resulting in reduced feed intake, increased water intake, and altered physiological functions such as reproductive and productive efficiency and changes in respiration rate (Nardone et al. 2010; Thornton et al. 2009). In cattle, for example, one of the major causes of reduced production in the dairy and beef industry is heat stress (Nardone et al. 2010), and significant economic losses in the US livestock industry ($1.69 to 2.36 billion) have been attributed to heat stress, of which 50% occurs in the dairy industry (St-Pierre et al. 2003; Havstad et al. 2016). In cattle, feed intake reduction leads to negative energy balance and reduced weight gain (Das et al. 2016). Heat stress also affects the reproduction efficiency of both livestock sexes. Effects in cows and pigs include impairment of embryo development and pregnancy rates (Nardone et al. 2010), and lower sperm concentration and quality in bulls, pigs, and poultry (Ayo et al. 2011; Naqvi et al. 2012). Other potential impacts on livestock production include health (Thornton et al. 2009; Nardone et al. 2010), animal growth, milk production, animal diseases, reproduction, biodiversity (Rojas-Downing et al. 2017), mortality (Howden et al. 2008), and biodiversity (Steinfeld et al. 2006; Thomas et al. 2004). In addition to body functions and metabolic rate impact, greater energy deficits affect cow fitness and longevity (Thornton 2010; Roland et al. 2016). Climate change is one of the direct drivers of the biodiversity loss (Armenteras and Finlayson 2012). It is projected that climate change may eliminate 15% to 37% of all species in the world (Thomas et al. 2004; IPCC 2014a). Animal production systems will also be affected by climate change policy and the national targets to address GHG emissions, since the livestock production sector is estimated to contribute significantly to agricultural GHG emissions.

14.5.3 ROLE OF LIVESTOCK SECTOR ON CLIMATE CHANGE

Being unique among economic sectors, agriculture and its production processes involve direct weather inputs—solar radiation available to plants, temperature, and precipitation. Climate change alters the weather patterns such as frequency of extreme temperatures and precipitation events

(Porter et al. 2014), and therefore has a direct biophysical effect on agricultural productivity. In addition to temperature, precipitation, and transpiration regimes, plant development, growth, and productivity of crop and pasture species will also respond to increases in atmospheric CO_2 concentration. Other effects of climate change on crop and pasture growth and yield include weed, pests, and pathogen pressure. Free-air CO_2 enrichment (FACE) experiments have shown that elevated CO_2 concentrations stimulate photosynthesis leading to increased plant productivity under optimum temperature, nutrients, and water regimes (Kimball et al. 2002; Kimball 2016; Vanuytrecht and Thorburn 2017). However, increased temperature during growing periods, as well as impacts of variable climate and increased frequency of extreme events, often limit the direct effects of CO_2 on crops and pastures (Caldwell et al. 2005; Baker 2004; Xiao et al. 2005; Bishop et al. 2014).

Traditional livestock production systems are resource-driven such that they use locally available resources with limited alternative uses—such as crop residues and extensive grazing lands that are not suitable for crop production or other uses. These systems convert waste material and other resources of limited alternative use into edible products and goods and services. Its size and environmental impacts are relatively limited. A large part of the livestock sector remains resource-driven where millions of pastoralists and smallholders depend on livestock for their daily survival, income, and food. The traditional systems are often difficult to intensify and typically suffer from lack of infrastructures and market barriers.

The rising demand for livestock products is changing the relationship between livestock and natural resources. As a result, over time the livestock sector has become increasingly demand-driven, its growth has been faster, and it now competes for natural resources with other sectors, and therefore causes widespread environmental pressures in many global areas (Gerber et al. 2013). Given the current and future projected increased consumption of animal source protein—meat, milk and their products, and eggs, resulting from human population growth, increasing incomes, and urbanization during the second half of the 20th and first half of 21st centuries (Alexandratos and Bruinsma 2012), the livestock sector has been growing at an accelerated rate, and is expected to continue growing in a phenomenon which has been termed as the livestock revolution (Delgado et al. 1999). As a result, environmental impacts associated with livestock production have become greater, and the sector is currently known to be resource-hungry. Intensive livestock production tends to cluster in locations with cost advantage—often close to cities or ports where insufficient land is available for the recycling of animal waste, leading to nutrient overload and pollution. While livestock production has been able to keep pace with human demand and address some of the food security concerns, its expansion has significant implications for the environment. Of the grazing areas, 2300 Mha (67%) are in the developing world, and over the last 40 years, land under pasture or grazing lands has expanded by 330 Mha (FAO 2015), mostly in Latin America, and it is projected to increase by 100 to 120 Mha by 2050 (Smith et al. 2010). Croplands are also projected to expand by 190 Mha by 2050 to supply additional feed for monogastric livestock and also intensive ruminant production to meet the demand for changing diet (Smith et al. 2010). Most of the expansion has occurred at the expense of forests (i.e., deforestation). Land use and land cover change have significant impacts on environment, biodiversity, biogeochemical cycles, hydrological processes, land dynamics as well as regional climate patterns and land dynamics, and by using the land, humans alter many patterns and processes.

Between 1980 and 2000, 83% of agriculture land expansion in the tropics occurred at the expense of forests, and livestock were a major contributor (Gibbs et al. 2010). Livestock contribute 80% of all agricultural non-CO_2 emissions (Tubiello et al. 2013), and this makes the sector responsible for about 12% of all anthropogenic GHG emissions (Westhoek et al. 2011). Satisfying the future food demand using a business-as-usual model of production will likely lead to serious environmental impacts. Therefore, either consumption has to be reduced or considerable productivity gains must be achieved if the sustainability of the global food system is to be increased (Bouwman et al. 2005; Wirsenius et al. 2010). Quantitative assessments of the effects of land use and anthropogenic GHG emissions caused by human diets with reduced meat content agree on the expected benefits from such changes in consumer preferences (Popp et al. 2010; Wirsenius et al. 2010), although productivity improvements rather than reduced consumption are preferred options which historically have led to equilibrium in agricultural markets (Evenson and Rosegrant 2003).

The impacts of climate change on livestock systems have not received as much attention as those of crops. Significant changes in physical and biological systems associated with climate change have already occurred on all continents and oceans as a result of warming global temperature (Hansen and Stone 2016; Rosenzweig et al. 2008). The impacts of global temperature increase on food production will be far-reaching. For example, climate change will alter the regional distribution of hungry people and cause significant impact on food security, with large negative impacts in sub-Saharan Africa. Smallholder and subsistence farmers, pastoralists, and artisanals will suffer complex, localized impacts of climate change, due to both constrained adaptive capability in many places and to other climate-related processes. Some of the impacts on the livestock system include changes in productivity, quality and quantity of rain-fed forage, reduced water availability and associated widespread water shortages, heat stress, changes in severity and distribution of livestock diseases and vectors, and biodiversity losses (Thornton et al. 2009). For example, climate change will influence the productivity of rangelands and yields of feed crops (Thornton and Gerber 2010). Also, heat stress directly impairs production—meat, milk, egg yield and quality, as well as reproductive performance and animal health and welfare (Thornton et al. 2009). In addition, climate change may affect the quantity of crop residues which are key dry-season feed resources for ruminants, providing as much as 60% of the diet of ruminants under mixed crop–livestock systems in developing countries (Thornton and Herrero 2014). Climate change will have severe negative impacts in many parts of the tropics and subtropics even with small increases in temperature. Therefore, more extensive adaptation is needed to reduce the vulnerability to future climate changes.

14.5.4 GREENHOUSE GAS EMISSIONS

Using the current 100-year global warming potential (GWP-100), the annual GHG emissions for the year 2010 were estimated at 49.5 Pg CO_2 eq yr^{-1}, and the shares of CO_2, CH_4, N_2O, and combined fluorinated gases (F-gases) are 76, 16, 6, and 2%, respectively (Victor et al. 2014). Agriculture, forestry and land use (AFOLU) accounts for 24% (11.9 Pg CO_2 eq yr^{-1}) of global anthropogenic GHG emissions (Victor et al. 2014) and the share of the sector's total emission is predicted to grow to 36% by 2030 (USEPA 2012). The agricultural sector is the largest contributor to the global anthropogenic non-CO_2 GHG emissions, accounting for 54% of non-CO_2 emissions globally (USEPA 2012).

The livestock sector has come into focus in recent years because of its large environmental footprint. The sector is also growing faster and increasingly competing for natural resources such as land and water, while causing widespread environmental pressures in many parts of the world (De Haan et al. 1997; Gerber et al. 2013). A study of the United Nations Food and Agriculture Organization (FAO) which assessed the full impact of the global livestock sector on the environment by taking into account the entire production chain provided aggregate perspectives on the role of livestock on environment. It revealed significant impacts on all aspects of environment, including GHG emission and climate change, air pollution, water supply and water pollution, land degradation, and loss of biodiversity, and indicated that the environmental footprint of the global livestock industry is much larger than commonly thought and its potential climate change mitigation contribution is also large (Steinfeld et al. 2006). The global meat and milk consumptions are projected to double by 2050 compared to 2000 (FAO 2006; Alexandratos and Bruinsma 2012). As the global demand for livestock products increases, the emissions from the livestock sector will also increase at much higher rates. Because of its considerably larger carbon footprint, livestock GHG emissions as well as the sector's mitigation potential, the livestock sector has gained more attention in recent studies (Steinfeld et al. 2006; Westhoek et al. 2011; Gerber et al. 2013; Weindl et al. 2015; Herrero et al. 2016; Rojas-Downing et al. 2017). The general conclusions from these studies are that the livestock sector is a large user of natural resources and a significant contributor to GHG emissions and climate change. It also faces the difficult challenge of reducing GHG emissions while also responding to significant demand growth for livestock products.

Identifying sources of GHG emissions associated with different livestock production chains, the major contributing gases, and their baseline emissions is essential for evaluating the GHG mitigation potential of the livestock sector. The existing estimates of GHG emissions from livestock have been undertaken following two main approaches: (1) IPCC GHG emission guidelines which comprise direct non-CO_2 emissions of CH_4 from enteric fermentation and manure and N_2O from manure management and utilization) (IPCC 2006), and (2) lifecycle analysis (LCA) which involves the systematic analysis of livestock production systems to account for all inputs and output with defined boundary depending on the goal of the analysis (FAO 2010). The LCA takes into account the entire livestock commodity chain—from land use and feed production, livestock farming, and waste management (i.e., cradle to farm gate), to product processing and transportation (farm-gate to retail) (FAO 2010; Gerber et al. 2013). Therefore, LCA incudes extra GHG emission sources in supply chain emissions arising from feed production as well as those from processing and transportation of livestock commodities to markets that are reported under other sectors (e.g., fuels to transport products in the transport sector, energy used in processing industry sector and land use change in IPCC GHG inventories).

Several estimates of livestock contribution to global anthropogenic GHG emission based on both IPCC inventory method and LCA exist. Livestock GHG emissions under IPCC emissions guidelines comprise direct non-CO_2 enteric CH_4 and manure CH_4 and N_2O. At the farm level, ruminants and monogastrics to some extent emit CH_4 as part of their digestive process. Microbial fermentation by symbiotic microorganisms inhabiting the rumen of cattle, buffalo, goats, and sheep convert fiber and celluloses into products that can be digested by ruminant animals. In this process, CH_4 is produced and exhaled as a by-product of the fermentation process. Poorly digestible food rations cause higher CH_4 emissions per unit of ingested energy. Non-ruminants such as pigs and horses also produce CH_4 during digestion process depending on the feed but the amounts are much lower compared to those from ruminants (Gerber et al. 2013). Cattle and buffaloes are an important source of CH_4 because of their large population and high emissions from their ruminant digestive systems. The enteric CH_4 is the largest emission source in the livestock sector, which is estimated to contribute 1.6 to 2.7 Pg CO_2 eq yr^{-1} (Table 14.7; USEPA 2012; Tubiello et al. 2013; Popp et al. 2010; Gerber et al. 2013; Herrero et al. 2013b, 2016). Between 1961 and 2010, global enteric emission increased from 1.4 to 2.2 Pg CO_2 eq. yr^{-1}, with an average annual growth rate of 0.70% yr^{-1} (Smith et al. 2014; USEPA 2012; Tubiello et al. 2015; Gerber et al. 2013) mainly due to the increase in the global livestock population (USEPA 2012). In 2010, 1.0–1.5 Pg CO_2 eq yr^{-1} or 75% of the emissions came from developing countries. During the 2000–2010 decade, Asia and Central and South America contributed most of the enteric CH_4 emissions, followed by Africa. The decadal growth rates were 2.4, 2.0, 1.1, and -1.7% yr^{-1} for Africa, Asia, Central and South America, and Europe, respectively (Smith et al. 2014). Cattle contributed the largest share of enteric CH_4 emissions, accounting for 75% of the total, followed by buffalo, sheep, and goats in that order.

Manure handling, storage, and application release CH_4, N_2O, and NH_3. CH_4 is released from the anaerobic decomposition of OM in manures and slurries, mostly when manure is managed in liquid form such as in deep lagoons or holding tanks. N_2O is released from nitrification and/or denitrification of the mineralized N. Direct and indirect N_2O emissions can vary significantly between the type of management system used as well as the temperature and humidity at the time of application of manure in the field. The number of animals and the type and amount of feed consumed are the primary factors that determine the type and amount of GHG emissions. The emissions from manure management contribute 0.2 to 0.4 and 0.2 to 0.5 Pg CO_2 eq. yr^{-1} of CH_4 and N_2O, respectively (Table 14.7; Gerber et al. 2013; Bodirsky et al. 2012; FAO 2013b). The CH_4 emissions from manure are generally low, except under CAFO where manure is handled in liquid-base systems (IPCC 2006). The NH_3 released into the atmosphere during storage and processing of manure can later be transformed

TABLE 14.7
Global Greenhouse Gas Emissions from Livestock

Emission Type and Source	Emissions (Pg CO_2 eq.)	Reference
IPCC emissions guidelines		
• CH_4 Enteric fermentation	1.6–2.7	USEPA (2012); Tubiello et al. (2013);
• CH_4 Manure management	0.2–0.4	Herrero et al. (2013b); Bodirsky et al.
• N_2O Manure management	0.2–0.5	(2012); Popp et al. (2010);
Total IPCC guidelines	2.0–3.6	Tubiello et al. (2015)
LCA approach		
• Feed production		
- CO_2 Land use change for soybean	0.23	Steinfeld et al. (2006); Gerber et al. (2013);
- CO_2 Land use change - pasture expansion	0.43	FAO (2013a,b); Herrero et al. (2016)
- CO_2 Direct energy	0.11	
- CO_2 Feed processing	0.92	
- N_2O from fertilizers and crop residue	0.92–2.0	
- N_2O Leguminous feed crops	0.20	
- N_2O Legume pasture	0.1	
- CH_4 from rice feed	0.03	
• Animal management and production		
- N_2O Indirect manure emissions	0.62	
- N_2O manure deposited on pastures	0.16	
- CO_2 indirect embedded energy	0.02–0.15	
• Post-farm gate		
- CO_2 from post-farm gate	0.02–0.1	
Total CO_2	2.0	
Total CH_4	1.8–3.1	
Total N_2O	1.5–2.5	
Total LCA approach emissions	5.6–7.6	
Total global livestock sector	2.5–7.6	

into N_2O in soils. Global manure management emissions increased from 0.73 to ~1.0 Pg CO_2 eq. yr^{-1} between 1961 and 2010, or an average annual rate of 1.1% yr^{-1} (Smith et al. 2014; Tubiello et al. 2013). During the 2000–2010 decade, Central and South America, Asia, and Africa were the largest source, with the growth rates of 1.5% yr^{-1} for both Asia and Central and South America, and 2.4% yr^{-1} for Africa (Tubiello et al. 2013).

In addition to enteric and manure management emissions, LCA also accounts for other GHG sources in livestock production processes, including upstream livestock feed production, animal management and production, and post-farm gate emissions. Emissions from upstream feed production results from the interaction of diverse drivers which may be direct or indirect, and can involve impact on C balance of land used for pasture and/or feed-crops, grazing, cultivation, application of fertilizers and pesticides, SOC losses from land use change, and transport of feed (Gerber et al. 2013; FAO 2013a,b; Herrero et al. 2016). For example, when forest is cleared for pasture and feed crops, large SOC pools stored in vegetation and soil are released into the atmosphere. CO_2 emissions generally result from SOC decomposition when natural habitats are disturbed by the expansion of feed crops and pasture. CO_2 also originates from the use of fuel to synthesize fertilizer, processing and transporting of feed and livestock products. Energy consumption, either direct or indirect, and the associated CO_2 emissions occur along the entire livestock production chain. Other GHG sources estimated by LCA include energy CO_2 emissions estimated at 0.13 Pg CO_2 eq. yr^{-1}, CO_2 emissions from land use change (0.66 Pg CO_2 eq. yr^{-1}), CO_2 from feed processing and production (0.92 Pg CO_2 eq. yr^{-1}), indirect embedded energy post-farm gate CO_2 (0.04 Pg CO_2 eq. yr^{-1}), and feed rice CH_4 emissions (0.03 Pg CO_2 eq. yr^{-1}) (Table 14.7; Bodirsky et al. 2012; FAO 2013b; Gerber et al. 2013). In addition to manure management, N_2O also comes from the use of synthetic fertilizers for feed production and from direct deposition of animal excreta on pastures during grazing and high N crop residues decomposition. The N_2O emissions attributed to feed production are estimated at 1.3–2.0 Pg CO_2 eq. yr^{-1} (Bodirsky et al. 2012; Herrero et al. 2013b; Gerber et al. 2013; FAO 2013a, 2013b; Rojas-Downing et al. 2017), which make feed production the largest source of N_2O under livestock. It includes N_2O emissions from manures applied to feed crops and pasture, fertilizer, leguminous feed crops, and indirect manure emissions (Table 14.7). Overall, enteric fermentation remains the largest source of GHG emissions, accounting for about 40%. Emissions associated with energy consumption, direct or indirect, are related to fossil fuel use during feed production, and fertilizer manufacturing. In total, the processing chain energy use contributes about 20% of total emissions (Gerber et al. 2013). Overall, most of CH_4 and N_2O emissions occur during production and waste management, while CO_2 emission occurs during feed production, processing, and transportation of livestock products.

The estimated global GHG emissions from the livestock production vary from 8% to 18% of global anthropogenic GHG emissions (Steinfeld et al. 2006; O'Mara 2011; USEPA 2012; Gerber et al. 2013). The large range reflects methodological difference—IPCC emissions inventories versus LCA. The animal agricultural is responsible for 8%–11% based on IPCC emissions inventories, whereas livestock contribution based on LCA is as high as 18% of global GHG emissions. The global total GHG emissions from the livestock sector ranges from 5.6 to 7.5 Pg CO_2 eq yr^{-1} (Table 14.7; Gerber et al. 2013; Herrero et al. 2016). About 44% of the sector's emissions are in the form of CH_4, while the remaining is shared almost equally between N_2O (29%) and CO_2 (27%) (Gerber et al. 2013). The livestock sector accounts for about 37% and 65% of global CH_4 and N_2O emissions (Steinfeld et al. 2006). The global non-CO_2 emission intensity of livestock products is estimated at 44 kg CO_2 eq. kg^{-1} protein and ranges from 9 to 500 kg CO_2 eq. kg^{-1} protein (Herrero et al. 2016). Pigs and poultry (monogastrics) have lower emission intensity (Table 14.8) and hence are at the lower end of the range. Based on IPCC guideline estimates, the CH_4 from enteric fermentation is

TABLE 14.8
Global Greenhouse Gas Emissions and Emission Intensity by Animal Types

Animal	Emission	Emission Intensity (kg CO_2 eq. kg^{-1} Product)
Dairy cattle	1.8	18.2
Beef cattle	2.3	67.6
Cattle	4.3	46.2
Buffaloes	0.6	53.4
Sheep and goats	0.4	23.4
Pigs	0.7	6.1
Poultry	0.6	5.4

Sources: Herrero, M., P. Havlik, H. Valin, A. Notenbaert, M.C. Rufino, P.K. Thornton, M. Blummel, F. Weiss, D. Grace, and M. Obersteiner, *Proc. Natl. Acad. Sci. U.S.A.*, 110 (52), 20888–20893, 2013; Gerber, P.J., H. Steinfeld, B. Henderson, A. Mottet, C. Opio, J. Dijkman, A. Falcucci, and G. Tempio, Food and Agriculture Organization of the United Nations (FAO), Rome, Italy, 2013.

projected to increase by 22% by 2030, and its relative share to agricultural GHG emissions will increase by 33% (USEPA 2012). Global emissions from enteric fermentation are driven by the size of the livestock population and the management practices in use. Therefore, the quantity, quality, and type of feed will influence CH_4 emissions.

The projections based on the current growth of the livestock sector by 2050 suggest that CH_4 emissions from enteric fermentation, CH_4 and N_2O emissions from manure management will likely grow at 0.9%–5%, 0.9%–4% and 1.2%–3% yr^{-1}, respectively. The current trends suggest that emissions from livestock will increase by 1%–1.5% yr^{-1} across all sources other than land use change (USEPA 2012; Havlik et al. 2013; Herrero et al. 2016). Emissions from deforestation over the same period are projected to increase at a slower rate ~0.8% yr^{-1} (Havlik et al. 2013). Overall, cropland expansion is growing at a faster rate than pasture expansion at 5% yr^{-1}, primarily due to accelerated global growth of pork and poultry production (Herrero et al. 2016).

Cattle production is the main contributor to the sector's emissions contributing 64% to 78% of emissions, representing as much as 4.6 Pg CO_2 eq. yr^{-1} (Table 14.8; Herrero et al. 2013b; Gerber et al. 2013; FAO 2013b), of which beef cattle and dairy that produced milk and meat accounted for 2.5 and 2.1 Pg CO_2 eq. yr^{-1}, respectively (FAO 2013b). Other species—buffalo, small ruminants (goats and sheep), pigs, poultry—contributed much lower and nearly similar emissions estimated at 0.6, 0.5, 0.7, and 0.7 Pg CO_2 eq. yr^{-1}, respectively (Table 14.8). Pigs, poultry, buffalos, and small ruminants have lower emission levels, representing 7% and 10% of livestock sector emissions (Figure 14.3; Gerber et al. 2013). In addition, mixed crop–livestock system dominates livestock emissions at 58% of the global livestock sector emissions, while grazing systems contribute 19% (Herrero et al. 2013b). The developing countries contribute 70% and 53% of non-CO_2 livestock sector's emission from ruminants and monogastrics, respectively (Herrero et al. 2013a). This share is projected to increase as livestock production increases to meet the demand in the developing world. Fifty-eight percent of the total emissions from livestock come from mixed crop-livestock systems, mainly due to their prevalence.

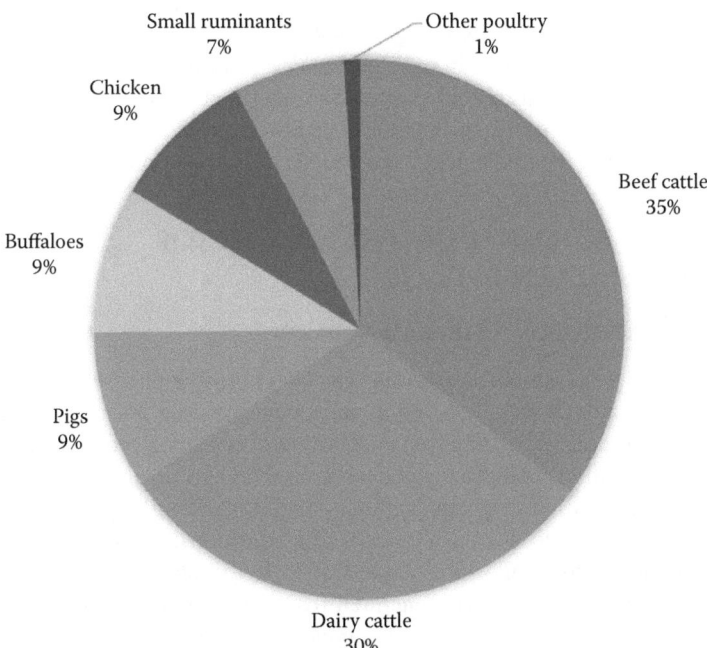

FIGURE 14.3 **(See color insert.)** Global estimates of greenhouse gases emission by species. (Based on data from Gerber, P.J., H. Steinfeld, B. Henderson, A. Mottet, C. Opio, J. Dijkman, A. Falcucci, and G. Tempio, Food and Agriculture Organization of the United Nations (FAO), Rome, Italy, 2013.)

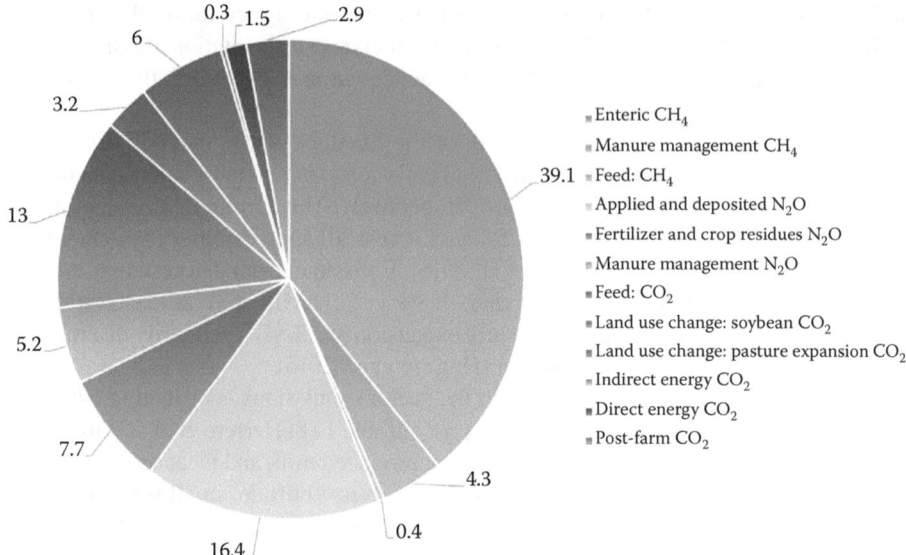

FIGURE 14.4 (See color insert.) Global livestock sector supply chain categories greenhouse gases emissions percentage. (Data from Gerber, P.J., H. Steinfeld, B. Henderson, A. Mottet, C. Opio, J. Dijkman, A. Falcucci, and G. Tempio, Food and Agriculture Organization of the United Nations (FAO), Rome, Italy, 2013.)

Grazing-based systems contribute 19% of global livestock emissions (Herrero et al. 2013a), while industrial and other systems comprise the rest. Emissions from production, processing, and transportation of feed account for 45% of the livestock sector, while fertilization of feed crops and deposition of manure on pastures during grazing generate N_2O emissions representing ~50% of feed emissions or 25% of livestock sector N_2O emissions (Figure 14.4). Increasing population, incomes, and urbanization are projected to drive increases in the consumption of milk and meat during the first half of the 21st century, at least at previous observed rates (Steinfeld et al. 2006; Rosegrant et al. 2009). This could increase the contribution of livestock to GHG emissions by nearly 50% under the business-as-usual scenario, necessitating the implementation of GHG emissions mitigation.

14.6 MITIGATION POTENTIAL OF LIVESTOCK SECTOR GREENHOUSE GAS EMISSIONS

14.6.1 LIVESTOCK PRODUCTION SYSTEMS MANAGEMENT

There are various effective options for mitigating GHG emissions from livestock production. These can be grouped into: (i) technical and management interventions; (ii) increased livestock, pasture, and feed crop productivity; and (iii) moderation of livestock products demand. Technical and management interventions comprise practices that could be implemented within the livestock sector supply chain such as manure management—including handling, storage and application phases, use of feed additives to decrease enteric CH_4 emissions, improved feed digestibility to reduce enteric CH_4, and increasing soil C sequestration in grasslands through best pasture management practices. Because enteric CH_4 generated from the gastrointestinal tracts of livestock is the single largest source of anthropogenic CH_4, most of the GHG emissions mitigation research for livestock has focused on options for reducing enteric CH_4 produced by ruminants (Martin et al. 2010; Cottle et al. 2011; Boadi et al. 2004; Knapp et al. 2014). Enteric CH_4 contributes 17% of the global CH_4 emission and 3.3% of global GHG emissions (Knapp et al. 2014). Practices that increase livestock, pasture, and feed crop productivity include management of animal health and productivity and avoiding deforestation and pastures encroachment. Moderation of demand for livestock will require lifestyle changes to rely less on food products with a high C cost associated with their production (i.e., animal based food) in favor of plant-based healthy diets (Popp et al. 2010; Bajželj et al. 2014).

Management of livestock manure includes activities associated with handling, storage, and application of manures in field in situations where manure can be collected and stored as well as the direct deposition of manure on pasture. CH_4, N_2O and NH_3 are emitted throughout these processes. Although NH_3 is not a GHG, its transformation can potentially lead to N_2O emissions. Livestock manure contributes 2% of global CH_4 emissions and accounts for 0.4% of global GHG emissions (Knapp et al. 2014). Most CH_4 emissions resulting from manure are produced under anaerobic conditions during storage and only little following land application. Droppings from ruminants in grazing land do not produce significant quantities of CH_4 because they remain largely aerobic. Therefore, strategies for reducing CH_4 emission centers around preventing anaerobic conditions during storage or capturing and either utilizing or transforming CH_4 produced under anaerobic conditions during storage. N_2O is produced as a product of nitrification and denitrification, and manure contains the necessary stimulators of soil nitrification and denitrification processes. The N_2O emissions can be reduced by reducing N losses to the environment through the storage of manure or slurries appropriately and that minimize losses from volatilization, runoff, and leaching. Application of manures to match plant demands while also avoiding heavy rains reduces N_2O emissions. Nitrification inhibitors are also effective in reducing N_2O emissions, and they are suitable for cropland and grassland in different climates. Practices such as compacting and covering farmyard manure, as well as anaerobic digestion of manure slurries prior to application to soils have been suggested as effective approaches for reducing the environmental impacts of manures and slurries. The highest mitigation potential for manure (0.01 to 0.075 Pg CO_2 eq. yr^{-1}) is achieved from lower N losses to the environment (Smith et al. 2008).

Feed additives are generally used to reduce CH_4 through manipulation of rumen microbial processes. They include chemical compounds which act as alternative electron receptors, ionophoric antibiotics, enzymes, and probiotic cultures (Hristov et al. 2013b). The ability of feed additives to reduce CH_4 emissions have mainly been tested in short-term experiments (Hristov et al. 2013a). However, their effects are often much reduced in the long-term due to adaptation of rumen microbial ecosystems. Additionally, there are environmental concerns associated with some of the compounds, as well as issues associated with public acceptance which constrain the adoption of these techniques on a large scale. Recent experiments with CH_4 inhibitors show promise for future however (Hristov et al. 2013a). Overall, dietary lipids and nitrates (electron receptor) appear to be the most promising feed additives (Hristov et al. 2013b). The estimated potential for this option can be as high as 0.68 Pg CO_2 eq. yr^{-1} under the assumption that up to 10% improvement in the digestibility of animal feed rations can be attained and widespread application can occur among developing countries.

Management practices that increase livestock, crop, and pasture productivity improve GHG emission intensities, spare land use, and promote structural change in the livestock sector. Some approaches for increasing livestock productivity include improving the animal genetic potential for production through selection for high production traits, breeding and/or biotechnology, improving animal reproductive performance and efficiency, live-weight gain, and animal health reducing mortality rates of calves and adult animals (Hristov et al. 2013a). These practices reduce GHG emissions per unit of product (Gill et al. 2009). For example, in extensive livestock production systems, reducing the herd size increases the availability of feed and productivity of both the total herd and individual animals while also lowering CH_4 and overall GHG emissions per unit product. Furthermore, increasing live-weight gain rates reduce slaughtering age and decrease GHG emission per unit product in beef and meat production systems in general. Deforestation and grasslands encroachment due to intensification can also be avoided by addressing the unsustainable practices in the global food production system (Smith 2013). In contrast, intensification of livestock production avoids arable and pasture land expansion and the associated deforestation (Pretty and Bharucha 2014; Smith 2013). Estimated improved animal management could mitigate as much as 0.2 Pg CO_2 eq. yr^{-1} by 2050.

Intensification also minimizes GHG emissions by avoiding deforestation. Options for sustainable intensification include adapting new technologies such as genetic modification, cloning, and nanotechnology to improve the efficiency of existing food production. There are ranges of other sustainable intensification options that can also address widespread unsustainable practices within global food. For example, substantial crop yield increases between 1961 and the present associated with the Green Revolution have led to a more than 200% increase in crop production, with only 11% expansion in the global cropland area (Hertel et al. 2014). An estimated emission of ~1.3 Pg CO_2 was avoided from land use change through intensification during this period (Hertel et al. 2014).

Demand for livestock products can be moderated by reducing consumption of livestock products, especially for societies where the consumption is high. Because resource use efficiency of livestock production is much lower compared to crops, and nearly one-third of cereals are fed to animals (Steinfeld et al. 2006), reducing the consumption of livestock products in regions with overconsumption will lessen the need for more livestock products. It is estimated that the production of beef protein requires about 50 times more land than the production of vegetative proteins (Nijdam et al. 2012). Even though meat contributes only 15% of the the total energy in global human diet, nearly 80% of the global arable land is used for grazing or feed and fodder production for livestock (Steinfeld et al. 2006). Currently, about 80% of agricultural land is used for animal grazing or feed and fodder production. Several researchers have assessed the hypothetical mitigation potential of different dietary scenarios and demonstrate substantial mitigation effect at relatively lower cost compared to alternative mitigation strategies (Smith et al. 2013b; Stehfest et al. 2009). A shift from livestock consumption to more crop-oriented human diets reduces global resources requirements substantially. Positive effects of reducing animal protein intake have also been reported among populations consuming high levels of animal source products (Stehfest et al. 2009). Removal of ruminant meat from diets and the adoption of a healthy diet results in potential mitigation benefits of 5.8 and 4.3 Pg CO_2 eq. yr^{-1}, respectively (Stehfest et al. 2009; Smith et al. 2013b). Because livestock resource use efficiency is low compared to that of crops, reducing the consumption of livestock products and consuming healthy diets, especially in countries where overconsumption of animal products is widespread, could also reduce the need for more animal-source food. It is estimated that switching to low animal product diets that maintain global average energy demand at 2800 kcal capita^{-1} day^{-1} compared to global mean of 3100 kcal capita^{-1} day^{-1} and the spared land is used for bioenergy or afforestation, an estimated GHG emissions reduction of 0.7–7.3 and 4.6 Pg CO_2 eq yr^{-1}, respectively, can be achieved (Herrero et al. 2016). Theoretically, therefore, reducing the demand for livestock remains the most effective GHG emission mitigation option, with additional potential benefits of increasing food security and sparing land that can be used for afforestation or energy crops production.

14.6.2 Soil Carbon Sequestration in Grasslands

The large proportion of global grassland area has been stressed due to intensive grazing in order to produce more livestock, especially Africa's grasslands (Reid et al. 2004) which are also vulnerable to climate change. Improved grazing management increases forage production, enhances efficient use of land resources, increases profitability, and rehabilitates the degraded land. A conceptual SOC cycling in grasslands is presented in Figure 14.5. SOC losses occur as a result of overgrazing and OM decomposition due to grassland soil disturbance. Despite historic loss, global grasslands still contain substantial SOC content and high natural soil fertility. About 343 Pg C is stored in the top 1-m soil depth of the global grassland (Conant et al. 2017). However, these pools are susceptible to loss when grasslands are converted to other land uses or overgrazed. It is estimated that about 20% of the global native grazing lands have been converted to croplands (Ramankutty et al. 2008)

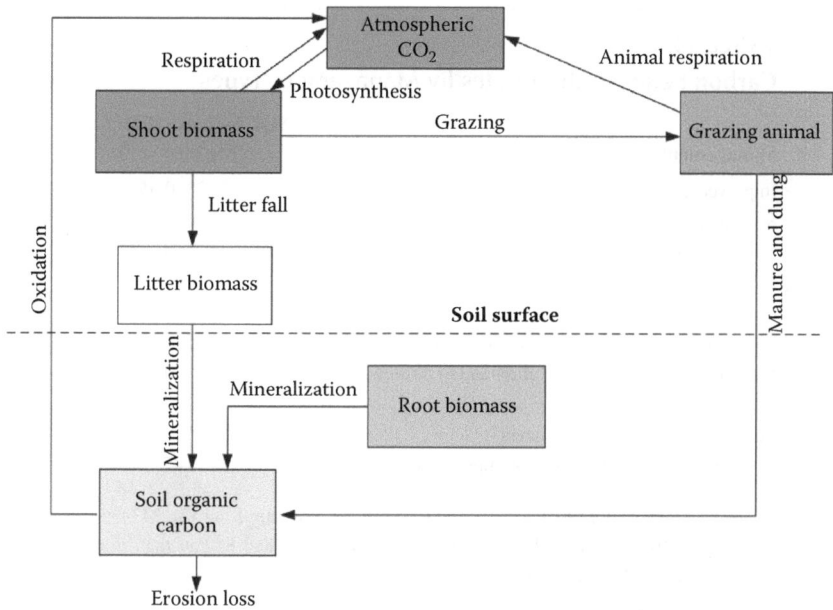

FIGURE 14.5 Conceptual C cycling in the grazed grassland.

leading to the loss of about 60% of SOC stocks (Guo and Gifford 2002). Other disturbances that deplete grassland systems C stocks include fire, invasive species, management practices that affect species composition, mismatch between forage supply and animal demands, nutrients, and water input (Smith et al. 2008; Herrero et al. 2016). Grassland ecosystem SOC stocks can be increased by changing grassland management practices that have led to grazing land degradation and by adopting BMPs proven to increase C sequestration in grazing land soils.

Best grassland management practices offer a significant potential for C sequestration in grasslands soils. Grazing land and pasture management practices that increase C stocks will also mitigate CO_2 emissions while also providing opportunities for profitable investment. Similar to other land uses, increasing C inputs to grazing land soil increases C sequestration. Synthesis of global grassland management indicated that improved grazing management, fertilization of pastures, sowing legumes, and improved grass species, irrigation of pastures, and conversion from cultivation increased SOC at rates ranging from 0.105 to 3.0 Mg ha^{-1} yr^{-1}, and an average of 0.47 Mg ha^{-1} yr^{-1} across all improved grassland management (Table 14.9, Conant et al. 2001, 2017). Assuming that recommended grassland management practices are implemented globally, an estimated 1.59 Pg C yr^{-1} (Table 14.10) or 5.82 Pg CO_2 yr^{-1} will be sequestered for the next 20–30 years. Higher sequestration potential exists in Africa (0.43 Pg C yr^{-1}), Asia (0.39 Pg C yr^{-1}), and South America (0.26 Pg C yr^{-1}) (Table 14.10).

Other published global management-based sequestration estimates show that improved grazing management practices could sequester 0.11 to 0.15 Pg CO_2 yr^{-1} over a 20-yr period globally (Smith et al. 2014; Henderson et al. 2015). Also, sowing legumes in some grassland areas could sequester 0.18 Pg CO_2 eq yr^{-1}. Recommended pasture and grazing management could therefore reduce GHG emissions by 0.29 Pg CO_2 eq. yr^{-1} (Table 14.11; Gerber et al. 2013). Nearly 50% of leguminous pasture sequestration is in developing countries (Henderson et al. 2015).

Grassland soils have large potential reservoir for storing CO_2, but this potential depends on how grasslands are managed for ruminants grazing. About 80% of this potential is in the developing countries where overall yield gap is large. Management practices that can increase grassland soil C stocks in grazing land include managing forage consumption, nutrient and water

TABLE 14.9
Carbon Sequestration Rates by Management Types

Management	SOC Sequestration Rate (Mg ha^{-1} yr^{-1})
Improved grazing	0.28–0.35
Sowing legumes	0.66–0.75
Sowing grass species	3.04
Earthworm introduction	2.55
Irrigation	0.11
Land use conversion: cultivated cropland to pasture	0.85–1.01
Land use conversion: native grassland to pasture	0.02–0.35
Fertilization: organic manure	0.82
Fertilization: inorganic fertilizers	0.54
Average: management practices improvement	0.47

Sources: Conant, R.T., K. Paustian, and E.T. Elliott, *Ecol. Appl.*, 11 (2), 343–355, 2001; Conant, R.T., C.E.P. Cerri, B.B. Osborne, and K. Paustian, *Ecol. Appl.*, 27 (2), 662–668, 2017.

TABLE 14.10
Estimated Technical Potential of the Global Grasslands Carbon Sequestration by Continents

Region	Grassland Area (Mha)[a]	SOC Sequestration Potential (Pg C yr^{-1})[b]	Degraded Grassland	
			%	Area (Mha)
Europe	437.3	0.21	35	153.1
Developing Asia	832.8	0.39	20	166.6
Africa	911	0.43	31	282.4
Latin America and Caribbean	550.1	0.26	13	71.5
North America	253.7	0.12	11	27.9
Oceania	393	0.18	19	74.7
World	3377.9	1.59	21	709.4

[a] Grassland area from Table 14.2. Source: FAO, Food and Agriculture Organization of the United Nations (FAO), Rome, Italy, 2009.
[b] SOC sequestration rate from Table 14.9.

management, sowing improved deep-rooted tropical grass species, fire control and management, and controlled removal of aboveground biomass. A large proportion of global grazing land is currently facing pressures to produce more livestock through intensive grazing, especially in Africa and other tropical regions (Reid et al. 2004). Better grassland management practices can reverse historical soil C losses by increasing SOC stocks in grazing land soils. Likewise, predictions have indicated that net primary production (NPP) can recover from overstocking after a period of destocking (Henderson et al. 2015). For example, a meta-analysis of and synthesis of data from multiple studies on grazed ecosystems of the Northern Great Plains showed that grazing regimes implemented after a "dust bowl" enhanced C storage in grazed soils in the past 70–80 years. The data indicate that 1.6 Mg C ha^{-1} (5.84 Mg CO$_2$-eq ha^{-1}) has been restored in the top 0–15 cm soil depth as a result of grazing management and recover soil C lost during the period of widespread grassland deterioration in the beginning of the 20th century (Wang et al. 2016). Practices that enhance SOC sequestration have economic benefits also arising from increased forage production. In addition

TABLE 14.11
Technical Potential for Livestock Sector GHG Emissions Mitigation

Practice	Mitigation Potential (Pg CO$_2$ eq yr^{-1})	References
Animal management	0.01–0.5	Herrero et al. (2016);
Use of feed additives	0.2–0.3	Havlik et al. (2014);
Improvement of digestibility	0.68	Hristov et al. (2013a,b);
Avoiding LUC by intensification	0.15	Smith et al. (2008);
Manure management	0.05	Herrero et al. (2016);
Rangeland restoration and rehabilitation	0.1–0.2	Gerber et al. (2013)
Carbon sequestration by pasture improvement	0.2	
Carbon sequestration by legume sowing	0.13	
Improved grazing management	0.15–0.7	
Livestock technical mitigation potential	4–7.5	

to GHG mitigation benefits, grazing land management also improves productivity and other ecosystem services. Despite C sequestration benefits, a potential increase in N$_2$O emissions by as much as 0.06 Pg CO$_2$ eq. yr^{-1} by leguminous pasture could offset up to 28% of SOC sequestration benefits (Herrero et al. 2016).

The technical mitigation potential of the livestock production sector is large and represents as much as 50% of the global mitigation potential of agriculture, forest, and land use (AFOLU) (i.e., 4.0 to 7.5 Pg CO$_2$ eq. yr^{-1}) (Table 14.11) compared to 4.6–12.9 Pg CO$_2$ eq yr^{-1} (Smith et al. 2014), but most of this potential has not been realized due to low adoption rate of BMPs in livestock sector and trade-offs associated with reducing the consumption of livestock products. Similarly, the societal impacts of opportunities for land sparing on socio-economic and gender equity are not well understood and require attention due to the relevance of mitigation policies. In addition, investment is needed for the transformation of the developing world livestock sector so that it becomes market-driven, which could catalyze the adoption of sustainable intensification practices. Finally, global trends of increasing demand and consumption of meat and other livestock products is incompatible with reducing GHG missions from agriculture. Reducing the global consumption of livestock products can bring benefits in reducing GHG emissions; however, more research is needed to understand how this goal can be achieved without negative trade-offs in some regions and nutrition security goals. Therefore, there is a need for research to determine the required interventions for limiting global growth in livestock products consumption. Overall, mitigating livestock sector emissions remains challenging, as it is for other sectors such as energy and food crops production.

14.6.3 INORGANIC CARBON AND ITS ROLE IN CARBON SEQUESTRATION IN GRASSLANDS

Although most C in the soil is associated with OM, inorganic C (SIC) pool is a principal component of arid and semiarid region soils around the world. The amount of SIC that forms in these soils depends on soil age. Calcic and petro-calcic horizons are generally found in older arid and semiarid soils. In soils, the SIC occurs largely as carbonate minerals—calcite (CaCO$_3$), dolomite [CaMg(CO$_3$)$_2$], or precipitated pedogenically. In arid and semiarid regions that cover as much as one-third of the surface of the planet, the SIC pool is approximately 2 to 10 times larger than SOC storage (Schlesinger 1982; Eswaran et al. 2000) while the SIC rate of accumulation is generally higher than that of SOC in these soils (Landi et al. 2003). However, quantifying carbonate C sequestration rates remains a challenge. Many soils contain both lithogenic carbonate (i.e., primary or geogenic carbonate derived from bedrock) and pedogenic carbonate (i.e., secondary or authigenic carbonate formed by CO$_2$ reactions and precipitated in place) and only pedogenic carbonate

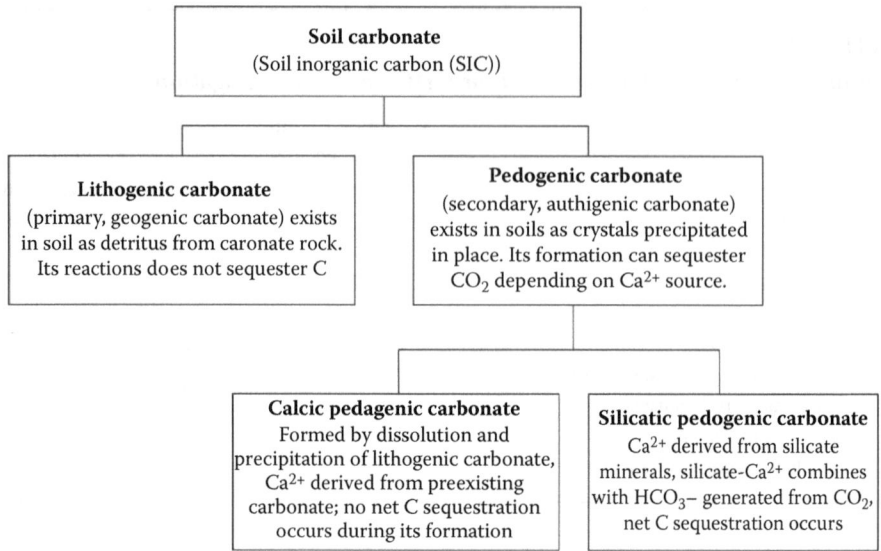

FIGURE 14.6 Classification of soil carbonates based on origin and calcium sources (Monger et al. 2015).

formation can potentially sequester atmospheric CO_2 depending on the Ca^{2+} source. Pedogenic carbonate is formed in soil through the dissolution and precipitation of carbonate material. This process is common in arid, semiarid, and some sub-humid climates (Monger et al. 2015). Classification of soil carbonates is outlined in Figure 14.6. Although active C sequestration occurs during the formation of silicatic pedogenic carbonates, tracing and distinguishing Ca^{2+} that originated from silicates versus carbonate mineral is difficult. Strontium (Sr) isotopes can be used as proxy indicator for Ca (Capo and Chadwick 1999; Van der Hoven and Quade 2002), however, many routine soil analytical laboratories lack the capacity for this kind of specialized analysis. Another possibility is to use $\delta^{13}C$ or $\delta^{18}O$ isotope signature to track C source. However, this technique is not yet well developed. A cost-effective routine analysis is needed for reliable estimate of the SIC sequestration rates.

In addition to silicatic carbonate formation, groundwater also acts as a large sink for C. Groundwater stores C as HCO_3^-, and dissolved CO_2 leaches into groundwater on all continents regardless of climate. The global volume of groundwater is 23.4×10^6 km^3, and HCO_3^- concentration up to 5 mmol L^{-1} suggests inorganic C (IC) pool of nearly 1400 Pg C in groundwater (Oki et al. 2003; Oki and Kanae 2006; Zhu and Schwartz 2011; Monger et al. 2015). Ma et al. (2014) estimated an average net ecosystem CO_2 exchange of -25.0 ± 12.7 g C m$_2$ yr^{-1}, with most of C uptake leaching into the groundwater. Similarly, estimated C sequestration rate by the downward flux of HCO_3^- into groundwater in steppes of Russia range from 21 to 74 kg C ha^{-1} yr^{-1} (Lapenis et al. 2008) and 9.0 to 71.5 Mg C ha^{-1} yr^{-1} in irrigated agriculture region in California (Eshel et al. 2007). The global downward fluxes of IC into the groundwater range from 0.2 to 0.36 Pg C yr^{-1} as HCO_3^- (Kessler and Harvey 2001) and a residence time equivalent to groundwater that can be up to thousands of years.

14.7 STRATEGIES FOR ADAPTATION IN THE LIVESTOCK PRODUCTION SECTOR

Climate change poses challenges for growth and development, particularly in low-income countries. Current and projected future climate change poses an unprecedented challenge to the adaptive capacity of agriculture globally. Climate change effects are increasing the complexity and uncertainty of agricultural management. Therefore, projected climate changes over the next century may require major adjustments to production practices, particularly for production systems operating at their marginal limits of climate. Agricultural systems are human-dominated ecosystems, and the vulnerability of agriculture to climatic change is strongly dependent on the biophysical effects of climate change and the responses taken by humans to moderate those effects at the farm, regional, and global scales. Effective adaptive action undertaken by the multiple dimensions offers potential for capitalizing on the opportunities presented by climate change and minimizing the costs through reduction or avoiding the severity of detrimental effects from changing climate.

Some of the adaptation measures for livestock production include developing drought, pest, and heat stress resistance in animals through breeding, integrating livestock systems with crop production and forest systems, improving soil quality, minimizing off-farm flow of nutrients and pesticides, changing the timing and location of farm operations, diversifying livestock to increase drought and heat wave tolerance, and taking actions that increase the capacity of the livestock production system to minimize the effects of climate change on productivity such as sustainable livestock production systems. There are significant synergies between adaptation and mitigation actions, such as improved rangeland management, that can both sequester C and improve grassland productivity.

The adaptation capacity of traditional livestock producers in extensive systems can be increased through number of ways, including: (i) promoting production adjustments through (a) diversification, intensification, integration of pasture management, livestock and crop production, changing land use and irrigation alteration of timing of operations conservation of nature and ecosystems, and (b) introduction of mixed livestock farming systems—such as stall feeding and grazing; (ii) breeding strategies—such as strengthening local breeds which are adapted to local climate stress and feed sources, and improving local breeds through cross-breeding with heat- and disease-tolerant breeds; (iii) promoting international trade, credit schemes and market access; (iv) adopting institutional and policy changes through introduction of livestock early warning systems and other forecasting and crisis preparedness systems; (v) improving science and technology research to provide greater understanding of causes of climate change and its impact on livestock, to facilitate development of new breeds and genetic types, to improve animal health and to improve water and soil management; and (vi) developing livestock management systems that allow efficient and affordable adaptation practices for rural poor populations that are unable to purchase expensive adaptation technologies. Such systems include providing shade and water to reduce heat stress from increased temperature, reducing livestock numbers, using more productive animals to increase efficiency of production while reducing GHG emissions, and adjusting the livestock numbers and head composition to optimize the use of feed resources (Sidahmed et al. 2008).

There is an urgent need for deliberate and effective strategies for adapting climate change, since it is occurring much faster than the natural ecological adaptation, which can exacerbate already existing vulnerabilities and increase the impacts of other stresses such as poverty, unequal access to resources, natural disasters, food insecurity, and incidences of animal diseases. Although livestock producers have traditionally adapted to environmental and climate change, increased human population, urbanization, economic growth, growing consumption of foods of animal origin, and commercialization have made the coping mechanisms less effective (Sidahmed et al. 2008). Since livestock are key assets held by rural poor communities, particularly in pastoral and agro-pastoral systems where they fulfil multiple economic, social, and risk management functions, the loss of livestock assets will lead to chronic poverty with long-term effects on their livelihoods. Research is needed to (a) improve projections of future climate conditions at different timescales ranging from season to multi-decades, (b) understand the sensitivity of livestock production systems to direct and indirect effects of climate change such as changes in temperature, pathogens, soil and water components, and (c) develop and extend the knowledge, management strategies and tools needed by stakeholders to enhance the adaptive capacity of livestock production systems to climate variability and extreme events.

14.8   CONCLUSIONS

Livestock are global resources of significant benefits to society, especially for food, income, nutrients, employment, security, traction, and others. In the process of providing these benefits, livestock use significant amount of land resources and cause significant soil degradation. The livestock sector uses more than 26% of terrestrial land for grazing, and an additional one-third of the global arable land is devoted to animal feed production. There is a growing concern about the increasing environmental footprint of the global livestock production systems resulting from increasing global demand for livestock products. In addition, due to projected climate change impacts, more people are at risk of hunger in this century. Capitalizing on the positive aspects of livestock systems (i.e., sustainable intensification and incorporating mixed crop–livestock systems under properly regulated industrial livestock and crop production) may offer the potential of providing ecosystem services payments in the long-run. These systems might also help in achieving the goals of balancing livestock production, livelihoods, and environmental protection. Climate change will have significant impact on livestock performance and productivity and food security in many regions, and predictive models indicate that the impact will be detrimental. In addition to global temperature change, livestock production will be negatively impacted due to feed and fodder availability, diseases, heat stress, and water availability especially in arid and semiarid regions. During the 21st century, warming is projected to decrease livestock productivity as well as increase the number of days when animals will be experiencing heat stress. Conversely, livestock production also influences climate change. The livestock production chain contributes 11% to 18% of all anthropogenic GHG emissions. Enteric fermentation in ruminant animals is the largest GHG contributor in the animal production stage. It is expected that the contribution of livestock to the global food supply chain will continue to grow as a result of growing global population, income growth, urbanization, and dietary shifts toward livestock products from emerging middle class in developing countries. As a result, multiple existing trade-offs and competing demands for natural resources will intensify, but reducing livestock production is not a viable option because of the sector's role in attaining balanced nutrition. As a result, the contribution of livestock production to anthropogenic GHG emissions will continue to increase. Livestock production also contributes to other environmental quality deterioration, including degradation of water quality, land degradation, air quality pollution, and loss of biodiversity. Evidence suggests that livestock and environmental trade-offs are currently substantial, and this will increase significantly in the future as a result of increased demand for livestock products. Some of the most important impacts are those associated with increases in anthropogenic GHG emissions, soil degradation, and land use change for feed production for both ruminants and monogastrics as the livestock sector continues to expand to meet the demand for increasing animal sourced food. As a result, livestock production will continue to have significant impacts on a wide range of environmental dimensions—ranging from land use change, GHG emissions, water cycle and quality deterioration, nutrient balance,s and biodiversity. Therefore, there is a need for a fundamental shift in the way demands for livestock products are met, both at the local, national, and international levels. Governments and institutions need to develop and enact appropriate policies at the local, national, and international levels that account for livestock–environment interactions and emphasize on sustainability of both crop and livestock production systems. At the same time, there are significant opportunities for the livestock sector to contribute in improving the livelihoods of smallholder farming systems and improving environmental management through sustainable land management. Sustainable intensification of crop and livestock systems through integration of crop and livestock production systems offer promising alternatives in the developing world as well as rural areas of the developed countries. There is strong evidence that better management practices can sequester significant amounts of C while also minimizing CH_4 and N_2O emissions through nutrient recycling, and can play an important role in improving the water productivity of the whole ecosystems as well as other environmental, socioeconomic, and ecosystem benefits. Many impacts of climate change on livestock production can be avoided, reduced, or delayed by implementing mitigation and adaptation strategies. The livestock sector has a large untapped potential to contribute to climate change mitigation and adaptation while also addressing food insecurity. Livestock account for as much as half of the technical mitigation potential of the AFOLU sector. Livestock GHG emissions can be mitigated through management options that promote sustainable intensification of livestock production, promote C sequestration in grazing land and pastures, reduce emissions from manures, reduction in the demand for livestock products and interventions such as use of feed additives and improvement of feed digestibility. Realizing this potential will require new and extensive initiatives at the national and international levels including the promotion of research and development of new mitigation technologies, enhancing capabilities to monitor, report and verify emissions from livestock production. There is a significant knowledge gap about the impact of climate change on livestock production, since most of the studies have been performed at the continental and regional levels. Among the areas that need to be addressed include:

- Enhancing understanding of how climate change affects pasture and rangeland composition and the consequences it will have on livestock productivity.
- Identifying and implementing managements required to achieve economic, environmental, and ecosystem benefits.
- Evaluating the costs and benefits of adopting more environmentally friendly livestock production practices, especially in extensive production systems.
- Assessing the impacts of livestock on biodiversity remains a missing link for the adoption of systems such as mixed crop-livestock systems.
- Evaluating the impact of extreme weather events on livestock production systems is still lacking.
- Understanding to what degree heat affects the biology of animals and promotion of new diseases.
- Understanding the impacts of climate change at local levels on localities of poor households and the way climate change alters the fragile relationship between livelihoods and production dependent on natural resources remains particularly fraught with uncertainty.

REFERENCES

Alexandratos, N., and J. Bruinsma. 2012. World Agriculture towards 2030/2050: The 2012 Revision. ESA Working Paper No. 12-03. Agricultural Development Economics Division, Food and Agricultural Organization of the United Nations (FAO), Rome, Italy. 154 p. Available at: http://www.fao.org /docrep/016/ap106e/ap106e.pdf.

Arima, E.Y., P. Richards, R. Walker, and M.M. Caldas. 2011. Statistical confirmation of indirect land use change in the Brazilian Amazon. *Environ. Res. Lett.* 6 (2), doi: 10.1088/1748-9326/6/2/024010.

Armenteras, D., and C.M. Finlayson. 2012. Biodiversity. In *Global Environment Outlook 5*, edited by UNEP, pp. 133–166. Valetta, Malta: Progress Press Ltd.

Asner, G.P., A.J. Elmore, L.P. Olander, R.E. Martin, and A.T. Harris. 2004. Grazing systems, ecosystem responses, and global change. *Annu. Rev. Environ. Resour.* 29:261–299, doi: 10.1146/annurev .energy.29.062403.102142.

Ayo, J.O., J.A. Obidi, and P.I. Rekwot. 2011. Effects of heat stress on the well-being, fertility, and hatchability of chickens in the Northern Guinea Savannah zone of Nigeria: A review. *ISRN Veterinary Science* 2011:10, doi: 10.5402/2011/838606.

Bajželj, B., K.S. Richards, J.M. Allwood, P. Smith, J.S. Dennis, E. Curmi, and C.A. Gilligan. 2014. Importance of food-demand management for climate mitigation. *Nature Clim. Change* 4:924, doi: 10.1038 /nclimate2353.

Baker, J.T. 2004. Yield responses of Southern US rice cultivars to CO_2 and temperature. *Agric. For. Meteorol.* 122 (3–4):129–137, doi: 10.1016/j.agrformet.2003.09.012.

Bishop, K.A., A.D.B. Leakey, and E.A. Ainsworth. 2014. How seasonal temperature or water inputs affect the relative response of C_3 crops to elevated [CO_2]: A global analysis of open top chamber and free air CO_2 enrichment studies. *Food Energy Secur.* 3 (1):33–45, doi: 10.1002/fes3.44.

Blaikie, P., and H. Brookfield. 2015. *Land Degradation and Society*. New York: Routledge, Taylor & Francis Group.

Boadi, D., C. Benchaar, J. Chiquette, and D. Masse. 2004. Mitigation strategies to reduce enteric methane emissions from dairy cows: Update review. *Can. J. Anim. Sci.* 84 (3):319–335.

Bodirsky, B.L., A. Popp, I. Weindl, J.P. Dietrich, S. Rolinski, L. Scheiffele, C. Schmitz, and H. Lotze-Campen. 2012. N_2O emissions from the global agricultural nitrogen cycle—Current state and future scenarios. *Biogeosciences* 9 (10):4169–4197, doi: 10.5194/bg-9-4169-2012.

Bouwman, A.F., A.H.W. Beusen, and G. Billen. 2009. Human alteration of the global nitrogen and phosphorus soil balances for the period 1970–2050. *Global Biogeochem. Cycles* 23 (4), doi: 10.1029/2009GB003576.

Bouwman, A.F., K.W. Van der Hoek, B. Eickhout, and I. Soenario. 2005. Exploring changes in world ruminant production systems. *Agric. Syst.* 84 (2):121–153, doi: 10.1016/j.agsy.2004.05.006.

Bouwman, L., K.K. Goldewijk, K.W. Van Der Hoek, A.H.W. Beusen, D.P. Van Vuuren, J. Willems, M.C. Rufino, and E. Stehfest. 2013. Exploring global changes in nitrogen and phosphorus cycles in agriculture induced by livestock production over the 1900–2050 period. *Proc. Natl. Acad. Sci. U.S.A.* 110 (52):20882–20887, doi: 10.1073/pnas.1012878108.

Bradford, G.E. 1999. Contributions of animal agriculture to meeting global human food demand. *Livestock Production Science* 59 (2–3):95–112, doi: 10.1016/s0301-6226(99)00019-6.

Bruinsma, J. 2009. The resource outlook to 2050: By how much do land, water and crop yields need to increase by 2050? In *Proceeding of the Expert Meeting on How to Feed the World in 2050, June 24–26, 2009.* FAO headquarters, Rome, Italy: Food an Agriculture Organization of the United Nations, pp. 1–33.

Burkholder, J., B. Libra, P. Weyer, S. Heathcote, D. Kolpin, P.S. Thome, and M. Wichman. 2007. Impacts of waste from concentrated animal feeding operations on water quality. *Environ. Health Perspect.* 115 (2):308–312.

Caldwell, C.R., S.J. Britz, and R.M. Mirecki. 2005. Effect of temperature, elevated carbon dioxide, and drought during seed development on the isoflavone content of dwarf soybean Glycine max (L.) Merrill grown in controlled environments. *J. Agric. Food. Chem.* 53 (4):1125–1129, doi: 10.1021/jf0355351.

Capo, R.C., and O.A. Chadwick. 1999. Sources of strontium and calcium in desert soil and calcrete. *Earth Planet. Sci. Lett.* 170 (1):61–72, doi: https://doi.org/10.1016/S0012-821X(99)00090-4.

Capper, J.L., R.A. Cady, and D.E. Bauman. 2009. The environmental impact of dairy production: 1944 compared with 2007. *J. Anim. Sci.* 87:2160–2167.

Chapagain, A.K., and A.Y. Hoekstra. 2003. Virtual water flows between nations in relation to trade in livestock and livestock products. Value of Water Research Report Series No. 13. UNESCO-IHE, Institute of Water Education, The Netherlands. 59 p.

Chapman, S.C., S. Chakraborty, M.F. Dreccer, and S.M. Howden. 2012. Plant adaptation to climate change—Opportunities and priorities in breeding. *Crop Pasture Sci.* 63 (3):251–268, doi: https://doi.org/10.1071/CP11303.

Conant, R.T., C.E.P. Cerri, B.B. Osborne, and K. Paustian. 2017. Grassland management impacts on soil carbon stocks: A new synthesis. *Ecol. Appl.* 27 (2):662–668, doi: 10.1002/eap.1473.

Conant, R.T., K. Paustian, and E.T. Elliott. 2001. Grassland management and conversion into grassland: Effects on soil carbon. *Ecol. Appl.* 11 (2):343–355, doi: 10.1890/1051-0761(2001)011[0343:GMACIG]2.0.CO;2.

Cottle, D.J., J.V. Nolan, and S.G. Wiedemann. 2011. Ruminant enteric methane mitigation: A review. *Anim. Prod. Sci.* 51 (6):491–514, doi: 10.1071/an10163.

Das, R., L. Sailo, N. Verma, P. Bharti, J. Saikia, Imtiwati, and R. Kumar. 2016. Impact of heat stress on health and performance of dairy animals: A review. *Veterinary World* 9 (3):260–268, doi: 10.14202/vetworld.2016.260-268.

De Haan, C., H. Steinfeld, and H. Blackburn. 1997. *Livestock and the Environment: Finding a Balance*. Rome, Italy: Food and Agriculture Organozation of the United Nations (FAO).

de Toledo, P.M., E. Dalla-Nora, I.C.G. Vieira, A.P.D. Aguiar, and R. Araújo. 2017. Development paradigms contributing to the transformation of the Brazilian Amazon: Do people matter? *Curr. Opin. Environ. Sustain.* 26–27:77–83, doi: https://doi.org/10.1016/j.cosust.2017.01.009.

Delgado, C., M. Rosegrant, H.S.I. Feld, S. Ehui, and C. Courbois. 1999. *Livestock to 2020: The Next Food Revolution. Food, Agriculture, and the Environment Discussion Paper 28.* International Food Policy Research Institute, Washington, D.C. 83 p.

Delgado, C.L. 2003. Rising consumption of meat and milk in developing countries has created a new food revolution. *J. Nutr.* 133 (11):3907S–3910S.

Diaz, R., and R. Rosenberg. 2008. Spreading dead zones and consequences for marine ecosystems. *Science* 321:926–929.

Doreau, M., M.S. Corson, and S.G. Wiedemann. 2012. Water use by livestock: A global perspective for a regional issue? *Animal Frontiers* 2:9–16, doi: 10.2527/af.2012-0036.

Entz, M., W. Bellotti, J. Powell, S. Angadi, W. Chen, K. Ominski, and B. Boelt. 2005. Evolution of integrated crop–livestock production systems. In *Grassland: A Global Resource*, edited by D. A. McGilloway, pp. 137–148. Wageningen, the Netherlands: Wageningen Academic Publications.

Erb, K.-H., V. Gaube, F. Krausmann, C. Plutzar, A. Bondeau, and H. Haberl. 2007. A comprehensive global 5 min resolution land-use data set for the year 2000 consistent with national census data. *J. Land Use Sci.* 2 (3):191–224.

Ercin, A.E., and A.Y. Hoekstra. 2016. European water footprint scenarios for 2050. *Water* 8 (6):226.

Ericksen, P.J., J.S.I. Ingram, and D.M. Liverman. 2009. Food security and global environmental change: Emerging challenges. *Environ. Sci. Policy* 12 (4):373–377, doi: https://doi.org/10.1016/j.envsci.2009.04.007.

Eshel, G., P. Fine, and M.J. Singer. 2007. Total soil carbon and water quality: An implication for carbon sequestration. *Soil Sci. Soc. Am. J.* 71 (2):397–405, doi: 10.2136/sssaj2006.0061.

Eswaran, H., P.F. Reich, J.M. Kimble, F.H. Beinroth, E. Padmanabhan, and P. Mocharoen. 2000. Global carbon stocks. In *Global Climate Change and Pedogenic Carbonates*, edited by R. Lal, J.M. Kimble, H. Eswaran, and B.A. Stewart, pp. 15–27. Boca Raton, FL: Lewis Publishers.

Evenson, R.E., and M. Rosegrant. 2003. The economic consequences of crop genetic improvement programmes. In *Crop Variety Improvement and Its Effect on Productivity: The Impact of International Agricultural Research*, edited by R.E. Evenson and D. Gollin, pp. 473–497. Willington, U.K.: CABI Publishing.

FAO. 2006. *World agriculture: Towards 2030/2050: Prospects for food, nutrition, agriculture and major commodity groups*. Food and Agriculture Organization of the United Nations (FAO), Rome, Italy. 71 p.

FAO. 2009. *The state of food and agriculture: Livestock in the balance*. Food and Agriculture Organization of the United Nations (FAO), Rome, Italy. 166 p.

FAO. 2010. *Greenhouse gas emissions from the dairy sector: A life cycle assessment*. Food and Agriculture Organization of the United Nations (FAO), Rome, Italy. 94 p.

FAO. 2012. *Livestock sector development for poverty reduction: An economic and policy perspective—Livestock's many virtues*. J. Otte, A. Costales, J. Dijkman, U. Pica-Ciamarra, T. Robinson, V. Ahuja, C. Ly, and D. Roland-Holst (eds.). Food and Agriculture Organization of the United Nations (FAO), Rome, Italy. 161 p.

FAO. 2013a. *Greenhouse gas emissions from pig and chicken supply chains: A global life cycle assessment*. Food and Agriculture Organization of the United Nations (FAO), Rome. 83 p. Available at: http://www.fao.org/docrep/018/i3460e/i3460e.pdf.

FAO. 2013b. *Greenhouse gas emissions from ruminant supply chains: A global life cycle assessment*. Food and Agriculture Organization of the United Nations (FAO), Rome, Italy. 151 p. Available at: http://www.fao.org/docrep/018/i3461e/i3461e.pdf.

FAO. 2015. *The state of food and agriculture. Social protection and agriculture: Breaking the cycle of rural poverty*. Food and Agriculture Organization of the United Nations (FAO), Rome, Italy. 129 p.

FAOSTAT. 2017. *FAO Statistical Databases*. Food and Agriculture Organization of the United Nations (FAO) [cited November, 2017 2017]. Available from http://faostat.fao.org.

Fearnside, P. 2008. The roles and movements of actors in the deforestation of Brazilian Amazonia. *Ecol. Soc.* 13 (1):23.

Fearnside, P.M. 2005. Deforestation in Brazilian Amazonia: History, rates, and consequences (Deforestación en la Amazonía Brasileña: Historia, tasas y consecuencias). *Conserv. Biol.* 19 (3):680–688, doi: 10.1111/j.1523-1739.2005.00697.x.

Fehlenberg, V., M. Baumann, N.I. Gasparri, M. Piquer-Rodriguez, G. Gavier-Pizarro, and T. Kuemmerle. 2017. The role of soybean production as an underlying driver of deforestation in the South American Chaco. *Global Environ. Change* 45 (Supplement C):24–34, doi: https://doi.org/10.1016/j.gloenvcha.2017.05.001.

Foley, J.A., N. Ramankutty, K.A. Brauman, E.S. Cassidy, J.S. Gerber, J. Johnston, N.D. Mueller, C. O/'Connell, D.K. Ray, P.C. West, C. Balzer, E.M. Bennett, S.R. Carpenter, J. Hill, C. Monfreda, S. Polasky, J. Rockstrom, J. Sheehan, S. Siebert, D. Tilman, and D.P.M. Zaks. 2011. Solutions for a cultivated planet. *Nature* 478 (7369):337–342, doi: 10.1038/nature10452.

Garnett, T., M.C. Appleby, A. Balmford, I.J. Bateman, T.G. Benton, P. Bloomer, B. Burlingame, M. Dawkins, L. Dolan, D. Fraser, M. Herrero, I. Hoffmann, P. Smith, P.K. Thornton, C. Toulmin, S.J. Vermeulen, and H.C.J. Godfray. 2013. Sustainable intensification in agriculture: Premises and policies. *Science* 341 (6141):33–34, doi: 10.1126/science.1234485.

Geist, H.J., and E.F. Lambin. 2002. Proximate causes and underlying driving forces of tropical deforestation. *Bioscience* 52 (2):143–150, doi: 10.1641/0006-3568(2002)052[0143:pcaudf]2.0.co;2.

Gerber, P.J., H. Steinfeld, B. Henderson, A. Mottet, C. Opio, J. Dijkman, A. Falcucci, and G. Tempio. 2013. *Tackling climate change through livestock—A global assessment of emissions and mitigation opportunities.* Food and Agriculture Organization of the United Nations (FAO), Rome, Italy. 115 p.

Gibbs, H.K., A.S. Ruesch, F. Achard, M.K. Clayton, P. Holmgren, N. Ramankutty, and J.A. Foley. 2010. Tropical forests were the primary sources of new agricultural land in the 1980s and 1990s. *Proc. Natl. Acad. Sci. U.S.A.* 107 (38):16732–16737, doi: 10.1073/pnas.0910275107.

Gill, M., P. Smith, and J.M. Wilkinson. 2009. Mitigating climate change: The role of domestic livestock. *Animal* 4 (3):323–333, doi: 10.1017/S1751731109004662.

Gollehon, N., M. Caswell, M. Ribaudo, R. Kellogg, C. Lander, and D. Letson. 2001. Confined animal production and manure nutrients. Agriculture Information Bulletin No. 771. Resource Economics Division, Economic Research Service, U.S. Department of Agriculture (USDA), 771 Washington, D.C. 35 p.

Gornall, J., R. Betts, E. Burke, R. Clark, J. Camp, K. Willett, and A. Wiltshire. 2010. Implications of climate change for agricultural productivity in the early twenty-first century. *Philos. Trans. R. Soc. London, Ser. B* 365 (1554):2973–2989, doi: 10.1098/rstb.2010.0158.

Guo, L.B., and R.M. Gifford. 2002. Soil carbon stocks and land use change: A meta analysis. *Global Change Biol.* 8 (4):345–360, doi: 10.1046/j.1354-1013.2002.00486.x.

Hansen, G., and D. Stone. 2016. Assessing the observed impact of anthropogenic climate change. *Nature Clim. Change* 6 (5):532–537, doi: 10.1038/nclimate2896.

Hansen, M.C., S.V. Stehman, P.V. Potapov, B. Arunarwati, F. Stolle, and K. Pittman. 2009. Quantifying changes in the rates of forest clearing in Indonesia from 1990 to 2005 using remotely sensed data sets. *Environ. Res. Lett.* 4 (3):034001.

Hatfield, J.L., K.J. Boote, B.A. Kimball, L.H. Ziska, R.C. Izaurralde, D. Ort, A.M. Thomson, and D. Wolfe. 2011. Climate impacts on agriculture: Implications for crop production. *Agron. J.* 103 (2):351–370, doi: 10.2134/agronj2010.0303.

Havlik, P., H. Valin, M. Herrero, M. Obersteiner, E. Schmid, M.C. Rufino, A. Mosnier, P. Thornton, H. Bottcher, R.T. Conant, S. Frank, S. Fritza, S. Fuss, F. Kraxner, and A. Notenbaert. 2014. Climate change mitigation through livestock system transitions. *Proc. Natl. Acad. Sci. U.S.A.* 111:3709–3714.

Havlik, P., H. Valin, A. Mosnier, M. Obersteiner, J.S. Baker, M. Herrero, M.C. Rufino, and E. Schmid. 2013. Crop productivity and the global livestock sector: Implications for land use change and greenhouse gas emissions. *Am. J. Agric. Econ.* 95:442–448.

Havstad, K.M., J.R. Brown, R. Estell, E. Elias, A. Rango, and C. Steele. 2016. Vulnerabilities of southwestern U.S. rangeland-based animal agriculture to climate change. *Clim. Change*: 1–16, doi: 10.1007/s10584-016-1834-7.

Henderson, B.B., P.J. Gerber, T.E. Hilinski, A. Falcucci, D.S. Ojima, M. Salvatore, and R.T. Conant. 2015. Greenhouse gas mitigation potential of the world's grazing lands: Modeling soil carbon and nitrogen fluxes of mitigation practices. *Agric. Ecosyst. Environ.* 207 (Supplement C):91–100, doi: 10.1016/j.agee.2015.03.029.

Hendrickson, J., G.F. Sassenrath, D. Archer, J. Hanson, and J. Halloran. 2008. Interactions in integrated US agricultural systems: The past, present and future. *Renew. Agr. Food Syst.* 23 (4):314–324, doi: 10.1017/s1742170507001998.

Henry, B., E. Charmley, R. Eckard, J.B. Gaughan, and R. Hegarty. 2012. Livestock production in a changing climate: Adaptation and mitigation research in Australia. *Crop Pasture Sci.* 63 (3):191–202, doi: https://doi.org/10.1071/CP11169.

Herrero, M., P. Gerber, T. Vellinga, T. Garnett, A. Leip, C. Opio, H.J. Westhoek, P.K. Thornton, J. Olesen, N. Hutchings, H. Montgomery, J.F. Soussana, H. Steinfeld, and T.A. McAllister. 2011. Livestock and greenhouse gas emissions: The importance of getting the numbers right. *Anim. Feed Sci. Technol.* 166–67:779–782, doi: 10.1016/j.anifeedsci.2011.04.083.

Herrero, M., D. Grace, J. Njuki, N. Johnson, D. Enahoro, S. Silvestri, and M.C. Rufino. 2013a. The roles of livestock in developing countries. *Animal* 7:3–18, doi: 10.1017/s1751731112001954.

Herrero, M., P. Havlik, H. Valin, A. Notenbaert, M.C. Rufino, P.K. Thornton, M. Blummel, F. Weiss, D. Grace, and M. Obersteiner. 2013b. Biomass use, production, feed efficiencies, and greenhouse gas emissions from global livestock systems. *Proc. Natl. Acad. Sci. U.S.A.* 110 (52):20888–20893, doi: 10.1073/pnas.1308149110.

Herrero, M., B. Henderson, P. Havlik, P.K. Thornton, R.T. Conant, P. Smith, S. Wirsenius, A.N. Hristov, P. Gerber, M. Gill, K. Butterbach-Bahl, H. Valin, T. Garnett, and E. Stehfest. 2016. Greenhouse gas mitigation potentials in the livestock sector. *Nature Clim. Change* 6 (5):452–461, doi: 10.1038/nclimate2925.

Herrero, M., and P.K. Thornton. 2013. Livestock and global change: Emerging issues for sustainable food systems. *Proc. Natl. Acad. Sci. U.S.A.* 110 (52):20878–20881, doi: 10.1073/pnas.1321844111.

Herrero, M., P.K. Thornton, P. Gerber, and R.S. Reid. 2009. Livestock, livelihoods and the environment: Understanding the trade-offs. *Curr. Opin. Environ. Sustain.* 1 (2):111–120, doi: 10.1016/j.cosust.2009.10.003.

Herrero, M., P.K. Thornton, A. Notenbaert, S. Msangi, S. Wood, R. Kruska, J. Dixon, D. Bossio, J. van de Steeg, H.A. Freeman, X. Li, and P.P. Rao. 2012. Drivers of change in crop–livestock systems and their potential impacts on agro-ecosystems services and human wellbeing to 2030: A study commissioned by the CGIAR Systemwide Livestock Programme. International Livestock Research Institute (ILRI), Nairobi, Kenya. 101 p.

Herrero, M., S. Wirsenius, B. Henderson, C. Rigolot, P. Thornton, P. Havlik, I. de Boer, and P. Gerber. 2015. Livestock and the environment: What have we learned in the past decade? *Annu. Rev. Environ. Resour.* 40:177–202, doi: 10.1146/annurev-environ-031113-093503.

Hertel, T.W., N. Ramankutty, and U.L.C. Baldos. 2014. Global market integration increases likelihood that a future African Green Revolution could increase crop land use and CO2 emissions. *Proc. Natl. Acad. Sci. U.S.A.* 111 (38):13799–13804, doi: 10.1073/pnas.1403543111.

Hoekstra, A.Y. 2009. Human appropriation of natural capital: A comparison of ecological footprint and water footprint analysis. *Ecol. Econ.* 68 (7):1963–1974, doi: 10.1016/j.ecolecon.2008.06.021.

Hopkins, A., and A. Del Prado. 2007. Implications of climate change for grassland in Europe: Impacts, adaptations and mitigation options: A review. *Grass Forage Sci.* 62 (2):118–126, doi: 10.1111/j.1365-2494.2007 .00575.x.

Houghton, R.A., J.I. House, J. Pongratz, G.R. van der Werf, R.S. DeFries, M.C. Hansen, C. Le Quéré, and N. Ramankutty. 2012. Carbon emissions from land use and land-cover change. *Biogeosciences* 9 (12):5125–5142, doi: 10.5194/bg-9-5125-2012.

Howden, S., S. Crimp, and C. Stokes. 2008. Climate change and Australian livestock systems: Impacts, research and policy issues. *Anim. Prod. Sci.* 48 (7):780–788.

Hristov, A.N., J. Oh, J.L. Firkins, J. Dijkstra, E. Kebreab, G. Waghorn, H.P.S. Makkar, A.T. Adesogan, W. Yang, C. Lee, P.J. Gerber, B. Henderson, and J.M. Tricarico. 2013a. Mitigation of methane and nitrous oxide emissions from animal operations: I. A review of enteric methane mitigation options. *J. Anim. Sci.* 91 (11):5045–5069, doi: 10.2527/jas.2013-6583.

Hristov, A.N., J. Oh, C. Lee, R. Meinen, F. Montes, T. Ott, J. Firkins, A. Rotz, C. Dell, A. Adesogan, W. Yang, J. Tricarico, E. Kebreab, G. Waghorn, J. Dijkstra, and S. Oosting. 2013b. Mitigation of greenhouse gas emissions in livestock production: A review of technical options for non-CO$_2$ emissions. FAO Animal Production and Health Paper No. 177. P. J. Gerber, B. Henderson, and H.P.S. Makkar. Food and Agriculture Organization of the United Nations (FAO), Rome, Italy. 206 p.

IAASTD. 2009. Agriculture at Crossroads: A Global Report. B. D. McIntyre, H. R. Herren, J. Wakhungu, and R. T. Watson. International Assessment of Agricultural Knowledge, Science and Technology for Development (IAASTD), Washington, D.C. 608 p.

Ingram, J., P.J. Ericksen, and D.M. Liverman, eds. 2010. *Food Security and global Climate Change.* London, U.K.: Earthscan

INPE. 2017. *Deforestation estimates in Brazilian Amazon.* Instituto Nacional de Pesquisas Espaciais (INPE) [cited May 2017]. Available from http://www.obt.inpe.br/degrad/.

IPCC. 2006. *IPCC Guidelines for National Greenhouse Gas Inventories, Vol. 4: Agriculture, Forestry and Other Land Use. Chapter 10: Emissions from Livestock and Manure Management,* edited by H. S. Eggleston, L. Buendia, K. Miwa, T. Ngara, and K. Tanabe. Hayama, Japan: Institute for Global Environmental Strategies (IGES).

IPCC. 2013. *Climate change 2013: The physical science basis. Contribution of Working Group I to the Fifth Assessment Report of the Intergovernmental Panel on Climate Change (IPCC),* edited by T. F. Stocker, D. Qin, G.-K. Plattner, M. Tignor, S. K. Allen, J. Boschung, A. Nauels, Y. Xia, V. Bex, and P. M. Midgley. Cambridge, United Kingdom and New York, NY, USA: Cambridge University Press. 1535 pp.

IPCC. 2014a. *Climate Change 2014: Synthesis Report. Contribution of Working Groups I, II and III to the Fifth Assessment Report of the Intergovernmental Panel on Climate Change,* edited by C.W. Team, R.K. Pachauri, and L.A. Meyer. Geneva, Switzerland: Intergovernmental Panel on Climate Change. 151 pp.

IPCC. 2014b. *Climate Change 2014: Impacts, Adaptation, and Vulnerability. Part A: Global and Sectoral Aspects. Contribution of Working Group II to the Fifth Assessment Report of the Intergovernmental Panel on Climate Change,* edited by C.B. Field, V.R. Barros, D.J. Dokken, K.J. Mach, M.D. Mastrandrea, T.E. Bilir, M. Chatterjee, K.L. Ebi, Y.O. Estrada, R.C. Genova, B. Girma, E.S. Kissel, A.N. Levy, S. MacCracken, P. R. Mastrandrea, and L.L.White. Cambridge, U.K. and New York: Cambridge University Press. 1132 pp.

IPCC. 2015. Meeting Report of the Intergovernmental Panel on Climate Change Expert Meeting on Climate Change, Food, and Agriculture, edited by M.D. Mastrandrea, K.J. Mach, V.R. Barros, T.E. Bilir, D.J. Dokken, O. Edenhofer, C.B. Field, T. Hiraishi, S. Kadner, T. Krug, J.C. Minx, R. Pichs-Madruga, G.-K. Plattner, D. Qin, Y. Sokona, T.F. Stocker, and M. Tignor World Meteorological Organization, Geneva, Switzerland. 68 pp.

Kessler, T.J., and C.F. Harvey. 2001. The global flux of carbon dioxide into groundwater. *Geophys. Res. Lett.* 28 (2):279–282, doi: 10.1029/2000GL011505.

Kimball, B.A. 2016. Crop responses to elevated CO_2 and interactions with H_2O, N, and temperature. *Curr. Opin. Plant Biol.* 31:36–43, doi: https://doi.org/10.1016/j.pbi.2016.03.006.

Kimball, B.A., K. Kobayashi, and M. Bindi. 2002. Responses of agricultural crops to free-air CO_2 enrichment. *Adv. Agron.* 77:293–368, doi: 10.1016/s0065-2113(02)77017-x.

Kirschenmann, F.L. 2007. Potential for a new generation of biodiversity in agroecosystems of the future. *Agron. J.* 99 (2):373–376, doi: 10.2134/agronj2006.0104.

Knapp, J.R., G.L. Laur, P.A. Vadas, W.P. Weiss, and J.M. Tricarico. 2014. Enteric methane in dairy cattle production: Quantifying the opportunities and impact of reducing emissions. *J. Dairy Sci.* 97 (6):3231–3261, doi: https://doi.org/10.3168/jds.2013-7234.

Kraham, S.J. 2017. Environmental impacts of industrial livestock production. In *International Farm Animal, Wildlife and Food Safety Law*, edited by G. Steier, and K. K. Patel, pp. 3–40. Cham: Springer International Publishing.

Lal, R. 1990. Soil erosion and land degradation: The global risks. In *Advances in Soil Science: Soil Degradation*, edited by R. Lal, and B. A. Stewart, pp. 129–172. New York: Springer.

Landi, A., A.R. Mermut, and D.W. Anderson. 2003. Origin and rate of pedogenic carbonate accumulation in Saskatchewan soils, Canada. *Geoderma* 117 (1):143–156, doi: https://doi.org/10.1016/S0016-7061(03)00161-7.

Lapenis, A.G., G.B. Lawrence, S.W. Bailey, B.F. Aparin, A.I. Shiklomanov, N.A. Speranskaya, M.S. Torn, and M. Calef. 2008. Climatically driven loss of calcium in steppe soil as a sink for atmospheric carbon. *Global Biogeochem. Cycles* 22 (2), doi: 10.1029/2007GB003077.

Larson, S.C., and A. Wolk. 2012. Red and processed meat consumption and risk of pancreatic cancer: Meta-analysis of perspective studies. *Br. J. Cancer* 106:603–607.

Legesse, G., K.H. Ominski, K.A. Beauchemin, S. Pfister, M. Martel, E.J. McGeough, A.Y. Hoekstra, R. Kroebel, M.R.C. Cordeiro, and T.A. McAllister. 2017. Quantifying water use in ruminant production. *J. Anim. Sci.* 95 (5):2001–2018, doi: 10.2527/jas.2017.1439.

Lemaire, G., R. Wilkins, and J. Hodgson. 2005. Challenges for grassland science: Managing research priorities. *Agric. Ecosyst. Environ.* 108 (2):99–108, doi: 10.1016/j.agee.2005.01.003.

Lobell, D.B., M. Bänziger, C. Magorokosho, and B. Vivek. 2011. Nonlinear heat effects on African maize as evidenced by historical yield trials. *Nature Clim. Change* 1 (1):42–45.

Ma, J., R. Liu, L.-S. Tang, Z.-D. Lan, and Y. Li. 2014. A downward CO_2 flux seems to have nowhere to go. *Biogeosciences* 11, 6251–6262, doi: 10.5194/bg-11-6251-2014.

MacDonald, J.M., and W.D. McBride. 2009. The transformation of U.S. Livestock Agriculture Scale, Efficiency, and Risks: A Report from the Economic Research Service. Economic Research Service, United States Department of Agriculture (USDA), 43 Washington, D.C. 40 p.

Martin, C., D.P. Morgavi, and M. Doreau. 2010. Methane mitigation in ruminants: From microbe to the farm scale. *Animal* 4 (3):351–365, doi: 10.1017/s1751731109990620.

Matson, P.A., W.J. Parton, A.G. Power, and M.J. Swift. 1997. Agricultural intensification and ecosystem properties. *Science* 277 (5325):504–509, doi: 10.1126/science.277.5325.504.

McDermott, J.J., S.J. Staal, H.A. Freeman, M. Herrero, and J.A. Van de Steeg. 2010. Sustaining intensification of smallholder livestock systems in the tropics. *Livest. Sci.* 130 (1–3):95–109, doi: https://doi.org/10.1016/j.livsci.2010.02.014.

McMichael, A.J., J.W. Powles, C.D. Butler, and R. Uauy. 2007. Food, livestock production, energy, climate change, and health. *Lancet* 370 (9594):1253–1263, doi: 10.1016/S0140-6736(07)61256-2.

MEA. 2005. *Millennium Ecosystem Assesment: Ecosystems and Human Well-Being: Synthesis.* Washington D.C.: Island Press. 137 pp.

Mekonnen, M.M., and A.Y. Hoekstra. 2012. A global assessment of the water footprint of farm animal products. *Ecosystems* 15 (3):401–415, doi: 10.1007/s10021-011-9517-8.

Monger, H.C., R.A. Kraimer, S.E. Khresat, D.R. Cole, X. Wang, and J. Wang. 2015. Sequestration of inorganic carbon in soil and groundwater. *Geology* 43 (5):375–378, doi: https://doi.org/10.1130/G36449.1.

Morton, D.C., R.S. DeFries, Y.E. Shimabukuro, L.O. Anderson, E. Arai, F. del Bon Espirito-Santo, R. Freitas, and J. Morisette. 2006. Cropland expansion changes deforestation dynamics in the southern Brazilian Amazon. *Proc. Natl. Acad. Sci. U.S.A.* 103 (39):14637–14641, doi: 10.1073/pnas.0606377103.

Moyo, S., and F. Swanepoel. 2010. Multifunctionality of livestock in developing communities. In *The Role of Livestock in Developing Communities: Enhancing Multifunctionality*, edited by F. Swanepoel, A. Stroebel, and S. Moyo, pp. 1–11. Bloemfontein, South Africa: The Technical Centre for Agricultural (UFC) and Rural Cooperation (CTA).

Naqvi, S.M.K., D. Kumar, R.K. Paul, and V. Sejian. 2012. Environmental stresses and livestock reproduction. In *Environmental Stress and Amelioration in Livestock Production*, pp. 97–128. New York: Springer.

Nardone, A., B. Ronchi, N. Lacetera, M.S. Ranieri, and U. Bernabucci. 2010. Effects of climate changes on animal production and sustainability of livestock systems. *Livest. Sci.* 130 (1–3):57–69, doi: https://doi.org/10.1016/j.livsci.2010.02.011.

Nepstad, D., D. McGrath, C. Stickler, A. Alencar, A. Azevedo, B. Swette, T. Bezerra, M. DiGiano, J. Shimada, R. Seroa da Motta, E. Armijo, L. Castello, P. Brando, M.C. Hansen, M. McGrath-Horn, O. Carvalho, and L. Hess. 2014. Slowing Amazon deforestation through public policy and interventions in beef and soy supply chains. *Science* 344 (6188):1118–1123, doi: 10.1126/science.1248525.

Nepstad, D.C., C.M. Stickler, and O.T. Almeida. 2006. Globalization of the Amazon soy and beef industries: Opportunities for conservation (Globalización de las industrias de soya y Ganado del Amazonas: Oportunidades para la conservación). *Conserv. Biol.* 20 (6):1595–1603, doi: 10.1111/j.1523-1739.2006.00510.x.

Nijdam, D., T. Rood, and H. Westhoek. 2012. The price of protein: Review of land use and carbon footprints from life cycle assessments of animal food products and their substitutes. *Food Policy* 37 (6):760–770, doi: https://doi.org/10.1016/j.foodpol.2012.08.002.

O'Higgins, T.G., and A.J. Gilbert. 2014. Embedding ecosystem services into the Marine Strategy Framework Directive: Illustrated by eutrophication in the North Sea. *Estuarine Coastal Shelf Sci.* 140 (Supplement C):146–152, doi: https://doi.org/10.1016/j.ecss.2013.10.005.

O'Mara, F.P. 2012. The role of grasslands in food security and climate change. *Ann. Bot.* 110 (6):1263–1270, doi: 10.1093/aob/mcs209.

O'Mara, F.P. 2011. The significance of livestock as a contributor to global greenhouse gas emissions today and in the near future. *Anim. Feed Sci. Technol.* 166–167:7–15, doi: https://doi.org/10.1016/j.anifeedsci.2011.04.074.

Oki, T., and S. Kanae. 2006. Global hydrological cycles and world water resources. *Science* 313 (5790):1068–1072, doi: 10.1126/science.1128845.

Oki, T., S. Kanae, and K. Musiake. 2003. Global hydrological cycle and world water resources. *Membrane* 28 (5):206–214, doi: 10.5360/membrane.28.206.

Oldeman, R.L. 1994. The global extent of soil degradation. In *Soil Resilience and Sustainable Land Use*, edited by D. J. Greenland and I. Szabolcs, pp. 99–118. Willingford: CAB International.

Olesen, J.E., and M. Bindi. 2002. Consequences of climate change for European agricultural productivity, land use and policy. *Eur. J. Agron.* 16 (4):239–262, doi: https://doi.org/10.1016/S1161-0301(02)00004-7.

Pan, W., D. Carr, A. Barbieri, R. Bilsborrow, and C. Suchindran. 2007. Forest clearing in the Ecuadorian Amazon: A study of patterns over space and time. *Population Res. and Policy Rev.* 26 (5–6):635–659, doi: 10.1007/s11113-007-9045-6.

Peralta, J.M., J. Reynolds, and C.V. Kerr. 2014. Sustainability and animal agriculture. In *Encyclopedia of Food and Agricultural Ethics*, pp. 1673–1679.

Pilon, C., P. Moore, D. Pote, J. Pennington, J. Martin, D. Brauer, R. Raper, S. Dabney, and J. Lee. 2017. Long-term effects of grazing management and buffer strips on soil erosion from pastures. *J. Environ. Qual.* 46 (2):364–372.

Polley, H.W., D.D. Briske, J.A. Morgan, K. Wolter, D.W. Bailey, and J.R. Brown. 2013. Climate change and North American rangelands: Trends, projections, and implications. *Rangeland Ecol. Manage.* 66 (5):493–511, doi: 10.2111/REM-D-12-00068.1.

Popp, A., H. Lotze-Campen, and B. Bodirsky. 2010. Food consumption, diet shifts and associated non-CO_2 greenhouse gases from agricultural production. *Global Environ. Change* 20 (3):451–462, doi: 10.1016/j.gloenvcha.2010.02.001.

Porter, J.R., L. Xie, A.J. Challinor, K. Cochrane, S.M. Howden, M.M. Iqbal, D.B. Lobell, and M.I. Travasso. 2014. Food security and food production systems. In *Climate Change 2014: Impacts, Adaptation, and Vulnerability. Part A: Global and Sectoral Aspects. Contribution of Working Group II to the Fifth*

Assessment Report of the Intergovernmental Panel on Climate Change, edited by C.B. Field, V.R. Barros, D.J. Dokken, K.J. Mach, M.D. Mastrandrea, T.E. Bilir, M. Chatterjee, K.L. Ebi, Y.O. Estrada, R.C. Genova, B. Girma, E.S. Kissel, A.N. Levy, S. MacCracken, P.R. Mastrandrea, and L.L. White, pp. 485–533. Cambridge, U.K. and New York: Cambridge University Press.

Potter, P., N. Ramankutty, E.M. Bennett, and S.D. Donner. 2010. Characterizing the spatial patterns of global fertilizer application and manure production. *Earth Interact.* 14 (2):1–22.

Pretty, J., and Z.P. Bharucha. 2014. Sustainable intensification in agricultural systems. *Ann. Bot.* 114 (8):1571–1596, doi: 10.1093/aob/mcu205.

Ramankutty, N., A.T. Evan, C. Monfreda, and J.A. Foley. 2008. Farming the planet: 1. Geographic distribution of global agricultural lands in the year 2000. *Global Biogeochem. Cycles* 22 (1), doi: 10.1029/2007gb002952.

Randolph, T.F., E. Schelling, D. Grace, C.F. Nicholson, J.L. Leroy, D.C. Cole, M.W. Demment, A. Omore, J. Zinsstag, and M. Ruel. 2007. Invited review: Role of livestock in human nutrition and health for poverty reduction in developing countries. *J. Anim. Sci.* 85:2788–2800, doi: 10.2527/jas.2007-0467.

Ravi, S., D.D. Breshears, T.E. Huxman, and P. D'Odorico. 2010. Land degradation in drylands: Interactions among hydrologic–aeolian erosion and vegetation dynamics. *Geomorphology* 116 (3–4):236–245, doi: https://doi.org/10.1016/j.geomorph.2009.11.023.

Reid, R.S., K.A. Galvin, and R.S. Kruska. 2008. Global significance of extensive grazing lands and pastoral societies: An introduction. In *Fragmentation in Semi-Arid and Arid Landscapes: Consequences for Human and Natural Systems,* edited by K.A. Galvin, R.S. Reid, R.H. Behnke Jr, and N.T. Hobbs, pp. 1–24. Dordrecht: Springer.

Reid, R.S., P.K. Thornton, G.J. McCrabb, R.L. Kruska, F. Atieno, and P.G. Jones. 2004. Is it possible to mitigate greenhouse gas emissions in pastoral Ecosystems of the Tropics? In *Tropical Agriculture in Transition—Opportunities for Mitigating Greenhouse Gas Emissions?,* edited by R. Wassmann and P.L.G. Vlek, pp. 91–109. Dordrecht: Springer.

Ribaudo, M., N. Gollehon, M. Aillery, J. Kaplan, R. Johansson, J. Agapoff, L. Christensen, V. Breneman, and M. Peters. 2003. Manure Management for Water Quality: Costs to Animal Feeding Operations of Applying Manure Nutrients to Land. Agricultural Economic Report No. AER-824 U.S. Department of Agriculture (USDA), AER-824 Wahington, D.C. 97 pp.

Robinson, T.P., P.K. Thornton, G. Franceschini, R.L. Kruska, F. Chiozza, A. Notenbaert, G. Cecchi, M. Herrero, M. Epprecht, S. Fritz, L. You, G. Conchedda, and L. See. 2011. Global livestock production systems. Food and Agriculture Organization of the United Nations (FAO) and International Livestock Research Institute (ILRI), Rome, Italy. 152 pp.

Rojas-Downing, M.M., A.P. Nejadhashemi, T. Harrigan, and S.A. Woznicki. 2017. Climate change and livestock: Impacts, adaptation, and mitigation. *Clim. Risk Manage.* 16:145–163, doi: https://doi.org/10.1016/j.crm.2017.02.001.

Roland, L., M. Drillich, D. Klein-Jöbstl, and M. Iwersen. 2016. Invited review: Influence of climatic conditions on the development, performance, and health of calves. *J. Dairy Sci.* 99 (4):2438–2452, doi: https://doi.org/10.3168/jds.2015-9901.

Rosegrant, M.W., M. Fernandez, A. Sinha, J. Alder, H. Ahammad, C. Fraiture, B. Eickhour, J. Fonseca, J. Huang, O. Koyama, A.M. Omezzine, P. Pingali, R. Ramirez, C. Ringler, S. Robinson, P. Thornton, D. van Vuuren, and H. Yana-Shapiro. 2009. Looking into the future for agriculture and AKST. In *Agriculture at a Crossroads: Global report.,* edited by B.D. McIntyre, H.R. Herren, J. Wakhungu, and R.T. Watson, pp. 307–376. Washington, D.C.: Island Press.

Rosenzweig, C., D. Karoly, M. Vicarelli, P. Neofotis, Q.G. Wu, G. Casassa, A. Menzel, T.L. Root, N. Estrella, B. Seguin, P. Tryjanowski, C.Z. Liu, S. Rawlins, and A. Imeson. 2008. Attributing physical and biological impacts to anthropogenic climate change. *Nature* 453 (7193):353–U320, doi: 10.1038/nature06937.

Russelle, M.P., M.H. Entz, and A.J. Franzluebbers. 2007. Reconsidering integrated crop–livestock systems in North America. *Agron. J.* 99 (2):325–334, doi: 10.2134/agronj2006.0139.

Sakadevan, K., and M.-L. Nguyen. 2017. Livestock production and its impact on nutrient pollution and greenhouse gas emissions. *Adv. Agron.* 141:147–184.

Schauberger, B., S. Archontoulis, A. Arneth, J. Balkovic, P. Ciais, D. Deryng, J. Elliott, C. Folberth, N. Khabarov, and C. Müller. 2017. Consistent negative response of US crops to high temperatures in observations and crop models. *Nat. Commun.* 8:13931.

Schlesinger, W.H. 1982. Carbon storage in the caliche of arid soils: A case study from Arizona. *Soil Sci.* 133 (4):247–255.

Selman, M., and S. Greenhalgh. 2010. Eutrophication: Sources and drivers of nutrient pollution. *Renewable Resour. J.* 26 (4):19–26.

Sere, C., and H. Steinfeld. 1996. World Livestock Production Systems: Current Status, Issues and Trends. FAO Animal Production Paper No. 127. Food and Agriculture Organization (FAO), Rome, Italy. 51 pp.

Sidahmed, A.E., A. Nefzaoui, and M. El-Mourid. 2008. Livestock and climate change: Coping and risk management strategies for a sustainable future. In *Livestock and Global Change. Proceedings of International Conference Livestock and Global Climate Change 2008, 17–20 May, 2008 Hammamet, Tunisia*, edited by P. Rowlinson, M. Steele, and A. Nefzaoui, pp. 27–28. Penicuik, U.K.: Cambridge University Press.

Smith, J., K. Sones, D. Grace, S. MacMillan, S. Tarawali, and M. Herrero. 2013a. Beyond milk, meat, and eggs: Role of livestock in food and nutrition security. *Animal Frontiers* 3 (1):6–13, doi: 10.2527/af.2013-0002.

Smith, P. 2013. Delivering food security without increasing pressure on land. *Glob Food Secur.* 2 (1):18–23, doi: https://doi.org/10.1016/j.gfs.2012.11.008.

Smith, P., M. Bustamante, H. Ahammad, H. Clark, H. Dong, E.A. Elsiddig, H. Haberl, R. Harper, J. House, M. Jafari, O. Masera, C. Mbow, N.H. Ravindranath, C.W. Rice, C.R. Abad, A. Romanovskaya, F. Sperling, and F. Tubiello. 2014. Agriculture, Forestry and Other Land Use (AFOLU). In *Climate Change 2014: Mitigation of Climate Change. Contribution of Working Group III to the Fifth Assessment Report of the Intergovernmental Panel on Climate Change*, edited by O. Edenhofer, R. Pichs-Madruga, Y. Sokona, E. Farahani, S. Kadner, K. Seyboth, A. Adler, I. Baum, S. Brunner, P. Eickemeier, B. Kriemann, J. Savolainen, S. Schlömer, C.V. Stechow, T. Zwickel, and J. C. Minx, pp. 811–922. Cambridge, U.K. and New York: Cambridge University Press.

Smith, P., P.J. Gregory, D. van Vuuren, M. Obersteiner, P. Havlik, M. Rounsevell, J. Woods, E. Stehfest, and J. Bellarby. 2010. Competition for land. *Philos. Trans. R. Soc. London, Ser. B* 365 (1554):2941–2957, doi: 10.1098/rstb.2010.0127.

Smith, P., H. Haberl, A. Popp, K.H. Erb, C. Lauk, R. Harper, F.N. Tubiello, A.D. Pinto, M. Jafari, S. Sohi, O. Masera, H. Bottcher, G. Berndes, M. Bustamante, H. Ahammad, H. Clark, H.M. Dong, E.A. Elsiddig, C. Mbow, N.H. Ravindranath, C.W. Rice, C.R. Abad, A. Romanovskaya, F. Sperling, M. Herrero, J.I. House, and S. Rose. 2013b. How much land-based greenhouse gas mitigation can be achieved without compromising food security and environmental goals? *Global Change Biol.* 19 (8):2285–2302, doi: 10.1111/gcb.12160.

Smith, P., D. Martino, Z. Cai, D. Gwary, H. Janzen, P. Kumar, B. McCarl, S. Ogle, F. O'Mara, C. Rice, B. Scholes, O. Sirotenko, M. Howden, T. McAllister, G. Pan, V. Romanenkov, U. Schneider, S. Towprayoon, M. Wattenbach, and J. Smith. 2008. Greenhouse gas mitigation in agriculture. *Philos. Trans. R. Soc. London, Ser. B* 363 (1492):789–813, doi: 10.1098/rstb.2007.2184.

St-Pierre, N.R., B. Cobanov, and G. Schnitkey. 2003. Economic losses from heat stress by US livestock industries. *J. Dairy Sci.* 86, Supplement:E52–E77, doi: https://doi.org/10.3168/jds.S0022-0302(03)74040-5.

Stehfest, E., L. Bouwman, D.P. van Vuuren, M.G.J. den Elzen, B. Eickhout, and P. Kabat. 2009. Climate benefits of changing diet. *Clim. Change* 95 (1–2):83–102, doi: 10.1007/s10584-008-9534-6.

Steinfeld, H., P.J. Gerber, T. Wassenaar, V. Castel, M. Rosales, and C. de Haan. 2006. *Livestock's Long Shadow: Environmental Issues and Options*. Rome, Italy: Food and Agriculture Organization of the United Nations (FAO).

Thomas, C.D., A. Cameron, R.E. Green, M. Bakkenes, L.J. Beaumont, Y.C. Collingham, B.F.N. Erasmus, M.F. de Siqueira, A. Grainger, L. Hannah, L. Hughes, B. Huntley, A.S. van Jaarsveld, G.F. Midgley, L. Miles, M.A. Ortega-Huerta, A. Townsend Peterson, O.L. Phillips, and S.E. Williams. 2004. Extinction risk from climate change. *Nature* 427 (6970):145–148, doi: http://www.nature.com/nature/journal/v427/n6970/suppinfo/nature02121_S1.html.

Thornton, P., and M. Herrero. 2010. The Inter-Linkages between Rapid Growth in Livestock Production, Climate Change, and the Impacts on Water Resources, Land Use, and Deforestation. World Bank Policy Research Working Paper No. WSP 5178. World Bank Washington, D.C. 82 p. Available at: https://ssrn.com/abstract=1536991.

Thornton, P.K. 2010. Livestock production: recent trends, future prospects. *Philos. Trans. R. Soc. London, Ser. B* 365 (1554):2853–2867, doi: 10.1098/rstb.2010.0134.

Thornton, P.K., R.B. Boone, and J. Ramirez-Villegas. 2015. Climate Change Impacts on Livestock. Climate Change, Agriculture and Food Security (CCAFS), Working Paper No. 120. CGIAR Research Program on Climate Change, Agriculture and Food Security (CCAFS), Copenhagen, Denmark. 19 p. Available at: www.ccafs.cgiar.org.

Thornton, P.K., and P.J. Gerber. 2010. Climate change and the growth of the livestock sector in developing countries. *Mitig. Adapt. Strat. Glob. Change* 15 (2):169–184, doi: 10.1007/s11027-009-9210-9.

Thornton, P.K., and M. Herrero. 2014. Climate change adaptation in mixed crop–livestock systems in developing countries. *Glob. Food Secur.-Agr.* 3 (2):99–107, doi: 10.1016/j.gfs.2014.02.002.

Thornton, P.K., J. van de Steeg, A. Notenbaert, and M. Herrero. 2009. The impacts of climate change on livestock and livestock systems in developing countries: A review of what we know and what we need to know. *Agric. Syst.* 101 (3):113–127, doi: 10.1016/j.agsy.2009.05.002.

Tubiello, F.N., M. Salvatore, A.F. Ferrara, J. House, S. Federici, S. Rossi, R. Biancalani, R.D. Condor Golec, H. Jacobs, A. Flammini, P. Prosperi, P. Cardenas-Galindo, J. Schmidhuber, M.J. Sanz Sanchez, N. Srivastava, and P. Smith. 2015. The contribution of agriculture, forestry and other land use activities to global warming, 1990–2012. *Global Change Biol.* 21 (7):2655–2660, doi: 10.1111/gcb.12865.

Tubiello, F.N., M. Salvatore, S. Rossi, A. Ferrara, N. Fitton, and P. Smith. 2013. The FAOSTAT database of greenhouse gas emissions from agriculture. *Environ. Res. Lett.* 8 (1), doi: 10.1088/1748-9326/8/1/015009.

U.N. 2007. The State of World Population: Unleashing the Potential of Urban Growth. United Nations Population Fund, New York. 99 pp. Available at: http://www.unfpa.org/sites/default/files/pub-pdf/695_filename_sowp2007_eng.pdf.

U.N. 2017. World Population Prospects: The 2017 Revision. Working Paper No. ESA/P/WP/248. United Nations (U.N), Department of Economic and Social Affairs, Population Division.

UF/IFAS. Range Cattle Education Center. University of Florida Range Cattle Research Center. Available from http://rcrec-ona.ifas.ufl.edu/in-focus/if7-21-06.shtml.

USEPA. 2012. Global Anthropogenic Non-CO_2 Greenhouse Gas Emissions: 1990–2030. Office of Atmospheric Programs, Climate Change Division, U.S. Environmental Protection Agency (USEPA), EPA 430-R-12-006 Washington, D.C. 176 pp. Available at: https://www.epa.gov/sites/production/files/2016-08/documents/epa_global_nonco2_projections_dec2012.pdf.

Van der Hoven, S.J., and J. Quade. 2002. Tracing spatial and temporal variations in the sources of calcium in pedogenic carbonates in a semiarid environment. *Geoderma* 108 (3):259–276, doi: https://doi.org/10.1016/S0016-7061(02)00134-9.

van Horn, H.H., G.L. Newton, and W.E. Kunkle. 1996. Ruminant nutrition from an environmental perspective: Factors affecting whole-farm nutrient balance. *J. Anim. Sci.* 74 (12):3082–3102.

Vanuytrecht, E., and P.J. Thorburn. 2017. Responses to atmospheric CO2 concentrations in crop simulation models: A review of current simple and semicomplex representations and options for model development. *Global Change Biol.* 23 (5):1806–1820, doi: 10.1111/gcb.13600.

Victor, D.G., D. Zhou, E.H.M. Ahmed, P.K. Dadhich, J.G.J. Olivier, H.-H. Rogner, K. Sheikho, and M. Yamaguchi. 2014. Introductory Chapter. In *Climate Change 2014: Mitigation of Climate Change. Contribution of Working Group III to the Fifth Assessment Report of the Intergovernmental Panel on Climate Change*, edited by O. Edenhofer, R. Pichs-Madruga, Y. Sokona, E. Farahani, S. Kadner, K. Seyboth, A. Adler, I. Baum, S. Brunner, P. Eickemeier, B. Kriemann, J. Savolainen, S. Schlömer, C. von Stechow, T. Zwickel, and J. C. Minx, pp. 111–150. Cambridge, U.K. and New York: Cambridge University Press.

Wall, E., A. Wreford, K. Topp, and D. Moran. 2010. Biological and economic consequences heat stress due to a changing climate on UK livestock. *Adv. Anim. Biosci.* 1 (1):53.

Wang, X., B.G. McConkey, A.J. VandenBygaart, J. Fan, A. Iwaasa, and M. Schellenberg. 2016. Grazing improves C and N cycling in the Northern Great Plains: A meta-analysis. *Sci. Rep.* 6:33190, doi: 10.1038/srep33190.

Wassenaar, T., P. Gerber, P.H. Verburg, M. Rosales, M. Ibrahim, and H. Steinfeld. 2007. Projecting land use changes in the Neotropics: The geography of pasture expansion into forest. *Global Environ. Change* 17 (1):86–104, doi: https://doi.org/10.1016/j.gloenvcha.2006.03.007.

Weindl, I., H. Lotze-Campen, A. Popp, C. Mueller, P. Havlik, M. Herrero, C. Schmitz, and S. Rolinski. 2015. Livestock in a changing climate: Production system transitions as an adaptation strategy for agriculture. *Environ. Res. Lett.* 10 (9), doi: 10.1088/1748-9326/10/9/094021.

Westhoek, H., T. Rood, M.V.d. Berg, J. Janse, D. Nijdam, M. Reudink, and E. Stehfest. 2011. The Protein Puzzle: The consumption and production of meat, dairy and fish in the European Union. PBL Netherlands Environmental Assessment Agency, The Hague, Netherland. 218 pp.

WHO/FAO. 2003. Diet, nutrition, and prevention of chronic disease. Report of a joint WHO/FAO Expert Consultation. WHO Technical Series No. 916. World Health Organization (WHO), Geneva, Switzerland.

Wilkerson, V.A., D.R. Mertens, and D.P. Casper. 1997. Prediction of excretion of manure and nitrogen by Holstein dairy cattle. *J. Dairy Sci.* 80 (12):3193–3204.

Wilkins, R.J. 2008. Eco-efficient approaches to land management: A case for increased integration of crop and animal production systems. *Philos. Trans. R. Soc. London, Ser. B* 363 (1491):517–525, doi: 10.1098/rstb.2007.2167.

Wirsenius, S. 2003. The biomass metabolism of the food system: A model-based survey of the global and regional turnover of food biomass. *J. Ind. Ecol.* 7 (1):47–80.

Wirsenius, S., C. Azar, and G. Berndes. 2010. How much land is needed for global food production under scenarios of dietary changes and livestock productivity increases in 2030? *Agric. Syst.* 103 (9):621–638, doi: 10.1016/j.agsy.2010.07.005.

World_Bank. 2009. Minding the stock: Bringing public policy to bear on livestock sector development. Report No.44010-GLB. The World Bank, Washington, D.C. 74 pp.

Xiao, G.J., W.X. Liu, Q. Xu, Z.J. Sun, and J. Wang. 2005. Effects of temperature increase and elevated CO_2 concentration, with supplemental irrigation, on the yield of rain-fed spring wheat in a semiarid region of China. *Agric. Water Manage.* 74 (3):243–255, doi: 10.1016/j.agwat.2004.11.006.

Zhu, C., and F.W. Schwartz. 2011. Hydrogeochemical processes and controls on water quality and water management. *Elements* 7:169–174, doi: 10.2113/gselements.7.3.169.

15 Soil and Human Health in a Changing Climate

Kathi J. Kemper, Jeffery Lakritz, and Rattan Lal

CONTENTS

15.1 INTRODUCTION

Climate change involves increases in air, sea, and terrestrial temperatures; severity and frequency of extreme weather events; and decline in quality of air, water, and soils. The attendant adverse effects of the projected climate change on human health are also widely recognized (McMichael et al. 2003; Luber and Lemery 2015). Incidences of extreme climate events have increased the incidence of human diseases including malaia, dengue fever, cholera, and leptospirosis (Epstein 2005) (Table 15.1). Climate change may aggravate the incidence of excessive heat-related illnesses, vector and water-borne diseases, and increase in susceptibility to cardiovascular and respiratory diseases (Luber and Prudent 2009). The adverse effects of climate on human health may be relatively more severe in developing countries but can also jeopardize human health in economically and scientifically advanced countries as well (CDC 2015; Clayton et al. 2014; Fann et al. 2015; Schwartz et al. 2015).

Change in climate, being an active factor of soil formation (Jenny 1941), also changes soil properties, soil health, and its functionality. Soil health is defined as the soil's fitness to function as an ecosystem that sustains productive populations of soil organisms which support environmental, plant, and animal health. Healthy soil is associated with more productive croplands and indirectly by more productive animals. Soil health indicates "the capacity of soil to function as a vital living system to sustain biological productivity, promote environmental quality, and maintain plant and animal health" (Doran and Zeiss 2000). Healthy soil contains nutrients, water, microbial agents, algae, protozoa, nematodes, mites, insects, earthworm's, and plant roots which function to reduce erosion, recycle nutrients, and detoxify and reduce pests which may alter the production of viable biomass. Arable land fertilized with animal manure supports the growth of plants and maintains organic matter and microorganisms that maintain soil ecosystems.

As modern agricultural practices and demand for high quality animal sourced foods have increased around the globe, a shift from extensive, mixed crop and animal rearing to monocultures and intensive livestock production systems (generally concentrated into defined geographic locations) which have both resulted in the loss of fertile soil components through erosion, loss of organic matter, reductions of the numbers and species of soil organisms and increased release of gases which may alter the environment as well as concentrated sources of animal wastes, gases, and concentration of minerals, drugs, and other agents in the environment.

TABLE 15.1

Link between Climate Change, Environment Change, Soil Health, and Human Health

Disease	Agricultural/Environmental Changes	Incidence and Pathways
Malaria, Rift Valley fever	Deforestation, expansion, and intensification of agriculture	Increased breeding sites and vector resistance to pesticides
Schistosomiasis	Expansion of rice paddies, flood irrigation, increase in irrigated lands	Increase in human–snail contact
Lyme	Afforestation of marginal land	Incidence of tick host
Red Tide	Non-point source pollution, nutrient (N, P) runoff from agricultural lands	Increase in incidence of algal bloom by warm water
Dengue	Urban slums, high population density	Waterborne breeding ground for mosquito (*Aedes aegypti*)
Pulmonary diseases	Unhygienic urban environments	Increased incidence of rodents

Climate warming can reduce plants' available water capacity (PAWC) of the root zone and exacerbate drought. Soil functionality will also be adversely affected with increased risks of accelerated erosion, decomposition of soil organic matter (SOM), salinization, and exacerbation of other degradative processes by the current and projected climate change through increase in biotic and abiotic stresses (Frazer 2009).

There are several processes by which soil health affects human health: physical, biochemical, biological, and ecological (Figure 15.1). In addition, human and animal activities affect soil health. For example, sewage and animal waste containing antibiotics runs into water and eventually into

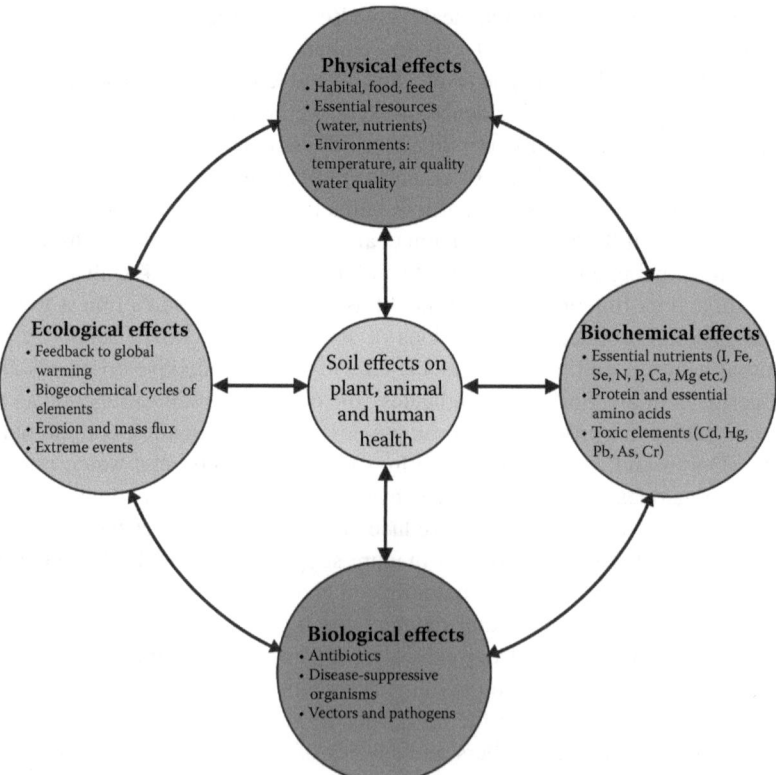

FIGURE 15.1   Direct and indirect effects of soil properties and processes on health of plants, animals, and humans.

soil, exerting selective pressure for soil microorganisms to develop antibiotic resistance (Diaz-Cruz 2003). Washing synthetic clothes disperses plastic microfibers into water and soil (Napper et al. 2016). Industrial activities such as mining can lead to massive soil and water contamination with heavy metals. Expanding farm land by cutting down forests and expanding human habitation into ever-growing cities and suburbs diminishes the availability of soil-restoring forests and perennial prairies (Seto 2012). The objective of this chapter is to describe the inter-connectivity between the health of soil, plants, animals, and people and provide specific examples of the adverse effects on health of plants, animals, and people with the degradation of soil quality and its functionality because of adverse changes in soil's physical, chemical, biological, and ecological properties.

15.2 FACTORS AFFECTING SOIL HEALTH IN A CHANGING CLIMATE

Soil health is affected by physical, chemical, biological, and ecological properties. Soil chemical properties, in relation to plant and animal health, include its capacity to retain and supply essential macro- and micronutrients. Positive and negative effects of soil health on plant and animal health are shown in Figure 15.2. There are about 20 nutrients needed by humans and these must be supplied through plants and animal products consumed by humans (Vogt et al. 2010). For example, bone and teeth formation in humans requires Ca, Mg, F, Cu, and Mn. The formation of protein, amino acids, and hormones depends on the availability of N. Processes governing energy transformation in biochemical pathways (ADP, ATP) require P, and cardiac functions need adequate supplies of K, Mg, Cu, and Se. Availability of several micronutrients is essential for several physiological functions. For example, Zn is needed for healthy skin, K for kidney functions and sugar metabolism, and asthma and bronchiolitis control needed adequate supplies of Zn, Cu, and Se (Vogt et al. 2010). Input of Na is needed for the development of the nervous system, I for goiter and thyroid, and Mn for immune system development. However, some nutrients needed by humans (i.e., Na, I, Se) are not essential to plant growth and development. Thus, humans must obtain these nutrients through animal-based food.

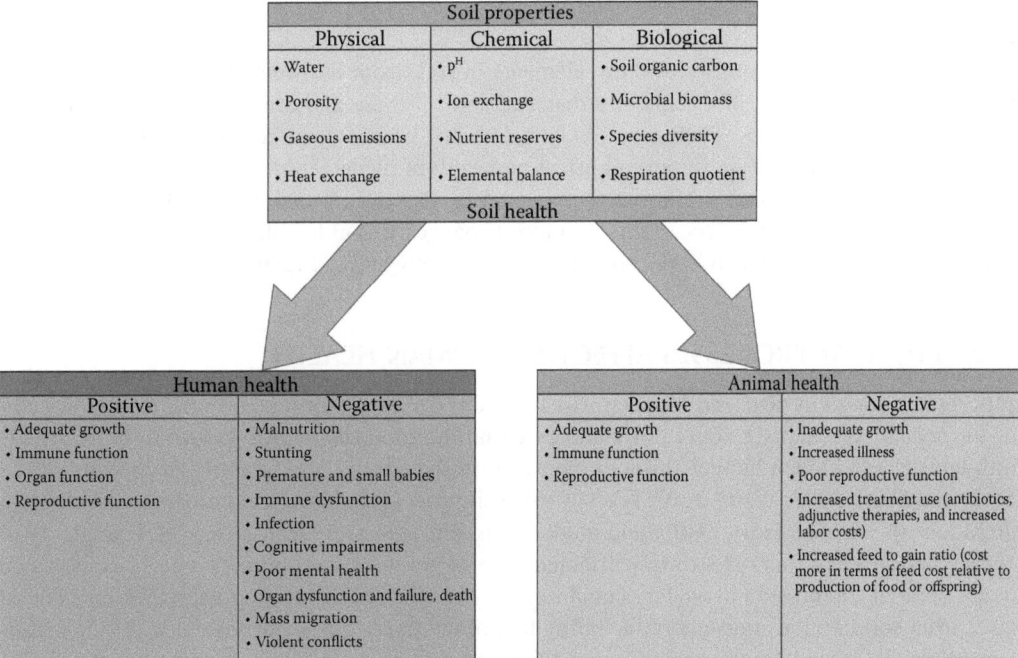

FIGURE 15.2 Positive and negative effects of soil properties on human and animal health.

TABLE 15.2

Climate Change and Soil Processes in Relation to Human and Animal Health

Climate Change	Soil	Effects on Health of Humans	Effects on Health of Animals
Elevated temperature	Soil desiccation	• Airborne particulates and pathogens • Heat stress	Reduced forage/grain production; reduced quality of feed produced
Elevated temperature	Reduced water availability	• Reduced potable water • Waterborne diseases	Reduced forage/grain production
Elevated temperature	Reduced arable land	• Limited resources for production of food • Limited resources for production of animal feed • Reduced soil quality	Competition between humans and animals for food stuffs
Reduced water availability	Poor quality water used to irrigate arable land	• Accumulation of heavy metals in soil (groundwater) • Accumulation of persistent xenobiotics in water • Accumulation of pathogens in water	Increased exposure to potential toxicants and pathogens through application of contaminated water
Increased water (floods, regional occurrences)	Contamination of soil with microbial pathogens	• Contamination of water with heavy metals, xenobiotics, pathogens	Increased Leptospirosis, Campylobacteriosis, Salmonellosisk, Listeriosis, Cryptosporidiosis
Increased disease in non-endemic areas	Warming environments of traditionally colder climates	• Increased blood feeding vectors in environment • Transmission of virus, bacteria, hemoparasites	Tick borne diseases (*Babesia bovis*, anaplasmosis), viral agents (Bluetongue virus)

In addition to the availability of these elements in soil, from the weathering of parent rock or input of soil amendments, biochemical transformations of these and other elements are governed by soil microbial processes. Therefore, biotic and abiotic processes affecting activity and species diversity of soil biota have an important effect on health of plants, animals and human dependent on the soil. These processes are strongly impacted by the current and projected climate change (Tables 15.1 and 15.2). Therefore, climate-induced changes in soil health (i.e., physical, chemical, biological, and ecological) can impact human health directly and indirectly.

15.3 PHYSICAL PROCESSES AFFECTING HUMAN HEALTH

Important among physical and biochemical functions are provisioning of food, feed, and habitat for people, and being a source of some essential micronutrients. Soil health directly affects both hunger and hidden hunger through its effect on the quality and quantity of food produced (Stein and Qaim 2007; WHO 2007; FAO 2016). Soil processes also affect human health through alterations of water and air quality, and moderation of micro- and meso-climates. The quality of surface and groundwaters is strongly affected by non-point source pollution and is exacerbated by accelerated soil erosion by water. The air quality is affected by wind erosion and emission of gases from soil into the atmosphere including that of methane (CH_4), nitrous oxide (N_2O), hydrogen sulfide (H_2S), and others depending on soil temperature and moisture regime and the degree of anaerobiosis.

15.3.1 CHEMICAL AND NUTRITIONAL EFFECTS OF SOIL ON HUMAN HEALTH

Healthy soil contains the proper balance of macronutrients and micronutrients (Figure 15.1) to support the optimal growth of humans and livestock that consume plants and fungi (Lal 2009). Globally, 815 million people (mostly in sub-Saharan Africa and Asia) are prone to food insecurity (FAO et al. 2017), and as many as 2 billion to hidden hunger and malnutrition. Soil and hence dietary deficiencies of iodine, iron, selenium, and zinc, have well-described health consequences for millions more (Golden 1982). For example, iodine deficiency leads to hypothyroidism, fatigue, mental slowing, goiter, stunting, and cretinism; iron deficiency leads to anemia and fatigue; selenium deficiency leads to hypothyroidism; and zinc deficiency increases the risks and severity of gastrointestinal and respiratory diseases, skin rashes (acrodermatitis enteropathica), and lack of appetite. Because of of the degradation of soil quality in the past 50 years of modern agricultural methods, fruits and vegetables grown just a few decades ago were richer in vitamins and minerals than most varieties grown today, particularly for calcium, iron, riboflavin, and ascorbic acid in the US (Davis et al. 2004); similar research in the UK showed marked reductions in the mineral content of commonly grown fruits and vegetables (Mayer 1997). Improved soil health is essential to ensuring humans can obtain adequate micronutrients from crops.

Chronic protein-calorie malnutrition from inadequate macronutrients (protein, fat, and carbohydrates) can also arise from poor soils, crops, and livestock production. Malnutrition from insufficient caloric intake leads to stunting, frailty, impaired immunity, increased susceptibility to infections, low birth weight in infants, reduced fertility, impaired intellectual development and function, apathetic or angry behavior, depression, poor coping, and reduced work capacity (Gabr 1987). Chronic food insecurity from poor soils secondary to climate change also leads to famine, mass migration, and violent conflicts (Reuveny 2007). Current examples can already be found in sub-Saharan Africa and central Asia. For example, from 2006 to 2011, up to 60% of Syria experienced the worst long-term drought and crop failures in hundreds of years, contributing to civil war and mass migration of refugees (Fischetti 2015).

As climate change accelerates, soil health and human nutrition are threatened by rising temperatures which can lead to soil drying, desertification, dustbowls, erosion, and loss of topsoil leading to abandonment of farms such as what occurred in the 1930s in the US (Tables 15.1 and 15.2; McLeman 2014); major weather events such as hurricanes, tornadoes, and floods which can wash away soil and crops and destroy food stores; and salination of crop-growing soils as sea levels rise. Most crop plants are sensitive to soil salinity and the land area affected by increased salinization is growing as climate change advances (Shrivastava and Kumar 2015). Thus, climate change increases the risk of human malnutrition and disease as well as mass migration and violent conflict, all jeopardizing human health.

Furthermore, contaminated soil, containing residues of toxic heavy metals such as lead, mercury, and cadmium, and naturally occurring hazards such as radon gas, as well as neuro-toxic and endocrine-disrupting chemicals, threaten human health (Table 15.3; Zhuang et al. 2009). For example, radon gas exposure increases the risk of lung cancer (Bissett et al. 2010). Hazardous waste sites expose tens of millions of people to toxic heavy metals (Ericson et al. 2013). Lead poisoning leads to loss of intellectual capacity. Endocrine-disrupting compounds such as polychlorinated biphenyls (PCBs), polybrominated biphenyls (PBBs), bisphenol A (BPA), and persistent pesticides such as DDT, dioxin, and phthalates have leached into air, water, and soil and are present in human tissue in early childhood, credibly contributing to early thelarche, disrupted male reproductive development, thyroid dysfunction, neurodevelopmental dysfunction, and obesity (Meeker 2012). Overall, soil contamination adversely affects human health (Table 15.3; European Commission of Science for Environmental Policy 2013).

Routes of exposure from soils to humans are diverse and include eating soil (geophagia, which may be a cultural practice, a childhood behavior, or a medical symptom); inhalation of airborne soil (dust); skin contact; and contact through run-off into water supplies or being taken up into plants

TABLE 15.3
Soil Toxicants and Human Health Effects

Contaminant	Human Health Effects	Animal Health Effects
Heavy metals		
Arsenic	Damage to GI track, skin, heart, liver, nerves, brain. Miscarriage, pre-term birth. Cancer. Death	Acute toxic doses: Gastrointestinal, respiratory, hepatic, renal, cardiovascular, neural, epithelial. Generally associated with exposure to ashes of burnt lumber (chromated, copper arsenate) or insecticides, herbicides or coccidiostats.
Asbestos	Pulmonary asbestosis; lung cancer	
Cadmium	Bone, liver, and kidney damage; cancer; itai-itai disease	Growth retardation, unthrifty, dry skin, hair loss, reproductive failure, sore/enlarged joints, visual defects, hepatitis, nephritis, anemia, Associated with super-phosphate fertilization and in some regions accumulation in sewage sludge
Chromium, hexavalent	Cancer	Dermatitis, oxidative stress, neurologic and chromosomal damage
Lead	Neurodevelopmental delay; lower intelligence; bone damage; high blood pressure; kidney disease	Anemia, myocarditis, impaired neurological function, blindness, colic, nephropathy, abnormal bone formation or dysregulation of calcium homeostasis, decreased fertility
Mercury	Brain and gastrointestinal track damage; lower IQ; poor coordination Liver, heart, kidney damage Teratogenic	Oral ulcers, gastroenteritis, renal failure, neurologic dysfunction
Endocrine-disrupting compounds and persistent organic pollutants (Dioxin, PCB, PBDEs, etc.)	Early thelarche, disrupted male reproductive development, thyroid dysfunction, neurodevelopmental delay and dysfunction, obesity, immune system damage, skin damage, liver damage, cancer	Hormonally active agents: DDT, PCB, PBDE, growth promotants (trenbolone, melengesterol, estrogens), plant toxicants; reduced fertility, congenital defects
Pesticides and herbicides	Non-Hodgkin lymphoma, Parkinson's disease, cognitive and psychomotor dysfunction, neurodegenerative and neurodevelopmental effects, leukemia, birth defects, pediatric brain tumors, cancer of the kidney, prostate, and pancreas	May be associated with heavy metal or trace metal accumulation; neurological disease, congenital deformities, death

Source: Modified and adapted from European Commission, Soil contamination: Impacts on human health, Science for Environment Policy In-Depth Report, Science Community Unit, University of West England, Bristol, England, Issue 5, September 2013, 29 pp; http://ec.europa.eu/scienc-environment-policy, 2013.

which are consumed by humans or animals that are consumed by humans. Soil toxin levels can be increased to dangerous levels by mining, smelting, pressure treated wood products, poor waste disposal, industry, agriculture, and burning fossil fuels whose mineral and particulate matter settles on soil (Wuana and Okieimen 2001). Overall, a variety of modern industrial and agricultural processes can lead to increased levels of biochemical and physical toxins in soil that adversely affect human health.

Climate change compounds these problems. As 2017 Hurricane Harvey showed, increasingly severe storms can lead to flooding of storage facilities of petrochemicals and other toxic compounds, releasing them to the flood waters where they eventually settle on land, including soil used to grow crops for humans and forage for livestock. Also, some human activities that lead to climate change, such as burning fossil fuels, release mercury into the atmosphere, eventually settling in water and soil, where it is taken up by trees and crops (Risch 2017).

15.3.2 BIOLOGICAL AND INFECTIOUS EFFECTS OF SOIL ON HUMAN HEALTH

Healthy soil limits levels of pathogenic helminths, protozoa, fungi, bacteria, and viruses, while supporting healthy populations of earthworms, fungi, and health-promoting microbes, including those that provide natural sources of antibiotic and antifungal medications (Table 15.4, Figure 15.1). About 25% of Earth's species find their home in the soil, and the vast majority of species are non-pathogenic to humans. In fact, many life-saving antibiotics were first identified in and derived from soil organisms. For example, the *Streptomyces* genus of bacteria is responsible for over half the clinically useful antibiotics of natural origin (Watve et al. 2001; Rosin Cerate 2014).

Soilborne infectious diseases of humans include infections by helminths, protozoa, fungi, bacteria, and viruses (Table 15.5) (Jeffrey and van der Putten 2011; Baumgardner 2012). Factors that affect the survival of soilborne viruses and other infectious agents include pH level, organic matter content, moisture content, soil particle size, sunlight exposure, and temperature. It is also notable that the use of pesticides such as glyphosate can reduce levels of neutral or beneficial microbes and theoretically allow overgrowth of pathogenic microbes. Veterinary antibiotics excreted in urine and manure may also place selective pressure on soil microbial communities and allow the spread of resistant traits amongst soil-dwelling microbes.

As climate change progresses, ecological processes are also affected (Figure 15.1). Soils may become warmer and drier (desertification) leading to higher amounts of airborne dust particles carrying microbes. This increases the risk of respiratory infections, particularly in vulnerable populations such as children who have higher relative respiratory rates compared with healthy adults (Mirsaeidi 2016). For example, the marked increase in symptomatic cases of coccidioidomycosis in Arizona and California between 2000 and 2010 has been attributed in part to climate change (Ampel 2010). Similarly, increasing temperatures exacerbate increases in Hantavirus Cardiopulmonary Syndrome associated with the conversion of native vegetation to sugarcane production and in fact may exceed the increases due to expansion of sugarcane-based agriculture (Prist 2017); increases in another hantavirus infection, hemorrhagic fever with renal syndrome (HFRS) or nephropathica epidemica (NE), have also been observed with increased temperature in Europe (Garcia-Menedez et al. 2015).

Finally, healthy soils contribute in two more notable ways to human health and well-being. First, soil-dwelling *Streptomyces* bacteria create an organic chemical compound, geosmin, which lends a distinct earthy taste to beets and contributes to the beautiful scent (petrichor) that occurs in the air when rain falls after a spell of dry weather. Second, the growing interest in the health effects of the microbiome and improved understanding of the gut–brain connection has turned research attention to the question of how soil micro-organisms might affect human mental health. Emerging research suggests that soil bacteria can improve mood and decrease anxiety, perhaps explaining the long-standing clinical observation that gardening improves mental health (Bested et al. 2013; Wood et al. 2016).

The impact of climate change—drought, flooding, storms, increased temperature, increased salinity, increased variability—likely has mixed and largely unknown effects on soil microorganisms. Drought decreases overall microbial biomass and diversity (Bastida 2017), but how this reduction affects the balance of beneficial to pathogenic microbes is uncertain. Effects of increased temperature and CO_2 levels are likely dependent on other factors such as baseline water stress and species composition (De Kauwe et al. 2017). Additional research is necessary to better understand the relationship between climate change and its effects on the soil microbiome and human health. In the meantime, it is clear

TABLE 15.4
Plant Mineral Nutrients in Soil and Their Effects on Human and Animal Health

Plant Nutrients	Human Health	Animal Health
B	Osteoporosis, poor memory and concentration, weak muscles, skin aging.	Seizure-like activity in goats dosed with 2 grams of borate fertilizer; nephritis in other species
Ca	Osteomalacia, rickets	Hypercalcemia is uncommon and generally only observed with iatrogenic administration of high doses (plasma iCa2+ > 3 mM) or animals consuming plants containing calcinogenic compounds (1,25-$(OH)_2$D-glycosides or toxic doses of vitamin D; low calcium diets or diets containing excessive phosphorous may develop rickets (calves), or osteomalacia (adults).
Chromium	Weakness; fatigue; impaired glucose tolerance; high blood pressure; high cholesterol	Glucose tolerance, immune responses, reduced disease incidence (calves), muscle growth, increased number of offspring (swine) in animals supplemented with organic chromium
Cu	Anemia; numbness or tingling of hands or feet; cardiac arrhythmia; osteoporosis; hypothyroidism; hypopigmentation	Excess soil/plant sulfate, molybdenum, zinc, iron reduces availability of copper; copper deficiency results in poor collagen function, faded coat color, anemia, reduced growth, enzootic ataxia, reduced immune functions. In excess, copper accumulates in liver and red cells. Hepatocellular necrosis, hemolytic anemia with methemoglobinemia and renal failure result in death
Fe	Anemia; fatigue; impaired immunity; inattention; prematurity	Iron deficiency anemia commonly observed in neonates associated with low iron stores at birth and milk diet
I	Hypothyroidism, cretinism; fatigue, mental slowing, goiter, stunting; stillbirths; increased infant mortality	Hypothyroidism, goiter, small birthweights, stillbirths and higher mortality
Mg	Constipation; irritability; anxiety; fatigue; weakness; loss of appetite; nausea; muscle spasms; tremors; apathy; ataxia; hypertension (over 300 different metabolic reactions)	Hypomagnesemia associated with altered behavior, CNS disorders (tetany, convulsions, death), altered function of PTH and thus hypocalcemia (weakness, open mouth, gasping)
Mn	Impaired carbohydrate metabolism; poor wound healing	Perosis, carbohydrate, lipid metabolism, growth retardation, dermatitis, reduced reproductive function
Mo	Intellectual deficiency; seizures; opisthotonus, coma; renal failure	Excess molybdenum interacts with copper
S	Weak or brittle connective tissue; impaired detoxification; fatigue; impaired glucose metabolism	Excess sulfur interacts with copper; Excess sulfur associated with H_2S formation in rumen, cerbrocortico necrosis, death
Se	Muscle weakness; myalgia; heart failure; cardiac arrhythmia; Keshan disease; hypothyroidism; increased susceptibility to infection	Selenium deficiency results in white muscle disease (skeletal or cardiac muscle mineralization), reduced thyroid hormone synthesis
Zn	Acrodermatitis enteropathica; diarrhea; respiratory infections; weakened immune system; growth retardation; delayed sexual maturation; alopecia; poor appetite; skin rashes; increased susceptibility to infections	Reduced growth rate, feed intake, reduced testicular size, dermatitis (parakeratosis), poor wound healing, thymic atrophy, impaired immunologic function

TABLE 15.5

Soil-Based Microbes and Their Antibiotic Derivatives

Soil-Based Microbe	Examples of Antimicrobial Compounds
Acremonium chrysogenum	Cephalosporins
Actinomycetes family of bacteria, including *Streptomyces*	Actinomysin, Amphotericin B, Chloramphenicol, Cycloserine, Daptomycin, Erythromycin, Imipenem, Kanamycin, Lincomycin, Neomycin, Nystatin, Rifamycin, Spectinomycin, Streptomycin, Tetracycline, Vancomycin, Virginiamycin
Aspergillus family	Fumagillin
Bacillus subtilis	Bacitracin
Micromonospora echinospora	Gentamicin
Paenibacillus polymyxa	Colistin (Polymyxin B)
Penicillium	Penicillin
Saccharopolyspora erythraea	Erythromycin
Streptoalloteichus tenebrarius	Tobramycin

Source: Aminov R, *Biochem Pharmacol*, 133, 4–19, 2017.

that increasing temperatures in the permafrost enable soil microbes to turn once-frozen vegetation into methane, nitrous oxide, and other greenhouse gases, further increasing climate change (Voigt 2017).

In summary, soil health affects human health physically, chemically, biologically, and ecologically (Figure 15.1). Climate change increasingly adversely affects how soil health impacts human health. Thus, sustainable management of soil health is essential to improve human health and advance Sustainable Development Goals (Agenda 2030) of the United Nations.

15.4 SOIL AND ANIMAL HEALTH

Globally, cattle population numbers approximate 1.49 billion head (FAOSTAT 2017; Wu et al. 2014). These animals are managed in a variety of production systems (grazing ~34%; mixed ~63%, and feedlot 2%; Gerber et al. 2015). Most of the world's livestock are managed under low intensity conditions (98%), in comparison to highly intensive, feedlot-type operations (2%) in developed countries (Gerber et al. 2015). Productivity of cattle fed in these three major systems is directly related to feed quality and thus by the quality of the feed produced from soil (Wu et al. 2014).

Livestock play key roles in the development of food security, through conversion of inedible by-products into meat and dairy foods, production of manure that can improve soil health for future production of feedstuffs, use as draft animals, and income for the owners (Randolph et al. 2007). A common theme among all cattle production is the animal's absolute need to consume forage, grains, or by-products of crop production. As such, all livestock production relies on soil to grow feedstuffs (Doran and Zeiss 2000).

The production of forage, crops, and crop products used to feed cattle is often associated with the use of fertilizers, water, and herbicides and pesticides to optimize production (Tilman et al. 2002). The manure produced by cattle consuming forage, crops, and by-products increases the nutrient value of soil if applied appropriately (Raup 1998). However, manure from livestock supplemented with minerals, when applied to forage or crops, may result in environmental contaminants inadvertently present in these soil treatments (Lopez et al. 2000; Lopez 2012). Some amendments contain heavy metals; drugs and other natural and synthetic compounds may alter soil health, crop growth, and disease resistance and limit productivity and (over time) may reduce animal productivity and health (Zhuang et al. 2009). Supplementation of animal diets with minerals (Table 15.6) to prevent deficiencies or balance the minerals based upon consumed forage, grains, and by-product mineral status is commonplace. Cattle feed supplements that are commonly used provide minerals close

TABLE 15.6

Soilborne Infectious Diseases of Humans and Animals

Category	Examples in Humans	Examples in Animals
Helminths	Ascariasis (roundworms), Echinococcosis, Hookworm, Strongyloidosis (threadworms), Trichuriasis (whipworms), Trichinellosis (Trichinosis)	Trichostrongyles, Monezia, Fascioliasis, Trichuris
Protozoa	*Entamoeba histolytica* (Amebiasis), Cryptosporidiosis, Cylcosporiasis, Giardiasis, Toxoplasmosis	Cryptosporidiosis, Girardiasis, Sarcocystis, Toxoplasmosis, Neospora
Fungi	Aspergillosis, Blastomycosis, Coccidioidomycosis (Valley fever), Histoplasmosis, Sporotrichosis, Mucormycosis, Nocardia	Blastomycosis, Histoplasmosis, Coccidiomycosis, Aspergillosis
Bacteria	Actinomycosis, Anthrax, *Bacillus cereus*, *Campylobacter*, *Clostridium botulinum* (Botulism), *C. perfringens* (gas gangrene), *Clostridium tetanus*, *Coxiella burnetti* (Q fever), *Escherichia coli* (Enterohemorrhagic, Enterotoxigenic, Verotoxigenic, and Enteropathogenic types), Legionella, Leptospirosis, *Listeria monocytogenes*, Lyme disease, Melioidosis, *Pseudomonas*, Salmonellosis (including typhoid fever and non-typhoid salmonella infections), Shigellosis, Tularemia, *Yersinia pestis* (plague)	Actinomycosis, Anthrax, *Bacillus cereus*, *Campylobacter*, *Clostridium botulinum* (Botulism), *C. perfringens* (enteric and histotoxic), *Clostridium tetani*, *Coxiella burnetti* (Q fever), *Escherichia coli* (Enterotoxigenic), Leptospirosis, *Listeria monocytogenes*, Salmonellosis (Sero-group B, C and D)
Viruses	Enteroviruses (poliovirus, coxsackieviruses, and echoviruses), Hantavirus (hantavirus pulmonary syndrome or hemorrhagic fever with renal syndrome)	Bovine viral diarrhea, Bovine herpes virus, Respiratory syncytial viruses, Parainfluenza virus, Coronaviruses, Bluetongue virus, Epizootic hemorrhagic disease virus, Ovine herpes virus type 2 (Malignant catarrhal fever), Rabies virus

to the maximum "tolerable" or "permissible" limits with the goal of maximizing the effects of the supplemented mineral on productivity while minimizing the possibility for toxicity (as manifested by reduced productivity). It is well known that the variety of trace- and macrominerals added to diets can optimize productivity, fertility, and immunologic competence and prevent nutritionally related disorders. It is also known that manure from animals supplemented with minerals can result in these minerals accumulating within the soil and run-off and eventually impact the productivity of the soil and animals consuming crops (Lopez et al. 2000; Lopez 2012). Minerals are commonly added to feed to ensure optimal health and productivity. In the past, with less intensive management systems, trace mineral deficiencies were addressed by simply adding these elements to the diet. Geographically, it was not uncommon for human deficiencies in trace minerals to be observed in farm animals (Lopez 2012). Because significant amounts of micro- and macrominerals added to feed are excreted in the animals' manure and urine, soil health may decline due to the accumulation of these minerals in soil and water (Lopez 2012). Application of excreta slurry to crop land can be expected to result in high soil trace mineral levels, plant accumulation, and potential toxicity to plants and animals consuming forage, concentrates, or mixed diets (Bengtsson et al. 2003; Coppenet et al. 1993; Lopez et al. 2000).

Current rates of mineral supplementation of animal feeds are based upon animal health and productivity. Establishment of the maximum tolerable limits by governmental advisory boards (National Research Council) provides a logical means with which to prevent toxicity. These standardized trace mineral supplements for cattle are commonly used because of the animals' ability to utilize what they need to minimize clinical deficiencies. The ability to "utilize what it needs" results in the addition of minerals in supra-physiologic concentrations based upon the relatively wide safety ranges for important trace minerals added to the diet, the documentation that supra-physiologic rates of specific minerals consumed are associated with animal growth promotion, improved animal health, productivity, and welfare. These rates are often in excess of the minimal animal requirements and while generally associated with limited risk for animal health, the excess minerals voided into the environment are of concern. Thus, it is well known that "A significant proportion of the minerals consumed are excreted in urine or manure leading to elevated concentration of minerals in soil, the crops grown in soil and eventually fed to animals and receiving water" (Suttle 2010). Depending upon the form and source of the minerals and processing of these commodities into animal supplements, contaminating substances including heavy metals such as arsenic, cadmium, mercury, and lead may occur (National Research Council 2005). Manure, manure slurries, and municipal wastes are commonly used in some areas to fertilize pastures (Poppenga 2000). Animal wastes contain salts, trace minerals, and microbes which may alter the soil microbiome and alter the health of cultivated plants. Accumulation of salt (NaCl) in the soil may cause salinization which can alter the uptake of metals by plants and soil compaction (Moral et al. 2007). In addition, copper and zinc are most likely to accumulate within soil when manure is applied to crop lands. Heavy metal accumulation within soil is also observed (Moral et al. 2007). Of significant importance, soils accumulating copper and zinc did so in proportion to the intensity of manure application rates to grazing lands and hepatic accumulation of copper and zinc in cattle (6–10 months of age; Lopez et al. 2000) Hepatic copper concentrations (but not zinc) in a significant proportion of the calves exceeded levels associated with toxicity in some species (Lopez et al. 2000).

Animals themselves may benefit or harm the soil through soil disturbances (overgrazing, consumption of one or few species of plants) and excretion of nitrogen, potassium, and phosphorous in excreta (Gerber et al. 2015). Digestion of forage, grains, and by-products by the rumen microflora is relatively inefficient with losses of consumed nitrogen and phosphorous of >50% into urine or feces (Oenema 2006). Excess nitrogen applied to crop land may reduce soil plant and microbial diversity and when washed into creeks, streams, and rivers may produce hypoxia that can make water bodies incapable of supporting aquatic life (Horrigan et al. 2002; Rabalais et al. 1996). Excretion of phosphate may lead to significant changes in the soil, water, and accumulation of wastes over time. Water-born algal blooms are associated with production of toxins that may accumulate in crops and aquatic organisms, has increased in occurrence in association with the use of fertilizers applied to soil, and may impact cattle health through consumption of forage grown on land irrigated with contaminated water (Corbel et al. 2014).

Cattle production systems also result in greenhouse gas emissions directly through rumen gas production, manure management, or changes in the use of land housing cattle (deforestation). Whereas crops and trees utilize carbon dioxide as a substrate, cattle release CO_2 and CH_4 into the environment in excess of utilization and sequestration by the soil biomass (Menzi and Gerber 2010).

Beef and dairy cattle production is estimated to generate approximately 61% of the greenhouse gases produced by agricultural industries (Gerber et al. 2013). Normal rumen function is based upon anaerobic fermentation of substrates consumed leading to the production of volatile fatty acids that can be used by the host to produce energy (O'Mara 2004). In the process of fermentation of plant substrates, production of carbon dioxide and hydrogen occurs which is further reduced to methane (O'Mara 2004). Rumen VFA may be converted to CO_2 or utilized to produce propionate (O'Mara 2004). Dietary changes can modify the rumen microbiome, thereby altering the production of methane (Vogels et al. 1980; O'Mara 2004). Strategies incorporating the feeding of more concentrates to cows (grain, grain by-products) results in a reduction of rumen pH, rumen microbial production of additional propionate that could be associated with reduced methane production (O'Mara 2004). Similar results can be anticipated when animal feed contain ionophores in their diet (increased production of propionate). Increases in biodiversity and soil health associated with managing cattle on grass over long periods and in feedlots under specific management practices also have been observed (Gerber et al. 2015). Thus, depending upon production management, there are associated environmental, animal, and human effects of cattle production. Some of these may reduce the greenhouse gas footprints of livestock. However, others suggest livestock consuming grains edible by humans also contributes to food insecurity (Horrigan et al. 2002).

Water use in cattle production varies tremendously depending upon the production system evaluated. Some estimate that cattle production is responsible for 10% of the global water footprint of agriculture (Hoekstra and Mekonnen 2012). Some of this water (rainwater) has less impact on total water use, whereas water from other sources (runoff, wastewater) results in degradation of water supplies (Mekonnen and Hoekstra 2012). Associated with the high levels of minerals provided to animals to sustain productivity and health, the application of manure, manure slurries, and municipal wastes are likely to contribute to increased soil concentrations of cadmium, lead, molybdenum, copper, and zinc (National Research Council 2005). If this soil is irrigated or drains into other water sources, excreted minerals and heavy metals may eventually contribute to lower water quality for animals and humans.

Soil and water contamination by pharmacologic agents may also alter soil health. Some agents may have substantial impact on plant growth, others on animal or human health. Several classes of therapeutic drugs are commonly used in conjunction with animal health prevention programs as well as therapeutically in cattle. Parasiticides (endo- and ecto-parasiticides), anti-bacterial agents, antifungal agents, growth promoting hormones, and non-medically important antimicrobials (monensin, lasalocid, decoquinate) are used to limit animal pain and suffering, production losses associated with diseases, and to promote growth. Many have been shown to alter the health of a soil biome the animals are maintained on through selection against susceptible species present in the environment. Pyrethrins, organophosphates, macrocyclic lactones, tetracyclines, beta lactams, aminoglycosides, monensin, progesterones, and estrogens are commonly used to treat disease, prevent infestation or promote growth of cattle (Boxall et al. 2002; McEachran et al. 2015; Noyes et al. 2016).

Some parasiticides were commonly used on cropland to control plant/soil parasites. Aldrin and dieldrin are organochlorine compounds that persist in the soil, sediments (lack bottoms), plants, and a variety of crops and trees (corn, hay fields, small grains and orchards, citrus, cotton, tobacco and vegetable crops; Jorgenson 2001). While most agents of this class are no longer marketed, the agricultural footprint of these compounds remains present. Further, these compounds are associated with endocrine disruption which may have impacts on a variety of soil borne species and has been associated with a variety of human disorders (Jorgenson 2001).

Animals maintained in confinement or grazed on pasture receive hormonal implants to improve their rate of gain. Hormonal implants are commonly used in animals introduced into feedlots as well as in various grazing schemes. The kinetics of hormonal release are reproducible and increased weight gain is very consistent in most animals. These hormone-containing implants may alter soil and water quality because the drugs, after absorption, are released into the environment by deposit with feces or excreted into the urine. Growth promoting steroids have been demonstrated in liquid run-off, surface soil, and manure samples obtained from animal containment facilities housing animals receiving implants (trenbolone, melengesterol acetate and their metabolites; Bartelt-Hunt et al. 2012; Jones et al. 2014). Further, untreated control animals produced natural steroids which were also detected in runoff, soil, and manure present in their containment facilities (Bartlet et al. 2012; Jones et al. 2014). Impacts of these compounds on aquatic organisms (fish) have been observed (Leet et al. 2009). Impacts upon terrestrial organisms have not been addressed as well.

The use of antimicrobial agents as growth promotants for all antibiotics (with the exception of those deemed non-medically important by the FDA; monensin and related compounds) was deemed inappropriate and are no longer available for such use in the US as of January 1, 2017. In hopes of reducing selection pressure in animal and environmental microbes, the sub-therapeutic levels of antibiotics for growth promotion is no longer legal in the US. In spite of this, therapeutic use of antimicrobials may also contribute to the development of resistant bacterial organisms in the environment and water. Particulate matter (wind driven dusts) from animal enclosures was shown to contain antimicrobial agents commonly used in feed prepared for housed animals (Tylosin, monensin, tetracycline, chrlotetracycline, oxytetracycline; McEachran et al. 2015). Further, microbes which commonly occur in rumainants were present and antimicrobial resistance genes were over-expressed in these agents. It is likely, although not directly examined in this study, that soil and water microbes would be altered as well. While some demonstrate the complexity of soil microbiomes (their ability to degrade antimicrobial agents), the ability of antimicrobial agents to alter soil microbiomes and expression of resistance genes requires additional study (Surette and Wright 2017). Similar comments could be made for water-borne microbes (Surette and Wright 2017). Other studies have evaluated the potential for soilborne organisms to contribute to disease spread with crop fertilization by spreading manure. As this may be animal growth stage related, management factors could be initiated to curtail this dissemination (Toth et al. 2013). Other studies evaluating the impact of antimicrobial agents present in manure from animals treated with these agents document alterations in the soil microbiome, altered soil nutrients, and impact the sensitivity of necessary organisms to other soil contaminants (Molael et al. 2017; Li et al. 2016).

15.5 CONCLUSIONS

Soil health is affected by biotic and abiotic factors. Climate, moisture, salinity, and temperature are abiotic factors. Temperature and moisture regimes affect soil health through alterations of biogeochemical and biogeophysical processes. The coupled cycling of water with those of elements (C, N, P, S, etc.) is also governed by climate change. In addition to the availability of water, climate change can also affect the amount and supply of essential macro- and micronutrients. Most macro- and micronutrients needed by plants are also needed by animals and humans. Human health is also affected by the concentration of contaminants (organic and inorganic) and organisms (pests and pathogens). Soil and dust are important sources of these pathogens. Whether pasture for grazing or crop land for production of high quality feedstuffs has a direct impact upon the health and productivity of cattle. Management of soil health is critical to maintain cattle for the production of high quality protein for human consumption as well as for work the animals do for their owners. Improvements in cattle production practices should center on soil health practices globally.

REFERENCES

Aminov R. 2017. History of antimicrobial drug discovery. *Biochem Pharmacol* 133:4–19.

Ampel NM. 2010. What's behind the increasing rates of coccidioidomycosis in Arizona and California? *Curr Infect Dis Rep* 12(3):211–6.

Bartelt-Hunt SL, Snow DD, Kranz WL, Mader TL, Shapiro CA, van Donk SJ, Shelton DP, Tarkalson DD, Zhang TC. 2012. Effect of growth promotants on the occurrence of endogenous and synthetic steroid hormones on feedlot soils and in runoff from beef cattle feeding operations. *Environ. Sci. and Technol.* 46:1352–1360.

Bastida F, Torres IF, Andrés-Abellán M, Baldrian P et. al. 2017. Differential sensitivity of total and active soil microbial communities to drought and forest management. *Glob Change Biol* 23(10):4185–4203.

Baumgardner DJ. 2012. Soil-related bacterial and fungal infections. *J Am Board Fam Med* 25(5):734–744.

Bengtsson H, Oborn I, Jonsson S, Nilsson I, Andersson A. 2003. Field balances of some mineral nutrients and trace elements in organic and conventional dairy farming—A case study at Ojebyn, Sweden. *Eur. J Agronomy* 20:101–116.

Bested AC et al. 2013. Intestinal microbiota, probiotics and mental health: From Metchnikoff to modern advances: Part III – Convergence toward clinical trials. *Gut Pathol* 5:4; https://www.ncbi.nlm.nih.gov/pmc/articles/PMC3605358/.

Bissett RJ, McLaughlin JF. 2010. Radon. *Chronic Dis Can* 29:38–50.

Boxall AB, Sarmah AK, Meyer MT. 2002. *Review of veterinary medicines in the environment.* UK: Environment Agency Bristol; www.environment-agency.gov.uk; sp6-012-8-tr-e-e.pdf.

CDC. 2015. Lyme disease: Data and statistics: Maps-reported eases of lyme disease, USA: 2001–2016. Center for Disease Control and Prevention. Atlanta, GA.

Clayton S, Manning CM, Hodge C. 2014. *Beyond storms and droughts: The psychological impacts of climate change.* Am. Psychological Assoc. and EcoAmerica, Washington, D.C.

Coppenet M, Golven J, Simon J, Corre LL, Roy ML. 1993. Chemical evolution of soils in intensive animal-rearing farms: The example of Finistere. *Agronomie* 13:77–83.

Corbel S, Mougin C, Bouaicha N. 2014. Cyanobacterial toxins: Modes of action, fate in aquatic and soil ecosystems, phytotoxicity and bioaccumulation in agriculture crops. *Chemosphere* 96:1–15.

Davis DR, MD Epp, HD Riordan. 2004. Changes in USDA food composition data for 43 garden crops, 1950 to 1999. *J Am Coll Nutr* 23(6):669–82.

De Kauwe MG, Medlyn BE, Walker AP, Zaehle S, Asao S, Guenet B, Harper AB, Hickler T, Jain AK, Luo Y, Lu X, Luus L, Parton WJ, Shu S, Wang YP, Werner C, Xia J, Pendall E, Morgon JA, Ryan EM, Carrillo Y, Dijkstra FA, Zelikova TJ, Norby RJ. 2017. Challenging terrestrial biosphere models with data from the long-term multifactor Prairie Heating and CO_2 Enrichment experiment. *Glob Change Biol* 23(9): 3623–3645.

Diaz-Cruz MS, Lopez de Alda MJ, Barcelo D. 2003. Environmental behavior and analysis of veterinary and human drugs in soils, sediments and sludge. *TrACT Trends in Analyt Chem* 22(6):340–351.

Doran JW, Zeiss MR. 2000. Soil health and sustainability: Managing the biotic component of soil quality. *Appl Soil Ecol* 15:3–11.

Epstein, PR. 2005. Climate change and human health. *N Eng J Med* 353:1433–1436.

Ericson B, Caravanos J, Chatham-Stephens K, Landrigan P, Fuller R. 2013. Approaches to systematic assessment of environmental exposures posed at hazardous waste sites in the developing world: The Toxic Sites Identification Program. *Environ Monit Assess* 185:1755–1766.

European Commission. 2013. Soil contamination: Impacts on human health. Science for Environment Policy In-Depth Report, Science Community Unit, University of West England, Bristol, England, Issue 5, September 2013, 29 pp; http://ec.europa.eu/scienc-environment-policy.

Fann N, Nolte CG, Dolwick P, Spero TL, Brown AC, Phillips S, Anenberg S. 2015. The geographic distribution and economic value of climate change-related ozone health impacts in the United States in 2030. *J Air and Waste Manag Assoc* 65:570–580.

FAO. 2016. Regional overview of food insecurity, Asia and the Pacific: Investing in zero hunger generation. FAO, Rome, Italy.

FAO, IFAD, UNICEF, WFP, WHO. 2017. The State of Food Security and Nutrition in the World 2017. Building Reliance for Peace and Food Security. ISBN: 978-92-5-109888-2, FAO, Rome, Italy, 112 pp.

FAOSTAT. 2017. FAO statistical database, accessed July 2017.

Fischetti M. 2015. Climate change hastened Syria's Civil War. *Scientific American* 2(March); retrieved at https://www.scientificamerican.com/article/climate-change-hastened-the-syrian-war/.

Frazer L. 2009. Climate change: Will warmer soil be as fertile? *Env Health Perspect* 117:A59.

Gabr M. 1987. Undernutrition and quality of life. *World Rev Nutr Diet* 49:1–21.

Garcia-Menedez F, Saari RK, Monier E, Selin NE. 2015. U.S. air quality and health benefits from avoided climate change under greenhouse gas mitigation. *Env Sci Technol*. 48:7580–7588.

Gerber PJ, Henderson B, Opio C, Mottet A, Steinfeld H. 2013. Tackling climate change through livestock—A global assessment of emissions and mitigation opportunities. FAO, Rome Italy; http://www.fao.org/3/i3437e.pdf (accessed 9/20/2017).

Gerber PJ, Mottet A, Opio CI, Falcucci A, Teillard F. 2015. Environmental impacts of beef production: Review of challenges and perspectives for durability. *Meat Sci* 109:2–12.

Golden MH. 1982. Trace elements in human nutrition. *Hum Nutr Clin Nutr* l36(3):185–202.

Hoekstra AY and Mekonnen MM. 2012. The water footprint of humanity. *PNAS* 109:3232–3237.

Horrigan L, Lawrence RS, Walker P. 2002. How sustainable agriculture can address the environmental and human health harms of industrial agriculture. *Environ Health Perspect* 110:445–456.

Jeffrey S, van der Putten WH. 2011. *Soil Borne Diseases of Humans*. JRC Publications Repository. Retrieved from: http://publications.jrc.ec.europa.eu/repository/bitstream/111111111/22432/2/lbna24893enn.pdf.

Jenny H. 1941. *Factors of Soil Formation*. McGraw Hill, New York.

Jones GD, Benchetler PV, Tate KW, Kolodziej EP. 2014. Trenbolone acetate metabolite transport in rangelands and irrigated pasture: Observations and conceptual approaches for agro-ecosystems. *Environ Sci and Technol* 48:12569–12576.

Jorgenson JL. 2001. Aldrin and Dieldrin: A review of research on their production, environmental deposition and fate, biotransformation, toxicology and epidemiology in the US. *Environ Health Perspect* 109:113–139.

Lal R. 2009. Soil degradation as a reason for inadequate human nutrition. *Food Secur* 1(1):45–57.

Leet JK, Lee LS, Gall HE, Goforth RR, Sassman S, Gordon DA, Lazorchak JM, Smith ME, Jafvert CT, Sepulveda MS. 2009. Assessing impacts of land applied manure from concentrated animal feeding operations on fish populations and communities. *Environ Sci and Technol* 46:13440–13447.

Li Y, Tang H, Hu Y, Wang X, Ai X, Tang L, Matthew C, Cavanagh J, Qiu J. 2016. Enrofloxacin at environmentally relevant concentrations enhances uptake and toxicity of cadmium in the earthworm Eisenia Fetida in farm soils. *J Hazard Mat* 308:312–320.

Lopez Alonso M, Benedito JL, Miranda M, Castillo C, Hernandez J, Shore RF. 2000. The effect of pig farming on copper and zinc accumulation in cattle in Galicia (North-Western Spain). *Vet J* 160:259–266.

Lopez-Alonso M. 2012. Trace minerals and livestock: Not too much, not too little. International Scholarly Research Network. ISRN Veterinary Science, Article ID 704825; doi:10.5402/2012/704825.

Luber G, N Prudent. 2009. Climate change and human health. *Trans Am Clin Climate Assoc* 120:113–117.

Luber G, Lemery J (Eds). 2015. *Global Climate Change and Human Health: From Science to Practice*. Wiley, New York, 672 pp. (ISBN: 978-1-118-5057-1).

Mayer AM. 1997. Historical changes in the mineral content of fruits and vegetables. 99:207–211.

McEachran AD, Blackwell BR, Hanson JD, Wooten KJ, Mayer GD, Cox SB, Smith PN. 2015, Antibiotics, bacteria and antibiotic resistance genes: Aerial transport from cattle feedyards via particulate matter. *Environ Health Persp* 123:337–343.

McLeman RA. 2014. What we learned from the dustbowl. *Popul Environ* 35(4): 417–40.

McMichael AJ, Campbell-Lendrum DH, Corvalán CF, Ebi KL, Githeko AK, Scheraga JD, Woodward A. (Eds). 2003. Climate Change and Human Health: Risks and Responses. WHO, Geneva, p. 322.

Meeker JD. 2012. Exposure to environmental endocrine disruptors and child development. A*rch Pediatr Adolesc Med* 166(6):E1–E7.

Mekonnen MM, Hoekstra AY. 2012. A global assessment of the water footprint of farm animal products. *Ecosystems* 15:401–415.

Menzi H, Gerber PJ. 2010. Nutrient balances for improving the use-efficiency of non-renewable resources: Experiences from Switzerland and Southeast Asia. Geological Society, London, Special publications, 266:171–181.

Mirsaeidi M. 2016. Climate change and respiratory infections. *Ann Am Thoracic Soc* 13(8):1223–1230.

Molael A, Lakzian A, Haghnia G, Astaraei A, Rasouli-Sadaghiani M, Ceccherini MT, Datta R. 2017. Assessment of some cultural experimental methods to study the effects of antibiotics on microbial activities in soil: An incubation study. *PLoS One* 127:e0180663. https://doi.org/10.1371/journal.pone.0180663

Moral R, Perez-Murcia MD, Perez-Espinosa A, Moreno-Caselles J, Paredes C, Rufete B. 2007. Salinity, organic content, micronutrients and heavy metals in pig slurries from South-eastern Spain. *Waste Manag* 28:367–371.

Napper IE, Thompson RC. 2016. Release of synthetic microplastic plastic fibres from domestic washing machines: Effects of fabric type and washing conditions. *Mar Pollut Bull* 112(1–2):39–45.

National Research Council 2005. *Mineral Tolerance of Animals*, 2nd Rev. Ed. The National Academies Press, Washington, D.C. https://doi.org/10.17226/11309.

Noyes NR, Yang X, Linke LM, Magnuson RJ, Cook SR, Zaheer R, Yang H, Woerner DR, Geornaras I, McArt JA, Gow SP, Ruiz J, Jones KL, Boucher CA, McAllister TA, Belk KE, Morley PS. 2016, Characterization of the resistome in manure, soil and wastewater from dairy and beef production systems. *Nature Scientific Rep* 6:24645. DOI:10.1038/srep24645.

O'Mara F. 2004. Greenhouse gas production from dairying: Reducing methane production. *Adv Dairy Technol* 16:295–309.

Oenema O. 2006. Nitrogen budgets and losses in livestock systems. Greenhouse gases in animal agriculture. Update *Proceedings 2nd International Conference Greenhouse Gases in Animal Agriculture*. Zurich Switzerland, 1293:262–271.

Poppenga RH. 2000. Current environmental threats to animal health and productivity. *Vet Clinics N Am: Food Animal Practice* 16:545–558.

Prist PR, Uriarte M, Fernandes K, Metzger JP. 2017. Climate change and sugarcane expansion increase Hantavirus infection risk. *PLoS Negl Trop Dis* 11(7):e0005705.

Rabalais NN, Turner RE, Justic D, Dortch Q, Wiseman WJ, Gupta BKS. 1996. Nutrient changes in the Mississippi River and system responses on the adjacent continental shelf. *Estuaries* 19:386–407.

Randolph TF, Schelling E, Grace D, Nicholson CF, Leroy JL, Cole DC, Demment MW, Omore A, Zinsstag J, Ruel M. 2007. Role of livestock in human nutrition and health for poverty reduction in developing countries. *J Animal Sci* 85:2788–2800.

Raup J. 1998. Yield, product quality and soil life after long term organic or mineral fertilization. In: *Proceedings of an International Conference*, Tufts University, Medford MA, 197:61–69.

Reuveny R. 2007. Climate change-induced migration and violent conflict. *Political Geo* 6: 656–673; http://www.csun.edu/~dtf46560/630/Misc/Reuveny-ClimateChangeMigration-2007.pdf.

Risch MR, DeWild JF, Gay DA, Zhang L, Boyer EW, Kraddenhoft DP. 2017. Atmospheric mercury deposition to forests in the eastern USA. *Environ Pollut* 228:8–18.

Rosin C. 2014. Mining drugs from microbes all over the world 20 September 2014, Retrieved from: http://www.rosincerate.com/2014/09/where-drugs-come-from-microbe-edition.html.

Schwartz JD, Lee M, Kinney PL, Yang S, Mills D, Sarofim MC, Jones R, Streeter R, Juliana A, Peers J, Horton RM. 2015. Projections of temperature-attributable premature deaths in 209 U.S. cities using a cluster-based poison approach. *Env Health* 14:85.

Seto KC, Güneralp B, Hutyra LR. 2012. Global forecasts of urban expansion to 2030 and direct impacts on biodiversity and carbon pools. *PNAS* 109(40):16083–16088.

Shrivastava P, Kumar R. 2015. Soil salinity: A serious environmental issue and plant growth promoting bacteria as one of the tools for its alleviation. *Saudi J Biol Sci* 22(2):123–131. PMC:4336437.

Stein, A.J., Qaim, M. 2007. The human and economic cost of hidden hunger. *Food and Nutr Bull* 28:125–134.

Surette M, Wright GD. 2017, Lessons from the environmental antibiotic resistome. *Annu Rev Microbiol* 71:309–329.

Suttle N. 2010. *Mineral Nutrition of Livestock*, 4th Ed. CABI Wallingford, Oxfordshire, UK, pp. 540–554.

Tilman D, Cassman KG, Matson PA, Naylor R, Polasky S. 2002. Agriculture sustainability and intensive production practices. *Nature* 418:671–677.

Toth JD, Aceto HW, Rankin SC, Dou Z. 2013. Short Communication: Survey of animal-borne pathogens in the farm environment of 13 dairy operations. *J Dairy Sci*: 965756–5761.

Trujillo-Gonzalez JM, Torres-Mora MA, Keesstra S, Brevik EC, Jimenez-Ballesta R. 2016. Heavy metal accumulation related to population density in road dust samples taken from urban sites under different land uses. *Sci Total Environ* 553:636–642.

Vogels GD, Hoppe WF, Stumm CK. 1980. Association of methanogenic bacteria with rumen ciliates. *Appl Evironmen Microbiol* 40:608–612.

Vogt KA, Patel-Weynand T, Shelton M, Vogt DJ, Gordon JC, Mukumoto CT, Suntana AS, Roads PA. 2010. *Sustainability Unpacked: Food, Energy, Water for Resilient Environments and Societies*. Earthscan, Washington, D.C., 305 pp.

Voigt C, Lamprecht RE, Marushchak ME, Lind SE, Novakovskiy A, Aurela M, Martikainen PJ, Biasi C. 2017. *Glob Chang Biol* 23(8):3121–3138.

Watve M, Tickoo R, Jog M, Bhole B. 2001. How many antibiotics are produced by the genus Streptomyces? *Arch Microbiol* 176(5):386–390.

WHO. 2007. Micronutrient deficiencies: Iron deficiency, anemia. WHO, Geneva, Switzerland.

Wood CJ, Pretty J, Griffen M. 2016. A case-control study of the health and well-being benefits of allotment gardening. *J Public Health* 38(3):e336–e344.

Wu G, Bazer FW, Cross HR. 2014. Land-based production of animal protein: Impacts, efficiency, and sustainability. *Ann NY Acad Sci* 1328:18–28.

Wuana RA, Okieimen FE. 2001. Heavy metals in contaminated soils: A review of sources, chemistry, risks, and best available strategies for remediation. *ISRN Ecol*, Article ID 402647.

Zhuang P, McBride MB, Xia H, Li N, Li Z. 2009. Health risk from heavy metals via consumption of food crops in the vicinity of Dabaoshan mine, South China. *Sci Total Environ* 407(5):1551–1561.

16 Climate Change and the Global Soil Carbon Stocks

Rattan Lal

CONTENTS

16.1 INTRODUCTION

The awareness about the importance of soil, land use and agricultural management practices in sequestering atmospheric CO_2 has been greatly enhanced since the Climate Summit of COP 22 in Paris in November 2015. Policymakers and private sectors are increasingly interested in making agriculture and soil as integral components of strategies for mitigation of and adaptation to climate change. Restoration of soil carbon (C) stock is of a specific interest for climate, food, and other ecosystem services. Soil C stock has two distinct but related components: soil organic C (SOC) and soil inorganic C (SIC) (Figure 16.1). The SOC stock is comprised of the remains of plants, animals, and microbes at various stages of decomposition and of the microbial byproducts. The SIC stock consists of primary or lithogenic carbonates and secondary or pedogenic carbonates (Monger et al. 2015; Zamanian et al. 2016). In general, SOC concentration is high in soils of the humid climates and that of SIC in soils of arid and semi-arid regions. To 1-m depth, total C stock is estimated at ~1500 Pg for SOC and 750 Pg for SIC (Batjes 1996, 2016).

The objective of this chapter is to describe the strong effects of soil C stock and its dynamics on the climate change, and vice versa, and managing SOC stocks for mitigation and adaption to the changing climate.

16.2 LONG-TERM VS. SHORT-TERM CARBON CYCLE

There are four principal reservoirs of C: geologic, oceanic, terrestrial (soil and plants), and atmospheric. Estimates of C stock in these reservoirs (Eg = 10^{18}g) include 60,000 as carbonates and 15,000 as organics in rocks, 42 in ocean, 4 in soils, 0.8 in atmosphere, and 0.6 in the biosphere (plants). Carbon cycles through these reservoirs following the long-term (10^3–10^7 yr at multi-million year scale) and the short-term (10^0–10^3 yr at decadal scale). Principal processes of the long-term cycling are (i) uptake of atmospheric CO_2 by weathering of Ca and Mg silicates, and (ii) the thermal decomposition of carbonate minerals and organic matter at depths by magmatism and diagenesis. Magmatism involves the formation of igneous rocks from magma. Diagenesis refers to the change

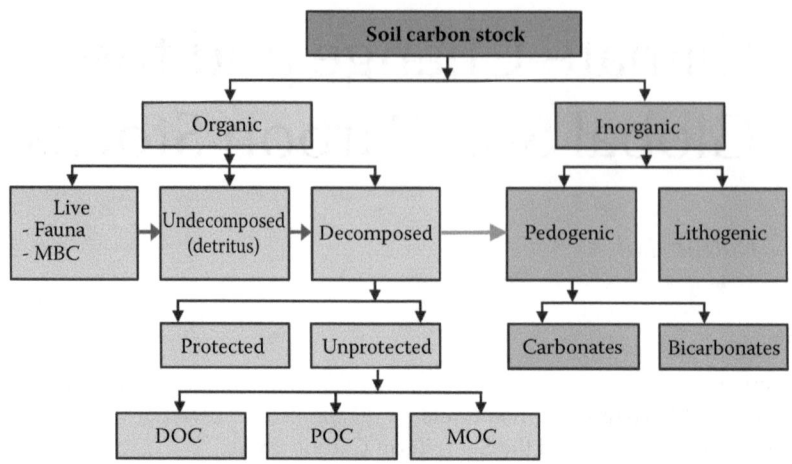

FIGURE 16.1 Components of total soil carbon stock: soil organic carbon and soil organic carbon.

of sediments into sedimentary rocks by lithification. Weathering of silicates and uptake of atmospheric CO_2 on a multi-million year-scale are described in Section 16.5.

The short-term (decadal C cycle) involves all but the geologic reservoirs. Principal reserves ($Eg = 10^{18}$ g) involved in the short-term cycle are oceanic (42), fossil fuel (5), soils (42), atmosphere (0.8), and vegetation (0.6). Fossil fuel combustion by humans is a special case of greatly accelerated weathering of the organic matter in the rocks.

There is an extremely little CO_2 in the atmospheric reservoir compared to that in the others. Thus, if inputs and outputs are not closely balanced, the atmospheric reservoir can be easily overwhelmed. This is exactly what happened because of the fossil fuel combustion since circa 1750 but especially since the 1950s. It is the overwhelming of the atmospheric reservoir that is the cause of global warming and necessitates an urgent action.

16.3 ANTHROPOGENIC ACTIVITIES AND THE SHORT-TERM CARBON CYCLE

Anthropogenic activities have drastically altered the global C cycle with the attendant increase in atmospheric CO_2 concentration from land use conversion since the dawn of settled agriculture and onset of the Industrial Revolution since circa 1750. For example, the atmospheric concentration of gases has increased since circa 1750 by 145% for CO_2 to reach global abundance of 403.3 ppmv, and by 257% for CH_4 with concentrations of 185.3 ppbv in 2016 (WMO 2017). However, a widespread awareness about the risks of the current and projected climate change have sparked only since 2015. The Climate Summit (COP21) in Paris highlighted the importance of exploiting the potential of the soil of managed and natural ecosystems to store atmospheric CO_2 and reverse the warming trends. Over and above the focus on managing the SOC stock for improving availability of plant nutrients and enhancing crop growth to meet the food demands of the world's growing population, the focus has also shifted to realizing other ecosystem services provisioned by the SOC stock such as being a repository for the excess atmospheric CO_2. Information about the SOC stock and its dynamics are good indicators of climate change. Janzen (2005) suggested that SOC and its dynamics across four dimensions can be used as an indicator of changes in environment across time.

Strongly coupled with the global C cycle is the global N cycle, which has also been altered since the use of nitrogenous fertilizers after the World War II. Changes in the global N cycles are evident by increases in the concentrations of nitrates in soil and waters (surface and groundwater), and in the ever-increasing concentration of N_2O and NO_x in the atmosphere (WMO 2017). For example, the global abundance of N_2O in the atmosphere has increased by 122% since circa 1750 and reached the concentration of 328.9 ppbv in 2016 (WMO 2017). Similar to the focus on increasing global SOC stocks, there is a strong need for emphasizing the role of biological N fixation (BNF) to improving agronomic production (Olivares et al. 2013). In addition to legumes, cereals also have the capacity to fix N.

Global cycles of C and N are linked with the hydrological (H_2O) cycle. Soil is the major reservoir of fresh water within terrestrial ecosystems and has a strong impact on the hydrological cycle. The global aridity may be aggravated by the projected global warming because of the positive feedbacks of the projected soil moisture decrease on land temperature, relative humidity, and precipitation (Berg et al. 2016). Soil moisture storage is closely linked with SOC concentration and its effects on total and retention porosity (Hudson 1994; Huntington 2003).

16.4 CLIMATE CHANGE AND SOIL ORGANIC CARBON STOCK

The SOC stock to 3-m depth is estimated at ~3819 Pg C (Table 16.1). In comparison, C stock in the vegetation is about 570 Pg (live) (Eglin et al. 2010) and another 60 Pg in the detritus material (Lal 2004). Thus, total terrestrial stock of organic C is about 4450 Pg C (Table 16.1). The SOC stock to 3-m depth is 4.65 times the atmospheric stock (~820 Pg in 2017) and 6.06 times the vegetation stock (630 Pg). The combined SOC stocks in permafrost (1024 Pg) and peatlands (450 Pg) at 1474 Pg is huge and is by itself 1.8 times that of the atmospheric stock and 2.3 times the biotic stock. However, the SOC stock in permafrost and peatlands are also vulnerable to decomposition with the projected climate change. Indeed, the SOC stocks of the terrestrial biosphere are sensitive to the climate change and must be protected against inappropriate land use, safeguarded against global warming, and restored in degraded soils and denuded landscapes.

The terrestrial biosphere (vegetation and soil) has been an important residual sink for anthropogenic emissions (Le Quéré et al. 2009), along with a significant impact on the magnitude of the anthropogenic emissions retained in the atmosphere (Keenan et al. 2016). The magnitude of the terrestrial sink has relatively changed over time and has a trend of increase with increase in total anthropogenic emissions. The magnitude of the annual terrestrial sink was 1.7 Pg in the 1960s, 1.7 Pg in the 1970s, 1.6 Pg in the 1980s, 2.6 Pg in the 1990s, 2.6 Pg in the 2000s, 3.1 Pg from 2006 to 2015, 1.9 Pg in 2015 (Le Quéré et al. 2016), and 2.7 Pg in 2017 (Le Quéré et al. 2017). The residual terrestrial sink absorbed ~17% of the total anthropogenic emissions in 2015 and 25.2% in 2016.

TABLE 16.1
Estimate of Soil Organic C Stock to 3-m Depth

Ecoregion	Area (10⁶ km²)	SOC Stock (Pg C)	Vegetation Stock (Pg C)
Boreal forests	12.00	150	72.5
Croplands	14.00	248	3.5
Deserts	26.50	332	9.0
Permafrosts	18.78	1024	–
Peatlands	3.50	450	15.0
Temperate forests	12.00	262	99.0
Temperate grasslands	9.00	172	16.0
Tropical forests	24.50	692	276.0
Tropical savannas	15.00	345	7.25
Tundra	8.00	144	4.0
Total	**143.28**	**3819**	

Source: Adapted from Eglin, T., P. Ciais, S. Piao, P. Barre, V. Bellassen, P. Cadule, C. Chenu, T. Gasser, C. Koven, M. Reichstein and P. Smith. *Tellus Series B—Chemical and Physical Meteorology* 62:700–718, 2010.

The terrestrial sink is affected by changes in net primary productivity (NPP) and the attendant C input into the soil. It is argued that the terrestrial biosphere may cease to be a sink by 2050 and may even become a net source (Heinemeyer et al. 2010).

The mechanization of agriculture, especially plowing and other farm operations, has accelerated the depletion of SOC since the 1950s, and conversion of forest to managed ecosystems can reduce the SOC stock by as much as 15 Mg C/ha (Eglin et al. 2010). The historic loss of global SOC stock to land use conversion may be as much as 130 Pg C (Lal 2018; Sanderman et al. 2017). The SOC stock is prone to decomposition with the projected climate change due to decline in input of biomass C from NPP and to increase in decomposition with increase in temperature. However, there are numerous uncertainties regarding the future changes in SOC stock (Kruse et al. 2013). The global SOC stock may be vulnerable to both climate change and land use conversion. Drought-induced dieback of tropical forests may reduce the SOC stocks by 170 Pg C between 2000 and 2100 (Elgin et al. 2010). Based on the study of two plots on the Russian–Mongolian border, measurable change in the SOC stick due to climate change may take 43 and 26 years for 20- and 30-cm depths, respectively (Conen et al. 2003). Understanding of the soil biogeochemical cycling to climate change may be improved by studying/modeling aggregate size as a function of SOC decomposition/mineralization (Nie et al. 2014).

16.5   CLIMATE CHANGE AND SILICATE WEATHERING

Weathering of calcite, dolomite, and calcium and magnesium containing silicates can strongly impact the atmospheric concentration of CO_2 over the geological scale of hundreds of millions years. The classic research by Urey (1952) demonstrated that, over the geological timescale, the atmospheric concentration of CO_2 is impacted by the transformation of silicate rocks into carbonate rocks by weathering and metamorphism (Berner et al. 1983).

$$\left.\begin{array}{l} CO_2 + CaSio_3 \underset{\text{Metamorphism}}{\overset{\text{Weathering}}{\rightleftharpoons}} CaCO_3 + SiO_2 \\[2ex] CO_2 + MgSiO_3 \underset{\text{Metamorphism}}{\overset{\text{Weathering}}{\rightleftharpoons}} MgCO_3 + SiO_2 \end{array}\right\} \text{Urey Reactions}$$

The atmospheric concentration of CO_2 is affected by a balance between uptake by weathering and release by metamorphism-magmatism. Uptake of CO_2 also occurs by wreathing of carbonates (Chamberlin 1989), and as stated below by Berner et al. (1983):

$$CO_2 + CaCO_3 + H_2O \underset{\text{Precipitation}}{\overset{\text{Weathering}}{\rightleftharpoons}} Ca^{++} + 2HCO_3^-$$

$$2CO_2 + CaMg(CO_3)_2 + 2H_2O \underset{\text{Precipitation}}{\overset{\text{Weathering}}{\longrightarrow}} Ca^{++} + Mg^{++} + 4HCO_3^-$$

Thus, there is a strong coupling between chemical weathering and climate by consuming CO_2 by the dissolution of Ca and Mg silicates (Brady 1991). Weathering of silicate minerals removes CO_2 from the atmosphere. In the following reactions, Brady (1999) argued that two moles of CO_2 are removed, but one mole of CO_2 is re-emitted to the atmosphere after formation of marine carbonates.

$$2\,CO_2 + 4H_2O + CaAl_2Si_2O_8 \overset{\text{Weathering}}{\longrightarrow} Ca^{++} + 2Al(OH)_3 + 2SiO_2^{aq} + 2HCO_3^- \quad \frac{\partial^2\Omega}{\partial u \partial v}$$

It is also argued that carbonate weathering and other oceanic processes may be significant in controlling atmospheric CO_2 concentration in both the short-term (<3ka) and long-term scales

(Plummer et al. 1978). Liu et al. (2011) emphasized the importance of the "biological carbon pump" in natural aquatic ecosystems which transports C into the lithosphere by sedimentation and burial of photosynthates as follows:

$$Ca^{++} + 2HCO_3^- \longrightarrow CaCO_3 + \psi(CO_2 + H_2O) + (1 - \psi)(CH_2O + O_2)$$

Liu and colleagues argued that the importance of the "biological pump" at the short-timescale has been neglected.

Limiting global warming to <2°C necessitates a thorough understanding of the role of soil as a source and sink of atmospheric CO_2 (and CH_4). In this context, the potential of accelerated weathering of powdered silicate minerals (e.g., olivine, basalt) needs to be explored, both for potential benefits and possible pitfalls (Edwards et al. 2017). Mimicking the long-term C cycle has led to the development of several technologies toward an attempt to remove excess CO_2 from the atmosphere. Important among these are (i) enhancing oil recovery, (ii) mining the coalbed methane, (iii) oceanic and geologic sequestration, and (iv) carbonation. An important question that remains to be addressed is, will the reactions which occur over hundreds of millions of years over the geologic timescale be effective over the human timescale of decades? The significance of these chemical reactions is highlighted in Chapter 9 of this volume by Daryl Siemer, especially with reference to the weathering of olivine.

16.6   SOIL MOISTURE REGIME AND CLIMATE CHANGE

The projected climate change will also cause shifts in the precipitation regime along with the attendant changes in soil moisture reserves. The SOC stock increases with increasing precipitation because of the increase in soil moisture regime, the NPP, and input of biomass C into the soil. Any decline in soil moisture regime, because of increase in frequency and intensity of drought, may reduce the SOC stock. The soil moisture regime is also strongly related to the community composition, biodiversity, or species richness, and the patterns of occurrence of species within a community (Le Roux et al. 2013). Over and above the impact of change in temperature with climate change on SOC stock, change in soil moisture regime because of change in climate may strongly impact the SOC stock. Drought-related decline in SOC stock can reduce CO_2 flux and decrease soil C cycling of the labile pool (Garten et al. 2009). Changes in climate change may also impact soil fertility and crop nutrition, thereby altering the SOC stock (St. Clair and Lynch 2010). Therefore, these factors must be considered in the identification and design of appropriate farming systems (Altieri et al. 2015).

16.7   IMPACTS OF CLIMATE CHANGE ON PEATLANDS AND PERMAFROST

Peatlands, covering only 350 million hectare (Mha), contain 450 Pg SOC of the stock. Combined with SOC stocks contained in soils of the permafrost (1024 Pg), which are also mostly these two ecosystems, the total SOC stock of these two ecosystems is 1474 Pg or 38.9% of the global SOC stock to 3-m depth (Table 16.1). The SOC stocks in these ecosystems (peatland and permafrost soils) may be strongly vulnerable to climate exchange. Even short-term changes in climate may change the quality of peat in bogs and fens (Keller et al. 2004). Increases in the release of CO_2 to the atmosphere from increases in warming-induced mineralization may lead to a strong positive feedback to the climate change (Jones et al. 2005).

16.8   SOIL EROSION AND CLIMATE CHANGE

With the intensification of the hydrological cycle, and the increasing frequency of extreme events, the projected climate change can increase the soil erosion risks by both water and wind. Nearing et al.

(2004) observed that the complex processes of the impacts of climate change on erosion are moderated by changes in rainfall amount and intensity, number of days with precipitation events, ratio of rain:snow, NPP and the groundcover, the rate of biomass decomposition, activity and species diversity of soil biota (including the macro-, meso-, and microbiota), evapotranspiration, and change in land use and farming systems. Risks of soil erosion are also impacted by the increase in demands for multiple uses of soil resources. Therefore, future crop management changes, because of the changing socio-economic conditions of the growing population, can strongly impact the soil erosion hazard (O'Neal et al. 2005). On-site, accelerated erosion reduces use efficiency of inputs and agronomic productivity. Decline in NPP and production also reduces the amount to biomass C returned to the soil, aggravating the erosion hazard. Zhang and Nearing (2005) observed the greater magnitude of runoff and soil loss with adverse impacts on NPP and the agronomic yield, especially in the yield of winter wheat in Oklahoma. Climate change has severe impacts on the yield of spring wheat in the Indo-Gangetic Plains. Similar to the adverse impacts on cropland, climate change can also adversely impact the productivity of rangelands. Zhang et al. (2012) observed a dramatic increase in runoff and soil loss from rangelands in Arizona due to the increase in frequency and intensity of extreme events. Therefore, recommended rangelands management and policy interventions must consider these risks. Off-site, increases in erosion can aggravate sediment transport and adversely impact water quality, increase eutrophication, and damage property and infrastructure by "muddy flooding" (Mullan 2013). Excessive runoff can also impact the soil moisture regime, decreasing it on-site but increasing it in the flood-prone foot-slopes. Changes in soil moisture regime can impact CO_2 flux and SOC cycling and can have relatively more impact than that of the elevated temperature and the atmospheric concentration of CO_2.

16.9 CONCLUSION

The global soil C stock consists of SOC and SIC. The total C stock affects and is affected by the climate change over arange of temporal scales. Over the long timescale of hundreds of millions years, weathering of silicate minerals leads to removal of atmospheric CO_2. Over the short timescale of decades and centuries, soil erosion, soil moisture, and soil temperature regimes affect the SOC stock. The projected climate change can accelerate decomposition of SOC stock and the large SOC stocks in peatlands and permafrost (Cryosols) are vulnerable to global warming. Potential and limitations of accelerated weathering of powered silicates for limiting global warming, and vulnerability of the large SOC stocks in peatlands and permafrost must be critically assessed.

The SOC stock, an important component of global C cycle, has been strongly perturbed by anthropogenic activities including land use and land use change, drainage of wetlands soil tillage, accelerated soil erosion, and inputs of agrochemicals. There is a strong coupling of the C cycle with those of the water, N, P, and S. Climate change affects and is affected by perturbations of the biogeochemical and biogeophysical cycles, and the climate change may also uncouple these cycles. Understanding the impacts of weathering and "biological C pumps" on long-term and the short-term C cycles is important to identifying technological options of mitigating climate change. Thus, credible measurements of the soil C stocks (both organic and inorganic) and identification of processes, practices, and policies to enhance it by sequestering atmospheric CO_2 as SOC is important to addressing numerous global issues of the 21st century.

Thus, the importance of climate change related alterations in soil moisture reserves on SOC stocks cannot b]e overemphasized.

REFERENCES

Altieri, M., C. Nicholls, A. Henao and M. Lana. 2015. Agroecology and the design of climate change-resilient farming systems. *Agronomy for Sustainable Development* 35:869–890.

Batjes, N.H. 1996. Total carbon and nitrogen in the soils of the world. *European Journal of Soil Science* 47:151–163.

Batjes, N.H. 2016. Harmonized soil property values for broad-scale modelling (WISE 30 sec) with estimates of global soil carbon stock. *Geoderma* 269:61–68.

Berg, A., K. Findell, B. Lintner, A. Giannini, S.I. Seneviratne, B. van den Hurk, R. Lorenz, A. Pitman, S. Hagemann, A. Meier, F. Cheruy, A. Ducharne, S. Malyshev and P.C.D. Milly. 2016. Land–atmosphere feedbacks amplify aridity increase over land under global warming. *Nature Climate Change* 6:869–874.

Berner, R., A. Lasaga and R. Garrels. 1983. The carbonate-silicate geochemical cycle and its effect on atmospheric carbon-dioxide over the past 100 million years. *American Journal of Science* 283:641–683.

Brady, P. 1991. The effect of silicate weathering on global temperature and atmospheric co2. *Journal of Geophysical Research-Solid Earth* 96:18101–18106.

Chamberlin, G.S. 1898. The influence of great epochs of limestone formation upon the constitution of the atmosphere. *Journal of Geology* 6:609–621.

Conen, F., M. Yakutin and A. Sambuu. 2003. Potential for detecting changes in soil organic carbon concentrations resulting from climate change. *Global Change Biology* 9:1515–1520.

Edwards, D.P., F. Lim, R.H. James, C.R. Pearce, J. Scholes, R.P. Freckleton and D.J. Beerling. 2017. Climate change mitigation: Potential benefits and pitfalls of enhanced rock weathering in tropical agriculture. *Biology Letters* 13: doi:10.1098/rsbl.2016.0715.

Eglin, T., P. Ciais, S. Piao, P. Barre, V. Bellassen, P. Cadule, C. Chenu, T. Gasser, C. Koven, M. Reichstein and P. Smith. 2010. Historical and future perspectives of global soil carbon response to climate and land-use changes. *Tellus Series B—Chemical and Physical Meteorology* 62:700–718.

Garten, C., A. Classen and R. Norby. 2009. Soil moisture surpasses elevated co2 and temperature as a control on soil carbon dynamics in a multi-factor climate change experiment. *Plant and Soil* 319:85–94.

Heinemeyer, A., S. Croft, M. Garnett, E. Gloor, J. Holden, M. Lomas and P. Ineson. 2010. The millennia peat cohort model: Predicting past, present and future soil carbon budgets and fluxes under changing climates in peatlands. *Climate Research* 45:207–226.

Hudson, B.D. 1994. Soil organic matter and available water capacity. *Journal of Soil and Water Conservation* 49:189–193.

Huntington, T.G. 2003. Available water capacity and soil organic matter. In R. Lal (Ed) *Encyclopedia of Soil Science*, 2nd Edition, Taylor and Francis, 139–143.

Janzen, H. 2005. The soil carbon dilemma: Shall we hoard it or use it? *Canadian Journal of Soil Science* 85:467–480.

Jones, C., C. Mcconnell, K. Coleman, P. Cox, P. Falloon, D. Jenkinson and D. Powlson. 2005. Global climate change and soil carbon stocks; predictions from two contrasting models for the turnover of organic carbon in soil. *Global Change Biology* 11:154–166.

Keenan, T., I. Prentice, J. Canadell, C. Williams, H. Wang, M. Raupach and G. Collatz. 2016. Recent pause in the growth rate of atmospheric co2 due to enhanced terrestrial carbon uptake. *Nature Communications* 7.

Keller, J., J. White, S. Bridgham and J. Pastor. 2004. Climate change effects on carbon and nitrogen mineralization in peatlands through changes in soil quality. *Global Change Biology* 10:1053–1064.

Kruse, J., J. Simon, H. Rennenberg, R. Matyssek, N. Clarke, P. Cudlin, T. Mikkelsen, J. Tuovinen, G. Wieser and E. Paoletti. 2013. Soil respiration and soil organic matter decomposition in response to climate change. *Climate Change, Air Pollution and Global Challenges: Understanding and Perspectives From Forest Research* 13:131–49.

Lal, R. 2018. *Digging Deeper: A Holistic Perspective of Factors Affecting SOC Sequestration. Global Change Biology* (In press)

Le Quere, C., R. Andrew, J. Canadell, S. Sitch, J. Korsbakken, G. Peters, A. Manning, T. Boden, P. Tans, R. Houghton, R. Keeling, S. Alin, O. Andrews, P. Anthoni, L. Barbero, L. Bopp, F. Chevallier, L. Chini, P. Ciais, K. Currie, C. Delire, S. Doney, P. Friedlingstein, T. Gkritzalis, I. Harris, J. Hauck, V. Haverd, M. Hoppema, K. Goldewijk, A. Jain, E. Kato, A. Kortzinger, P. Landschutzer, N. Lefevre, A. Lenton, S. Lienert, D. Lombardozzi, J. Melton, N. Metzl, F. Millero, P. Monteiro, D. Munro, J. Nabel, S. Nakaoka, K. O'brien, A. Olsen, A. Omar, T. Ono, D. Pierrot, B. Poulter, C. Rodenbeck, J. Salisbury, U. Schuster, J. Schwinger, R. Seferian, I. Skjelvan, B. Stocker, A. Sutton, T. Takahashi, H. Tian, B. Tilbrook, I. Van Der Laan-Luijkx, G. Van Der Werf, N. Viovy, A. Walker, A. Wiltshire and S. Zaehle. 2016. Global carbon budget 2016. *Earth System Science Data* 8:605–649.

Le Quere, C., R. Andrew, J. Canadell, S. Sitch, J. Korsbakken, G. Peters, A. Manning, T. Boden, P. Tans, R. Houghton, R. Keeling, S. Alin, O. Andrews, P. Anthoni, L. Barbero, L. Bopp, F. Chevallier, L. Chini, P. Ciais, K. Currie, C. Delire, S. Doney, P. Friedlingstein, T. Gkritzalis, I. Harris, J. Hauck, V. Haverd, M. Hoppema, K. Goldewijk, A. Jain, E. Kato, A. Kortzinger, P. Landschutzer, N. Lefevre, A. Lenton,

S. Lienert, D. Lombardozzi, J. Melton, N. Metzl, F. Millero, P. Monteiro, D. Munro, J. Nabel, S. Nakaoka, K. O'brien, A. Olsen, A. Omar, T. Ono, D. Pierrot, B. Poulter, C. Rodenbeck, J. Salisbury, U. Schuster, J. Schwinger, R. Seferian, I. Skjelvan, B. Stocker, A. Sutton, T. Takahashi, H. Tian, B. Tilbrook, I. Van Der Laan-Luijkx, G. Van Der Werf, N. Viovy, A. Walker, A. Wiltshire and S. Zaehle. 2017. Global carbon budget 2017. *Earth System Science Data* (In Review).

Le Quere, C., M. Raupach, J. Canadell, G. Marland, L. Bopp, P. Ciais, T. Conway, S. Doney, R. Feely, P. Foster, P. Friedlingstein, K. Gurney, R. Houghton, J. House, C. Huntingford, P. Levy, M. Lomas, J. Majkut, N. Metzl, J. Ometto, G. Peters, I. Prentice, J. Randerson, S. Running, J. Sarmiento, U. Schuster, S. Sitch, T. Takahashi, N. Viovy, G. Van Der Werf and F. Woodward. 2009. Trends in the sources and sinks of carbon dioxide. *Nature Geoscience* 2:831–836.

Le Roux, P., J. Aalto and M. Luoto. 2013. Soil moisture's underestimated role in climate change impact modelling in low-energy systems. *Global Change Biology* 19, no 10.

Liu, Z., W. Dreybrodt and H. Liu. 2011. Atmospheric co2 sink: Silicate weathering or carbonate weathering? *Applied Geochemistry* 26:S292–S94.

Monger, H.C., R.A. Kraimer, S. Khresat, D.R. Cole, X. Wang, and J. Wang. 2015. Sequestration of inorganic carbon in soil and groundwater. *Geology* 43:375–378.

Mullan, D. 2013. Soil erosion under the impacts of future climate change: Assessing the statistical significance of future changes and the potential on-site and off-site problems. *Catena* 109:234–246.

Nearing, M., F. Pruski and M. O'neal. 2004. Expected climate change impacts on soil erosion rates: A review. *Journal of Soil and Water Conservation* 59:43–50.

Nie, M., E. Pendall, C. Bell and M. Wallenstein. 2014. Soil aggregate size distribution mediates microbial climate change feedbacks. *Soil Biology & Biochemistry* 68:357–365.

Olivares, J., E.J. Bedmar and J. Sanjuán. 2013. Biological nitrogen fixation in the context of global change. *Molecular Plant-Microbe Interactions* 26(5):1591–1601.

O'Neal, M., M. Nearing, R. Vining, J. Southworth and R. Pfeifer. 2005. Climate change impacts on soil erosion in Midwest United States with changes in crop management. *Catena* 61:165–184.

Plummer, L., T. Wigley and D. Parkhurst. 1978. Kinetics of calcite dissolution in CO2-water systems at 5-degrees-C to 60-degrees-C and 0.0 to 1.0 atm CO2. *American Journal of Science* 278:179–216.

Sanderman, J., T. Hengl, and G.J. Fiska. 2017. Soil carbon debt of 12,000 years of human land use. *PNAS* 134:9575–9580.

St Clair, S. and J. Lynch. 2010. The opening of Pandora's box: Climate change impacts on soil fertility and crop nutrition in developing countries. *Plant and Soil* 335, no 1–2: 101–115.

Urey, H.C. 1952. *The Planets, Their Origin and Development*. Yale University Press, New Haven, 245 p.

WMO. 2017. Greenhouse gas bulletin: The state of greenhouse gases in the atmosphere based on global observations through 2016. World Meteorological Organization. Geneva, Switzerland, 8 pp.

Zamanian, K., K. Pustovytov and Y. Kuzyakov. 2016. Pedogenic carbonate: Forms and formation processes. *Earth Science Processes* 157:1–17.

Zhang, X.C., M.A. Nearing. 2005. Impact of climate change on soil erosion, runoff, and wheat productivity in central Oklahoma. *Catena* 61:185–195.

Zhang, Y., M. Hernandez, E. Anson, M. Nearing, H. Wei, J. Stone and R. Heilman. 2012. Modeling climate change effects on runoff and soil erosion in southeastern Arizona rangelands and implications for mitigation with conservation practices. *Journal of Soil and Water Conservation* 67:390–405.

Index

Page numbers followed by f and t indicate figures and tables, respectively.